The Atlas of Breeding Birds in New York State

THE ATLAS OF BREEDING BIRDS
IN NEW YORK STATE

CORNELL UNIVERSITY PRESS

ITHACA AND LONDON

EDITED BY **Robert F. Andrle**

AND **Janet R. Carroll**

A project of the
Federation of New York State Bird Clubs, Inc.
New York State Department of Environmental Conservation
and Cornell University Laboratory of Ornithology

The preparation of this publication was funded by the Division of Fish and Wildlife, New York State Department of Environmental Conservation, and the "Return a Gift to Wildlife" program.

Cover illustration: Drawing of the Great Blue Heron by Karen L. Allaben-Confer. Courtesy of Berthold Seeholzer.
Title page illustration: Drawing of the Canada Goose by Cynthia J. Page.

First published 1988 by Cornell University Press.

Library of Congress Cataloging-in-Publication Data

The Atlas of breeding birds in New York State.

"A project of the Federation of New York State Bird Clubs, Inc., New York State Department of Environmental Conservation, and Cornell University Laboratory of Ornithology."
  Bibliography: p.
  Includes index.
1. Birds—New York (State) I. Andrle, Robert F. II. Carroll, Janet R. III. Federation of New York State Bird Clubs. IV. New York (State). Dept. of Environmental Conservation. V. Cornell University. Laboratory of Ornithology.
QL684.N7A84   1988   598.29747   87-47867
ISBN 0-8014-1691-4 (alk. paper)

Printed in the United States of America
The paper in this book is acid-free and meets the guidelines for permanence and durability of the Committee on Production Guidelines for Book Longevity of the Council on Library Resources.

To Gordon M. Meade, M.D.

His forty years of vision and leadership of the Federation of New York
State Bird Clubs are an inspiration to all and an example for the
more than four thousand volunteers who helped make this Atlas possible.

To the reader:

It is with great pride and pleasure that the Department of Environmental Conservation welcomes the publication of *The Atlas of Breeding Birds in New York State*. As participants, technical advisers, funding supporters, and users of the information contained in the Atlas, the Department salutes this landmark accomplishment and the efforts of the thousands of volunteers who contributed their time, energy, and skills to this important conservation program. Atlas data have already been used to identify areas of significant wildlife habitat and will continue to provide, for years to come, a database of bird distribution in New York that has never previously been available to researchers, wildlife managers, and naturalists.

I was fortunate enough to participate personally in the collection of Atlas data. I enjoyed the opportunity to join some Atlas workers in exploring woods, marshes, and fields in search of breeding birds. This experience will remain with me, as I know it will with others, to increase my future enjoyment and knowledge of the state's abundant natural resources.

A major source of funding to support the publication of the Atlas was provided by New York's "Return a Gift to Wildlife" program.

I look forward to continuation of breeding bird studies and hope that they will reveal an even greater variety of New York birdlife than appears in the Atlas. Then we will know that the environment is improving, both for wildlife and for the citizens of this great state.

HENRY G. WILLIAMS
Commissioner, New York State
Department of Environmental Conservation

31 May 1987

CONTENTS

ILLUSTRATIONS

FOREWORD

The production of this atlas demonstrates that quite disparate organizations can come together and, in a common effort, bring forth an outstanding contribution to science. It is also a striking testimony of how the amateur can play an essential role in such an endeavor. Joined in this project have been a volunteer organization of amateur and professional ornithologists, a large governmental agency, and a public membership research and educational institution. The three have different modes of operation, but despite these differences, they share a common objective, the study and protection of our avifauna. Within this mutual purpose they came together and worked amicably and productively.

Borrowing the concept of an atlas from the pioneering work of the British, the Federation of New York State Bird Clubs organized the Atlas structure, provided the survey personnel, carried out the censusing work, and gave overall guidance; the New York State Department of Environmental Conservation furnished a project coordinator, computerized and analyzed the data, created distributional maps, and provided the major financial support for operation of the project and preparation of the Atlas publication; the Cornell University Laboratory of Ornithology contributed invaluable advice, secretarial service, and financial management. The entire operation has been an exhilarating and stimulating experience for all involved.

Not only has the Atlas made a significant contribution to our knowledge of New York's birds, it has made an equally important one to the individuals involved, especially the surveyors. They have learned to be more observant of actions and activities they probably never noticed before and to understand what those things mean in the life of the bird. Their birding has taken on a new excitement and a meaning most had never before experienced; a pleasant hobby has become an absorbing intellectual and sensory pursuit. Their lives are better and fuller for it.

Many Atlas workers, enthralled by the sheer enjoyment of atlasing and having had their curiosity piqued by the unanswered questions they encountered, may well go on to explore new fields. Thus, the Atlas may bring fresh talents and avenues of investigation to New York ornithology.

And finally, the experience gained in the methods of wildlife surveying can serve as a guide for similar research on other wildlife, as is now occurring in Great Britain and other countries.

It all began in July 1978 at an Executive Committee meeting when the late Robert Arbib proposed that the Federation of New York State Bird Clubs undertake an atlas of the breeding birds of New York. Bob gave us an inspiration; we owe him our sincere appreciation. Unfortunately, Bob did not live to see his ideas come to fruition in this atlas. He will be remembered and missed by all who knew him.

Without our Regional Coordinators we would have been figuratively working without a head, hands, or feet. They recruited our fieldworkers—the Atlasers—instructed and advised them, encouraged the disheartened and the laggards, and spent endless hours checking and correcting data, all with enthusiasm and energy. They gave their time and energy, but through it all had fun and like the rest of us became better birders. To all of them our heartfelt thanks. By regions they were (1) Robert F. Andrle, (2) Robert G. Spahn, (3) Dorothy W. McIlroy, (4) Jay G. Lehman, (5) Dorothy W. Crumb, (6) Robert and June Walker, (7) John M. C. Peterson, (8) Mark Fitzsimmons, (9) Berna B. Weissman, and (10) Gilbert S. Raynor. Very special appreciation is owed to Gilbert S. Raynor, who did a superlatively imaginative job in Region 10, setting a high standard for everyone, until stricken by an incapacitating illness. Not to be omitted were former Regional Coordinators whose places were taken by others: Kate Dunham, Richard Guthrie, and Roger Robb.

The Regional Coordinators found many dedicated persons who helped with recruiting and organizing Atlasers, making block assignments and telephone calls, checking and correlating field reports, and distributing supplies. Those outstanding by regions were (1) Robert Brock, Elizabeth Brooks, William and Doris Vaughan; (3) Margaret Shepard, Walter Benning, Wilifred Howard; (4) Robert Daigle, Christine Lehman, Harriet Marsi; (5) Paul A. DeBenedictis, Belle Peebles; (6) Lee Chamberlaine, Bruce Beattie; (7) Elsbeth Johnson, Daniel Nickerson; (8)

Kate Dunham; (9) Barbara Butler, Janet Fitzpatrick, Valerie Freer, Stanley Lincoln, Ev Rifenburg; (10) Berna B. Weissman, Robert Adamo, James Clinton, John P. McNeil, Eric Salzman, John Ruscica.

Probably our greatest obligation in terms of steady, devoted day-after-day work is owed to the more than 4000 volunteers and block busters who gathered the data, without which there would be no Atlas. They rose early, walked and drove many miles, sloshed through marshes, slashed through briers and tangles, met interested and helpful property owners, were chased and threatened by irate and suspicious misanthropes, and cajoled and infected others to join in the "crusade." They did a magnificent piece of work—the base on which this monograph stands. If a name is missing from the list of Atlasers at the back of the book, it is not by intent but by misadventure. We are grateful to each and every one.

The Federation embarked on the Atlas project with high enthusiasm but little knowledge of what lay ahead. The NYSDEC, Division of Fish and Wildlife, soon joined us to provide financial support and many services. They furnished us with a Project Coordinator, Natasha Atkins, who later left and was replaced by the capable, indefatigable Janet R. Carroll. Others from the NYSDEC who worked diligently to make the Atlas a reality were Eugene R. McCaffrey and Robert L. Miller, who served as advisers and provided administrative expertise; Walton B. Sabin, who helped with many aspects of the book; Vincent Grilli, computer programmer; Barton Guetti, Mary Kate Cannistra, Kelly Mogiliski, and Katherine Barnes, who were responsible for production of species maps and overlays; Charles Dente, who managed the mailing lists; Katherine Goodrich and Sandra Miller, who prepared Atlas block maps; Karin O'Sullivan, who helped compile the observer list; and Tammy Madalla, who typed much of the manuscript. The enthusiastic support of Commissioner Henry G. Williams led to financial assistance from "Return a Gift to Wildlife."

Cornell's Laboratory of Ornithology was an official sponsor of the Atlas from the beginning. Believing that the Atlas project was consonant with the objectives of the Laboratory, its director, Charles Walcott, made available the good counsel and wise advice of Charles R. Smith, who served as Assistant to the Chairman and as a member of the Atlas Editorial Advisory Board. The Laboratory's facilities were frequently used for committee meetings, and the Atlas was widely publicized in its publications. Diane Johnson, Nancy Schrempf, and Maureene Stangle, of the Laboratory's staff, were invaluable in managing accounts and contracts, paying bills, and doing secretarial work.

Generous financial contributions to the Federation were given by the following:

Individuals: John W. Brown, Ezra A. Hale, Lewis Kibler, T. Spencer Knight, Mary Anne Sunderlin, Joseph W. Taylor, Kathryn Whitney, Clarence and Alice Wynd.

Bird clubs and Audubon societies: Alan Devoe Bird Club, Bedford Audubon Society, Buffalo Ornithological Society, Cayuga Bird Club, Hudson-Mohawk Bird Club, Hudson Valley Audubon Society, Linnaean Society of New York, Moriches Bay Audubon Society, Northern Adirondack Audubon Society, Onondaga Audubon Society, Ralph T. Waterman Bird Club, Sawmill River Audubon Society, Scarsdale Audubon Society, Sullivan County Audubon Society.

Corporations and foundations: American Conservation Association, American Sponge and Chamois Company, Consolidated Edison Company of New York, Long Island Lighting Company, New York Telephone Company, Niagara Mohawk Power Corporation, NL Industries Foundation, and Rochester Gas and Electric Corporation.

Some of the contributions were made to the Federation; others were given directly to the Regional Coordinators to assist the work in their Atlas region. The Buffalo Museum of Science provided work space and the services of their business office to Robert F. Andrle. Assisting him in typing and computer work were Betty Blakely, Marcya Foster, and Ruby Robinson.

After the species accounts were in final form, they were submitted to individuals with special knowledge of the species or family described, who read them critically for errors and omissions. These experts were Kenneth Able, Dean Amadon, David Austin, Jon Barlow, Paul Bishop, Hans Blokpoel, Robin Bouta, John Bull, Nan Chadwick, Scott Crocoll, Phyllis Dague, Robert Dickerman, William Dilger, James Glidden, Peter Gradoni, George Hall, Douglas Kibbe, Lawrence Kilham, Wesley Lanyon, Sarah Laughlin, Barbara Loucks, Richard Malecki, Nancy Martin, Robert L. Miller, Jack Moser, Paul Novak, Peter Nye, Karl Parker, Kenneth Parkes, David Peterson, John M. C. Peterson, Edgar Reilly, Michael Scheibel, Donald Slingerland, Charles R. Smith, Bryan Swift, and D. Vaughn Weseloh.

The selection of the artists who gave us the pen-and-ink drawings we will enjoy for years to come was made by Dorothy W. Crumb, who also supervised the production of the illustrations. Thanks go to Karen L. Allaben-Confer for contributing the logo for our stationery and our masthead on the newsletter, and to Robert F. Andrle, Paul A. DeBenedictis, and Francis G. Scheider for technical review of all illustrations.

To S. D. McDonald of the National Museum of Natural Sciences, Ottawa, Ontario, we express appreciation for the use of his Spruce Grouse vocalization cassette.

And to the people of New York who contributed to "Return a Gift to Wildlife," we give special thanks.

We are profoundly grateful to all who gave of their thought, their time, and their talents. And I personally extend my gratitude. You have made my task as chairman a much easier one.

GORDON M. MEADE
Chairman, New York State
Breeding Bird Atlas Project

A NOTE TO BIRDERS

The publication of this handsome and information-rich volume marks a milestone in New York State ornithology and is an achievement in which all who contributed can rightfully take pride. As you study the wealth of information gathered here, remember that this is a report on a discrete slice of time: the distribution of birds that have been found breeding in New York during the years 1980–85. Thus, this atlas is already a historical document, comprising for the first time significant baseline data against which future comparable studies can be measured. Those of us who labored on the text soon discovered how fragmentary, vague, or nonexistent is meaningful information about past distribution of the birds. Often we were forced to interpret past distribution from what we inferred or knew about the state's environmental history, or from the casual or ambiguous comments of earlier authors. We may have had to make educated guesses about where the Eastern Meadowlark bred in 1485, 1685, or even 1885, but thanks to this project we have a very good picture of where it bred in 1985! The Atlas nails it down for us now, and for all those Atlasers—and their readers—yet to come.

Preparation of the Atlas was indeed a daunting prospect. So many Atlas blocks to investigate! So much of it in mountainous, remote, near-wilderness areas! This book is proof that it was done. Almost all of our 5335 blocks were surveyed. Not all completely, some only minimally—but enough to provide us with surprisingly accurate and detailed distribution maps.

To be of value as an environmental monitor, of course, this project must be repeated at reasonable intervals. So, finally, a word to the Atlasers of the future. Study our work, learn from it, improve on it—but keep it comparable. Start planning early. And may you have as much fascination and as great a sense of achievement in the years 2030–35, or 2080–85, as we have had in the 1980s. And may you still have birds to survey!

ROBERT ARBIB
Chairman, Committee for
Publications and Research,
Federation of New York State Bird Clubs

Atlas Working Group

Chairman, New York State Breeding Bird Atlas Project   Gordon M. Meade
Assistant to the Chairman   Charles R. Smith
Project Coordinator   Janet R. Carroll
Regional Coordinators
    Region 1   Robert F. Andrle
    Region 2   Robert G. Spahn
    Region 3   Dorothy W. McIlroy
    Region 4   Jay G. Lehman
    Region 5   Dorothy W. Crumb
    Region 6   Robert and June Walker
    Region 7   John M. C. Peterson
    Region 8   Mark Fitzsimmons
    Region 9   Berna B. Weissman
    Region 10   Gilbert S. Raynor

Atlas Publication Staff

Editors   Robert F. Andrle, Janet R. Carroll
Editorial Assistance   Dorothy W. Crumb, Robert W. Darrow, Mark Fitzsimmons, Robert L. Miller, Edgar M. Reilly
Reference Editor   Walton B. Sabin
Art Committee   Dorothy W. Crumb, Chairwoman; Robert F. Andrle, Paul A. DeBenedictis, Francis G. Scheider

Business Manager   Charles R. Smith
Editorial Advisory Board   Gordon M. Meade, Chairman; Robert F. Andrle, Robert Arbib, Janet R. Carroll, Charles R. Smith
Map and Overlay Production   Habitat Inventory Unit, New York State Department of Environmental Conservation

Authors of Species Accounts

Robert F. Andrle
Robert Arbib
Richard E. Bonney, Jr.
Judith L. Burrill
Janet R. Carroll
John L. Confer
Paul F. Connor
Stephen W. Eaton
Mark Gretch
Richard A. Lent

Emanuel Levine
Gordon M. Meade
Paul G. Novak
David M. Peterson
John M. C. Peterson
Steven C. Sibley
Charles R. Smith
Gerald A. Smith
Robert G. Spahn

Artists

Karen L. Allaben-Confer
William C. Dilger

Cynthia J. Page
David A. Sibley

ABBREVIATIONS

a	acre, acres
AOU	American Ornithologists' Union
BBS	United States Fish and Wildlife Service Breeding Bird Survey
BP	before the present (1950 = present)
CBC	National Audubon Society Christmas Bird Count
cm	centimeter, centimeters
dbh	diameter at breast height (of tree)
ft	foot, feet
ha	hectare, hectares
in	inch, inches
km	kilometer, kilometers
LUNR	Land Use and Natural Resource Inventory of New York State
m	meter, meters
mi	mile, miles
NWR	United States National Wildlife Refuge
NYSDEC	New York State Department of Environmental Conservation
NYSDOT	New York State Department of Transportation
pers. comm.	personal communication
pers. obs.	personal observation
sq	square
USFWS	United States Fish and Wildlife Service
USGS	United States Geological Survey
v.r.	verification report
WMA	New York State Wildlife Management Area

The Atlas of Breeding Birds in New York State

INTRODUCTION

The goal of the Atlas project was to determine which species of birds breed in New York State and where they do so. Thousands of birders conducted extensive field surveying from 1980 through 1985, and three organizations—the Federation of New York State Bird Clubs, the New York State Department of Environmental Conservation, and the Cornell University Laboratory of Ornithology—worked together to oversee the fieldwork and compile the results. This book shows where breeding birds were found and discusses their status, distribution, and breeding habitats. Additional perspective on the state's avifauna is given in a chapter on the prehistory of New York's birds, as gleaned from paleontological research in the state.

The information contained herein should prove valuable to ornithologists, other scientists, government agencies, commercial interests, and the public for education, planning, and conservation. To help the reader fully understand and interpret this information, we present below an account of the planning done, the methods used for field surveys and for assembling the data, and some of the results attained.

Planning and Management

The Atlas Steering Committee decided to use the already established ten regions employed for bird sightings reported to the Federation's publication *The Kingbird* (see Map 1). To administer the project statewide, a Project Coordinator was hired by the New York State Department of Environmental Conservation, and ten Regional Coordinators were appointed to assist the Project Coordinator by managing Atlas work in each of the regions.

A metric-based coordinate system known as the New York Transverse Mercator Grid, a variation of the Universal Transverse Mercator Grid, was the basis for the grid developed for reporting all data collected. A 5 × 5

km grid unit or "block" was decided upon as the basic unit for Atlas field surveying because it was manageable in terms of the ability to obtain coverage, to present the species distributions in reasonable detail, and to make the data usable for further research.

This grid contains approximately 1300 10 × 10 km (38.6 sq mi) "squares," each subdivided for Atlas fieldwork into four 5 × 5 km (9.65 sq mi) "blocks" covering the 128,402 sq km (49,576 sq mi) area of the state. Each block was assigned a letter (A, B, C, D), and each square had coordinates numbered sequentially from south to north and west to east (see Fig. 1). There were 5335 blocks statewide. Sets of map squares for Atlas fieldwork were prepared by the NYSDEC using the latest NYSDOT planimetric-topographic and USGS topographic maps (Miller 1981). The

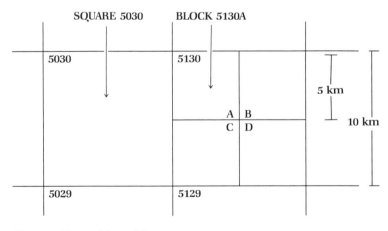

Figure 1. Key to Atlas grid system.

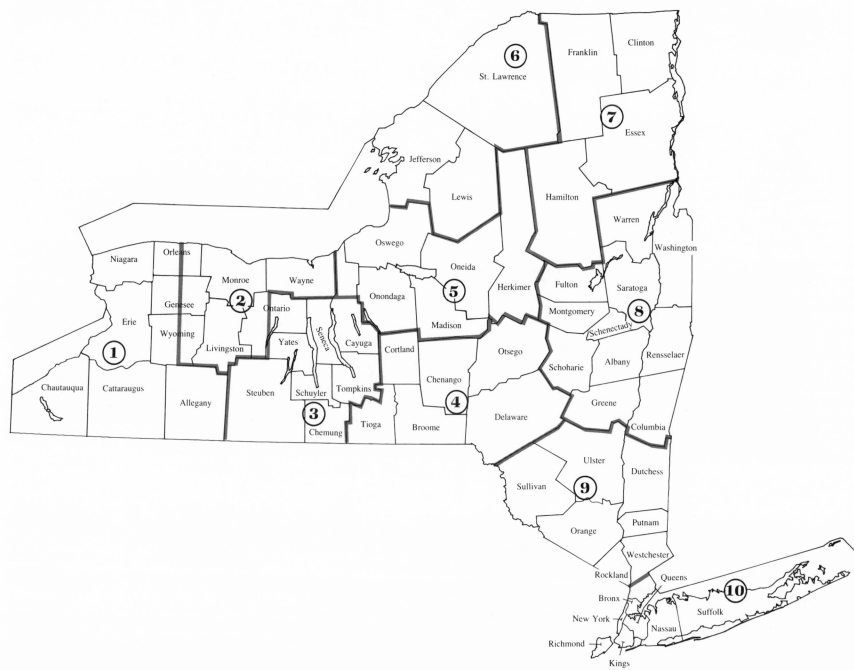

Map 1. *The Kingbird* regions of New York. (Source of data: Federation of New York State Bird Clubs.)

1a Tonawanda Wildlife Management Area
1b Iroquois National Wildlife Refuge
1c Oak Orchard Wildlife Management Area
2 Braddock Bay Wildlife Management Area
3 High Tor Wildlife Management Area
4 Montezuma National Wildlife Refuge
5 Howland Island Wildlife Management Area
6 Three Rivers Wildlife Management Area
7 Pharsalia Wildlife Management Area
8 Perch River Wildlife Management Area
9 Wilson Hill Wildlife Management Area
10 Jamaica Bay Wildlife Refuge

Map 2. The counties, national wildlife refuges, and state wildlife management areas of New York. (Source of data: NYSDEC.)

scale of the Atlas maps prepared for use by the surveyors is the same as that of USGS quadrangles—1:24,000. This grid system was developed in anticipation of the eventual conversion, especially of topographic maps, to the metric system, and because of its potential relation with an existing New York land use and natural resources (LUNR) survey (Miller 1981). (Currently designated wildlife areas are shown in Map 2.)

Recording Methods and Data Collection

A code system containing 16 types of breeding evidence was developed. It represents a synthesis of the British Trust for Ornithology, European Ornithological Atlas Committee, and Massachusetts Breeding Bird Atlas codes (see Fig. 2).

Data were recorded for each block each year on a field card and then transferred to a data sheet, which was designed to facilitate keypunching. Both the field card and data sheet contained spaces for names of observers, dates of fieldwork, hours spent atlasing, and comments (see Figs. 3, 4).

The volunteers were also provided with a "Handbook for Workers," which contained information on basic objectives, an explanation of the grid system, survey techniques, the breeding evidence codes, a sample verification report, and reporting instructions. Also, two other references were developed to assist Atlasers: a "Breeding Season Information" table (see Appendix C), which provided information on nidiology, and "Hints on Haunts," which gave habitat requirements for selected species. Newsletters were published three times a year during the Atlas period; these reported on progress, gave information on survey techniques, advised of administrative matters, and addressed problems encountered in the field.

Written details were required for 61 species identified as rare in the state, as well as for any species not previously known to breed in New York. These were each marked with an asterisk on the field card and data sheet. A verification report form was provided for this purpose, and all such reports were carefully reviewed. All verification reports are in the files of the NYSDEC (Wildlife Resources Center, Delmar, New York 12054).

Field Surveying and Coverage

The objective of field surveying was to record evidence of breeding on all species in each block, "confirming" as many as possible. This objective was best approached by surveying all habitat types occurring in each block. Appropriate codes were recorded for each species, with the codes upgraded during subsequent visits to a block.

Adequate coverage was defined as a "total list of at least 76 species identified from each block, with half (38) confirmed as breeders" (Smith 1982). These numbers were based on the experience of the Vermont Breeding Bird Atlas, which, after two years of field surveying, set a similar goal. Setting these numbers let the Atlaser know what was expected. In a few cases, however, it discouraged some from continued participation.

Criteria are listed in order of increasing certainty of breeding

POSSIBLE BREEDING—Enter in the "PO" column of data form
X   Species observed in breeding season in possible nesting habitat but no other indication of breeding noted.
     Singing male(s) present (or breeding calls heard) in breeding season.

PROBABLE BREEDING—Enter in the "PR" column of data form
P   Pair observed in suitable habitat in breeding season.
S   Singing male present (or breeding calls heard) on more than one date in the same place. This is a good indication that a bird has taken up residence if the dates are a week or more apart.
T   Bird (or pair) apparently holding territory. In addition to territorial singing, chasing of other individuals of same species often marks a territory.
D   Courtship and display, agitated behavior or anxiety calls from adults suggesting probable presence nearby of a nest or young; well-developed brood-patch or cloacal protuberance on trapped adult. Includes copulation.
N   Visiting probable nest site. Nest building by wrens and woodpeckers. Wrens may build many nests. Woodpeckers, although they usually drill only one nesting cavity, also drill holes just for roosting.
B   Nest building or excavation of a nest hole.

CONFIRMED BREEDING—Enter in the "CO" column of data form.
DD   Distraction display or injury feigning. Agitated behavior and/or anxiety calls are only "D," under Probable Breeding.
UN   Used nest found. *Caution:* These must be carefully identified if they are to be counted as evidence. Some nests (e.g., Northern Oriole) are persistent and very characteristic. Most are difficult to identify correctly.
FE   Female with egg in the oviduct.
FL   *Recently* fledged young (including downy young of precocious species—waterfowl, shorebirds). This code should be used with caution for species such as blackbirds and swallows, which may move some distance soon after fledging. Recently fledged passerines are still dependent on their parents and are fed by them.
ON   Adult(s) entering or leaving nest site in circumstances indicating occupied nest. *NOT generally used for open nesting birds.* It should be used for hole nesters only when a bird enters a hole and remains inside, makes a change-over at a hole, or leaves a hole after having been inside for some time. If you simply see a bird fly into or out of a bush or tree and do not find a nest, the correct code would be "N" under "Probable."
FS   Adult carrying fecal sac.
FY   Adult(s) with food for young. Some birds (gulls, terns, and raptors) continue to feed their young long after they are fledged, and even after they have moved considerable distances. Also, some birds (e.g., terns) may carry food long distances to young in a neighboring block. Be especially careful on the edge of a block. Care should be taken to avoid confusion with courtship feeding (D under Probable).
NE   Identifiable nest and eggs, bird setting on nest or eggs, identifiable eggshells found beneath nest, or identifiable dead nestling(s). If you find a cowbird egg in a nest, it is NE for Cowbird, and NE for the identified nest's owner.
NY   Nest with young. If you find a young cowbird with another species' young, it is NY for cowbird and NY for identified nest owner.

Figure 2. Atlas codes and criteria for breeding evidence.

After the first field season it was decided that the time spent trying to "confirm" species would be better used obtaining additional coverage and more species.

Smith (1982) presented a method of graphically illustrating the point at which an Atlaser's time in the field no longer represented "measurable dividends" in terms of additional species. Raynor (1983) developed a method for the Long Island–New York City region which estimated the number of species expected to be found in each block there.

82-14-25 (2/82)

New York Breeding Bird Atlas Project

Block ⬚⬚⬚⬚⬚ Year 19⬚⬚⬚

BREEDING CRITERIA CODES

PO
X Species seen in possible nesting habitat, or singing male(s) present.

PR
P Pair in suitable habitat.
S Singing male present on more than one date.
T Bird or pair on territory.
D Courtship and display.
N Repeatedly visiting probable nest site, or nest building by wrens and woodpeckers.
B Nest building or excavation of nest hole.

CO
DD Distraction display.
UN Used nest.
FE Female with egg in oviduct.
FL Recently fledged young.
ON Adult entering or leaving site indicating occupied nest (hole nesters only).
FS Adult with fecal sac.
FY Adult with food for young or feeding young.
NE Nest and eggs, adult incubating, identifiable dead nestlings, egg shells beneath nest.
NY Nest with young.

*Written details required except for well known, previously established breeding localities.

NAME	PO	PR	CO
Loon, Common *			
Grebe, Pied-billed *	X		
Cormorant, Double-crested *			
Heron, Great Blue			
Green			NE
Little Blue *			
Egret, Cattle *			
Great *			
Snowy *			

NAME	PO	PR	CO
Heron, Louisiana *			
Black-crowned Night			
Yellow-crowned Night *			
Bittern, Least			
American	X		
Ibis, Glossy *			
Swan, Mute			
Goose, Canada			
Mallard			FL
Mallard x Black Duck *			
Duck, Black			
Gadwall			
Pintail			
Teal, Green-winged			
Blue-winged		P	
Wigeon, American			
Shoveler			
Duck, Wood			FL
Redhead			
Duck, Ring-necked *			
Scaup, Lesser *			
Goldeneye, Common *			
Duck, Ruddy			
Merganser, Hooded			
Common			
Red-breasted			
Vulture, Turkey	X		
Goshawk			
Hawk, Sharp-shinned			
Cooper's			
Red-tailed			UN
Red-shouldered			
Broad-winged			
Eagle, Golden *			
Bald *			

NAME	PO	PR	CO
Hawk, Marsh			
Osprey			
Falcon, Peregrine *			
Kestrel, American		P	
Grouse, Spruce *			
Ruffed			NE
Bobwhite			
Partridge, Gray			
Pheasant, Ring-necked		P	
Turkey			FL
Rail, King *			
Clapper			
Virginia			
Sora			
Rail, Black *			
Gallinule, Common	X		
Coot, American			
Oystercatcher, American *			
Plover, Piping *			
Killdeer			DD
Woodcock, American	X		
Snipe, Common			
Sandpiper, Upland			
Spotted	X		
Willet *			
Gull, Great Black-backed *			
Herring			
Gull, Ring-billed			
Laughing *			
Tern, Gull-billed *			
Common			
Roseate *			
Least *			
Black			
Skimmer, Black *			

NAME	PO	PR	CO
Dove, Rock			FY
Mourning		P	
Parakeet, Monk *			
Cuckoo, Yellow-billed			
Black-billed			
Owl, Barn			
Screech			
Great Horned		S	
Barred			
Long-eared			
Short-eared			
Saw-whet			
Chuck-will's-widow *			
Whip-poor-will			
Nighthawk, Common			
Swift, Chimney			
Hummingbird, Ruby-throated		P	
Kingfisher, Belted		P	
Flicker, Common			ON
Woodpecker, Pileated			
Red-bellied	X		
Red-headed	X		
Sapsucker, Yellow-bellied			
Woodpecker, Hairy	X		
Downy	X		
Black-backed Three-toed *			
Northern Three-toed *			
Kingbird, Eastern		P	
Flycatcher, Great Crested		S	
Phoebe, Eastern			NE
Flycatcher, Yellow-bellied *			
Acadian			
Alder		S	
Willow		S	
Least		S	

NAME	PO	PR	CO
Pewee, Eastern Wood		S	
Flycatcher, Olive-sided *			
Lark, Horned	X		
Swallow, Tree	X		
Bank	X		
Rough-winged			
Barn			NE
Cliff			
Martin, Purple			ON
Jay, Gray *			
Blue	X		
Raven, Common *		P	
Crow, Common			
Fish			
Chickadee, Black-capped		P	
Boreal *			
Titmouse, Tufted			
Nuthatch, White-breasted	X		
Red-breasted			
Creeper, Brown			
Wren, House			ON
Winter			
Bewick's *			
Carolina			
Long-billed Marsh			
Short-billed Marsh			
Mockingbird			
Catbird, Gray			NE
Thrasher, Brown	X		
Robin, American			NE
Thrush, Wood			FY
Hermit			
Swainson's			
Gray-cheeked			
Veery		S	

NAME	PO	PR	CO
Bluebird, Eastern			
Gnatcatcher, Blue-gray			
Kinglet, Golden-crowned			
Ruby-crowned *			
Waxwing, Cedar			FL
Shrike, Loggerhead *			
Starling			ON
Vireo, White-eyed *			
Yellow-throated			
Solitary			
Red-eyed			NE
Philadelphia			
Warbling		S	
Warbler, Black-and-white			
Prothonotary			
Worm-eating			
Golden-winged		S	
Blue-winged		S	
Warbler, Brewster's *			
Warbler, Lawrence's *			
Tennessee			
Nashville			
Parula, Northern			
Warbler, Yellow			NE
Magnolia			
Cape May *			
Black-throated Blue			
Yellow-rumped			
Black-throated Green			
Cerulean			
Blackburnian			
Chestnut-sided			
Bay-breasted *			
Blackpoll *			
Pine			

NAME	PO	PR	CO
Warbler, Prairie		S	
Ovenbird		S	
Waterthrush, Northern			
Louisiana			
Warbler, Kentucky *			
Mourning		S	
Yellowthroat, Common		S	
Chat, Yellow-breasted		S	
Warbler, Hooded		S	
Warbler, Wilson's *			
Canada			
Redstart, American			NE
Sparrow, House			ON
Bobolink	X		
Meadowlark, Eastern		S	
Western *			
Blackbird, Red-winged			DD
Oriole, Orchard			
Northern			UN
Blackbird, Rusty			
Grackle, Common			FY
Cowbird, Brown-headed			NY
Tanager, Scarlet		S	
Cardinal			NE
Grosbeak, Rose-breasted		S	
Bunting, Indigo		S	
Dickcissel *			
Grosbeak, Evening *			
Finch, Purple			
House			
Siskin, Pine			
Goldfinch, American			FL
Crossbill, Red			
White-winged *			
Towhee, Rufous-sided		S	

NAME	PO	PR	CO
Sparrow, Savannah		S	
Grasshopper			
Henslow's			
Sharp-tailed *			
Seaside *			
Vesper			
Junco, Dark-eyed			
Sparrow, Chipping		S	
Clay-colored *			
Field		S	
White-throated			
Lincoln's *			
Swamp		S	
Song		S	

DATE	HOURS	OBSERVERS
6/7/83	4	R. Schrader
6/15/83	5	"
6/24/83	3	"
7/4/83	4	
		D. Cook

Figure 3. An Atlas field card.

MAP NAME **Orleans : West Gaines**
(county; and largest town, village landmark, or road junction in block)

NEW YORK BREEDING BIRD ATLAS PROJECT

PRINCIPAL OBSERVER

Name **Robert Schrader**

Address _____

Phone _____

RETURN TO:

IMPORTANT – PLEASE READ

1. Transfer data from small field record card to this form.
2. Use pencil when filling out this form.
3. Before submitting this form to the regional coordinator, sum the number of PO, PR and CO Species and enter at bottom of sheet.
4. Fill in the time spent in this block on the back of the form.

NAME					PO	PR	CO	NAME					PO	PR	CO	NAME					PO	PR	CO	NAME					PO	PR	CO
Loon, Common	C	O	L	O		1		2	Gallinule, Common	C	O	G	A	X	1		2	Swallow, Bank	B	K	S	W	X	1		2	Warbler, Black-throated Green	B	T	G	W		1		2
Grebe, Pied-billed	P	B	G	R		1		2	Coot, American	A	M	C	O		1		2	Rough-winged	R	W	S	W		1		2	Cerulean	C	R	W	A		1		2
Cormorant, Double-crested	D	C	C	O	X	1		2	Oystercatcher, American	A	M	O	Y		1		2	Barn	B	A	S	W		1	N E	2	Blackburnian	B	L	W	A		1		2
Heron, Great Blue	G	B	H	E		1		2	Plover, Piping	P	I	P	L		1		2	Cliff	C	L	S	W		1		2	Chestnut-sided	C	S	W	A		1		2
Green	G	R	H	E		1	N E	2	Killdeer	K	I	L	L		1	D D	2	Martin, Purple	P	U	M	A		1	O N	2	Bay-breasted	B	B	W	A		1		2
Little Blue	L	B	H	E		1		2	Woodcock, American	A	M	W	O	X	1		2	Jay, Gray	G	R	J	A		1		2	Blackpoll	B	P	W	A		1		2
Egret, Cattle	C	A	E	G		1		2	Snipe, Common	C	O	S	N		1		2	Blue	B	L	J	A	X	1		2	Pine	P	I	W	A		1		2
Great	G	R	E	G		1		2	Sandpiper, Upland	U	P	S	A		1		2	Raven, Common	C	O	R	A		1		2	Prairie	P	R	W	A		1		2
Snowy	S	N	E	G		1		2	Spotted	S	P	S	A	X	1		2	Crow, Common	C	O	C	R	1	P	2	Ovenbird	O	V	E	N		1	S	2
Heron, Louisiana	L	O	H	E		1		2	Willet	W	I	L	L		1		2	Fish	F	I	C	R		1		2	Waterthrush, Northern	N	O	W	A		1		2
Black-crowned Night	B	C	N	H		1		2	Gull, Great Black-backed	G	B	B	G		1		2	Chickadee, Black-capped	B	C	C	H	1	P	2	Louisiana	L	O	W	A		1		2
Yellow-crowned Night	Y	C	N	H		1		2	Herring	H	E	G	U		1		2	Boreal	B	O	C	H		1		2	Warbler, Kentucky	K	E	W	A		1		2
Bittern, Least	L	E	B	I		1		2	Gull, Ring-billed	R	B	G	U		1		2	Titmouse, Tufted	T	U	T	I		1		2	Mourning	M	O	W	A		1	S	2
American	A	M	B	I	X	1		2	Laughing	L	A	G	U		1		2	Nuthatch, White-breasted	W	B	N	U	X	1		2	Yellowthroat, Common	C	O	Y	E		1	S	2
Ibis, Glossy	G	L	I	B		1		2	Tern, Gull-billed	G	B	T	E		1		2	Red-breasted	R	B	N	U		1		2	Chat, Yellow-breasted	Y	B	C	H		1	S	2
Swan, Mute	M	U	S	W		1		2	Common	C	O	T	E		1		2	Creeper, Brown	B	R	C	R		1		2	Warbler, Hooded	H	O	W	A		1		2
Goose, Canada	C	A	G	O		1		2	Roseate	R	S	T	E		1		2	Wren, House	H	O	W	R		1	O N	2	Warbler, Wilson's	W	I	W	A		1		2
Mallard	M	A	L	L		1	F L	2	Least	L	E	T	E		1		2	Winter	W	I	W	R		1		2	Canada	C	A	W	A		1		2
Mallard x Black Duck	M	B	D	H		1		2	Black	B	L	T	E		1		2	Bewick's	B	E	W	R		1		2	Redstart, American	A	M	R	E		1	N E	2
Duck, Black	B	L	D	U		1		2	Skimmer, Black	B	L	S	K		1		2	Carolina	C	A	W	R		1		2	Sparrow, House	H	O	S	P		1	O N	2
Gadwall	G	A	D	W		1		2	Dove, Rock	R	O	D	O		1		2 F Y		Long-billed Marsh	L	B	M	W		1		2	Bobolink	B	O	B	O	X	1		2
Pintail	P	I	N	T		1		2	Mourning	M	O	D	O		1	P	2	Short-billed Marsh	S	B	M	W		1		2	Meadowlark, Eastern	E	A	M	E		1	S	2
Teal, Green-winged	G	W	T	E		1		2	Parakeet, Monk	M	O	P	A		1		2	Mockingbird	M	O	C	K		1		2	Western	W	E	M	E		1		2
Blue-winged	B	W	T	E		1	P	2	Cuckoo, Yellow-billed	Y	B	C	U		1		2	Catbird, Gray	G	R	C	A		1	N E	2	Blackbird, Red-winged	R	W	B	L		1	D D	2
Wigeon, American	A	M	W	I		1		2	Black-billed	B	B	C	U		1		2	Thrasher, Brown	B	R	T	H	X	1		2	Oriole, Orchard	O	R	O	R		1		2
Shoveler	S	H	O	V		1		2	Owl, Barn	B	A	O	W		1		2	Robin, American	A	M	R	O		1	N E	2	Northern	N	O	O	R		1	U N	2
Duck, Wood	W	O	D	U		1	F L	2	Screech	S	C	O	W		1		2	Thrush, Wood	W	O	T	H		1	F Y	2	Blackbird, Rusty	R	U	B	L		1		2
Redhead	R	E	D	H		1		2	Great Horned	G	H	O	W		1	S	2	Hermit	H	E	T	H		1		2	Grackle, Common	C	O	G	R		1	F Y	2
Duck, Ring-necked	R	N	D	U		1		2	Barred	B	D	O	W		1		2	Swainson's	S	W	T	H		1		2	Cowbird, Brown-headed	B	H	C	O		1	N Y	2
Scaup, Lesser	L	E	S	C		1		2	Long-eared	L	E	O	W		1		2	Grey-cheeked	G	C	T	H		1		2	Tanager, Scarlet	S	C	T	A		1	S	2
Goldeneye, Common	C	O	G	O		1		2	Short-eared	S	E	O	W		1		2	Veery	V	E	E	R		1	S	2	Cardinal	C	A	R	D		1	N E	2
Duck, Ruddy	R	U	D	U		1		2	Saw-whet	S	W	O	W		1		2	Bluebird, Eastern	E	A	B	L		1		2	Grosbeak, Rose-breasted	R	B	G	R		1	S	2
Merganser, Hooded	H	O	M	E		1		2	Chuck-will's-widow	C	W	W	I		1		2	Gnatcatcher, Blue-gray	B	G	G	N		1		2	Bunting, Indigo	I	N	B	U		1	S	2
Common	C	O	M	E		1		2	Whip-poor-will	W	P	W	I		1		2	Kinglet, Golden-crowned	G	C	K	I		1		2	Dickcissel	D	I	C	K		1		2
Red-breasted	R	B	M	E		1		2	Nighthawk, Common	C	O	N	I		1		2	Ruby-crowned	R	C	K	I		1		2	Grosbeak, Evening	E	V	G	R		1		2
Vulture, Turkey	T	U	V	U	X	1		2	Swift, Chimney	C	H	S	W		1		2	Waxwing, Cedar	C	E	W	A		1	F L	2	Finch, Purple	P	U	F	I		1		2
Goshawk	G	O	S	H		1		2	Hummingbird, Ruby-th.	R	T	H	U		1	P	2	Shrike, Loggerhead	L	O	S	H		1		2	House	H	O	F	I		1		2
Hawk, Sharp-shinned	S	S	H	A		1		2	Kingfisher, Belted	B	E	K	I		1	P	2	Starling	S	T	A	R		1	O N	2	Siskin, Pine	P	I	S	I		1		2
Cooper's	C	O	H	A		1		2	Flicker, Common	C	O	F	L		1		2 O N	Vireo, White-eyed	W	E	V	I		1		2	Goldfinch, American	A	M	G	O		1	F L	2
Red-tailed	R	T	H	A		1	U N	2	Woodpecker, Pileated	P	I	W	O		1		2	Yellow-throated	Y	T	V	I		1		2	Crossbill, Red	R	E	C	R		1		2
Red-shouldered	R	S	H	A		1		2	Red-bellied	R	B	W	O	X	1		2	Solitary	S	O	V	I		1		2	White-winged	W	W	C	R		1		2
Broad-winged	B	W	H	A		1		2	Red-headed	R	H	W	O	X	1		2	Red-eyed	R	E	V	I		1	N E	2	Towhee, Rufous-sided	R	S	T	O		1	S	2
Eagle, Golden	G	O	E	A		1		2	Sapsucker, Yellow-bellied	Y	B	S	A		1		2	Philadelphia	P	H	V	I		1		2	Sparrow, Savannah	S	A	S	P		1		2
Bald	B	A	E	A		1		2	Woodpecker, Hairy	H	A	W	O	X	1		2	Warbling	W	A	V	I		1	S	2	Grasshopper	G	R	S	P		1		2
Hawk, Marsh	M	A	H	A		1		2	Downy	D	O	W	O	X	1		2	Warbler, Black-and-White	B	A	W	A		1		2	Henslow's	H	E	S	P		1		2
Osprey	O	S	P	R		1		2	Black-backed Three-toed	B	T	W	O		1		2	Prothonotary	P	O	W	A		1		2	Sharp-tailed	S	T	S	P		1		2
Falcon, Peregrine	P	E	F	A		1		2	Northern Three-toed	N	T	T	W		1		2	Worm-eating	W	E	W	A		1		2	Seaside	S	E	S	P		1		2
Kestrel, American	A	M	K	E		1	P	2	Kingbird, Eastern	E	A	K	I		1	P	2	Golden-winged	G	W	W	A		1	S	2	Vesper	V	E	S	P		1		2
Grouse, Spruce	S	P	G	R		1		2	Flycatcher, Great Crested	G	C	F	L		1	S	2	Blue-winged	B	W	W	A		1	S	2	Junco, Dark-eyed	D	E	J	U		1		2
Ruffed	R	U	G	R		1	N E	2	Phoebe, Eastern	E	A	P	H		1		2 N E	Brewster's	B	R	W	A		1		2	Sparrow, Chipping	C	H	S	P		1	S	2
Bobwhite	B	O	B	W		1		2	Flycatcher, Yellow-bellied	Y	B	F	L		1		2	Lawrence's	L	A	W	A		1		2	Clay-colored	C	C	S	P		1		2
Partridge, Gray	G	R	P	A		1		2	Acadian	A	C	F	L		1		2	Tennessee	T	E	W	A		1		2	Field	F	I	S	P		1	S	2
Pheasant, Ring-necked	R	N	P	H		1	P	2	Alder	A	L	F	L		1	S	2	Nashville	N	A	W	A		1		2	White-throated	W	T	S	P		1		2
Turkey	T	U	R	K		1	F L	2	Willow	W	I	F	L		1	S	2	Parula, Northern	N	O	P	A		1		2	Lincoln's	L	I	S	P		1		2
Rail, King	K	I	R	A		1		2	Least	L	E	F	L		1	S	2	Warbler, Yellow	Y	E	W	A		1	N E	2	Swamp	S	W	S	P		1	S	2
Clapper	C	L	R	A		1		2	Pewee, Eastern Wood	E	W	P	E		1		2	Magnolia	M	A	W	A		1		2	Song	S	O	S	P		1	S	2
Virginia	V	I	R	A		1		2	Flycatcher, Olive-sided	O	S	F	L		1		2	Cape May	C	M	W	A		1		2									
Sora	S	O	R	A		1		2	Lark, Horned	H	O	L	A	X	1		2	Black-throated Blue	B	T	B	W		1		2									
Rail, Black	B	L	R	A		1		2	Swallow, Tree	T	R	S	W	X	1		2	Yellow-rumped	Y	R	W	A		1		2									

*written details (Verification Report) required except for well-known, previously established breeding localities.

PO = POSSIBLE
PR = PROBABLE
CO = CONFIRMED

NO. OF POSSIBLE: 0 1 7
NO. OF PROBABLE: 0 3 4
NO. OF CONFIRMED: 0 2 8
TOTAL: 0 7 9

RETURN THIS DATA SHEET TO YOUR REGIONAL COORDINATOR
BY SEPTEMBER 15

Figure 4. An Atlas data sheet.

Block Busting

Approximately 25% of Atlas coverage was conducted by an intense survey method termed block busting. In 1982 individuals were hired to do intensive fieldwork during June and July, with two goals. One was to cover a large number of blocks; the other was to determine how much time would be necessary to spend in a block in order to attain a desired level of coverage. Based on these results and the need to make the best use of paid survey teams, two-person teams spent one day in each block and camped there to assure an opportunity for recording nocturnal species.

Survey teams were hired each year from 1982 through 1985, and each team spent 16–20 person-hours surveying a block. Block busting by these teams was the principal method of coverage in the western Adirondacks, Tug Hill Plateau, Schoharie Hills, Catskill Peaks, Delaware Hills, Taconic Mountains, and the Mohawk Valley (see Map 3). The average number of species per block recorded by block-busting teams was just slightly lower than the overall average for the project.

An additional block-busting effort was made by individuals in various regions who were willing to travel away from their home areas to do intense atlasing during weekends and holidays. Travel expenses were reimbursed when requested by these individuals.

Data Checking, Processing, and Presentation

Regional Coordinators were responsible for review of all Atlas data submitted by field surveyors before forwarding the data to the Project Coordinator for further review. The data were then keypunched and edited, and reports were prepared. Both the Regional Coordinator and Atlaser were given copies of the block reports to check for errors. In addition to cumulative data for each block, nine other computer reports were generated, presenting the data in various ways. Computer-generated maps were also produced annually, showing block coverage and the distribution of each breeding species.

The Species Accounts

The discussions that accompany each distribution map consist essentially of several topics: abundance, history, distribution interpretation, atlasing methods and effectiveness, habitat, and nest description and location. Because there has been no statewide study of abundance for New York's breeding species, each author made an individual judgment based on the following abundance terms: abundant, very common, common, fairly common, rather uncommon, uncommon, very uncommon, rare, very rare, casual, accidental. Where abundance data were available, such as from colonial waterbird surveys and from Breeding Bird Surveys (Robbins et al. 1986), they were included.

The history of each species in New York is intended to present information on changes in distribution and/or abundance and to facilitate comparison with Atlas data. Although historic information from the literature is sometimes questionable, in most cases it provided some idea of the species' status and was usually the only available source of information.

Each author describes and interprets the current distribution of each species. Some factors taken into account are land use, forest cover, vegetation types, water areas, temperature, precipitation, elevation, and human influence (see Maps 4–9). Overlays for most of these factors are available from Cornell University Press for use with each species' distribution map.

Other subjects discussed were the codes used most frequently to record the species, the ease or difficulty in locating or "confirming" the species, and whether the map accurately depicts the species' range. If records from other research were included in the Atlas data for a species, it was noted.

Habitat is critical to any discussion of bird distribution. Atlasers collected habitat information usually only on rare species. Therefore, in most accounts information about habitat was gleaned from the literature, from relevant material in recent and ongoing research, and from the author's personal knowledge. Brief data on nest type, construction, and location, which are intimately related to the habitat and distribution of a species, are also included.

The total number of blocks in which the species was recorded, as well as a breakdown of this total into the numbers of blocks in which the species was recorded as "Possible," "Probable," and "Confirmed," is presented with each account. The total number of blocks in which a species was recorded is also expressed as a percentage of the total number of blocks surveyed (5323) in the state. In addition, for each species the number of blocks in which "Possible," "Probable," and "Confirmed" breeding was reported is expressed as a percentage of the total number of blocks in which the species was found. Since these percentages are rounded to the nearest whole number, their sum often does not total exactly 100%.

The American Ornithologists' Union *Check-list of North American Birds* (1983) is followed for nomenclature and sequence of bird species. Plant, other animal, and place names follow Mitchell (1986), NYSDEC (1985), and NYSDOT (1983), respectively.

Maps

The map with each species account shows the distribution to block level as found by Atlasers. Each symbol on the map represents a 5 × 5 km block in its correct location. The meanings of the symbols are as follows:

☒ Possible breeding
⊞ Probable breeding
■ Confirmed breeding

Where there is no symbol, the species was not observed.

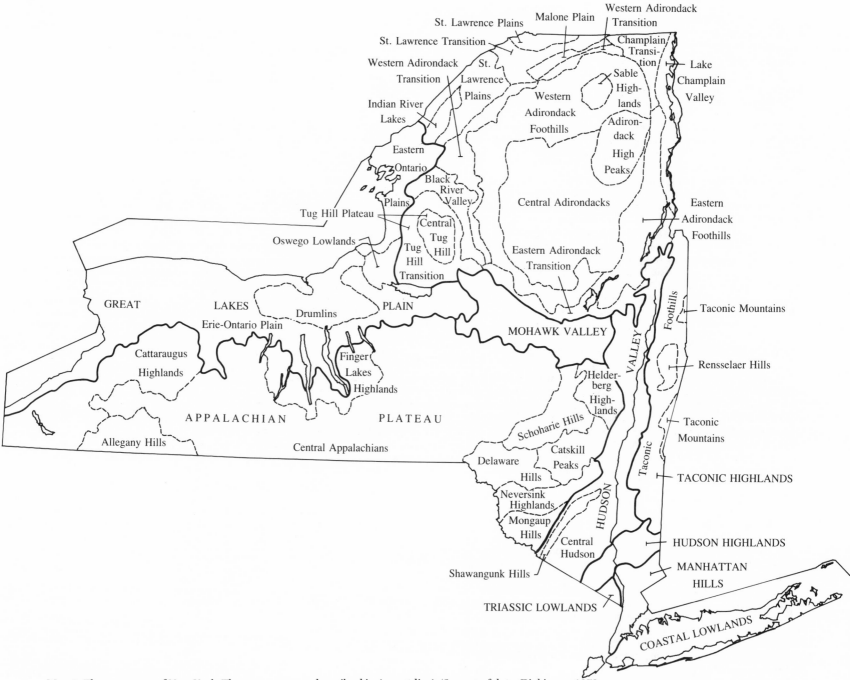

Map 3. The ecozones of New York. The ecozones are described in Appendix A. (Source of data: Dickinson 1979, 1983; Will et al. 1982; NYSDEC.)

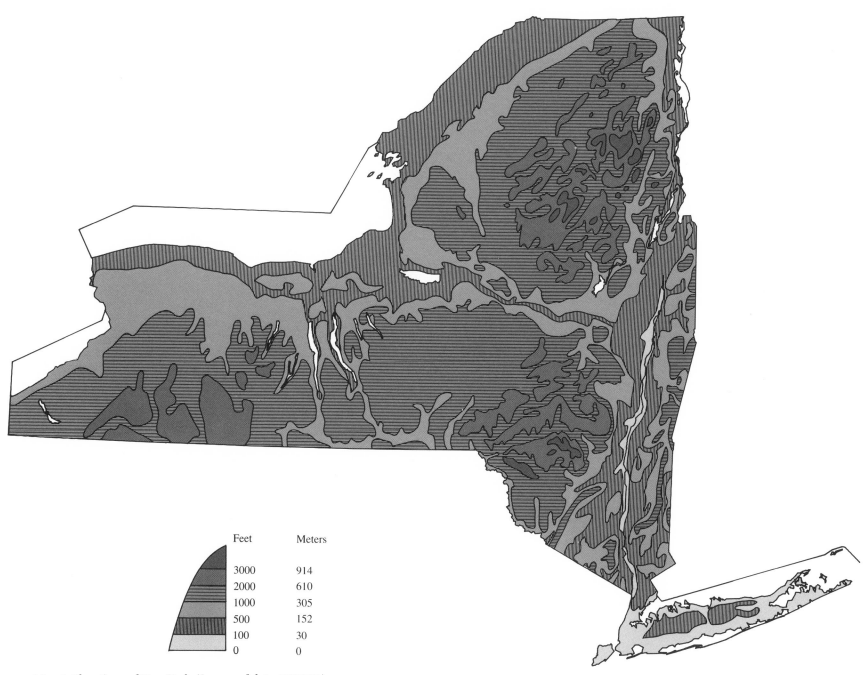

Feet	Meters
3000	914
2000	610
1000	305
500	152
100	30
0	0

Map 4. Elevations of New York. (Source of data: NYSDEC.)

Map 5. River systems of New York. (Source of data: NYSDEC.)

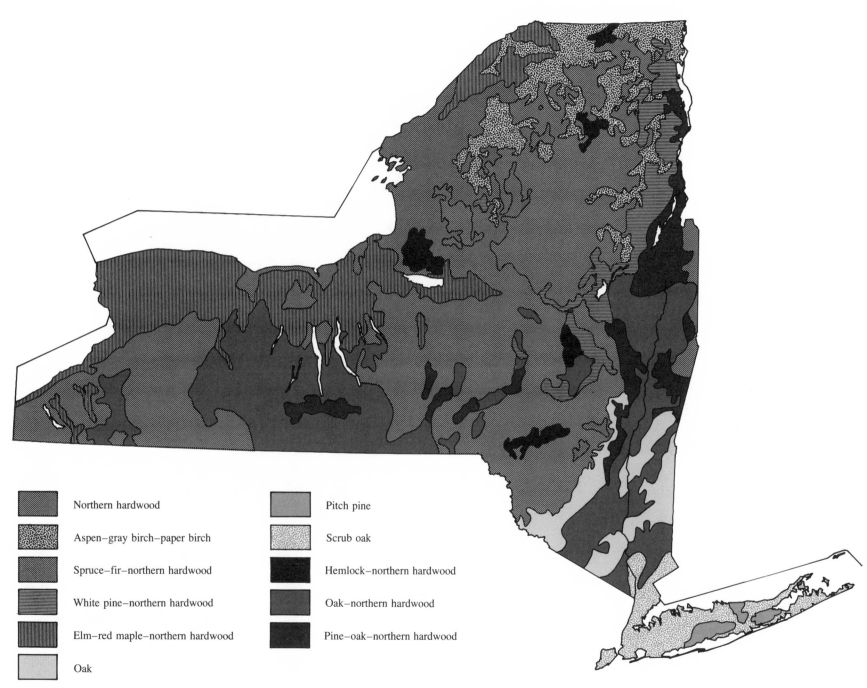

Map 6. Forest types of New York. (Source of data: Adapted by Forest Resources Planning, NYSDEC, from Stout 1958.)

Legend:

- Northern hardwood
- Aspen–gray birch–paper birch
- Spruce–fir–northern hardwood
- White pine–northern hardwood
- Elm–red maple–northern hardwood
- Oak
- Pitch pine
- Scrub oak
- Hemlock–northern hardwood
- Oak–northern hardwood
- Pine–oak–northern hardwood

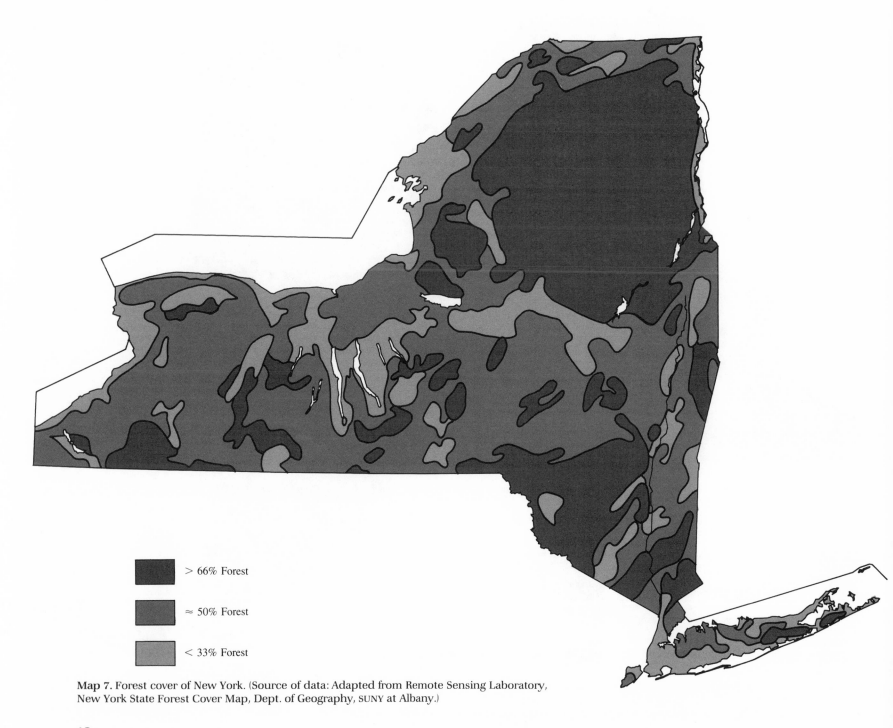

> 66% Forest

≈ 50% Forest

< 33% Forest

Map 7. Forest cover of New York. (Source of data: Adapted from Remote Sensing Laboratory, New York State Forest Cover Map, Dept. of Geography, SUNY at Albany.)

12

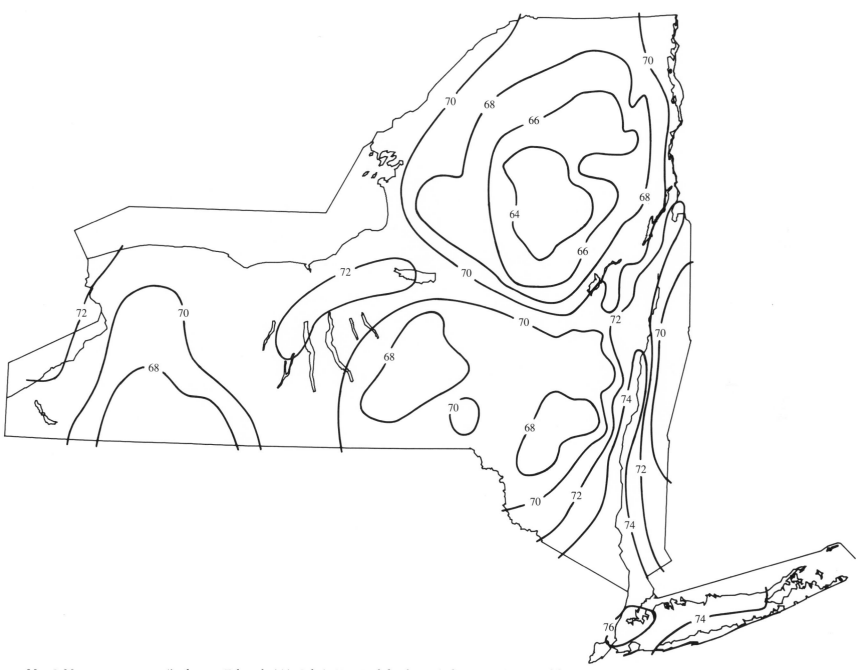

Map 8. Mean temperatures (in degrees Fahrenheit) in July in New York for the period 1931–64. (Source of data: Northeast Regional Climate Center, Cornell University.)

13

Map 9. Mean annual precipitation (in inches) in New York for the period 1931–64. (Source of data: Northeast Regional Climate Center, Cornell University.)

14

The map of coverage (see Map 10) indicates the number of species found for the entire Atlas period in each block. Where there is no symbol, the block was not surveyed.

Biases and Limitations

In few blocks can it be assumed that the species recorded constitute the total number actually breeding there during the Atlas survey period. The number of species recorded in a block was affected by the amount of time spent surveying a block, difficulty of access, weather, habitat type and variability, and observer's ability. Crepuscular and nocturnal species were often under-recorded because they required special efforts to find. The uneven distribution and inadequate numbers of observers were other factors affecting results. Where block busting was the major form of coverage, inherent deficiencies were found: "confirmed" records were very difficult to obtain; species such as waterfowl and raptors could easily be missed; when a block was entered only early in the breeding season, secretive cavity-nesting species already breeding were difficult to locate; species that cease singing later in the breeding season were also hard to record on late visits to a block. Bias was introduced into the data by the inclusion of information from other sources, such as the Long Island colonial waterbird survey, USFWS American Woodcock survey, and the Spruce Grouse research in the Adirondack region.

Results and Discussion

Over 4300 volunteers spent more than 200,000 hours in the field during the six years of the project. All but 12 of the state's 5335 blocks were worked by surveyors. Of the 12 uncovered, one was in the main artillery impact area of Fort Drum, six were in the Adirondacks where access was extremely difficult, and four were blocks with little land mass—two along the St. Lawrence River, two along Lake Ontario, and one along the Lake Erie shore. Of the 5323 blocks surveyed, only 337 had fewer than 50 species reported. The average number of species per block was 68. The Atlas data file contains 361,582 records, of which 39% were "confirmed" and 70% were either "probable" or "confirmed." A summary of atlasing progress is contained in Table 1.

The map illustrating the number of species recorded statewide reflects some of the biases and limiting factors previously mentioned. One very important factor affecting the potential for recording a variety of species among the regions was habitat—type, quality, and variability. For example, in many blocks in parts of Long Island habitat variability is limited and the average number of species recorded was quite low. In some areas, such as portions of central New York, habitat was diverse, and little effort was required to tally 76 or more species. Habitat types and quality at high elevations, in conjunction with rather severe climatic conditions, do not support a large number of species. Parts of the Great Lakes Plain contain much land used for fruit and row crops, and these areas, most

Table 1. The cumulative progress of Atlas surveying

	1980	1981	1982	1983	1984	1985
Species recorded	228	237	239	243	245	245
Blocks surveyed	909	1931	2901	4061	4964	5323
(% of total)	(17.0)	(36.2)	(54.4)	(76.1)	(93.0)	(99.8)

with small, disjunct woodlots, contributed to lower numbers of species in some blocks. Areas heavily developed with a high human population also yielded lower numbers of species (see Map 11).

Several other factors affected the number of species recorded. Concentrated efforts and considerable time spent by good birders, some in areas of less diverse habitat, could produce 76 or more species. The degree to which blocks and even certain ecozones permitted access, particularly in the Adirondacks and Catskills, also influenced coverage. The uneven distribution of observers and lack of them in some areas had a definite impact.

These and other factors are reflected in the summary of coverage by regions in Table 2. Generally, the coverage of blocks statewide was excellent, considering the large number of blocks to be surveyed, the available personnel, and the variety of obstacles encountered.

An overall analysis of the species distribution maps shows that they depict the ranges of New York's breeding birds to a degree of detail that can be used profitably by future investigators. The use of the small block (5 km sq) tends to minimize the effects of coverage biases, except, for example, where a species possesses a limited range in very mountainous terrain, or there is a disjunct distribution in few habitats over a large portion of the state. These maps reveal patterns of distribution of some species that were unknown before, including expansions, contractions, and hiatuses, some for known reasons, others for reasons not readily explained. The authors have mentioned many of these changes and

Table 2. Atlas coverage of New York, by regions

Region	Blocks surveyed	Species recorded	Species confirmed	Average species/block
1	634	178	157	62
2	338	176	159	76
3	448	179	167	74
4	539	168	150	76
5	550	188	174	72
6	578	198	177	63
7	690	202	179	61
8	755	182	160	65
9	509	176	159	74
10	282	187	169	58

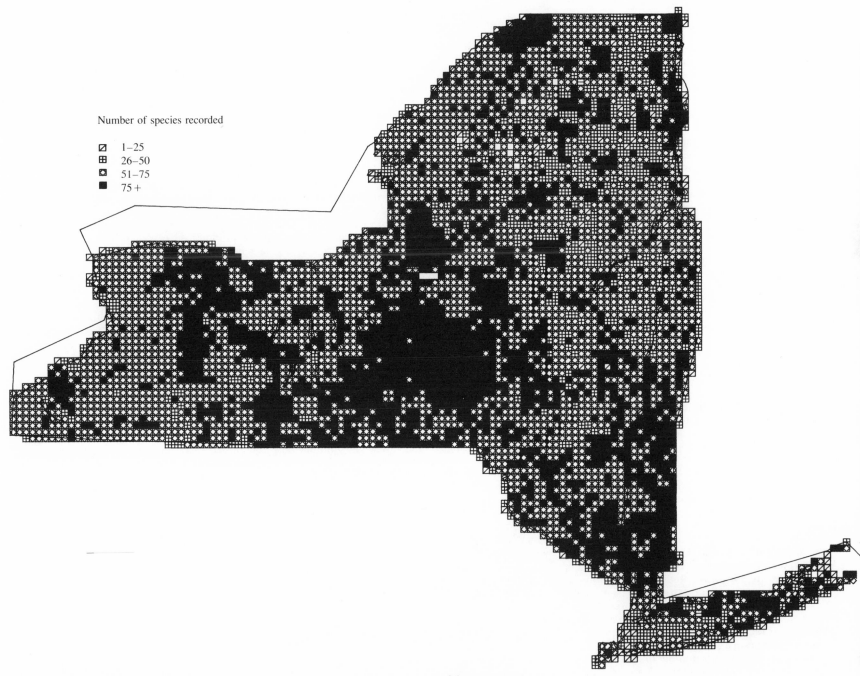

Map 10. Block coverage of New York by the Atlas project, 1980–85. (Source of data: NYSDEC.)

16

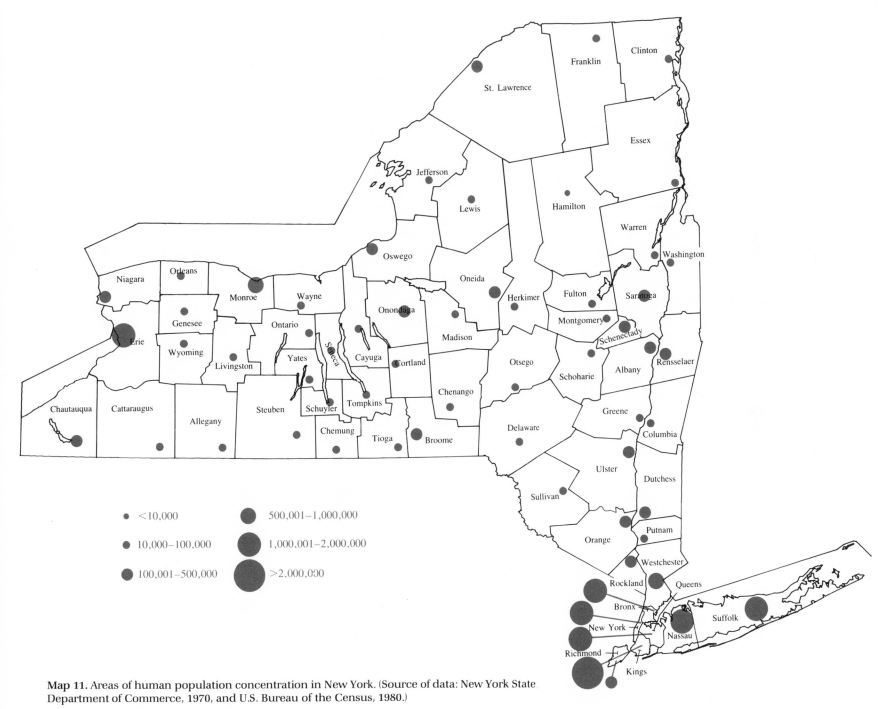

Map 11. Areas of human population concentration in New York. (Source of data: New York State Department of Commerce, 1970, and U.S. Bureau of the Census, 1980.)

Legend:
- ● <10,000
- ● 10,000–100,000
- ● 100,001–500,000
- ● 500,001–1,000,000
- ● 1,000,001–2,000,000
- ● >2,000,000

placed them in historical perspective when information warranted. Here is a wealth of material which provides a foundation for further research. Readers of this volume are urged to examine the distribution maps carefully, and to use the overlay maps if they possess them, to analyze distribution in light of what the authors discuss.

Breeding evidence was obtained for 242 species and 3 hybrids (Mallard × American Black Duck, Brewster's Warbler, and Lawrence's Warbler), and 230 of these were "confirmed." Five species were found breeding in New York for the first time: Forster's Tern, Yellow-throated Warbler, Palm Warbler, Blue Grosbeak, and Boat-tailed Grackle. Ten species known to have bred in the state previously but which were not "confirmed" during the Atlas period are Lesser Scaup, Golden Eagle, Black Rail, King Rail, Monk Parakeet, Bewick's Wren, Cape May Warbler, Wilson's Warbler, Western Meadowlark, and Dickcissel. The only one of these species that was unrecorded during the Atlas period was Bewick's Wren. Species accounts are not included for Lesser Scaup (reports of individual birds in three blocks) and Dickcissel (one bird recorded in one location in Wyoming County only in 1980).

Species that were recorded but not "confirmed" and not previously known to breed in the state are White-faced Ibis, Greater Scaup, Bufflehead, Caspian Tern, Summer Tanager, and Brewer's Blackbird. A single White-faced Ibis was observed in 1980 and 1984 at Jamaica Bay Wildlife Refuge. Both the Caspian Tern and Brewer's Blackbird have been expanding their eastern North American ranges. The Caspian Tern was found breeding on an eastern Lake Ontario island in 1986, and the Brewer's Blackbird has been observed in small numbers for several years in Region 2. The Summer Tanager was reported in two blocks on Long Island and one on Staten Island. Both the Greater Scaup and Bufflehead were probably summering individuals, a not-infrequent occurrence in various waters of the state. Of the foregoing species not known to breed previously, a species account is included only for Caspian Tern.

The results of this work give an insight into the distribution of New York's breeding birds never before achieved. This Atlas provides a unique base from which the distributional status of each breeding species can be projected in the future, and it also establishes a resource from which other kinds of ornithological research can be conducted.

Robert F. Andrle

Janet R. Carroll

PREHISTORIC BIRDS OF NEW YORK STATE

The Atlas of Breeding Birds in New York State contains a tremendous amount of uniformly compiled data on the distribution of breeding birds during the years 1980–85. The evaluation of these data depends partially upon chronological perspective. For example, if similar surveys are undertaken several decades from now, the systematic nature of the data in the current Atlas will permit accurate assessments of changes in the distribution of resident species of birds. Some assessment of changes in abundance may also be possible. Going back in time, comparisons of current (1980–85) data with past information are not nearly as quantifiable because previous attempts to portray the statewide distribution of birds were not done in such an organized and detailed manner. Nevertheless, much historical information exists about the birds of New York. Works by De Kay (1844), Eaton (1910, 1914), and Beardslee and Mitchell (1965) contain many observations on the distribution and abundance of New York birds during the first few centuries of European occupation. This essay expands the chronological perspective into prehistoric times, i.e., the period preceding European influence. A look at the past allows assessment of the nature of New York's birdlife during times of less habitat disturbance.

The prehistoric record of birds is based mainly upon bones excavated from archeological and paleontological sites. An archeological site is primarily cultural in origin, whereas a paleontological site is formed exclusively or primarily by natural, nonhuman processes. Of the sites listed below, only the Hiscock Site is regarded as paleontological, although even at Hiscock there is evidence of human presence.

The geology of New York restricts the geological time intervals during

I thank Janet Carroll for the idea of writing this paper. For their assistance, I thank William Ehlers, Jr., Robert Funk, Richard Futyma, George Hamell, Roxie Laybourne, Richard Laub, Norton Miller, Dominique Pahlavan, Susan Schubel, and Marie Zarriello. The manuscript was improved through comments by Robert Andrle, Janet Carroll, Robert Funk, George Hamell, Richard Laub, and Norton Miller. Contribution number 517 of the New York State Science Service.

which fossils of birds or other vertebrates might be found. Most of the state consists of Paleozoic rocks that predate the evolutionary origin of birds. Southeastern New York has exposures of Mesozoic rocks (roughly, the "Age of Dinosaurs") that are contemporary with fossils of early birds. No Mesozoic birds have been recorded, however, in New York. From the end of the Mesozoic, 65 million years ago, there is a huge chronological gap in the rock record of the state. The next youngest deposits are those of the Pleistocene ice ages. Although the cycles of Pleistocene glaciation began approximately 2 million years ago, most glacial deposits in New York are less than 50,000 years old. The majority of these sediments are from 18,000 to 10,000 years old, which represents the time of retreat of the last continental ice sheets. During this period a variety of deposits were left directly or indirectly by the melting glaciers, including the sediments of lakes, ponds, and bogs (usually silts and clays, often overlain by peats), gravels of many compositions (drumlins, kames, eskers, terminal moraines, and all sorts of tills), and the sedimentary infillings of caves. The period from 10,000 years ago to the present is known as the Holocene. Like other interglacial intervals, the Holocene is characterized by warmer climates than during glacial intervals.

At the worldwide peak of glaciation 18,000 years ago all of New York was covered by ice except the notch in southwestern New York that lies south of the Allegheny River (the Salamanca Re-entrant) and the seaward half of Long Island. The retreat of ice from New York was not uniform. Low-lying regions such as the Hudson and Mohawk river valleys became ice-free sooner than the adjacent uplands. Also, there were localized southward resurgences of ice lobes, even though the general picture was one of northward retreat. By 14,000–13,000 years ago most of the state was ice-free.

Major changes in habitat occurred in New York during and after the retreat of the glaciers (Miller 1973a, 1973b, in press; Davis 1983; Watts 1983). The newly deglaciated terrain was first occupied by a tundra-like community dominated by sedges, grasses, and other herbaceous plants, with few or no trees and shrubs. By about 12,000 years ago this community was colonized by boreal conifers (spruce and jack pine) to form an open woodland that persisted until about 10,000 years ago, with increases in the density of trees and the relative abundance of jack pines. At about 10,000 years ago various deciduous trees, temperate species of pines, and hemlocks arrived from the south and eventually displaced most of the boreal conifers. This mixed deciduous-coniferous forest has persisted into modern times in spite of fluctuations, such as a gradual increase in deciduous elements from 10,000 to 5000 years ago, a drastic reduction in hemlock about 5000 years ago (followed by a gradual increase), and a slight increase in boreal species during the past 2000 years.

Fossils of extinct mammals have been found in New York in late Pleistocene sediments (Fisher 1955; Drumm 1963). These include a ground sloth (*Megalonyx* sp.), giant beaver (*Castoroides ohioensis*), American mastodont (*Mammut americanum*), woolly mammoth (*Mammuthus primigenius*), Jefferson's mammoth (*Mammuthus jeffersonii*), long-nosed peccary (*Platygonus compressus*), moose-elk (*Cervalces scotti*), Vero tapir (*Tapirus veroensis*), and a zebra-like horse (*Equus* sp.). These species became extinct in New York, as well as throughout North America, from 11,000 to 10,000 years ago. The cause of their extinction is a subject of controversy. The major theories favor either the changing climates and habitats or the arrival of North America's first people. Other mammals recorded from glacial deposits in New York include species that survive outside of the state, such as the American elk (*Cervus canadensis*), caribou (*Rangifer tarandus*), and musk ox (*Ovibos moschatus*).

Until only two years ago there was no record in New York of the birds that lived alongside the extinct mammals. Although late Pleistocene fossils of birds were known from approximately 100 sites elsewhere in North America (Lundelius et al. 1983), the prehistoric record of birds in New York had been confined to species recorded from Holocene archeological sites. Recent work in Genesee County by the Buffalo Museum of Science in collaboration with the New York State Museum has yielded the first glimpse of New York's ice age birds. From 11,000-year-old sediments at the Hiscock Site, associated with bones of the extinct American mastodont, three bones of the California Condor have been found (Steadman and Miller, in press). Historic records of the California Condor are confined to the west coast of North America. Late Pleistocene fossils of the California Condor have been found across the southern United States and northern Mexico, from Florida to California. The fossils from the Hiscock Site represent a northward range extension of 1600 km. Pollen and plant macrofossils indicate that the habitat 11,000 years ago in western New York was an open woodland of spruce and jack pine (Miller, in press). The California Condor was able to live in a boreal environment when its food (large mammal carrion) was abundant. An ample food supply was provided by the herds of large mammals, mostly now extinct, that grazed and browsed on the late Pleistocene vegetation of North America. With the loss of most of these large mammals from 11,000 to 10,000 years ago, the range of the California Condor retreated across the continent. Although the California Condor barely survived the massive extinction of large mammals, many other large carrion-feeding birds,

including Old World vultures, storks, eagles, and other condors, became extinct in North America at that time.

Feathers are another unexpected discovery in the late Pleistocene sediments of the Hiscock Site. The feathers are from uncontaminated sediment samples (2–3 liters in volume) collected from freshly exposed walls of the excavations. These samples were packaged in the field in aluminum foil or plastic bags and then frozen. Laboratory analysis consisted of sieving the sediments through very fine-mesh screens, followed by sorting the particulate residue under a microscope. This work, conducted largely by Marie Zarriello and Richard Futyma, has yielded about 50 whole or partial feathers from the seven sediment samples examined thus far. The fossil feathers are beautifully preserved, retaining color as well as microstructure. The only other late Pleistocene feathers are from a few extremely arid caves in southwestern United States. The identification of feathers is a highly specialized process that often involves scanning electron microscopy. Three feathers from the Hiscock Site have been identified microscopically and macroscopically by Roxie Laybourne of the USFWS. A lower breast feather of the Pied-billed Grebe was recovered from sediments dated at 11,000 years. Two upper back or lower neck feathers of a male Northern Oriole are from sediments that are 9500 to 10,200 years old. Further details of the fossil feathers from the Hiscock Site await additional studies.

The record of Holocene birds in New York is more complete than that of late Pleistocene birds, yet it still is far from a thorough picture. For example, there are few sites in extreme western New York and none in the Adirondack region. Except for fossils from the Hiscock Site, the Holocene birds of New York are known exclusively from archeological sites. Aquatic birds and land birds both are represented, best typified by the Lamoka Lake and Riverhaven No. 2 sites (aquatic birds) and the Dutchess Quarry Cave and Hansen Rockshelter sites (land birds). The richest sites include species from both groups. One of the best samples of both aquatic birds and land birds is from Holocene peats of the Hiscock Site, derived from a small freshwater wetland surrounded by forest until 180 years ago.

The archeological record of birds includes some important biases. First, the sediments of most archeological sites have been sifted through screens with meshes of 0.25 or 0.5 inch or even larger. Thus the bones of small species, especially passerines, are not recovered. The fossils of small birds collected at the Hiscock Site demonstrate that a much more diverse fauna can be obtained when fine-mesh screens are used. Another bias is that the birds recorded from archeological sites are present at the sites because Amerindians brought them there. If most of the bones found are assumed to represent food items, a potential prejudice in the record is that birds which were hunted regularly by Amerindians, such as Ruffed Grouse, Wild Turkey, and Passenger Pigeons, might be represented in the sites in greater than natural relative abundances. Other birds, such as hawks and eagles, may have been captured or killed for their feathers or claws. Bird bones were used as tools (awls, fishhooks) and ornaments (beads, pendants), although such artifacts often are overlooked in faunal analyses. Various Amerindian rituals involved birds, such as the Great Blue Heron buried with an adult prehistoric Iroquois woman (perhaps Seneca) along the Genesee River near Avon (Wray 1964). The Great Blue Heron and other birds had spiritual value for certain Amerindian groups. For example, three of the eight clans of the Senecas were named after birds—heron, snipe, and hawk (Wray 1964). To summarize, Amerindians did not sample their local avifaunas uniformly. The uses of birds also varied among different groups of Amerindians. Furthermore, standard archeological techniques usually do not recover all of the bones buried at archeological sites.

In spite of these limitations, the archeological record of birds in New York provides significant clues to the prehistoric status of certain species. Ruffed Grouse, Wild Turkey, and Passenger Pigeons occur more frequently in the sites than any other species. That these three species occur as well in the nonarcheological Hiscock Site shows that their abundance in archeological sites is not due solely to human factors, but reflects the heavily forested habitats of pre-European times. Archeological records of Red-shouldered Hawks and Barred Owls outnumber those of Red-tailed Hawks and Great Horned Owls. This may be another reflection of the predominance of forest before the arrival of Europeans, as the former two species tend to prefer forested areas more than do the latter two species. Except for its wetlands, most of New York was covered prehistorically by deciduous and mixed deciduous-coniferous forests. Although Amerindians influenced the Holocene forests and wildlife of New York, their impact was not nearly so great as that of the Europeans.

The prospects for expanding the prehistoric record of New York's birds are very good. Much has been learned during the past several years, particularly about the late Pleistocene avifauna. Knowing that the California Condor once lived here, perhaps we will discover other unexpected species. The continued study of fossil feathers is another exciting new approach. The season during which the feather was lost can be determined by the stage of molt of the feathers. This information in turn will help to determine the seasonality of prehistoric migrations. Several new projects in caves and rockshelters of eastern New York show promise for yielding the remains of birds. Working with archeologists in the field as well as in the laboratory, and using the most modern techniques of excavation, fossil preparation, and radiocarbon dating, we can expand greatly the late Pleistocene and Holocene record of birds in New York over the next decade. By the time another Atlas is compiled the prehistoric birds of New York, like their modern counterparts, will be much better known.

Birds Recorded from Archeological and Paleontological Sites

The sites are listed first; birds are listed second, keyed by letter to the list of sites. Map 12 shows the locations of the sites. Sites from which only "unidentified birds" have been recorded are not listed. The sites are grouped by age, as determined by cultural associations and radiocarbon dating. Years BP = years before present ("present" = AD 1950). Site A and perhaps part of site B = late Pleistocene (Paleoindian cultural stage; ca.

11,000–10,000 years BP). Parts or all of sites B and C through site M = early through mid-Holocene (Archaic stage; ca. 10,000–3300 years BP). Site N = late Holocene (Transitional stage; ca. 3300–3000 years BP). Sites O–RR = latest Holocene (Woodland stage; ca. 3000–300 years BP). For brevity, radiocarbon ages are given without standard deviations. All ages presented here are approximations. ? = uncertain identification.

The Sites

A, Hiscock Site, Genesee County. Late Pleistocene strata only; radiocarbon dated at 11,000 years BP; Holocene strata considered separately below. (Steadman et al. 1986; Laub, in press; Miller, in press; Steadman, in press; Steadman and Miller, in press a, in press b)

B, Dutchess Quarry Cave No. 1, Orange County. Stratum 2 only; radiocarbon and culturally dated at 12,500–6500 years BP; stratum 1 considered separately below. (Funk et al. 1969; Guilday 1969)

C, Dutchess Quarry Cave No. 1, Orange County. Stratum 1 only; culturally dated at 6500–1000 years BP; stratum 2 considered separately above. (Funk et al. 1969; Guilday 1969)

D, Sylvan Lake Rockshelter, Dutchess County. Stratum 3 only; strata 1 and 2 considered separately below; radiocarbon dated at greater than 6560 years BP. (Funk 1976)

E, Sylvan Lake Rockshelter, Dutchess County. Stratum 2 only; strata 1 and 3 considered separately above and below; radiocarbon and culturally dated at 6560–3000 years BP. (Funk 1976)

F, Rabuilt Cave, Dutchess County. Culturally dated as middle to late Archaic. (Vargo and Vargo 1983)

G, Lamoka Lake Site, Schuyler County. Radiocarbon dated at 4500–4400 years BP. (Guilday 1980; Ritchie 1980)

H, Frontenac Island Site, Cayuga County. Radiocarbon dated at 4500–4000 years BP. (Ritchie 1980)

I, Bronck House Rockshelter, Greene County. Stratum 4 only; strata 1–3 considered separately below; culturally dated as late Archaic. (Funk 1976)

J, Zimmermann Rockshelter, Greene County. Stratum 2 only; stratum 1 considered separately below; culturally dated as late Archaic. (Funk 1976)

K, Moonshine Rockshelter, Greene County. Culturally dated as late Archaic. (Weinman and Weinman 1969)

L, Claverack Rockshelter, Columbia County. Most of the bone is from strata culturally dated as late Archaic. (Funk 1976)

M, Cole Gravel Pit Site, Livingston County. Culturally dated as late Archaic. (Brown et al. 1973)

N, Stoney Brook Site, Suffolk County. Radiocarbon dated at 2900 years BP. (Ritchie 1980)

O, Riverhaven No. 2 Site, Erie County. Culturally dated as early Woodland. (Granger 1978)

P, Scaccia Site, Livingston County. Radiocarbon dated at 2820 years BP. (Funk 1973a)

Q, Kipp Island Site, Seneca County. Radiocarbon and culturally dated at 1450–1320 years BP. (Ritchie 1980)

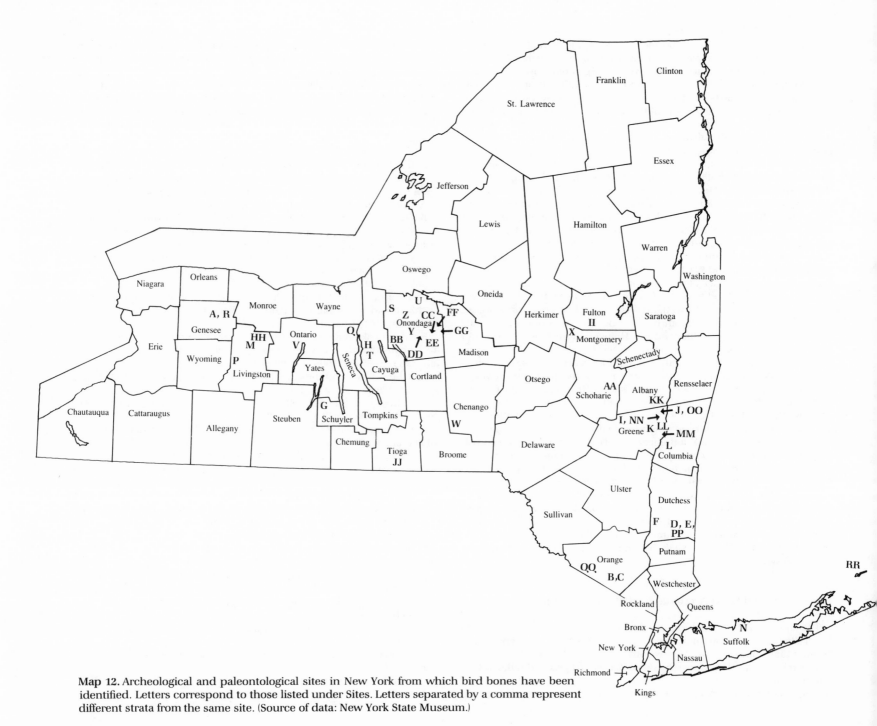

Map 12. Archeological and paleontological sites in New York from which bird bones have been identified. Letters correspond to those listed under Sites. Letters separated by a comma represent different strata from the same site. (Source of data: New York State Museum.)

22

R, Hiscock Site, Genesee County. Late Holocene strata only; radiocarbon dated at 950–250 years BP; early Holocene peat radiocarbon dated at 8500 years BP, although no bird bones are certainly from this stratum; late Pleistocene strata considered separately above. (Steadman et al. 1986; Laub, in press; Miller, in press; Steadman, in press; Steadman and Miller, in press a)

S, Carpenter Brook Site, Onondaga County. Culturally dated as middle to late Woodland. (Ritchie 1947)

T, Levanna Site, Cayuga County. Culturally dated as middle to late Woodland. (Ritchie 1928)

U, Wickham Site, Onondaga County. Culturally dated as middle to late Woodland. (Ritchie 1946)

V, Sackett or Canandaigua Site, Ontario County. Radiocarbon dated at 820 years BP. (Ritchie 1973a)

W, Bates Site, Chenango County. Radiocarbon dated at 825–760 years BP. (Ritchie 1973b)

X, Snell Site, Montgomery County. Radiocarbon dated at 800 years BP. (Ritchie et al. 1953; Ritchie and Funk 1973a)

Y, Cabin Site, Onondaga County. Culturally dated at 700 years BP. (Tuck 1971)

Z, Furnace Brook Site, Onondaga County. Radiocarbon dated at 650–580 years BP. (Tuck 1971)

AA, Nahrwold No. 1 Site, Schoharie County. Radiocarbon dated at 640–500 years BP. (Guilday 1973a; Ritchie 1973c)

BB, Kelso Site, Onondaga County. Radiocarbon dated at 560 years BP. (Ritchie and Funk 1973b)

CC, Bloody Hill Site, Onondaga County. Radiocarbon dated at 530 years BP. (Tuck 1971)

DD, Keough Site, Onondaga County. Culturally dated at 530 years BP. (Tuck 1971)

EE, Burke Site, Onondaga County. Radiocarbon dated at 470 years BP. (Tuck 1971)

FF, Cemetery Site, Onondaga County. Culturally dated at 430 years BP. (Tuck 1971)

GG, Barnes Site, Onondaga County. Culturally dated at 400 years BP. (Gibson 1968; Tuck 1971)

HH, Markham and Puffer estate site, Livingston County. Culturally dated as late Woodland. (Wray 1964)

II, Garoga Site, Fulton County. Culturally dated at 400–350 years BP. (Funk 1973b; Guilday 1973b)

JJ, Owego Sewage Plant, Tioga County. Culturally dated as late Woodland. (Versaggi et al. 1982)

KK, Fish Club Cave, Albany County. Culturally dated as Woodland, with some Archaic material in lowest levels; exact cultural association of bones uncertain. (Funk 1976)

LL, Tufano Site, Greene County. Radiocarbon dated 1250 years BP. (Funk 1976)

MM, Black Rock Site, Greene County. Radiocarbon dated at 1100 years BP. (Funk 1976)

NN, Bronck House Rockshelter, Greene County. Strata 1–3 only; stratum 4 considered separately above; culturally dated as middle and late Woodland. (Funk 1976)

OO, Zimmermann Rockshelter, Greene County. Stratum 1 only; stratum 2 considered separately above; culturally dated as middle and late Woodland. (Funk 1976)

PP, Sylvan Lake Rockshelter, Dutchess County. Stratum 1 only; strata 2 and 3 considered separately above; culturally dated as middle to late Woodland. (Funk 1976)

QQ, Hansen Rockshelter, Orange County. Culturally dated as Woodland. (Anonymous 1985)

RR, Hawks Nest Point Site, Fishers Island, Suffolk County. Culturally dated as late Woodland. (R. E. Funk and J. Pfeiffer, pers. comm.)

The Birds

Family Gaviidae—loons
  Common Loon (*Gavia immer*) M
Family Podicipedidae—grebes
  Pied-billed Grebe (*Podilymbus podiceps*) A
  Horned Grebe (*Podiceps auritus*) G
Family Ardeidae—bitterns and herons
  American Bittern (*Botaurus lentiginosus*) C
  Great Blue Heron (*Ardea herodias*) G, M, O, EE
Family Anatidae—swans, geese, and ducks
  Tundra Swan (*Cygnus columbianus*) G
  Snow Goose (*Chen caerulescens*) P
  Canada Goose (*Branta canadensis*) B, C, D, G, O, Q, U, AA, II, RR
  Wood Duck (*Aix sponsa*) G, R
  Green-winged Teal (*Anas crecca*) V?, R
  Mallard/Black Duck (*Anas platyrhynchos/rubripes*) M, O, R, Z
  Blue-winged Teal (*Anas discors*) Y
  Gadwall (*Anas strepera*) M
  Unidentified dabbling duck (*Anas* sp.) M, O, V
  Canvasback (*Aythya valisineria*) O
  Unidentified pochard (*Aythya* sp.) O, II
  White-winged Scoter (*Melanitta fusca*) RR
  Common Goldeneye (*Bucephala clangula*) O
  Bufflehead (*Bucephala albeola*) G
  Common Merganser (*Mergus merganser*) O
  Red-breasted Merganser (*Mergus serrator*) O
  Unidentified merganser (*Mergus* sp.) Q
  Unidentified duck (Anatidae sp.) C, H, N, O, P, X, QQ, RR
Family Cathartidae—American vultures
  California Condor (*Gymnogyps californianus*) A
  Turkey Vulture (*Cathartes aura*) B, C
Family Accipitridae—kites, eagles, hawks, and allies
  Bald Eagle (*Haliaeetus leucocephalus*) H, O
  Cooper's Hawk (*Accipiter cooperii*) R
  Northern Goshawk (*Accipiter gentilis*) G
  Red-shouldered Hawk (*Buteo lineatus*) G, R, QQ

Red-tailed Hawk (*Buteo jamaicensis*) C
Unidentified hawk (*Buteo* sp.) M, II
Golden Eagle (*Aquila chrysaetos*) G
Unidentified hawk (Accipitridae sp.) F
Family Phasianidae—partridges, grouse, turkeys, and quail
Ruffed Grouse (*Bonasa umbellus*) C, D, E, G, H, J, M, O, R, T, W, Z, AA, CC, EE, II, QQ
Wild Turkey (*Meleagris gallopavo*) C, D, E, F, G, H, I, K, L, M, N, O, P, Q, R, S, T, U, V, W, AA, BB, DD?, FF, GG, II, JJ, KK, LL, MM, NN, PP, QQ
Northern Bobwhite (*Colinus virginianus*) D, M
Family Rallidae—rails, gallinules, and coots
Sora/Virginia Rail (*Porzana carolina/Rallus limicola*) R
Family Gruidae—cranes
Sandhill Crane (*Grus canadensis*) II
Family Scolopacidae—sandpipers, phalaropes, and allies
Solitary Sandpiper (*Tringa solitaria*) R
American Woodcock (*Scolopax minor*) QQ
Family Columbidae—pigeons and doves
Mourning Dove (*Zenaida macroura*) M
Passenger Pigeon (*Ectopistes migratorius*) B, C, E, F, G, H, J, M, R, Y, Z, AA, BB, II, MM, OO, QQ
Family Cuculidae—cuckoos, roadrunners, and anis
Unidentified cuckoo (*Coccyzus* sp.) R
Family Strigidae—typical owls
Eastern Screech-Owl (*Otus asio*) M

Great Horned Owl (*Bubo virginianus*) G, M
Barred Owl (*Strix varia*) C, E, M, R, QQ
Family Picidae—woodpeckers and allies
Yellow-bellied Sapsucker (*Sphyrapicus varius*) R
Downy Woodpecker (*Picoides pubescens*) R
Northern Flicker (*Colaptes auratus*) R
Family Corvidae—jays, magpies, and crows
Blue Jay (*Cyanocitta cristata*) F?, R, QQ
American Crow (*Corvus brachyrhynchos*) E, M, QQ
Common Raven (*Corvus corax*) R
Family Muscicapidae—Muscicapids
Subfamily Turdinae—solitaires, thrushes, and allies
American Robin (*Turdus migratorius*) M, R
Family Mimidae—mockingbirds, thrashers, and allies
Gray Catbird (*Dumetella carolinensis*) R
Family Emberizidae—Emberizids
Subfamily Icterinae—blackbirds and allies
Common Grackle (*Quiscalus quiscula*) R
Northern Oriole (*Icterus galbula*) A
Family undetermined
Unidentified passerine (Passeriformes sp.) R

David W. Steadman

24

SPECIES ACCOUNTS

Common Loon *Gavia immer*

Possessed of a form both lovely and functional, a voice that haunts, and a need for quiet spaces, the Common Loon has taken on meaning beyond itself in recent years. It has become one of the most admired of New York's birds and is a subject of intense study and concern. Fortunately, this great northern diver is still common on lakes throughout the Adirondacks, although along the St. Lawrence River it is uncommon, and elsewhere it is rare.

The history of the Common Loon in New York is imperfectly known. As Arbib (1963) suggested, "Undoubtedly the loon was far more widespread as a breeding bird in pre-Columbian and early colonial days, . . . for the state was then largely forested." Within the Adirondacks nesting was reported by De Kay on Raquette Lake, Hamilton County, as early as 1844 (Eaton 1910). Then, in the spring of 1869, *Adventures in the Wilderness; or, Camp-Life in the Adirondacks* by W. H. H. Murray appeared, "and by June there was an influx of tourists such as the Adirondacks had never seen before" (Donaldson 1921). The great rush of "Murray's Fools" eagerly devoured his sporting adventures, including an account of "Loon-Shooting in a Thunder-Storm," in which Murray cheerfully limited himself to 50 rifle bullets to shoot a loon from a guideboat (Murray 1869). The impact of this publicity on the loon is unknown, but it does appear that loon numbers decreased in the years that followed. Eaton (1910) found only a few in Franklin, Hamilton, Herkimer, and St. Lawrence counties, but none in Essex County during 1905, reporting: "As a summer resident it is now confined to the secluded ponds and lakes of the Adirondacks, being most numerous in the western and southwestern parts of that region."

Arbib (1963) sent questionnaires to ornithologists, bird clubs, and game protectors, and placed a request for breeding site information in the *New York State Conservationist*. On the basis of the replies, he reported 120 pairs nesting at more than 90 locations and estimated the state's total breeding population to be about 240 pairs. More recently there have been two concerted efforts to survey New York's Common Loons. A field survey of 420 lakes larger than 10 ha (24.7 a) conducted by the NYSDEC from 1977 to 1979 counted 114 breeding pairs occupying 91 lakes. The total New York population was estimated at fewer than 200 pairs of loons (Trivelpiece et al. 1979). Then in the summers of 1984 and 1985 NYSDEC teams surveyed 557 lakes, ponds, and reservoirs. This intensive effort resulted in a count of 157 pairs on 128 lakes. An additional 247 adults not known to breed were also found. Between 1983 and 1985 volunteers monitored 165 lakes and six sites on the St. Lawrence River that were not visited by NYSDEC teams, locating 17 breeding loon pairs and 87 nonbreeding adults (Parker et al. 1986). These results are included on the accompanying Atlas map, which shows the Common Loon in a total of 369 blocks, "confirmed" in 152, some blocks having more than 1 pair, and some lakes falling into several blocks. The recent survey found no significant increase in the number of breeding loons, number of chicks produced, or fledging rate, but there was an increase in the number of lakes with adults presumed to be nonbreeders and an overall increase in numbers of this group (Parker et al. 1986).

Nesting loons require a body of water large enough for displays, feeding, and nesting, with sufficient length to clear surrounding trees on takeoff and deep enough to provide escape cover. The location should also offer a suitable nest site, generally an island or low hummock, rarely a sheltered shoreline location, as close as possible to water deep enough for escape (Palmer 1962). Parker (1985), in a 1983–84 study of 24 Adirondack lakes with pH levels ranging from 4.5 to 6.7, found no significant difference in reproductive success between loons on highly acidic and less acidic lakes. One lake in Hamilton County contained no fish, tadpoles, or crayfish, yet the pair of Common Loons successfully raised two chicks during 1985 (Parker and Brocke 1985). Common Loons on highly acidic lakes seem able to adapt, for now at least, to food declines, but Parker wondered whether they could cope with acidification in the long term.

The eggs (usually 2, rarely 1) are laid in a nest of dead vegetation, mud, sticks, grasses, and aquatic debris close to the water's edge. Most Adirondack nests are built on islands, islets, hummocks, or floating bog mats and only rarely are mainland nests employed (pers. obs.). Boaters and hikers can greatly assist New York's loons by giving the birds a wide berth, especially when disturbed calls or distraction displays by adults suggest the presence of a nearby nest or concealed chicks that could be lost to predators or abandoned.

John M. C. Peterson

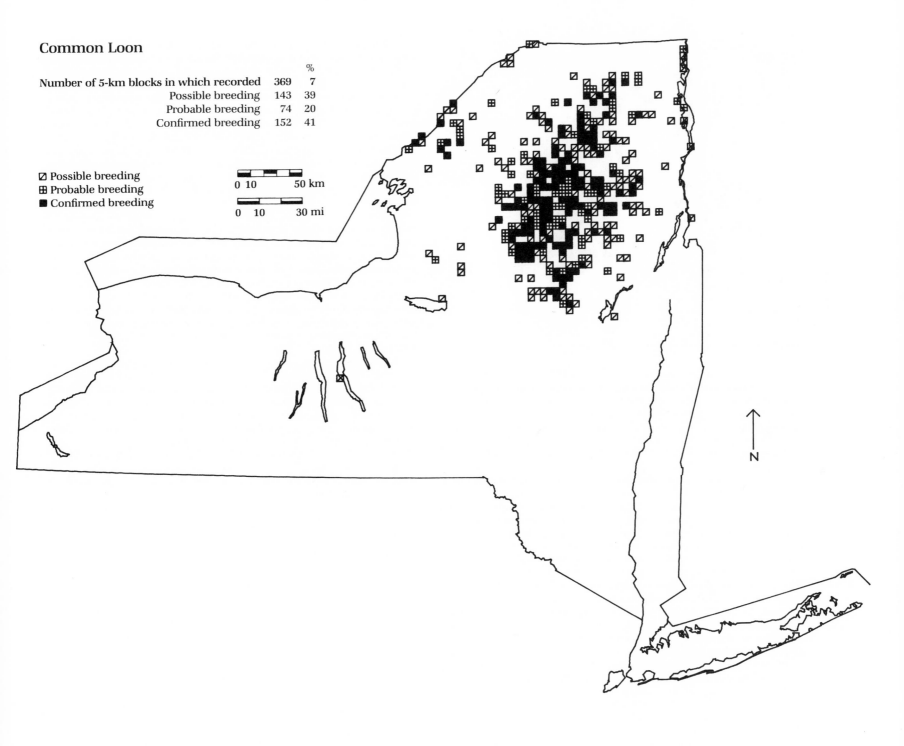

Common Loon

Number of 5-km blocks in which recorded		%
	369	7
Possible breeding	143	39
Probable breeding	74	20
Confirmed breeding	152	41

◩ Possible breeding
⊞ Probable breeding
■ Confirmed breeding

0 10   50 km

0 10   30 mi

N

©KLA-C 1986

Pied-billed Grebe *Podilymbus podiceps*

A small diving bird whose body is highly modified for an aquatic life, the Pied-billed Grebe is a locally rare to uncommon breeder in New York. Its numbers have declined over the years. Eaton (1910) stated that it was found in undisturbed situations throughout the state during the breeding season but noted that outdoor recreational activities and the draining of marshy ponds and streams had reduced its numbers. Also, a common practice was to shoot this species, which presented an elusive but challenging target (Bent 1919).

Griscom (1923) and Cruickshank (1942) stated that this species bred only sporadically in the New York City region. Bull (1964), however, indicated that the breeding population had increased, despite reduced marshland acreage, adding that the creation of waterfowl refuges was probably one reason. At Jamaica Bay Wildlife Refuge there were an estimated 40 or more pairs with young present in 1961.

Bull (1974) noted that this species was fairly well distributed statewide, although least plentiful in the higher mountains apparently as a result of lack of proper habitat. More recently, observers have reported a widespread decrease in numbers of nesting Pied-billed Grebes in New York (Kaufman 1984) and adjacent states (Leck 1984; Kibbe 1985a). The reasons for this are not clear, although Cattaraugus County suffered a great reduction in numbers of Pied-billed Grebes and other marsh birds as wetlands were filled (Eaton 1981).

Atlas surveying found evidence of breeding in relatively few yet widely distributed blocks in most regions of the state. There were no records for the Adirondack High Peaks, Catskill Peaks, and Allegany Hills, and very few in most adjacent upland ecozones. Except for the Hudson Valley, most of the records of this grebe are from the many state and national wildlife refuges across New York. In some areas of sparse distribution or brief field coverage this wary breeder may have been overlooked.

During the breeding season the Pied-billed Grebe becomes so secretive, especially during incubation, that its presence may be undetected unless its loud, discordant springtime call, which ends in a series of cuckoolike notes, is heard. The call, however, can be mistaken for that of the Common Moorhen (Kibbe 1985a). Also, this species is seen in flight less often than other grebes and usually escapes by diving or skulking in surrounding vegetation. In addition, it quietly leaves the nest when approached, even at a distance, and usually remains out of sight (Bent 1919).

Habitat requirements for the Pied-billed Grebe include a combination of open water, usually fresh, along with an abundance of emergent aquatic vegetation. These conditions may be found along marshy shorelines of ponds, shallow lakes, or marshy bays, and slow-moving streams with sedgy banks. Extensive marshes containing areas of open water or channels of sufficient depth (Pough 1951) may also be suitable. Rarely, it nests in brackish water with limited tidal fluctuation (Chabreck 1963). The species of marsh plants may vary; the grebe's primary need is for vegetation that will furnish adequate cover and nesting material, and for enough open water to allow it to become airborne (Eaton 1910; Bent 1919).

The nest, built in water by both members of the pair, usually floats and is anchored to or built around emergent dead or growing vegetation such as cat-tails, bur-reeds, sedges, rushes, and bushes. Water depth at the site may vary from 0.3 m to 0.9 m (1–3 ft). Most of the structure (up to a bushel or more in volume) is submerged; the top layer is smaller, somewhat more organized, and raised slightly above the water. The nest is constructed of available dead and decayed plant material, and occasionally algae or mud. Some green vegetation may be included, but the overall result is a rather soggy structure. When the incubating adult leaves the nest, it usually covers the eggs with the remains of wet plants from the rim of the nest (Bent 1919; Palmer 1962).

Paul F. Connor

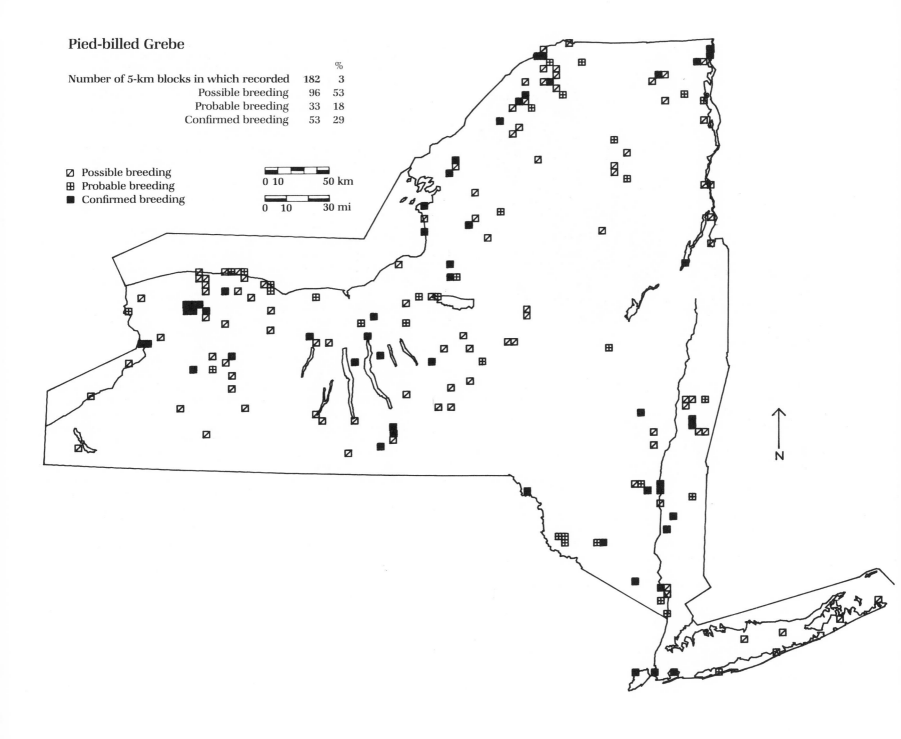

Pied-billed Grebe

Number of 5-km blocks in which recorded		%
Number of 5-km blocks in which recorded	182	3
Possible breeding	96	53
Probable breeding	33	18
Confirmed breeding	53	29

☑ Possible breeding
⊞ Probable breeding
■ Confirmed breeding

0 10   50 km

0 10   30 mi

N

Double-crested Cormorant *Phalacrocorax auritus*

A large, primitive-looking waterbird whose appearance and habits might fit nicely into a Cretaceous landscape with Hesperornis, the Double-crested Cormorant is a fairly recent and still local addition to the breeding avifauna of our state. Its population is increasing in New York as it finds new locations for its colonies.

In precolonial times the Double-crested Cormorant could have nested within the boundaries of the future state. We do know that elsewhere along the East Coast to the north the species abandoned colonies in a losing confrontation with advancing civilization; it has been persecuted by humans, who considered this bird a competitor for coastal food fish. I traced a decline in abundance to an eastern population low in the mid-1920s (Palmer 1962), but the decline and contraction of the breeding range had been documented earlier. Eaton (1910), Griscom (1923), and Cruickshank (1942) all considered the Double-crested Cormorant a migrant only, with scattered records of summer lingerers and winter survivors.

After 1930, however, the population began to burgeon, and breeding colonies moved south along the coast, west up the St. Lawrence River drainage, and east along the north shores of the Great Lakes. As a result of this combined surge, a colony was established in 1945 on Gull Island, 6.4 km (4 mi) west of Henderson Harbor, Jefferson County (Kutz and Allen 1947). Meanwhile, the coastal population was colonizing sites from the Maritimes to Maine, to Massachusetts and finally to Long Island (Fishers Island) in 1975 (Bull 1981).

As the map shows, several colonies followed. There are "confirmed" records in six blocks, representing five colonies. In 1986 the largest active colonies in New York were those at Little Galloo Island in eastern Lake Ontario (Jefferson County), on Long Island at Gardiners Island in Gardiners Bay (Suffolk County), and the three Hungry Point islands north of Fishers Island in Long Island Sound (Suffolk County). Censuses of nests at these colonies conducted in 1986 found approximately 1468 nests at Little Galloo (Weseloh, pers. comm.), more than 800 nests on Gardiners Island (Stoutenburgh, pers. comm.), and an estimated 669 at the Hungry Point islands (Horning, pers. comm.). Additionally, a colony is located on an island in the East River in New York City, and another on The Four Brothers in Lake Champlain, Essex County. There are promising records in Nassau County, at Stillwater Reservoir in northern Herkimer County, and Lake DeForest in Rockland County, which could represent potential breeding areas for this expanding species. A sixth colony was located in 1986 on Trinity Reservoir in Westchester County (Askildsen 1986).

The Gull Island colony appears to have been abandoned in favor of Little Galloo during the 1970s. This may have been the result of deteriorating nesting habitat at Gull Island, repeated human intrusion, nest predation by rats, foxes, or other mammals, or it may have been the result of a shift in the occurrence of food fish species during this cormorant's breeding season (Chamberlaine 1986). The colony at Little Galloo is now much bigger than it ever was at Gull Island. In 1979, 275 nests were counted on Little Galloo, 595 in 1982 (Weseloh, pers. comm.), and over 1400 in 1986, as indicated above.

The Double-crested Cormorant is a colonial nester, preferring isolated, undisturbed islands or other protected sites, even sea-facing promontories with coniferous or deciduous trees in which the crude stick platforms can be built. Repeated use and the effects of cormorant guano eventually kill the trees, which then fall. The fallen trees can be used as nest sites, or the colony may be abandoned or moved (Mendall 1936). Ground nesting has been frequently reported, and at Little Galloo Island in 1986, 63% of the nests were on the ground (Weseloh, pers. comm.). The territory of the pair is limited to the nest site and a perch for the off-duty mate. The nest is built by both sexes, although the male's role is limited to material gathering, and takes 4 days to build (Palmer 1962).

The colony at Little Galloo Island is located mainly around the perimeter of the island which is approximately 17.8 ha (44 a) in area. Nests are placed in almost every available spot on every standing and fallen tree and on the ground in slightly elevated locations. Guano has killed almost all of the trees on the island, and it will not be long before there will be none standing. This island haven is also the breeding site for the Cattle Egret, the Black-crowned Night-Heron, the Great Black-backed, Herring, and Ring-billed gulls, and the Caspian Tern (Weseloh, pers. comm.).

Robert Arbib

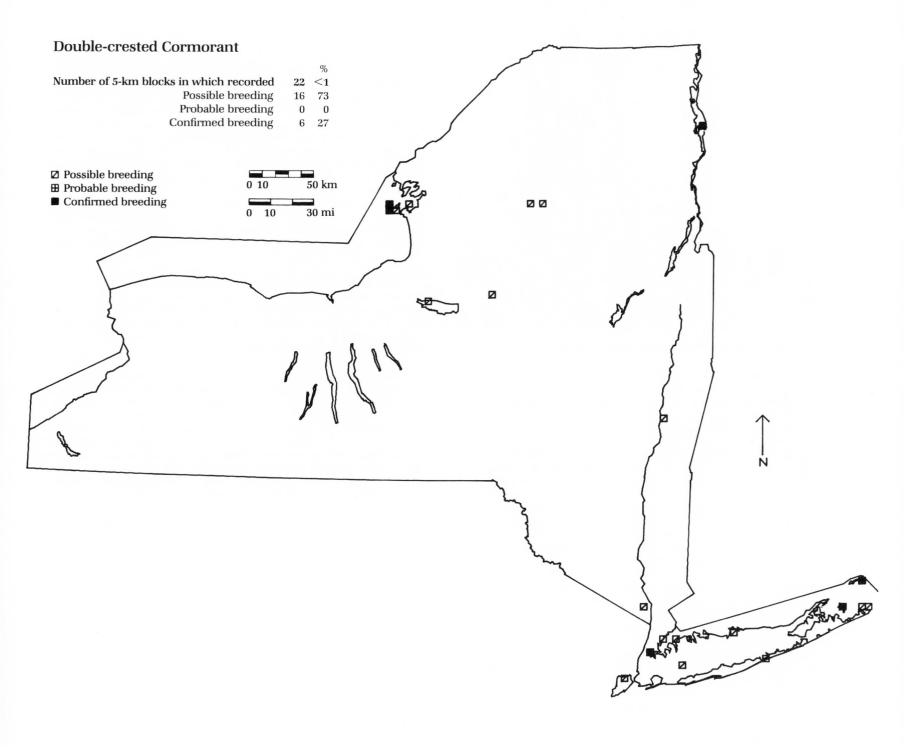

Double-crested Cormorant

		%
Number of 5-km blocks in which recorded	22	<1
Possible breeding	16	73
Probable breeding	0	0
Confirmed breeding	6	27

◩ Possible breeding
⊞ Probable breeding
■ Confirmed breeding

0 10   50 km

0 10   30 mi

N

©KLA·C 1987

American Bittern *Botaurus lentiginosus*

A denizen of marshes, the American Bittern emits a humorous mating call that resembles the call of a bullfrog and triggers a cacophony of sound from other males of the species. Although the American Bittern is a fairly common breeding species in the state, in New York and elsewhere this species has declined since the 1950s, especially on the Appalachian Plateau, in the Hudson Valley, and on Long Island.

The American Bittern was apparently once much more common. According to Eaton (1910), "This bird undoubtedly occurs, and probably breeds, in every county of the state. It is a fairly common summer resident on eastern Long island and on all the marshes of the interior. On the Montezuma marshes it is so common that a dozen birds may often be heard booming at the same time." He found it nesting as far north as Elk Lake in Essex County and within 10 miles of Mount Marcy, and noted that it probably bred in suitable localities throughout the Adirondack region wherever it was left undisturbed.

Bull (1974) called it a "widespread breeder." Although at the time a decline was apparently occurring, he did not note it. In 1976 the American Bittern was provisionally placed on the *American Birds'* Blue List of declining species, where it remains to this day (Arbib 1975; Tate 1986). This decline is apparently continentwide. In New York, Temple and Temple (1976) documented a significant population decline in the Cayuga Lake basin between 1940 and 1970, and attributed it to the reduction of marshes. A similar decline has undoubtedly occurred in other parts of the state where large areas of marshland have been drained at an acceler-

ated rate since World War II, the Freshwater Wetlands Act notwithstanding.

Atlas observers found the American Bittern on the Great Lakes Plain, mainly in Orleans, Genesee, Monroe, and Wayne counties in protected wetlands, and also south of the five major Finger Lakes and around Oneida Lake. It was also recorded in the wetlands bordering and near the St. Lawrence River, as well as in the Lake Champlain Valley. Many of these marshes are part of state wildlife management areas or wildlife refuges. Along the Hudson River this bittern was recorded in a few of the remaining large tidal marshes. In the central and northern Adirondacks it occurs in wetland areas along river systems. It was missing from most blocks in the eastern Appalachian Plateau and was scarce west of the Finger Lakes because of the paucity of wetlands.

Atlas observers had difficulty "confirming" this and other marsh birds; only 7 nests were found. Most "confirmed" records were of fledged young observed by chance. Any record of American Bitterns' calling from a marsh is probably good indication, however, that the species breeds there.

Inland, the American Bittern's preferred nesting habitat is cat-tail–bulrush marshes but it also nests in sedge meadows, beaver ponds and meadows, kettle-hole bogs, alder thickets, and shrub swamps. In estuarine communities it nests largely in salt marshes or freshwater tidal marshes. Eaton (1910) noted that "it seems to establish itself wherever there are weedy marshes 10 or more acres in extent."

The nest site is typically in a wet situation in fresh- or saltwater marshes but can be on dry land in grassy meadows or on a floating island in a lake (Palmer 1962). Materials used for the nest depend on local vegetation and may include cat-tail, cordgrass, bulrush, or sedge. The nest is 5.1–15.2 cm (2–6 in) deep; sometimes eggs are placed on the ground (Palmer 1962). The outside diameter of the nest ranges from 25.4 to 40.6 cm (10–16 in) (Harrison 1975). By bending and breaking vegetation with its bill the American Bittern constructs and uses definite paths to and from its nest (Palmer 1962). The young leave the nest at about 20 days and go to nearby new nest platforms built by the female. There is some evidence that the male may be polygamous (Palmer 1962).

Stephen W. Eaton

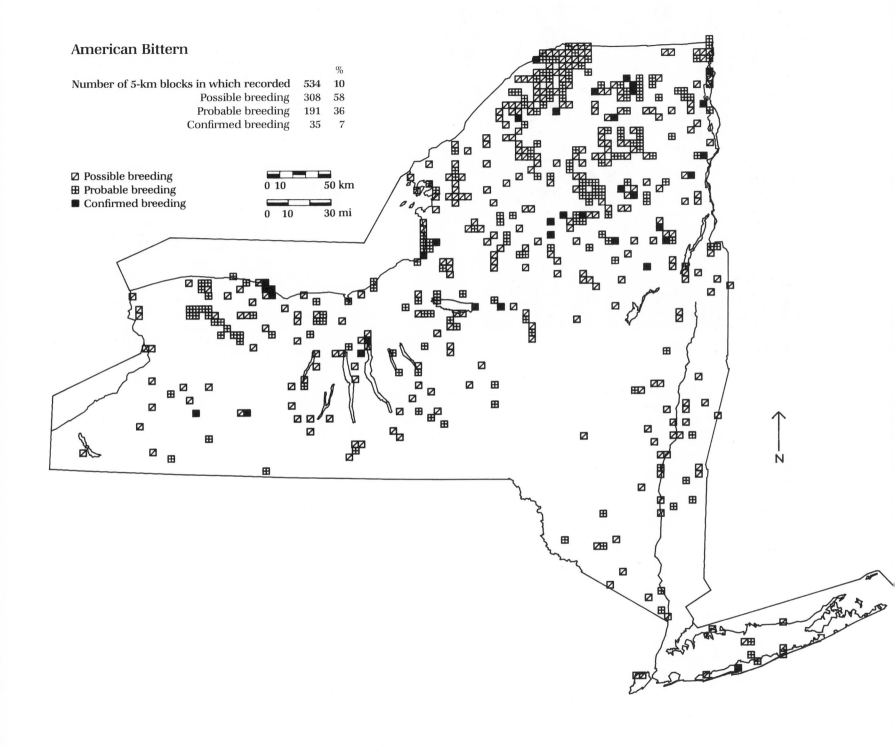

American Bittern

Number of 5-km blocks in which recorded	534	10
Possible breeding	308	58
Probable breeding	191	36
Confirmed breeding	35	7

%

◩ Possible breeding
⊞ Probable breeding
■ Confirmed breeding

0 10   50 km

0 10   30 mi

N

Least Bittern *Ixobrychus exilis*

A small, inconspicuous heron, the Least Bittern is an uncommon to rare nesting species in the state. The draining of wetlands has adversely affected the population and range of this species, and its spotty distribution is not surprising given the scattered occurrence of remaining wetland habitat. In addition, New York is near the northern periphery of the species' range (AOU 1983).

Eaton (1910) considered the Least Bittern to be locally common in parts of the state including the Great Lakes Plain, the Finger Lakes, the Hudson and St. Lawrence valleys, and Long Island. Local surveys by Saunders (1926) and Hyde (1939) indicated that the species was common in the marshes of central New York and along Lake Ontario, and Bull (1974) concurred. In his description of its breeding distribution, Bull did not include the St. Lawrence River area (Eastern Ontario and St. Lawrence plains and St. Lawrence Transition) in the species' New York range.

Atlas data correspond fairly well to the known historical range, with some exceptions. During the Atlas period, "confirmed" and "probable" breedings were most frequently recorded in central and western New York and along the Hudson Valley, the locations of many historical sightings. Scattered "probable" and "possible" records in the Black River Valley and the highland regions of the Central Adirondacks, on the Tug Hill Plateau, and on the Appalachian Plateau provide evidence of breeding in areas with few historical records. In addition, the species was found in the Lake Champlain Valley and in parts of the Adirondacks where populations were previously unknown. (It should be noted that there are few birders in some of these areas.) The Least Bittern continues to be absent from most of the Appalachian Plateau and the Catskills.

The inconspicuous habits (Eaton 1910; Saunders 1926) and specialized cat-tail habitat (Saunders 1926; Weller 1961) of this species make it difficult to detect. In some areas, such as the northeastern shore of Lake Ontario and western St. Lawrence County, visits to interiors of extensive marshlands were limited because of the small number of Atlas observers. Even in areas with greater numbers of observers, this species may have been overlooked.

The Least Bittern occupies a variety of wetland habitats including cat-tail and sedge marshes, salt marshes, and other areas with emergent vegetation (Palmer 1962). In upstate New York it prefers open-canopy wetlands, particularly cat-tail marshes. Large marshes, such as the barrier dune wetlands along the eastern end of Lake Ontario, are important habitat for this species. Salt marshes are utilized on Long Island. Eaton (1910) and Saunders (1926) indicated that areas where cat-tail or sedge border open water are favored habitat within a marsh.

Detailed studies of the breeding biology of the Least Bittern are few. Available information, primarily from Saunders (1926) and Weller (1961), indicates that the male builds much of the nest. It is a poorly woven mat of cat-tail, grass, sedge, and other herbaceous vegetation supported by the previous year's growth (Eaton 1910; Weller 1961). It is placed in dense stands of emergents, about 10 cm (4 in) to 1.2 m (4 ft) above the water and usually close to open water (Saunders 1926; Weller 1961; Palmer 1962).

Gerald A. Smith

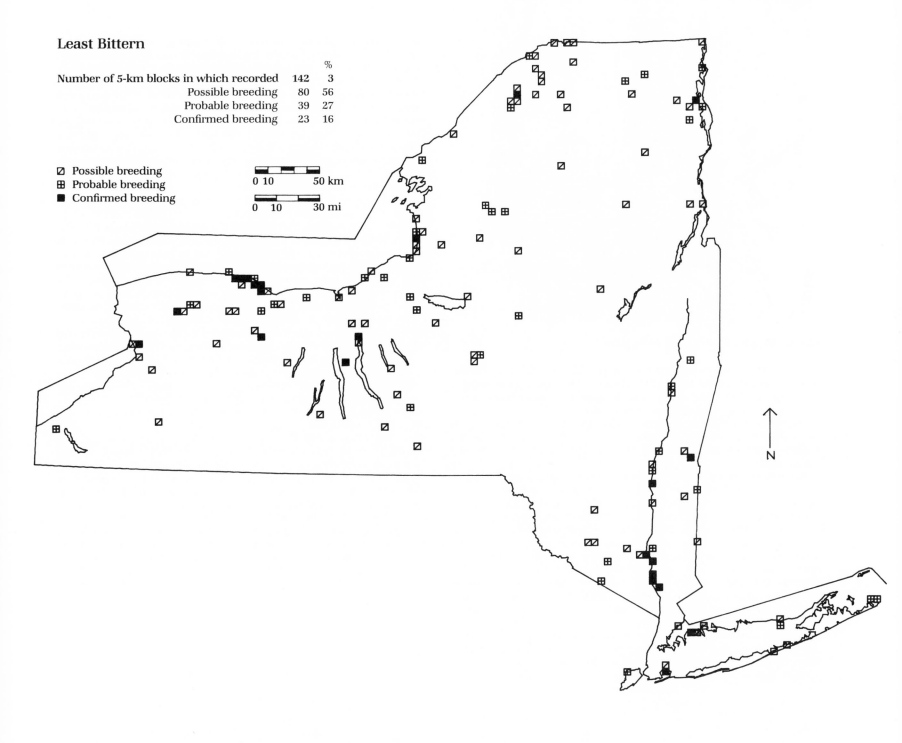

Least Bittern

Number of 5-km blocks in which recorded	142	%
Possible breeding	80	56
Probable breeding	39	27
Confirmed breeding	23	16

▨ Possible breeding
⊞ Probable breeding
■ Confirmed breeding

0 10   50 km

0 10   30 mi

N

©KLA-C 1986

Great Blue Heron *Ardea herodias*

The Great Blue Heron has clearly recovered from the killing and habitat loss that occurred at the turn of the century. Atlas data and other studies (Benning 1969; McCrimmon 1981, 1982) show both an increase in the number of heronries that are larger and a wider distribution of heronries than in the 1960s. This large heron is a local but fairly common species throughout most of upstate New York during the breeding season. On the Coastal Lowlands the Great Blue Heron is now only a nonbreeding summer visitor.

According to Eaton (1910) a heronry existed on Long Island at Jamaica Bay until the late 19th century, and there was still a breeding colony on Gardiners Island, Suffolk County, in 1900. He wrote, "Breeding colonies formerly existed in every large swamp in the State, but constant persecution and the destruction of the large trees which furnished their nesting sites have greatly reduced the number of heronries." He noted that the largest remaining colony, near Constantia, Oswego County, on the northwestern shore of Oneida Lake, still had about 500 pairs of Great Blue Herons early in this century. At that time other colonies were located in the Tonawanda Swamp, Orleans County; the Clyde River, Wayne County; and several Adirondack localities, but most others had "passed into history within the last two decades."

Between 1964 and 1968 Benning (1969), assisted by the Federation of New York State Bird Clubs, conducted a 5-year survey of great blue heronries and reported that 41 were active for at least 1 year during that period. The largest colony, in Marengo Swamp, Wayne County, had 175 active nests in 1967. About 40% of reported heronries were located in a narrow belt 64.4–80.5 km (40–50 mi) wide, running from Broome County north through Cortland and Onondaga counties into Jefferson County. Despite considerable annual fluctuations in numbers of nesting birds within heronries, Benning believed that the overall number of nests statewide did not change drastically from year to year. An even more rigorous survey was conducted from 1972 to 1981. Aerial surveys were made of almost all of the 110 heronries that had been reported by the Colonial Bird Register, the NYSDEC, biologists, refuge managers, and bird and garden clubs (McCrimmon 1981, 1982). McCrimmon found that the nesting abundance of the Great Blue Heron had probably increased since Benning's survey. The largest colony, at Seneca Pool, Montezuma NWR, had an estimated 456 active nests in 1980.

Atlas coverage was complicated by the large number of Great Blue Heron sightings in blocks where breeding could not be "confirmed." In addition, unconfirmed nesting almost certainly took place in some blocks. Many "possible" and "probable" records undoubtedly represent feeding or wandering birds. Currently, the largest colonies in the state are on Valcour Island in Clinton County waters of Lake Champlain, where at least 666 active nests were found in 426 trees in 1986 (Mitchell, pers. comm.) and on Ironsides Island in Jefferson County waters of the St. Lawrence River, where 621 active nests were found in 1985 (G. Maxwell, G. Smith, pers. comms.). Benning (1969) believed that a shortage of observers in the Adirondacks and Catskills might account for at least part of the seeming scarcity of herons in those regions. McCrimmon (1981, 1982) recorded only 3 great blue heronries in the northern and eastern regions of the state, including the Adirondacks, but also indicated that "this may reflect a relative scarcity of observers in those areas." The Atlas map, which reflects improved coverage, suggests that Great Blue Herons are scarce in the Catskill Peaks but as widespread in the Adirondack High Peaks and Central Adirondacks as elsewhere.

Nesting habitat is usually lowland swamp or upland hardwood forest; islands, forest-bordered lakes and ponds, and riparian woodlands, including conifers, are also used (Bent 1926; Palmer 1962). The phenomenal recovery of the beaver from near extirpation in New York at the end of the 19th century has provided additional nesting and foraging areas. Nests vary greatly in their dimensions from flimsy new platforms of sticks just 0.5 m (19 in) across to bulky older structures 0.9–1.2 m (3–4 ft) across. The nest is a shallow depression lined with fine twigs, moss, pine needles, or grass. Additions may be made throughout the incubation period and until the young are quite large. Nests, usually built in hardwoods, are placed from 7.6 to 30.5 m (25–100 ft) above the ground (Palmer 1962).

John M. C. Peterson

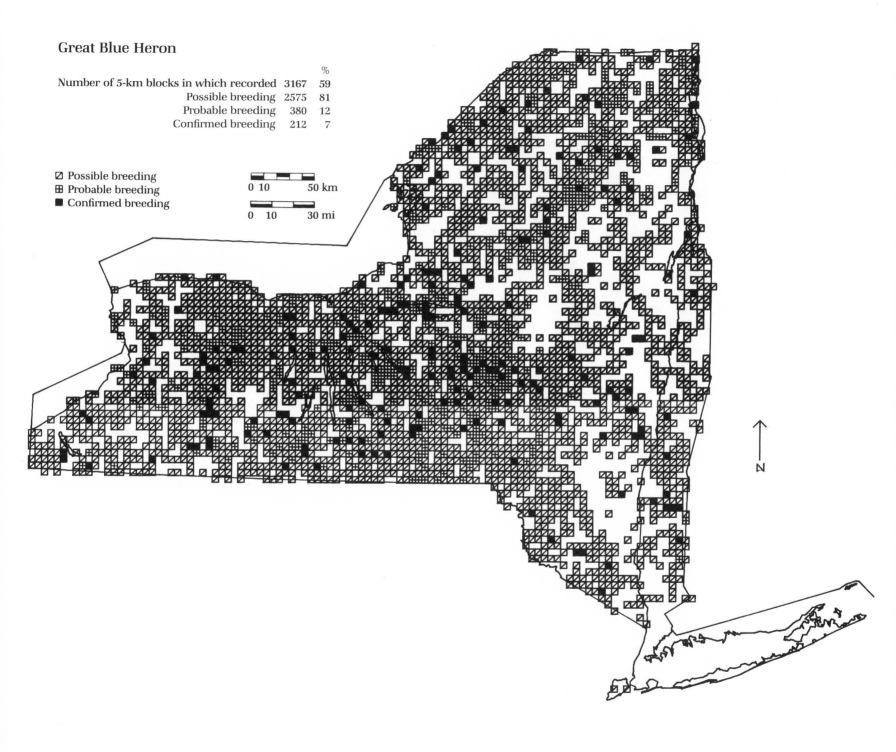

Great Blue Heron

		%
Number of 5-km blocks in which recorded	3167	59
Possible breeding	2575	81
Probable breeding	380	12
Confirmed breeding	212	7

◩ Possible breeding
⊞ Probable breeding
■ Confirmed breeding

0 10   50 km

0 10   30 mi

Great Egret *Casmerodius albus*

The Great Egret is the largest of the "southern" wading birds to occur in New York. In the early 1900s this egret was regarded as an occasional summer visitor; breeding occurred no farther north than Virginia (Eaton 1910). During the late 1800s and early 1900s the Great Egret was nearly driven to extinction because of the demand for its plumes. With protection granted under the Migratory Bird Treaty Act, it again became a regular summer visitor to New York. By the late 1940s hundreds of Great Egrets could be found along the coast in the late summer and early fall (Bull 1964).

The first recorded nesting of the Great Egret in New York occurred at Fishers Island, eastern Suffolk County, in 1953. By the early 1960s three additional colonies were established along Long Island's southwestern shore between Jamaica Bay Wildlife Refuge and southern Nassau County (Bull 1964). Between 1964 and 1974 the Great Egret continued to expand its breeding range along the south shore as far east as Moriches Bay, central Suffolk County. In eastern Suffolk County, Gardiners Island was colonized in 1967 (Bull 1974). The Great Egret established itself in existing wading bird colonies in association with the Snowy Egret, Black-crowned Night-Heron, and other species. Before 1973 no individual colony contained more than 50 pairs (Bull 1974). In 1975 the population reached its peak, when 410 pairs were found nesting at 13 colonies. Over half of these were at one colony at Jones Beach, Nassau County (Buckley and Buckley 1980). Other active colonies during the 1970s were on Huckleberry Island (Westchester County), Shooters Island (Richmond County), Fishers and Gardiners islands (eastern Suffolk County), Stony Brook Harbor (north

shore of western Suffolk County), and along the southwestern shore between Babylon (southeastern Suffolk County) and Jamaica Bay Wildlife Refuge (Erwin and Korschgen 1979; Buckley and Buckley 1980). After its peak in 1975, the population fluctuated between 216 and 311 pairs during 1976–78 (Buckley and Buckley 1980). In 1985, 296 pairs of Great Egrets were counted at 16 colonies (Peterson et al. 1985).

The Atlas has documented a distribution that has not changed greatly since the 1970s. In addition to the areas noted previously, breeding was "confirmed" on Plum Island at the eastern end of Long Island and at South Brothers Island in Bronx County. Although evidence of breeding was found on the north shore near Stony Harbor and on Fishers Island, breeding was not "confirmed" in those historical sites. The Great Egret record in central New York was of birds observed flying in and out of a Great Blue Heron colony. The Atlas also documents a record on Lake Champlain. However, breeding has never been confirmed there or on the Vermont side, where a pair was recorded by the Vermont Atlas at Missisquoi NWR in 1977 (Fichtel 1985). The Great Egret recently nested to the west of New York on the Canadian side of Lake Erie in 1981 and 1982 (Weir 1982), and to the north at Dundee, Quebec, in 1984 (J. Peterson 1984b). It may be just a matter of time before the Great Egret is found breeding either on Lake Champlain or the Great Lakes basin of New York.

The Great Egret is the largest species to inhabit New York's coastal mixed species wading bird colonies. At one colony the Great Egret was observed to out-compete all other wading birds for nest sites (Burger 1979b). The species tends to nest in the tallest vegetation available, usually 1–3.7 m (3.2–12 ft) off the ground (Bull 1974; Burger 1979b). At the large colony at Jones Beach the Great Egret nested in pines that were planted for erosion control (Bull 1974; Buckley and Buckley 1980). At most other colonies it is found nesting in dense greenbriar, bayberry, and poison ivy thickets, often with its nest built in the taller cherry, shadbush, or sumac trees found interspersed among the shrubs (Bull 1974; Peterson et al. 1985). When colony vegetation is very low, the Great Egret will sometimes nest on or near the ground (Bull 1974; Burger 1979b; Salzman 1985a). The Great Egret nest is usually a firm platform of sticks (Burger 1979b). No documented information is available concerning the breeding success of the Great Egret in New York colonies.

Although the Great Egret population in New York has fluctuated during the late 1970s and 1980s, this relatively recent colonizer appears to be maintaining a stable breeding population (Erwin and Korschgen 1979; Buckley and Buckley 1980; Peterson et al. 1985).

David M. Peterson

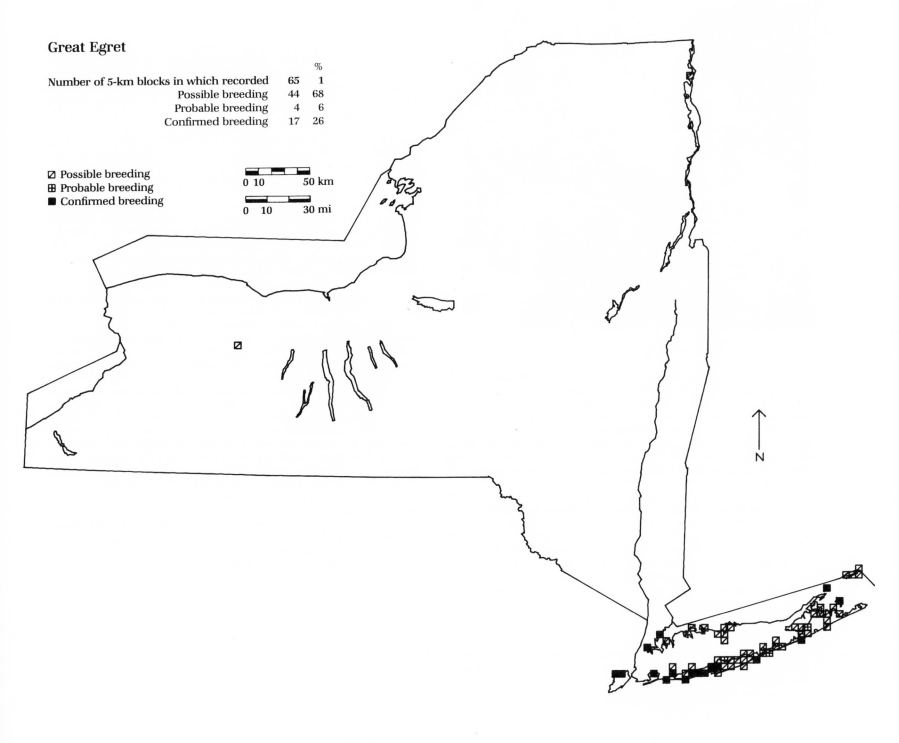

Great Egret

		%
Number of 5-km blocks in which recorded	65	1
Possible breeding	44	68
Probable breeding	4	6
Confirmed breeding	17	26

◪ Possible breeding
⊞ Probable breeding
■ Confirmed breeding

0 10   50 km

0 10   30 mi

N

Snowy Egret *Egretta thula*

The Snowy Egret is the common "white heron" seen along the coast of New York. Like the Great Egret, though, it has not always been a breeding species in the state. Giraud (1844) said the species was not uncommon on Long Island but did not document any nestings. The only confirmed nesting for the species during the 1800s was a single pair observed at Fire Island in 1885 (Bull 1964). The Snowy Egret suffered greatly from plume hunting beginning in the late 1800s and quickly disappeared from the state. In the 1920s Griscom (1923) listed the egret as extirpated as a breeding species in New York.

The Snowy Egret began to reappear in the 1930s and over the next 20 years became more abundant. New Jersey's first nesting occurred in 1939 (Bull 1964; Erwin 1979), and in 1949, 2 nests were discovered on Long Island at Oak Beach, Suffolk County (Bull 1964). During the 1950s additional sites were colonized along the coast in southern Nassau County and at Jamaica Bay Wildlife Refuge. The number of nesting birds also increased, with over 100 pairs counted at one colony on the Meadowbrook Causeway, Nassau County, in 1959 (Bull 1964). During the early 1960s the Snowy Egret again expanded its range north to southern Maine (Erwin 1979).

On Long Island it began nesting in colonies in Moriches Bay, on the central south shore of Suffolk County, and at Gardiners and Fishers islands off eastern Long Island, Suffolk County (Bull 1974). Between 1974 and 1977 the population along the coast increased from 730 to 1401 pairs. The number of active colonies increased from 9 in 1974 to 18 in 1976. The 1978 population dropped slightly to 1228 pairs, but the number of active

colonies rose to 21 (Buckley and Buckley 1980). When a Long Island–wide survey was again conducted in 1985, only 650 pairs of Snowy Egrets were counted in 15 colonies (Peterson et al. 1985). Local surveys conducted on Long Island documented the decline. For example, between 1982 and 1984 a wading-bird survey along 33.8 km (21 mi) of Long Island's south shore between Jones Inlet, Nassau County, and Fire Island Inlet, Suffolk County, was conducted (D. Peterson 1984). A 1977 census of that area revealed 509 pairs (Erwin 1979; Buckley and Buckley 1980), but in 1982 only 260 pairs were counted. Further decreases to 132 pairs in 1983 and 89 pairs in 1984 were observed. The number of colonies decreased from 5 in 1977, 1982, and 1983 to 4 in 1984 (D. Peterson 1984).

Despite local declines on Long Island, the Atlas documents the expansion of the Snowy Egret's range in the state since Bull's (1974) account. In the early 1970s breeding Snowy Egrets had been recorded only along the southwestern shore of Long Island; in Moriches Bay, south central Suffolk County; and on Gardiners and Fishers islands, eastern Suffolk County. Atlas surveyors "confirmed" additional nesting off Staten Island in Richmond County, on Fire Island, Shinnecock Bay, Stony Brook Harbor on Long Island Sound, and Plum Island off Orient Point (all in Suffolk County); on South Brother Island in Bronx County; and on Huckleberry Island off New Rochelle, Westchester County. The Snowy Egret has been recorded nesting on islands in Lake Champlain in Vermont, occasionally being observed on the New York side of the lake (J. Peterson 1983). In 1986 breeding was recorded in Hamilton Harbor on the Canadian side of Lake Ontario (Weseloh, pers. comm.). It is possible that upstate breeding may be imminent.

To date, except for the single 1889 record, this species has been found breeding in mixed colonies (Erwin 1979; Buckley and Buckley 1980; Peterson et al. 1985). When it first recolonized the state in the late 1940s and 1950s, the Snowy Egret nested at sites already colonized by the Black-crowned Night-Heron (Bull 1964). Currently, the species nests in colonies with as many as eight other species of wading birds. Among the nesting wading birds, the Snowy Egret generally nests in the middle levels of the vegetation, usually 1–3 m (3–10 ft) off the ground (Bull 1974). At sites where the vegetation is very low, nests have been found on or near the ground (Salzman 1985a).

The Snowy Egret has been very successful at reestablishing itself as a breeding species in New York. Although its numbers appear to have declined along the coast between the late 1970s and the mid-1980s, it still breeds over a wide area of the region.

David M. Peterson

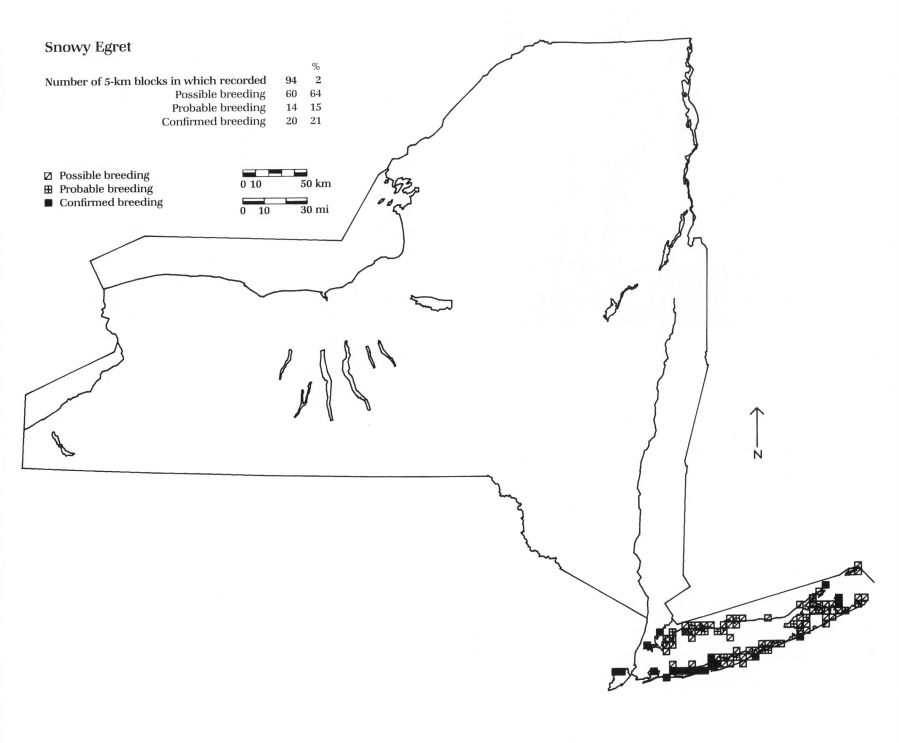

Snowy Egret

Number of 5-km blocks in which recorded	94	2
Possible breeding	60	64
Probable breeding	14	15
Confirmed breeding	20	21

%

◩ Possible breeding
⊞ Probable breeding
■ Confirmed breeding

0 10   50 km

0 10   30 mi

N

Little Blue Heron *Egretta caerulea*

The Little Blue Heron is one of a group of southern wading birds that first began breeding in New York in the 1950s. Eaton (1910) described the species as an accidental summer visitor. This heron had become a regular visitor to the coast by the 1920s and was occasionally observed inland. It was most numerous during the 1930s and 1940s when post-breeding congregations of as many as 100 birds were found along the coast. This period corresponded with its colonization of New Jersey (1935) and its establishment as a common nester in that state by the 1940s (Erwin 1979). For unknown reasons, sightings of the species in New York declined in the late 1940s and 1950s during a period when other southern herons were becoming more common in the state (Bull 1964).

Although the Little Blue Heron was well established as a nesting species in New Jersey by the 1940s (Bull 1964; Erwin 1979) and had even nested in Massachusetts during 1940 and 1941 (Erwin 1979), the first New York nest was not discovered until 1958 at Tobay Pond, Nassau County (Bull 1964). By 1967 small numbers were found nesting at three more heronries on the southwestern shore of Long Island between Jamaica Bay Wildlife Refuge and Oak Beach, western Suffolk County. The species also nested at a fifth heronry on Gardiners Island, eastern Suffolk County (Bull 1974). While continuing to spread to additional areas on Long Island during the late 1960s and 1970s, this heron in small numbers also colonized the New England coast from Connecticut to southern Maine (Erwin 1979).

Censuses conducted from 1974 to 1978 throughout Long Island and New York Harbor recorded from 9 to 34 pairs of Little Blue Herons nesting in 3–5 colonies (Buckley and Buckley 1980). In 1985, 68 pairs were surveyed in 8 colonies on Long Island and Staten Island (Peterson et al. 1985).

Although the Atlas documents breeding in wading-bird colonies off Staten Island and in eastern Long Island on Gardiners and Plum islands, most records are along the southwestern shore of Long Island. This area consists of barrier beach islands that protect shallow bays. The bays contain extensive salt marsh and islands of maritime shrubland that are ideal feeding and nesting areas for wading birds.

In New York the Little Blue Heron nests among other wading birds in mixed colonies. These are usually located in small trees and shrubs on isolated bay islands, although some have been discovered on barrier islands and causeways where human disturbance is greater. The Little Blue Heron generally nests in the middle levels of the vegetation above the Glossy Ibis, which nests on or near the ground; below the Great Egret, which breeds in the tops of the trees and shrubs; and among the similar sized Snowy Egret and Tricolored Heron (Peterson et al. 1985). Little information has been collected on the breeding biology of this species in New York.

Other southern wading birds, such as Snowy and Great Egrets and Glossy Ibis, have greatly increased their numbers since colonizing coastal New York during the 1950s and early 1960s, but Little Blue Heron numbers have not increased proportionally. Competition with other colonial species, limitations imposed by climate or habitat, or some unknown factor appears to be preventing the Little Blue Heron from becoming a more abundant member of New York's wading-bird community.

David M. Peterson

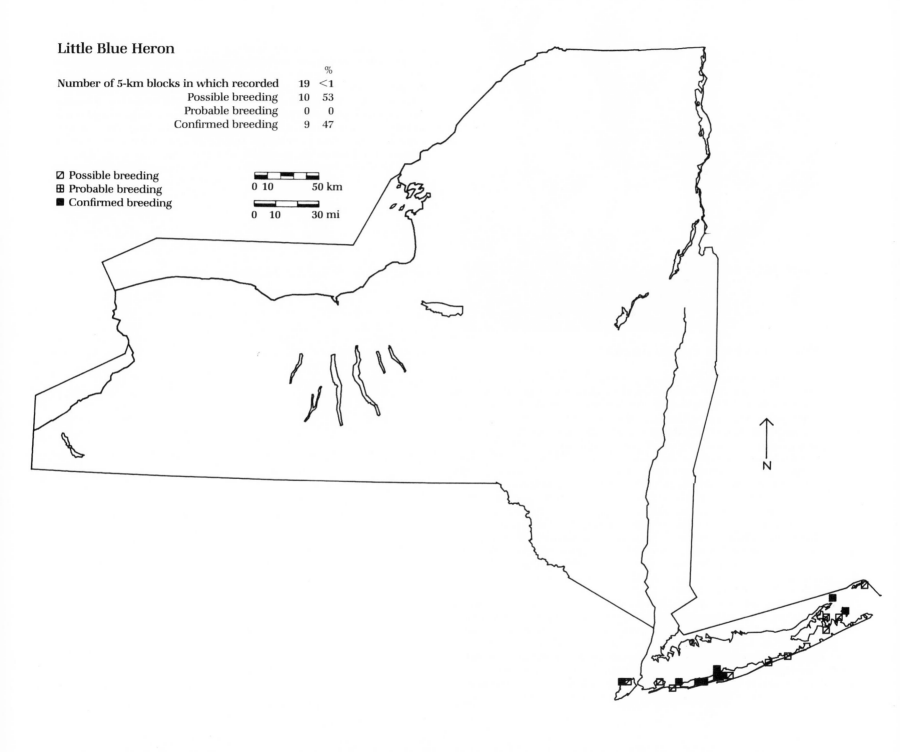

Little Blue Heron

◨ Possible breeding
⊞ Probable breeding
◼ Confirmed breeding

0 10   50 km

0 10   30 mi

N

©KLA-c 1986

Tricolored Heron *Egretta tricolor*

Formerly known as the Louisiana Heron, the Tricolored Heron is another of the group of southern wading birds that has only recently become established as a regular breeding species in New York. Griscom (1923) knew of only one Long Island record of this species from 1836. Between 1925 and 1941 Cruickshank (1942) listed 14 occurrences in the state, and by 1964 over 40 reports had been documented (Bull 1964). It was first found breeding at Jamaica Bay Wildlife Refuge in 1955 (Meyerriecks 1957). The state's second confirmed breeding occurred 11 years later, also at Jamaica Bay (Hays 1969). Since the early 1970s the Tricolored Heron has become a regular breeder in small numbers along the southwestern shore of Long Island. Between 1973 and 1985 the state's population appears to have stabilized, with 10–25 pairs being found in 3–5 heronries each year (Erwin 1979; Buckley and Buckley 1980; Peterson et al. 1985). The establishment of a stable population on Long Island has occurred during a period when the species established itself as a regular breeder in New Jersey and extended its breeding range into Connecticut, Massachusetts, and southern Maine (Erwin 1979).

The Tricolored Heron has been found breeding only in mixed-species heronries with the Snowy and Great Egrets, Black-crowned Night-Heron, Glossy Ibis, and other waders. In New York most of these heronries are located on the southwestern shore of Long Island, as reflected in the Atlas map. Wherever other mixed-species heronries are located along the coast of New York, however, there is a potential for this species to be found nesting. Outside of the southwestern shore Tricolored Herons had previously been observed in Long Island heronries in Stony Brook Harbor

and Plum Island, northwestern and eastern Suffolk County (Buckley and Buckley 1980), and Shinnecock Bay, southwestern Suffolk County (Raynor, pers. comm.). No evidence of breeding in these areas was reported during the period of Atlas fieldwork.

Little is known about the biology of this species in New York. Although its population has been monitored since its original colonization, it is not known why it began to breed in the state or what may be currently limiting its breeding population. Since there are many other coastal mixed species heronries, especially along the salt marshes of Long Island's southwestern shore, it would seem that a larger population of this species could exist (Erwin 1979; Buckley and Buckley 1980; Peterson et al. 1985).

David M. Peterson

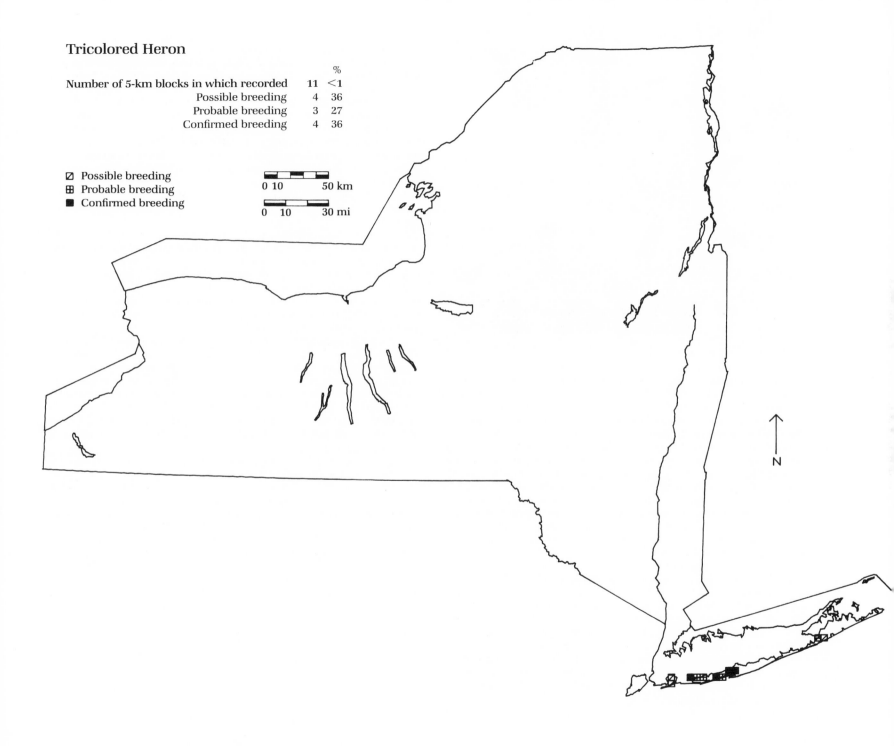

Tricolored Heron

		%
Number of 5-km blocks in which recorded	11	<1
Possible breeding	4	36
Probable breeding	3	27
Confirmed breeding	4	36

◪ Possible breeding
⊞ Probable breeding
■ Confirmed breeding

0 10   50 km

0 10   30 mi

N

Cattle Egret *Bubulcus ibis*

The Cattle Egret is an Old World species that crossed the Atlantic from Africa to South America during the 1870s. It moved north through northern South America and the Caribbean and first appeared in North America in the 1940s (AOU 1983). By 1958 it was found breeding as far north as southern New Jersey (Bull 1964). It was first observed in New York in 1954 at East Moriches, Suffolk County. By the 1960s it was established as a locally common visitor in the state, with scattered individuals and small flocks appearing along the coast and in upstate areas (Bull 1964, 1974).

The Cattle Egret began breeding in southern Ontario in 1960 (Bull 1964). By 1968 its range in Ontario extended to Pigeon Island on extreme eastern Lake Ontario, approximately 8 km (5 mi) from the New York border (Bull 1974). Meanwhile, in 1964, its breeding range along the Atlantic Coast expanded north from New Jersey into Rhode Island, bypassing New York (Erwin 1979).

Not until 1970 did the first confirmed nesting of the Cattle Egret in New York occur, at Gardiners Island off eastern Long Island (Bull 1974). In 1973 two additional and widely separated sites were colonized at Jones Beach, southern Nassau County (Bull 1974), and on Lake Champlain at The Four Brothers, Essex County. During its second year the latter colony contained 20 nesting pairs (Mack 1974). It can only be guessed whether the Lake Champlain birds originated from coastal populations to the south or Canadian birds from the west. More easily attributed to Canadian population expansion was a colony that appeared near Rochester at Braddock Bay WMA in 1974. The site contained 5 nests but occurred only one season (O'Hara 1974).

Through the mid-1970s the Cattle Egret population was centered along the coast and on Lake Champlain. The New York Lake Champlain population varied, from several pairs in 1975 to 8–10 pairs in 1976, to 3 pairs in 1979 (Mack 1975; Kibbe 1976; Mack 1979).

On the Coastal Lowlands the population increased from 16 pairs at three colonies in 1974 to 136 pairs at four colonies in 1978. Its breeding range shifted from Gardiners Island to the south shore from southern Nassau County to Jamaica Bay Wildlife Refuge. As further expansion occurred, the Cattle Egret was found nesting at Shooters Island, Richmond County, and South Brother Island, Bronx County (Erwin 1979; Buckley and Buckley 1980).

In 1977 a third breeding area containing 2 nests was discovered at Little Galloo Island, Jefferson County, in eastern Lake Ontario (Blokpoel and Weseloh 1982).

During the Atlas period eastern Lake Ontario, Lake Champlain, and the coastal region were the only areas in which the Cattle Egret was recorded breeding. The two upstate sites remained small. The Four Brothers had 8 active nests in 1980, and Cattle Egrets were observed on the island in 1985, but a nest count was not made (Mack 1980; Peterson 1985c). The maximum size documented for the Little Galloo Island colony was 6 nests in 1982 (Crowell 1982). It is believed that only 1 pair nested on the island in 1986 (Weseloh, pers. comm.)

The upstate population remains low, but the coastal population continues to grow. In 1985, 351 pairs were counted at four colonies in southwestern Long Island and Staten Island (the South Brother Island colony was not surveyed) (Peterson et al. 1985). The Cattle Egret along the coast has apparently become specialized at feeding in grassy upland areas of parkways, airports, and landfills (Buckley and Buckley 1980). Although its first breeding in the state occurred in eastern Long Island, this species has not begun to use the croplands and horse farms of eastern Long Island.

The Cattle Egret nests in association with other wading bird species. On Lakes Champlain and Ontario it has been found in colonies with Black-crowned Night-Herons, gulls, and Double-crested Cormorants (Mack 1974; Chamberlaine 1978; Blokpoel and Weseloh 1982; Ellison 1985a). On the coast it has nested in colonies with as many as 7 other species of wading birds (Erwin 1979; Buckley and Buckley 1980; Peterson et al. 1985). Although the Cattle Egret forages in upland areas where it does not compete with other wading birds for food, it has been suggested that this exotic species may be competing with native wading birds within the nesting colonies. The species is more aggressive than native wading birds and competes favorably with other species for nesting locations within a mixed species colony (Burger 1978).

David M. Peterson

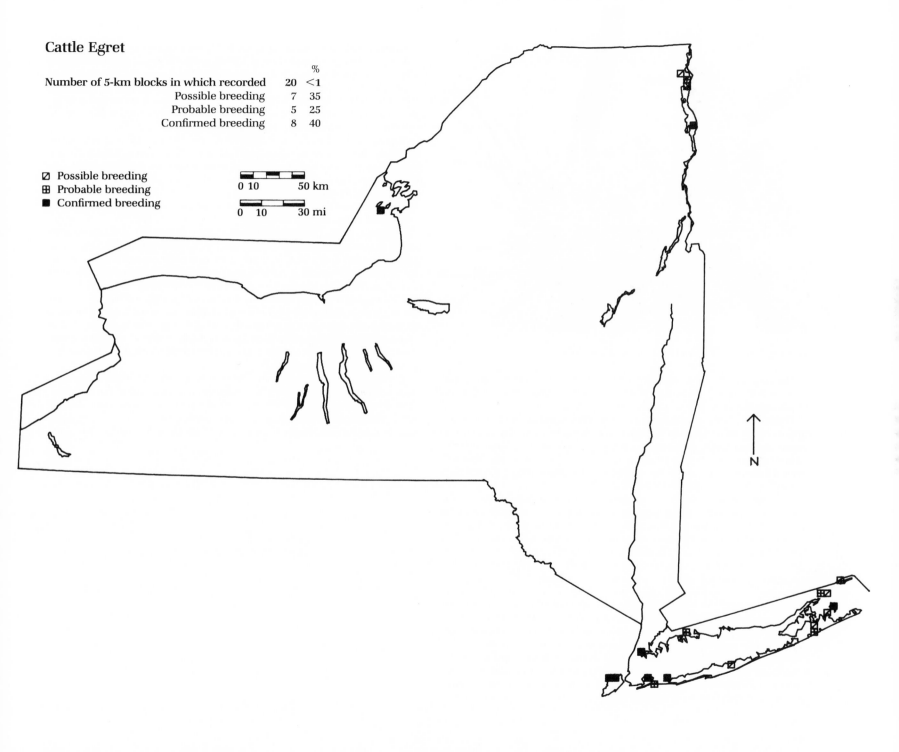

Cattle Egret

Number of 5-km blocks in which recorded		%
	20	<1
Possible breeding	7	35
Probable breeding	5	25
Confirmed breeding	8	40

☒ Possible breeding
⊞ Probable breeding
■ Confirmed breeding

0 10   50 km

0 10   30 mi

N

Green-backed Heron *Butorides striatus*

The Green-backed Heron is the most widespread and abundant heron in New York. It is a fairly common breeder throughout the state at low elevations but very uncommon in the mountains.

The status of this small heron, formerly known as the green heron, has apparently not changed in this century. Eaton (1910) stated that it bred in every county except Franklin and Hamilton and believed that with improved observation, it could have been found in those. Bull (1974) considered this species a widespread breeder in the state except for the high elevations of the Adirondacks and Catskills.

Atlas data show evidence of breeding by the Green-backed Heron in all counties, including Franklin and Hamilton. It was "confirmed" in all but Herkimer, Hamilton, Schoharie, and Greene counties, probably because of low population density in those areas and lack of intense coverage. The Green-backed Heron is present in relatively few blocks at elevations above 305 m (1000 ft). Almost entirely restricted to stream and river systems at higher elevations, it was absent from most of the Adirondacks, Catskills, and Tug Hill Plateau because of extensive forests and unsuitable stream and river habitats. In western New York this heron was found infrequently in Chautauqua, Cattaraugus, and Allegany counties, where there are few wetlands and many streams at high elevations are small and narrow with little suitable nesting habitat (Andrle, pers. comm.). Gaps in distribution in the Mohawk Valley and eastern end of the Appalachian Plateau can be attributed largely to a lack of suitable habitat, because coverage was thorough in those areas and almost all Atlas records in those regions coincided with the distribution of river systems.

Unlike most other heron species, the Green-backed Heron is usually a solitary nester, although it sometimes nests in small groups (Palmer 1962). Because it is often quite secretive around the nest site, obtaining "confirmed" records can be difficult and locating nests can be extremely difficult. Only 25% of the "confirmed" records were of active nests. However, the bird is relatively easy to spot as it flies from one feeding area to another, or to and from the nest site.

The Green-backed Heron is so widespread because of its ability to use many varieties of wetland habitats. In New York it frequents marshes, swamps, streams, rivers, ponds, and lakes of all sizes (Eaton 1910). It is as widespread and abundant in saltwater habitats as in freshwater, showing no preference for either (Palmer 1962).

It prefers nest sites near water, but the Green-backed Heron will readily nest in dry woods, orchards, or thickets some distance from water (Townsend 1926). On the outer south shore of Long Island this species often nests in low thickets of poison ivy (Cruickshank 1942). Nests are placed from on the ground to in the treetops but are usually 3–6.1 m (10–20 ft) above the ground. At Jamaica Bay Wildlife Refuge 6 of 136 nests found in 1955 were on the ground (Meyerriecks 1960).

Palmer (1962) described nest construction in this species as a three-stage process. The male begins construction, and once the pair-bond is formed, he brings twigs to the female and she adds them to the nest. After the eggs have been laid, both sexes add twigs to the outside of the nest and lining material to the nest cup. This behavior continues for as long as 2 weeks after the young hatch.

The nest varies from a flimsy platform to a bulky, tightly woven structure of many twigs (Palmer 1962). Because it is often added to in subsequent years, a reused nest tends to be more bulky than a new one (Palmer 1962). Occasionally, old nests of other species, such as the Snowy Egret and Black-crowned Night-Heron, are also used (Meyerriecks 1960).

Steven C. Sibley

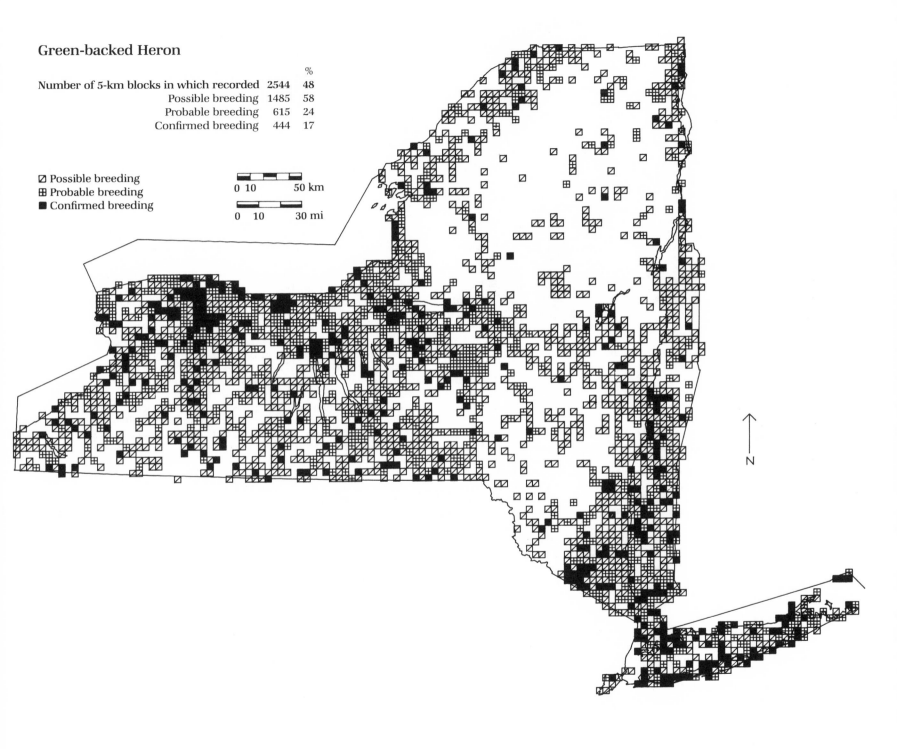

Green-backed Heron

		%
Number of 5-km blocks in which recorded	2544	48
Possible breeding	1485	58
Probable breeding	615	24
Confirmed breeding	444	17

☑ Possible breeding
⊞ Probable breeding
■ Confirmed breeding

0 10   50 km

0 10   30 mi

N

©KLA-C 1986

Black-crowned Night-Heron *Nycticorax nycticorax*

One of the better known night sounds on Long Island is the familiar *quawk* of the Black-crowned Night-Heron. It is a fairly common breeding species there, but in other parts of the state it is local and uncommon. The general trend over the last five decades has been one of fewer breeding locations and smaller colonies on Long Island (Bull 1974; Peterson et al. 1985), and a sharp decrease in the number of breeding locations in the rest of the state.

Eaton (1910) considered this species to be a common summer resident on Long Island. Griscom (1923) called the Black-crowned Night-Heron a common summer resident there, and Allen (1938) stated that more than 3000 pairs had been found on Long Island in the 1930s. Within the next several years, however, declines became evident. For example, Bull (1964) noted that the colony at Great Neck, Nassau County, had 1000 nests in 1934, but by 1951 only 600 pairs were counted. The major factor in the decline at most north and south shore colony sites was the development of these areas for homes (Allen 1938; Peterson et al. 1985). In addition, the egg-shell thinning effects of DDT, which Anderson and Hickey (1972) indicated were severe for this species in other areas, may have been a factor. During the period 1974–78 Buckley and Buckley (1980) found the Long Island Black-crowned Night-Heron population to be about 25% of its former numbers in the mid-1930s, with between 430 and 760 pairs recorded at a maximum of 23 colonies. They also noted that the average colony size was smaller. During their survey, numbers of breeding pairs and colony numbers increased. In 1985 a survey of colonial waterbirds on Long Island located 1430 pairs at 22 colonies, a significant increase in individuals but no increase in the number of colonies (Peterson et al. 1985).

In upstate New York, Judd (1907) indicated that this heron was an uncommon breeding species in Albany County in the upper Hudson Valley. Eaton (1910) disagreed with this assessment, indicating that this heron was common in the Hudson and Lake Champlain valleys as far north as Washington County. Eaton also mentioned heronries in Erie County, and many years later Beardslee and Mitchell (1965) stated that there were 2 colonies on Grand Island in Erie County and called the species fairly common. In the 1930s the Grand Island colonies had over 600 nests; however, Beardslee and Mitchell indicated that the numbers at those colonies had decreased. At one time there were apparently many colonies in upstate New York, as Bull (1974) reported colony locations along the upper and lower Hudson Valley, the Lake Champlain Valley, eastern Lake Ontario, in central New York, and along the Niagara River. Bull noted, however, that by the early 1970s some of the colonies were no longer active.

The Atlas shows that on Long Island the Black-crowned Night-Heron is now established along both the north and south shores. Colonies were found on Gardiners, Plum, and Fishers islands off the eastern end of Long Island. However, the Fishers Island colony was no longer active in 1985 (Peterson et al. 1985). Heronries were also located off Staten Island on Shooters and Pralls islands, on South Brother Island in Bronx County, and on Huckleberry Island in Westchester County. Locating a colony is apparently not difficult. Not only is it fairly obvious visually, but the stench of rotting fledgling corpses, which seems common at large heronries, and regurgitated fish is revealing.

Loss of wetland habitat was probably the major factor in the apparent further loss of colonies in upstate New York documented by the Atlas. Few colonies remain, with "confirmed" breeding documented only on Lake Champlain, at Little Galloo Island in eastern Lake Ontario, at Montezuma NWR, Perch River and Braddock Bay WMAs, and Lakeview Marsh WMA on the eastern shore of Lake Ontario. The colonies along the Niagara River in New York, the Hudson Valley, and in central New York appear to be gone. The other Atlas records may represent single nests but may also be of post-breeding individuals that wandered away from their natal areas. According to Judd (1907), the Black-crowned Night-Herons in Albany County did not breed in colonies.

On Long Island the Black-crowned Night-Heron nests in mixed-species heronries in red cedar or pitch pine–scrub oak woodlands, or in maritime shrubland of mixed poison ivy, greenbrier, sumac, and cherry (Bull 1974). In upstate New York this heron nests in forested wetlands, mainly in red maple–northern hardwood swamps or in mixed deciduous and coniferous forested uplands near streams or ponds (Bull 1974). The nests vary from crudely to solidly built platforms of reeds and branches (Gross 1923) and can be placed on the ground (Salzman 1985a) to well up into the middle and upper levels of 12.2 m (40 ft) tall trees (Bull 1974; Peterson et al. 1985).

Emanuel Levine

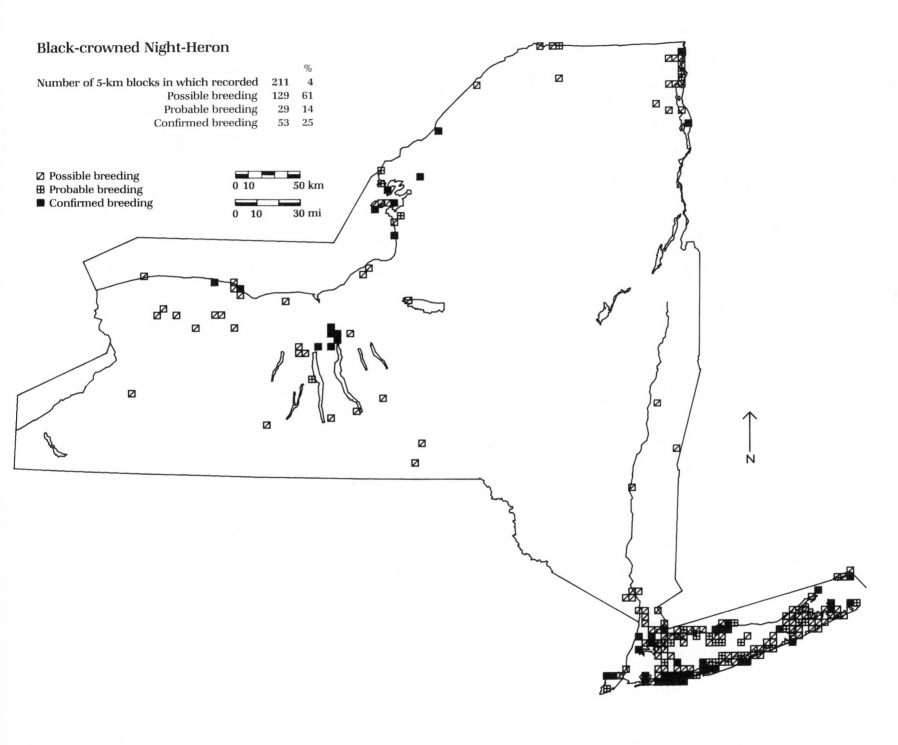

Black-crowned Night-Heron

Number of 5-km blocks in which recorded	211	% 4
Possible breeding	129	61
Probable breeding	29	14
Confirmed breeding	53	25

▨ Possible breeding
⊞ Probable breeding
■ Confirmed breeding

0 10   50 km

0 10   30 mi

N

©KIA-C 1987

wet or dry woodlands near swamps, rivers, and harbors (Bull 1974; Peterson et al. 1985). The nest consists of a platform of sticks and is similar to that made by the Black-crowned Night-Heron. Nest height ranges from 1.5 to 3.7 m (5–12 ft) in the coastal shrubland to 4.6–18.3 m (15–60 ft) in mature woodlands (Bull 1974).

In New York the Yellow-crowned Night-Heron is restricted to coastal areas, in large part probably because its primary prey is crustaceans, particularly fiddler crabs, which live in coastal salt marshes (Riegner 1982). The highest densities of nesting Yellow-crowned Night-Herons occur in areas with extensive salt marshes. Small colonies or individual nests have been found in areas with more limited salt-marsh habitat.

David M. Peterson

Yellow-crowned Night-Heron *Nycticorax violaceus*

The Yellow-crowned Night-Heron is one of several southern herons that have established themselves as breeding species in New York in the last 50 years. Before 1938, when the first nesting was reported in Massapequa, Nassau County (Bull 1964), it was regarded as a rare (Eaton 1910) to casual (Griscom 1923) visitor to Long Island and the New York City area. Since 1938 it has become established as a regular but uncommon nesting species on Long Island, especially along the southwestern and north central shores. Fewer than 50 pairs have been located breeding in the state during any one year (Bull 1974; Erwin 1979; Buckley and Buckley 1980; Peterson et al. 1985). The species no longer nests at several historical mainland nesting areas in western Long Island as a result of development and habitat destruction (Allen 1938; Bull 1964).

The Atlas has not located any significant new breeding areas for this species beyond those identified by Bull (1974). The Yellow-crowned Night-Heron is still found breeding in the Stony Brook–Mount Sinai area of northwestern Suffolk County, in scattered localities on the eastern end of Long Island, in the extensive salt marshes along Long Island's southwestern shore, and in Westchester County on the northwestern shore of Long Island Sound. In Nassau County, however, nesting was only "confirmed" at two out of seven historical sites.

The Yellow-crowned Night-Heron nests in mixed-species heronries or in association with Black-crowned Night-Heron colonies. It may also breed singly away from other nesting herons; such isolated nesting pairs are often difficult to detect. The species nests both in low coastal shrubland, primarily on isolated dredge spoil and salt-marsh islands, and in

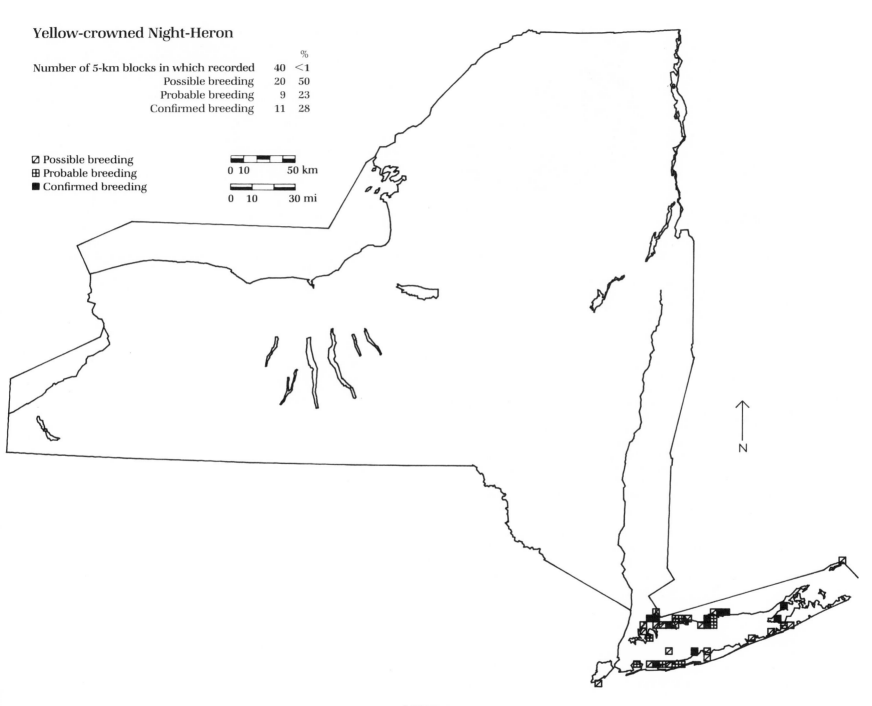

Yellow-crowned Night-Heron

		%
Number of 5-km blocks in which recorded	40	<1
Possible breeding	20	50
Probable breeding	9	23
Confirmed breeding	11	28

◩ Possible breeding
⊞ Probable breeding
■ Confirmed breeding

0 10   50 km

0 10   30 mi

N

© KLA-C 1986

the late 1970s, it appears that New York's Glossy Ibis population may still be increasing (Erwin 1979; Peterson et al. 1985).

As the Atlas map indicates, the nesting range of the Glossy Ibis has not expanded greatly in New York. The spread onto islands off Staten Island and the discovery of a small colony in Shinnecock Bay on the eastern south shore of Long Island have been the only changes. The Glossy Ibis is most abundant along Long Island's southwestern shore from Jamaica Bay to Babylon, western Suffolk County. A much smaller number nests in eastern Long Island on Plum Island and Gardiners Island (Erwin 1979; Buckley and Buckley 1980; Peterson et al. 1985).

Most nesting Glossy Ibis in the state have been found in mixed-species colonies. It tends to nest low in the vegetation, sometimes even building its nest on the ground (Hays 1977; Burger 1979b). The nest may consist of a bulky pile of common reed or a platform of sticks lined with a few common reed stems. The Glossy Ibis also has been found to reuse the old nests of the Snowy Egret and Black-crowned Night-Heron (Bull 1974). This species has been very successful in expanding its range into the northeastern United States, establishing a large, stable population in New York in less than 25 years.

David M. Peterson

Glossy Ibis *Plegadis falcinellus*

Early in this century the Glossy Ibis occurred as a rare vagrant in upstate and coastal New York (Eaton 1910). Until the 1940s Cruickshank (1942) considered the bird an accidental visitor. The only known breeding population in the United States was in southern Florida (Hays 1977). Between 1944 and 1959 the Glossy Ibis became an annual visitor to the state. It began to be seen along the coast regularly, and in 1961 the first breeding Glossy Ibis was discovered at Jamaica Bay Wildlife Refuge (Bull 1964). Considering that the species first nested in New Jersey in 1955, Maryland in 1956, and Virginia in 1956, the growth of the Glossy Ibis population was quite dramatic (Erwin 1979). By 1967 over 100 pairs were nesting on Long Island at Jamaica Bay; Lawrence Marsh, Nassau County; and Oak Beach, western Suffolk County (Bull 1974; Erwin 1979). Glossy Ibis numbers increased, and it spread into several other wading-bird colonies along the southwestern shore of Long Island. In 1970 the first recorded nesting occurred in eastern Long Island at Gardiners Island (Bull 1974). That same year another range expansion occurred, and Glossy Ibis moved up the New England coast as far north as southern Maine (Erwin 1979).

Between 1974 and 1977 the New York population grew from 465 pairs in 11 colonies to 842 pairs in 17 colonies. The number dropped to 574 pairs in 18 colonies in 1978 (Buckley and Buckley 1980). In 1985 a survey of all wading-bird colonies on Long Island and Staten Island counted 888 pairs in 15 colonies (Peterson et al. 1985). Although population declines were observed throughout the Northeast from Virginia northward during

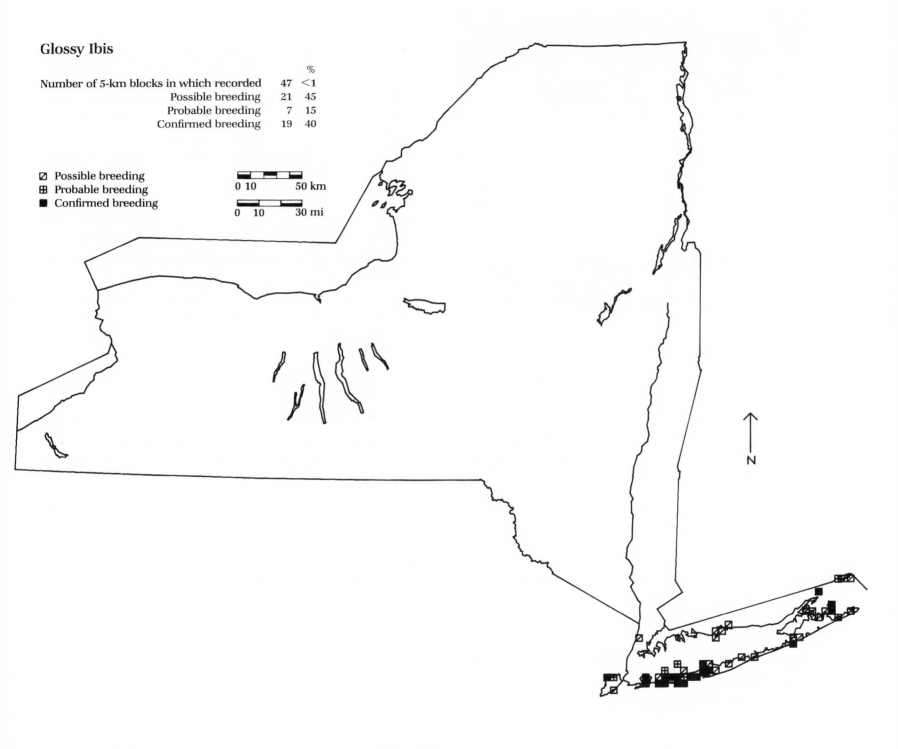

Glossy Ibis

		%
Number of 5-km blocks in which recorded	47	<1
Possible breeding	21	45
Probable breeding	7	15
Confirmed breeding	19	40

◱ Possible breeding
⊞ Probable breeding
◼ Confirmed breeding

0 10   50 km

0 10   30 mi

N

CJ Page
© 1987

Mute Swan *Cygnus olor*

The Mute Swan is an introduced species now locally common on Long Island and in the lower Hudson Valley but only a casual breeder elsewhere in the state. Native to Eurasia, the Mute Swan was first released in small numbers in New York in the late 1800s by private individuals (Bump 1941). The species did not become established as a feral breeder, however, until after 1912 (Bull 1974). In 1910, 216 Mute Swans, probably all semidomesticated (Delacour 1954), were imported and released in the lower Hudson Valley, chiefly near Rhinebeck, Dutchess County (Bull 1974). In 1912 another 328 Mute Swans (Delacour 1954) were released on the south shore of Long Island at Southampton and Oakdale, Suffolk County (Bull 1974). Both of these populations soon became established and slowly increased; by 1967 the Oakdale population was estimated at 700 birds (Palmer 1976a).

The stronghold of this species in New York continues to be the bays and ponds of Suffolk County at the eastern end of Long Island. Although the Long Island population seems to have stabilized, the Mute Swan is still increasing in numbers in the lower Hudson Valley and gradually extending its range northward. It can probably be expected to continue a slow but steady expansion up the valley (Weissman, pers. comm.).

The Mute Swan has been seen at some time in virtually every part of the state. Almost all of these sightings appear to be of semidomesticated birds that have escaped from private waterfowl collections, parks, or zoos. Feral pairs have occasionally bred in upstate New York, but these have not become permanently established. There is reason to believe, however, that feral populations will eventually become established throughout New York because the Hudson Valley population is still expanding and feral populations in Toronto, Ontario, and Traverse City, Michigan (McKee 1984), are thriving in conditions harsher than those found in portions of upstate New York.

During the Atlas Survey the Mute Swan was reported in two blocks away from Long Island and the Hudson Valley. Breeding was "confirmed" at Perch River WMA in Jefferson County, where a pair raised broods in 1983, 1984, 1985, and 1986. Pairs and single adults had been seen there for a number of years. The "possible" record in western New York was of a single individual seen in only one year. In both 1986 and 1987 a feral pair of Mute Swans nested along the Cayuga Lake inlet in Ithaca. The nest was built adjacent to a fence surrounding a waterfowl pond that contained a pinioned pair of Mute Swans. The origin of this feral pair is unknown.

Because of its large size, pure white coloration, and conspicuous habits, the Mute Swan is virtually impossible to overlook. In the Atlas blocks in which it was "confirmed" half the records were of adults with dependent young.

The Mute Swan can be found in almost any open-water habitat, both fresh- and saltwater, but it shows a strong preference for small ponds and streams of 4.1 ha (10 a) or less (Minton 1968). On Long Island the preferred nesting areas are the extensive bays and marshes along the north and south shores (Cruickshank 1942). In the lower Hudson Valley the Mute Swan nests in the ponds, lakes, and New York City reservoirs throughout the valley rather than along the Hudson River (Weissman, pers. comm.). Perch River WMA is a 2833-ha (7000-a) expanse of cat-tail marsh and ponds dotted with numerous islands (R. and J. Walker, v.r.).

There has been some concern about the effects the Mute Swan may have on native waterfowl (Reese 1975). Because of the swan's long neck, large appetite, and tendency to remain in one area, it was feared that it would deplete the aquatic vegetation normally available to native duck species. A recent study in Connecticut, however, found no competition between swans and native waterfowl: they feed on different vegetation and at different water depths (O'Brien and Askins 1985). The Mute Swan has been known to kill waterfowl that invade its nesting area and will readily attack people who get too close to its young or nest (Reese 1975).

The Mute Swan usually nests on small islands, along isolated shorelines, or on platforms it builds in shallow water within its defended territory (Johnsgard 1975). Minton (1968) found that established breeders tend to use the same nest site year after year. The nest itself consists of a large mound of vegetation gathered from nearby in which the female forms a depression for the eggs (Palmer 1976a).

Steven C. Sibley

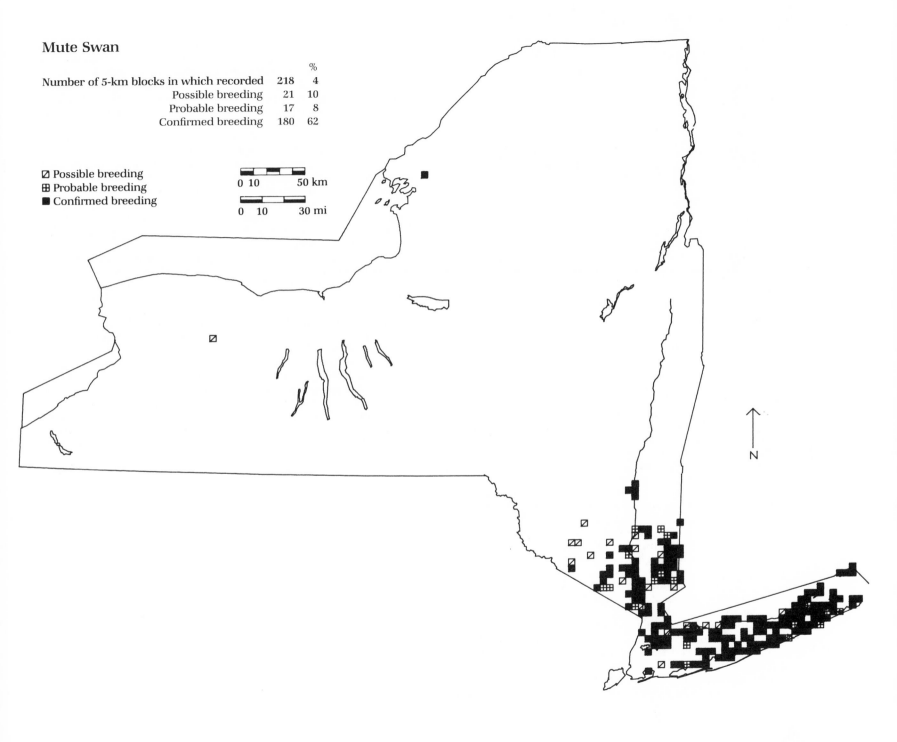

Mute Swan

		%
Number of 5-km blocks in which recorded	218	4
Possible breeding	21	10
Probable breeding	17	8
Confirmed breeding	180	62

◩ Possible breeding
⊞ Probable breeding
■ Confirmed breeding

0 10   50 km

0 10   30 mi

N

C J Page © 1987

Canada Goose *Branta canadensis*

The Canada Goose, one of the state's most familiar waterfowl, was not recorded by early naturalists as a breeding bird in New York. It has been introduced into various wildlife management areas and national wildlife refuges in the state. Today it is common on Long Island and in the lower Hudson Valley and is spreading from wildlife management areas into other parts of New York.

The Atlantic Flyway population of the Canada Goose, which includes New York, has experienced a rather dramatic increase in the past three decades. In the late 1940s winter bird counts estimated the population to be less than 300,000 birds; today it is conservatively estimated to be 850,000. This growth seems to be related to the creation of several waterfowl refuges and an increased food supply made available by changes in agricultural practices (Purdy and Malecki 1984) In addition, local introductions may have added to the population. New York's breeding population is estimated to be about 24,000 (Moser, pers. comm.), and it continues to grow.

The Canada Goose was introduced into New York in the 1930s, but early documentation of breeding is scarce. The first record of breeding at Montezuma NWR was in 1949 (Gingrich, pers. comm.). In 1961 at least 50 pairs bred there and at nearby Howland Island WMA (Bull 1974). From 1980 to 1985 an annual average of 230 young were raised at Montezuma. The Canada Goose was first recorded nesting at Oak Orchard WMA in 1950 (D. Carroll, pers. comm.); in 1963 over 20 pairs raised broods there (Beardslee and Mitchell 1965); and in 1980, 300 young were produced. This number had increased to 450 young by 1985 (D. Carroll, pers. comm.)

On Long Island in 1970, 100 pairs were estimated to nest on Gardiners Island (Bull 1974), and nesting has been documented at Jamaica Bay Wildlife Refuge since about 1976. From 1980 to 1985 an annual average of 60 young were raised there (Riepe, pers. comm.) Other national wildlife refuges and wildlife management areas around the state have also been centers for the spread of this resident population, including Wilson Hill, Perch River, and Three Rivers WMAs. Introductions in southern Ontario, Canada, may also be supplementing the St. Lawrence River population (Ontario Breeding Bird Atlas, preliminary map).

Atlas workers found the Canada Goose nesting over much of Long Island, the Central Hudson Valley, Taconic and Hudson highlands, the Manhattan Hills, and the Triassic Lowlands. The largest gaps in its distribution occurred in the Central Appalachians, east and west of the Finger Lakes; and the Adirondacks. These areas are farthest from the introduction sites and, except for the Adirondacks, are generally deficient in wetland habitat.

In spring the Canada Goose seeks areas free from disturbance for its nest site. In general, it prefers to nest on small islands within impounded areas which command a clear view in all directions. Permanent water must be nearby to which its young can be led shortly after hatching (Palmer 1976a). In New York the Canada Goose favors cat-tail marshes. Muskrat houses located in these marshes act as platforms and loafing areas. Nests are sometimes placed on old tree stumps (Bull 1974), on islands in lakes, and on beaver dams.

The nest is typically a depression lined with sticks, cat-tails, reeds, and grasses gathered from nearby (Harrison 1975). The nest site is selected and the nest built principally by the female. As incubation proceeds, the female plucks increasing amounts of down from her breast and places it around the eggs. The goose and gander are monogamous and form a lifelong pair-bond. The gander does not incubate, but he is always nearby, guarding and defending the nest and territory against intruders (Palmer 1976a).

Stephen W. Eaton

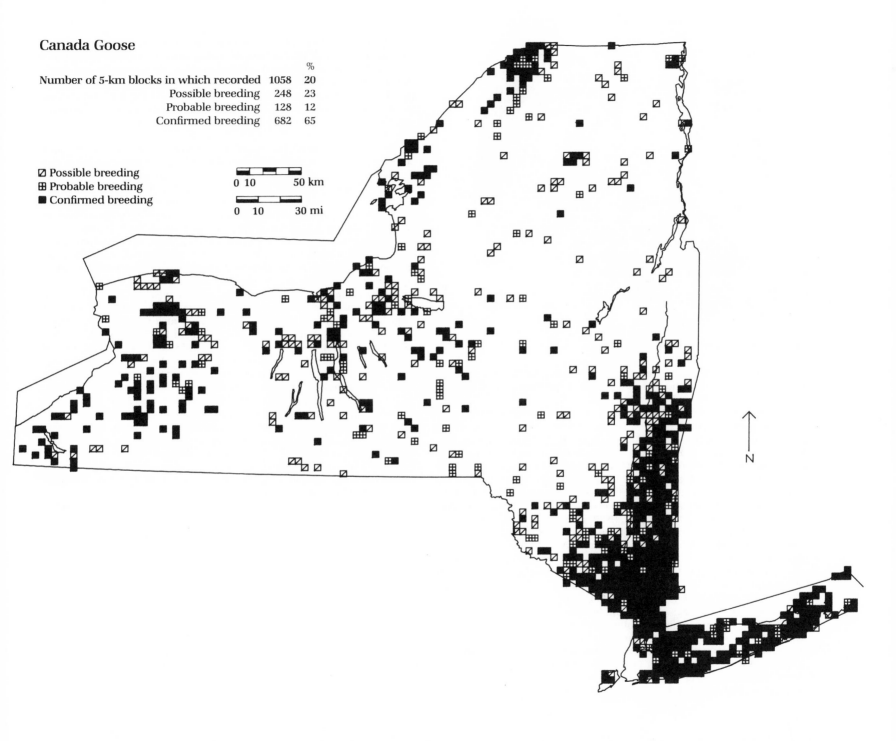

Canada Goose

		%
Number of 5-km blocks in which recorded	1058	20
Possible breeding	248	23
Probable breeding	128	12
Confirmed breeding	682	65

☒ Possible breeding
⊞ Probable breeding
■ Confirmed breeding

0 10   50 km

0 10   30 mi

N

Wood Duck *Aix sponsa*

Certainly one of the most beautiful North American ducks, the Wood Duck is once again a common summer resident throughout the state, with breeding "confirmed" in every county.

Of its historical status Eaton (1910) stated it was "formerly a common summer resident," but by the early 1900s it was only fairly common in the marshes bordering the Seneca River and along the shores of Lake Ontario. In an address to the American Ornithologists' Union in 1930, he expressed concern about the toll that hunting was taking on the Wood Duck in western and central New York (Eaton 1953).

The pressure from hunters on the species was such that in 1918 Canada and the United States in concert declared a closed season that continued through 1941 (Kortright 1942). In New York, however, there was never a closed season on Wood Duck, and prior to 1911 there was not even any bag limit. In 1912 a limit was set at 25 Wood Ducks a day; in 1931 that was cut to 15, then to 12 in 1933, 10 in 1935, and finally to 1 a day in 1942. It remained at that figure until 1952 when the bag limit was raised to its present 2 birds a day. The population is apparently at a high, stable level (Moser, pers. comm.). During the 1980s the Wood Duck has ranked first or second in the bags of New York hunters.

The reduced pressure from hunters in the 1930s and 1940s obviously helped the Wood Duck population to grow (Bellrose 1978). However, additional factors have facilitated the increase in New York. According to Ermer (1984), many biologists believe that the increase in the state's beaver population over the past 40 years is largely responsible for the increase in the Wood Duck population. He conservatively estimated that at current population levels beavers have created 24,291–32,389 ha (60,000–80,000 a) of shallow-water impoundments in New York. Additionally the abandonment of open farmland in the state in the last three decades and the decrease of logging in areas contiguous to swampland, ponds, and streams have greatly increased Wood Duck nesting areas. Conservation organizations, assisted by the state, have also campaigned vigorously for the building and placing of nesting boxes in suitable areas (USFWS 1976).

The greatest concentrations of Atlas records occur in central and western New York, as well as in the Hudson Valley. There were a surprising number of records in the heavily populated Coastal Lowlands. Fewer records were found in the Mohawk Valley and to the north, and in the Catskill Mountains. Most of the "confirmed" records were at elevations below 610 m (2000 ft). Although this species apparently tolerates a wide range of habitats, it was not found in areas of heavy forest cover.

The Wood Duck is a vocal species, announcing its arrival and tending its young with distinctive calls. These calls and its beautiful plumage allowed Atlas observers to identify it readily. Most "confirmed" records were of broods. Nests were found only infrequently.

The Wood Duck is the only dabbling duck to nest in a hole. It normally chooses natural cavities or abandoned Pileated Woodpecker holes in large branches, tree trunks, or stumps over or near water (Palmer 1976b). Bull (1974), however, reported an example of a stick nest, Carleton (1971) a possible cliff nest, and Bent (1923) mentioned nests in barns. These are extreme departures from the norm. The Wood Duck does not bring any material to the cavity or hole, using its own down plus whatever materials it finds in the hole to build the nest (Kortright 1942).

The height of the nest may vary from almost ground level to 18.3 m (60 ft). Although there are accounts of the young being carried from the nest hole to the water one at a time, either in the female's bill or on her back, the fledgling usually makes the trip on its own, simply by stepping out into space (Kortright 1942). The one time I observed this phenomenon, a brood of 10 fluttered and fell from a nest hole about 6.1 m (20 ft) high. The young seemed none the worse for the experience, and the female subsequently shepherded them to a pond about 12.2 m (40 ft) away.

Emanuel Levine

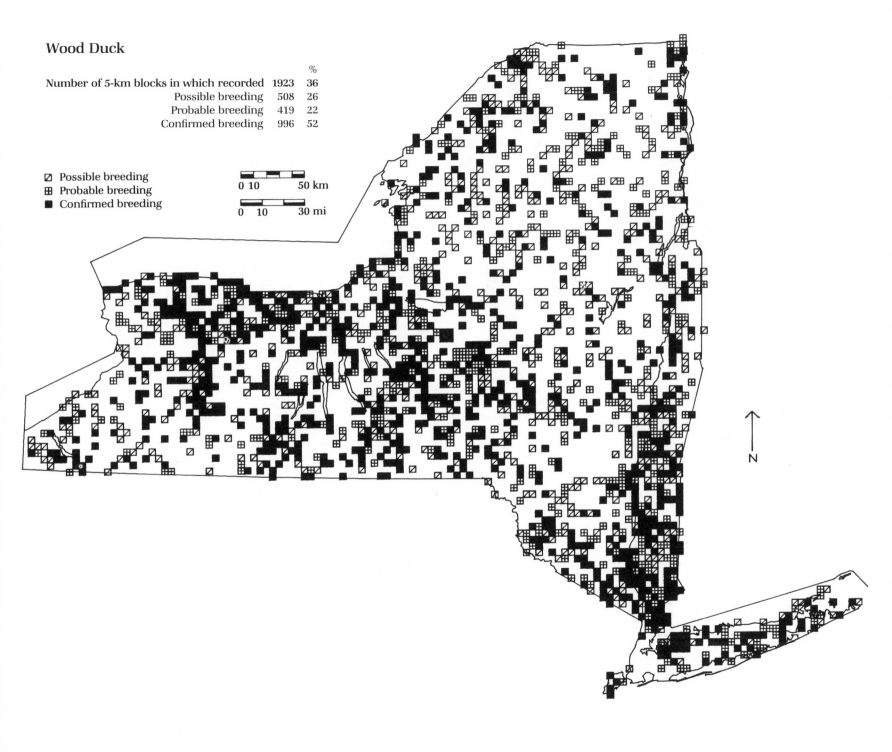

Wood Duck

Number of 5-km blocks in which recorded		%
	1923	36
Possible breeding	508	26
Probable breeding	419	22
Confirmed breeding	996	52

◩ Possible breeding
⊞ Probable breeding
■ Confirmed breeding

0 10   50 km

0   10   30 mi

N

CJPage
©1987

Green-winged Teal *Anas crecca*

The Green-winged Teal, a handsome speedster, is an uncommon breeder in widely scattered localities throughout the state. The southeastern edge of its range, which during the first half of this century extended from north central and northwestern United States throughout most of Canada, apparently changed in the 1950s when the Atlantic Flyway population was increasing (Bellrose 1978). The range descriptions in the fifth edition of the AOU Check-list documented breeding as far south only as western New York, but in the sixth edition sporadic breeding was described as far south as Delaware (AOU 1957, 1983).

Two factors appear to have contributed to the expansion of this species' range. Early in this century spring hunting was prohibited, and the population grew dramatically (Cruickshank 1942). This increase may have led to competition for nest sites, forcing some individuals to move away from the main range. At the same time refuges created in New York and other states provided suitable nesting sites (Palmer 1976a).

The population now appears to have stablized. In a study of harvest and banding data to determine whether a suspected recent decline had occurred along the Atlantic Flyway, Reynolds (1985) found "no evidence of a decline in survival or recruitment." In New York the Green-winged Teal remains a sporadic breeder, typical of a species at the edge of its range.

Eaton (1910) reported only two breeding records in the state, one at Montezuma NWR and another at Strawberry Island in the Niagara River. Summer records of the Green-winged Teal in New York were scarce in the 1950s, and breeding was documented only at Howland Island WMA in

1957 (Scheider 1957) and Stephens Pond, Oswego County, in 1958 (Scheider 1958). Breeding was suspected at Montezuma in 1959, but the first brood was not reported until 1962 (Hoyt 1962). Broods were first reported on Long Island in 1960 at Easthampton, Nassau County (Bull 1964), and at Jamaica Bay Wildlife Refuge in 1961 (Elliott 1961).

During the 1960s summer sightings continued, but breeding was consistent only at Montezuma, Howland Island, and Jamaica Bay, with single year reports at Coxsackie, Greene County, in 1962 (Wickham 1962); at Amherst, Erie County, in 1965; at Riverside Marsh, Chautauqua County, in 1967 (Sundell 1967); and at Cuba Marsh, Allegany County, in 1969 (Sundell 1969). Broods were observed throughout 1970 and the early 1980s at Montezuma, but breeding was not documented after 1972 at Howland Island or at Jamaica Bay after the early 1970s (Riepe, pers. comm.). During this period summer sightings continued to be sporadic: a brood reported at Rome Marsh, Oneida County (Rusk and Scheider 1970), provided the only record in *The Kingbird* other than at Montezuma. Bull (1974) listed 22 breeding localities statewide.

The Atlas map documents "confirmed" breeding in 32 blocks. The records from new upstate sites in New York probably reflect the intensity of Atlas coverage. The records from Long Island, particularly along the eastern tip, are surprising, even though there was at least one record of a summering bird from Gardiners Island in 1967 (Davis and Heath 1967). Many of the "possible" and "probable" records may be of nonbreeding summering birds.

The Green-winged Teal nests in upland locations in dense stands of grass, weeds, and brush. A ground nester, it places its nest near the edges of ponds and lakes, or at some distance from the water in dense grass or at the base of a shrub (Kortright 1942). Like most dabblers, this teal nests in a hollowed-out area containing some vegetation. The nest is concealed by surrounding vegetation that sometimes forms a canopy. After laying has begun, down is added to the nest and continues to be added throughout incubation (Palmer 1976a).

Emanuel Levine

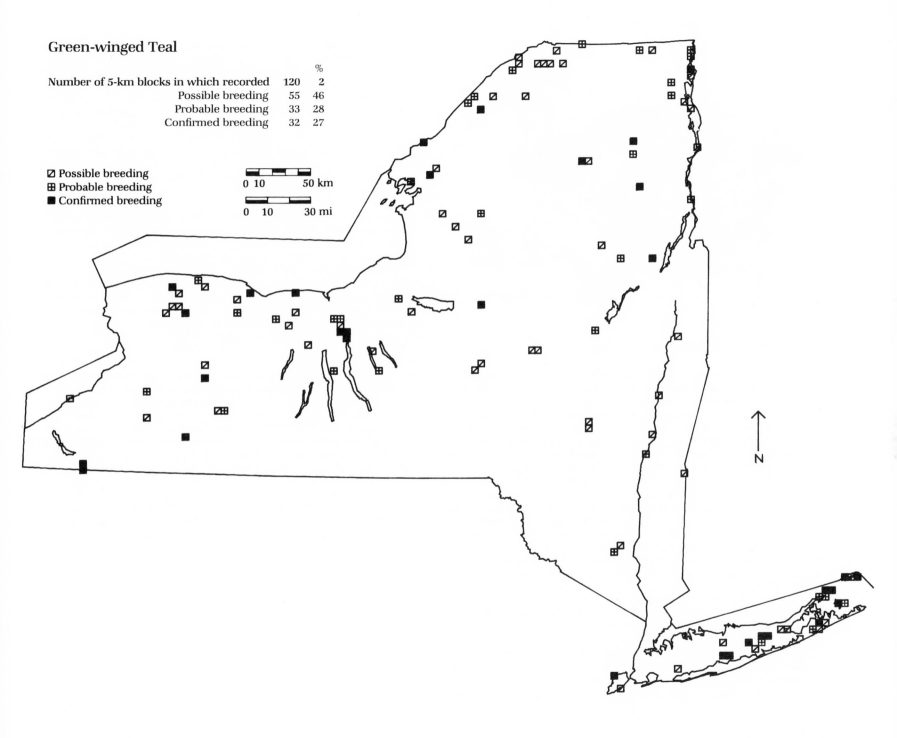

Green-winged Teal

		%
Number of 5-km blocks in which recorded	120	2
Possible breeding	55	46
Probable breeding	33	28
Confirmed breeding	32	27

☑ Possible breeding
⊞ Probable breeding
■ Confirmed breeding

0 10   50 km

0 10   30 mi

N

American Black Duck *Anas rubripes*

Long prized by duck hunters throughout the Northeast, the American Black Duck is common in wetland habitats mainly in the eastern part of the state. Throughout the remainder of New York this species is an uncommon breeder. Historically, it was the duck species that most commonly bred in New York (Eaton 1910). In the 1950s, however, the population began to decline (Benson 1966), and from the mid-1950s to early 1980 a decline of almost 60% was calculated (Grandy 1983).

Much of the black duck decline has been blamed on increasing Mallard populations (Smith and Ryan 1978). At Montezuma NWR breeding Mallards outnumbered American Black Ducks 4:1 in censuses taken in the early 1960s (Cowardin et al. 1967), yet the Mallard did not begin to breed there until the early 1900s (Benton 1949). Speiser (1982) stated that in the Hudson Highlands the black duck had "greatly decreased in numbers as a breeder since about 1975, as Mallards continue to prosper." There is some evidence that hybridization between Mallards and American Black Ducks is at least partially responsible for the latter's decline. "The primary zone of contact between Mallards and American Black Ducks has moved considerably eastward during the past half century, and current evidence indicates that hybridization between them will continue to increase" (Johnsgard 1975).

Habitat loss, pollution (from toxicants), excessive hunting, and acid rain have probably had a much greater effect than hybridization on American Black Duck populations (Moser, pers. comm). Their decline coincides closely with an increase in acid deposition in many lakes within the American Black Duck's range, including the Adirondacks.

There is growing evidence that this acidification is contributing to the decline of many waterfowl species (Hansen 1987).

On the positive side, the coastal salt-marsh population in New York has remained stable (Bull 1974). In addition, the beaver has created a wealth of excellent new black duck habitat in upstate New York in recent years.

The Atlas map documents breeding mainly in the Coastal Lowlands, in the Hudson Valley and to the east, and in the Adirondack region, with concentrations in the Central Adirondacks and the Western Adirondack Foothills. The American Black Duck is also found in the Catskill region, especially in the Neversink Highlands and the Mongaup Hills, in the many small lakes and ponds in southeastern Madison County, and along Lake Champlain. Breeding was localized in the marshes of western and central New York.

Although the black duck is more wary of humans than is the Mallard, Atlas coverage was very good because it is a conspicuous bird. Breeding is easily "confirmed" by observation of females with their young (83% of all "confirmed" records).

The American Black Duck breeds in a wide variety of habitats, ranging from coastal salt marshes and brackish tidal marshes to the freshwater marshes, bogs, lakes, ponds, and streams of the inland hemlock–northern hardwood forest (Stewart 1958). Specific habitat requirements are open water with thick vegetation such as reeds or shrubs somewhere along the water's edge (Saunders 1926).

The American Black Duck usually nests on the ground, placing its nest in dense vegetation along the edge of a stream, river, lake, or pond, or in the marsh vegetation no farther than 1.6 km (1 mi) from open water (Cowardin et al. 1967). It is often placed near a break or change in the vegetation; a path may lead from the nest to open water (Coulter and Miller 1968). Nests are occasionally found in elevated sites such as stumps, snags, crotches of trees, and artificial nest platforms (Cowardin et al. 1967), as well as old hawk and crow nests (Bent 1923; Bull 1974). In New York a pair nested in an abandoned Osprey nest at Orient Point in eastern Suffolk County, 6.1 m (20 ft) above the ground (Bull 1974).

Ground nests are generally shallow depressions kicked out by the hens in the soil or plant litter. Grasses, leaves, ferns, and down are all used to line the nest (Palmer 1976a). Down from the hen's breast and belly is placed in the nest after the first eggs are laid and continues to be added to the nest lining throughout incubation (Palmer 1976a).

Steven C. Sibley

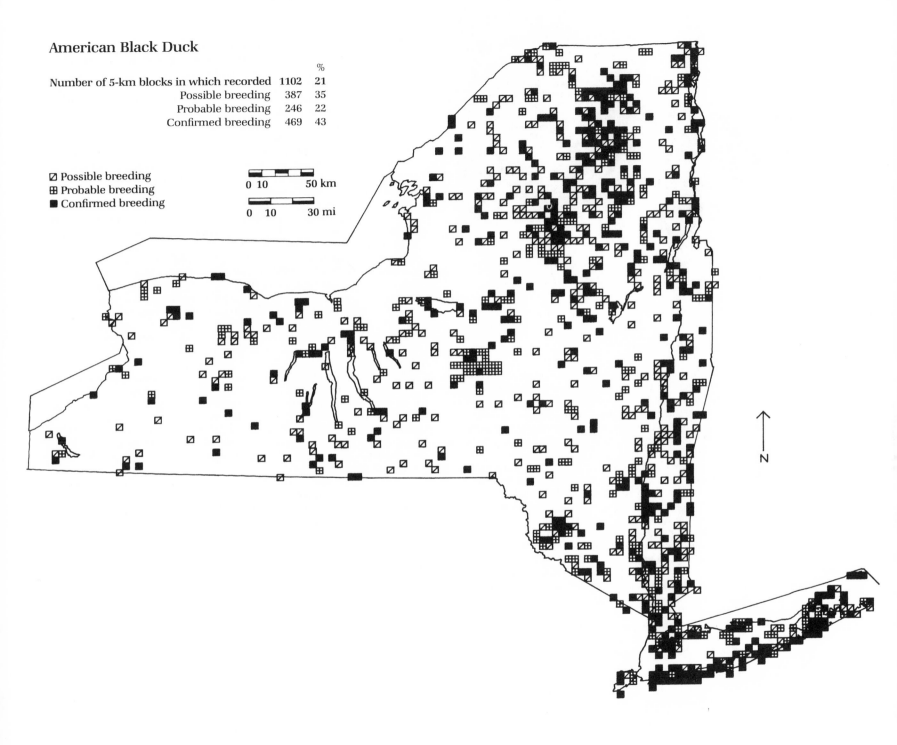

American Black Duck

Number of 5-km blocks in which recorded		%
	1102	21
Possible breeding	387	35
Probable breeding	246	22
Confirmed breeding	469	43

▨ Possible breeding
⊞ Probable breeding
■ Confirmed breeding

0 10   50 km

0 10   30 mi

N

CJ Page © 1987

Mallard *Anas platyrhynchos*

The Mallard, probably our best-known duck, is a common breeder in wetland habitats throughout the state. It was, however, virtually unknown as a breeder in New York as recently as 1900; the species has become common only in the past 30 years. Eaton (1910) stated that it bred "very rarely in the counties of Cayuga, Livingston, Monroe, Ontario, Orleans, Oswego, Seneca, and perhaps in Washington," but nowhere else in New York.

The Mallard spread eastward across the state during the early and mid-1900s. This spread was aided by the creation or improvement of almost 900 marshes by the New York State Conservation Department (Cook 1957). From 1952 to 1956 the department also released more than 20,000 pen-reared adult and duckling Mallards, many into these newly created marsh habitats (Foley 1954, 1956). In western New York and along the St. Lawrence River, breeding Mallards settled into the improved marshes at a greater rate than any other duck (Foley 1956).

Also during the early 1900s, semidomesticated Mallards regularly escaped from parks, sanctuaries, hunting clubs, and estates, and established feral breeding populations around New York City and on Long Island (Cruickshank 1942). On Staten Island the first confirmed nesting record occurred in 1950, and the Mallard is now the most common breeding duck there (Siebenheller 1981). This population increase is blamed, in part, for the decline of the American Black Duck over the past 50 years.

Atlas data show the centers of Mallard distribution to be in the Great Lakes Plain, Coastal Lowlands, and Hudson Valley. Atlas breeding rec-

ords are sparse above 305 m (1000 ft) in the forested Adirondack, on the Tug Hill Plateau, and in the Catskill regions, where less suitable habitat exists.

Atlas coverage was excellent for this species because of its conspicuousness and the ease with which breeding could be "confirmed" by locating hens with broods. Blocks where none were sighted, therefore, usually reflect a true absence of the Mallard and suitable wetland habitat.

Although the Mallard breeds in virtually all types of wetlands, it is mostly a bird of freshwater marshes. In the Coastal Lowlands, where freshwater marshes are few, it regularly nests in brackish tidal and salt marshes (Bull 1974). Other habitats used for breeding include rivers, streams, lakes, wooded swamps, and farm ponds (Bent 1923).

Semidomesticated and feral individuals still form a large percentage of the state's population, especially in urban and suburban shore-front parks. These sites often contain very high concentrations of Mallards, many of which are hybrids between wild Mallards, domesticated Muscovy Ducks, and domesticated forms of the Mallard (e.g., Pekin duck).

The Mallard usually nests on the ground, along or near wetland habitat where it feeds. The nest is commonly found in or under bushes, in open grassy or weedy areas, in cat-tails or in common reed (Bent 1923). It is placed where the ground is dry or only slightly marshy, or on higher land, especially islands, not far from water. Nests have been found, however, as far as 0.5 km (0.3 mi) from the nearest water (Bent 1923). Tall vegetation ranging from 15.2 to 127 cm (6–50 in) high often conceals the nest (Johnsgard 1975).

The Mallard will also nest in elevated sites such as stumps, dead snags, or in the crotches of living or dead trees (Cowardin et al. 1967). During a study at Montezuma NWR at the north end of Cayuga Lake, many Mallards were found nesting in elevated sites from 0.5 to 4.6 m (1.5–15 ft) above the ground, many of them in artificial wire mesh nesting platforms set out for the ducks in the crotches of trees (Cowardin et al. 1967). A nest in Bronx Park, New York City, was located in a hole 6.1 m (20 ft) high in a dead oak (Bull 1974).

Nests typically consist of a shallow depression which the hen forms by kicking her feet and rotating her body (Palmer 1976a). The nest is well lined with vegetation and down feathers (Bent 1923). Instead of gathering plants for the nest, the hen picks them from the immediate vicinity while she is sitting in the nest depression.

Steven C. Sibley

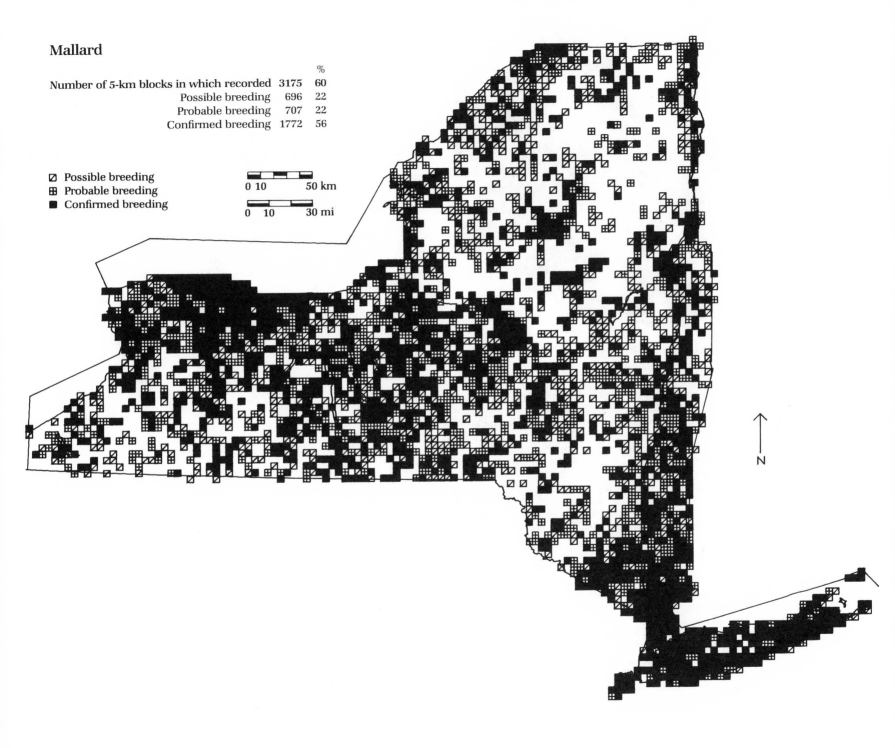

Mallard

		%
Number of 5-km blocks in which recorded	3175	60
Possible breeding	696	22
Probable breeding	707	22
Confirmed breeding	1772	56

◩ Possible breeding
⊞ Probable breeding
■ Confirmed breeding

0 10   50 km

0 10   30 mi

N

Mallard × American Black Duck
Anas platyrhynchos × rubripes

The Mallard hybridizes with at least 40 duck species (Heusmann 1974). In New York it most often interbreeds with the American Black Duck. Unlike many hybrids, Mallard × American Black Duck offspring are fertile, as are their backcrosses (Heusmann 1974). Although little is known about the abundance of these hybrids in their breeding range, Johnsgard (1967) estimated that hybrids made up 1.8% of the total Atlantic Flyway population of the Mallard, the American Black Duck, and their hybrids. By the 1970s estimates were much higher (Heusmann 1974).

The Mallard became established as a breeder in New York in the mid-1900s. Until that time the American Black Duck was the most common duck species in the state. As the Mallard increased, the American Black Duck declined. A similar decline occurred in the eastern United States, where by the late 1950s the Mallard outnumbered the American Black Duck 6:1 (Johnsgard 1961). The first record of a Mallard × American Black Duck hybrid on the state waterfowl count was in 1960 (Jones 1980). As the range of both species became more and more sympatric—particularly the winter range, where most pair formation occurs—hybridization increased (Johnsgard 1961).

A large proportion of New York's American Black Ducks winters along the East Coast from New York to Chesapeake Bay and south (Bystrak, pers. comm.). Inland they are found in fewer numbers in the agricultural areas of central and western New York where both Mallards and Black Ducks feed on waste grain in farm fields (Moser, pers. comm.). The Mallard is most commonly found inland. It would seem, then, that most hybrid pair formation occurs at inland locations.

Many ponds in urban areas in the state support mixed Mallard and American Black Duck flocks, and duck feeding by local residents encourages the birds to stay year round. This is especially true from southern Westchester County to the tip of Long Island, where, according to Atlas data, the greatest concentration of hybrids is found. Other locations on the map also reflect this phenomenon. For example, Andrle (pers. comm.) reports hybrids in ponds in Buffalo at Forest Lawn Cemetery and Delaware Park Zoo, where duck feeding is common and the water is open in the winter. Some of the Atlas records, however, are undoubtedly of birds that paired in winter ranges and then moved away to breed. In many cases ducks return to the natal areas of the hens to breed. Whether this is true in the case of hybrids is unknown.

It is very likely that Mallard × American Black Duck hybrids were under-reported. Not only is identification difficult, but a second-generation hybrid resulting from a cross between a hybrid and a typical Mallard or American Black Duck shows only slight deviation from the pure species (Johnsgard 1961).

Although the Mallard prefers open-marsh habitat for breeding and the American Black Duck prefers forests and large bodies of water, intermediate situations provide breeding grounds for both.

Janet R. Carroll

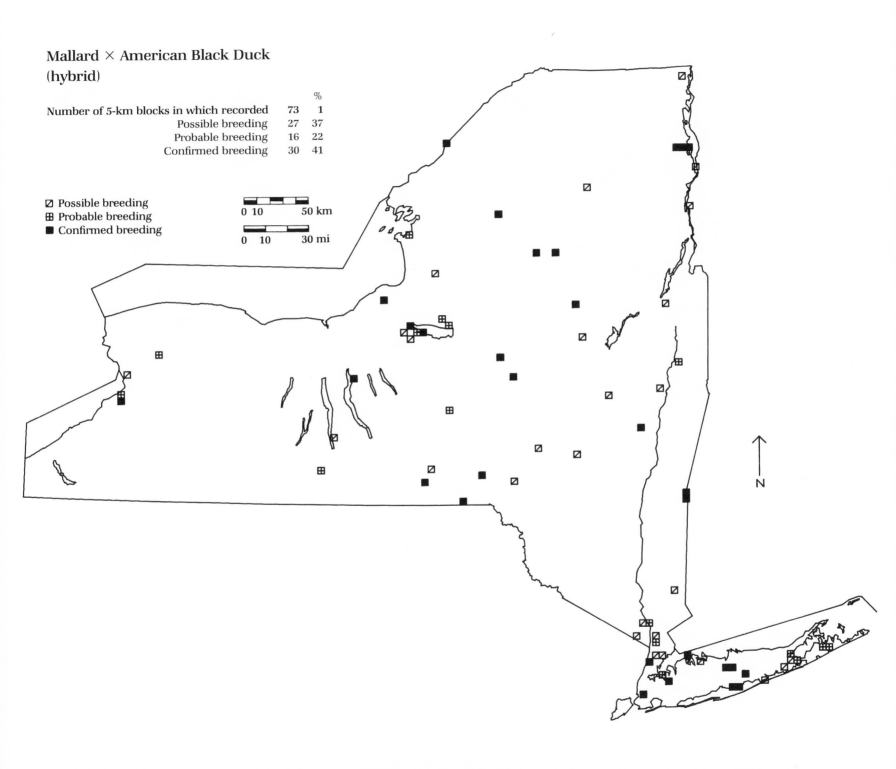

Mallard × American Black Duck
(hybrid)

		%
Number of 5-km blocks in which recorded	73	1
Possible breeding	27	37
Probable breeding	16	22
Confirmed breeding	30	41

☑ Possible breeding
⊞ Probable breeding
■ Confirmed breeding

0 10   50 km

0 10   30 mi

N

C.Page ©1987

Northern Pintail *Anas acuta*

The Northern Pintail, a slender, graceful dabbling duck, is one of the more widely distributed species of waterfowl in North America. A relatively recent breeder in New York, it is presently rare and local. De Kay (1844) did not consider the Northern Pintail a nesting species in the state, and Eaton (1910) indicated that it bred only in Orleans County, probably at Oak Orchard Swamp (Beardslee and Mitchell 1965). Although it has nested since 1934 at Pymatuning State Park in Pennsylvania, near New York's southwestern border (Todd 1940), and also in 1938 at Mohawk Island in the Canadian waters of Lake Ontario (Beardslee and Mitchell 1965), a New York nesting was not documented until 1945 when a nest was found at Perch River WMA (Kutz and Allen 1946).

Since 1947 the Northern Pintail has been known to breed on Little Galloo Island, Jefferson County (Bull 1974), although it was not "confirmed" there during Atlas fieldwork. In 1959 the New York State Conservation Department released the Northern Pintail at Wilson Hill WMA, Montezuma NWR, and Howland Island WMA. Broods of young were raised successfully that year at all three locations (Bull 1974). In 1967, 3 broods were reported at Oak Orchard WMA. The only coastal record came from Jamaica Bay Wildlife Refuge where a pair with downy young was reported in 1962 (Bull 1974).

The Northern Pintail has been "confirmed" in only three blocks, all in upstate areas: Chazy Landing on Lake Champlain, Clinton County (Peterson 1984b); St. Regis Lakes area, Franklin County; and Montezuma. Pairs were observed in the large wetlands at Rome WMA, Lake Alice WMA, and along Lake Champlain, north of Goose Bay on the St. Lawrence River, Jefferson County; and on Long Island at Hempstead Lake.

The Northern Pintail is fairly easily identified. Although the map reflects a localized distribution pattern, it may have been missed in large marshes where habitat was not readily accessible.

Local populations of the Northern Pintail can fluctuate dramatically. When water levels are low and drought occurs, emigration or nonbreeding is common; when water levels are high and flooding occurs, this duck is pioneering and high numbers are reported (Derrickson 1978). Krapu suggested (1974) that its ability to occupy recently flooded areas in order to exploit the available invertebrate food supply is a major factor in its widespread distribution.

Where the Northern Pintail is an established breeder, it typically selects open habitat with many scattered bodies of water. The nest is located in grasslands, barrens, dry tundra, dry meadows near water, farmland with streams, or grain stubble (Palmer 1976a). In New York, Bull (1974) flushed a female from a nest of 10 eggs under some bushes near the shore of Little Galloo Island. Kutz and Allen (1946) found a nest on the Perch River Flat, Jefferson County, which they described as "dense to open swamp, as well as extensive marsh land and low lying pasture." The species is not colonial, but in optimum habitat several pairs may nest together.

The Northern Pintail uses a natural depression or makes one, and adds down during laying and incubation. Intermingled feathers are also present, and as incubation proceeds the bowl deepens and dry vegetation may be added to the down and feathers (Palmer 1976a). Both sexes nest at the age of 1 year; pair formation begins on the winter range so that many spring migrants are paired.

Stephen W. Eaton

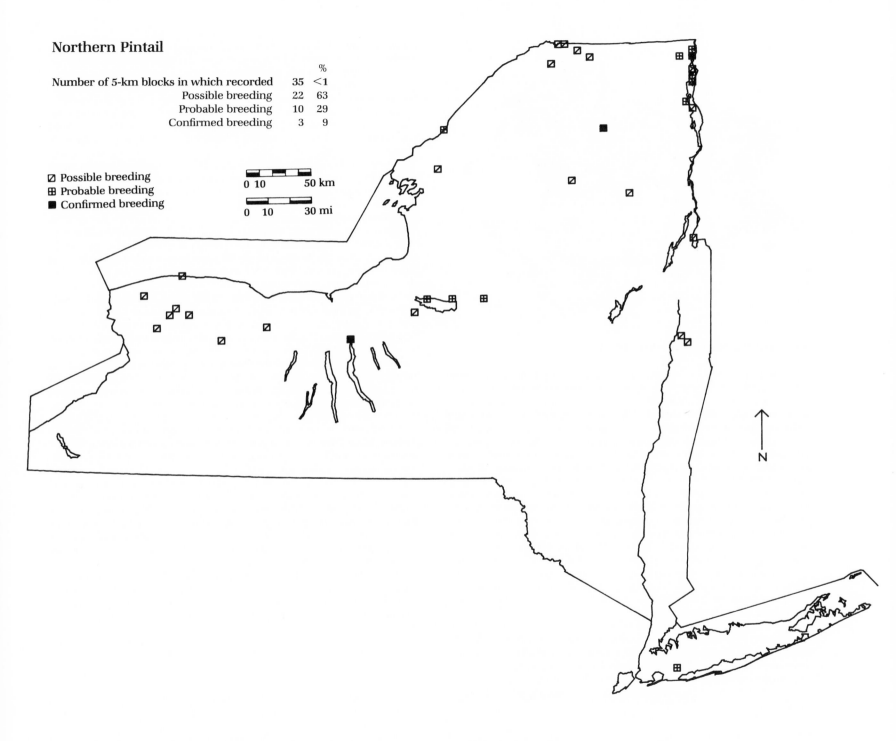

Northern Pintail

		%
Number of 5-km blocks in which recorded	35	<1
Possible breeding	22	63
Probable breeding	10	29
Confirmed breeding	3	9

☑ Possible breeding
⊞ Probable breeding
■ Confirmed breeding

0 10   50 km

0 10   30 mi

N

CJ Page
©1987

Blue-winged Teal *Anas discors*

The highly migratory Blue-winged Teal is a fairly common but local breeder in freshwater marshes throughout the state and a very uncommon breeder in salt and brackish tidal marshes on the coast. Populations of this small duck have changed little in this century in New York. Eaton (1910) said that it bred rarely on Long Island but more commonly in the Finger Lakes and other regions, a statement that still holds true today.

Populations have, however, disappeared from several historical nesting locations because of destruction of local marsh habitat. At Montezuma NWR there were over 40 breeding pairs of Blue-winged Teal in 1962 (Bull 1974), 15 in 1970, 25 in 1971, 2 in 1984 and 1985, and none in 1986 (Gingrich, pers. comm.). The Atlas project has documented these disappearances and also greatly increased our knowledge of this species' statewide distribution. The discovery of widespread breeding in the St. Lawrence Plains was especially important, as Bull (1974) listed no breeding records at all for that region. This discovery is probably a result of lack of knowledge of that region prior to the Atlas rather than a sudden increase in Blue-winged Teal populations. Blue-winged Teal were "confirmed" in the majority of the adjacent blocks in Ontario (Ontario Breeding Bird Atlas, preliminary map).

In New York the Blue-winged Teal was "confirmed" as a breeder in many parts of the state. However, suitable breeding habitat is very localized, and the species is virtually absent from the Catskills, Adirondacks, and higher elevations of the Appalachian Plateau.

Atlas data show most "confirmed" breeding in the large freshwater marshes of western New York and in the marshes adjoining the Great Lakes, St. Lawrence River, and Finger Lakes. Breeding was also "confirmed" in the Hudson, Mohawk, and Lake Champlain valleys and on the coasts of Long and Staten islands.

This distribution pattern roughly matches that mapped by Bull (1974) except for the discovery of breeding in many blocks in the St. Lawrence Plains. In addition, although breeding was "confirmed" in far fewer Long Island and Staten Island localities than cited by Bull, the Blue-winged Teal was recorded as a "possible" or "probable" breeder in virtually every location Bull listed.

The Blue-winged Teal prefers wetlands such as marshes and pond edges with protective vegetative cover (Palmer 1976a) and rarely ventures onto open water away from such cover (Johnsgard 1975). It also nests in salt and brackish water cordgrass meadows interspersed with tidal ponds and creeks (Stewart 1962).

This species was easily "confirmed" as a breeder by observing a female with a brood of young (86% of all "confirmed" records) but was not as easy to "confirm" as other duck species, such as Mallard, because of its more secretive nature. The fairly high number of "possible" and "probable" breeding records compared with "confirmed" records reflects this bird's habit of breeding in dense herbage about marshes and ponds rather than in areas of open water. Also, it usually conceals its nest better than other duck species. It is unlikely, however, that the Blue-winged Teal was missed in blocks where it was present because it can often be seen flying over the marshes during morning and evening hours.

The Blue-winged Teal usually nests on dry ground in dense vegetation within 91.4 m (300 ft) of water (Glover 1956). Nest sites include meadows and pastures, dry sedge, hay fields, and cat-tails (Palmer 1976a), often along the edges of paths, roads, or railroad tracks (Bent 1923). The nest is usually built amidst lush plant growth 20.3–61 cm (8–24 in) high (Johnsgard 1975), with the vegetation forming a concealing canopy over the nest (Palmer 1976a). There are apparently no records of Blue-winged Teal nesting in elevated sites such as trees. However, they have nested both in and on muskrat houses (Bent 1923).

The nest is a shallow depression in the ground lined with grasses and down. The female may scrape out several nest depressions before selecting the one it will use (Palmer 1976a). Nest depressions average 20.3 cm (8 in) in diameter; the bowl of the nest, inside the lining, measures about 12.7 cm (5 in) across and 5.1 cm (2 in) deep (Glover 1956). A small amount of dry grasses and other vegetation surrounding the nest is used initially to line it, with down feathers added later. Large amounts of down from the hen's breast and belly are used to line the sides and rim of the nest but are not added until after egg laying has begun, usually after at least 4 of the eggs have been laid (Glover 1956). Down is used to help insulate the eggs and occasionally to conceal them when the female leaves the nest (Bent 1923).

Steven C. Sibley

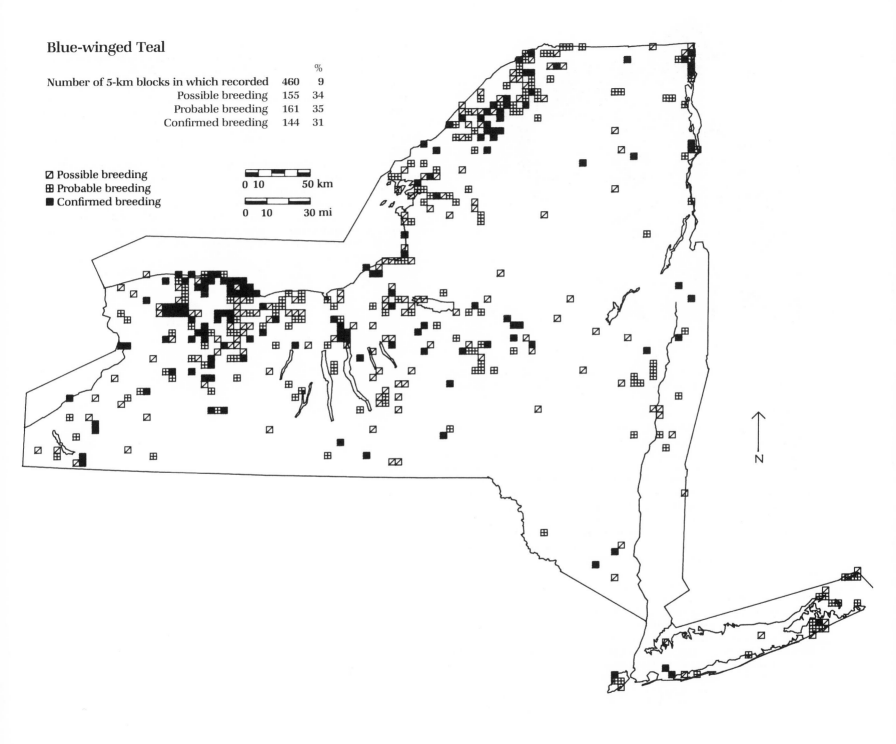

Blue-winged Teal

		%
Number of 5-km blocks in which recorded	460	9
Possible breeding	155	34
Probable breeding	161	35
Confirmed breeding	144	31

◨ Possible breeding
⊞ Probable breeding
■ Confirmed breeding

0 10   50 km

0 10   30 mi

N

CJ Page
© 1987

marshy areas with shallow waterways and abundant vegetation which surround dry meadows for nesting. In upstate New York at Oak Orchard Swamp it nests in grasslands (D. Carroll, pers. comm.), and at Montezuma it favors grassy dike margins (Gingrich, pers. comm.) adjacent to cat-tail–bulrush marshes and sedge meadows. At Jamaica Bay it nests beneath upland shrubs and in grass thickets (Riepe, pers. comm.). The Northern Shoveler feeds in the open water of the marsh using its spatulate bill to filter feed in the areas just below and at the surface of the water (Palmer 1976a).

Preferred nest sites are close to water, in low grass. There may be little cover early in the season. The duck shapes a nest bowl, lines it with vegetation gathered near the nest site, and lays a clutch at the rate of 1 egg a day. During egg laying it adds nest down. Feathers are interspersed with down, a characteristic that can help an observer distinguish this species' nests from those of related species. During incubation the hen usually feeds twice a day, covering her eggs carefully before she leaves the nest (Palmer 1976a).

Stephen W. Eaton

Northern Shoveler *Anas clypeata*

The Northern Shoveler, a strikingly colored and specialized puddle duck, is a rare breeder in the state. With its uniquely adapted bill, it moves in a circular pattern, skimming the water for seeds or small insects (Palmer 1976a).

Eaton (1910) reported that the Northern Shoveler first nested in New York at Montezuma NWR early in the century. It continued to nest there at irregular intervals until 1982 (Gingrich, pers. comm.); as many as 8 broods were raised there in 1962. In 1931 nesting was first recorded at Oak Orchard Swamp when a hen with young was observed, and the following year a nest was found (Beardslee and Mitchell 1965). Bull (1974) reported nesting at Howland Island WMA in 1961, Wilson Hill WMA in 1959, Jamaica Bay Wildlife Refuge in 1956, and Tobay Pond, Nassau County, in 1958.

The majority of Atlas records were from wildlife management and refuge areas as well. Large, protected marshes seem to attract these waterfowl. Fledglings were observed not only at the historical locations at Montezuma and Jamaica Bay but also at Lawrence Marsh, Nassau County. Pairs were seen at Oak Orchard WMA; Braddock Bay WMA; Galen Marsh WMA, Wayne County; Pralls Creek Marsh, Staten Island; and Kings Bay and Montys Bay WMAs in Lake Champlain. The Northern Shoveler was reported in fewer blocks than any of the other puddle ducks.

All New York records represent eastern outlying areas away from the Northern Shoveler's main range in the prairie pothole country of the northwestern and north central United States and Canada (Palmer 1976a). Similar casual records occur in Ohio, Pennsylvania, and Delaware (AOU 1983). In its main North American breeding range it frequents open

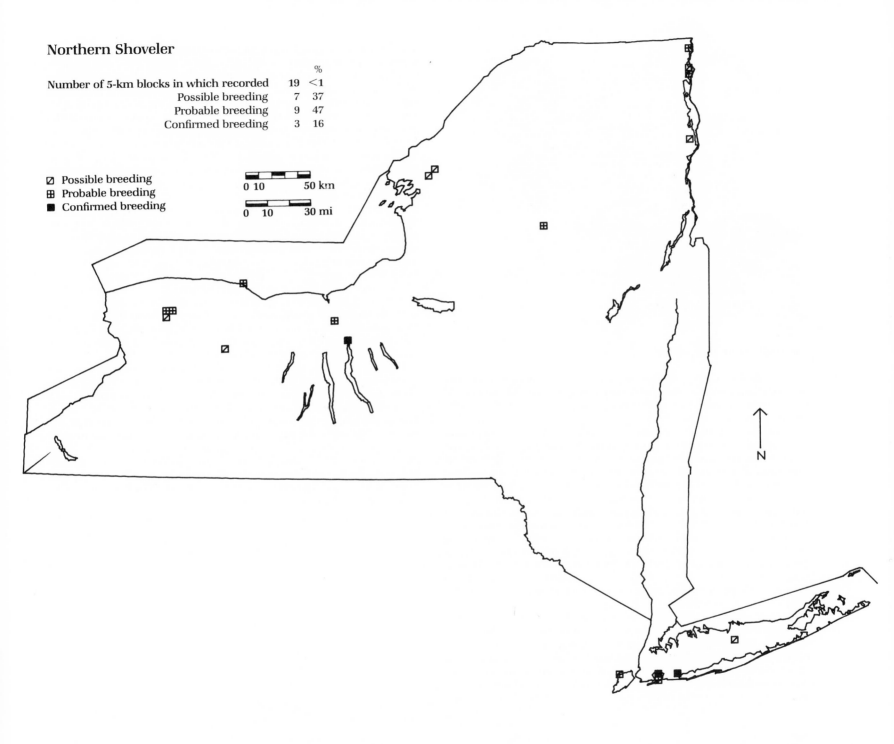

Northern Shoveler

		%
Number of 5-km blocks in which recorded	19	<1
Possible breeding	7	37
Probable breeding	9	47
Confirmed breeding	3	16

◩ Possible breeding
⊞ Probable breeding
■ Confirmed breeding

0 10   50 km

0 10   30 mi

N

C.J. Page ©1987

Gadwall *Anas strepera*

The Gadwall, a swift-flying gray duck, has recently become a common breeder along the south shore of Long Island and a regular but uncommon breeder in and near wildlife management areas and national wildlife refuges in upstate New York.

De Kay (1844) stated that the Gadwall bred in New York; however, Eaton (1910) was unable to find evidence that it bred nearer to New York than the Gulf of St. Lawrence. By 1934 it was breeding at Pymatuning State Park just south of New York in western Pennsylvania (Todd 1940). Breeding was first documented in New York in 1947 (Sedwitz et al. 1948) at Tobay Pond near Jones Beach State Park, Long Island; the species had apparently spread from other recently established localities along the Atlantic Coast. In 1950, 85 adults and 118 young were reported at Tobay Pond; numbers have fluctuated in subsequent years (Bull 1964). Bull noted that the species has decreased at Tobay since 1952 but mentioned that in 1961, 3 pairs nested at Jamaica Bay Wildlife Refuge. It still nested there in 1986 (Riepe, pers. comm.). By 1974 the Gadwall had been documented as breeding at eight localities along the south shore of Long Island (Bull 1974).

Upstate, it was first recorded nesting at Montezuma NWR in 1950 (Parkes 1952a). It has nested there continuously since, producing 153 young in 1980 and from 1981 to 1986 an annual average of 27 young (Gingrich, pers. comm.). The Gadwall was first observed nesting at Oak Orchard and Tonawanda WMAs in 1971 (Bull 1974), and from 1980 to 1985 an average of 26 young were produced annually (D. Carroll, pers. comm.). Bull (1976) first reported it nesting along the St. Lawrence River at Wilson Hill WMA in 1970, and Crowell and Smith (1985) said that it was "widely distributed as a breeder" there.

Atlas workers found that the Gadwall had increased its range along the south shore of Long Island and still nested at Montezuma and at Oak Orchard and Tonawanda WMAs. In addition to the areas shown on Bull's (1974) map of breeding locations, the Gadwall was "confirmed" at Wilson Hill and other marshes along the St. Lawrence River; on Stony Island in Lake Ontario; at Chazy Landing, Lake Champlain; and in a few other blocks scattered over the rest of the state. The areas along the St. Lawrence River where this duck was "confirmed" are across from a population located by Atlas workers in Ontario (Ontario Breeding Bird Atlas, preliminary map).

The Gadwall prefers open areas and is found in lacustrine communities such as ponds, cat-tail–bulrush marsh, sedge meadow, managed waterfowl wetland; in estuarine communities such as salt marsh, freshwater tidal marsh (lower Hudson River); and in terrestrial communities including maritime dune system (sand and beachgrass) and maritime heathland (bayberry).

It often nests in brackish or alkaline locations. Cooch (1964) found that waterbirds that inhabit areas where water alkalinity is high possess a well-developed salt gland. Like a similar gland in marine birds, it may help maintain the proper osmotic balance of body fluids by excreting salt. This may explain in part the Gadwall's penchant for the brackish waters of the coastal marshes of Long Island.

Located on islands and in meadows, nests are well concealed in thick grasses or tall reeds (Harrison 1975). At Jamaica Bay Wildlife Refuge the Gadwall nests in upland shrubs and grassy thickets (Riepe, pers. comm.); at Montezuma, along grassy dike edges (Gingrich, pers. comm.); and at Oak Orchard WMA in grasslands (D. Carroll, pers. comm.) The nest itself is merely a hollow in the substrate, lined with material gathered from the immediate vicinity, usually reeds, grasses, stems, and roots mixed with down from the bird's breast. As with most waterfowl, the Gadwall adds down during incubation (Harrison 1975). An average of 6–8 young were counted in broods at Jamaica Bay, Montezuma, and Oak Orchard (D. Carroll, Gingrich, Riepe, pers. comms.).

Stephen W. Eaton

Gadwall

Number of 5-km blocks in which recorded	114	% 2
Possible breeding	29	25
Probable breeding	38	33
Confirmed breeding	47	41

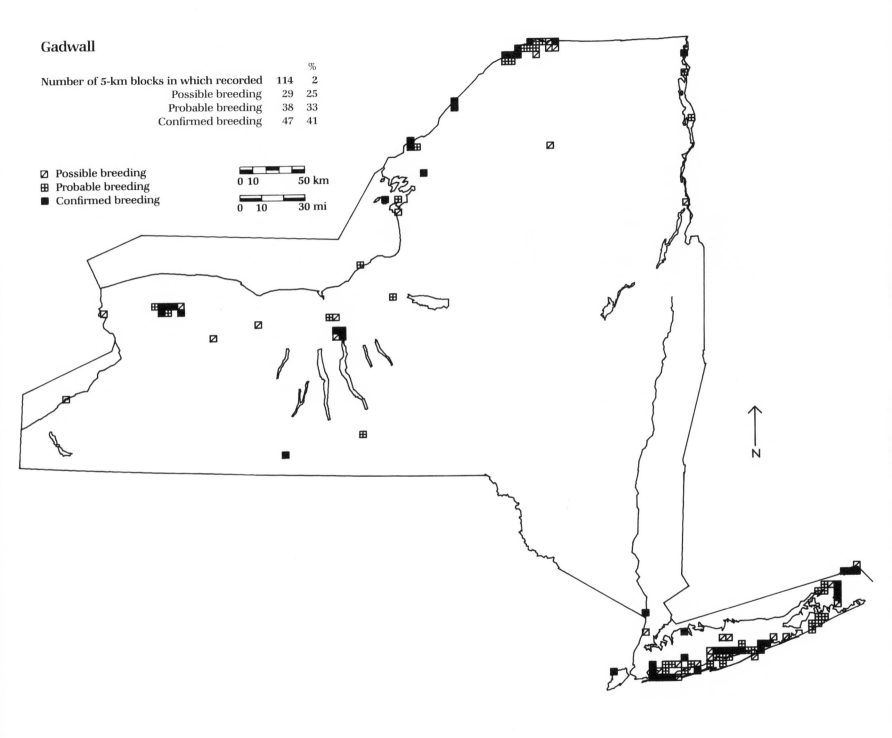

Ⓩ Possible breeding
⊞ Probable breeding
■ Confirmed breeding

0 10   50 km

0 10   30 mi

N

American Wigeon *Anas americana*

A delicacy of the table and a very uncommon breeder in the state, the American Wigeon has recently been expanding its range southeastward from Canada and is becoming established in western New York and along the St. Lawrence River. The most terrestrial of the genus *Anas*, the American Wigeon inhabits large marshes and grazes along their exposed shorelines.

In the early 1900s this species occurred only as a migrant in New York (Eaton 1910). Breeding in the Northeast was first documented in 1936 when it was found nesting at Pymatuning State Park in western Pennsylvania (Todd 1940). It first nested in New York in 1954 at Montezuma NWR (Bull 1974; Gingrich, pers. comm.). Additional nestings were later found at Wilson Hill WMA in 1959 (Lamendola, pers. comm.); at Oak Orchard WMA in 1961 but apparently not since (D. Carroll, pers. comm.), although pairs were observed there during the Atlas survey; Jamaica Bay Wildlife Refuge and Flushing Meadows, Queens County, both in 1961. Howland Island WMA first recorded this species nesting in 1968 (Bull 1974). In 1974 breeding was documented in Rockland County, which at the time was only the seventh breeding locality for the state and the first for the Hudson Valley (Bull 1976).

During the period of Atlas fieldwork, evidence of breeding was recorded at most historical sites except for Rockland County, Jamaica Bay, and Flushing Meadows, which have been changed considerably by development in recent years. Breeding was "confirmed" at Times Beach in Buffalo Harbor, Erie County, where 5 females were observed with at least 30 young (Andrle, pers. comm.); at the Perch River WMA; at Chazy Landing

along Lake Champlain in Clinton County; and at North Sea in the Great Peconic Bay, Suffolk County.

This increase in the number of breeding areas seems to be part of a southeastward extension of the American Wigeon's North American breeding range, which began in the early 1950s. Formerly, the eastern edge of the range extended from western Minnesota north through Manitoba to Hudson Bay and eastward into northern Ontario and Quebec along Hudson Bay. The current movement occurred from eastern Manitoba southward through Ontario into New York (Palmer 1976a). The American Wigeon was found on the Canadian side of the St. Lawrence River in several blocks by Ontario Atlas workers (Ontario Breeding Bird Atlas, preliminary map).

Atlasers observed broods in nine blocks and found only 1 nest. Most of the "probable" records were of pairs. Because it seems to be more secretive than other duck species, the American Wigeon is less easily located.

Little information is available on nest construction and nesting habitat in New York, but in other parts of North America the American Wigeon favors large lakes and marshes characterized by abundant submerged vegetation (DeGraaf et al. 1980) and waterways with exposed shoreline (Palmer 1976a) for feeding. Unlike the Mallard and the Northern Pintail, it is not usually associated with small, temporary ponds. The nest is usually built in a dry place and generally contains bits of grass, weed stems, and down. The American Wigeon's nest can be identified by the nest down, which has light centers with very conspicuous white tips and is intermingled with molted breast feathers (Palmer 1976a). This duck tends to nest away from other American Wigeons, a single pair inhabiting a marsh-bordered pond. The drake departs early, perhaps a week after incubation begins.

Stephen W. Eaton

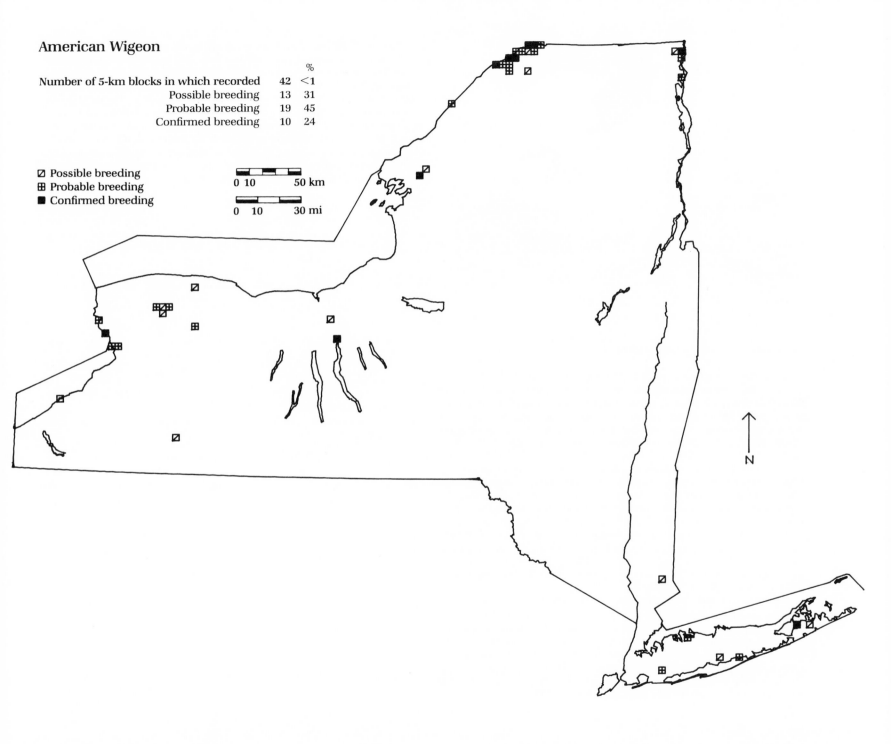

American Wigeon

		%
Number of 5-km blocks in which recorded	42	<1
Possible breeding	13	31
Probable breeding	19	45
Confirmed breeding	10	24

▨ Possible breeding
⊞ Probable breeding
■ Confirmed breeding

0 10   50 km

0 10   30 mi

N

Breeding habitats of the Canvasback typically consist of shallow prairie marshes surrounded by cat-tail, bulrush, and similar emergent vegetation. They must be permanent and large enough to have sufficient open water for easy landings and takeoffs. Such marshes also usually have an abundance of submerged aquatic vegetation such as pondweed, the most important group of food plants for the species (Johnsgard 1978).

The nest is built in bulrush, reeds, or cat-tail over shallow water 1–18.3 m (3–60 ft) from the edge of open water. Constructed of bulrush, reeds, sedge, and other water plants, it is often anchored to surrounding plants or placed on floating mats of dead plants and built up above high water, and lined with down (Bent 1923; Terres 1980).

Richard E. Bonney, Jr.
Judith L. Burrill

Canvasback *Aythya valisineria*

The Canvasback is a harbinger of winter, for it generally breeds in central and western Canada and visits New York only as a migrant or wintering bird. Eaton (1914) referred to this colorful duck with the ski-slope-shaped bill as a rare migrant or occasional winter visitor, and Bull (1974) stated that winter concentrations had increased greatly since the 1940s, making the species a common to locally very abundant winter visitant.

Nevertheless, there are some summer records. One bird was seen on Chautauqua Lake outlet on 19 and 21 June 1948 (Beardslee and Mitchell 1965). Two individuals were observed from 19 June to 18 July at the Jamaica Bay Wildlife Refuge, Queens County, in 1956, and a nonbreeding pair summered there in 1964 (Bull 1974). Even more interesting are a pair and brood of 6 young reported from Montezuma NWR in 1962 (Hoyt 1962). Breeding Canvasbacks also were reported, without details, from the same locality in 1965 (Bull 1974). Bull believed that these birds were of introduced stock and that there was no evidence that the species was an established breeder in New York.

But 1980 and 1981 brought still more breeding records from Montezuma. On 3 July 1980, the first year of the Atlas, a Canvasback brood of 5 young was discovered at the west side of the Main Pool in an area known as Black Lake, a shallow cat-tail marsh. The birds were estimated to be 5 weeks old. And on 7 July 1981 a brood of 9 1-week-old young was reported at the Main Pool. These represent the only "confirmed" records for the entire Atlas period. No Canvasbacks are known to have bred at Montezuma or anywhere else in New York since 1981.

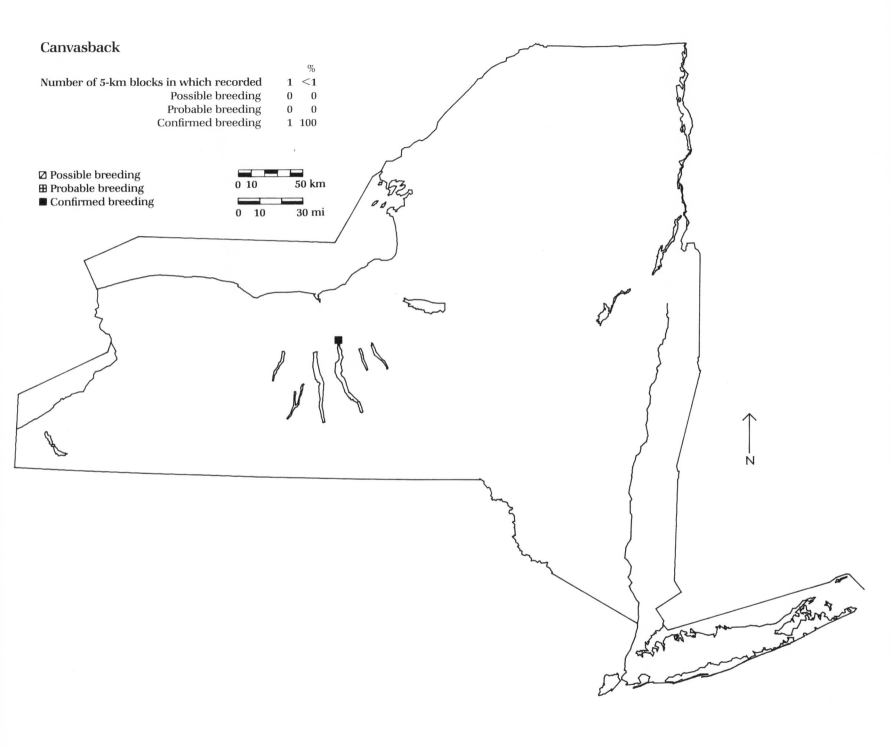

Canvasback

Number of 5-km blocks in which recorded		%
	1	<1
Possible breeding	0	0
Probable breeding	0	0
Confirmed breeding	1	100

◨ Possible breeding
⊞ Probable breeding
■ Confirmed breeding

0 10   50 km

0 10   30 mi

N

Redhead *Aythya americana*

The Redhead is a rare breeder in New York. It was introduced into the state by the NYSDEC in 1952 in order to increase the species diversity in the state's wetlands and provide additional hunting opportunity. Between 1952 and 1963 the NYSDEC released ducklings and adults at ten refuges and waterfowl preserves into large and small wetlands including marsh, swamp, open water, and bog habitats. Only Montezuma NWR, with its extensive areas of cat-tail and bulrush, and Jamaica Bay Wildlife Refuge, with its tidal marshes and open water, met the Redhead's needs. The initial goal was to establish breeding colonies; however, the attempt was not very successful. Today only a few remnant populations exist.

At Montezuma productivity has declined since 1967, when 9 broods were reported (Benson and Browne 1972). During the period from 1955 to 1963 the number of broods varied from 2 to 17. However, from 1970 to the mid-1980s the breeding population had only 1 or 2 broods a year. One of the factors responsible for this decline has been the crowding out of cat-tails by purple loosestrife (Benson and Browne 1972).

Atlas surveyors found evidence of breeding at six widely scattered localities. "Confirmed" breeding was observed at two Adirondack sites: Upper Chateaugay Lake, Clinton County (Gretch et al. 1981), and along the Saranac River, Franklin County (Dudones 1983). Breeding was also "confirmed" at Montezuma and at Jamaica Bay. Management personnel at Jamaica Bay reported a single brood in 1980 and another in 1983, but all young perished from predation.

The species was also recorded, though not "confirmed," in the vicinity of Wilson Hill WMA along the St. Lawrence River (Van Riet, pers. comm.), as well as at Ausable Marsh WMA, Clinton County (Lacombe, pers. comm.). Management personnel at Wilson Hill reported that the Redhead was not encountered during summer duck-banding operations; however, the species is occasionally observed there in August (Van Riet, pers. comm.). Although Redheads observed at this time are possible breeders, postbreeding dispersal from other areas, such as colonies in Quebec at Lake St. Francis, Lake St. Peter (Bentley 1981), or from more southern and western breeding areas cannot be dismissed. It is well documented that some young wander north and east in late summer (Weller 1964).

During the 1930s the Redhead's primary breeding area in central and western North America was hard hit by severe drought. As a result its breeding range shifted eastward. It was during this time that the appearance of what was suspected to be a natural brood was first noted at Montezuma in 1939 (Weller 1964). In the same year a lone male was seen on 9 July at Oak Orchard Swamp (Beardslee and Mitchell 1965).

The Redhead typically nests in emergent vegetation of large marshes; however, many nest sites have been reported over open water in dense vegetation, and occasionally on islands or over dry land. Nesting studies revealed that it has a strong preference for bulrush beds over other types of vegetation, with cat-tail and sedge its second and third choices, respectively (Bellrose 1978). The Redhead begins to nest late in the season. Benson and Browne (1972) stated that in captivity it did not start laying until the middle of May, peaking in early June. Gretch et al. (1981) calculated that the Redhead at Upper Chateaugay Lake began laying the first or second week in May. Similar egg dates have been obtained in the southeast part of the breeding range. It is difficult to determine normal clutch size for this species because hens may lay eggs in the nests of other Redheads and even in the nests of other species. Weller (1959) calculated that Redheads which parasitize the nests of other species lay an average of 10.8 eggs.

Mark Gretch

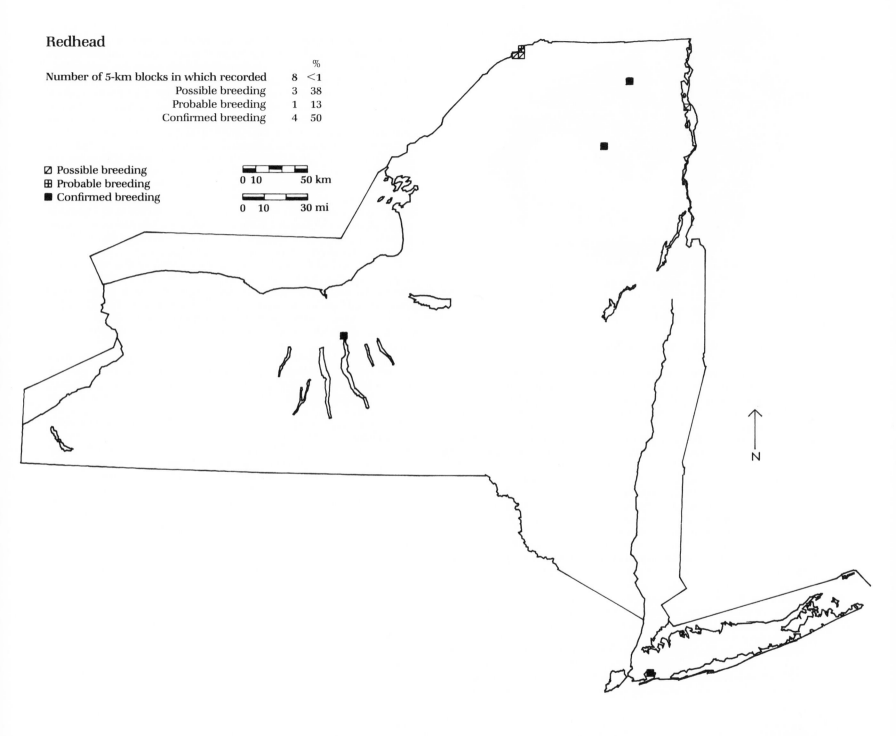

Redhead

		%
Number of 5-km blocks in which recorded	8	<1
Possible breeding	3	38
Probable breeding	1	13
Confirmed breeding	4	50

◨ Possible breeding
⊞ Probable breeding
■ Confirmed breeding

0 10   50 km

0 10   30 mi

N

C.J.Page
© 1987

Ring-necked Duck *Aythya collaris*

Even within southern Franklin County, where most Atlas records were documented, this northern diving duck is a rather uncommon breeder. Elsewhere in the Adirondacks and northern New York the species is locally rare to uncommon, and the population within the state may be experiencing a decline.

The Ring-necked Duck is a relatively new addition to the breeding birds of the state. It was not until 1946 that C. William Severinghaus of the New York Conservation Department spotted a brood on Jones Pond in southern Franklin County (Severinghaus and Benson 1947) for the first confirmed nesting record for the state.

A North American species, the Ring-necked Duck breeds from Alaska eastward and reaches its greatest density in the boreal forests of western Canada, nesting south into the northern Midwest. During the 1930s many ring-necks wandered or lingered east of this historical range, expanding into Quebec and east to Newfoundland, and south into Maine (Bellrose 1978). The Ring-necked Duck is more pioneering than most species (Mendall 1958), and the New York population might have been established by chance, perhaps even an overflow from earlier colonizations in Maine, Quebec, or elsewhere east of the traditional range. Yet if the Ring-necked Duck began an eastward expansion during the 1930s and became established in New York during the 1940s, it seemingly reached its population peak in the state during the 1960s. The Adirondacks were surveyed specifically for Ring-necked Duck broods from 1955, when 17 broods were counted, until this NYSDEC project was terminated in 1970 with a count of 29 broods, down from the high of 59 broods in 1966 and 1967 (Moser 1982).

By 1974 the Ring-necked Duck had been found breeding on over 20 bodies of water in six northern counties: Clinton, Essex, Franklin, Hamilton, Jefferson, and St. Lawrence (Bull 1974). Atlas observers "confirmed" breeding in 23 blocks, including the first records for Herkimer, Warren, and Washington counties. In 1984 a drake summered on Smith Pond on the Merwin Preserve, an area of small ponds and marshy wetlands in Columbia County (K. Dunham, v.r.); another extralimital Atlas record was obtained in Sullivan County. The Atlas map shows the Ring-necked Duck scattered over a somewhat wider area around the core population of southern Franklin County than Bull's (1974) map indicated, but observers did not record this duck near Perch Lake, Jefferson County, or on some of the Adirondack lakes where the species previously was known to nest.

Coverage was generally good and was enhanced by the efforts of the NYSDEC Adirondack Loon Survey teams, who covered the shorelines of almost all lakes and ponds during 1984 and 1985. The Atlas map can be considered a reasonable indication of the range in New York, though it remains incomplete.

The Ring-necked Duck nests in sedge meadows or bog lakes, where seasons of drought cause declines in brood production or force pairs to relocate (Moser 1982). Degradation of habitat on the Dodge Flow reduced brood production in 1962 (Moser 1982). Conversely, when new vegetative growth of aquatic plants occurred at Lake Alice WMA, Clinton County, after a barrier dam broke, a pair of Ring-necked Ducks bred there in 1984, followed by 2 breeding pairs in 1985 (Gretch, pers. comm.). Declines may also be linked to acid rain, or perhaps habitat may not be extensive enough to hold populations through poor seasons. Also, some lakes are increasingly subject to disturbance by boaters, fishermen, and campers, with a direct negative effect upon the Ring-necked Duck (Benson, pers. comm.).

This is among the least abundant of the North American diving ducks (an estimated 550,000 breeding birds), yet it is the most heavily harvested diver, 477,600 in the 1984–85 hunting season. Ingestion of lead shot and high crippling rates are also consequences of hunting. Meanwhile, wintering habitat is being lost through drainage and channelization throughout the Southeast and along the Atlantic Coast (25,000 acres of marsh each year in Louisiana alone). The national objective is to maintain the breeding population at or above 666,500 birds, and to sustain an annual harvest of 400,000 birds in the United States and 110,000 in Canada (Office of Migratory Bird Management 1985).

The Adirondack lake areas most extensively used by the Ring-necked Duck have some cat-tail and northern white cedar and edible seed- and tuber-producing aquatic plants (Benson, pers. comm.). Nests are built on islets of floating marsh plants, in clumps of vegetation, or on islands, usually in an association of leatherleaf, sedge, and sweet-gale. Egg laying and nest building are concurrent, the nest assuming a cup shape as eggs are laid, down then being added, and surrounding plants being woven into an overhead canopy. A single nest with eggs was located in the state during the Atlas period.

John M. C. Peterson

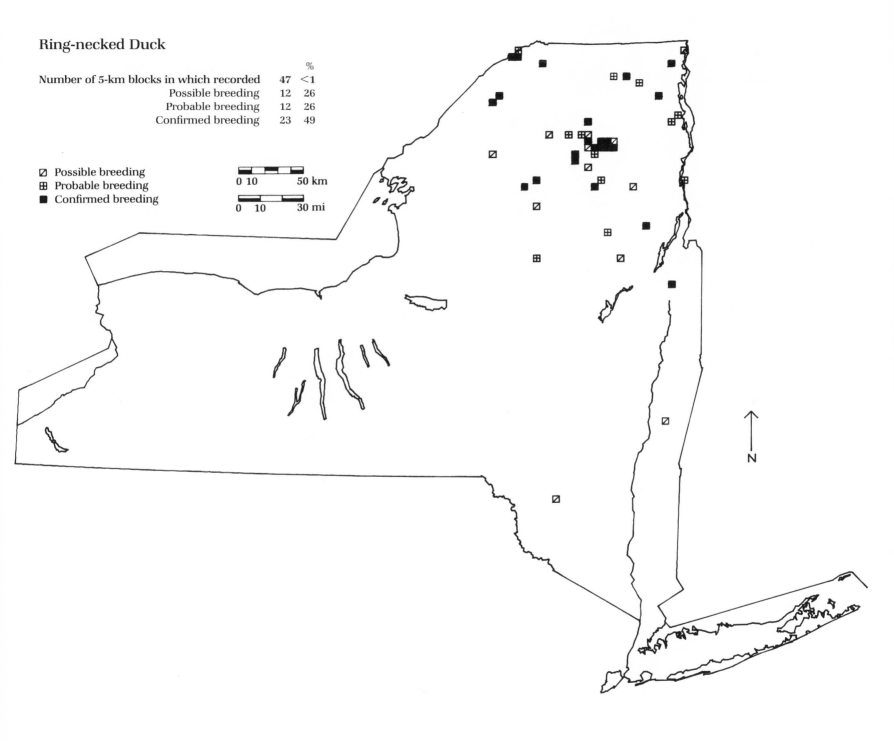

Ring-necked Duck

Number of 5-km blocks in which recorded		%
	47	<1
Possible breeding	12	26
Probable breeding	12	26
Confirmed breeding	23	49

◻ Possible breeding
⊞ Probable breeding
■ Confirmed breeding

0 10   50 km

0 10   30 mi

N

C.J.Page
©1987

Common Goldeneye *Bucephala clangula*

The Common Goldeneye, a hardy, cavity-nesting duck, is a rare to very uncommon breeding species in New York. Two discrete breeding populations exist in the state: one lies entirely in the Lake Champlain Valley, elevation 30.5–61 m (100–200 ft); the other straddles the Western Adirondack Foothills and the Central Adirondacks, elevation 305–762 m (1000–2500 ft). New York is along the southern periphery of the Common Goldeneye's primary breeding ground.

The earliest records, from 1878 and 1879, are of young found in Hamilton County during June (Eaton 1910). Although Merriam (1881) recorded it as a summer resident and breeder in the Adirondacks, Roosevelt and Minot (1929) did not report it in their surveys of 1874, 1875, and 1877; neither did Saunders (1929a) in the High Peaks of Essex County and the Lake Clear–St. Regis Lakes regions of Franklin County or Silloway (1923) at Cranberry Lake, St. Lawrence County. Except for the mention in Barnum's (1886) list of the birds of Onondaga County, the Common Goldeneye was not known to nest in the state outside of the Adirondack region. Evidently it was a rare breeder a century ago, as it is today.

The first record of a Common Goldeneye breeding on Lake Champlain was from Milton, Vermont, in 1928 (Ellison 1985b). More recently, Bull (1974) reported eight breeding localities; three on Lake Champlain and five in the Central Adirondacks.

As the Atlas map indicates, the Adirondack population is clustered primarily in the southern half of Franklin County, with the exception of reports from northern Herkimer County (Fischer Pond) and western Essex County. The Lake Champlain Valley population is confined to a narrow stretch along the lake in Clinton County and Wickham Marsh, Essex County.

The prospect of increasing marina development may threaten the Lake Champlain population. Another factor affecting this duck is lake acidification. According to a study done in Quebec, the Common Goldeneye "actually seems to prefer some acidic lakes where fish are totally absent," perhaps because of the lack of competition from predatory fish. Whether this competitive advantage will last as acidification increases has been questioned (Hansen 1987).

The two New York populations appear to have different habitat preferences. The Lake Champlain Valley population breeds in open, wooded areas adjacent to extensive marshes that provide cover for the young (pers. obs.). The Adirondack population prefers small, isolated ponds, particularly beaver ponds where northern hardwoods predominate (C. Delehanty, T. Mack, M. Milligan, v.r.).

Flocks of Common Goldeneye appear on Lake Champlain soon after the ice begins to break up in mid-March. The male initiates courtship displays. Pairing occurs thereafter in late March and early April. This duck uses bucket-shaped holes or cavities in trees for nest sites (Bellrose 1978). Because adequately sized cavities in trees always seem to be in short supply, the Common Goldeneye has benefited from the installation of artificial nest boxes. Bellrose (1978) reported that the species favors nest boxes placed 5.5–6.1 m (18–20 ft) above the water, but boxes placed about 1.5 m (5 ft) above the water are also frequently used. Lake shore locations are preferred to those more distant from the shore. The female adds down to the wood chips that have been placed in the box. Occupied nests often have down clinging to the entrance hole. Once incubation starts, the female usually covers the eggs with down when leaving the nest. The pale green color of the eggs is unlike that of the eggs of other cavity-nesting ducks. Drakes leave the nesting area soon after incubation begins (Bellrose 1978).

Mark Gretch

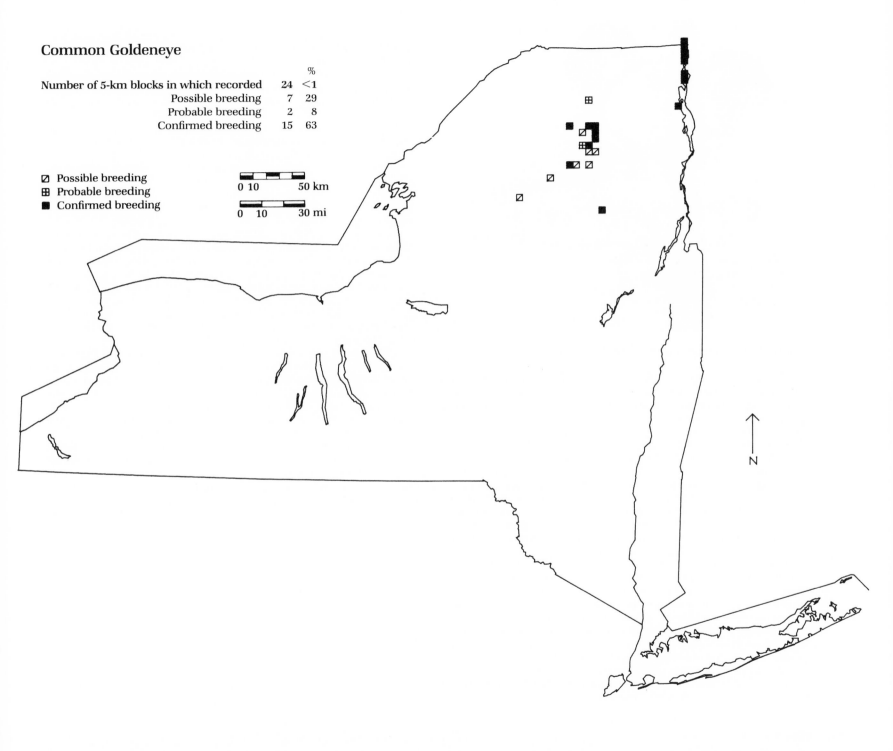

Common Goldeneye

Number of 5-km blocks in which recorded	24	% <1
Possible breeding	7	29
Probable breeding	2	8
Confirmed breeding	15	63

◩ Possible breeding
⊞ Probable breeding
■ Confirmed breeding

0 10   50 km

0 10   30 mi

N

C.J.Page
©1987

Hooded Merganser *Lophodytes cucullatus*

Smallest of the mergansers, the Hooded Merganser is a widespread but uncommon breeder in New York. It is curious that one of its rivals in beauty among North American ducks, the Wood Duck, is also sometimes a rival for a nesting cavity.

Eaton (1910) mentioned that the Hooded Merganser bred in scattered counties in western New York, as well as in the Adirondacks and the Catskills. He also stated that "it is one of the four species of wild ducks which breed to any extent in the State as a whole." Fifty years later, according to Foley (1960), there were still only four species that made up 95% of the breeding ducks in the state: the American Black Duck, the Mallard, the Wood Duck, and the Blue-winged Teal. The Hooded Merganser was included in the dozen species that made up the additional 5%. Bull (1974) showed a concentration of breeding records in the Adirondacks and none in the Catskills. The balance of the records are scattered throughout the state.

Palmer (1976b) pointed out that until the turn of the century the combination of fewer nesting trees and unrestricted hunting accounted for the scarcity of the species throughout its range, but increases have been noted since the 1930s. The Hooded Merganser, like the Wood Duck, has benefited from the reforestation of New York and also from the resurgence of the beaver. Browne (1975) summarized all known nesting occurrences in New York between 1941 and 1973. There were a total of 229, distributed over 26 counties, with the greatest number occurring in Franklin, Essex, and Genesee. A comparison of the Atlas map to Browne's map of the nest locations shows documented nestings in 17 additional counties during the Atlas period; active nests were unknown in these areas between 1941 and 1973.

Atlas work shows further concentration in the Adirondacks, branching out into neighboring Lake Champlain Valley, where breeding had gone unrecorded previously. A nesting in Westchester County (Treacy 1968) and one in Schoharie County (Riexinger et al. 1978), each a first for the county, were not repeated. Before 1980 no breeding had taken place on Long Island; however, in that year there was a successful nesting on Shelter Island, Suffolk County (Scheibel 1981).

The Hooded Merganser is a striking and familiar duck to most observers; however, since it nests in wooded swamps that are often inaccessible, it may have been missed in some blocks. More than half of the Atlas records are "confirmed"; most of these represent broods. Only 20 nests were found.

In addition to flooded woodlands, two other factors affect the breeding of this species: the amount of forest cover and stream quality. In areas of the state with heavy agricultural land use, few Hooded Mergansers breed. Johnsgard (1978) stressed the dependence of the species on "clear and unpolluted streams" that provide foraging areas for the small fish and invertebrates that are its main food during the breeding season. In New York such streams mainly occur in those forested areas that are best protected from the inroads of civilization. The dearth of records from the Catskills is therefore puzzling.

Although this duck nests primarily in cavities in trees and stumps, it will also nest in fallen logs and in holes in banks. Nests have been recorded as high as 30.5 m (100 ft) above the ground. The young descend shortly after hatching by stepping out into space and falling to the ground (Bent 1923). As in other cavity-nesting duck species, the female brings no nesting material with her into the hole but adds some down from her breast to whatever happens to be there (Kortright 1942).

The literature abounds with reports of competition between the Wood Duck and Hooded Merganser for nest sites, and there are a number of accounts of both ducks' laying eggs in the same nest at the same period. One account in Bent (1923) told of 30 Wood Duck eggs and 5 Hooded Merganser eggs placed several layers deep in 1 small nest.

Emanuel Levine

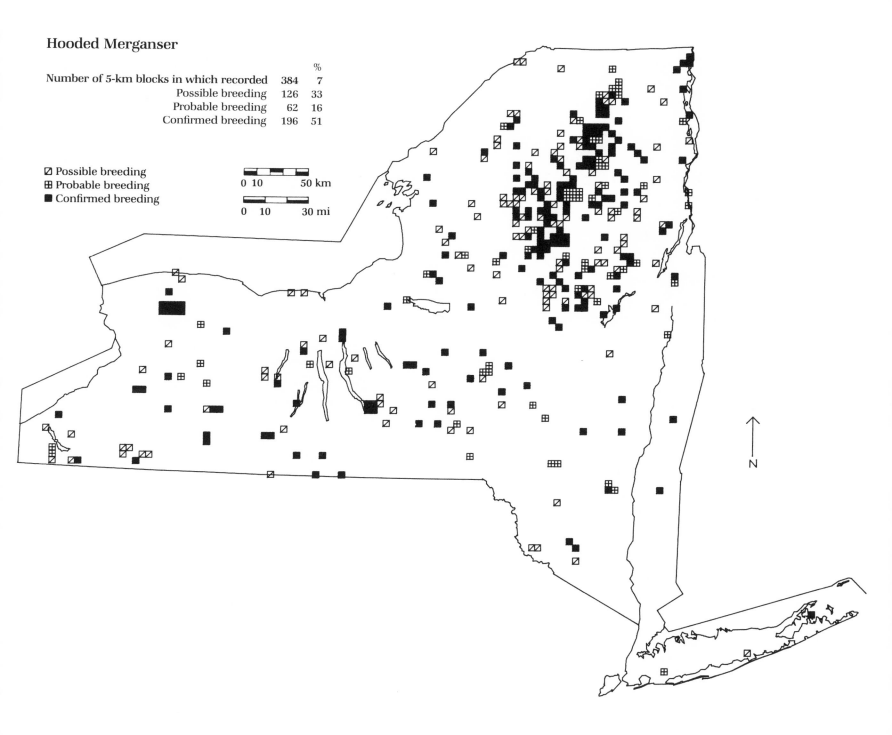

Hooded Merganser

Number of 5-km blocks in which recorded		%
	384	7
Possible breeding	126	33
Probable breeding	62	16
Confirmed breeding	196	51

◨ Possible breeding
⊞ Probable breeding
■ Confirmed breeding

0 10   50 km

0 10   30 mi

N

© 1987

Common Merganser *Mergus merganser*

The diving, fish-eating Common Merganser is one of 3 species of mergansers which nest in New York, and it is the largest of 18 species of ducks known to breed in the state. It is fairly common in northern New York, particularly in the Central Adirondacks, and uncommon and local elsewhere.

De Kay (1844) indicated that the Common Merganser bred from Pennsylvania northward into New York. Eaton (1910) reported that it nested near Buffalo in Erie County, at Montezuma NWR, and near Little Sodus Bay on Lake Ontario, as well as in the Adirondacks, where he said that summer visitors and fishermen disturbed this species to such an extent that it was fast disappearing from lakes and rivers. However, Bull (1974) still considered it one of the characteristic breeding birds of lakes in the forested areas of the Adirondacks.

Bull's map of the Common Merganser's breeding distribution showed the largest concentration in the Central Adirondacks, but it also plotted a restricted and more recent (since 1954) clustering of records along tributaries of the Delaware River, Delaware County. An isolated nesting in 1968 in southern Allegany County, adjacent to the Genesee River, was also shown. Bull (1974) listed 32 breeding records in the Adirondacks, 6 along the Delaware River, and the Allegany County record.

Atlas observers "confirmed" the species in 20 counties, most of which are in the Central Adirondacks; however, it was also "confirmed" along the Salmon River on the Tug Hill Plateau, the Schoharie and Esopus creeks in the Catskills, and the Delaware River and several of its tributaries in southeastern New York. Some of the records came from reser-

voirs around the state, including the Tomhannock in Rensselaer County, Gilboa in Schoharie County, and Ashokan in Ulster County. One female with young was seen in a tributary of Lake Erie near Silver Creek. In addition, broods were observed on Lake Champlain. The breeding distribution of this species is restricted by the availability of forested streams, lakes, and reservoirs.

Most "confirmed" records were of fledged young; only 20 nests were found. What appears to be an expansion of this species' range may in this case reflect the intensity of Atlas coverage and other survey work. Common Loon surveyors in the Adirondacks in 1984 and 1985 recorded observations of the Common Merganser and other waterfowl and provided these records to the Atlas project.

The Common Merganser usually nests near relatively cool, clear, medium-gradient streams, seldom in muddy or weed-choked waters. This duck needs good visibility to locate prey. Although the nest is usually placed in hollow trees, it is sometimes found in cavities in cliffs or on the ground. The species will use nest boxes (Palmer 1976b). The nest down is light colored in contrast to that of the Red-breasted Merganser, which is dark. Larson (1982) reported observing 5 fledgings leaving a chimney in Wellsville in Allegany County, where this species had apparently nested for 14 or 15 years. Most females and young spend the summer on lakes and rivers, gradually moving toward the river mouths as migration time approaches.

Ontario researchers who have studied the effects of acid rain on waterfowl indicate that because the Common Merganser lives "on waters that are both sensitive to acid rain and that are receiving levels of acid rain known to cause acidification," it is one of the species that may suffer serious declines in the future (Hansen 1987).

Stephen W. Eaton

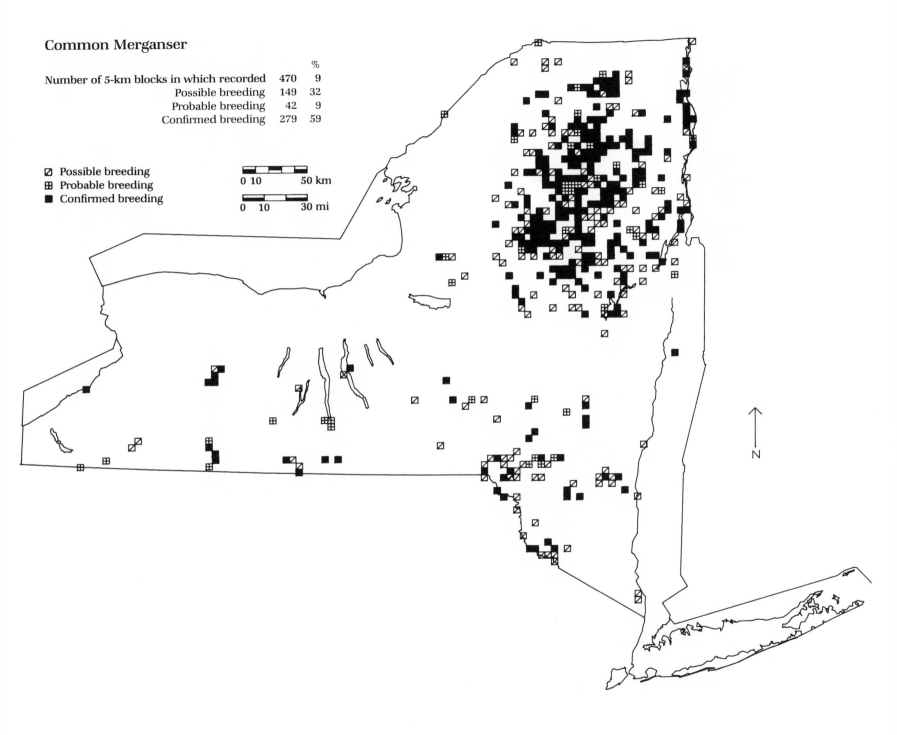

Common Merganser

		%
Number of 5-km blocks in which recorded	470	9
Possible breeding	149	32
Probable breeding	42	9
Confirmed breeding	279	59

◩ Possible breeding
⊞ Probable breeding
■ Confirmed breeding

0 10   50 km

0 10   30 mi

N

Red-breasted Merganser *Mergus serrator*

The Red-breasted Merganser, which feeds almost exclusively on fish, is a very rare breeder in the state. Although its breeding status has apparently changed very little in this century, breeding records have come from various locations, and there is a great deal of confusion over which records are reliable. Much of the confusion stems from the difficulty of distinguishing female Red-breasted Mergansers from Common Mergansers in the field. Bull (1974) questioned Eaton's (1910) statement that "a few are known to nest in the Adirondacks" and even went so far as to say that Bagg (1911) had completely confused the two species.

According to Bull (1974), nesting was verified in New York only on Little Galloo and Gull islands at the eastern end of Lake Ontario (Hyde 1939), and at seven Long Island localities, including islands at the eastern end and beaches on the south shore. Coulter and Miller (1968), however, banded a female Red-breasted Merganser at The Four Brothers in Lake Champlain in June 1957 and then found the same individual on a nest there in June 1958. This species was also "confirmed" as breeding on The Four Brothers during the Atlas survey. At least one pair may breed there every year (Peterson, pers. comm.).

Although there were Atlas sightings throughout the eastern end of Long Island, breeding was "confirmed" in only one block. This result is not unexpected; historically there has been about one confirmed breeding a decade on Long Island (Bull 1974). Also, although individuals of this species regularly linger through the summer on Oneida Lake and Lake Ontario (Smith and Ryan 1978), there was only one sighting from these two areas during the Atlas survey. This dearth of records is most likely a sign of the rare and sporadic occurrence of this species as a breeder at the southern edge of its range rather than any change in its abundance. The Long Island records are disjunct from the main range, which is principally to the north of New York in Canada (Bellrose 1978).

There were no Atlas records for Little Galloo and Gull islands in eastern Lake Ontario, although breeding was verified in 1936 on Little Galloo and Gull islands (Hyde 1939) and in 1967 on Little Galloo Island (Bull 1974). This species may have stopped breeding on these formerly infrequently visited islands. Today ornithologists regularly survey Little Galloo Island, and it is unlikely this merganser would have been overlooked.

The Atlas project also reported five other "possible" and "probable" breeding records at sites that Bull (1974) did not mention. "Possible" breeding records in St. Lawrence County and at two locations in Steuben County were of females sighted on rivers. Two "probable" breeding records at unusual locations were of a pair observed on the Hudson River in Greene County and pairs seen at two places on the north shore of Long Island in Nassau County.

Most Atlas sightings were of single adults, and all "confirmed" records were of females with broods of young on the water. No nests were found during the Atlas survey. Although this species is easy to detect in the open-water habitats it frequents, identification can be difficult, and all Atlas sightings were closely scrutinized.

The Red-breasted Merganser breeds in a variety of wetland habitats. Preferred areas are near the coast (Johnsgard 1975), along saltwater shorelines, bays, tidal channels, and, most often, around islands. It regularly nests inland near clear rivers, lakes, ponds, and streams, where its preferred nesting habitat is on small islands with dense low growth (Palmer 1976b). In New Brunswick the Red-breasted Merganser regularly nests amid tern and gull colonies, perhaps because association with a more aggressive species provides protection from avian predators (Young and Titman 1986).

The Red-breasted Merganser nests on or very close to the ground. Almost always placed under some protective cover, such as a pile of driftwood or thick shrubbery, the nest is usually within a few yards of water but sometimes as far as 45.7 m (150 ft) inland (Bellrose 1978). The nest site is selected by the female, usually 2–3 weeks before egg laying begins (Johnsgard 1975). The nest itself consists of a shallow depression in the ground heavily lined with gray down feathers and a few white breast feathers (Bent 1923).

Steven C. Sibley

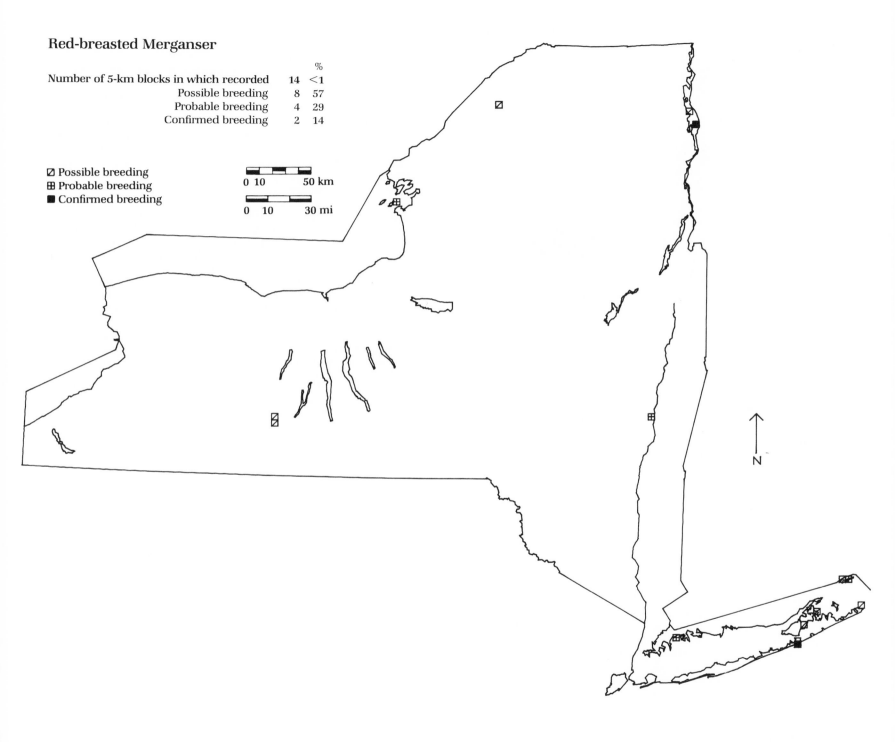

Red-breasted Merganser

Number of 5-km blocks in which recorded	14	% <1
Possible breeding	8	57
Probable breeding	4	29
Confirmed breeding	2	14

☑ Possible breeding
⊞ Probable breeding
■ Confirmed breeding

0 10   50 km

0 10   30 mi

N

C J Page
© 1987

Ruddy Duck *Oxyura jamaicensis*

The Ruddy Duck nests in small colonies between the Atlantic Coast and its main breeding range in western states and provinces. In New York it is a rare and local breeder.

The Ruddy Duck was first reported breeding in New York along the Lake Ontario shore at Sandy Creek, Monroe County, on 30 May 1891. A pair with 5 or 6 young, apparently about 2 weeks old, were observed (Eaton 1910). There have been no records from Sandy Creek since that time. On 1 September 1907 an adult duck with a brood of unfledged young were observed in the Seneca River marshes, presumably at either Montezuma NWR or Howland Island WMA (Eaton 1910). Bull (1974) reported nesting at Oak Orchard Swamp, Genesee County, in 1961, and 3 broods were reported at Montezuma in 1963. D. Carroll (pers. comm.) reported one nesting of the Ruddy Duck at Oak Orchard WMA in 1977 and none since. Although no broods have been observed at Montezuma since 1977 (Gingrich, pers. comm), a nest with eggs was found in 1981 in one of the far corners of the refuge (Dewey, pers. comm.). The Ruddy Duck was first observed nesting at Jamaica Bay Wildlife Refuge in 1955, and by the 1960s as many as 40 broods were observed there (Bull 1974). This duck does not appear to be well established anywhere in the state except at Jamaica Bay, where it produced about 20 young annually from 1980 to 1985 (Riepe, pers. comm).

An additional Atlas report of fledglings was recorded at Mud Creek on Patchogue Bay on the south shore of Long Island. The only record in northern New York was of a female observed in July 1982 at Wilson Hill WMA. The Ruddy Duck was observed there during the breeding season

previous to the start of the Atlas, but breeding has never been confirmed (Van Riet, pers. comm.).

This duck is quite familiar to birders, for whom it is often a favorite. It is unlikely that Atlasers missed this species, because nearly all potential breeding areas are much-used birding spots and have well-monitored waterfowl populations.

The Ruddy Duck breeds in cat-tail–bulrush marshes. It builds a nest in dense stands of emergent aquatics such as cat-tail, bulrush, and the common reed, and it uses adjoining areas of open water for display, feeding, and resting. The duck first constructs an anchored floating platform of plant materials; there it lays its eggs. When the clutch is completed, the duck adds a rim, and incubation begins. If the water table rises, the duck uses additional material at the site to build up the nest and keep the eggs dry. It adds down during the initial days of incubation. The down feathers are light colored, small, and when flattened show a dense white center. Breast feathers, which may be mingled with the down, are brown with white tips; the darkened portions of their shafts are darker than the adjoining web (Palmer 1976b).

The behavior of the Ruddy Duck is quite like the grebe's. If disturbed, it generally dives instead of taking wing; ashore it moves awkwardly because the hind limbs, as in the grebes, are positioned rather far back on its body. When leaving the nest the female slips into the water, submerges, then surfaces a considerable distance away. At about the time the pair-bond is established, long before actual nest building, both sexes build platforms that have no apparent relation to the later nest (Palmer 1976b).

Stephen W. Eaton

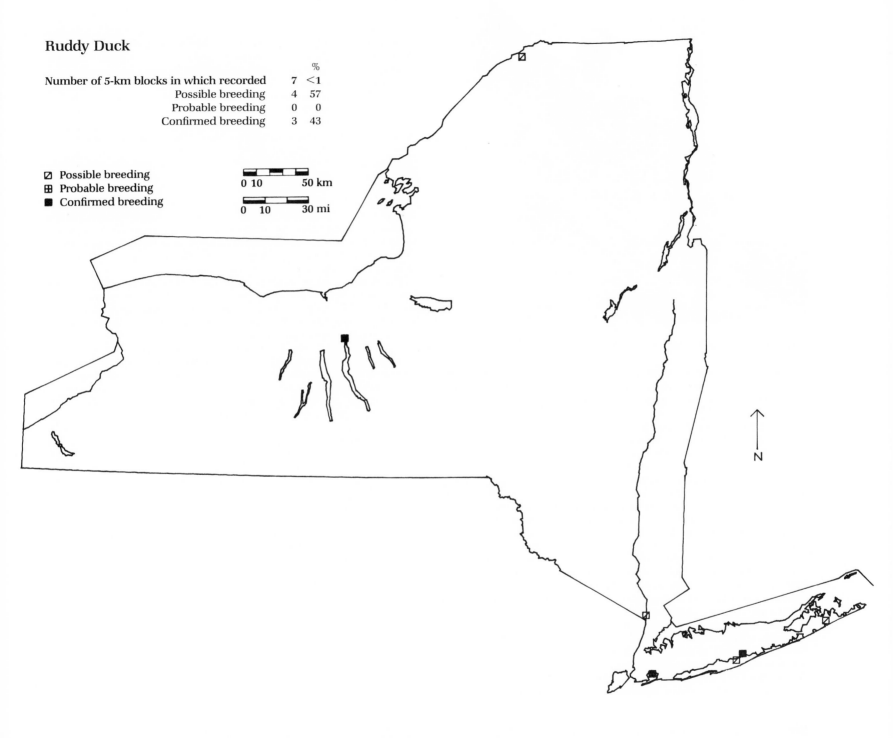

Ruddy Duck

Number of 5-km blocks in which recorded   7   <1

		%
Possible breeding	4	57
Probable breeding	0	0
Confirmed breeding	3	43

☑ Possible breeding
⊞ Probable breeding
■ Confirmed breeding

0 10   50 km

0 10   30 mi

N

© KLA-C 1987

Turkey Vulture *Cathartes aura*

Repulsive to look at, offensive in habits, clumsy and awkward on the ground but the personification of grace and skill in the air, the Turkey Vulture is steadily becoming a more common sight in New York's skies. Because of its breeding season habit of foraging over miles of territory and migrating in often large groups, it may seem to be a fairly common species; in fact, it is generally a rather uncommon breeding bird, still only fairly common locally despite its increasing numbers.

De Kay (1844) considered the turkey buzzard, as it was previously called, common in southern cities but rare in New York State. Northern New Jersey was believed to be the northern edge of its breeding range. Eaton (1914) thought of it as only a summer visitant on Long Island, in the Hudson Valley, and in the warmer portions of western New York. During the late 1800s and early 1900s occasional scattered sightings were reported in the interior of New York. Specimens were taken in May 1884 in Orleans County and in July 1891 in Niagara County (Beardslee and Mitchell 1965). The Turkey Vulture was seen in Ontario County in July 1908 (Wetmore 1930), in Steuben County in 1909 (Burtch 1910), and in the Cayuga Lake basin of Tompkins, Seneca, and Cayuga counties in 1910 (Wright and Allen 1910). Increasing reports from this time on indicated this species' widespread infiltration in the state, even into the Western Adirondack Foothills (Belknap 1967). Nesting was first confirmed in New York in June 1925 at Lewisboro, Westchester County (Howes 1926); a second nest was found in Orleans County in 1927 (Tyler 1937). Bull (1974) documented 15 nesting locations in the state.

Several factors have been suggested to account for the spread of the Turkey Vulture in the Northeast (Bagg and Parker 1951). Amelioration of the climate may have been a cause for the northward spread of several southern species of birds since the end of the Ice Age. Land development may be forcing the Turkey Vulture to utilize more of its potential range. In addition, a marked increase in the deer populations, resulting in higher mortality through starvation, and an increased number of road-killed animals have provided an additional food supply. Similarly, the great bison massacre of 1883 attracted thousands of Turkey Vultures into eastern Montana for a short time (Cameron 1907).

Although the Atlas map shows the Turkey Vulture to be widespread throughout most of New York, the "possible" and most of the "probable" records are undoubtedly sightings of foraging or wandering breeders or unmated birds, and these records cannot be linked to breeding in particular blocks. The Atlas records do seem to indicate that the influx of this species has been along the Great Lakes Plain from the west and via the valleys of the Genesee, Susquehanna, and Hudson rivers from the south. Atlas observers "confirmed" breeding in only 27 blocks. It is extremely difficult to find the nest of this species.

The Turkey Vulture exhibits a strong attachment to old nest sites, returning to nest in the same place year after year (Ritter 1983). The breeding season begins with a behavior called "dual nest sitting." The vulture pair perches for long periods near the nest for several days to several weeks before egg laying. In eastern North America the Turkey Vulture frequently nests in forests, mainly bottomland hardwoods, although nesting on rocky cliffs is common. The nest site is a dark recess: a cave, hollow tree, brush pile, old building, rock crevice, vine tangle, or thicket (Jackson 1983). Jackson (1983) thought that loss of forest habitat during the 19th and early 20th centuries in the eastern United States may have been responsible for a change in the type of nest site used, from tree cavities to thickets.

A nest was located in 1986 at an Orange County town park about 500 yards from a nature center. An adult with 2 chicks was found in a cave formed by a natural rock outcropping and rock wall, with small entrances on both sides. The cave was about 1.4 m (4.5 ft) deep and 0.5 m (1.5 ft) high. The park is in a mature oak–northern hardwood forest, and the area around the cave was strewn with boulders (Seymour, pers. comm.). Tyler (1937) reported nesting in abandoned henhouses, and an Atlas observer found a nest in a barn (Olds, pers. comm.). The Turkey Vulture makes no nest preparation (Ritter 1983).

With rare exceptions the Turkey Vulture's diet is animal carcasses. At the nest decayed flesh is regurgitated and fed to the young (Terres 1980). The odor produced by the putrid flesh led some Atlas observers with good olfactory senses to nest sites.

Gordon M. Meade

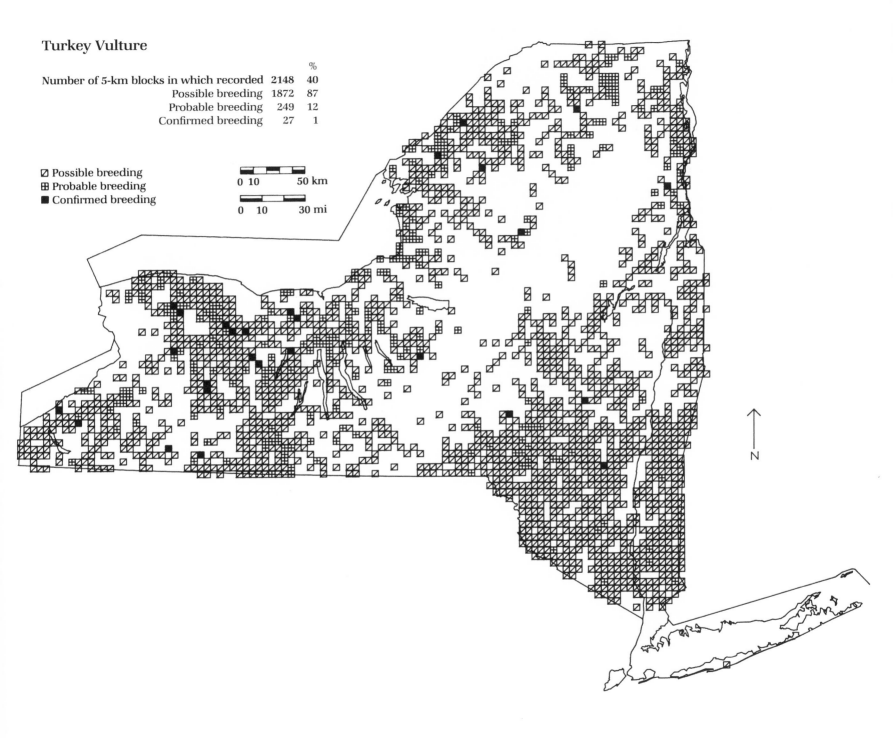

Turkey Vulture

		%
Number of 5-km blocks in which recorded	2148	40
Possible breeding	1872	87
Probable breeding	249	12
Confirmed breeding	27	1

◪ Possible breeding
⊞ Probable breeding
■ Confirmed breeding

0 10   50 km

0 10   30 mi

N

©KLA·C 1987

Osprey *Pandion haliaetus*

The Osprey was one of the birds of prey which experienced reproductive failure in the 1950s and 1960s when high concentrations of DDT caused thinning of its eggshells (Spitzer 1978). Osprey productivity throughout the New York–Boston region and elsewhere reached very low levels. Recently in many places, including New York, the population is recovering. The Osprey now is a common breeding species on eastern Long Island; however, it is still uncommon in the Adirondacks and St. Lawrence River Valley, and rare elsewhere in the state. It was initially listed as Endangered in New York in 1978, but its status was changed to Threatened in 1983.

A century ago there was a large population of Ospreys on Long Island. Allen (1892) indicated that during the late 1870s an estimated 1000–2000 adults roosted on Plum Island at the eastern tip of Long Island. Eaton (1914) spoke of its breeding on eastern Long Island: he noted more than 100 nests on Gardiners Island in 1910. By 1930 the number of nests on Gardiners Island was reported to be over 300 (Knight 1932). Cruickshank (1942) called it "common to locally abundant" on eastern Long Island. The decline in productivity on Long Island began in 1948 and continued until the late 1960s. In 1966 only 4 young were produced from an estimated 55–60 active nests on Gardiners Island (Puleston 1977).

In 1977 the NYSDEC began actively to monitor Osprey nests on Long Island using ground and aerial surveys, as well as reports by local observers. Signs were posted warning individuals to stay away from nesting areas, artificial nest platforms were built, and predator guards were placed around nest trees and poles supporting nest platforms. In addi-

tion, the plight of the Osprey on Long Island was highly publicized. As a result of the ban on the use of DDT and these management efforts, the recovery of the Osprey population on Long Island has been dramatic. In 1980, 101 young fledged from 87 active nests on Long Island and in 1986 there were 144 active nests and 186 fledglings (NYSDEC files).

Away from Long Island, by the early 1900s the Osprey was a summer resident only in the Adirondacks where, according to Eaton (1914), it was becoming more rare each year. During a study conducted in the Adirondack Park between 1970 and 1972 by the NYSDEC, only 15 out of 60 nesting attempts were successful. Most of the unsuccessful attempts were a failure of the eggs to hatch (Singer 1974). More recently the NYSDEC conducted annual surveys of both the Adirondack and St. Lawrence River Valley populations which indicated that productivity has improved; for example, 39 young were produced in 1986 (NYSDEC files). Clum (1986) found that during the period 1981–84 prey quantity appeared to be influencing the productivity of the Adirondack Osprey population. She also found that "the pH of foraging sites . . . was positively correlated with the number of young produced per active nest."

Osprey hacking, using the technique described in accounts for the Peregrine Falcon and Bald Eagle, was begun in 1980 at Allegheny Reservoir, Cattaraugus County. Young Ospreys were taken from nests on Long Island and released at the reservoir. From 1980 to 1986, 30 birds were released. Although hacked Ospreys have returned to the reservoir and were even observed building a nest and copulating, no successful nesting in the area has been documented (Loucks 1986a).

The Atlas map identifies the two major breeding concentrations of Osprey in the state and also indicates a nest that has been active at Montezuma NWR since 1979 (Loucks 1986a). The record near Oneida Lake was an observation of two food-begging young with an adult on 20 July 1985. Attempts to find a nest there were unsuccessful. (In 1986 a pair nested at Three Rivers WMA and produced at least 1 young, but the young did not fledge [NYSDEC files].) Most of the "possible" and "probable" records elsewhere in the state probably represent either late migrants or wandering birds.

On Long Island the Osprey usually nests on large tracts of undeveloped land near major estuaries, on tributaries, or along the many harbors or bays; or on islands, such as the Nature Conservancy's Mashomack Preserve on Shelter Island, and Gardiners Island. One nest, however, was located in the pine barrens several miles from water. On Long Island the Osprey appears to tolerate human disturbance. In the Adirondacks the Osprey nests near lakes, ponds, and rivers, and on small islands, usually near a mature forest. Although the nest is most frequently built in a tall tree, nests at Gardiners Island were found on the ground (NYSDEC files).

Janet R. Carroll

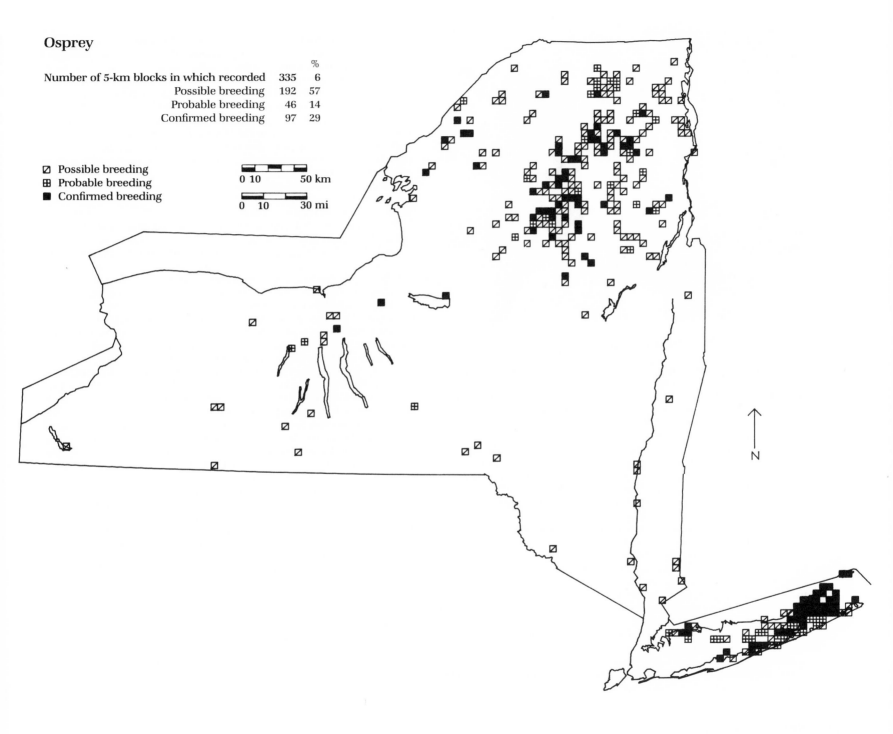

Osprey

		%
Number of 5-km blocks in which recorded	335	6
Possible breeding	192	57
Probable breeding	46	14
Confirmed breeding	97	29

Possible breeding
Probable breeding
Confirmed breeding

0 10   50 km

0 10   30 mi

N

Bald Eagle *Haliaeetus leucocephalus*

The endangered Bald Eagle is a rare breeding species in New York. The NYSDEC identified 72 verifiable nesting locations used by Bald Eagles during the period 1860–1960 (Nye 1979). Some of these eyries were alternate nest sites for the same pairs of birds, and all were not occupied at once. There are now 4 pairs holding territories in the state.

The Bald Eagle has probably nested in New York for centuries, but the earliest records are from the late 1800s, in Essex, Erie, Jefferson, and Dutchess counties. Of the 72 known eyries, 26 were in the Central Adirondacks and the Eastern and Western Adirondack Foothills; 22 were in the Eastern Ontario Plains, Oswego Lowlands, and Oneida Lake area; and the rest were scattered among the Great Lakes Plain, Lake Champlain Valley, Dutchess County (1), and Long Island (2) (NYSDEC files).

The Atlas map shows "confirmed" breeding in Livingston and Jefferson counties. The territorial adults indicated as "probable" in Genesee County successfully bred in 1986. The "probable" record in Sullivan County was of an adult male that had been defending a territory at a reservoir for 3 years but in 1986 was observed at Montezuma NWR with a mate. Many of the "possible" records represent adult eagles from southern states which breed early and wander into New York in late May and June to spend the summer (Spofford 1959).

Although the Bald Eagle has always been persecuted, pesticides were the direct cause of its extirpation as a breeder in the state. Productivity at eagle eyries along the Atlantic Coast and in the Great Lakes region drastically declined after World War II when DDT and other pesticides began to take their toll on many raptor populations (Graham 1976). The decline of the Bald Eagle was rapid, and by 1960 most nests in New York were vacant. The Livingston County nest was the only one that remained active; it remains so today.

In 1976 the NYSDEC contracted with Cornell University's Peregrine Fund to test techniques for releasing Bald Eagles into the wild in an attempt to restore the population. Using a method called hacking, researchers took 2 young eagles from a nest in Wisconsin and brought them to Montezuma NWR. They were held there until they were able to fly and then released (Milburn and Cade 1977). Hacking has been used by the NYSDEC in subsequent years at four sites around the state. A total of 166 birds have been released at Montezuma; Oak Orchard WMA; Alcove Reservoir, Albany County; and near Tupper Lake, Franklin County. The 4 Bald Eagle territories in New York are occupied by 7 hacked eagles and 1 eagle of unknown origin.

The history of the Livingston County eyrie is remarkable. A pair of adult-plumaged eagles were in residence at the nest in 1965, making them at least 4 or 5 years old at that time. Only once, in 1973, did they successfully raise young, although an egg was laid each year until 1979 (Rauber 1976). Manipulative experiments were begun to try to obtain production from this pair. In 1978 and 1980 a dummy egg and later a captive-raised eagle chick were placed in the nest. The adults incubated the egg and successfully reared the chick to fledging both years. In 1981 an adult male eagle believed to be one of the Livingston County pair was shot to death near the nest site. The female, however, appeared at the nest in 1981 with a new mate: an eagle hacked in 1977 at Montezuma (Nye and Allen 1983). No eggs were laid again, so a dummy egg was put in the nest and then replaced by 2 eaglets, both of which successfully fledged. In 1982 and 1983 the fostering continued successfully. In 1984 the male came back to the eyrie with a new female. The old female was presumed dead. Although the pair was observed building a nest, no successful nesting took place that year. In 1985, however, the pair produced 1 young: the first natural production at this site in 13 years. At the times of the disappearance of the male and the female of the original pair, they would have been at least 20 and 23 years old, respectively.

The Bald Eagle nests near undisturbed large lakes or reservoirs, in marshes and swamps, or along rivers. The Livingston County eyrie is typical, located on the western slope of an undeveloped lake that is 12.8 km (8 mi) long and 0.5–0.8 km (0.3–0.5 mi) wide. At the southern end of the lake is an open cat-tail marsh. The lake is surrounded by a mature forest of oak, beech, hickory, white pine, and hemlock (Rauber 1976). Although many of the historic eyries are now unsuitable for habitation by eagles because of development, the NYSDEC has estimated that at least 30 of the known historic sites are still potential eagle nest sites (Nye and Peterson 1980). Eagles usually nest in trees but may also nest on cliffs. The large stick nests are added to year after year, sometimes becoming mammoth structures.

Janet R. Carroll

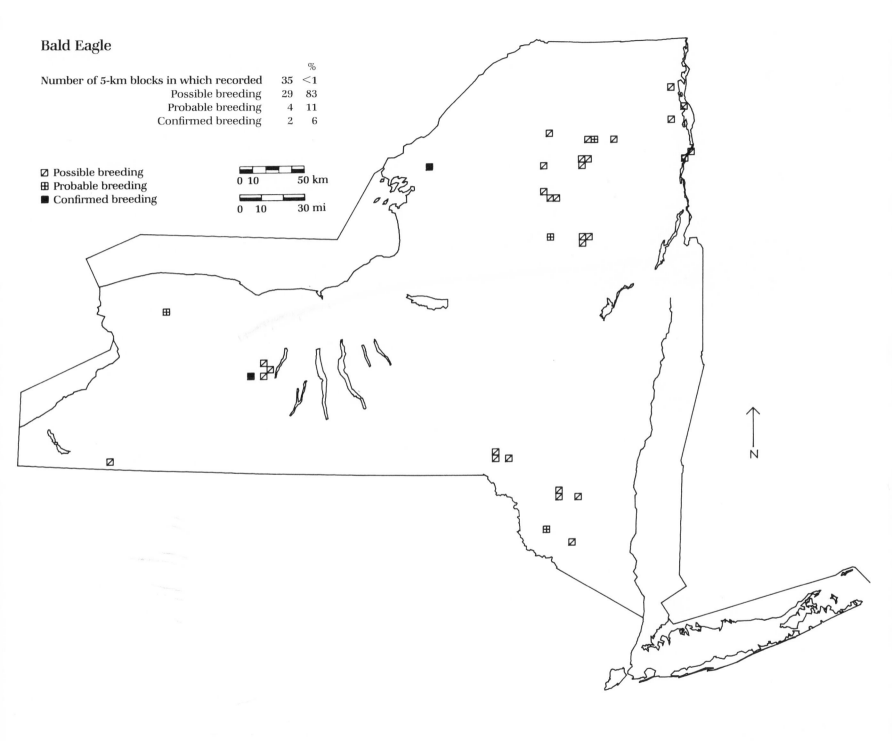

Bald Eagle

		%
Number of 5-km blocks in which recorded	35	<1
Possible breeding	29	83
Probable breeding	4	11
Confirmed breeding	2	6

☑ Possible breeding
⊞ Probable breeding
■ Confirmed breeding

0 10   50 km

0 10   30 mi

N

Northern Harrier *Circus cyaneus*

The Northern Harrier, a graceful bird, is an uncommon, scattered breeder throughout New York. In the early 1900s this species bred throughout the state, from the highlands of the western Appalachian Plateau and the Adirondacks to the tidal reaches of Long Island (Eaton 1914). Saunders (1926) considered it to be fairly common in and near central New York's extensive marshes, and Hyde (1939) found it common in wetland habitats and upland agricultural areas near Lake Ontario and the St. Lawrence River. But by the mid-1900s the species had declined (Bull 1974).

This decline continued throughout central New York during the 1950s and 1960s (Scheider, pers. comm.) and in western New York (Beardslee and Mitchell 1965). A similar decline was occurring in many parts of the Northern Harrier's range, including nearby New Jersey (Dunne 1984), and the species was placed on the *American Birds'* Blue List (Tate 1986). It is listed as Threatened in New York. The reasons for its decline are unclear, but the period during which it occurred matches that of other raptor populations, reduced as a result of pesticide contamination.

The Atlas map identifies "confirmed" and "probable" records in central, western, and northern portions of upstate New York, including the western Great Lakes Plain and Appalachian Plateau regions west of Seneca Lake, and in northeastern Lake Ontario in Jefferson County through the St. Lawrence Plains to the northern Lake Champlain Valley. Breeding was also documented along the south shore of Long Island and in the interior areas of eastern Suffolk County. Records were scarce in much of the Adirondacks, eastern Appalachian Plateau, Catskills, Hudson Valley, and southeastern New York, as well as on most of western Long Island.

The breeding range of the Northern Harrier in New York has apparently changed locally as a result of regional declines. Comparison of Atlas data with historical data suggests its current distribution is more fragmented. Although the harrier is still present in the state, in the southeastern Lake Ontario region extending to the Syracuse area in Onondaga County and in the lower Hudson Valley there are few records of it. In contrast, its distribution in some areas, particularly northern New York and Long Island, has changed little. This pattern probably reflects the types of human activities occurring in these regions and differing rates of the harrier's decline and recovery. Pressures on upland nesting pairs, including the change to a corn monoculture and a decrease in the amount of pasture and grassland habitat, vary among areas. The effect of wetland drainage and changes in vegetation caused by alterations in the water level of eastern Lake Ontario marshes are important local variables. The prognosis for the harrier's recovery, such as is being observed in southern New Jersey (Dunne 1984), is uncertain for many parts of New York.

The Northern Harrier is a highly visible species, but finding its well-hidden nest on the ground is difficult, and Atlas observers were warned to avoid disturbing the nest of this Threatened species. Pairs were reported quite often, and observations of fledglings were the most frequently used "confirmed" code. The species' large hunting ranges and the presence of "brown" individuals, some of them nonbreeding yearlings, may explain many "possible" breeding records where breeding does not occur.

This raptor uses various wetland and upland habitats for breeding throughout its range (Apfelbaum and Seelbach 1983). A 1979 survey of nesting harriers found that in central and western New York most nests were in hay fields; along Lake Ontario and the St. Lawrence River most were in cat-tail marshes and wet meadows. In the Adirondacks bogs and other wetland sites provided nesting habitat (Kogut 1979). On Long Island nest sites are above the tidal salt marsh (England 1985) in stands of pure common reed, and mixed common reed and poison ivy; in mixed shrub vegetation along the barrier beach; and in bearberry in the pitch pine–scrub oak barrens.

The Northern Harrier is one of only a few New York raptors that nest on the ground. The nest, a substantial structure especially when built in wet areas, is made of grasses and sticks woven into a loose mat of herbaceous vegetation (Brown and Amadon 1968). It is usually placed on a hummock or directly on the ground (Bull 1974). Often only a small proportion of the young fledge because of heavy predation on the eggs and young (Munoff 1963). On Long Island in 1985, 9 of 16 nesting attempts were successful in fledging at least 1 young (England 1985).

Gerald A. Smith

102

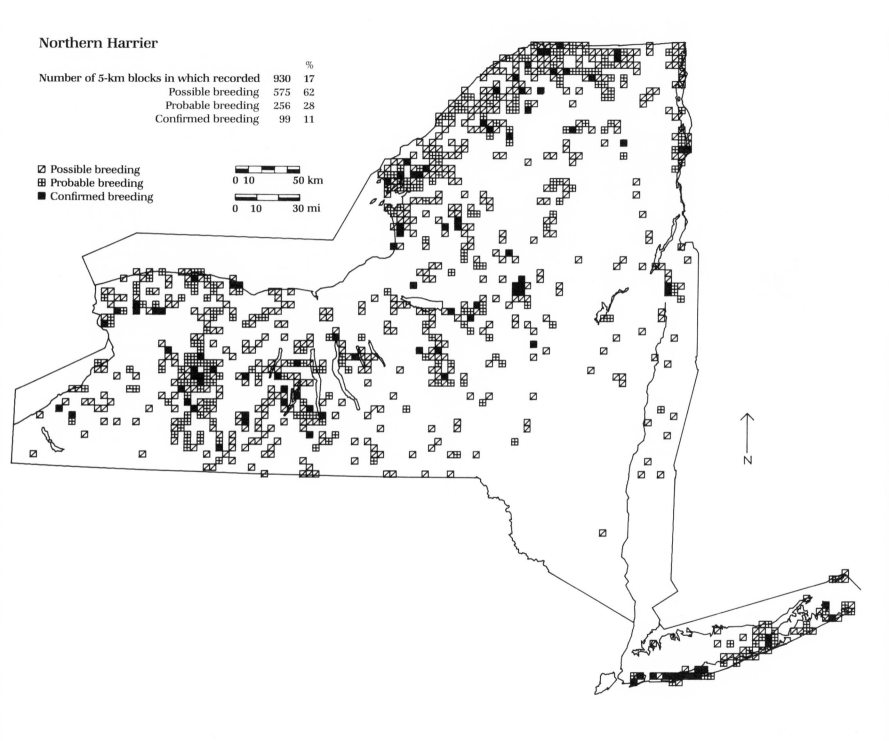

Northern Harrier

Number of 5-km blocks in which recorded	930	% 17
Possible breeding	575	62
Probable breeding	256	28
Confirmed breeding	99	11

◨ Possible breeding
⊞ Probable breeding
■ Confirmed breeding

0 10　　　50 km

0 10　　　30 mi

N

Sharp-shinned Hawk *Accipiter striatus*

The Sharp-shinned Hawk, a small, bold accipiter, is an uncommon to very uncommon and unevenly distributed breeding species. The immediate future of this species seems secure, but with rapid changes occurring in woodlands on both breeding and wintering grounds, the long-term future of this and other migrant forest-dwelling birds is uncertain.

Eaton (1914) stated that the Sharp-shinned Hawk nested in wooded country throughout the state, especially in heavily forested regions. Reflecting the views of many naturalists of his era and most of the general public, he recommended "they . . . be destroyed wherever the more desirable song and game birds are to be preserved." Bull (1974) indicated that, although secretive and difficult to detect, this raptor was fairly common, and he called it a widespread breeder in wooded country throughout the state. Its secretive nesting habits and the lack of systematic surveys of breeding populations muddle the picture of its past distribution. It is not clear whether the breeding population declined substantially during the mid-1940s through early 1970s when DDT was commonly used. Such was the case for many other birds of prey. However, a decline seems likely, based on evidence from migration counts.

The Sharp-shinned Hawk has a wide though scattered distribution throughout much of New York. Breeding was "confirmed" most frequently in the heavily forested areas of the Adirondacks and Appalachian Plateau. This hawk was noticeably scarce in the fragmented forests of the eastern Great Lakes Plain, St. Lawrence Plains, Hudson Valley, and Coastal Lowlands.

Comparison of Atlas and historical data suggests this species still nests in most areas of the state where it bred during Eaton's (1914) time. The current epicenter of breeding appears still to lie within heavily forested upland areas. However, comparison of data from the intense, short-term Atlas project and the more long-term but fragmentary historical records is risky at best, and particularly so for the Sharp-shinned Hawk, which has a patchy distribution, a large home range, and a nest that is extremely difficult to find.

Of the 3 accipiters in the state, the Sharp-shinned Hawk was found in the most blocks, yet fewer nests were found for this species than for any other of the more common diurnal raptors. Only 16 active nests were found during the period of Atlas fieldwork. Most of the "probable" records are of pairs or single birds exhibiting territorial behavior. A calling and circling Sharp-shinned Hawk announces that a human or other intruder has entered its territory; however, unlike the Northern Goshawk, it seldom attacks. Fledglings or adults feeding young provided the most common "confirmed" records. To confirm presence or absence of the Sharp-shinned Hawk, intense coverage is necessary, particularly in heavily forested blocks. The need for special surveys targeted at this and other uncommon, difficult-to-detect species is critical.

Breeding Sharp-shinned Hawks use a variety of mixed and coniferous woodlands for nesting. This species typically breeds in heavily forested areas of the state. Areas of extensive mixed forest provide optimal habitat in many regions; coniferous forests are widely used in the Adirondacks and other northern sectors. Forests in which it breeds must contain evergreens to serve as nest trees (Apfelbaum and Seelbach 1983). Bull (1974) stated that New York nests have been found only in conifers, with 80% located in hemlocks.

The Sharp-shinned Hawk builds a stick nest that is comparatively bulky for its size (Eaton 1914). Materials include large sticks interwoven with small twigs and strips of bark. Nests are extremely inconspicuous and difficult to find, which undoubtedly contributed to the low percentage of "confirmed" breeding records.

Gerald A. Smith

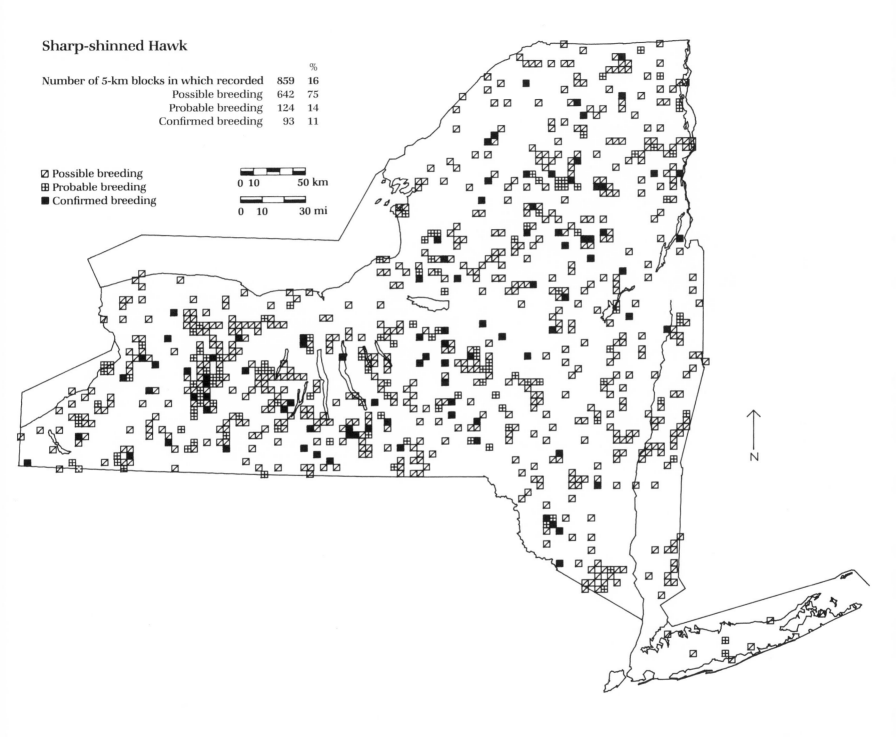

Sharp-shinned Hawk

		%
Number of 5-km blocks in which recorded	859	16
Possible breeding	642	75
Probable breeding	124	14
Confirmed breeding	93	11

◩ Possible breeding
⊞ Probable breeding
■ Confirmed breeding

0 10   50 km

0 10   30 mi

N

Cooper's Hawk *Accipiter cooperii*

The Cooper's Hawk, a medium-sized accipiter, is a fairly common breeder in much of the state. The Cooper's Hawk was long persecuted by humans because it preyed on chickens that roamed unprotected on small farms. Although such persecution no longer persists, the population of Cooper's Hawks in New York has continued to decline; it was listed as a Species of Special Concern in 1983 by the NYSDEC. Recently, however, increases have been noted.

Eaton (1914) found that in the wilder and more wooded portions of the state the Cooper's Hawk was one of the most common breeding species. He said, however, "In the more thickly settled districts it is much less common than formerly, the nesting birds having been killed off on account of their destructiveness to poultry and game." Temple and Temple (1976), describing the abundance of this hawk in central New York, noted a sudden decline in the population beginning in the late 1950s. By 1965 its numbers had drastically reduced. Schriver (1969) has shown that this decline coincided with the heavy use of the pesticide DDT after World War II. Metabolites of the DDT accumulated in the tissues of many species high on the food chain, including the Cooper's Hawk, causing eggshell thinning. Spofford (1969) documented a continuous decline in migrant Cooper's Hawks passing over Hawk Mountain, in Pennsylvania, from 1950 to 1964. The species was put on the *American Birds'* Blue List in 1971 and was still there in 1986. Its numbers appear to be recovering (Tate 1986), probably as a result of a cleaner environment.

The Cooper's Hawk nests in wooded bottomlands in most parts of the state. Atlas observers found it in most counties except Niagara, Rockland,

Nassau, and Suffolk, and the five counties that make up New York City. It is best represented on the Appalachian Plateau, particularly in the Finger Lakes Highlands and along the great river valleys such as the Allegheny, Genesee, Chemung, and the east branch of the Susquehanna. Farther east, it occurs along tributaries of the Delaware and Hudson rivers, and in the Taconic Highlands. The Cooper's Hawk was found in the Adirondacks but infrequently on the Central Tug Hill and Transition, where the Northern Goshawk was better established. The Cooper's Hawk was also scarce on the St. Lawrence Plains but occasionally "confirmed" on the Great Lakes Plain, where the Northern Goshawk was scarce. Perhaps most of the woodlots there are not large enough to attract this accipiter. The general absence of the Cooper's Hawk on Long Island is probably a result in part of a lack of proper forest habitat. It generally avoids large urban centers at nesting time. However, it may nest in a patch of woods near a smaller city, relying on birds of the city for prey (pers. obs.). Like other woodland raptors, the Cooper's Hawk is probably under-represented on the Atlas map.

Meng (1951) found it nesting in hemlock–northern hardwood and oak–northern hardwood forests of the Finger Lakes Highlands. In the Adirondacks it nests in spruce–fir–northern hardwoods; on the Great Lakes Plain in elm–red maple–northern hardwoods. In Bull's (1974) summary of trees most often used for nesting, out of 57 nests, 21 were in beech, 12 in maple, 7 in oak, 6 in ash, 4 in pine, 2 in birch, and 1 each in elm, cherry, hickory, and basswood.

Nest heights range from 9.5 to 22 m (30–70 ft). The nest is placed in an upright crotch of a deciduous tree or next to the trunk on a horizontal limb of a conifer, often a white pine. A substantial structure of sticks and twigs, the nest is lined with chips of the outer bark of oak or pine. The nest is not decorated with greenery as often as nests of buteos are. The Cooper's Hawk usually builds a new nest annually but may repair an old one or build on an old squirrel's or crow's nest (Harrison 1975).

The Cooper's Hawk nest is less easily located than that of the Northern Goshawk and often is not as well defended. Only 36 nests were located, in contrast to 84 reported for the Northern Goshawk. It should be noted that many of the nest records of the Northern Goshawk were obtained from a researcher doing nest studies on that species.

Stephen W. Eaton

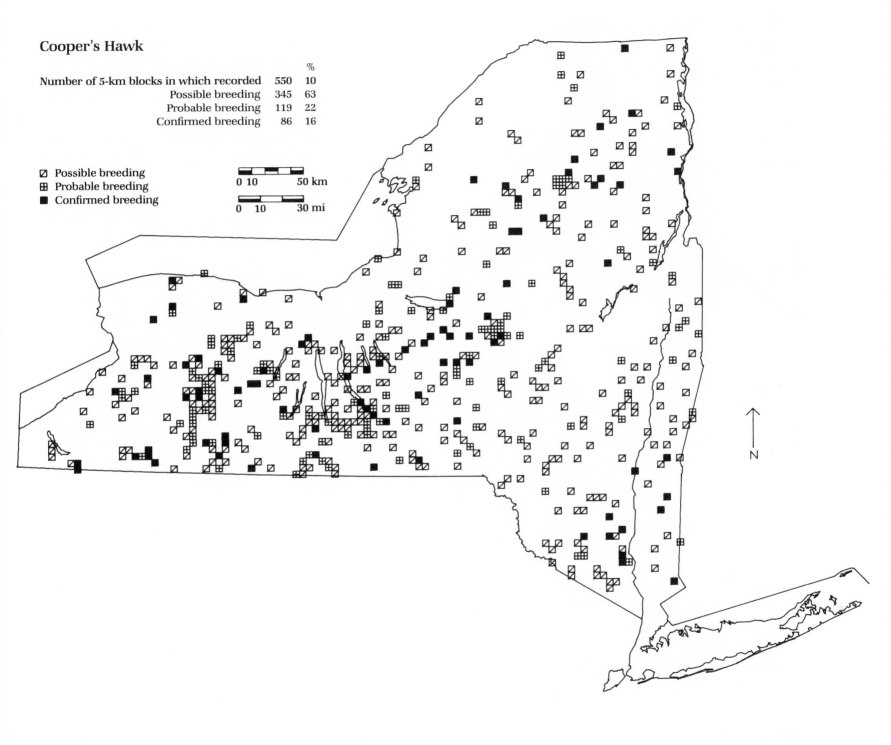

Cooper's Hawk

		%
Number of 5-km blocks in which recorded	550	10
Possible breeding	345	63
Probable breeding	119	22
Confirmed breeding	86	16

☑ Possible breeding
⊞ Probable breeding
■ Confirmed breeding

0 10 50 km

0 10 30 mi

N

©KLA-C 1987

Northern Goshawk *Accipiter gentilis*

The Northern Goshawk, a large, fierce, boreal accipiter of the taiga and northern hardwoods, is an uncommon breeder in New York. Before the demise of the Passenger Pigeon around 1900, the Northern Goshawk was believed to have been common in the Northeast (Bent 1937). But by the early 1900s Eaton (1914) called it a rare summer resident "even in the wildest portions of the Adirondack forest." However, from the 1950s onward, the species' range appears to have expanded and its numbers to have increased in the state. In central New York, Temple and Temple (1976) documented an increase in numbers during the period 1960–71, and Bull (1974) recorded 48 breeding sites in the state since 1952, most in the High Peaks and Central Adirondacks but others at higher elevations throughout New York. The Atlas documents further expansion.

The movement into the state appears to be from two sources: from the north into the Adirondacks (Bull 1974), and from Pennsylvania into the Catskills and the Appalachian Plateau (Todd 1940; Eaton 1981). Reforestation of New York has provided additional nesting habitat for this species (Todd 1940; Bull 1974). On the Appalachian Plateau at least two other species are following a similar pattern: the Wild Turkey and the Common Raven.

The Northern Goshawk was found by Atlas observers throughout most of New York, and the species was recorded in all but 11 counties. It is most notably absent from Long Island and the New York City area north into Rockland County, as well as in the farm country of western New York. The woodlands in these areas are fragmented and do not provide good nesting habitat for this raptor. In comparing the breeding distribu-

tion as illustrated in Bull (1974) to the Atlas distribution, the species appears to be spreading out, particularly from Chautauqua County through the Finger Lakes, along the Mohawk Valley, and the area east of the Hudson River.

This accipiter was "confirmed" in 128 blocks in the state during Atlas fieldwork, exceeding the number of "confirmed" records of all but the two most common diurnal raptors, the Red-tailed Hawk and American Kestrel. The goshawk is much more territorial and vocal near its nest site than are other hawks. In addition, a researcher studying the Northern Goshawk in central New York and banding young in the nest provided almost half of the "confirmed" records. When the results of this special effort in Madison and Chenango counties are compared with records from the rest of the state, it is apparent that this species was under-recorded. In addition, some observers were reticent about reporting localities of nest sites because of the goshawk's popularity among falconers.

The Northern Goshawk prefers dense, mature, continuous forest, either coniferous or mixed. The nest tree is usually near a small break in the forest, such as a logging road (Reynolds et al. 1982). This hawk generally returns each year to the same wooded area to nest. The nesting area may contain several old nests, one of which the bird may use; or it may build a new nest (Brown and Amadon 1968). Of 40 nests reported by Bull (1974), 34 were in deciduous trees, 6 were in pines. Beech was most often used, followed by maple and birch. I have observed that in the Allegany Hills it also frequently nests in poplar. The nest is usually in the middle or upper level of the tree; the average nest is from 9.1 to 12.2 m (30–40 ft) off the ground. The nest is placed in a crotch of a tree where large branches diverge and is a bulky structure of large sticks, slightly hollowed and lined with bark chips and evergreen sprigs. Plucking sites, termed "butcher blocks," can be found near the nest on the ground beside horizontal logs. The Northern Goshawk carries its prey to this area and plucks it before feeding its young at the nest (Jones 1979).

Stephen W. Eaton

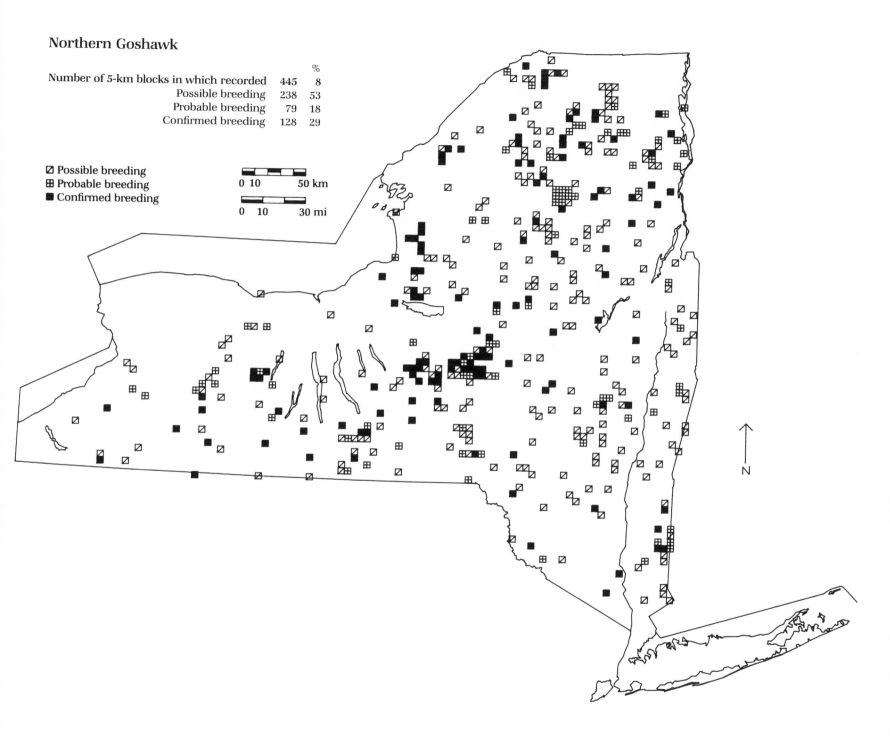

Northern Goshawk

		%
Number of 5-km blocks in which recorded	445	8
Possible breeding	238	53
Probable breeding	79	18
Confirmed breeding	128	29

◨ Possible breeding
⊞ Probable breeding
■ Confirmed breeding

0 10   50 km

0 10   30 mi

N

©KLA-C 1987

Red-shouldered Hawk *Buteo lineatus*

No longer does the piercing cry of the Red-shouldered Hawk echo through the many areas that once held the swamps where it bred. One after another of these breeding haunts have been drained for housing developments or agricultural land. This woodland buteo, once described by observers of the early 19th century as a common breeder in the state, is now uncommon.

Eaton (1914) considered the Red-shouldered Hawk to be the most common large hawk of the central and western counties; it apparently outnumbered the Red-tailed Hawk in many areas. Hyde (1939), while surveying the riparian areas of Lake Ontario and the St. Lawrence River, noted the Red-shouldered Hawk more frequently than the Red-tailed Hawk. By the 1970s Bull (1974) still considered it to be a fairly common breeder in the interior lowlands of the state but rare in the mountains and in areas of patchy forests. He cited Meng's 1951 report of 18 nests within 17 miles of Ithaca, Tompkins County, as evidence of the species' abundance in that area and era. Crocoll (1984) considered the Red-shouldered Hawk a common breeding hawk in Chautauqua County. Some observers noted declines, however, including Rusk and Scheider (1969), who stated that the species was "now unreported from all the wooded swamps of the Ontario lake plain." This statement pertained only to the lake plain in Region 5 (the Great Lakes Plain, including the Oswego Lowlands and Eastern Ontario Plains).

Atlas observers found the Red-shouldered Hawk to be sporadically distributed. "Confirmed" and "probable" breeding was reported most frequently from heavily forested sectors of the Appalachian Plateau, Cats-

kill Peaks, the Delaware, Mongaup, and Rensselaer hills, the Tug Hill Plateau, and Lake Champlain Valley. It was also widely scattered throughout much of the Adirondacks, particularly near river systems. Few records of Red-shouldered Hawks were reported from the lowland, lightly forested areas of the Great Lakes Plain and Long Island. The lack of records in some areas, including the Western Adirondack Transition and St. Lawrence Transition, probably reflects inadequate coverage rather than an absence of breeding Red-shouldered Hawks.

Comparison of Atlas data and the historical record strongly suggests that the breeding distribution of the Red-shouldered Hawk has changed substantially. Whether this change reflects an overall population decline statewide or a shift of breeding areas is difficult to determine. Breeding was recorded infrequently by Atlas observers in lowland areas of the central and western counties, where the Red-tailed Hawk has replaced the Red-shouldered Hawk as the common large breeding hawk. The virtual absence of the Red-shouldered Hawk from red maple–hardwood swamps along Lake Ontario and the scarcity of records in riparian situations on the Great Lakes Plain is striking. Equally striking is the large number of records from the Adirondack and Tug Hill Plateau forests. This is in sharp contrast to distribution described earlier in this century. The future of New York breeding populations of this hawk now seems closely tied to the heavily forested higher elevations of the state. Reforestation should provide additional nesting habitat for this species elsewhere as well (Crocoll 1984).

The Red-shouldered Hawk uses a variety of woodland habitats, including lower wetlands and mesic upland forests (Portnoy and Dodge 1979; Bednarz and Dinsmore 1981, 1982; Morris and Lemon 1983). In Chautauqua County it nested on level terrain, close to some form of surface water (Crocoll 1984). Included in this hawk's hunting areas are nearby openings such as pastures, fallow fields, and wood edges (Bednarz and Dinsmore 1981; Apfelbaum and Seelbach 1983). Bednarz and Dinsmore (1981, 1982) believed that reductions of forest cover through development and timber harvest expose breeding Red-shouldered Hawks to competition from the more aggressive Red-tailed Hawk. Other factors that may adversely affect breeding include increased firewood cutting, development of second homes, and recreational activity in New York forests, as well as the substantial reduction in wetland habitats and increased acid precipitation.

The Red-shouldered Hawk builds a bulky stick nest close to the trunk in large deciduous trees (Apfelbaum and Seelbach 1983). American beech was the favored tree in western New York (Crocoll 1984) and elsewhere in the state (Bull 1974). The hawk's habit of building its nest close to the trunk of large, mature trees may increase its stability. Nest sites for the more agile flying Red-shouldered Hawk are usually lower in the tree canopy than those of Red-tailed Hawks. Cutting that opens the canopy and permits easier access to Red-tailed Hawk may work to the disadvantage of the Red-shouldered Hawk (Bull 1974; Bednarz and Dinsmore 1982).

Gerald A. Smith

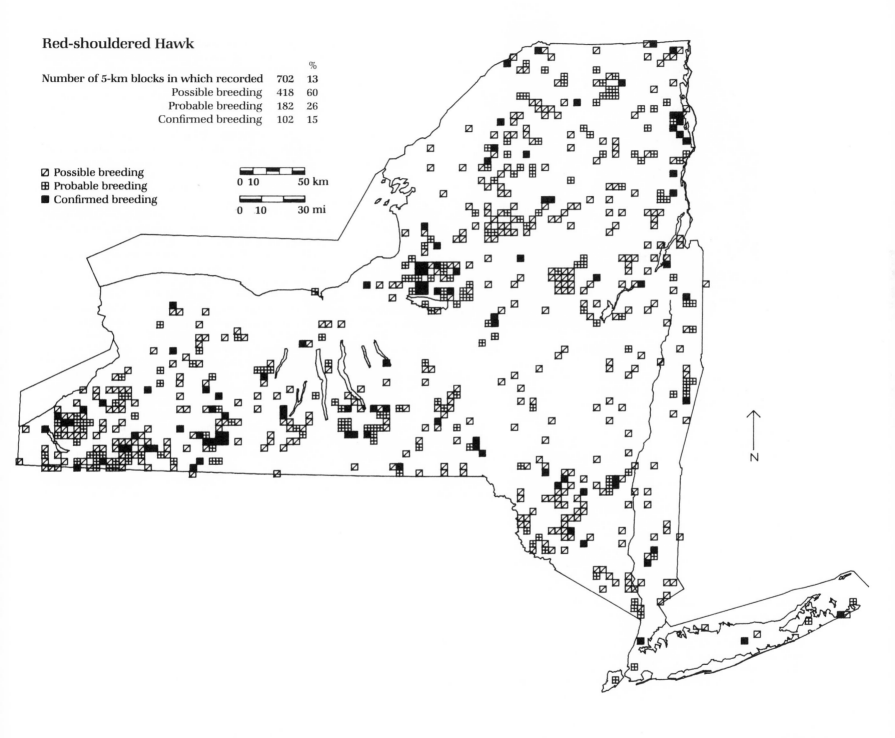

Red-shouldered Hawk

		%
Number of 5-km blocks in which recorded	702	13
Possible breeding	418	60
Probable breeding	182	26
Confirmed breeding	102	15

◪ Possible breeding
⊞ Probable breeding
■ Confirmed breeding

0 10   50 km

0 10   30 mi

N

Broad-winged Hawk *Buteo platypterus*

The smallest of New York's buteos, the Broad-winged Hawk is a fairly common breeder in the more heavily wooded areas of the state. It is widely distributed and found in many forest types, from mixed spruce-fir and northern hardwoods in the Adirondacks to pitch pine–scrub oak on Long Island.

Eaton (1914) described the Broad-winged Hawk as the most common hawk in the Adirondacks, yet irregularly distributed on Long Island and in eastern New York and "almost unknown as a breeding species" in western and central New York. Beehler (1978) also called it the most common raptor in the Adirondacks. More recently, the picture has changed in some parts of the state. For example, Beardslee and Mitchell (1965) rated it as "uncommon," rather than Eaton's "almost unknown," in the counties of southwestern New York.

This raptor has apparently expanded its range in the state, particularly south and west of the Mohawk Valley. The number of Atlas records from the Catskills and the Hudson Valley, in particular, was unexpected. The Broad-winged Hawk is now absent only from the agricultural areas of the Great Lakes and St. Lawrence plains and the Mohawk Valley, the less wooded areas along the Appalachian Plateau and Hudson Valley, and the urbanized western end of the Coastal Lowlands. The distribution of this hawk is determined by the existence of extensive forest cover, either primary or secondary woodland (Bull 1974). It is usually absent from small groves or farm woodlots (Brown and Amadon 1968). The recent reversion of farmland to forests in New York certainly must be a factor in its expansion.

Atlas observers did have difficulty locating the nests of this species. When found on its nest, it often sits tight until the observer is quite close. Some individuals will even allow themselves to be lifted off the nest by hand (Bent 1938). The Broad-winged Hawk is, however, highly visible as it soars searching for food. Its call is distinctive and is easily recognized. This small raptor is probably better represented on the Atlas map than most birds of prey.

The habitat of this species comprises fairly large tracts of forest, either mixed coniferous and deciduous, or deciduous. In a study of the Broad-winged Hawk in Chautauqua County from 1978 to 1980, Crocoll (1984) found that it nested in upland maple-beech-hemlock associations, and in an adjacent conifer plantation. Although the nest trees included a variety of deciduous and coniferous species, fully one-third of the nests were in tamarack. Crocoll also found that the Broad-winged Hawk tended to nest on slopes near a stream, pond, or swamp and about 26.5–223.7 m (87–734 ft) from the edge of the forest. In another study in the Central Adirondacks, of 14 nests found, 12 were in yellow birch. Evidently the structure of this tree provides ideal main crotches for supporting the stick nests built by this raptor (Matray 1974).

The nest is placed in the crotch of a main trunk, usually the first substantial crotch in the tree (Crocoll 1984) or on a horizontal branch against the trunk, never away from the trunk (Bent 1938). Nests have been found from 1 to 27.4 m (3–90 ft) above the ground (Bent 1938); in Chautauqua County the nests were 5.8–16.2 m (19–53 ft) high (Crocoll 1984). The nest is built of twigs and lined with bark and lichens. Both the male and the female construct the nest; however, only the female brings the bark that lines the nest cup (Matray 1974). This hawk, like many other raptor species, usually places small branches of green leaves under, over, or around the eggs or young. These are replaced regularly with fresh ones that the adult breaks from treetops with its bill (Bent 1938). This behavior is exhibited by many raptor species for reasons not satisfactorily understood (Matray 1974). Crocoll (1984) found that of 15 Broad-winged Hawk nests, 5 were old nests that the hawks had rebuilt.

Emanuel Levine

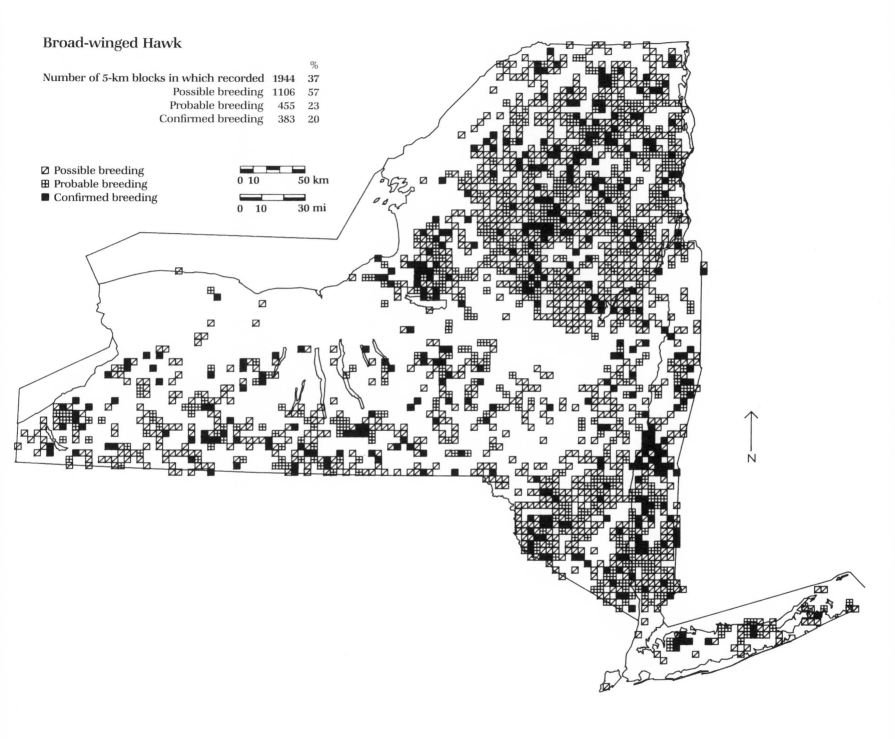

Broad-winged Hawk

		%
Number of 5-km blocks in which recorded	1944	37
Possible breeding	1106	57
Probable breeding	455	23
Confirmed breeding	383	20

◰ Possible breeding
⊞ Probable breeding
■ Confirmed breeding

0 10   50 km

0 10   30 mi

N

Red-tailed Hawk *Buteo jamaicensis*

The Red-tailed Hawk, a conspicuous buteo, is the most common and widely distributed large diurnal raptor in New York. Red-tailed Hawk numbers seem to have increased substantially in many areas during the last 30–60 years. In the Kingston area of nearby southeastern Ontario, there has been a great increase in breeding populations, particularly since the 1950s (Quilliam 1973). Hyde (1939), surveying the Lake Ontario shoreline during two summers, recorded a few scattered pairs or singles; today the species is common in that area.

Eaton (1914) noted the Red-tailed Hawk breeding in all parts of New York, primarily in upland areas. Bull (1974) also considered the species to be a widespread breeder, and he indicated it was present in remote areas of the Adirondacks as well as in the Long Island pitch pine–scrub oak barrens. Surveys by Hagar (1957) and Minor and Minor (1981) have found the species to be common in a variety of habitats.

The Atlas distribution corresponds well with the known historical range in the state. Atlas observers found the Red-tailed Hawk to be present throughout most regions. Evidence of breeding was absent or infrequent only in much of heavily urbanized New York City and Nassau County and parts of the well-forested mountain regions. Atlas data illustrate widespread breeding in the Central Adirondacks and Adirondack High Peaks, which was previously poorly documented, as was the bird's nesting in heavily developed areas. Minor and Minor (1981) first documented breeding in such densely populated counties as Erie, Monroe, and Onondaga.

Atlas observers documented breeding of the Red-tailed Hawk in more

blocks than were counted for any other raptor species. Because of its propensity to nest on the edges of woodlots overlooking open areas, nests were not difficult to find, and 457 active nests were located. Fledglings were recorded in more than 400 blocks. It is unlikely that this species is under-represented on the Atlas map.

The Red-tailed Hawk adapts to a wide variety of breeding habitats: from large city parks and cemeteries to woodlots mixed with agriculture to heavy forests with some open areas. As long as it is not harassed by humans and adequate nest sites and foraging areas exist (Minor and Minor 1981), coexistence with people even in areas with substantial environmental modification is possible. Although changes in land-use patterns may greatly affect productivity (Howell et al. 1978), the species' overall adaptability to human changes in the environment has permitted it to maintain or increase its population in many areas. Such success contrasts sharply with population declines noted in many raptor species, such as the Red-shouldered Hawk, which have been adversely affected by habitat alteration (Bednarz and Dinsmore 1982). Some modification of habitat, including increased forest fragmentation and agriculture, apparently even benefits this species by providing additional nesting habitat and permitting access to areas previously better suited to other raptor species (Bednarz and Dinsmore 1982).

The Red-tailed Hawk builds a large, conspicuous stick nest lined with twigs and bark. The nest is usually built on a branch of a large deciduous tree; conifers are rarely used (Eaton 1914; Bull 1974). The nest tree is often located along the edge of the woods or in the interior of small woodlots, but may be isolated in an open space. It is quite conspicuous during the leafless period. New York nests are most often placed in trees, but Andrle (1969) reported a nest on a cliff face along the Ontario side of the Niagara River gorge.

Gerald A. Smith

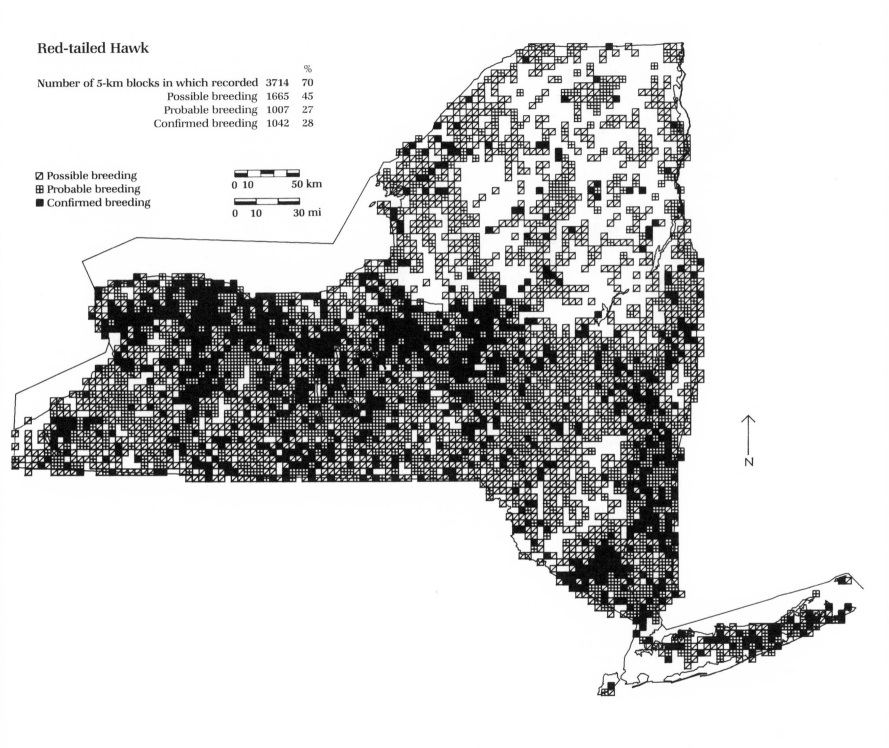

Red-tailed Hawk

		%
Number of 5-km blocks in which recorded	3714	70
Possible breeding	1665	45
Probable breeding	1007	27
Confirmed breeding	1042	28

◨ Possible breeding
⊞ Probable breeding
■ Confirmed breeding

0 10   50 km

0   10   30 mi

N

©KLA-C 1987

Golden Eagle *Aquila chrysaetos*

The Golden Eagle has never been common in the state, even though New York was formerly the eastern stronghold of its breeding range. In the Northeast only New York, New Hampshire, and Maine have had breeding Golden Eagles (Spofford 1971). Unfortunately this magnificent raptor has been extirpated as a breeding species in the state and now breeds only in Maine, where 4 active nests were reported in 1985 and 2 in 1986 (Nye, pers. comm.).

According to Spofford (NYSDEC files), only six areas in the state offered solid evidence of breeding: five in the Adirondacks and one in the lower Hudson Valley. Although Eaton (1914) stated that it was formerly an uncommon breeder in the Hudson Highlands, Catskills, and Adirondacks, there was no evidence it was still breeding there by the early 1900s.

The first verified nest for New York was found in 1957 by G. T. Chase on an escarpment in St. Lawrence County. A young bird was successfully fledged that year. Although subsequent nesting attempts were made, no other young were known to be raised at that site. The last known successful New York nest produced 1 eaglet in 1970 (NYSDEC files). The Atlas map reflects Golden Eagles observed in the area of that eyrie in 1980, 1981, 1983, and 1984. In March 1981 a pair was observed repairing and adding to the nest, but no eggs were laid.

Golden Eagles in the East appear commonly to experience nesting failure. According to Spofford's calculations (1971), annual productivity over the 20-year period 1951–70 was very low: an average of 0.15 eaglets fledged per site. He speculated that an uncertain food supply may have been one cause of irregular breeding.

Mortality from other than natural causes may have further reduced this small population. From mid-1946 to 1965, 10 Golden Eagles were found either dead or injured in the Adirondacks: 6 were caught in traps, 2 were shot, 1 was injured on a high tension line, and 1 died of unknown causes (NYSDEC files).

The NYSDEC makes annual aerial surveys of all historic and many potential Golden Eagle nest sites in the Adirondacks in search of nesting eagles, but like the Atlas survey, these surveys have located no nesting pairs. The Atlas record in Cattaraugus County was of a lone adult observed in June 1984 and probably represented an unmated wandering bird.

The Golden Eagle prefers wild areas with open and edge habitat where small game is abundant. An inventory to measure changes in vegetative cover types from 1942 to 1968 around known historic Golden Eagle eyries was conducted by the Habitat Inventory Unit of the NYSDEC. At all but one eyrie, the amount of open areas had significantly decreased. Many believe the loss of an already limited amount of open space may be the major reason for this eagle's extirpation as a breeding species in the state (NYSDEC files).

Most of the Golden Eagle nests in New York were placed on inaccessible cliff ledges overhung by a protective tree or rock. (The failure of nesting to succeed at one eyrie was believed to have been the result of exposure to drenching snow melt early in the season.) This species also nests in trees, with 1 New York nest located 27.4 m (90 ft) above the ground in a white pine. The nest was made of dead sticks and brush of all kinds, as well as boughs of mountain ash, American beech, white pine, and red pine. All former New York eyries were located at elevations between 457 m and 792 m (1500–2600 ft) (NYSDEC files).

In addition to annual surveys for breeding Golden Eagles, the NYSDEC has taken steps to minimize disturbance at all historical eyries in the hope that one day this eagle will nest again. Expectations are not high, however, given the limited number of individuals observed in the East and changes in habitat.

Janet R. Carroll

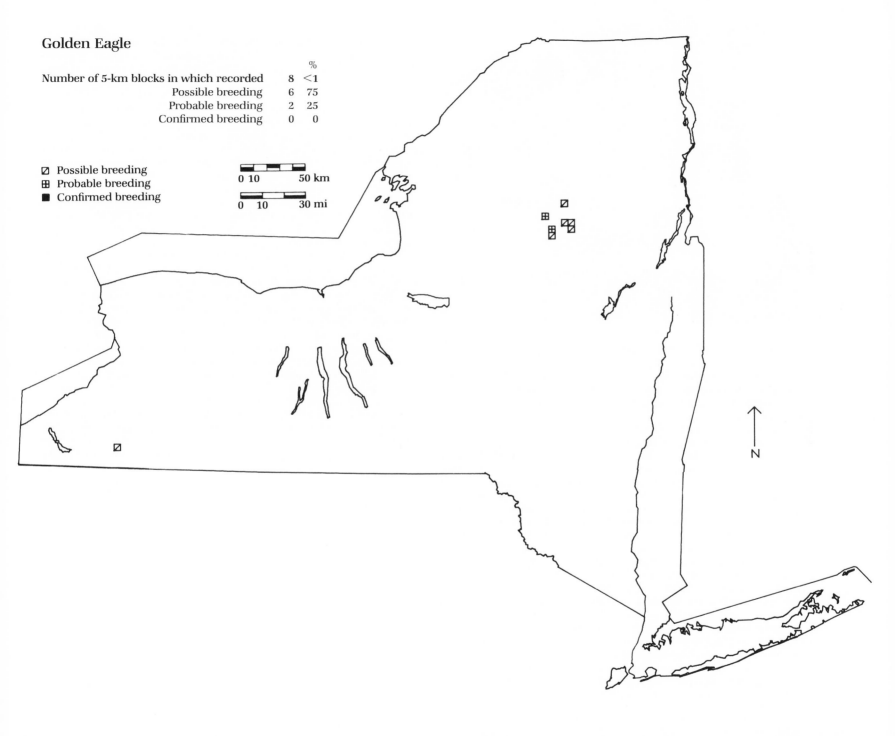

Golden Eagle

Number of 5-km blocks in which recorded	8	% <1
Possible breeding	6	75
Probable breeding	2	25
Confirmed breeding	0	0

☑ Possible breeding
⊞ Probable breeding
■ Confirmed breeding

0 10   50 km

0 10   30 mi

N

©KLA·C 1987

American Kestrel *Falco sparverius*

The American Kestrel, a small, colorful falcon, is the most common and widely distributed diurnal raptor in the state. With minimal human effort this adaptable species will likely remain a part of New York's avifauna for decades to come.

According to Eaton (1914), the American Kestrel was a common summer resident in all parts of the state. He noted that it was found most frequently in agricultural regions, less so in heavily forested areas. Hyde (1939) found it to be a common breeder along Lake Ontario and the St. Lawrence River. Bull (1974) noted that in addition to being a widespread breeder in farm country, it was also reported nesting in city parks, airports, and other open spaces in highly urbanized areas. It also breeds in woodlot edges, beaver meadows with standing snags, and borders of forest openings. The American Kestrel and the Red-tailed Hawk share the distinction of being the only New York breeding hawks that apparently did not suffer serious population declines during the first half of this century. Several studies of the American Kestrel, including those by Heintzelman and Nagy (1968), found that pesticides have not affected this species' reproductive success. Moreover, its ability to tolerate a wide range of breeding habitats, including heavily modified sites, has largely spared it from the negative effects of habitat alteration.

Atlas observers found this falcon to be widespread, except in the more extensively forested higher elevations of the Adirondacks, Tug Hill Plateau, Taconic Mountains and Rensselaer Hills, Catskill Peaks and adjacent hill and highland ecozones, and the Allegany Hills. Particularly striking is the absence of nesting kestrels in much of the Central Adirondacks and in the Eastern and Western Adirondack Foothills, probably as a result of the contiguous forest cover in most of the area. The American Kestrel was recorded, however, in the Adirondack High Peaks and other parts of the northern Adirondacks where open areas are available.

Atlas data are consistent with the species' known historical range. Comparison of these data with those of Eaton (1914) shows little change in the American Kestrel's New York breeding range. The Atlas survey illustrates well this species' tolerance for breeding in heavily urbanized situations where suitable habitat exists. The kestrel was found breeding within the boundaries of the state's largest cities.

The American Kestrel was reported in more Atlas blocks than all other raptors except the Red-tailed Hawk. Its distinct plumage and propensity to forage in open areas make it highly visible and easy to recognize. Observers were fairly successful at finding nests, locating 346. Observations of fledglings and adults feeding young, however, provided the most common "confirmations."

Where suitable nest cavities in trees, buildings, or nest boxes and sufficient nonforested foraging areas are present, this raptor is likely to occur. When not directly disturbed, breeding kestrels are quite tolerant of human activities, readily nesting near human habitation. In recent years increasing public appreciation of raptors and the resulting decreased mortality from shooting has undoubtedly benefited this tame and potentially easy target. Suitable nest cavities are harder to find as dead trees are removed for firewood, which may decrease the population in some local areas. Increased efforts to retain dead trees and to construct nest boxes would help solve these problems.

American Kestrel nest sites are usually in cavities in dead or injured trees or telephone poles, nest boxes, and crannies of old buildings (Bull 1974), and in deserted nests of other hawks (Eaton 1914). The eggs are laid directly on the cavity floor on whatever debris is there; no nesting material is added.

Gerald A. Smith

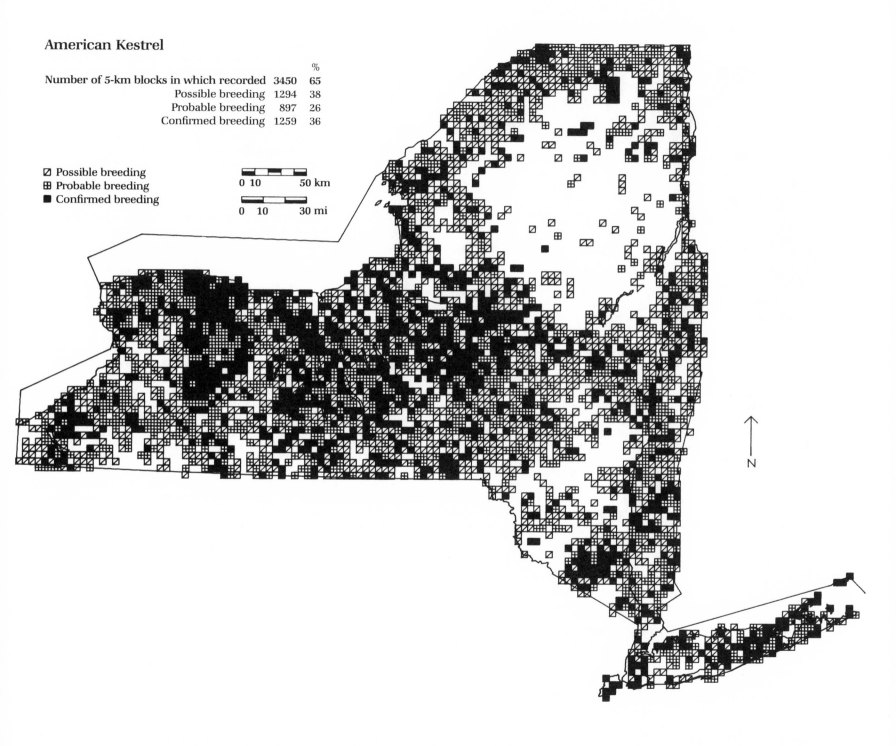

American Kestrel

Number of 5-km blocks in which recorded 3450   65

		%
Possible breeding	1294	38
Probable breeding	897	26
Confirmed breeding	1259	36

▨ Possible breeding
⊞ Probable breeding
■ Confirmed breeding

0 10   50 km

0 10   30 mi

N

Peregrine Falcon *Falco peregrinus*

The endangered Peregrine Falcon is now one of the better known raptors in the state. The reintroduction program run by Cornell University's Peregrine Fund and operated in New York with the support of the NYSDEC is highly publicized. As a result of reintroductions since 1974, New York now has 5 active eyries, 4 of which are shown on the Atlas map. The fifth was discovered in 1986 along the southern shore of Lake Champlain. A survey conducted in the spring of 1985 resulted in observations of 17 peregrines, including 7 pairs and 3 individuals at a total of 10 locations in the Adirondack and New York City areas (Gilroy 1985).

At one time there were thought to be 350 pairs of Peregrine Falcons in the eastern United States, occupying 275 eyries. According to Bull (1974), there was strong evidence that at least 40 eyries were occupied in New York during the late 1800s until 1940. These eyries were located within the Central Adirondacks, Lake Champlain Valley to Lake George, the Helderberg Highlands, Catskill Peaks, Shawangunk Hills, lower Hudson Valley, and Finger Lakes Highlands (Eastern Peregrine Falcon Recovery Team 1979). The earliest published report of nesting in the state came from the Helderbergs, where a clutch of 4 eggs was found in 1884 (Lintner 1884).

The extirpation of the Peregrine Falcon in the East was dramatic. Although some decline was noted before 1940, between the late 1940s and early 1950s observers began to notice serious problems at the eyries. Eggs disappeared, pairs attempted but failed to raise young from first and second clutches, and eggs were found broken. Occupancy at some eyries began to be sporadic, and finally eyries were abandoned. By 1957 only single birds were observed at the cliff sites, and by 1962 no peregrines were seen (Eastern Peregrine Falcon Recovery Team 1979). In New York the last reported successful nesting was in 1956 (Bull 1974). The cause, indiscriminate use of the pesticide DDT and its effect on reproduction, is well documented.

Successful captive breeding of the Peregrine Falcon by Heinz Meng in the early 1970s and later by the Peregrine Fund, under the direction of Tom Cade, raised hopes that a method could be developed by which this falcon could be reestablished in the East. The Peregrine Fund began an intensive breeding program and later began to release the captive-raised birds to the wild. As of 1986, 148 peregrines had been released in New York. From 1983 to 1986, 26 young were produced at the 5 eyries in the state. Two of these eyries are at sites formerly occupied by peregrines (Loucks 1986b). The goal of the Eastern Peregrine Falcon Recovery Plan of the USFWS is to establish a nesting population of approximately 20 pairs in New York (Cade and Dague 1985).

The continued presence of DDT and its metabolites in the environment has been cause for concern because of their potential effects on peregrine reproduction. The Peregrine Fund has been analyzing addled and infertile eggs and egg shell fragments collected from eyries in the East. Although these chemicals are still present, the levels do not appear to be high enough to impair reproduction (Cade and Dague 1985).

Peregrine eyries are commonly found on ledges on cliff faces or escarpments with a commanding view of the surrounding countryside. This falcon does not build a nest, but lays its eggs in a shallow scrape it makes on bare or grassy ledges. Lintner (1884) described one such eyrie: "The eggs . . . were placed upon the bare surface of a ledge in an extremely wild situation; there was no appearance of a nest, but the eggs were surrounded merely by a few bones and feathers." An Adirondack eyrie used in 1985 was located on a ledge facing southwest approximately 10.1 m (33 ft) from the top of a 45.7-m (150-ft) high cliff. The nest ledge was partially protected by a small white pine growing above it. The cliff overlooked a vast expanse of Adirondack wilderness (O'Brien, pers. comm.).

In New York City peregrines now nest on the superstructures beneath bridge roadways. They lay eggs on the gravel-like substance formed from tar and dirt that falls from the roadway above. The bridge nests are apparently well protected from the elements, and the falcons share the structures with nesting Rock Doves, on which they prey. Between 1943 and 1953 peregrines in the city also nested on skyscrapers, in one instance on a corner shield ornament between the 15th and 16th floors of the St. Regis Hotel (Herbert and Herbert 1965).

Janet R. Carroll

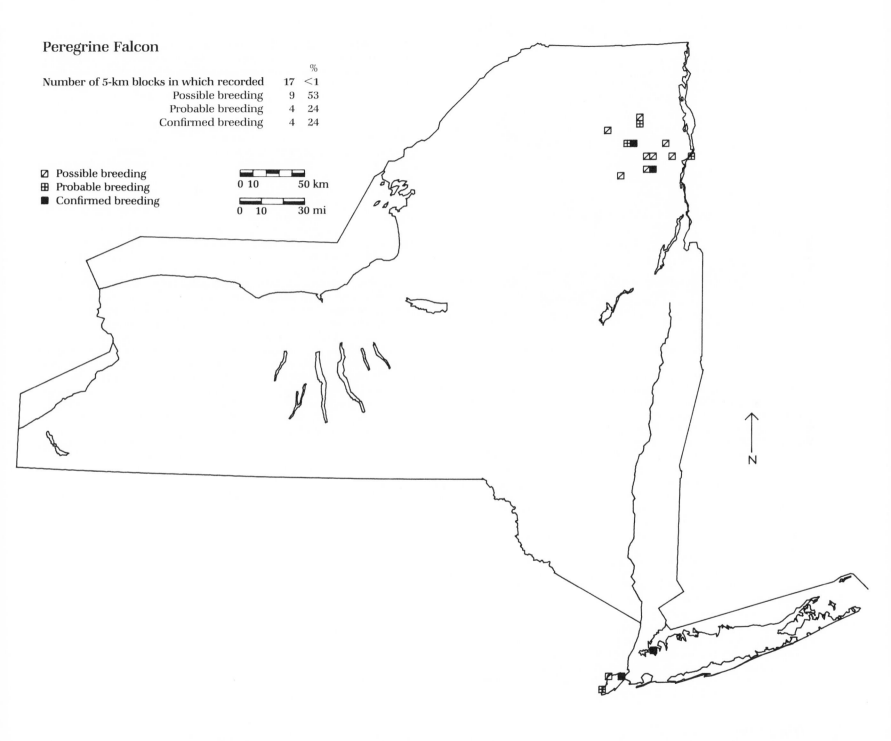

Peregrine Falcon

		%
Number of 5-km blocks in which recorded	17	<1
Possible breeding	9	53
Probable breeding	4	24
Confirmed breeding	4	24

◨ Possible breeding
⊞ Probable breeding
■ Confirmed breeding

0 10   50 km

0 10   30 mi

N

Gray Partridge *Perdix perdix*

The Gray Partridge is rather uncommon in New York. This introduced Eurasian species is found along a narrow belt that extends from Lake Ontario to Lake Champlain, following the northern border with Ontario and Quebec provinces and extending down the Lake Champlain Valley. The origin of the first introductions into New York is unknown, but the earliest releases were made at "several localities" in the state in 1909 (Eaton 1910). The first documented release site was Batavia, Genesee County, prior to 1917; these partridges bred, and a few may have survived until about 1940 (Bump 1941).

Czechoslovakia provided Gray Partridges for the initial major releases by the New York State Conservation Department from 1927 to 1932. As many as 27,750 adult birds were imported for statewide stocking (Brown 1954), although Bump (1941) noted the actual release of just 17,731 birds in "poor-good" condition. Transplant efforts were halted in 1932 because of poor results. By 1937 Jefferson, Onondaga, and St. Lawrence counties still had a few partridge (Austin 1980). After the end of World War II there were three or four colonies located in Allegany County and several east of Cayuga Lake (Wells 1951).

Northeastern New York has long been the stronghold of the Gray Partridge in the state. Atlas records in the four northeastern counties (Clinton, Franklin, Jefferson, and St. Lawrence) may derive from birds released by the Province of Ontario between 1909 and 1938 (Peck and James 1983). In 1937 Severinghaus found that Jefferson and St. Lawrence counties had only a few surviving Gray Partridge. "In the early 40's, he was convinced by surveys he made and contacts he had with Ontario provincial conservation officers, that the gray partridge population had increased greatly since 1937 as a result of an expanding population in Ontario, north of the St. Lawrence River extending south into New York" (Austin 1980).

In reality the Gray Partridge occurs in many more blocks than the Atlas map indicates. NYSDEC winter surveys made during the Atlas period recorded Gray Partridge in 10 towns in both Clinton and Franklin counties.

Since 1981 the NYSDEC has been trapping and transferring partridge from Jefferson and Clinton counties to areas in Cayuga and Ontario counties. Thus far 260 birds have been transferred to these two areas. An additional 180 have been transferred from Minnesota and the Province of Ontario. These releases have not succeeded in establishing new populations (Austin, pers. comm).

Gray Partridge in northern New York prefer farmland devoted to dairying. In this region from 1 to 6% of the land is planted in oats and from 10 to 18% in corn (Bureau of Census 1977). Much of the remaining open land within the Gray Partridge's range is pasture or hayfield with numerous hedgerows that provide adequate nesting and brood-rearing areas. In England, where the species is native, the greatest number of nests are in low, incomplete hedges (Southwood and Cross 1969; Sharrock 1976). A much larger portion of the partridge's range in the midwestern states, in the Canadian provinces, and in Europe is planted in cereal grains than is the case in New York. In this respect, the quality of New York's occupied range is marginal (Austin, pers. comm.). The nest is a simple depression, lined with a few dead leaves or woven dry grasses which sometimes form a canopy, and placed among bushes or in fields of clover, grain, or grass (Bent 1932; Peck and James 1983). Although a nest with eggs was reported in one Atlas block, no documented nest with eggs has ever been described in New York for this species.

John M. C. Peterson

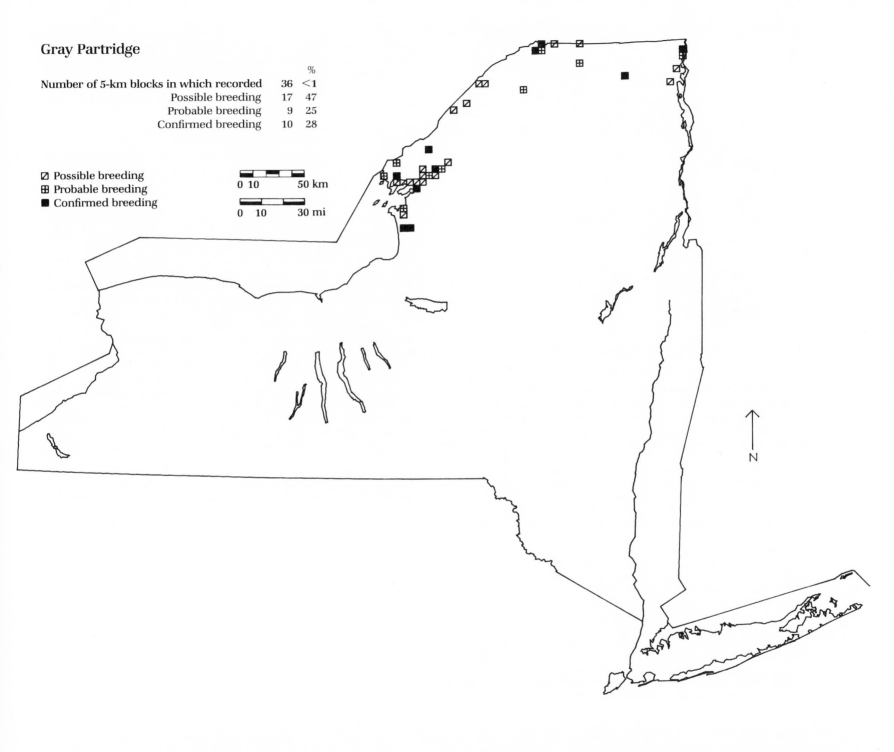

Gray Partridge

		%
Number of 5-km blocks in which recorded	36	<1
Possible breeding	17	47
Probable breeding	9	25
Confirmed breeding	10	28

☑ Possible breeding
⊞ Probable breeding
■ Confirmed breeding

0 10   50 km

0 10   30 mi

N

Ring-necked Pheasant *Phasianus colchicus*

The Ring-necked Pheasant, one of the true pheasants most characteristic of the eastern China subspecies (Brown and Robeson 1959), was introduced into the state and is now a common resident. DeGraff (1975), in reviewing the species' history in New York, noted that releases were first made in 1892 on Gardiners Island, off Long Island, followed by another release in 1903 on the Wadsworth estate in the Genesee River Valley, Livingston County. The first game farm in the state (the second in the United States) was established in Sherburne, Chenango County, in 1908 (Whipple 1935). By 1946 six more farms had been added, and releases of as many as 60,000 birds were being made annually (Brown and Robeson 1959). The first hunting season was initiated in 1908, and by 1926 as many as 200,000 pheasants were reported harvested each year (DeGraff 1975).

In 1942, however, the population began to decline, and for six years the trend continued as the hunting take dropped 20,000 birds annually. Hunting was then restricted, and releases were increased to 190,000 birds in 1947. In 1947 and 1948 almost all of the Great Lakes Plain was closed to hunting. Slowly, the pheasant population began to recover, until by the late 1960s population levels on the Great Lakes Plain reached an all-time high (DeGraff 1975).

Then, in the early 1970s, the population again declined, this time a result of changes in land use and severe winters. Loss of the Federal Soil Bank and Crop Set Aside programs was largely responsible for loss of habitat. Small general farms became large dairy farms. Fields were enlarged, hedgerows removed, and corn became the primary crop. Fallow grasslands were either plowed or reverted to seedling and sapling stage forests. These changes decreased the type of habitat needed to sustain high pheasant populations (Bureau of Wildlife 1979).

From 1979 to 1982 the NYSDEC monitored wild hens with transmitters to determine productivity, mortality, and habitat use. Nesting success and brood survival were poor, and mortality was high in winter and early spring. It was found that 94% of the monitored hens were killed by predators, mainly red and gray foxes, Great Horned Owls, and Red-tailed Hawks. This apparent increase in predation was attributed to loss of protective cover (Penrod and Austin 1985).

A pilot study by NYSDEC and Soil and Water Conservation District personnel will be conducted to attempt to establish and maintain grass cover in Genesee, Livingston, and Niagara counties on a cost-sharing basis with farmers and landowners. Such a project will benefit not only pheasants but also many other grassland birds such as Vesper, Grasshopper, and Henslow's sparrows (Penrod and Austin 1985).

The Atlas distribution identifies the primary, secondary, and marginal breeding ranges of the Ring-necked Pheasant (Bureau of Wildlife 1979). It is well established on the Great Lakes Plain, much of which is considered the primary range of this species. The populations along the Hudson Valley and on Long Island consist mainly of birds released at hunting preserves, although there are some areas where wild populations exist. Records on the map from elsewhere are probably of released birds. The species' distribution is tied to heavy agricultural use of land. As the Atlas map indicates, on the Appalachian Plateau, where farmland is broken up by large forested tracts, the pheasant does not do well.

The Ring-necked Pheasant frequents cropland and cultivated fields, undisturbed grassland, cat-tail–bulrush marsh, and maritime dune system (on Long Island), but except during the severest winter weather it avoids woodlands. Most of this pheasant's life is spent in grassy fields and, in the winter, along hedgerows and in cat-tail swamps (Penrod and Austin 1985).

The pheasant places its nest on the ground in open weedy fields, bushy pastures, or hay fields. It uses a natural hollow or one scraped by the female and lines this with weed stalks, grasses, and leaves (Harrison 1975).

Stephen W. Eaton

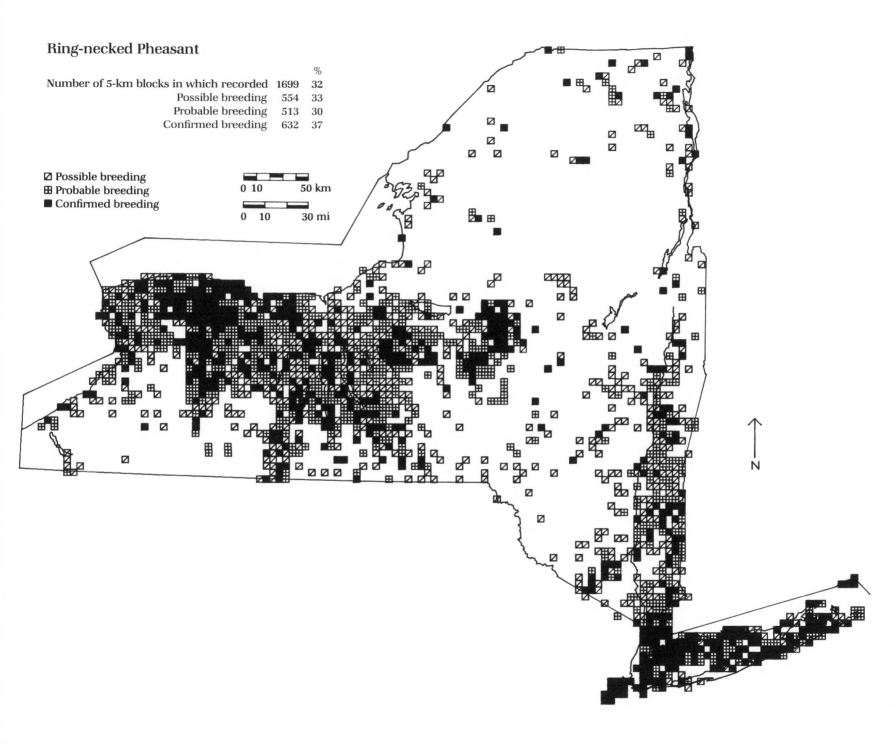

Ring-necked Pheasant

		%
Number of 5-km blocks in which recorded	1699	32
Possible breeding	554	33
Probable breeding	513	30
Confirmed breeding	632	37

◩ Possible breeding
⊞ Probable breeding
■ Confirmed breeding

0 10   50 km

0 10   30 mi

N

CJ Page © 1987

Spruce Grouse *Dendragapus canadensis*

An unobtrusive resident of the boreal forest, the Spruce Grouse is uncommon even in its New York stronghold, where Franklin and St. Lawrence counties meet in the Western Adirondack Foothills. South of this contiguous range, the Spruce Grouse becomes locally rare in the Central Adirondacks, and rare to casual in the Adirondack High Peaks.

The Spruce Grouse was more abundant and widespread before the Adirondacks were settled. Merriam (1881) still found this bird "tolerably common in certain localities," noting it as common near Brown's Tract Ponds in September 1879 and finding it near Big Moose in 1880 and 1882; the Atlas map suggests that only a relict population remains in this area near the Hamilton-Herkimer County line a century later. Between 1874 and 1877 Roosevelt found what was then known as the Canada grouse quite plentiful in some parts of Franklin County (Roosevelt and Minot 1929) near the nucleus of the present range. Some years later Eaton (1910) cautioned: "It was formerly common throughout the tamarack and spruce swamps of the North Woods but for many years it has become scarcer and scarcer, until it is now threatened with extermination in our State." Bull (1974) presented a map showing 16 known New York breeding localities in six counties. A survey conducted from 1976 to 1980 by Chambers (1980a, 1980b) found Spruce Grouse at 27 sites distributed through adjacent parts of Franklin, Hamilton, and St. Lawrence counties. The Spruce Grouse population apparently declined during the period 1977–79, but has increased since then (Bouta 1987).

Eaton (1910) noted that the unwary nature of the Spruce Grouse allowed whole flocks to be shot by hunters. Saunders (1929a), however, added: "Its tameness has always made it easy to kill, yet there is so much wild country still in the region, so many mountain slopes where no trail has yet been cut, that it hardly seems that man alone can have been responsible for its decrease." Neither Eaton nor Saunders called attention to habitat destruction brought about by logging of spruce forests and flooding of vast tracts of forest behind dams, both of which must have had an impact on the species.

The Atlas map shows Spruce Grouse recorded in 27 blocks in five counties. The range does not extend as far west as Bull's 1974 map, since no Atlas records were obtained in Fulton or Herkimer counties; it does extend farther east than the map presented by Fritz (1977). Although the Atlas map is generally representative of the 1980–85 range of Spruce Grouse, coverage was incomplete because of the difficulty of access into the large tracts of spruce–balsam fir habitat. Records from a continuing study of this grouse were included in the Atlas data.

The boreal acid bog forest that remains in the Adirondacks is patchy. According to Bouta (1987), "Small, isolated patches of conifers that are rapidly recolonized [by the grouse] after extirpation [of the birds] may exhibit more preferred habitat characteristics than larger contiguous areas where populations [of grouse] have become extirpated." The remnant Spruce Grouse live on islands within a sea of potentially encroaching hardwoods, pines, and open space. Even in such islands of suitable habitat, these isolated grouse populations can lose their viability and disappear without continued immigration of birds from other areas (Fritz 1979).

The Spruce Grouse is aptly named. The western Adirondack heartland is generally boreal acid bog forest where this grouse prefers immature or uneven-aged spruce-fir stands. The male commonly displays in areas where conifers are larger and more widely spaced with an open understory and roosts near the display areas but where the trees are smaller and spaced more closely (Bouta 1987). It may feed and nest in more open areas with a ground cover of mosses, lichens, and shrubs. Typical shrubs include young conifers, bog rosemary, blueberry, Labrador tea, leatherleaf, speckled alder, sheep laurel, and withe-rod (Gradoni and Bishop 1981). Habitat preference within the Adirondack High Peaks region is not as well known but appears to be similar. Drainage is generally poor in these areas, and during summer the grouse may move to neighboring upland spruce forest to dust and eat blueberries (Fritz 1977).

Only 3 nests have been found thus far in New York. The first was found by John Ozard on 6 June 1977, near the Jordan River, St. Lawrence County. The nest was located on a dry upland spruce site completely concealed under a young black spruce and contained 4 eggs (Ozard, pers. comm.). The grouse's nest is generally a slight depression in the moss, lined with dead grass and leaves, and placed under a spruce bough or within brushy vegetation (Townsend 1932).

John M. C. Peterson

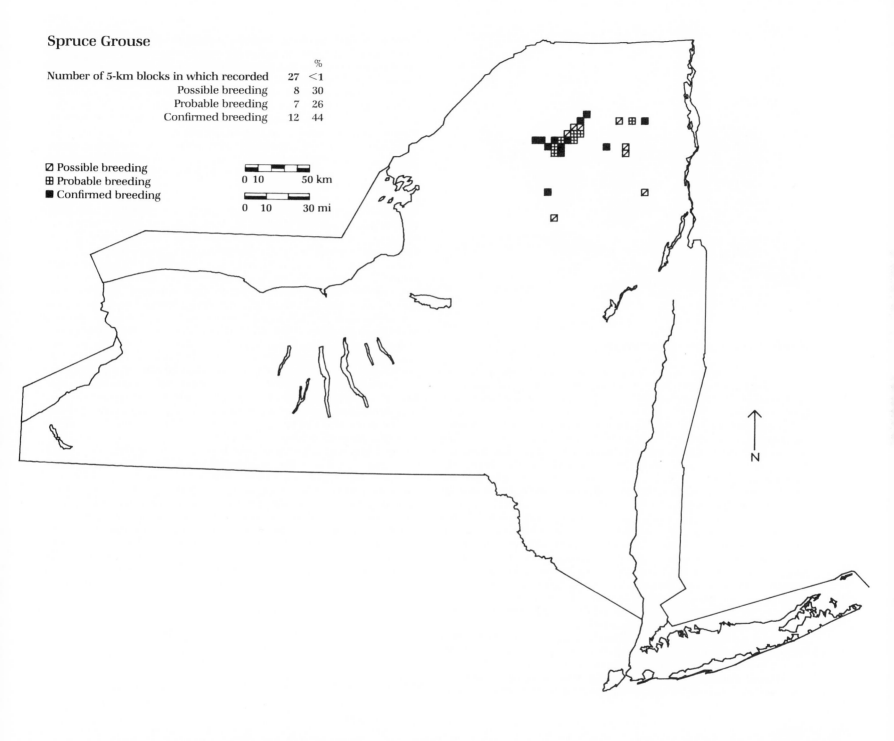

Spruce Grouse

		%
Number of 5-km blocks in which recorded	27	<1
Possible breeding	8	30
Probable breeding	7	26
Confirmed breeding	12	44

◩ Possible breeding
⊞ Probable breeding
■ Confirmed breeding

0 10   50 km

0 10   30 mi

N

Ruffed Grouse *Bonasa umbellus*

A favorite among birders, sportsmen, and gourmets, the Ruffed Grouse is still a common resident of the state, found wherever there is appropriate forest cover. Current populations have stabilized at levels very near historical highs.

Grouse numbers have not always been so good. During the late 1800s it was feared that the Ruffed Grouse would follow the Heath Hen to extinction. By 1880 the cumulative effects of land clearing, logging, and fire had taken their toll; only 25% of the state remained in forests. Fortunately, changes in agricultural practices and industrialization reversed this trend. A mere century later (1980), more than 60% of the state was again forested (Considine 1984).

This reforestation provided ideal habitat conditions for the Ruffed Grouse, which is especially well adapted to early successional stage forests, exactly the kind found from 10 to 40 years following land abandonment. Changes in habitat may still be affecting grouse populations today. Recent harvest data from New York indicate that fluctuations occur frequently and may be the result of the rather rapid local transition of young second-growth forest to maturity (Slingerland, pers. comm.). Gullion (1984) reported that alteration of successional stages can quite easily and quickly affect population numbers.

Bump et al. (1947) gave a thorough account of the history of the Ruffed Grouse in the state. They said, "With the exception of certain Adirondack mountain tops which rise to timberline or above, and the coastal marshes of Long Island, the entire area of the state originally constituted suitable Ruffed Grouse habitat." They stated that densities attained in some areas considerably surpassed those of primitive times, as settlement opened up the forest, and where mixed cover occurred. Some of the finest grouse coverts in the state are located on the Appalachian Plateau and in many of the valleys in the Catskills, particularly where abandoned farms have been reverting to edge and forest.

The Ruffed Grouse was formerly found in almost every county of the state except for those of New York City (Eaton 1910; Bump et al. 1947). Atlas workers did not find the Ruffed Grouse in the New York City area, and did not "confirm" breeding in either Nassau or Niagara counties where loss of forest habitat from urban sprawl has occurred. This species is poorly represented near cities and in areas with heavy agricultural land use. Atlas workers had little difficulty locating this bird. A drumming male grouse can be heard for a considerable distance, the drumming serving to maintain a territory approximately 3.2–4 ha (8–10 a) (Gullion and Martinson 1984). It was also relatively easy to "confirm" because it has a long period of unusually strong parental care; the female is particularly obvious when she and her brood are approached.

Bump et al. (1947) described grouse habitat in the state during different stages of the life cycle. Nesting occurs mainly in second-growth woodlands, less frequently in mature and "spot-lumbered" woodlands. If conifers are present in these types, usage increases. The grouse brood spends most of its time in slashings, overgrown lands, and second-growth hardwoods, which are characterized by diverse vegetation. Ruffed Grouse use aspen extensively when it is available, both as a food item and for cover. The very thick stem densities created by aspen root suckers offer the cover needed by grouse for protection from avian predators and provide buds that are eaten in winter and spring. Gullion (1984) believed that this tree was necessary to maintain high populations in Minnesota. In fact, under conditions of intensive aspen management, Gullion was able to virtually eliminate the cyclic population fluctuations for which the Ruffed Grouse is so well known.

The drumming of the male Ruffed Grouse attracts females, and after a brief courtship of but a few minutes, the female wanders away in search of a nest site, and there is no further association between the male and his mate. The nest may be placed a few hundred feet or half a mile away from the drumming log in an area where the hen has unrestricted vision for a distance of 15.2–18.3 m (50–60 ft) (Gullion and Martinson 1984). It is placed on the ground at the base of a tree or some upright object with protection on one side. The hen hollows a site and lines it with hardwood leaves, pine needles, and other available material, mixed eventually with a few grouse feathers (Harrison 1975).

Stephen W. Eaton

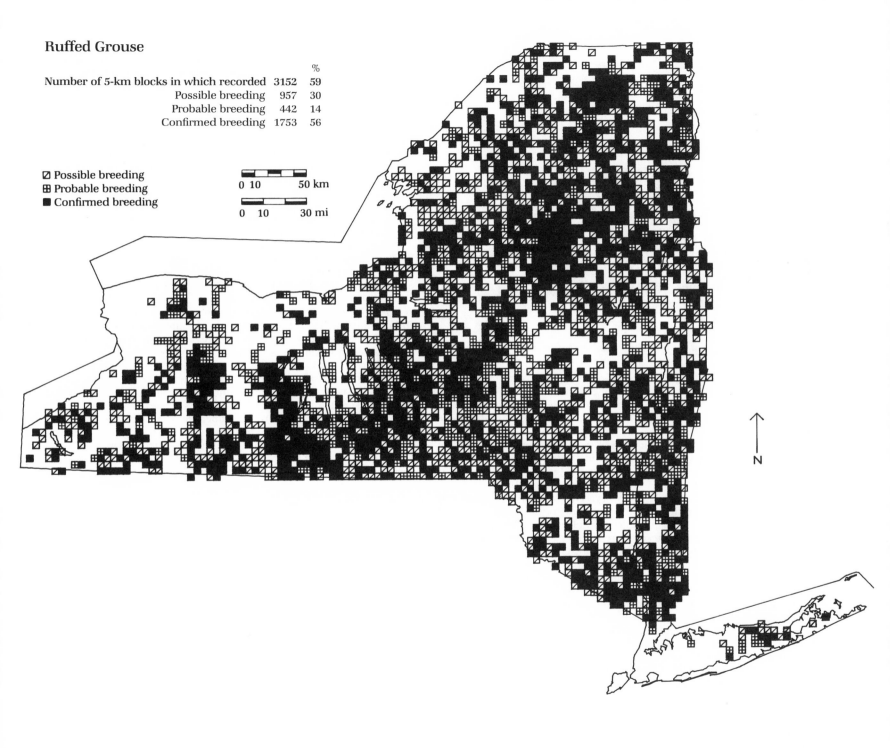

Ruffed Grouse

		%
Number of 5-km blocks in which recorded	3152	59
Possible breeding	957	30
Probable breeding	442	14
Confirmed breeding	1753	56

☑ Possible breeding
⊞ Probable breeding
■ Confirmed breeding

0 10   50 km

0 10   30 mi

N

Wild Turkey *Meleagris gallopavo*

The Wild Turkey, a distinctly American bird, is now common in most of southern New York and is managed as a game bird in the state. Based on data from banding, harvest, and distribution studies, the statewide population is estimated to be from 55,000 to 60,000 individuals (Glidden, pers. comm.).

By the mid-1800s the Wild Turkey was apparently near extirpation. De Kay (1844) stated that it was "only found in the counties of Sullivan, Rockland, Orange, Allegany and Cattaraugus." By the early 1900s it had disappeared from New York and New England (Eaton 1910). However, by the end of the 1940s the Wild Turkey began to reappear in Allegany and Cattaraugus counties as it expanded its range from McKean and Potter counties in Pennsylvania (Eaton 1953; DeGraff 1973). A small population had remained in the oak-hickory forests of central Pennsylvania, and then moved into the northern hardwood forests in the higher plateaus of northern Pennsylvania and into the Allegany Hills of New York. What started the resurgence is not completely understood, but the reversion of farmland to forest was probably the most important factor, in addition to the recovery of forests from excessive lumbering and a lessening of hunting pressure during World War II.

From 1952 to 1959 the New York State Conservation Department conducted a stocking program. More than 3000 game-farm birds were banded and released in 22 counties, but few survived (Foley 1963; Austin 1964). In the 1960s another attempt to increase the turkey population by trapping wild birds in Allegany State Park and transferring them to potentially good habitat elsewhere in the state was begun. As a result of this successful effort, between 1960 and 1986, 1540 wild birds were trapped and transferred to other parts of New York. In addition, Connecticut, Massachusetts, New Hampshire, New Jersey, Minnesota, Vermont, and Ontario have all established successful Wild Turkey populations using New York's wild-trapped stock (Glidden, pers. comm.).

Atlas workers "confirmed" this species in the forests of the Appalachian Plateau, Catskills, Taconic Highlands (probably birds that moved into the Taconics from Vermont, Massachusetts, and Connecticut), and the Lake Champlain Valley. It was also found in portions of the St. Lawrence and Great Lakes plains and in two blocks along the south shore of Long Island, where domesticated stock have been released (Riehlman, pers. comm.), as well as on Shelter, Gardiners, and Fishers islands. In southwestern New York, where the Wild Turkey population has been established the longest, the absence of turkey records in blocks where they undoubtedly occur reflects the lack of intense Atlas coverage.

The Wild Turkey frequents forested uplands, such as Appalachian oak-hickory, mixed mesophytic, maple-basswood rich mesic, hemlock–northern hardwood (now a favored woods in the northeastern states), and maple sugar bush. To a lesser extent it uses open-canopy uplands, such as successional old fields, cultivated fields, and recreation land. It seems most at home in mature stands, the understory of which is not too thick. In summer the diet of the Wild Turkey comprises mainly grass and sedge seeds. It switches to mast crops by mid-September and also relies on mast again when the snow melts in the spring (Eaton et al. 1970). Agricultural lands provide important brood range for this species and serve as alternate foraging areas during the winter (Porter 1980).

The Wild Turkey nests on the ground, usually under a downed treetop in the forest, often near an opening. It nests occasionally at the edge of the forest and rarely in open abandoned fields (Grzybowski 1974). The nest is a depression in the ground lined with whatever leaves or grasses are already present in the scrape. The average clutch size in New York is 12 (Glidden 1977), but I have found as many as 26 eggs in a single nest, the result of at least 2 birds laying in the same nest.

Stephen W. Eaton

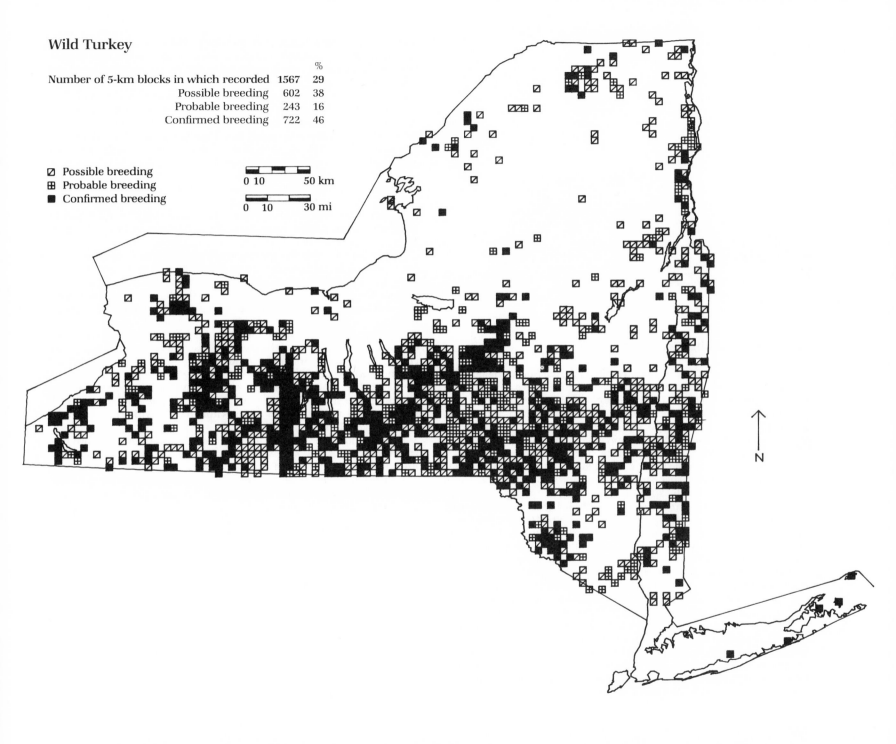

Wild Turkey

Number of 5-km blocks in which recorded	1567	% 29
Possible breeding	602	38
Probable breeding	243	16
Confirmed breeding	722	46

▨ Possible breeding
⊞ Probable breeding
■ Confirmed breeding

0 10   50 km

0 10   30 mi

N

Northern Bobwhite *Colinus virginianus*

The owner of one of the best-known bird calls, the Northern Bobwhite is described as an "uncommon to rare" breeding species by Bull (1964, 1974), an assessment that still seems to be accurate. As the Northern Bobwhite is a sedentary species, CBC numbers reported over the years can be used as an index of population fluctuation in the state. In 25 years of counts in Kings, Queens, Nassau, and Suffolk counties, where the only really viable wild population now exists, the maximum count was 828 birds in 1972 (American Birds 1973). The 1985 count turned up only 361 (American Birds 1986). Perhaps the word *uncommon* is too liberal a term to describe the status of the species.

Historically, the Northern Bobwhite was probably always absent from the Adirondacks and the higher Catskills, but Eaton (1910) wrote that it was fairly well distributed "as far north as the counties of Jefferson, Oneida, Saratoga and Washington." In an address to the American Ornithologists' Union 20 years later, he stated that he had seen the species "practically disappear from the interior of the state." In a list compiled by Frank Rathbun in 1879 for Cayuga County, the Northern Bobwhite was included as a common resident (Benton 1950); atlasing found "confirmed" breeding in only one block in the northern part of that county. In Westchester County, where there were few Atlas records, the bobwhite had once been resident in some numbers (Griscom 1923). The extensive quail population reported by Eaton (1910) was probably an artifact of exceptionally good habitat and mild weather. The small 19th-century farm with its diverse cereal crops, inefficient harvest techniques, weeds, large hedgerows, and limited predator populations must have presented perfect quail habitat. But as farming became cleaner, more mechanized, and more efficient, habitat quality declined. Moreover, the loss of farmland to forest succession and to development has only hastened the decline (Slingerland, pers. comm.).

In New York City the Northern Bobwhite bred in Central Park until 1893 (Griscom 1923); Cruickshank (1942) reported its gradual disappearance from the entire New York City area. In the 1950s and 1960s only 10% of the New York City region Christmas Bird Counts reported bobwhites, and rarely were more than a dozen individuals seen in any single reporting area (Audubon Field Notes 1950–1969).

The reasons for the decline of this species are many, but it should be pointed out that New York is the northern limit of its range (AOU 1983). Certainly, unregulated hunting early in this century was a factor, and hunting continues to take a toll: the estimated kill of this highly prized game bird in the United States in 1970 was *35 million* birds (Johnsgard 1973).

In New York the species is now virtually confined to Long Island. There are scattered records elsewhere in the state, but these are probably traceable to the releases of quail raised in captivity for hunting or dog training (Slingerland, pers. comm.). It is a difficult species to miss during the breeding season, as the males call constantly, often from exposed perches.

On Long Island it breeds in a number of habitats, including open fields of tall grass, in weedy and cultivated fields, along the edges of golf courses, and even in the open scrub pine forest. It requires nearly bare ground and associated herbaceous cover, with cultivated crops or a similar source of food and brushy cover nearby. It needs open country for breeding. The nest is a shallow depression on the ground lined with plant material and covered with plants from nearby, which it pulls over to hide the nest (Johnsgard 1973).

It is questionable which of the New York birds are of the native race *virginianus*. Beardslee and Mitchell (1965) described the unsuccessful introduction and stocking of non-native birds over many years, especially in western New York. Stocked birds of any species do not do as well as established native populations, but some may have persisted.

Emanuel Levine

Northern Bobwhite

		%
Number of 5-km blocks in which recorded	236	4
Possible breeding	50	21
Probable breeding	67	28
Confirmed breeding	119	50

◪ Possible breeding
⊞ Probable breeding
■ Confirmed breeding

0 10   50 km

0 10   30 mi

N

Black Rail *Laterallus jamaicensis*

The Black Rail's small size and secretive habits make it one of the most elusive of bird species, and in New York it is the rarest rail. Only a single Atlas record was documented, at Oak Beach Marsh, Long Island. New York is at the northern limit of its range, which along the Atlantic Coast is from New York to Florida (AOU 1983).

This rail was first found in the United States in 1836 (Allen 1900). Because of spring and fall records from several locations on Long Island and also in Yates and Schuyler counties, Eaton (1910) speculated that breeding occurred in New York. A nest was finally found by G. Carleton at Oak Beach Marsh on Long Island in May 1937; however, from the early 1940s until 1968 there were no records of this species at Oak Beach or elsewhere in the state (Post and Enders 1969). In the late 1930s measures were taken to control mosquitos on Long Island. These included ditching salt marshes and spraying with DDT. Although Oak Beach Marsh itself was not disturbed, habitat disturbance along the Great South Bay and contamination may have adversely affected the Black Rail (Post and Enders 1969).

In 1968 this rail was once again found at Oak Beach Marsh. Four individuals were trapped from 4 May to 20 June, although no nests were found (Post and Enders 1969). Records of the Black Rail from Oak Beach Marsh persisted until 1978. In 1980, the first year of the Atlas project, a Black Rail was heard in the marsh, and in 1983 a rail was observed in mid-May and then heard on several later dates. Records from the early 1970s were of an estimated 5 territorial males, but both Atlas records and records in 1977 and 1978 (Greenlaw and Miller 1982) appeared to represent only 1 bird.

The Black Rail typically nests on that part of the coastal salt marsh where only the highest tidal waters can reach. Inland, this rail chooses dense marshes with wet ground, preferring undisturbed marsh to disturbed areas (Todd 1977). The dominant vegetation at Oak Beach is cordgrass with common reed at the edges. The highest part of the marsh where the Black Rail was found by Post and Enders (1969) has patches of salt-meadowgrass and spikegrass interspersed with shallow pools. The rails were observed ranging into the cordgrass and common reed.

The nest of this rail is made of fine grasses loosely placed with a canopy of standing grasses; there is a passage to one side (Clark 1884). According to Carleton (pers. comm.), he and others found the nest at Oak Beach Marsh after flushing an adult and subsequently scouring the area. He noted, "The nest looked ordinary, I thought at the time, like a Red-winged Blackbird's nest." That nest is still the only Black Rail nest from New York in the collection of the American Museum of Natural History (Sloss, pers. comm.).

Janet R. Carroll

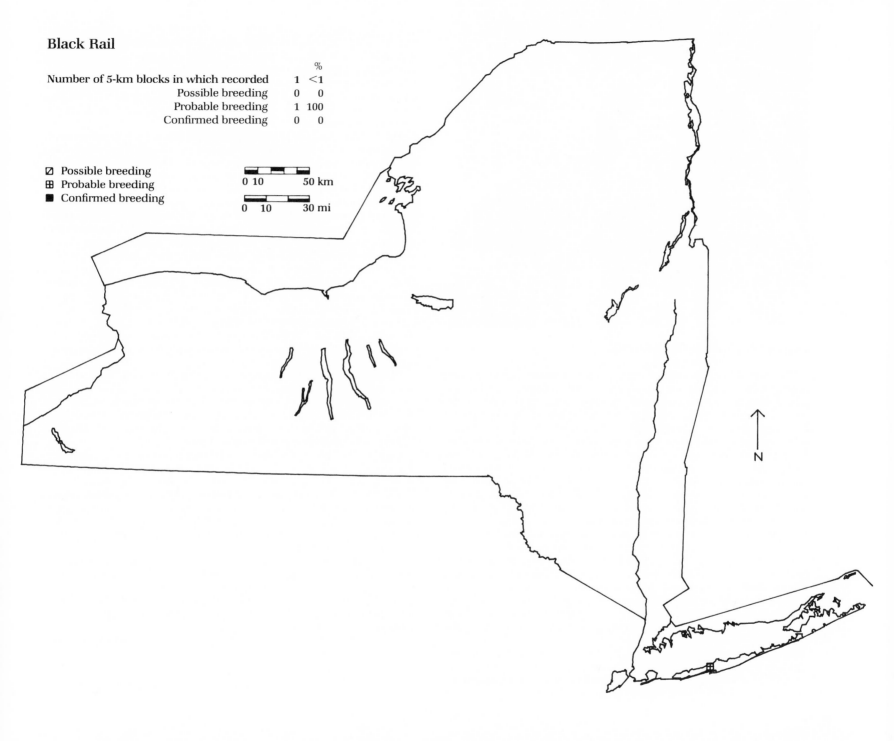

Black Rail

		%
Number of 5-km blocks in which recorded	1	<1
Possible breeding	0	0
Probable breeding	1	100
Confirmed breeding	0	0

▨ Possible breeding
⊞ Probable breeding
■ Confirmed breeding

0 10   50 km

0 10   30 mi

N

©KLA-C 1987

Clapper Rail *Rallus longirostris*

The Clapper Rail is a fairly common bird of New York's coastal salt marshes. In the mid-1800s it was considered to be abundant on Long Island (Giraud 1844) and was listed by Eaton (1910) as common on the coast, being particularly abundant along Long Island's south shore but uncommon in eastern Long Island. During the next 50 years the Clapper Rail was able to maintain a large population in the salt marshes along Long Island's southwestern shore. In 1950 at least 120 pairs were estimated along 7.2 km (4.5 mi) of salt marsh in the Town of Hempstead, southern Nassau County (Bull 1964). In the Hempstead marshes in 1965, 96 nests were located (MacNamara and Udell 1970). Another 60 pairs were found the same year on approximately 77 ha (190 a) of salt marsh to the east of Hempstead in the Town of Oyster Bay (Bull 1970).

Elsewhere on the coast the Clapper Rail was documented at salt marshes on Long Island's north shore at Lattingtown, Nassau County (12 pairs in 1954) and at Mount Sinai Harbor, Suffolk County (30 pairs in 1942). In eastern Long Island the species has always been an uncommon breeder with nesting records from East Hampton and Orient Point, Suffolk County (Eaton 1910; Bull 1964). The Clapper Rail also nested along the Harlem River in New York City until 1924 and was documented breeding in 1959 along the Hudson River at Piermont, Rockland County (Bull 1964).

During the 1970s Clapper Rail numbers apparently declined, particularly in the center of the species' range in the Town of Hempstead. A 14.6 ha (37 a) study area there contained 13 breeding pairs in 1966 but only 4 pairs by 1979. Similar declines were observed around the Hemp-

stead marshes (Spencer 1979; Zarudsky, pers. comm.). Potential causes of the decline included the effects of Hurricane Belle in 1976 and two successive hard winters that killed birds on the wintering grounds in the southeastern United States (Smith et al. 1978). Apparently the population up to the mid-1980s still had not recovered. Only 3 pairs of nesting Clapper Rails were located in the Hempstead study plot in 1985 (Zarudsky, pers. comm.).

The Atlas documents a shrinking range for this rail in comparison with that presented by Bull (1974). Breeding was not recorded in several of the former nesting areas. In Shinnecock Bay, southeastern Suffolk County, where the species was formerly a common breeder (Salzman 1985b), it was "confirmed" in only one Atlas block. Several areas of the Westchester County shoreline still support Clapper Rails, but on Staten Island, where the species was formerly common (Eaton 1910), no evidence of breeding was found. It formerly bred at Stony Brook Harbor, northwestern Suffolk County, but was recorded only as "possible" there during the Atlas period. In eastern Long Island, where it has always been uncommon, "confirmed" breeding was recorded in only two Atlas blocks in addition to the record at Shinnecock Bay. Along the north shore marshes evidence of breeding was recorded but not "confirmed." It is likely that most Atlas "possible" and "probable" records for this species represent actual breeding. The stronghold of the Clapper Rail on Long Island remains along the southwestern shore where it was frequently "confirmed." Loss of salt marshes along the coast may in part explain the contraction of the species' breeding range.

Of 137 nests located in Hempstead in 1965, 96 were located in either high or low marsh vegetation; 34 were found on slightly higher ground on levees and spoil deposits, and 7 nests were located in miscellaneous sites such as the top of a piling and on an abandoned cane chair (MacNamara and Udell 1970). This bird often nests near other species, including Willets and Common Terns (Zarudsky 1985b).

Clapper Rail nests tend to be elevated platforms in the vegetation. Runways are often visible around the nests (Johnson 1973). Of 137 Hempstead nests examined, 37 were domed by vegetation and the rest were open (MacNamara and Udell 1970; Johnson 1973). Increased tidal flooding that washed out nests and young is thought to be a possible cause of the decline in the Shinnecock Bay population (Salzman 1985b). Flooding may also have limited reproduction on the Hempstead marshes (MacNamara and Udell 1970).

If the factors limiting the state's Clapper Rail population are natural, then in time the population should return to its historical level, at least where extensive salt marsh habitat still remains. If human activity has caused the decline, the drop in Clapper Rail numbers may indicate widespread damage to New York's coastal environment, particularly its remaining salt marshes.

David M. Peterson

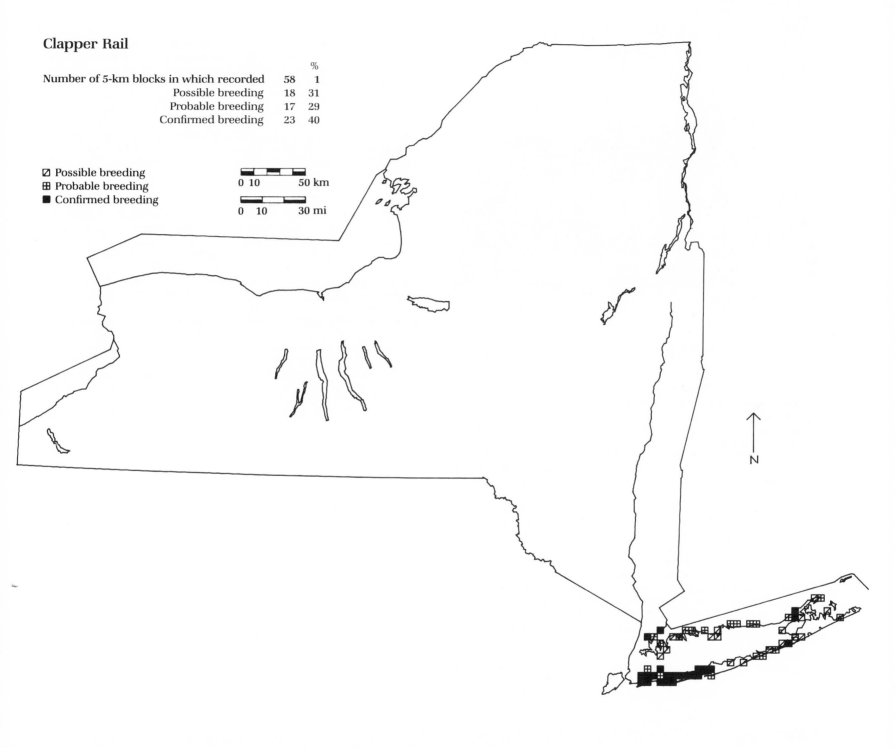

Clapper Rail

		%
Number of 5-km blocks in which recorded	58	1
Possible breeding	18	31
Probable breeding	17	29
Confirmed breeding	23	40

◩ Possible breeding
⊞ Probable breeding
■ Confirmed breeding

0 10   50 km

0 10   30 mi

N

©KLA-C 1987

King Rail *Rallus elegans*

The largest member of the North American rail family, the King Rail is now and has historically been a very rare nesting species in New York. Early accounts described its breeding range as the southern and middle parts of the country, excluding New York (Giraud 1844; Maynard 1881). The King Rail was apparently not observed in the state until the early 1900s. Both Eaton (1914) and Cooke (1914) described breeding in central and western New York, north to Tompkins and Erie counties. Although there was no early mention of the species on Long Island, Bull (1974) reported records from there, similar in frequency to those for other parts of the state.

According to Bateman (1977), its population stronghold was the coastal area from the Delaware Valley south to Florida. He also described an inland population in the Lake Erie marshes in Ohio. New York lies along the northern limit of this rail's range.

As the map of the King Rail's distribution illustrates, no evidence of breeding was documented on Long Island during the period of Atlas fieldwork, and few records were found elsewhere, none "confirmed." Of the five Atlas records, only two are actual observations of the rail. The others represent responses to tape-recorded calls. All records fall during the period 15 May to 2 July, well within the breeding season as described in Appendix C. A difficult species to locate, the King Rail was undoubtedly missed by Atlas workers. Its call is similar to that of other rail species, and its secretive habits make observation unusual. This description by an Atlas observer illustrates how one record was obtained. "We were first aware of the rail as it called steadily. . . . Because it was impossible to find in the marsh vegetation, we decided to tape its own voice and play it back, whereupon the rail became very excited and came toward the recorder. It was a very large chicken-sized rail, much larger than a Virginia Rail. The bird at one observation was in the tall grasses with neck outstretched trying to get a better view of us" (A. McHale, v.r.).

Not only is the species difficult to locate but its population, like that of other marsh birds, may have been affected by the loss of wetlands in the state. A 2-year study in Massachusetts during 1984 and 1985 failed to locate the species (Melvin, pers. comm.). Massachusetts, like New York, is at the edge of this rail's range and has never had a large population. Little is known about the population density of this bird in any of its breeding range.

Most rails are restricted to very specific breeding habitats; however, the King Rail seems to be less selective. Nests have been found in coastal salt marshes and brackish tidal marshes, freshwater marshes, shrub swamps, and upland fields associated with marshes. The Lake Erie population was found in extensive cat-tail marshes along the southwestern shore (Meanley 1969). All Atlas records are from freshwater marshes representing a variety of types of emergent vegetation particularly large, dense stands of cat-tails. The King Rail's habitat requirements, however, are not well known.

Meanley (1969) described the nest as "a round, elevated platform with a saucer shaped depression. It usually has a round or cone-shaped canopy and a ramp." Interbreeding of the King Rail and Clapper Rail has been documented in brackish tidal marshes where their breeding habitats overlap. Because hybridization does occur, some taxonomists question whether the Clapper Rail and King Rail are two distinct species (Meanley 1985).

Janet R. Carroll

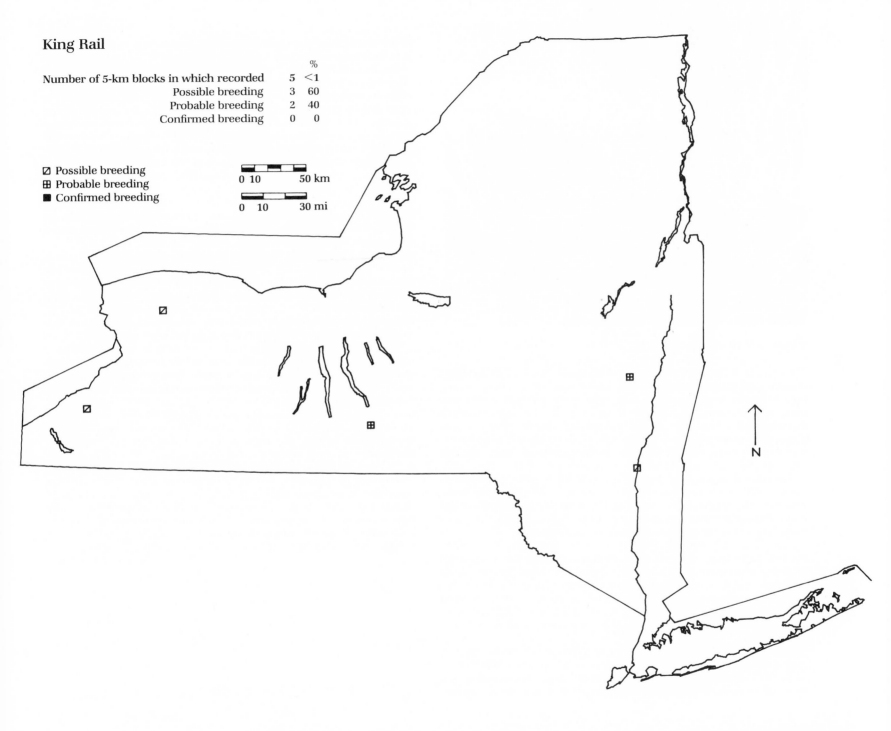

King Rail

Number of 5-km blocks in which recorded		%
	5	<1
Possible breeding	3	60
Probable breeding	2	40
Confirmed breeding	0	0

◩ Possible breeding
⊞ Probable breeding
■ Confirmed breeding

0 10   50 km

0 10   30 mi

Virginia Rail *Rallus limicola*

An elusive inhabitant of wetlands, the Virginia Rail may be the most common rail in the state today, although Eaton (1910) considered the Sora to be the most abundant. Zimmerman (1977) indicated that New York was in the area of highest breeding density for the Virginia Rail and considered it common as a breeding species. Concern has been expressed about the status of this species, and some observers believe the population has decreased since the 1940s because of loss of wetland nesting habitat (Post and Enders 1970; Zimmerman 1977).

Information on the historic distribution of the Virginia Rail is scarce. Eaton (1910) said that it was "a fairly common summer resident on the marshes of Long Island and central and western New York" and thought that it bred almost everywhere but the Adirondacks. It undoubtedly was found at most of the large cat-tail marshes in the state.

Atlas observers found the Virginia Rail to be concentrated in two areas: the marshes along the Great Lakes Plain (particularly the refuges and wildlife management areas in Orleans, Genesee, Monroe, and Onondaga counties); and the marshes along the St. Lawrence Plains and Transition. Elsewhere in the state, including the Adirondacks, it is widely scattered in various large and small marshes. A 1986 study of six tidal freshwater marshes along the Hudson River (Swift, pers. comm.) found only the Virginia Rail.

Atlas observers were encouraged to look diligently for rails. The Virginia Rail responds readily to recordings of its call and even calls of other rails, especially in the early breeding season. Because special efforts are required to find rails, however, it is likely that this species is under-represented on the Atlas map. It was extremely difficult to "confirm," and very few nests were found.

The Virginia Rail nests in freshwater marshes that have abundant emergent vegetation, especially cat-tail and sedge. It is also found in brackish tidal and salt marsh (DeGraaf et al. 1980). It usually selects the drier areas in the marsh in which to nest (Forbush 1912). Pospichal and Marshall (1954) found the water depth at pond nest sites in central Minnesota ranged between 12 and 44 cm (4.7–17.3 in). At areas where the vegetation seemed adequate but where the water exceeded that range, no nesting occurred. Lehman and Peterson (1983) advised Atlas observers that areas of even less than an acre in size that have "cattails with nearby shrubs such as willow, dogwood, and alder; also sedge meadows, marshes with mixed woody and herbaceous growth" were good habitats for the Virginia Rail. Swift (pers. comm.) found that along the Hudson River it nested in persistent cat-tail cover, which was often interspersed with purple loosestrife or scattered woody species.

Post and Enders (1970) found the Virginia Rail nesting in a salt marsh bordering Great South Bay at Oak Beach, Suffolk County. The standing vegetation of the unditched marsh was entirely cordgrass with an edge of common reed. Post and Enders said that the Virginia Rail was absent from most salt marshes in New York and suggested that ditching for mosquito control might be responsible, because it encouraged salt-meadowgrass, a species of salt marsh grass not used by this rail.

The nest is well concealed in marsh vegetation in drier areas or over water. It is built of cat-tails, coarse grass stems, or other available plant material; sometimes only rushes are used. The nest may be entirely of coarse material or have a lining of finer material. Over the nest there is usually a canopy of live sedge and rush (Harrison 1975). There is often, too, a loosely formed ramp leading up to the suspended nest (Post and Enders 1970). Post and Enders found that 6 of 8 nests in tidal salt marshes were flooded in May and June, and concluded that salt marsh may represent a marginal habitat for this species.

The Virginia Rail is listed as a migratory game bird. Except for Long Island, the hunting season on this species is open statewide; in 1985 this season ran from 1 September to 9 November. The daily bag limit was 25. Zimmerman (1977), in discussing management of this species as a game bird, recommended, among other things, protection of existing wetlands, restoration of marsh habitat throughout the breeding and wintering ranges, and measuring hunting pressure by permit-type hunting.

Stephen W. Eaton

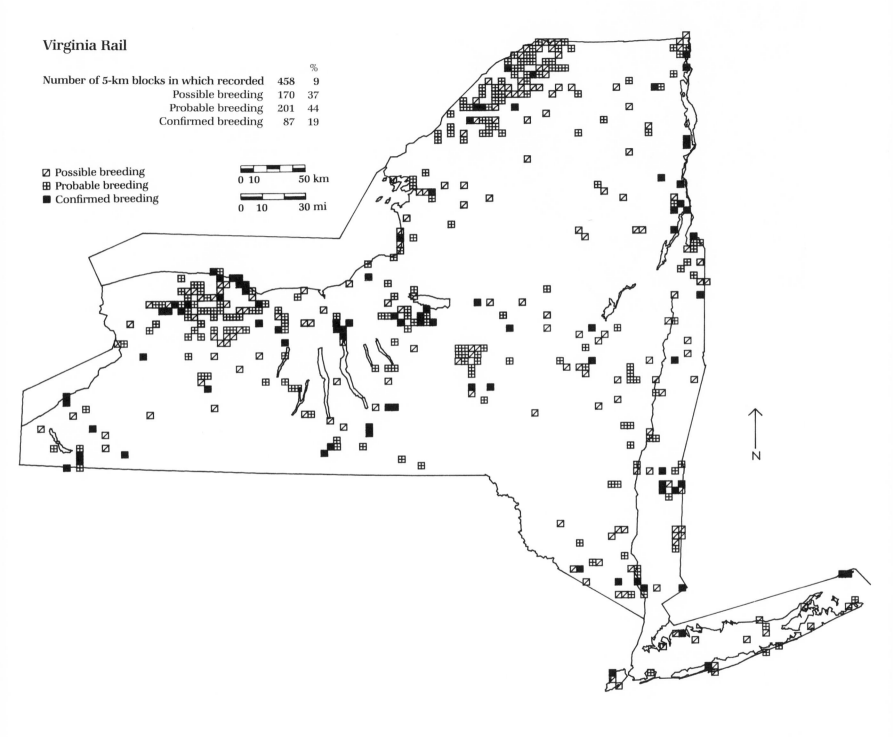

Virginia Rail

Number of 5-km blocks in which recorded	458	%
		9
Possible breeding	170	37
Probable breeding	201	44
Confirmed breeding	87	19

◪ Possible breeding
⊞ Probable breeding
■ Confirmed breeding

0 10   50 km

0 10   30 mi

N

Sora *Porzana carolina*

The Sora, a small, stocky rail once killed for the market by the thousands in the Delaware marshes of New Jersey (Stone 1937), is becoming scarcer in many parts of the state as wetlands are reduced or modified. Although Odom (1977) called the Sora the most common rail in North America, in New York today it appears to be uncommon.

Formerly the Sora was probably a common breeding species in the state. Eaton (1910) said that it was the most abundant species of rail, "being common in the marshes of central New York and the Great Lake region and probably breeding in nearly every county of the state, but . . . uncommon as a summer resident in our coastal district." The Sora has since decreased steadily in the northeastern United States (DeGraaf et al. 1980). Bull (1974) indicated that although this species was widely distributed in the rest of the state away from Long Island, it was not common. On Long Island he found only six breeding records—four on the north shore and two on the south shore—and noted that for more than 35 years up to 1974 there were no breeding records on Long Island.

The Sora continues to be widely distributed in large marshes in the state. Atlas records were concentrated at Iroquois NWR, Oak Orchard and Tonawanda WMAs, and at wildlife management areas on the St. Lawrence Plains. Sora records were absent from 12 counties in the state. There was only one record from Long Island, at the southern edge of Great South Bay in 1980. Greenlaw and Miller (1982) felt that this represented a "fugitive" population since the Sora was not found in 1977, 1978, or 1981. They also thought it conceivable that a population still occurred on Gardiners Island, although Atlas observers did not find the Sora there.

Atlas data appear to document a reduction in the range of this species. Because Atlas records of the Sora were hard to obtain, however, this rail may be under-recorded on the map. It does not seem to be as responsive to tape-recorded calls, as is the Virginia Rail, and it is difficult to flush (Walkinshaw 1940).

The Sora nests in freshwater marshes, ponds, swamps, bogs, grassy meadows, and sloughs that have abundant and dense vegetation. It prefers sedge, cat-tail, or bulrush, the mud and water in which are deep (DeGraaf et al. 1980). It nests more rarely in salt marshes where cordgrass and the common reed occur (Greenlaw and Miller 1982).

The nest is a loosely built structure usually about 15.2 cm (6 in) above the water level, with the base sometimes at or below it; it is located close to open water (Pospichal and Marshall 1954; Harrison 1975). It is supported and arched over by cat-tail or bulrush. Typically made of dead cat-tail blades, bulrush, or grass, it is difficult to see even at very close range. Sometimes the bird constructs a runway leading to the nest (Harrison 1975). Pospichal and Marshall (1954) found that the Sora's optimum breeding habitat in central Minnesota was ponds where cat-tail was abundant. Nesting densities increased with the increase in the amount of cat-tail.

Two of the nests found in salt marshes by Greenlaw and Miller (1982) were in erect green and persistent cordgrass, 9 and 15 m (29.5 and 49.2 ft), respectively, from a stand of common reed; another was built in cordgrass within a small, isolated clump of common reed. Two of the 3 nests were successful; the unsuccessful one was in the clump of common reed. The 3 nests were placed on a platform of material built directly on the mud and lacked the ramp that may be found leading to nests in freshwater marshes.

According to Odom (1977), "There were no accurate harvest figures for Soras but throughout many areas of the country, the potential harvest is high." If this bird is to be a game species, data on its abundance should be obtained. Population studies on this species and the Virginia Rail are urgently needed.

Stephen W. Eaton

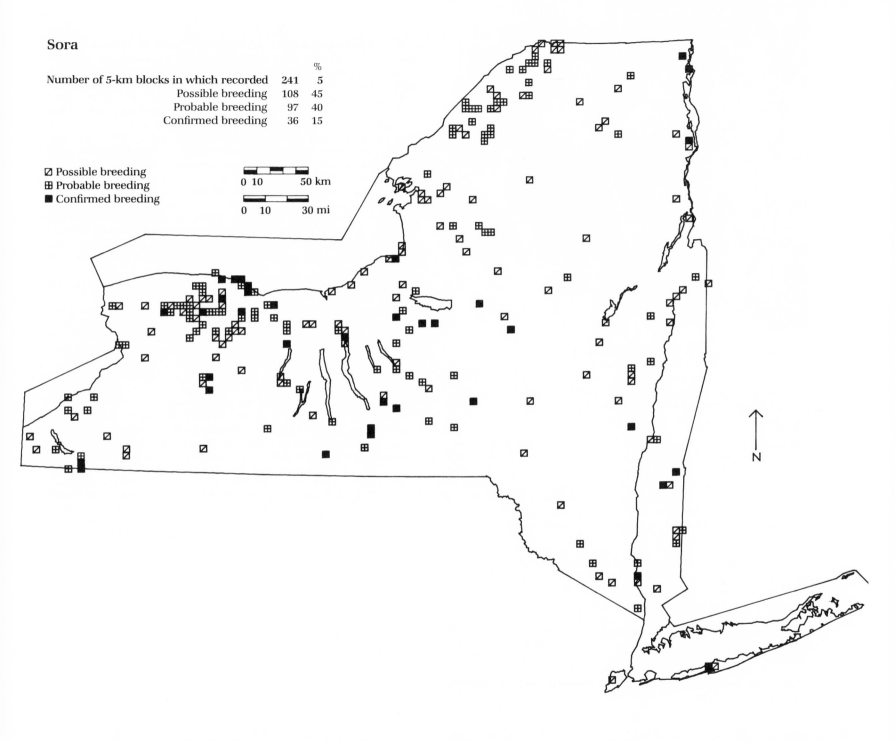

Sora

Number of 5-km blocks in which recorded	241	5 %
Possible breeding	108	45
Probable breeding	97	40
Confirmed breeding	36	15

◪ Possible breeding
⊞ Probable breeding
■ Confirmed breeding

0 10   50 km

0 10   30 mi

N

Common Moorhen *Gallinula chloropus*

The Common Moorhen is a fairly common breeder in freshwater marshes throughout the state. Formerly called the common gallinule, it is now less abundant and less widespread than it was once, probably because of the draining and filling of its marsh habitats (Strohmeyer 1977).

Cruickshank (1942) described the species as a formerly common summer resident around New York City but said it was almost entirely extirpated from Brooklyn, Queens, and Nassau counties and present only at one or two localities in Suffolk, Bronx, Westchester, and Rockland counties. Its status on Long Island has changed little since 1942, although the destruction of marshes there continues.

Within the past decade the Common Moorhen has been extirpated from a number of other local areas in the state; a dramatic example is Siebenheller's (1981) account of three Staten Island sites used in the late 1970s but destroyed by 1980. It may have declined even in the larger upstate marshes if the Montezuma NWR marsh complex at the north end of Cayuga Lake is representative. From 1955 to 1970 the estimated average annual number of Common Moorhens on the refuge was 162, but after 1970 the averages steadily declined. In 1985, 2 pairs raised 13 young at Montezuma, and in 1986 there were no broods observed (Montezuma NWR files). Montezuma staff mentioned the invasion and replacement of cat-tails by purple loosestrife and the destruction of habitat adjacent to marshes as factors that may be contributing to the decline. It is apparent that the Common Moorhen population now depends on protection of marshland habitat.

Freshwater marsh habitat can be found in all of the state's ecozones, but the Common Moorhen is virtually absent from areas above 305 m (1000 ft) in elevation. This may be because there are fewer marshes, especially large cat-tail marshes, at higher elevations.

The moorhen is rather secretive; adults spend most of their time amidst the emergent vegetation of the marsh rather than in open water (Strohmeyer 1977). During the breeding season this species is somewhat easier to locate, as courtship calls can be heard at dawn and dusk. Later in the season the young often frequent open water, where they are easily seen. Although this species is easier to locate than other rail species, some were likely missed by Atlas observers. The vast majority (90%) of Atlas "confirmed" records were of fledged young.

Bull (1974) and Cruickshank (1942) both describe the Common Moorhen's preferred breeding habitat in the state as well-vegetated freshwater marshes containing numerous small patches of open water. They list characteristic plants of these marshes as cat-tail, pickerel-weed, bur-reed, water arum, buttonbush, and often duckweed on the surface of all open water areas.

The Common Moorhen nests in most types of emergent vegetation but usually selects well-concealed spots in cat-tails (Strohmeyer 1977). It often places the nest where it can have immediate access to open water (Bent 1926), although 19 nests found in Iowa from 1963 to 1966 were placed 2–28 m (6.6–92.4 ft) from open water (Frederickson 1971). The moorhen builds the nest at the water's surface where the water is 0.3 to 0.9 m (1–3 ft) deep (Strohmeyer 1977); nests reach heights of 7.6–30.5 cm (3–12 in) above the water (Bull 1974). The nest itself is constructed of dead aquatic vegetation (Bent 1926) or leaf blades bent over in a criss-cross pattern to form a foundation, with nearby vegetation then molded to form the nest cup (Helm et al. 1987). During incubation, the bird often bends growing leaf blades over the nest to form a canopy. It continues to add nest material during incubation; one pair was observed raising the nest a total of 25.4 cm (10 in) by adding new vegetation to prevent it from being inundated by rising water during a flood (Bent 1926). The foundation is usually well anchored to surrounding emergent plants; however, a free-floating nest was found in New York (Bull 1974). Many nests have a sloping ramp constructed from the nest's rim to the water (Strohmeyer 1977). The Common Moorhen also constructs not only egg nests, but trial nests, brood nests, and elevated platforms (Helm et al. 1987).

Steven C. Sibley

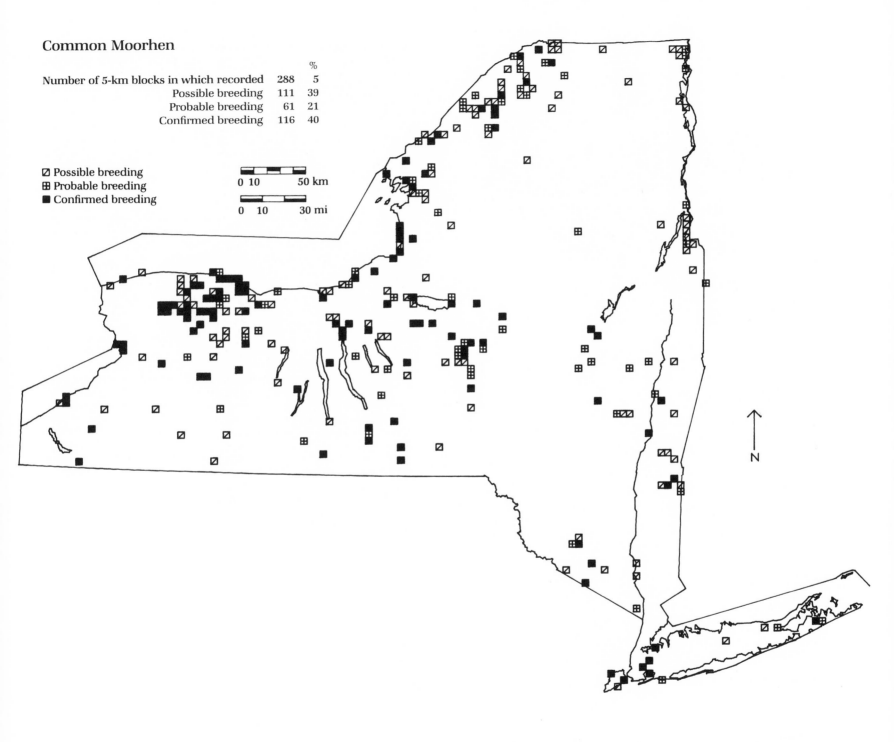

Common Moorhen

		%
Number of 5-km blocks in which recorded	288	5
Possible breeding	111	39
Probable breeding	61	21
Confirmed breeding	116	40

☒ Possible breeding
⊞ Probable breeding
■ Confirmed breeding

0 10   50 km

0 10   30 mi

N

American Coot *Fulica americana*

The American Coot, the most aquatic representative of the rail family, is a locally common breeding species in New York. Like several other wetland birds, it is well established mainly in the wildlife management areas and national wildlife refuges in the state. The American Coot is considered a migratory game bird. The daily bag limit in New York is 15, with a possession limit of 30 individuals. This bird is not, however, a popular target because it is not very palatable. In 1985 the total American Coot harvest in New York, as estimated by the USFWS, was less than 1% of the total waterfowl harvest for the state (Carney et al. 1986).

Forbush (1912) said, "This species was formerly one of the most abundant waterfowl on the freshwaters of North America." A severe decline occurred in the late 1800s over much of its range, which Frederickson et al. (1977) theorized was a result of the loss of wetlands that occurred between 1870 and 1930 and after World War II in its main breeding range in the Midwest and Canada, and also from overharvesting. Populations also fluctuate according to water and marsh conditions. During the drought of the early and middle 1960s, breeding ground surveys conducted by the USFWS showed a severe drop in the numbers of this species.

Like other waterfowl, the American Coot is an opportunist and can be lured to nest in many scattered locations between traditional wintering and summer breeding areas. Eaton (1910) reported that "a few remain to breed in the Montezuma marshes and about the eastern end of Lake Ontario." Bull (1974) also noted its presence at Montezuma NWR, indicat-

ing that "at least *200* broods were produced." He stated that about 50 pairs bred at Jamaica Bay Wildlife Refuge in 1961; elsewhere in the state, "only a single pair, or at the most, a few pairs have nested."

From 1980 to 1985 an average of 33 young were produced at Montezuma (Gingrich, pers. comm.). At Jamaica Bay only a few young were raised in 1985 and none in 1986. Riepe (pers. comm.) thought that the freshwater pond at Jamaica Bay where the coot had previously nested had become too brackish. Another major breeding ground in New York is at the Oak Orchard WMA–Iroquois NWR–Tonawanda WMA complex. The American Coot began nesting at Oak Orchard WMA in 1932 (Beardslee and Mitchell 1965), and D. Carroll (pers. comm.) reported that the number of young raised there from 1980 to 1986 was "in the hundreds."

Atlas observers also "confirmed" breeding at Perch River WMA, Jefferson County, and Fish Creek and Upper and Lower Lakes WMAs, St. Lawrence County. The coot was reported nesting in other widely scattered areas, including Tifft Farm Nature Preserve at Buffalo, Erie County, and on Staten Island. It was also "confirmed" near Oswego, Oswego County, in 1984 (DeBenedictis 1984). Most of the "confirmed" records were observations of broods; only 9 nests were found. The majority of "probable" records were of pairs. Because the adults of this species stay paired during the entire nesting season (Ward 1953), these records most likely indicate that breeding did take place.

The American Coot nests in freshwater marshes, along lakes, ponds, streams, and rivers, and in wet meadows. It prefers areas with abundant emergent vegetation and shallow water, 0.3–1.2 m (1–4 ft) deep (DeGraaf et al. 1980). In a study in Colorado the coot did not seem to prefer any specific plant species as nesting cover; the structure and location of the vegetation were more important (Gorenzel et al. 1982). In Iowa the most numerous breeding populations occurred where cat-tail marshes had equal areas of open water and emergent vegetation (Frederickson et al. 1977).

The American Coot is highly territorial, defending its nest against other coots and other species of waterfowl (Gullion 1952). It builds a platform—usually floating, but sometimes on top of a muskrat house—for copulation, brood rearing, nesting, and probably night roosting. The nest is placed over water and attached to emergent vegetation. It is made of dead or rooted bulrush, cat-tail, reed, or grass (Harrison 1975). Coots that nest early in the season use available dead vegetation; those that nest later use new emergent vegetation (Gorenzel et al. 1982).

Stephen W. Eaton

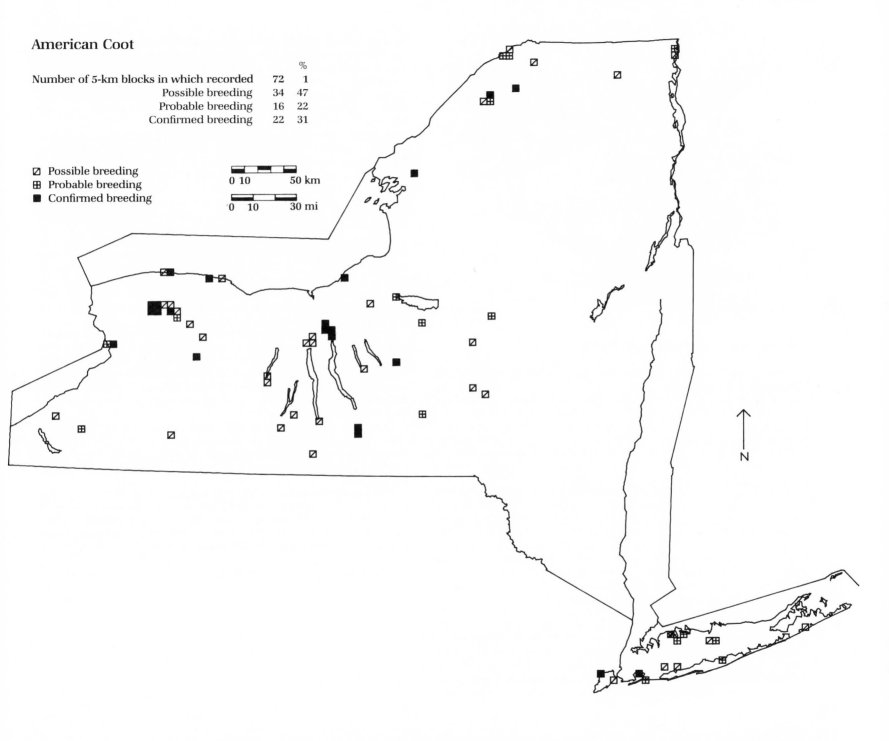

American Coot

Number of 5-km blocks in which recorded	72	% 1
Possible breeding	34	47
Probable breeding	16	22
Confirmed breeding	22	31

◩ Possible breeding
⊞ Probable breeding
■ Confirmed breeding

0 10   50 km

0 10   30 mi

N

©KLA-C 1987

Piping Plover *Charadrius melodus*

The Piping Plover, a small, pale shorebird, was declared federally an Endangered Species in the Great Lakes basin and a Threatened Species along the Atlantic Coast in 1986 (Sidle 1985), and is listed as Endangered by the NYSDEC. Historically, along the Great Lakes in New York there were three documented nesting sites for Piping Plover, all on the eastern shore of Lake Ontario. At its known peak population from 1935 to 1936, approximately 26 pairs were located in northwestern Oswego County and northern Cayuga County (Bull 1974). The population quickly declined, in part because of increased shoreline recreation and development. The last breeding on the New York shore of Lake Ontario before Atlas fieldwork occurred at Sandy Pond, Oswego County, in 1955 (Bull 1974). In 1984, however, a pair reappeared at Sandy Pond and fledged at least 2 young (1 of which was later found dead at the site) (Spahn 1984). The Great Lakes population has dropped from an estimated historical population of 500 pairs to only 17 pairs in 1984; the Sandy Pond nesting is the only breeding site east of Michigan (Russell 1983; Sidle 1985).

On the coast of Long Island the Piping Plover is still fairly common along stretches of suitable habitat. The species was regarded as common before the 1840s (Giraud 1844). Heavy hunting greatly reduced numbers, and by 1910 breeding was limited to Gardiners Island and the east end of Long Island (Eaton 1910). The species recovered quickly after protection was granted in 1913 and reached a population of over 500 pairs in the 1930s (Wilcox 1939). With increased development along the coast and greatly increased recreational and vehicular traffic on the species' nesting beaches, the population again declined. During the 1970s there were only an estimated 80 to 100 nesting pairs on Long Island (Cairns and McLaren 1980). Subsequent coastal surveys during 1984 and 1985 have recorded 110 and 114 pairs of Piping Plovers at 52 and 57 sites along the coast (Peterson et al. 1985). In 1985 Long Island's breeding Piping Plovers constituted approximately 13% of the entire federally designated Threatened coastal population. Only Massachusetts and Virginia had as many breeding Piping Plovers as New York (Sidle 1985).

In addition to recording the first breeding Piping Plover on Lake Ontario in more than 25 years, the Atlas has documented several changes in Piping Plover distribution since Bull's (1964, 1974) summary of breeding sites. During the Atlas period no Piping Plovers were "confirmed" on either Staten Island or in northern Nassau County along Long Island Sound. On a more positive note, the Piping Plover appears to be more widespread, or at least better surveyed, in the harbors of northern Suffolk County farther east along Long Island's north shore.

The breeding biology of the Piping Plover on Long Island has been well documented by Wilcox (1959). On Long Island it nests primarily along beaches, although it is also found nesting on dredge spoil. Although Wilcox documented nests on sandy areas, on Long Island the bird also nests on pebble substrates (Peterson et al. 1985). It nests singly and generally locates its nest on the beach above the high-tide line or up on the edge of the dunes (Wilcox 1959).

On the south shore beaches Wilcox (1959) found that this plover preferred open, washed-over areas as opposed to more vegetated sites. The Least Tern uses the same habitat. Between 1983 and 1985 approximately 90% of nesting Piping Plovers on Long Island were found in the vicinity of Least Tern colonies (Peterson et al. 1985). It is likely that the species are forced to share the same nesting areas owing to the relative scarcity of undisturbed beach sites.

The Sandy Pond area in Oswego County is a spit of land on the east shore of Lake Ontario. The beach where the Piping Plover bred is bordered by sparse grass and shrubs. Although the location of the nest was not known, the young were observed hiding in the beach grass (Crumb, pers. comm.).

The apparent stabilization of Long Island's Piping Plover population from the 1970s through the mid-1980s and the reappearance of the Piping Plover on Lake Ontario are favorable signs for the species in New York. The recent granting of Endangered and Threatened status by the USFWS for the inland and coastal populations should provide additional protection. The future of the Piping Plover in New York depends largely on the protection of remaining shoreline habitats from further development and recreational disturbance.

David M. Peterson

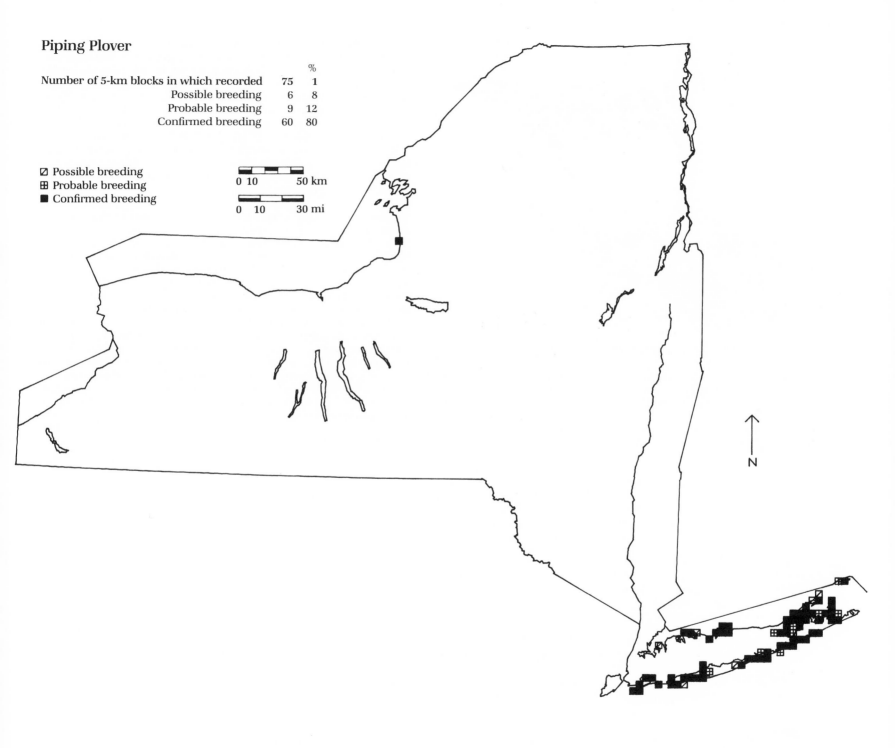

Piping Plover

Number of 5-km blocks in which recorded		%
Number of 5-km blocks in which recorded	75	1
Possible breeding	6	8
Probable breeding	9	12
Confirmed breeding	60	80

▨ Possible breeding
⊞ Probable breeding
■ Confirmed breeding

0 10   50 km

0 10   30 mi

N

Killdeer *Charadrius vociferus*

A familiar plover, the Killdeer is a common breeding bird in the state. Its early status in New York is vague, but it was apparently quite common early in the 1800s on the north shore of Long Island (Giraud 1844). By Eaton's time, however, it was very rare there (Eaton 1910); and by the mid-1900s it was again commonly seen during the breeding season (Cruickshank 1942) although concentrated on the Hempstead Plain in Nassau County.

Elsewhere in New York and in much of the Northeast a similar decline and then a subsequent increase occurred. In the 1800s the Killdeer population began to decline as a result of heavy hunting pressure during spring and fall migrations. In the early 1900s some states, including Massachusetts, closed the season on the Killdeer (McAtee 1911). In a 1907 survey conducted by Forbush, all 51 correspondents questioned said Killdeer numbers were decreasing (Forbush 1912). Within a decade after the passage of the Migratory Bird Treaty Act of 1918, it was illegal to hunt the Killdeer in the United States (Matthiessen 1967). Eaton (1953), in a review of the status of New York birds from 1910 to 1930, stated that "the most spectacular improvement [in the shorebirds] is that of the Killdeer which shows several hundred percent increase in the interior of New York and is extending its range." The protection of this species from hunting and the clearing of land for agricultural use (which provided excellent habitat for the Killdeer) were probably responsible for the increase.

The Atlas map shows the widespread distribution of this species. It is again found in the lower Hudson Valley, where it was unknown in Eaton's time (Eaton 1910), and on Long Island. Only in the heavily forested areas of the Central Adirondacks, Adirondack High Peaks, sections of the Eastern and Western Adirondack foothills, the Central Tug Hill, the Rensselaer Hills, the Catskill Peaks, the Hudson Highlands, and the Delaware, Mongaup, and Allegany hills was it absent entirely or recorded only rarely. This species has adapted fairly well to development in many areas. Palmer (1967) pointed out that it is "not particularly wary; adapts readily to man and his noises, as along roadsides and at airports."

The Killdeer was found in 74% of the Atlas blocks surveyed. Although it was most often "confirmed" by recording recently fledged young, 264 nests were located. Distraction displays are commonly observed for this species and were recorded in 25% of the "confirmed" blocks.

The Killdeer inhabits heavily grazed meadows and pastures, newly plowed fields, and the shores of lakes, ponds, and streams. Around human habitation it can be found on golf courses, lawns, parking lots, driveways, and at airports. Like most plovers it prefers closely cropped fields or areas of sparse vegetation and is therefore plainly visible. It places the nest in the open, allowing the incubating bird an extended view. The male makes various scrapes in the nesting area, and the female selects one for its eggs. The nest scrape may be lined with pebbles, wood chips, grass, or assorted debris, or it may be unlined (Harrison 1975).

The Killdeer nests on crushed gravel rooftops in western and central New York. Mitchell (1954) described a rooftop nest in downtown Buffalo which was used for at least 3 years, and Hoyt (1959) recorded several nests on the roofs of some Cornell University Veterinary College buildings. In 1976 a Killdeer nested on the roof of a building in Olean, Cattaraugus County (Forness, pers. comm.), and in 1977 one nested on the roof of the St. Bonaventure University Library, Cattaraugus County (Eaton 1981). Nesting on rooftops is apparently sporadic and seldom successful but is reported widely in such areas as California, Nebraska, and Ontario, Canada (Mitchell 1954).

Stephen W. Eaton

Killdeer

Number of 5-km blocks in which recorded 3939 74
Possible breeding 1014 26
Probable breeding 748 19
Confirmed breeding 2177 55

%

☒ Possible breeding
⊞ Probable breeding
■ Confirmed breeding

0 10   50 km

0 10   30 mi

N

American Oystercatcher *Haematopus palliatus*

Formerly occurring only casually in New York (Eaton 1910; Bent 1929), the American Oystercatcher has recently expanded its range northward into New Jersey and Long Island (Bull 1964). The first documented New York nest was found on Gardiners Island, at the east end of Long Island, in 1957, and the first nesting on the south shore apparently was at Moriches Inlet, Suffolk County, in 1960 (Post 1961).

Bull (1974) called the American Oystercatcher a "regular, but rare and local breeder on Long Island since 1957," with no evidence of breeding in New York before that date. Breeding densities cited by Bull (1974) were 14 pairs at Gardiners and Cartwright islands on the eastern end in 1966 and 15 pairs on Gardiners Island in 1970. More recently, Zarudsky (1985a) estimated the annual number of breeding pairs for Long Island "at well above one hundred, an increase not originally anticipated during the 1960's." Indeed, Post and Raynor (1964), discussing the recent range expansion of the American Oystercatcher into New York, said, "The outlook for future population increase on Long Island is unfavorable." Zarudsky's 1985 estimate of more than 100 breeding pairs (or 200 adults) compares well with unpublished 1986 data from a Long Island colonial waterbird survey, coordinated by Cornell University Laboratory of Ornithology's Seatuck Research Program, which found 230 adults at 38 breeding sites (MacLean, pers. comm.).

In the early 1800s the American Oystercatcher bred much farther north (e.g., New England) along the Atlantic Coast than at present (Post 1961). By the mid- to late 1800s its range had contracted and it was scarce on Long Island (Post 1961). The increase described above was in part due to termination of market hunting of this species. Another factor may have been an increase in numbers of its major prey, blue and ribbed mussels, which occurred as pollution of waters around Long Island abated (Lauro 1977). Tomkins (1947) stated that the American Oystercatcher fed "mainly on bivalves" and was "limited in range by the accessible supply of this sort of food." Availability of suitable undisturbed nesting sites and competition for nest sites with gulls and terns may also be limiting factors for this species (Harrison 1975; Zarudsky 1985a).

As the Atlas map indicates, the range of the American Oystercatcher is limited to the south shore and eastern end of Long Island. This distribution has not changed from historical data (Bull 1974), except for the records at Jamaica Bay Wildlife Refuge. However, an increased number of nesting locations were found by Atlas observers. Atlas data are probably quite accurate for this species because of its conspicuousness (Tomkins 1947) and the addition of data to the Atlas from several researchers.

On Long Island some American Oystercatchers nest on salt-marsh islands that lack the sand substrate usually associated with this species (Frohling 1965; Zarudsky 1985a). This is a departure from the southern breeding habitat described by Bent (1929) as "the higher parts of the dry, flat, sand beaches, well above the high water mark." Harrison (1975) suggested that discovery of American Oystercatcher nests on this type of salt-marsh island "may indicate that greater habitat flexibility is developing" in this species; Post and Raynor (1964) tempered their pessimism about a future population increase with the possibility that the species could adjust to disturbed habitats.

Zarudsky (1985a) presented evidence that this adjustment was occurring on Long Island. Of 45 nests studied in the Town of Hempstead in 1984, 33 were on dredge spoil, 10 on high sand bars, 1 on salt marsh, and 1 on construction material. However, while providing man-made nest sites, dredge spoil is also attractive to recreational boaters, dogs, and competing gulls, which may ultimately limit range expansion of the species on Long Island (Zarudsky 1985a).

The oystercatcher also nests in more typical habitat, described by Bull (1974) as "sandy islands with shells, pebbles, and sparse vegetation" with the additional requirement of nearby marine intertidal mud flats for feeding on mollusks. Long Island nests are usually placed in relative isolation, such as behind a barrier beach rather than on the beach itself, and away from human disturbance (Bull 1974; Gochfeld 1976).

The nest is usually a simple scrape or hollow in a sand or shell substrate and is often ringed with shell fragments (Tomkins 1954; Harrison 1975). Zarudsky (1985a) described Long Island nests as "usually a depression in the sand lined with shell fragments or pieces of littoral drift."

Richard A. Lent

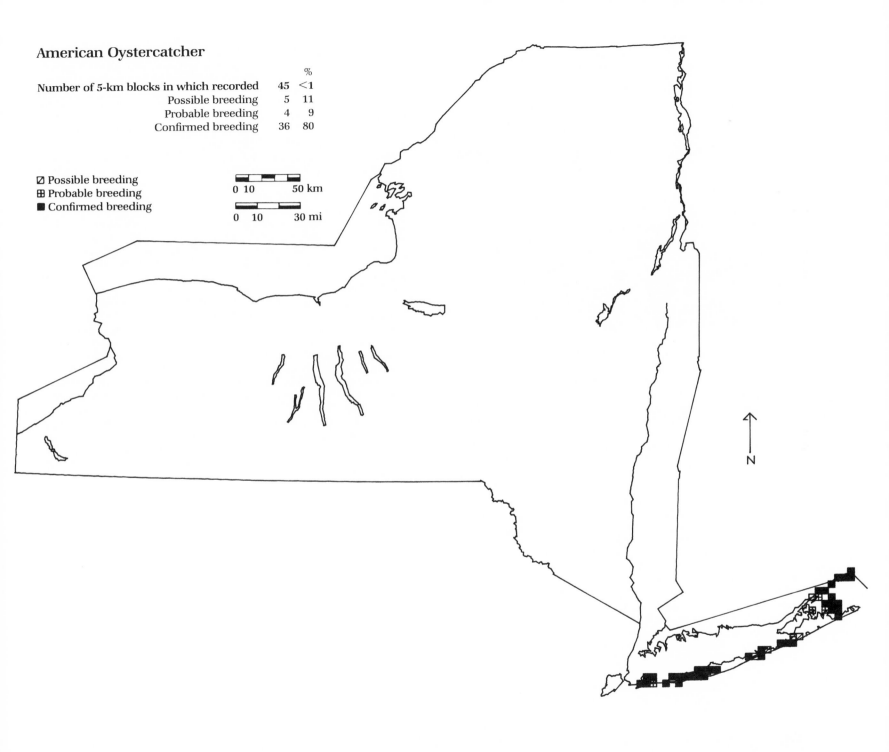

American Oystercatcher

		%
Number of 5-km blocks in which recorded	45	<1
Possible breeding	5	11
Probable breeding	4	9
Confirmed breeding	36	80

☑ Possible breeding
⊞ Probable breeding
■ Confirmed breeding

0 10   50 km

0   10   30 mi

N

© KLA-C 1987

Willet *Catoptrophorus semipalmatus*

The Willet is a common breeding species along the salt marshes of Long Island's south shore. Its current status is remarkable, given that it was first documented as breeding in the state in 1966 (Davis 1968). Before the 1960s the breeding range of the eastern subspecies of the Willet (*C.s. semipalmatus*) along the East Coast was split into two widely separated populations. One population nested along the south Atlantic Coast north to southern New Jersey, and the other nested in southern Nova Scotia (Bull 1974). Before 1887 the Willet also nested in Massachusetts and had been recorded in southern Connecticut (Bull 1964; Davis 1968).

The historical record for New York is unclear. Giraud (1844) listed the species as a fairly common summer resident but did not record any breeding. De Kay (1844) listed the species as occurring from May to November and indicated that "many remain to breed." Giraud (1844) also wrote, "Its flesh though palatable was not considered so great a delicacy as its eggs," implying that Long Islanders knew of nests in the region. However, no known specimens of nests or eggs have been documented in the state from the 1800s (Bull 1964). Whether or not breeding ever did occur in New York at that time, subsequent hunting and possible egg-collecting pressure would have greatly reduced Willet populations, as it did elsewhere along the East Coast (Forbush and May 1936). Through the 1960s the species continued to become more common in the spring and summer. Three nests were finally discovered in Hempstead, southern Nassau County, in 1966 (Davis 1968). The following year the Willet was found breeding at Oak Beach, southwestern Suffolk County. By 1968 it expanded west even farther and was found breeding at Tiana Beach in Shinnecock Bay, southcentral Suffolk County (Bull 1974; Wilcox 1980).

Although the Willet's breeding range has not greatly changed over the last 20 years, the population of nesting birds has increased dramatically. From the first 3 nests found in 1966, the population increased to at least 83 pairs that were observed during a coastal helicopter survey in 1975 (Lauro and Spencer 1975). Zarudsky (1980) listed the Willet as abundant in the salt marshes of the Town of Hempstead, southern Nassau County, with most islands containing at least 2 or more breeding pairs. Lauro (DiCostanzo 1983b) censused 150 pairs, along approximately 29 km (18 mi) of salt marsh between Jones Beach, eastern Nassau County, and the Robert Moses Causeway, western Suffolk County. An additional 10 pairs nested that year at Jamaica Bay Wildlife Refuge (DiCostanzo 1983b). Farther east along the salt marshes of eastern Great South Bay, Moriches Bay, and Shinnecock Bay the Willet has become a common nester (Salzman 1985b). In eastern Long Island, however, it has not attained the high breeding density that it has in the more extensive salt marshes of southwestern Suffolk County, southern Nassau County, and Jamaica Bay (Wilcox 1980).

As the Atlas map indicates, other than one "confirmed" block on Gardiners Bay, the Willet has not expanded its range beyond Shinnecock Bay. On the western south shore "confirmed" breeding has been recorded throughout southwestern Suffolk County, southern Nassau County, and west to Jamaica Bay. The Willet is always found close to salt marshes. It appears that it has colonized most of the remaining major salt marsh areas along the coast. Further expansion may occur into smaller marshes if the population continues to increase along Long Island's south shore.

Nests studied along Shinnecock Bay were concealed in short grass on the salt marsh barely above the high-tide mark (Wilcox 1980). In addition to nesting sites they found in salt marshes, Davis and Morgan (1971) and Salzman (1985b) cited nesting on the barrier beach, and Zarudsky (1980) noted nests in the low vegetation of spoil islands.

David M. Peterson

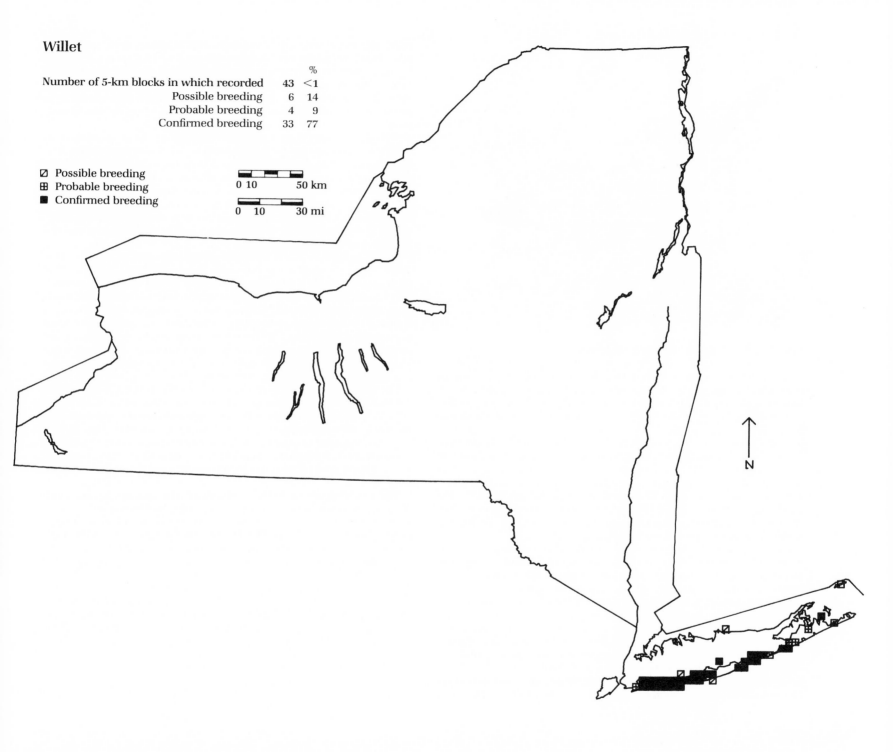

Willet

		%
Number of 5-km blocks in which recorded	43	<1
Possible breeding	6	14
Probable breeding	4	9
Confirmed breeding	33	77

◨ Possible breeding
⊞ Probable breeding
■ Confirmed breeding

0 10   50 km

0 10   30 mi

N

Spotted Sandpiper *Actitis macularia*

The Spotted Sandpiper, also called "teeter-tail" and "tip-up" for its characteristic bobbing behavior, is one of the best known shorebirds in New York. It is a common breeder throughout the state, and its status has changed little, if any, in the past century. Eaton (1914) stated that it was "common in every county."

In appropriate habitats this species can be abundant and sometimes nests in loose colonies (Terres 1980). On Great Gull Island at the mouth of Long Island Sound the breeding density of the Spotted Sandpiper has been almost as high as 1 pair per 0.4 ha (1 a) (Bull 1974). One factor affecting the bird's abundance is its ability to cope with human activities. Smith and Ryan (1978) stated that along the spits at Sandy Pond on Lake Ontario the Spotted Sandpiper seemed "able to tolerate the human activity which the Piping Plover could not."

Local populations of the Spotted Sandpiper can fluctuate. If open habitat becomes overgrown, flooding occurs, or predation of eggs and chicks is high, the sandpipers may leave the area. However, additional habitat is created by the "scouring action of water and ice" along shorelines and human creation of ponds, lakes, and other water bodies (Oring et al. 1983).

Of New York's eight breeding shorebird species, only the Killdeer was recorded in more blocks during Atlas fieldwork. The Spotted Sandpiper is scattered in the wooded highlands of the Adirondacks, Catskills, and Appalachian and Tug Hill plateaus, where suitable habitat is scarce. In all areas this species' distribution corresponds closely with rivers, lakes, and shorelines. It is, however, absent from long stretches of several rivers,

especially in the western Adirondacks, where forests grow to the river edge. However, methodical searches along other sections would probably be successful in locating Spotted Sandpipers. Reasons for the large gaps in distribution in the western Appalachian Plateau are also unclear, although almost all gaps correspond to areas above 457 m (1500 ft) in elevation where the land is largely wooded and there are few lakes or ponds.

The Spotted Sandpiper is an active, conspicuous bird not easily overlooked by Atlas observers. Nests, however, were extremely difficult to locate (only 10% of the "confirmed" records); most (72%) were of dependent fledged young. The large number of "confirmed" records in Orleans, Monroe, and Genesee counties were of dependent fledged young seen in July during return visits to blocks and block busting. Although the bird itself was not difficult to locate, the large percentage (51%) of "possible" breeding records was the result of sometimes lengthy observations of a "spotty" that did nothing but pace.

This sandpiper occurs in wetland habitats, both fresh and salt. Cruickshank (1942) stated that it can be seen "frequenting the shores of both fresh and salt-water bodies and even accepting low, wet meadows with only square yards of water." The Spotted Sandpiper is also often seen in upland habitats. Eaton (1914) stated that in addition to wetland areas, it could be "found in pastures, cultivated fields and meadows, sometimes at considerable distance from the water." Bogs, small fast-running mountain streams, and lakes and ponds "entirely wooded to the water's edge" are the only wetlands and waterways not frequented by this species during the breeding season (Eaton 1914).

The Spotted Sandpiper nests in a wide variety of situations. Tyler (1929) stated that "few birds show a greater variation in this respect and among the places which the bird selects to lay its eggs there is but one point in common—the proximity to water." Bull (1974) described typical New York nesting sites as "nests on the ground in open grassy fields or occasionally in cultivation where planted to grain and vegetable crops; also along ponds and streams, and in coastal sand dunes among beach grass."

The nest consists of a shallow depression in the ground. It is often well concealed among thick grass or weeds and sometimes under bushes, logs, driftwood piles, or rock ledges (Terres 1980). As is typical of other sandpipers, this bird uses only a few dry grasses, leaves, or weeds to line the nest hollow where it lays the eggs (Eaton 1914). Occasionally, a more bulky, deep cup nest is built, with mosses or bits of rotting wood used as lining (Tyler 1929).

This species is unique among New York's breeding birds in that whenever there is a surplus of males, the female is polyandrous (Hays 1972; Oring and Knudsen 1972). The female initiates courtship and leaves the male largely responsible for incubating the eggs and caring for the young.

Steven C. Sibley

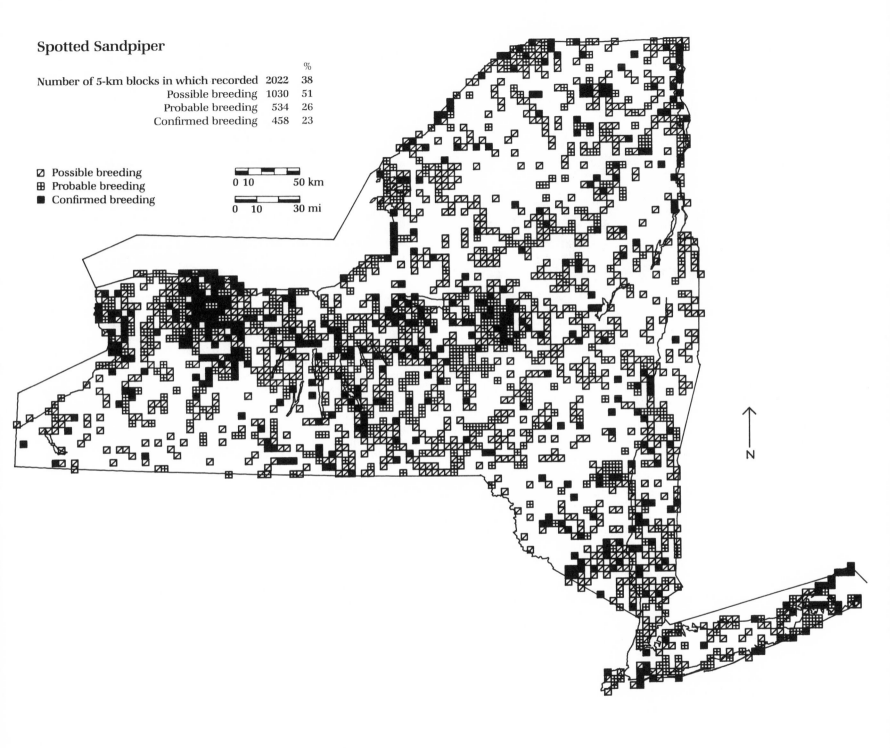

Spotted Sandpiper

Number of 5-km blocks in which recorded 2022   38

		%
Possible breeding	1030	51
Probable breeding	534	26
Confirmed breeding	458	23

◨ Possible breeding
⊞ Probable breeding
■ Confirmed breeding

0 10   50 km

0 10   30 mi

N

Upland Sandpiper *Bartramia longicauda*

The "notes rapidly rising and swelling, then slowly falling and dying away into a hollow windlike whistle" was the description used by Eaton (1910) for the call of the Upland Sandpiper, a call that in New York can be heard over rolling fields and flat plains. It is an uncommon species in the state, occurring mainly in farming country.

The Upland Sandpiper probably moved into the eastern United States from the western prairies as forests were cleared and pastures and meadows created by early settlers. Whether prairie habitats such as the Hempstead Plains on Long Island were already occupied by this species is not known. During the late 19th century market hunting began to take its toll. Forbush (1912) said that "about 1880 . . . the destruction of the Upland Plover began in earnest." Eaton (1910), apparently concerned about the hunting pressure, called for its protection, and as Aldo Leopold (1949) said, "the belated protection of the federal migratory bird laws came just in time."

From about 1940 to 1980, however, land use changes have adversely affected this species (White 1983). Prairies are being cultivated, wetlands drained, farms abandoned, and grasslands developed for residential or industrial purposes, making them unsuitable for this species. Also in New York farmers are planting less grain and more corn (Richmond and Nicholson 1985), further eliminating nesting habitat. Tate (1986) noted the decline in the Northeast due to these changes. This species has appeared on the *American Birds'* Blue List from 1975 to 1986. The Upland Sandpiper is listed as a Species of Special Concern by the NYSDEC.

Eaton (1910) stated that this sandpiper was "a summer resident of eastern Long Island," where he believed the population was decreasing. He also found it along the Great Lakes and St. Lawrence plains of inland New York, as well as in Clinton and Rensselaer counties. He said that it was "holding its own" in northern and western New York, with increases noted in northern Erie County and western Monroe County. Bull (1974) also indicated a continuing decline on Long Island and described it as "locally common" in the agricultural areas of central and western New York, and along the central Hudson and Mohawk river valleys.

Atlas workers found the Upland Sandpiper along the Great Lakes Plain, although with a decided shift to the south from the distribution shown on Eaton's 1910 map. The Mohawk Valley remains a stronghold, with some apparent expansion there. This species also occurs in scattered blocks across the Appalachian Plateau, with small concentrations in the farmlands of the eastern Finger Lakes Highlands and also the Cattaraugus Highlands. The eastern Great Lakes Plain, St. Lawrence Plains, and Lake Champlain Valley populations seem to have expanded, particularly in St. Lawrence and Franklin counties. The Upland Sandpiper still occurs in Orange County and a few other parts of the Central Hudson Valley, but it is gone from most of the Taconic Highlands. It was found locally on Long Island, including at John F. Kennedy International Airport where Davis (1982) reported an estimate of 12–20 pairs in 1981. These apparent local changes in distribution may be the result of habitat changes but more likely reflect the intense survey effort of the Atlas. New York's distribution is contiguous with that of Ontario, Canada, along the Great Lakes and St. Lawrence plains (Ontario Breeding Bird Atlas, preliminary map) and along the Champlain Valley in Vermont (Kibbe 1985b). The map gives a good picture of the bird's distribution in the rapidly decreasing grasslands of New York.

The species is not difficult to locate when the males are displaying but may become very hard to find during incubation. "Confirmed" records were mainly observations of recently fledged young; however, distraction displays were reported in several blocks.

The nesting habitats of this bird in agricultural areas upstate are old pastures and hay fields (Bull 1974). In addition, it can be found in mowed grass areas adjacent to expressways and airports. These areas most closely duplicate the shortgrass prairie that was its former nesting habitat. White (1983) found that certain other habitat features in addition to having the appropriate crops may be essential requirements for this species, including the presence of fenceposts, large and unbroken fields, little forest, and lack of rugged topography. It nests in loosely spaced colonies of from 0.6 to 6.1 ha (1.5–15 a) per nest. The nest is well hidden in a depression in a thick clump of grass that arches over the top. The grass is twisted in a circle to form a cup (Harrison 1975).

Stephen W. Eaton

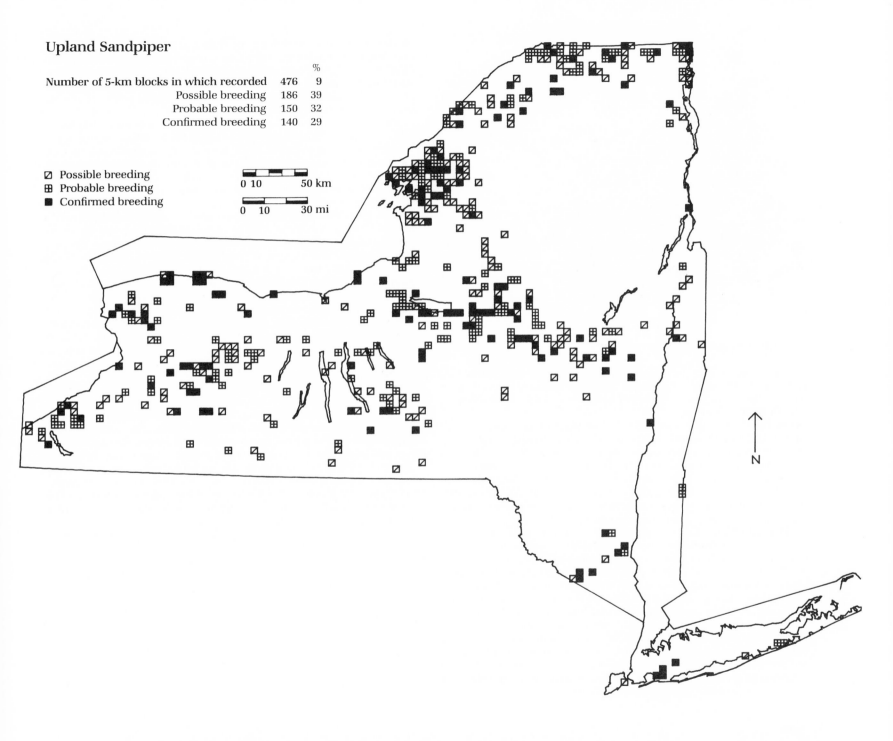

Upland Sandpiper

Number of 5-km blocks in which recorded	476	% 9
Possible breeding	186	39
Probable breeding	150	32
Confirmed breeding	140	29

☒ Possible breeding
⊞ Probable breeding
■ Confirmed breeding

0 10   50 km

0 10   30 mi

N

©KLA-C 1986

Common Snipe *Gallinago gallinago*

The *winnowing* of the Common Snipe, heard from marshes on late spring evenings, adds a mysterious quality to wetlands in the state. A common but local breeder in New York, the species is at the southern edge of its breeding range.

The Common Snipe population was reduced in the late 1800s and early 1900s, mainly because of excess harvesting (Forbush 1912). Spring and fall hunting reduced populations, but other stresses, such as drought on the breeding range and extended cold periods on the wintering grounds, caused a further decline; subsequently the hunting season was closed from 1941 to 1953. Although snipe hunting resumed in 1954, the tradition, passed down from father to son, had been broken. Today few people hunt the Common Snipe (Fogarty et al. 1977).

Eaton (1910) stated that this species was neither a common nor a regular breeder in wetlands where it nested but suggested that it would undoubtedly become a common breeder in the swamps of central and western New York and Lake Ontario if spring shooting and disturbance on its breeding ground were stopped. Bull (1974) called it fairly numerous in the largest marshes and wetlands in central and western New York but indicated that it was somewhat more local in the Adirondacks. He said it was quite rare in the Mohawk and Hudson valleys, showing the most southern breeding record to be near Warwick, Orange County. Loss of habitat from the draining of wetlands, which began after World War II, has probably adversely affected the Common Snipe population.

Atlas workers found the bird best represented on the Great Lakes and St. Lawrence plains and in the Lake Champlain Valley. These populations represent southern extensions into New York from a more widespread distribution in Ontario, Canada (Ontario Breeding Bird Atlas, preliminary map). It was also well represented in the Mohawk and Black River valleys and the Tug Hill region. In the Adirondacks and on the Appalachian Plateau, it is more scattered because both of these areas have fewer suitable wetlands.

Although the range of the Common Snipe is well defined, and although Atlas data were supplemented by records from USFWS, the species is probably under-represented on the map. It is found most easily early in the breeding season by listening at dusk for a *winnowing* male intent on finding a mate, but most Atlasers were not out early in the breeding season. When the eggs hatch, *winnowing* ceases and the Common Snipe is very difficult to locate.

Tuck (1965) found that a close relationship exists between the distribution of the Common Snipe and the distribution of tamarack. He found that isolated populations in areas outside the optimum snipe breeding range, such as New York, were in bogs and marshes. Along the Great Lakes Plain it is found in and around the margins of the large cat-tail–bulrush marshes, and on the Appalachian Plateau it is found in smaller sedge meadows (Rosche 1967; Eaton 1981). In the Adirondacks it breeds mainly in marshes and wet areas along rivers and streams and around beaver ponds and meadows (Peterson, pers. comm.). An exposed area within the habitat, such as matted sedge beds, mud banks, grazed meadows, or even paved highways, provides an area to perform copulatory displays. For other diurnal activities of the snipe, cover is a necessary component (Tuck 1965).

The male arrives on the breeding grounds 10–14 days before the female and establishes its territory. When the female arrives *winnowing* intensifies, and a spectacular courtship display begins. The eggs are laid in a scrape at a spot on the ground selected by the female. As laying progresses, the nest is added to, sometimes acquiring a canopy. It often becomes a structure that can be lifted out intact (Fogarty et al. 1977). The nest is placed in or at the edge of a marsh or bog, concealed in a tussock of grass projecting above the water (Harrison 1975). In a study in Newfoundland, Tuck (1965) found that nests were placed in a damp site usually among tall cover such as alder or grass. He found that if the nest was in an open situation, tall cover was nearby.

Stephen W. Eaton

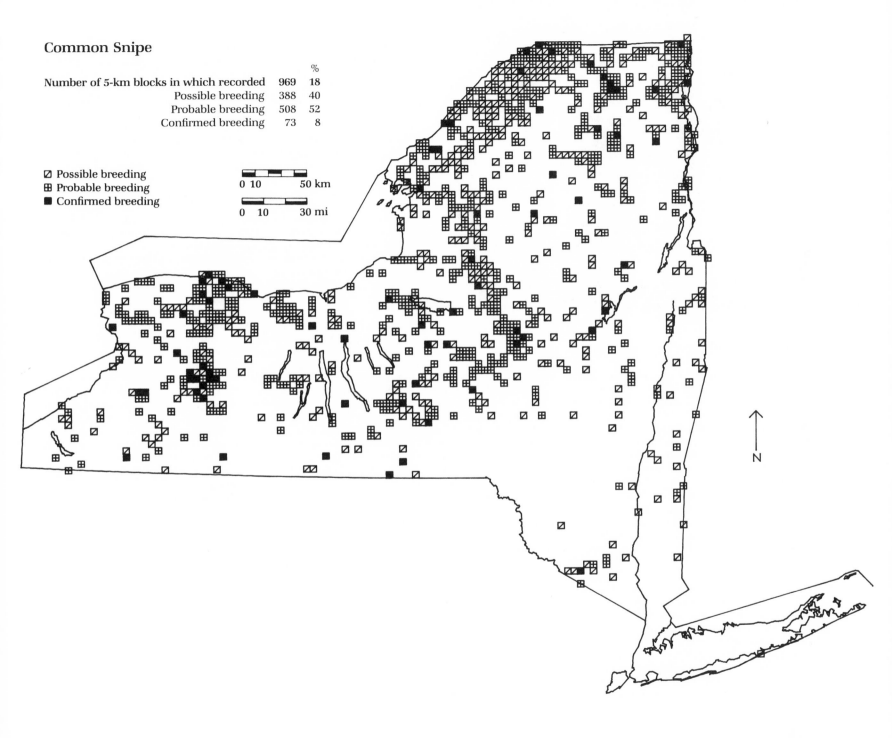

Common Snipe

Number of 5-km blocks in which recorded	969	% 18
Possible breeding	388	40
Probable breeding	508	52
Confirmed breeding	73	8

☑ Possible breeding
⊞ Probable breeding
■ Confirmed breeding

0 10   50 km

0 10   30 mi

N

©KLA·C 1986

American Woodcock *Scolopax minor*

An unusual shorebird of upland areas, the American Woodcock is fairly common throughout the state. Its abundance in early colonial days is unrecorded, but presumably it was present near the many undrained wetlands, beaver ponds, meadows, and forest openings of the time. The scattered clearings of early settlers created additional habitat.

The American Woodcock probably became a popular food and game bird in the last quarter of the 18th century. As early as 1791 New York passed a law protecting it in Queens, Kings, and New York counties from 20 February to 1 July (Palmer 1912; Pettingill 1936). Naturalists of the early 19th century wrote of great numbers of woodcock; De Kay (1844) stated that it bred in every part of the state and said it was abundant in New York City as late as 1814. However, by late in the century the years of habitat loss, long open hunting seasons, improved firearms, and market hunting had taken a serious toll on breeding populations throughout the Northeast (Sheldon 1971).

Eaton (1910) also described the American Woodcock as formerly a common breeder in every county and said that it still bred sparingly in all parts of the state but was fast disappearing from more inhabited areas because of overhunting, draining of swamps, and other habitat destruction. As increasingly restrictive hunting regulations were enacted in various states between 1900 and 1920, the species responded well to the protection, although it never regained the numbers known earlier (Sheldon 1971). Moreover, declining agriculture after 1900, with widespread land abandonment, improved habitat conditions for the American Woodcock. Nevertheless, a slow but significant decline in its eastern population has been detected over the past 20 years; the USFWS singing-ground survey documented a decrease at a mean annual rate of 2.6% from 1968 to 1986 (Kelly 1986).

Atlas surveying has revealed a fairly uniform distribution statewide in suitable habitats, minor variations corresponding partly with degree of coverage. Least favorable to the species are heavily urbanized areas and the highest elevations of the Adirondacks and Catskills. Yet, it still breeds in three boroughs of New York City and is unrecorded only in Manhattan (New York County) and Bronx County. It is still common locally on Long Island, where Cruickshank (1942) called it "astonishingly common" in some localities in the courtship season. Although the map includes records of the USFWS singing-ground survey, this species is under-represented, mainly because the birds sing primarily at dawn and dusk in late March, April, and May, when some observers were not yet atlasing.

The American Woodcock utilizes various habitats; the main requirement is an abundant supply of earthworms, which in one study constituted by volume over two-thirds of the diet, the remainder being mostly other invertebrates (Sperry 1940). The species seeks a favorable interspersion of habitats such as those provided by varied forestry practices on adjacent tracts and abandoned agricultural land in various stages of succession (Pough 1951). In the Northeast, breeding abundance is closely related to four types of habitats: (1) alder or young hardwoods on moist, rich soils, for daytime feeding cover; (2) forest openings or clearings, for springtime singing grounds; (3) young second-growth hardwoods, for nesting and brood habitat; and (4) large fields to roost in at night, beginning in early summer (Sepik et al. 1981). The nesting territory of the female is separate from the male's singing ground; nests are located from within 91 m (300 ft) of an occupied singing ground (Sepik et al. 1981) to as far as 274 m (900 ft) or more (Pettingill 1936). Nests are usually situated in woods but are also found in brushy cover, moist thickets, elevated spots in swamps, or in fields (Pettingill 1936; Pough 1951).

The very simple nest is placed where leaf litter or other plant matter is accumulated on the ground. The female shapes it into a cuplike depression, presumably by moving her body around, and probably does not add to material already there; possibly she moves pieces of material from the edge of the site to the nest with her bill (Pettingill 1936). The incubating hen is remarkably well camouflaged even in the open; her color pattern blends with the surrounding dead leaves and litter (Sheldon 1971). Some nests are near the base of a small tree or shrub or are partially concealed by logs, fallen branches, or ground vegetation (Bent 1927; Pettingill 1936). Pettingill described 16 nest sites in Tompkins County.

Paul F. Connor

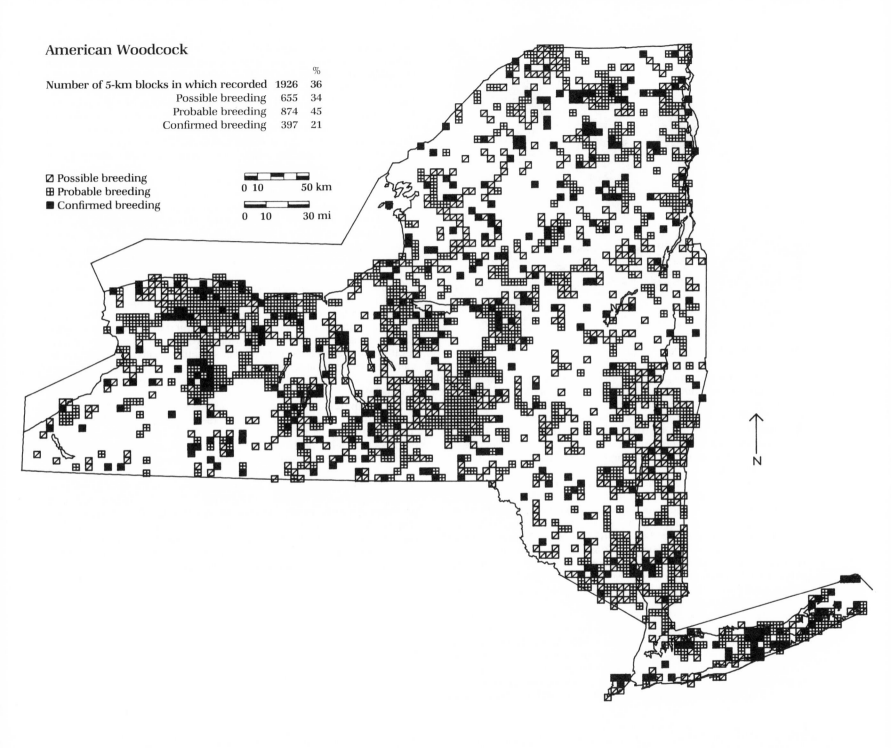

American Woodcock

		%
Number of 5-km blocks in which recorded	1926	36
Possible breeding	655	34
Probable breeding	874	45
Confirmed breeding	397	21

◩ Possible breeding
⊞ Probable breeding
■ Confirmed breeding

0   10   50 km

0   10   30 mi

N

Laughing Gull *Larus atricilla*

In the 1970's the Laughing Gull was discovered nesting on Long Island after having been extirpated as a breeding species in the state for almost 80 years (Bull 1974; Buckley et al. 1978). Bent (1921) noted that "in Giraud's day [1844] the Laughing Gull was a common summer resident" on the Long Island coast but doubted that any breeding colonies were left. Pressures from the millinery trade and commercial egg collecting had greatly reduced its numbers all along the East Coast by the late 1800s (Buckley et al. 1978).

Eaton (1910) called the species "rare, nesting only on the salt marshes of [Long Island's] Great South bay." Bull (1974) categorized the Laughing Gull as a "common to locally abundant coastal migrant and summer visitant" in New York, noting that although there had been a recent "great decrease" in the size of breeding colonies in the northeastern United States, "great numbers" of Laughing Gulls were still seen in New York each year during migration. The last known Long Island breeding records before the birds' reappearance were of a set of eggs found on Cedar Island, Great South Bay, on 14 June 1890 (Bull 1974), and a nesting pair at Orient Point State Park in 1900 (Buckley et al. 1978).

Recent searches of Laughing Gull habitat on Long Island (Buckley et al. 1978; Post and Riepe 1980) have produced proof that it has indeed reestablished itself as a breeding species in New York. Buckley et al. (1978) discovered "Long Island's first Laughing Gull nest in probably 78 years" in the Line Island complex of salt-marsh islands in Great South Bay. The single nest contained 1 warm egg. This breeding site has apparently not been sustained (Peterson et al. 1985); however, in 1979 a breeding colony of 12 to 15 pairs was discovered at Jo Co Marsh at Jamaica Bay Wildlife Refuge (Post and Riepe 1980). The Jamaica Bay colony has continued to be viable (Buckley and Buckley 1984a) and was the only Atlas record of "confirmed" breeding for the species; "probable" and "possible" breeding records were scattered along the coasts of Long Island and Staten Island but may at present just represent summering birds. The colony at Jamaica Bay has been steadily increasing since 1979, with 2802 breeding pairs in 1984 (Buckley and Buckley 1984a). As of 1985, 3 distinct colonies were recognized in the Jamaica Bay salt marshes, with an estimated 2741 breeding pairs (Peterson et al. 1985).

Currently, a major factor limiting expansion of Laughing Gull range is competition with larger gull species, notably, the Herring Gull, which preys on both eggs and young of the Laughing Gull (Nisbet 1971; Harrison 1975; Buckley et al. 1978; Burger 1979a). Quality of the preferred salt-marsh nesting habitat may also be a factor (Buckley et al. 1978). Nisbet (1971) believed that controlling the larger gull species that directly compete with the Laughing Gull could improve the chances of survival of the northeastern population. About the Laughing Gull in New Jersey, Burger (1979a) stated, "Unless long-used Laughing Gull nesting areas are protected from Herring Gulls . . . their numbers may decline dramatically as they have done farther north." Buckley et al. (1978), discussing the return of the Laughing Gull to Long Island, commented that "the almost exclusive breeding of Herring and Great Black-backed Gulls on high, dredge spoil or natural islands . . . may offer Laughing Gulls competition-free marshes where their breeding numbers might recover unimpeded. Its fate on Long Island over the next 10 or 20 years should be monitored closely."

The Laughing Gull breeds in salt marshes along the Atlantic Coast and in dry habitats (Burger and Gochfeld 1985) such as open sandy beach (Eaton 1910; Nisbet 1971; Buckley and Buckley 1984a). It may nest in colonies with terns or the Black Skimmer (Harrison 1975). It often builds the nest on the ground on coastal islands, in grass clumps or reeds in salt marshes, or on beaches. It prefers moderate to heavy vegetation in the nest area (Nisbet 1971; Burger and Gochfeld 1985), possibly to protect the nest from predators, storms, and high tides. Nests are usually large and well built of weeds, sedge, and grass to raise the eggs above tidal flooding (Terres 1980), a common cause of nest loss (Harrison 1975). Nests may be mere hollows in the sand, lined with grass and other stems, but they usually are bulky and substantially interwoven, built up several inches above ground level with a lining of grasses (Bent 1921; Harrison 1975).

Richard A. Lent

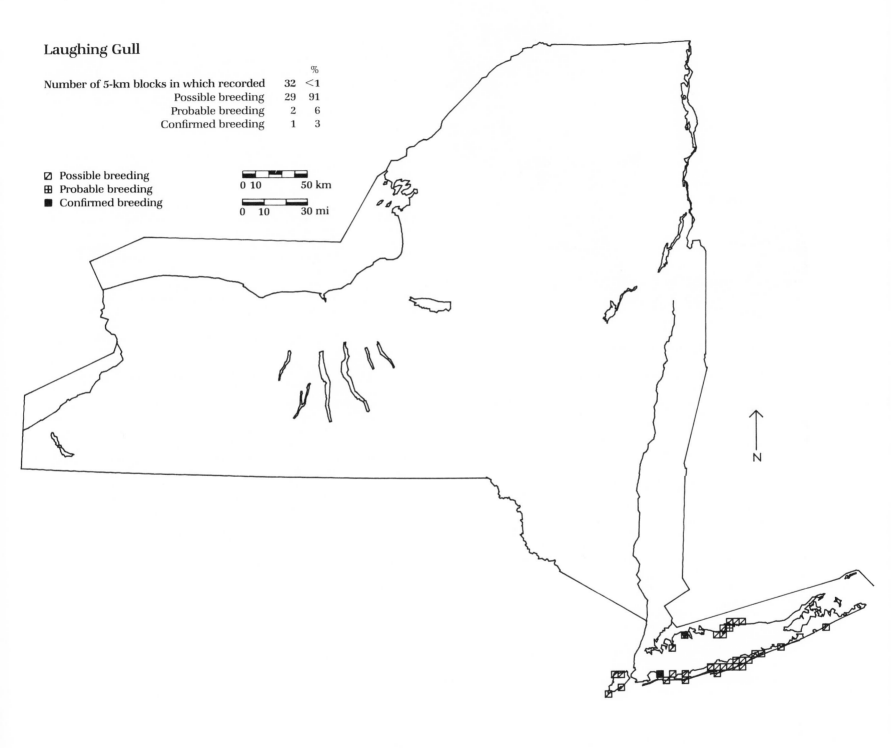

Laughing Gull

Number of 5-km blocks in which recorded		%
	32	<1
Possible breeding	29	91
Probable breeding	2	6
Confirmed breeding	1	3

▨ Possible breeding
⊞ Probable breeding
■ Confirmed breeding

0 10   50 km

0 10   30 mi

N

© KLA-C 1986

Ring-billed Gull *Larus delawarensis*

The Ring-billed Gull has become a common colonial breeder on natural islands and man-made structures in four local upstate areas: Buffalo and along the Niagara River, Oneida Lake, the eastern end of Lake Ontario and along the St. Lawrence River, and on The Four Brothers islands in Lake Champlain. The Ring-billed Gull was once considered a rare species in New York, even as a migrant (Eaton 1910) but is now commonly seen in large flocks in the spring (Eaton 1981) and is becoming a nuisance in many areas.

Audubon and Chevalier (1840–1844) referred to this species as "The Common American Gull" but said that in the East its population was small, and Bent (1921) attributed this to egg collecting. Barrows (1912) documented its disappearance as a breeding bird from much of the Great Lakes in the early 20th century. By the 1940s, however, the population there had recovered and was stable from 1940 to 1960, with an estimated 27,000 pairs (Ludwig 1974). In the 1960s the population began to increase and the total Great Lakes–St. Lawrence River nesting population in 1984 was estimated to be about 700,000 pairs (Blokpoel and Tessier 1986).

This gull was first found nesting in New York on Gull Island at the eastern end of Lake Ontario in 1936 (Hyde 1939). In 1938 it was found on Little Galloo Island, an island of about 17.4 ha (43 a) not far from Gull Island (Bull 1974). (The Ring-billed Gull no longer nests on Gull Island and has not for some time.) Kutz (1946) estimated that on 7 June 1945, there were 2000 breeding gulls at Little Galloo, most of them Ring-billed Gulls. Since that time the population on Little Galloo has increased dramatically. In 1981 a total of 73,780 nests were counted on a census conducted 11 and 12 May (Blockpoel and Weseloh 1982).

Another large colony was located in 1949 on Island "C" of The Four Brothers in Lake Champlain (Belknap 1955). In 1955 when Belknap visited the island, 2000 adults occupied an area of less than an acre. By 1967 the Ring-billed Gull colony was estimated to be 2500 pairs (Bull 1974); by 1980, 10,000 pairs (Mack 1980); and in 1985, 16,329 nests were counted (Peterson 1986a).

The third colony to be located in New York was on Long Island in Oneida Lake, discovered by L. J. Loomis in 1952 (Belknap 1955). In 1955 that population was estimated at 150 pairs. In 1957 the Ring-billed Gull nested on both Long and Wantry islands with a combined total of 700 pairs. In 1970 breeding was reported on Wantry but not on Long (Rusk and Scheider 1970), and a dramatic decline was noted the following year, with just 37 nests on Wantry Island only (Rusk and Spies 1971). The decline did not last. During the period 1979–85 the Ring-billed Gull population at Oneida Lake increased from 399 nests in 1979 to 2269 nests in 1985 on Little, Long, and Wantry islands (Bollinger 1985).

Nesting colonies were also located in Buffalo Harbor and in the upper Niagara River. In 1986 the total number of nests of Ring-billed Gulls at these colonies was estimated to be about 24,000 (Hotopp 1986). The Ring-billed Gull also nests along the St. Lawrence River, including at Eagle Wing Group islands, which had 135 nests in 1984 (Crowell and Smith 1985).

Belknap (1968) described the habitat on Little Galloo Island as rock covered by a thin layer of soil. Grass and weeds constitute much of the vegetation; the interior resembles a level, grassy meadow with a few trees around the perimeter. More recently the Ring-billed Gull has been found nesting on various artificial sites like those in Buffalo Harbor and along the Niagara River, including dredge spoil, gravel adjacent to a breakwater, and a water-control structure. In addition, it has been found nesting on harbor dikes, piles of rubble, and slag dumps. Nesting substrates vary greatly and include sand, earth, driftwood, concrete, slag, rocks, and boulders (Blokpoel and Tessier 1986). The nest is placed in a scrape on the ground, on matted vegetation, or at the upper edges of beaches among rocks. It is made of dried grasses, mosses, weeds, or rubbish and lined with finer grasses and feathers (Harrison 1975).

The Great Lakes population of the Ring-billed Gull has expanded so much that there now is interest expressed in controlling its numbers. Farmers complain that this gull damages crops and destroys soil fertility by eating earthworms; also it is an airport hazard and can transmit diseases to humans. This gull has taken over the nesting areas of other species, including the Common Tern (Blokpoel and Tessier 1986).

Stephen W. Eaton

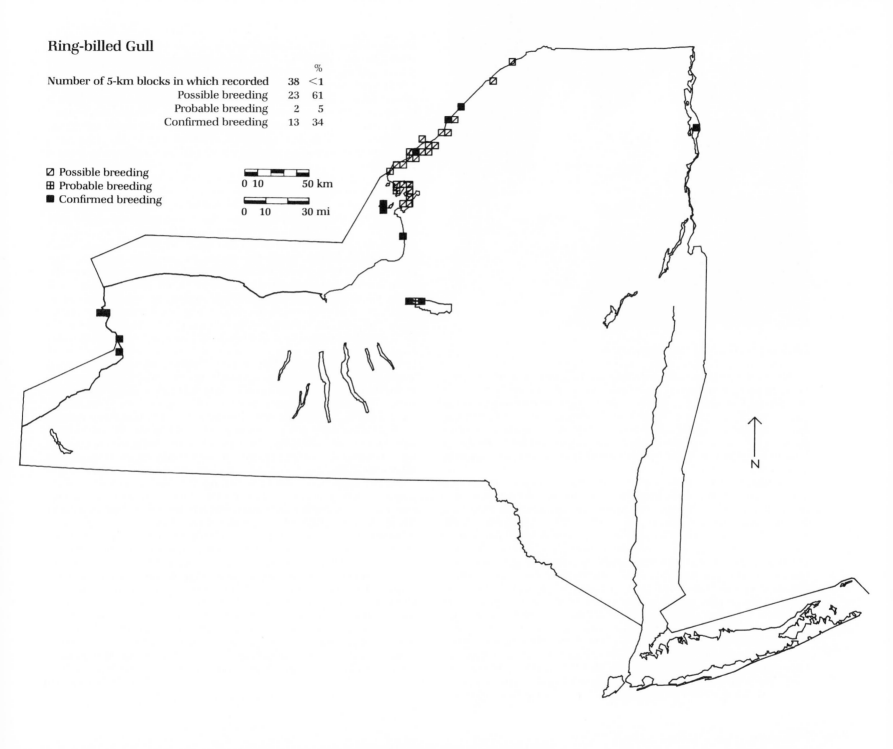

Ring-billed Gull

		%
Number of 5-km blocks in which recorded	38	<1
Possible breeding	23	61
Probable breeding	2	5
Confirmed breeding	13	34

◩ Possible breeding
⊞ Probable breeding
■ Confirmed breeding

0 10 50 km

0 10 30 mi

N

Herring Gull *Larus argentatus*

Once nearly extirpated as a breeding bird in New York, save for a small remnant population nesting on remote lakes in the Adirondack interior, the Herring Gull made a good recovery during this century. This gull is sometimes an abundant but quite local colonial nester on the Coastal Lowlands, and at most of the large lakes and rivers in upstate New York where it occurs. Within the Central Adirondacks and Sable Highlands, however, the Herring Gull is widely distributed, but it nests in single pairs rather than in colonies and is fairly uncommon.

The history of the persecution of the Herring Gull by eggers and gunners dates back at least to the 19th century when the Reverend G. Ingersoll was collecting eggs on "one of the islands called the Four Brothers," in Lake Champlain (Thompson 1853). Alvah Jordan (1888) visited these same islands in 1887 to collect adults and eggs, and lamented: "Eight or ten years ago a large colony of these Gulls used to breed here, but owing to the relentless persecution . . . the colony had been reduced to some fifty pairs and I do not think they raised a single brood last season." Eaton (1910) reported that it was said the Herring Gull formerly bred on the islands in Lake Ontario but had no evidence that they were still doing so. By the first decade of this century this gull was barely hanging on in the Central Adirondack counties of Franklin, Hamilton, and Herkimer, and was still persecuted in spite of the new migratory bird treaty and other protective efforts (Eaton 1910). Yet a change was taking place. The relict colony on the The Four Brothers islands was vigorously protected under the ownership of Edward Hatch. The coastal Herring Gull still nested only as far south as Maine at the turn of the century, but by 1931 it had arrived on eastern Long Island (Bull 1974). Within the next half century the Herring Gull had achieved the recovery documented on the Atlas map.

Today, the Herring Gull is most numerous on the Coastal Lowlands, where recent surveys found a population numbering over 24,000 pairs at 34 colonies. The largest single colony in the state was on Plum Island, Suffolk County, where 7000 pairs were found in 1985. As the Atlas map indicates, most of these colonies are on islands in bays along the south shore and the eastern tip of Long Island, with only a few records along the north shore.

In the Adirondacks the Herring Gull nests on rocks or islands in an unknown number of lakes, with "confirmed" records in only 35 blocks. Although noted as "possible" or "probable" on many other lakes, it is unlikely that this gull nested in all of the blocks indicated on the Atlas map.

All nesting sites in the state other than the Adirondacks and the Coastal Lowlands are shared with the smaller but more numerous Ring-billed Gull. At least a few Herring Gull colonies such as the one at The Four Brothers have experienced declines (Wolfe 1923; Belknap 1955; Peterson 1986a). The St. Lawrence River appears to have only about 28 pairs (18 in St. Lawrence County and just 10 in Jefferson County), occupying eight islands and a buoy, and the Herring Gull there is considered "in danger from chemical contamination of its food source" (Maxwell and Smith 1983). The sole colony in Lake Ontario is on Little Galloo Island, Jefferson County, where, Blokpoel and Weseloh (1982) stated, "It would appear that Herring Gulls are reproducing very well and increasing" to 350 nests in 1981. A NYSDEC survey of the upper Niagara River located two nests at Buckhorn Diversion Weir. In Buffalo Harbor, Erie County, 108 nests were found on Donnelly's Pier and at least two at Stony Point, a recently established gull colony in a dredged material disposal site in Lake Erie outside the harbor (Hotopp 1986). In 1985 a total of 60 nests were counted on Long, Wantry, and Little islands on Oneida Lake (Claypoole 1986).

The Herring Gull occupies a wide range of shoreline and insular habitats in New York. On Long Island it nests almost exclusively on islands, where it prefers dredge spoil areas and abandoned buildings and docks (Peterson et al. 1985). On freshwater Adirondack lakes and ponds it prefers rocks or islands, and on large freshwater lakes and rivers it also chooses islands, islets, or navigation buoys. In the Niagara River gorge it uses natural cliff ledges and talus slopes (Andrle 1976).

The nest of this species varies from a simple scrape or a few dry stems to more elaborate mounds of earth and mosses, built up over many years of repeated use (pers. obs.). Some Adirondack nests are especially lovely constructions of mosses and lichens, surrounded by native northern plants, and yet even in the same region some nests on emergent boulders are hardly more than a mound of organic duff (pers. obs.). Along the coast nests were placed on the sand or in a grassy area, and nests in small trees were found in at least two Long Island colonies (Peterson et al. 1985).

John M. C. Peterson

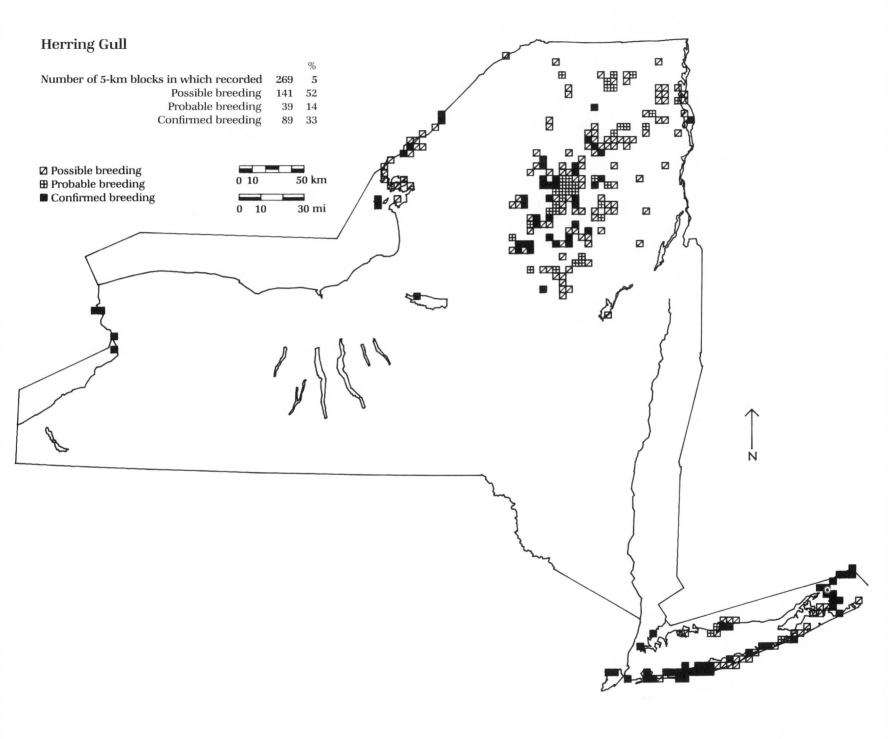

Herring Gull

Number of 5-km blocks in which recorded	269	% 5
Possible breeding	141	52
Probable breeding	39	14
Confirmed breeding	89	33

◪ Possible breeding
⊞ Probable breeding
■ Confirmed breeding

0 10   50 km

0 10   30 mi

N

©KLA-c 1986

Great Black-backed Gull *Larus marinus*

"Aggressive, predatory, merciless tyrant" are words sometimes used to describe the Great Black-backed Gull, a large gull known as a winter visitor to the Atlantic Coast since the time of De Kay (1844). Eaton (1910) said it was common in winter on the shores of Long Island and uncommon in the interior along Lakes Ontario, Cayuga, and Seneca. Today it is a fairly common breeding species only on New York's Atlantic Coast; elsewhere in New York there are only four known nesting areas.

CBC records for western New York show that the Great Black-backed Gull began an amazing increase in the 1940s. The combined totals for the Buffalo (Erie County), Rochester (Monroe County), and Geneva (Ontario County) counts go from no birds for the period 1930–34 to a 5-year average of 42 birds for 1940–44, gradually climbing to an average of 204 for the period 1960–64 (Peakall 1967), and averaging 2032 for the years 1981–85 (*American Birds* 1981–85).

As numbers of winter visitors in the interior as well as on the coast were burgeoning there was also a concomitant, steady movement of breeding south from Nova Scotia, which was the southernmost breeding location in 1921 (Bent 1921). In 1930 nesting was confirmed on the Isle of Shoals, New Hampshire (Jackson and Allen 1932), in Maine (Norton and Allen 1931), and Massachusetts (Eaton 1931). By 1944 the total New England population was not less than 3500 individuals, a remarkable growth from the early 1930s (Gross 1945).

New York was logically the next invasion stop; in July 1942 Wilcox (1944) banded young on Cartwright Island near the east end of Long Island. By 1958 there were 50 pairs on Long Island, and in 1966, 320–480

pairs were on Gardiners Island alone (Peakall 1967). In 1974, 1838 pairs were counted on a survey of islands around Long Island and in New York Harbor (Buckley and Buckley 1980). During another survey of Long Island's offshore islands and south shore beaches in 1985, 6948 pairs were estimated in 23 colonies (Peterson et al. 1985). The Atlas map indicates "confirmed" breeding in 35 blocks on Long Island proper and the Westchester County shore of Long Island Sound.

It was only a matter of time until the Great Black-backed Gull would be found nesting in the interior. In 1954 a pair nested on Lake Huron in Ontario (Krug 1956); nests were discovered on the north shore of Lake Ontario in 1962 (Woodford 1962) and in Quebec in 1963 (Carleton 1963). New York's first known upstate nesting was on The Four Brothers islands in Lake Champlain in 1975. This gull has continued to nest there through 1985, when there were 3 pairs with 2 active nests (Peterson 1986a, and pers. comm.). In 1981 4 nests were found at Little Galloo Island in northeastern Lake Ontario (Blokpoel and Weseloh 1982). The colony persisted, with 3 nests in 1984 and 4 in 1986 (Weseloh, pers. comm). Additionally, a nest with 2 eggs was found by Atlas observers on Eagle Wings Group in the St. Lawrence River; the eggs did not hatch. An unsuccessful nesting attempt was made on an island in Oneida Lake near Syracuse in 1983 (DeBenedictis 1983).

Wherever this gull nests, it usually finds a place near or among Herring Gulls. Of 23 colonies located on Long Island, 22 contained both Great Black-backed Gull and Herring Gull nests and 1 contained only Great Black-backed Gull nests (Peterson et al. 1985). The nest sites can be associated with fresh, brackish, or salt water. Atlas observers found nests on rocky beaches among large boulders, perhaps with scattered deciduous trees and underbrush, and on the tops of rocks, sandy mounds, or just on a sandbar. Most nests found by surveyors on Long Island were on islands, with a preference shown for man-made sites, including "dredge spoil areas and abandoned buildings and docks" (Peterson et al. 1985). The island locations of many nest sites may afford a degree of freedom from mammalian predators. The nest itself is simply a collection of moss, grass, twigs, rubbish, and a few feathers or perhaps no more than a depression in the sand. Nests were found in small trees in at least two Long Island colonies (Peterson et al. 1985).

The range expansion and growth in numbers of this striking and handsome gull has already created a problem in tern colonies. In Maine on Eastern Egg Rock, the Arctic and Common Tern colony contained at least 1000 birds in 1880; by 1936 all but a few pairs of terns had been supplanted by Herring Gulls and by 1974 by Great Black-backed Gulls. After several years of gull control the gulls ceased nesting there and the terns returned (Kress 1983).

Gordon M. Meade

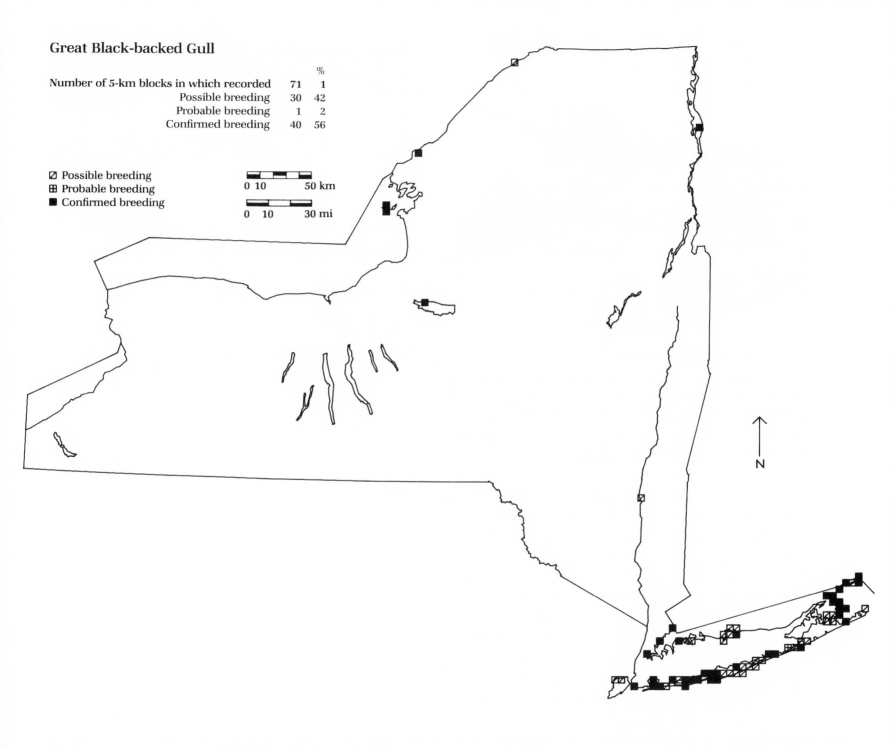

Great Black-backed Gull

Number of 5-km blocks in which recorded	71	% 1
Possible breeding	30	42
Probable breeding	1	2
Confirmed breeding	40	56

☑ Possible breeding
⊞ Probable breeding
■ Confirmed breeding

0 10   50 km

0 10   30 mi

N

interspersed among the Common Terns at any colony. The nest is similar to that of the Common Tern, consisting of a loose pile of whatever dead vegetation or debris is in the immediate area (Buckley et al. 1975; Peterson et al. 1985).

David M. Peterson

Gull-billed Tern *Sterna nilotica*

The Gull-billed Tern is a recent addition to the breeding birds of New York. Bull (1974) considered the species very rare in the state, occurring primarily along the coast after hurricanes. As a breeding species the Gull-billed Tern was known to nest in small numbers as far north as New Jersey (Buckley et al. 1975; Erwin 1979). In 1975, 2 pairs of Gull-billed Terns were found nesting in a Common Tern colony on a dredge spoil island in Hempstead, Nassau County (Buckley et al. 1975). One pair was again found breeding in the area in 1976, but no birds were located in 1977. Since 1978, 1–3 pairs have nested on salt-marsh and dredge spoil islands of Hempstead (Bull 1976; Scheibel and Morrow 1979; Buckley and Buckley 1980; Zarudsky 1980; Peterson et al. 1985). During 1983 and 1984, 1–2 pairs were also found nesting on the barrier beach at the large Common/Roseate Tern and Black Skimmer colony at Cedar Beach in Babylon, Suffolk County (Peterson et al. 1985). The records on the Atlas map are of the Hempstead and Cedar Beach colonies.

New York's breeding Gull-billed Terns constitute a relatively isolated outpost at the northern edge of the species' breeding range. Its breeding population to the south in New Jersey is small and scattered (Erwin 1979; Buckley and Buckley 1984b). As New Jersey's population has not expanded greatly, it appears unlikely that the number of breeding Gull-billed Terns in New York will increase greatly, either.

The Gull-billed Tern has displayed an ability to utilize a variety of nesting habitats, from barrier beach sites to marsh islands. Since its arrival as a breeding species in New York, it has always been found nesting in association with Common Terns. Normally, only 1–2 pairs nest

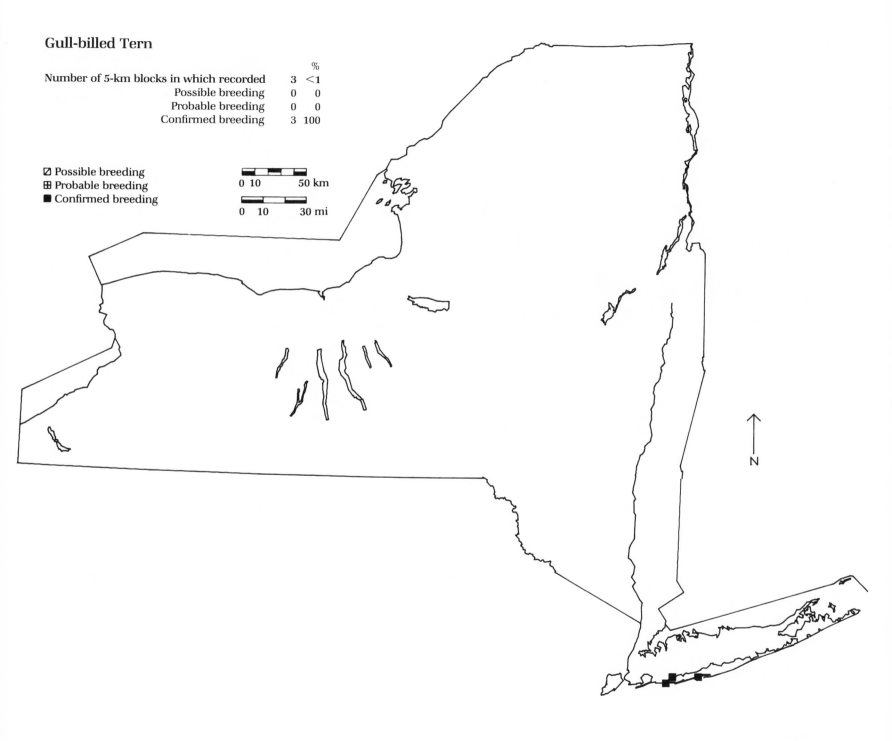

Gull-billed Tern

Number of 5-km blocks in which recorded		%
	3	<1
Possible breeding	0	0
Probable breeding	0	0
Confirmed breeding	3	100

▨ Possible breeding
⊞ Probable breeding
■ Confirmed breeding

0 10   50 km

0 10   30 mi

N

©KLA-C 1987

Caspian Tern *Sterna caspia*

Although this large, crested tern has long nested on the Great Lakes, not until after Atlas fieldwork had concluded was the Caspian Tern confirmed as a breeder in New York. Found in 1986 at a single colony on Little Galloo Island, in the Jefferson County waters of eastern Lake Ontario, this is quite obviously a rare nester within the state.

In North America the Caspian Tern has nested in at least ten states and seven Canadian provinces (Martin 1978). During the 1980–85 Atlas period there were two active colonies on the Canadian side of Lake Ontario: Pigeon Island on the eastern end (Martin 1978) and the Eastern Headland at the west end near Toronto (Fetterolf and Blokpoel 1983). Another Ontario colony was established at Hamilton Harbour during 1986 (Weseloh, pers. comm.).

Neither Giraud (1844) nor De Kay (1844) mentioned the Caspian Tern in New York. Eaton (1910) described this bird as "a regular but rather uncommon transient visitant both on the coast and the larger lakes of New York." Bull (1974) described the Caspian Tern as a variously uncommon to common migrant on the outer coast of Long Island, found in largest numbers after fall hurricanes. Upstate, he noted the greatest fall concentrations at Braddock Bay, Monroe County, and Sandy Pond, Oswego County, and that elsewhere the species was relatively rare to uncommon.

The nesting population on the Great Lakes declined between 1925 and 1960, then began increasing and apparently dispersing (Ludwig 1965). During the Atlas period there were several upstate reports of Caspian Terns during the months of June and July that probably represented summer stragglers or post-breeding wanderers. On Lake Champlain 1 was at Port Henry, Essex County, 13 June 1984, and 3 were seen at Treadwell Bay, Clinton County, 25 and 26 June 1984 (Peterson 1984b). A dozen were displaying among Ring-billed Gulls on a rocky islet several hundred meters offshore in Jefferson County waters of Lake Ontario on 9 June 1985 (R. and J. Walker, v.r.). Several were diving for fish in a lagoon on Calf Island, Jefferson County, the following day (R. and J. Walker, v.r.), and up to 10 were present in Jefferson County during the summer of 1985, including an immature among adults on a gravel beach 20 July (W. Hendrickson, v.r.). Finally, on 21 June 1986, Weseloh encountered over 100 Caspian Terns on the southwest side of Little Galloo Island and discovered that they were nesting on the receding shoreline of a small water impoundment or lagoon. Subsequent nest surveys provided a high count of 112 nests containing a total of 180 eggs, plus 2 chicks, on 7 July (Weseloh, pers. comm.). To date, this represents the first and only nesting record for the state.

Census efforts on the Canadian side of Lake Ontario during 1986 gave high counts of 330 nests on Pigeon Island, 202 nests on the Eastern Headland, and 48 nests at Hamilton Harbour (Blokpoel, Weseloh, pers. comms.). Including Little Galloo, the total Lake Ontario population was approximately 692 pairs. The most recent figures available for the entire population on the Great Lakes show just over 2500 pairs on the Canadian side (Scharf and Shugart 1983; Weseloh et al. 1986; Blokpoel, pers. comm.) and an additional 1700 pairs nesting on the U.S. Great Lakes (Shugart et al. 1978; Weseloh, pers. comm.), suggesting a total Great Lakes population of about 4200 pairs of Caspian Terns.

A number of colonial waterbirds nested on Little Galloo in 1986, the most numerous being the Ring-billed Gull (approximately 70,000 pairs). Although the Caspian Terns nested among the Ring-billed Gulls, placement and timing may have assisted the terns. They nested later than the gulls, settling where the waters of the lagoon had just receded, on land that was inundated when the gulls arrived. Most Ring-billed Gulls had hatched their young by the time the first tern eggs were laid, about 26 May, so that territorial defense by the gulls had lessened (Weseloh, pers. comm.).

The Caspian Tern has increased on Georgian Bay, Ontario, on Lake Huron and on Lake Ontario in recent years (Weseloh, pers. comm.), and further colonization in New York is possible. The limiting factor may be suitable nest sites, ideally islands, where mammalian predation and human disturbance are limited. Coyotes, raccoons, and striped skunks, as well as Herring and Ring-billed Gulls, are known predators of Caspian Terns (Shugart et al. 1978; Fetterolf and Blokpoel 1983).

The nests on Little Galloo Island were simple scrapes or depressions on the bare ground, usually lined with twigs and other dead vegetation, sometimes a few feathers or other litter (pers. obs.).

John M. C. Peterson

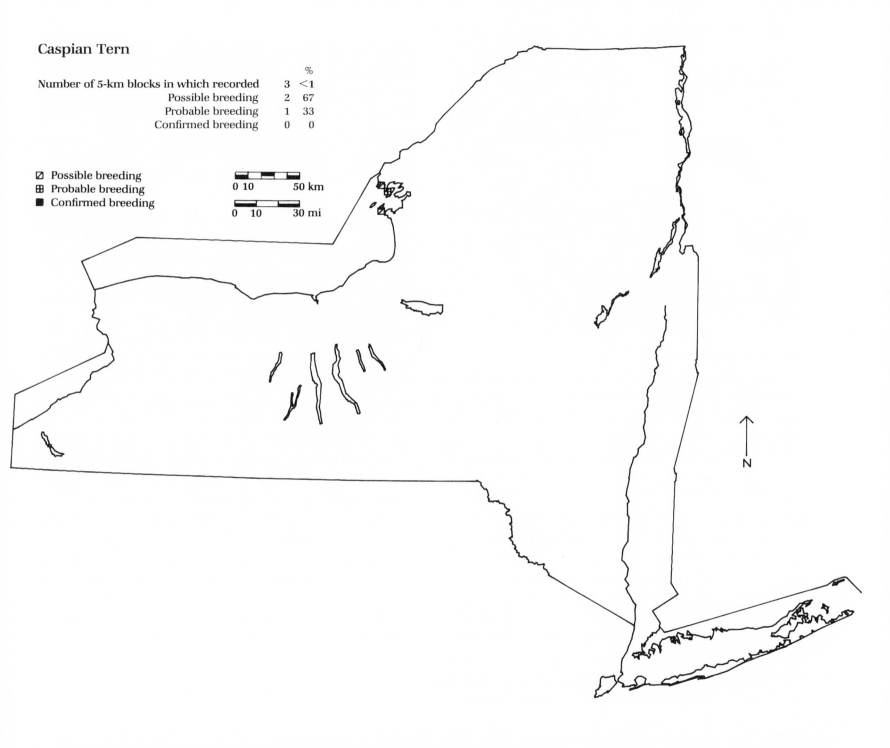

Caspian Tern

Number of 5-km blocks in which recorded   %
                                        3 <1
                    Possible breeding   2 67
                    Probable breeding   1 33
                   Confirmed breeding   0   0

☑ Possible breeding
⊞ Probable breeding
■ Confirmed breeding

0 10   50 km

0 10   30 mi

N

Roseate Tern *Sterna dougallii*

The Roseate Tern is one of the two tern species listed by the NYSDEC as Endangered. It breeds in a small number of colonies along the coast. During the plume-hunting era in the late 1800s and early 1900s the Roseate Tern was heavily hunted. Bull (1964) reported that only 10 pairs nested at Great Gull Island, Suffolk County, in 1889, and Erwin (1979) listed it extirpated as a breeding species in the state in 1900. Eaton (1910) called it an uncommon but regular summer resident on Long Island with a few pairs nesting in Common Tern colonies on Gardiners Island and neighboring islands. In 1914, 30 pairs were nesting at Gardiners Island (Bull 1964).

By 1934 the New York population had recovered to the point where 400 pairs were present at Orient Point in eastern Long Island (Bull 1964). For the next 20 years little information was documented about the species in New York, although the population in Connecticut had grown to 1600 pairs by 1941 (Bull 1964).

During the 1950s and early 1960s new sites were colonized. In 1951, 3 pairs nested at Short Beach in southern Nassau County, the westernmost nesting of the species in the state. The Common Tern colony at Cedar Beach near Fire Island Inlet, southwestern Suffolk County, was occupied by 15 pairs of Roseate Terns in 1958. By 1960, 350 pairs were breeding in the area (Bull 1964). Great Gull Island was recolonized after over 65 years of abandonment around 1955 (Hays 1984). During the 1960s and early 1970s Roseate Tern numbers continued to grow. By 1968 over 1000 pairs were nesting at Great Gull Island, and colonies containing up to 700 pairs were present near Fire Island, Moriches, and Shinnecock inlets along the south shore of Suffolk County. In addition, 20 pairs were located at Robins Island in Great Peconic Bay, and the first record of breeding on Long Island's north shore was documented with two pairs present at Stony Brook Harbor (Davis and Morgan 1968). The peak documented population of the Roseate Tern occurred in 1975 when 2254 pairs were recorded at 11 colonies. That year the Great Gull Island colony contained approximately 1500 pairs (Buckley and Buckley 1980; Hays 1984).

In 1976 the Great Gull Island population dropped to 700 pairs because of enroachment of vegetation on the island (Hays 1984). In 1978 New York's entire population decreased to 843 pairs in 7 colonies (Buckley and Buckley 1980; Hays 1984). By 1985 the population increased slightly to 967 pairs, but only 4 active colonies were recorded (Peterson et al. 1985).

Although the decline of the Roseate Tern can in part be explained by habitat loss due to vegetation changes, competition with gulls, coastal development, and predation (Buckley and Buckley 1981; Spendelow 1982; Peterson et al. 1985), some studies indicate that current productivity should be sufficient to maintain the population (Buckley and Buckley 1981; Safina 1985b). Market hunting of the Roseate Tern on its wintering ground in northeastern South America has been cited as another possible cause of the decline (Buckley and Buckley 1984b).

The Atlas documents only a slight change in the distribution of the Roseate Tern from the 1970s. The colonies at Short Beach and Stony Brook Harbor have been abandoned. Along the south shore the 3 remaining colonies are still located adjacent to major bay inlets. On the east end of Suffolk County the major colony is still at Great Gull Island, with smaller satellite colonies on islands in Peconic and Gardiners bays. During the Atlas period the satellite colonies shifted from year to year, with no sites being active every year (Peterson et al. 1985).

While the Common Tern on Long Island has greatly increased between the 1970s and 1980s, the Roseate Tern has not, perhaps because the Roseate Tern is more specialized both in its feeding and nesting habits. The Roseate Tern has been observed feeding almost exclusively along inshore ocean waters at tidal rips, sandbars, and bay inlets. This type of feeding habitat is more limited than the foraging areas used by the Common Tern (Safina 1985a). The Roseate Tern diet may also be more restricted than that of the Common Tern (Safina 1985b).

In New York the Roseate Tern has always been found nesting in Common Tern colonies (Bull 1974; Peterson et al. 1985); it usually places its nest in dense vegetation, whereas the Common Tern nests more in open areas with less vegetation. Roseate Tern nests are usually found under vegetation and debris (Bull 1974; Hays 1984). On Great Gull Island it also nests in crevices among boulders along the shore (Hays 1984). On Gardiners Island the Roseate Tern nests on open sand; on Lanes Island in Shinnecock Bay it nests on salt marsh (Buckley and Buckley 1980; Hays 1984).

David M. Peterson

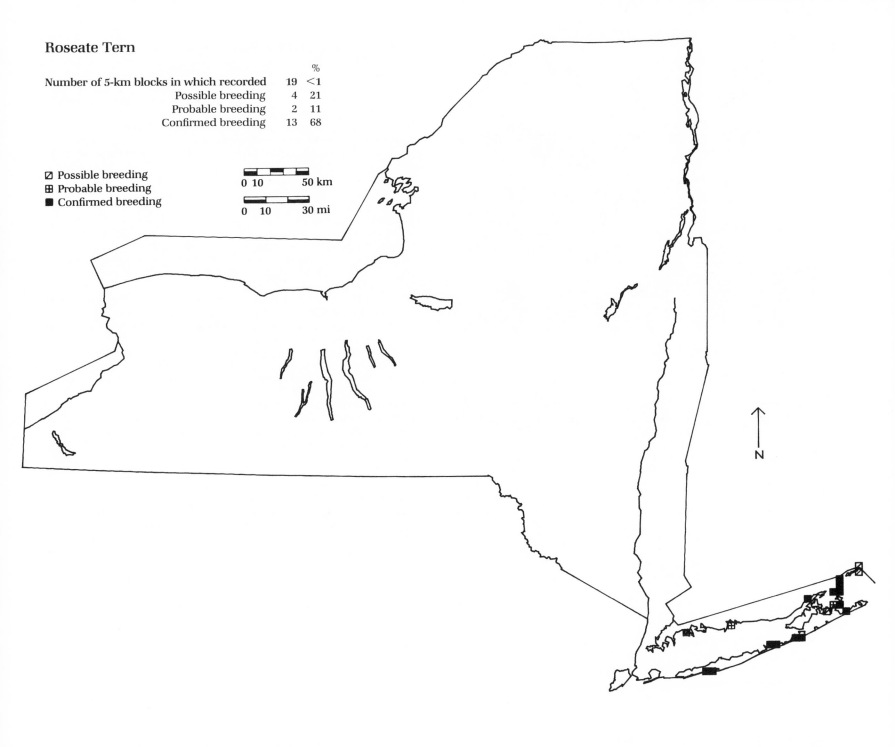

Roseate Tern

		%
Number of 5-km blocks in which recorded	19	<1
Possible breeding	4	21
Probable breeding	2	11
Confirmed breeding	13	68

◨ Possible breeding
⊞ Probable breeding
■ Confirmed breeding

0 10   50 km

0 10   30 mi

N

Common Tern *Sterna hirundo*

The Common Tern is New York's most abundant and widespread tern, occurring both along the coast and on major inland waters. It was listed as Threatened by the NYSDEC because of recent serious declines in the upstate population. Eaton (1910) listed the species as abundant on the coast but noted no breeding upstate.

During the late 1800s and early 1900s the Common Tern population was decimated by plume hunters. Most colonies that survived were at protected areas on Gardiners and Fishers islands off eastern Long Island (Eaton 1910). When protection from hunting was granted, the Common Tern population increased along the coast. By 1930 a colony at Orient Point in eastern Long Island contained 6000 nesting pairs (Bull 1964).

In upstate New York this tern was first documented breeding on the upper St. Lawrence River (1917), Oneida Lake (1929), and eastern Lake Ontario (1936). By 1952 the Little Galloo colony in eastern Lake Ontario contained 1000 pairs (Bull 1974). Nesting was recorded in the Buffalo area in 1944. Colonies of 1000–1400 pairs became established in eastern Lake Erie and on the Niagara River (Courtney and Blokpoel 1983). A small number of pairs also nested at Montezuma NWR at the north end of Cayuga Lake (Bull 1974).

It appears that Common Tern numbers in the entire eastern Great Lakes region peaked in the early 1960s and remained stable through the early 1970s (Courtney and Blokpoel 1983). The eastern Lake Ontario, Oneida Lake, and St. Lawrence River populations were estimated to be approximately 2500 pairs in the early 1970s (Smith et al. 1984). Data available for eastern Lake Erie and the Niagara River are less exact, but apparently 1000 pairs of Common Terns may have nested in the area (Courtney and Blokpoel 1983).

Since the early 1970s upstate Common Tern numbers have declined, dropping to an estimated 1000 pairs by 1982 on eastern Lake Ontario, Oneida Lake, and the St. Lawrence River (Smith et al. 1984). On Lake Erie and the Niagara River the population declined to approximately 800 pairs by 1983 (Hotopp 1986). Disturbance, flooding, predation, and pressure from nesting Ring-billed Gulls have forced out the Common Tern from most historical upstate sites (Courtney and Blokpoel 1983; Smith et al. 1984; Claypoole 1986).

Along the coast the Common Tern expanded from nesting primarily on eastern Long Island and adjacent islands to colonizing areas along Long Island's north and south shores. Sites near bay inlets along the south shore, in particular, became major colonies. During the 1950s and 1960s colonies of from 1000 to 6000 pairs became established in the vicinity of four major inlets (Bull 1964, 1974). In eastern Long Island, Great Gull Island was recolonized around 1955 after over 65 years of abandonment, and by 1964, 1500 pairs of Common Terns were breeding there (Hays 1984).

Surveys of the New York coast between 1974 and 1978 documented up to 14,972 pairs of Common Terns in 34–54 colonies (Buckley and Buckley 1980). By 1985 the coastal population had grown to an estimated 22,857 pairs occupying 41 colonies (Peterson et al. 1985).

The Atlas documented several changes in Common Tern distribution from Bull's (1974) range map. Upstate Common Terns were not found in several areas where they formerly bred, including Irondequoit Bay near Rochester, the south end of Cayuga Lake, and Onondaga Lake near Syracuse. Only one Atlas block along the eastern shore of Lake Ontario had "confirmed" breeding. New breeding areas were found by Atlas observers along sections of the St. Lawrence River. On the coast the Atlas has not documented any significant change in the species' breeding range.

The Common Tern nests in a variety of shoreline areas. Inland it has been found on small islands, navigation structures, and jetties (Courtney and Blokpoel 1983; Smith et al. 1984; Hotopp 1986). At Montezuma NWR it often nests on muskrat lodges and duck blinds (Bull 1974). On the coast it nests on offshore islands, barrier beach dunes, dredge spoil areas, and on salt marshes. In 1985 over 65% of all coastal colonies were located on salt-marsh islands (Peterson et al. 1985). That year, though, over half of the state's coastal population nested at only two colonies: Great Gull Island (5500 pairs) and Cedar Beach (6000 pairs) in southwestern Suffolk County (Peterson et al. 1985).

Common Tern nests are constructed of dead vegetation and debris collected in the vicinity of the nest. Nests may vary from scrapes in the ground to small piles of dead vegetation (Bull 1974). At some inland sites eggs have been laid on the bare concrete (Hotopp 1986).

David M. Peterson

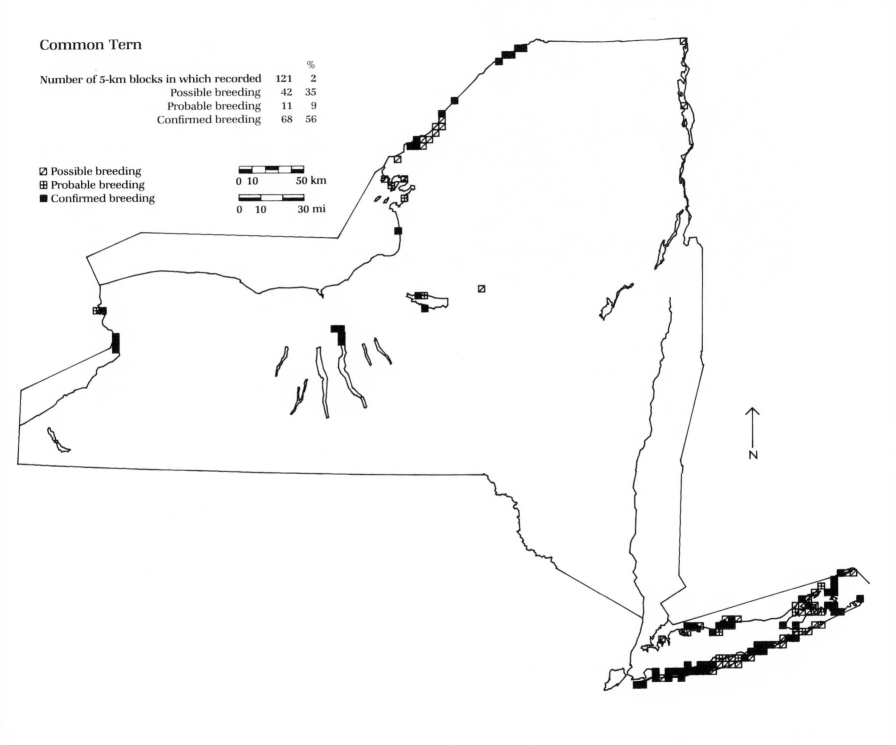

Common Tern

		%
Number of 5-km blocks in which recorded	121	2
Possible breeding	42	35
Probable breeding	11	9
Confirmed breeding	68	56

▨ Possible breeding
⊞ Probable breeding
■ Confirmed breeding

0 10   50 km

0 10   30 mi

N

Forster's Tern *Sterna forsteri*

The Forster's Tern, a species superficially similar to the Common Tern, was not known before the Atlas survey to nest in New York. Unlike the other terns breeding in the state, which also nest on other continents, the Forster's Tern is confined to North America. It was found breeding in New York, on Long Island, for the first time in 1981. Otherwise, it is known in the state chiefly as a fall migrant on the coast and the Great Lakes, but before 1924 this tern was very scarce, or possibly was overlooked (Bull 1974). Forster's Terns in New York also include Atlantic Coast birds that apparently disperse northward before fall migration (AOU 1983). Griscom (1923) knew of only five Long Island records, none since 1883, but he thought it may have been a visitor at an earlier period, when the species nested in New Jersey.

Major breeding grounds of the Forster's Tern are in the interior from the Great Lakes westward, but there are two coastal populations, one on the Gulf Coast and another on the East Coast. As a nesting bird, it has recently increased on the East Coast, spreading north from Virginia and Maryland, and in 1955–56 several pairs bred at Brigantine NWR in New Jersey, a state where it had nested in the 1800s (Leck 1984). Paxton et al. (1982, 1985) cited a steady increase in the number of Forster's Terns observed during the breeding season at various colonies in New Jersey, from 661 adults in 1977 to 2333 adults in 1985.

On 1 June 1981 a breeding pair was discovered in a colony of Common Terns on Hewlett Hassock, a salt-marsh island in Hewlett Bay on the southwestern shore of Long Island in the Town of Hempstead, Nassau County (Zarudsky 1981). Breeding was "confirmed" on 8 June when a nest with 3 eggs was located amid very similar Common Tern nests. Subsequently, a young bird was noted on the nest on 16 June, and later, on 23 June, 2 chicks were located. No further breeding was "confirmed" during the next two years, but there were occasional south shore reports in June, especially at Jamaica Bay Wildlife Refuge. In 1984 a pair nested again in the Hewlett Bay area, Hempstead, on North Green Sedge Island, but no birds were observed there in 1985 (Peterson et al. 1985). Paxton et al. (1985) reported that a few Forster's Terns were noted by A. J. Lauro and J. Zarudsky in the general area of their western Long Island beachhead in 1985.

In breeding plumage, this species closely resembles the Common Tern, although the primaries are pale silvery rather than dark above and the call notes are distinctive, with a nasal quality, including a diagnostic *tsaap*. Careful observation of adults is required where Forster's and Common Terns nest together, as at Hewlett Hassock, because of the similarity of their nests and eggs, and some may have been missed by Atlas observers for this reason. However, both tern colony monitoring and Atlas coverage have been thorough on the Coastal Lowlands.

Inland, the Forster's Tern has recently been nesting in Ontario on the north shore of Lake Erie after a long absence. Goodwin (1976) stated that breeding in Ontario was confirmed when a colony of some 50 pairs was found at Long Point 29 May to 6 June 1976. This locality is in the eastern part of Lake Erie not far from western New York. Breeding continued at Long Point after 1980, and also breeding was "probable" at a Lake Ontario north shore locality (Ontario Breeding Bird Atlas, preliminary map).

In the breeding season the Forster's Tern frequents large marshes and adjacent open water, nesting in extensive freshwater marshes or marshy lake shores in the interior, and salt marsh islands on the coast, such as those in bays behind the outer barrier beaches. Long Island's Hewlett Hassock is a salt-marsh island of about 10.9 ha (27 a) dominated by cordgrass (Zarudsky 1981).

Forster's Terns are relatively colonial and tend to nest very close together at favorable sites (Johnsgard 1979). On the East Coast the nests are usually large and well-built piles of dead grasses and sedges, the nest proper cup-shaped and lined with reeds and grasses; the nests are often placed on windrows of tide-drifted plant remains (Bent 1921). The Hewlett Hassock nest was situated in cordgrass about 38 cm (15 in) high, constructed of cordgrass drift, and located on a broad platform of wrack; Common Tern nests are of similar construction in this habitat (Zarudsky 1981). Away from the coast nests may be placed on floating masses of decaying cat-tails and composed of compact piles of cat-tail stalks with a neat central depression, or very often placed on top of muskrat houses, where the nest itself may be little more than a depression or hollow (Bent 1921).

Paul F. Connor

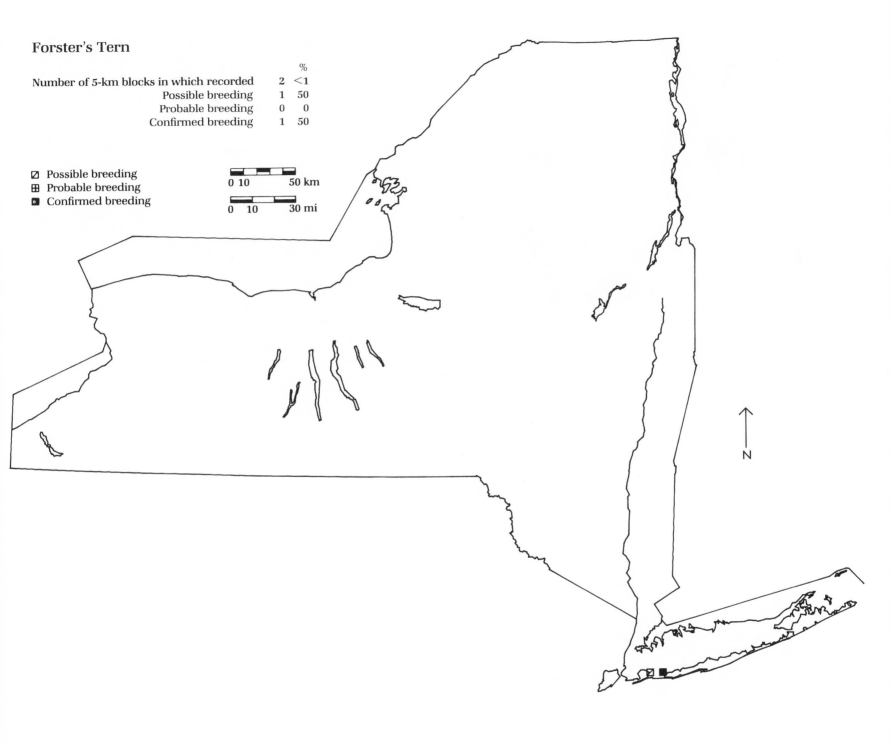

Forster's Tern

		%
Number of 5-km blocks in which recorded	2	<1
Possible breeding	1	50
Probable breeding	0	0
Confirmed breeding	1	50

◪ Possible breeding
⊞ Probable breeding
◪ Confirmed breeding

0 10   50 km

0 10   30 mi

N

©KLA-C 1986

Least Tern *Sterna antillarum*

The Least Tern, the smallest of the North American terns, is a locally common inhabitant of New York's coastal shores. Rarely seen away from salt water, it nests in small colonies along the shore (Bull 1974). In 1983 the NYSDEC listed the Least Tern as Endangered in the state because of threats to the species from increasing development and recreational pressures along the coast.

Historically, the Least Tern was a common summer resident during the mid-1800s on Long Island (Giraud 1844). Like the Common and Roseate terns, though, it was heavily hunted for its feathers in the late 1800s and early 1900s. Plume hunting resulted in its extirpation as a breeding species in the state in 1882 (Eaton 1910). With protection from hunting in the early 1900s, the Least Tern recovered to the point where a small colony reestablished itself in the state in 1926 (Bull 1964). Numbers increased, and by 1936 a colony of 200 pairs was found along Long Island's south shore in western Suffolk County (Bull 1974). In 1942, 6 south shore colonies containing over 300 pairs were recorded. By the 1950s the Least Tern was well established on Long Island's south shore, with over 200 pairs nesting at Short Beach, Nassau County, in 1951. Colonies also occurred on the north shore of Long Island. Twenty pairs nested at Port Jefferson, Suffolk County, in 1951 (Bull 1964).

By the early 1970s Least Tern colonies were found on both the north and south shores of Long Island and in the Peconic bays section in eastern Long Island. The Least Tern also nested at Great Kills, Staten Island (Bull 1974), and on Fishers Island off the eastern end of Long Island (Duffy 1977). During the mid-1970s surveys located and counted the Least Tern colonies in the state (Erwin and Korschgen 1979; Buckley and Buckley 1980). Although most Least Tern colonies were previously thought to be located on the beaches of Long Island's south shore, the region-wide survey found 55–70% of the Least Tern breeding population to be located at the mouths of the major harbors on Long Island's north shore and in small colonies in the Peconic bays area (Erwin and Korschgen 1979; Buckley and Buckley 1980). The presence of colonies in these areas had not been well documented in previous accounts (Bull 1964, 1974; Duffy 1977). A very large colony was documented at Eatons Neck in northwestern Suffolk County. As many as 851 pairs bred at the site in 1976. Between 1974 and 1978, from 1719 to 2628 pairs were surveyed at between 29 and 47 colonies on Long Island. Least Tern populations through the mid-1980s have remained stable and may have actually increased slightly. Surveys conducted between 1982 and 1985 recorded between 2536 and 3114 pairs breeding in 39–59 colonies along New York's coast (Scheibel 1982; Peterson et al. 1985).

Results from the Atlas document the relative stability in the species' distribution in the state between Bull's (1974) map of Least Tern colony sites and the Atlas period. Although no colonies were located by the Atlas on Staten Island or on the northern shore of Nassau County, breeding expansions onto Fishers Island and throughout eastern Long Island were documented.

The Least Tern nests on beaches, dredge spoil areas, and other open shoreline sites. Colonies tend to be small. In 1985, 64% of colonies contained 50 or fewer adults. Data from 1985 indicate that vegetation covers less than 10% of the colony area at over two-thirds of the sites. More vegetation was often present at older colonies, where plant succession was encroaching on formerly bare nesting areas (Peterson et al. 1985). The nests themselves are often no more than scrapes in the sand, which are sometimes lined with small shells or other debris (Bull 1974).

The areas where the Least Tern prefers to nest are heavily used for recreation and shoreline development. Although some terns have benefited from some types of human activity, primarily by nesting on spoil areas deposited from building projects or navigational maintenance activities, most Least Tern colonies in New York experience varying degrees of disturbance from people, pets, and four-wheel-drive vehicles. Natural factors such as predation and flooding also can greatly affect colony productivity (Peterson et al. 1985). It is not known whether the pressures of coastal development are forcing many terns into nesting at suboptimal sites that are more susceptible to natural disturbances.

David M. Peterson

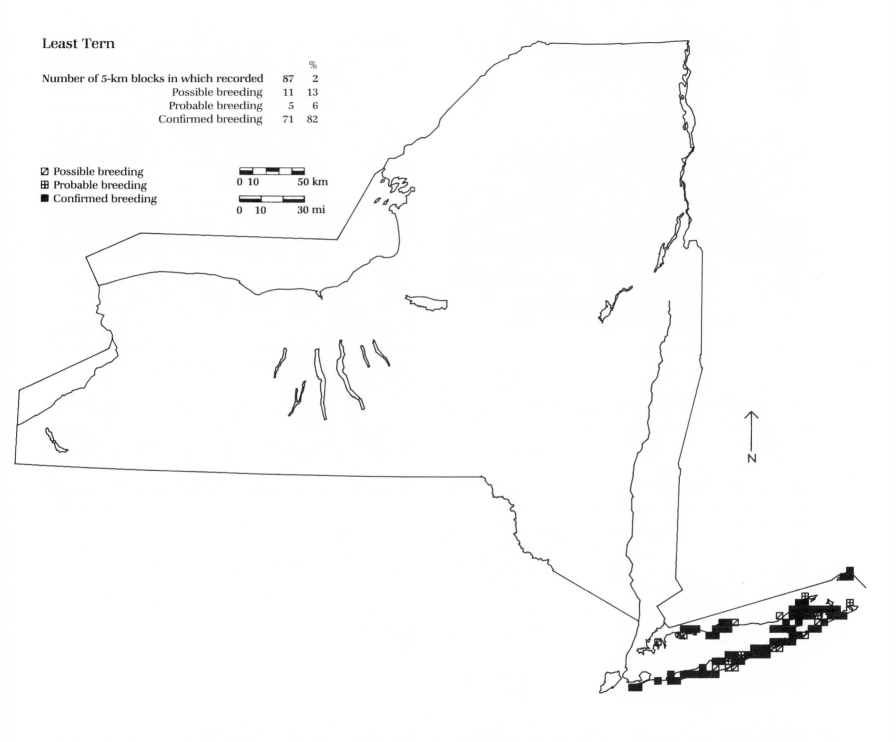

Least Tern

		%
Number of 5-km blocks in which recorded	87	2
Possible breeding	11	13
Probable breeding	5	6
Confirmed breeding	71	82

☑ Possible breeding
⊞ Probable breeding
■ Confirmed breeding

0 10   50 km

0 10   30 mi

N

©KLA-C 1987

Black Tern *Chlidonias niger*

The Black Tern is a conspicuous member of large, inland, freshwater marsh communities in the northern half of the state. It may be fairly common in some marshes in the state (about 50 pairs were recently observed at Wilson Bay, Jefferson County), but its numbers have decreased dramatically in recent years. New York is on the southeast edge of its widespread North American breeding range.

The Black Tern was reported in the state's early ornithological literature, when breeding was suspected but not proven (De Kay 1844). Breeding along the southern and eastern shores of Lake Ontario was first reported by Eaton (1910), with a colony estimated at 150 pairs located at the mouth of Sandy Creek, Jefferson County, in 1903. Through the first half of this century numbers increased, with colonies established in several places including Oak Orchard Swamp (now WMA) in 1937 and Tifft Farm in Buffalo, Erie County, in 1946 (Beardslee and Mitchell 1965). Bull (1974) reported three colonies known to have consisted of over 100 pairs: the Sandy Creek colony (now about 5 pairs), 200 pairs at Montezuma NWR (only about 10 pairs in 1985), and 100 pairs at the Perch River WMA (now only 20–30 pairs). The majority of others had at most a few dozen pairs.

More recently, Black Tern numbers have declined, presumably because of increased human disturbance and habitat loss due to either draining of marshes for agriculture or development, or raising of water levels, particularly in the marshes along Lake Ontario. In addition, Goodwin (1953) reported that some individuals are highly sensitive to human disturbance. The species has been on the *American Birds'* Blue List in other parts of its range since 1978 (Arbib 1977) and is currently listed as a Species of Special Concern in New York. As Beardslee and Mitchell (1965) stated, it is somewhat surprising that this tern has not colonized more of what appears to be suitable marsh habitat in the Adirondacks, Finger Lakes Highlands, or Appalachian Plateau, areas home to many other species with similar ranges and habitat requirements.

The Black Tern is found in the marshes of the southern and eastern shores of Lake Ontario, along the nearby St. Lawrence Plains, and in the extensive marshes in and adjacent to Montezuma NWR and the Oak Orchard WMA–Iroquois NWR complex. There were scattered Atlas records of small colonies away from these areas, including historically occupied breeding sites on the west end of Oneida Lake, Oswego County, and at Tupper Lake, Franklin County, in the Adirondacks. Newly reported sites were in Genesee and Tompkins counties, to the east of Oneida Lake, and at the south end of Lake Champlain. However, in none of these areas were there any "confirmed" records, and these observations may have been of birds that had already dispersed from nesting areas elsewhere. This is one of the earliest departing species following breeding.

Atlas coverage for the Black Tern was good. The bird spends much time throughout the breeding season in the air, coursing back and forth over marshes, nearby fields, or water, hawking for insects or occasionally picking minnows from near the surface of the water. Its flight is buoyant and nighthawk- or swallowlike, and it frequently utters a grating *kik-kik-kik* call while on the wing. This, along with aggressive defense of nest and young to the point of even striking a human intruder (Bent 1921), made it one of the easier marsh-breeding birds to locate and to "confirm," even with only brief Atlas forays.

The Black Tern typically arrives in early May to take up residence in large tracts of cat-tails or reeds, nesting in loose colonies. Nesting density is a function of the number of small openings in the emergent vegetation with suitable floating material to serve as nest bases (Goodwin 1953; Richardson 1967; Tilghman 1979). The specific nest base used appears to depend mainly on availability. In a study at three nest sites Bergman et al. (1970) found that floating, dead vegetation was used as a base for 29–92% of nests studied and old muskrat houses for 7–48%. In New York, Goodwin (1953) reported almost exclusive use of floating material as a nest base, with floating, partially anchored boards or old logs preferred. Floating mats of dead vegetation were used otherwise. Use of old muskrat houses was reported at Oak Orchard WMA (Beardslee and Mitchell 1965) and at 10 of 32 known nesting locations across the state (Bull 1974). Some nests are elaborately built, but most are just a few pieces of old weed stems or dead cat-tails formed into a loose cup shape (Bent 1921). The location is often such that the eggs may get wet, and storms occurring during the egg period may exact a heavy toll (Goodwin 1953). Bergman et al. (1970) reported an average of 197 nests only 3.3 ± 0.2 cm (1.3 ± 0.1 in) above water.

Robert Spahn

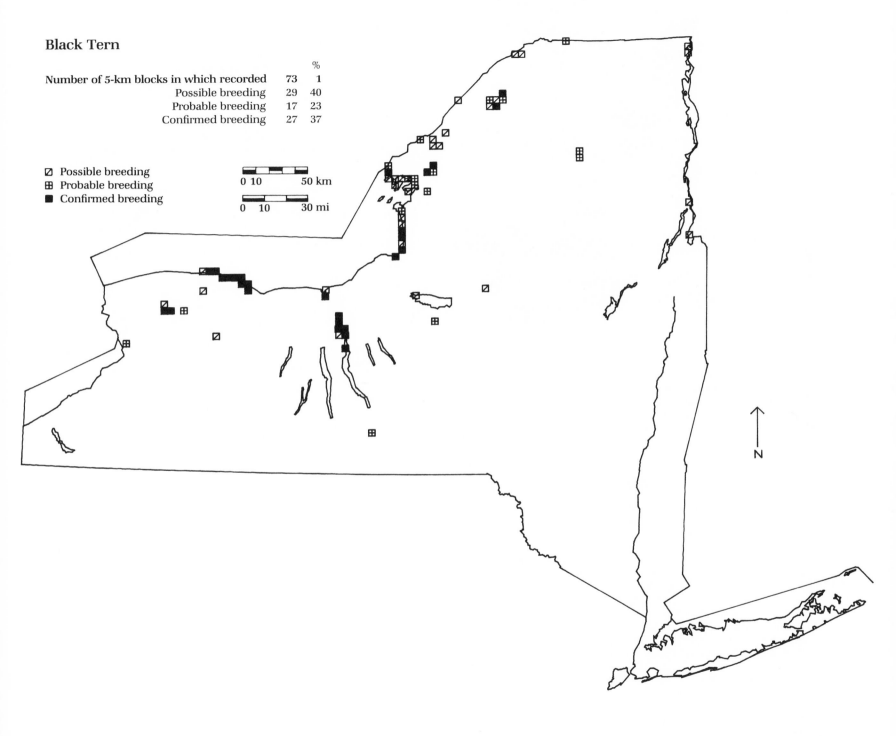

Black Tern

Number of 5-km blocks in which recorded	73	% 1
Possible breeding	29	40
Probable breeding	17	23
Confirmed breeding	27	37

Possible breeding
Probable breeding
Confirmed breeding

0 10   50 km

0 10   30 mi

N

©KLA-C 1986

Black Skimmer *Rynchops niger*

The black-and-white plumaged Black Skimmer is one of the most unusual looking of New York's birds. Its colorful red and black bill is uniquely designed, having a longer lower mandible than upper mandible for skimming across the water's surface to capture small fish. It is a fairly common breeding species in its restricted range along the coast of Long Island.

The Black Skimmer was first recorded breeding in New York at Gilgo Island, Suffolk County, in 1934 (Bull 1974). Within 20 years the Black Skimmer was found nesting in scattered locations from Jamaica Bay Wildlife Refuge to Orient Point, eastern Suffolk County. From the early 1960s on, some colonies numbered as many as 200 pairs (Bull 1974). No complete census of Black Skimmer colonies was made, however, until the early 1970s. Several surveys conducted during 1974–85 recorded between 339 and 495 pairs of Black Skimmers in 9–13 colonies (Erwin 1979; Buckley and Buckley 1980; Buckley and Buckley 1984b; Peterson et al. 1985). It appears that Black Skimmer populations have remained relatively stable since the early 1970s.

Except for occasional nesting in Massachusetts and recently in Connecticut (Buckley and Buckley 1984b), Long Island's population is at the northern edge of the species' range (Erwin 1979). As the Atlas map indicates, the Black Skimmer nests primarily on the southern shore of Long Island. Away from the south shore, it nests on Gardiners Island, Great Gull Island, and on Peconic Bay in eastern Suffolk County. This species has occasionally nested on the north shore at Huntington in western Suffolk County (Bull 1974; Peterson et al. 1985), but not during the period of Atlas fieldwork. The "possible" and "probable" Atlas records probably represent foraging birds.

In New York the Black Skimmer has always been found nesting in association with Common or Least Terns. Most colonies are located either on natural beaches or on dredge spoil islands. Small numbers of Black Skimmers also nest in several salt-marsh island Common Tern colonies (Bull 1974; Buckley and Buckley 1980; Peterson et al. 1985). On the south shore of Long Island the Black Skimmer nests near bay inlets. The largest of these colonies has contained as many as 200 pairs (Bull 1974; Peterson et al. 1985).

The Black Skimmer usually nests on sandy areas with little vegetation and is susceptible to predation, flooding, and human disturbance. Since most Black Skimmers breed at major Common Tern colonies, efforts being made to protect terns also benefit the Black Skimmer. Continued protection of Black Skimmer nesting sites should insure that this unique bird remains as a breeding species along New York's coast.

David M. Peterson

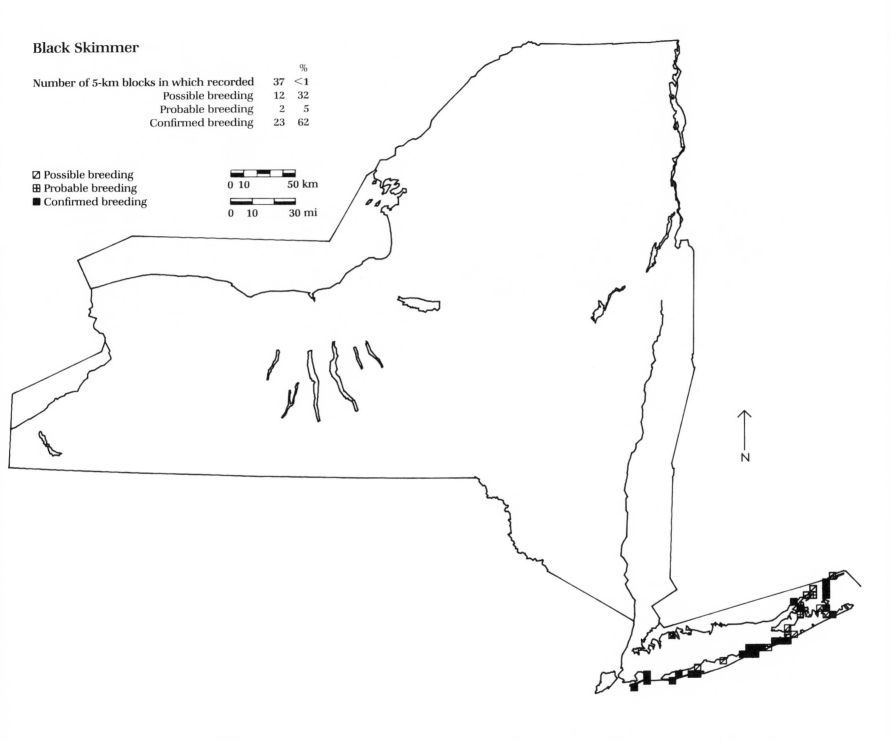

Black Skimmer

		%
Number of 5-km blocks in which recorded	37	<1
Possible breeding	12	32
Probable breeding	2	5
Confirmed breeding	23	62

◨ Possible breeding
⊞ Probable breeding
■ Confirmed breeding

0 10   50 km

0 10   30 mi

N

Rock Dove *Columba livia*

The hardy and successful Rock Dove is an abundant breeder throughout most of New York. This species, commonly known as the domestic or homing pigeon, is not native to North America, having been introduced from Europe by the French in the early 1600s (Schorger 1952). A large percentage of the Rock Doves in New York, however, exhibit color patterns that differ from those of the original wild stock. "Countless varieties resulting from the process of artificial selection encouraged by the pigeon fancier are at large" (Cruickshank 1942). Unlike the European Starling, House Sparrow, and other introduced species, the Rock Dove was first introduced so long ago that there is no accurate record of its introduction or subsequent range expansions (Long 1981). Although historically excluded from ornithological publications (e.g., Eaton's *Birds of New York* and Audubon Christmas Bird Counts from 1950 to 1974), the species has been abundant in urban and rural areas in New York since the 1800s.

Many attempts have been made to control populations of this species because accumulations of droppings around roosting and nesting sites are messy and a potential health threat. During the 6-year period from 1946 to 1951, the Buffalo Department of Public Works destroyed over 94,000 individuals "with little apparent effect on their overall abundance" (Beardslee and Mitchell 1965).

This dove is sparsely distributed in the highlands of the Adirondacks, Catskills, Delaware, Mongaup, and Allegany hills, and Tug Hill Plateau. It has expanded its range somewhat in the past decade in the Adirondacks, moving into hamlets and farmland. The first "confirmed" breeding at Indian Lake, Hamilton County, occurred during the Atlas period (Peterson, pers. comm.).

There was no difficulty "confirming" this species: nests were reported in more than 900 blocks, used nests in 95. This was one of only a few species where records of nests far outnumbered those of fledglings.

In most of the state the Rock Dove is found in agricultural areas in the vicinity of large barns, farmyards, and grain elevators; in urban areas it is found primarily in parks and around large buildings. It nests almost exclusively on man-made structures. "Nests are commonly placed on window sills and building cornices" (Bull 1974), as well as on beams under bridges and similar sites. However, this species got its name from the fact that the native European population originally nested on cliffs and rock ledges. It still nests occasionally on such places, such as the cliffs of Niagara River gorge (Beardslee and Mitchell 1965) and those in the Mohonk Lake area of Ulster County (Smiley 1964). In 1980 it was discovered breeding in a small natural cave with a 0.9-m (3-ft) outer opening, halfway up a 21-m (70-ft) cliff in Ausable River chasm in the Adirondacks (Peterson, pers. comm.).

The Rock Dove has been recorded nesting in every month of the year in Buffalo, New York City, and Oswego County (Bull 1974; Smith and Ryan 1978). The nest is usually only a small collection of twigs and grasses arranged on a ledge to form a hollow for the eggs. However, a nest in Buffalo was made entirely of large nails (Beardslee and Mitchell 1965). The male does most of the nest building.

Steven C. Sibley

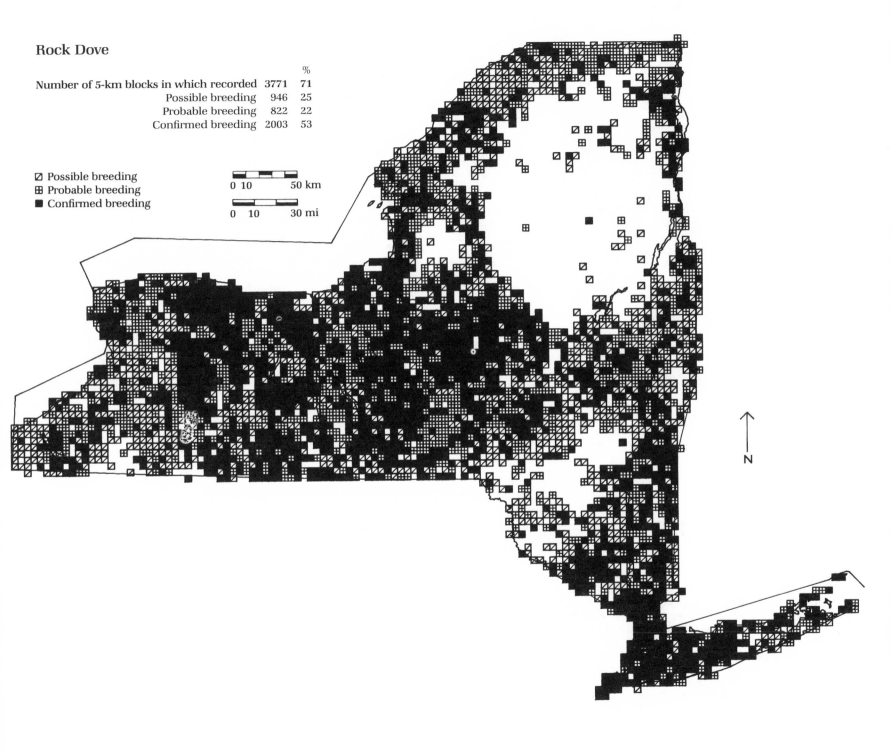

Rock Dove

		%
Number of 5-km blocks in which recorded	3771	71
Possible breeding	946	25
Probable breeding	822	22
Confirmed breeding	2003	53

☑ Possible breeding
⊞ Probable breeding
■ Confirmed breeding

0 10   50 km

0 10   30 mi

N

Mourning Dove *Zenaida macroura*

Although the Mourning Dove is now a common bird in New York, except where forest cover is heavy, before 1900 it was rare. It has become increasingly common as deciduous forests have been cleared and with passage of the Migratory Bird Treaty Act of 1918 (Slingerland 1983). During the 1960s and 1970s the Mourning Dove became a very common winter bird at feeders in New York (Peakall 1963; Scheider 1971).

DeGraff (1985) said, "The very activity that wiped out the turkey—the great land-clearing at the turn of the century—provided the habitat for dove [population] growth and [range] expansion. Even after much of the state had reverted to woodland, evolving agriculture ... stimulated a Mourning Dove population which now exceeds anything in the state's history." DeGraff described its adaptability using Long Island as example: "For the past 40 years truck farms, potato farms and small hamlets have slowly given way to the tidal wave of housing development. Yet wherever there is a surviving farm or a garden or a green space, doves continue to thrive."

The Mourning Dove is now one of the more abundant birds in the United States: of a population estimated at about 500 million, approximately 10 million are found in New York (Decker 1985). Although call-count surveys showed a decrease in dove populations from 1963 to 1973 (Keeler et al. 1977), numbers have remained stable or increased slightly during the past 10 years (Dolton 1986). The call-count survey is a roadside count taken on 900 randomly selected routes. Calling birds are counted for 3-minute periods at 20 stations along the 20-mile route.

The Atlas map shows the widespread distribution of this species, which is absent only from the heavily forested, mountainous regions of the state. Nests of this species were found in more than 700 blocks, but most "confirmed" records were of fledglings or adults feeding young, which were reported in more than 1300 blocks.

This bird inhabits croplands and cultivated fields, developed commercial, residential, or recreational land, gardens, arboretums, small conifer plantations, and orchards. It favors open areas with scattered trees and is frequently observed along roadsides looking for seeds or dusting (Edminster 1954). Unlike its relative the Passenger Pigeon, this dove has readily adapted to humans' use of the land.

The nest, essentially a platform of sticks, is surprisingly strong and lined with few grasses, weeds, or rootlets, if it is lined at all (Harrison 1975). It is usually placed 3–7.6 m (10–25 ft) from the ground, although it is sometimes built as high as 15.2 m (50 ft). The dove typically nests in evergreens or in tangles of shrubs or vines, although other nest sites reported in New York include a willow in a wetland (Stoner 1932), on a hummock of cutgrass in a swamp (Lesperance 1960), and a wheat field (Benton 1961b). The Mourning Dove sometimes uses the nest of other birds, including the American Robin, the Gray Catbird, Common Grackle, and the Black-crowned Night-Heron (Harrison 1975).

During incubation the male relieves the female at the nest from early morning until late afternoon (Moore and Pearson 1941). Newly hatched young are fed pigeon's milk, a sloughing from cells lining the crop of the adult bird. Within a few days of hatching, the milk is supplemented with various seeds and grains (Keeler et al. 1977).

Stephen W. Eaton

190

Mourning Dove

Number of 5-km blocks in which recorded 4402   83 %

Possible breeding   1085   25
Probable breeding   1202   27
Confirmed breeding   2115   48

◪ Possible breeding
⊞ Probable breeding
■ Confirmed breeding

0 10   50 km

0 10   30 mi

N

Monk Parakeet *Myiopsitta monachus*

The Monk Parakeet is an introduced species that is now only an accidental breeder in New York because of an active control and retrieval program by federal and state wildlife officials. This species is native to southern South America (Argentina, Uruguay, Bolivia, Paraguay) and was unrecorded in the state until 1967 (Niedermyer and Hickey 1977), when large numbers were imported into the United States for the pet trade (Bull 1974). In the period 1968–72, 64,225 Monk Parakeets were reported to have been imported into the United States (Banks 1976).

Some Monk Parakeets may have made their way into the wild from a broken crate at John F. Kennedy International Airport in Queens County, but most were likely from escapes and intentional releases from pet shops, aviaries, and private homes (Bull 1975). Owners were often disturbed by this bird's raucous calls and lack of speaking ability (Briggs and Haugh 1973). The Monk Parakeet builds, maintains, and roosts in its nest year-round, and nests were soon reported. By December 1970 there were 7 known nests in the New York City area (Bull 1971), and by late 1971 there were 19 in New York City and on Long Island (Bull 1974). These sightings did not yet represent breeding. The first successful breeding in New York took place at Valley Stream, Nassau County, in 1971 when two young were fledged (Niedermyer and Hickey 1977).

This population expanded rapidly, probably supplemented by continued releases and escapes. By late 1972 the Monk Parakeet had been reported away from New York City and Long Island in at least 20 locations, including Eden, Erie County (Sundell 1973); Elmira, Chemung County; Schenectady, Schenectady County; Watertown, Jefferson Coun-

ty; and at Derby Hill, Oswego County (Terres 1980; Long 1981). One was even seen on Great Gull Island, Suffolk County, on 4 June 1972 (Davis and Morgan 1972).

Numbers of Monk Parakeets peaked in 1972 and, as concern over its spread mounted, a control program was initiated in 1973. Federal and state officials began removing birds from the wild because of fears the species would become an agricultural pest, spread disease (psittacosis), and compete with native wildlife (Niedermyer and Hickey 1977). By 1975 almost all had been removed from the wild (Niedermyer and Hickey 1977). The species had obviously been doing well, as at least 34 of the 87 birds collected in 1973 were juveniles that most likely were hatched in the wild in the state (Niedermyer and Hickey 1977).

In its native South America the Monk Parakeet is a severe agricultural pest causing "crop losses . . . from 2%–15% with some as high as 45% annually" (Long 1981). According to Davis (1974), the species raids such crops as corn, sorghum, sunflower, millet, and other grains, and fruits such as pears, grapes, apples, and peaches. There is every reason to believe the Monk Parakeet could thrive and become a pest in New York, as the climate resembles that in its native Argentina (Briggs and Haugh 1973).

This species has only rarely been reported in New York since 1975. There were only three Atlas records, all from Long Island. There were two "possible" breeding records from Nassau and Suffolk counties on the north shore of Long Island and one "probable" breeding record in Richmond County on Staten Island. The Staten Island record represents several nests built under the eaves of houses, which were removed by homeowners when they reached huge proportions.

The Monk Parakeet seems to prefer suburban habitats. Of 367 observations in the United States from 1970 to 1975, 77% were in urban-suburban habitats; the remainder were in rural areas (Niedermyer and Hickey 1977). In South America it is also "particularly common in the vicinity of human habitation" (Long 1981). Most of the individuals around New York City fed regularly at bird feeders, eating "mixed wild bird seed, suet, bread, various fruits, and other staples" (Bull 1974).

The Monk Parakeet nest is a large bulky structure made of sticks, and in New York, Bull (1974) described nest sites "in trees and in a rosebush only seven feet up, but also on buildings, on telephone poles, in broken street lamps and lampposts, on the steel structures of an abandoned crane, and even on a U.S. Coast Guard microwave tower, nearly 100 feet up." The nest is round or dome-shaped, usually about 0.9 m (3 ft) in diameter (Terres 1980). Most nests contain entrances at the bottom and on the sides (Bull 1974). This species is often colonial, and nests have been known to contain as many as 20 compartments (Long 1981). All New York nests were used by only 1 or 2 pairs (Bull 1974).

Steven C. Sibley

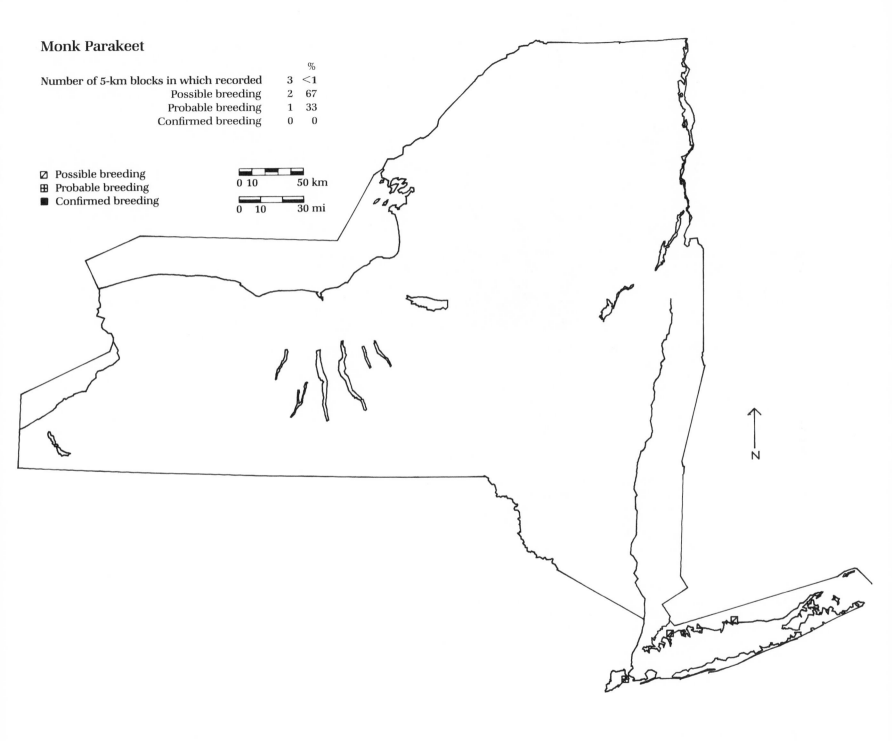

Monk Parakeet

		%
Number of 5-km blocks in which recorded	3	<1
Possible breeding	2	67
Probable breeding	1	33
Confirmed breeding	0	0

☒ Possible breeding
⊞ Probable breeding
■ Confirmed breeding

0 10   50 km

0 10   30 mi

N

Black-hooded Parakeet *Nandayus nenday*

The Black-hooded Parakeet is a common cage bird imported to the United States from central South America. Cage birds of this and other psittacines often escape or are released into the wild, but in New York they do not usually survive the winters.

In June 1980 adult and downy young Black-hooded Parakeets were observed in Westchester County, and later a group of 6 spent several months in the area of Indian Point in that same county foraging in a meadow and also being fed by residents. In December, during a particularly cold period, all but 1 bird disappeared. The remaining parakeet was livetrapped and found to have severe frostbite. Its upper mandible and both feet were nearly gone (Carroll 1982).

In 1984 New York City Atlas workers reported watching a pair enter and leave a nest hole several times; later local residents reported seeing fledglings. Also in 1984 a pair was observed in Nassau County on Long Island.

In its native South America this parakeet inhabits savannahs and palm groves (Forshaw 1978). In California, where a feral population exists, the birds forage in city gardens and rural areas on chinaberry seeds, palm fruits, sunflower seeds, and various nuts, and use old woodpecker holes and natural cavities as nest sites (Fisk and Crabtree 1974).

Janet R. Carroll

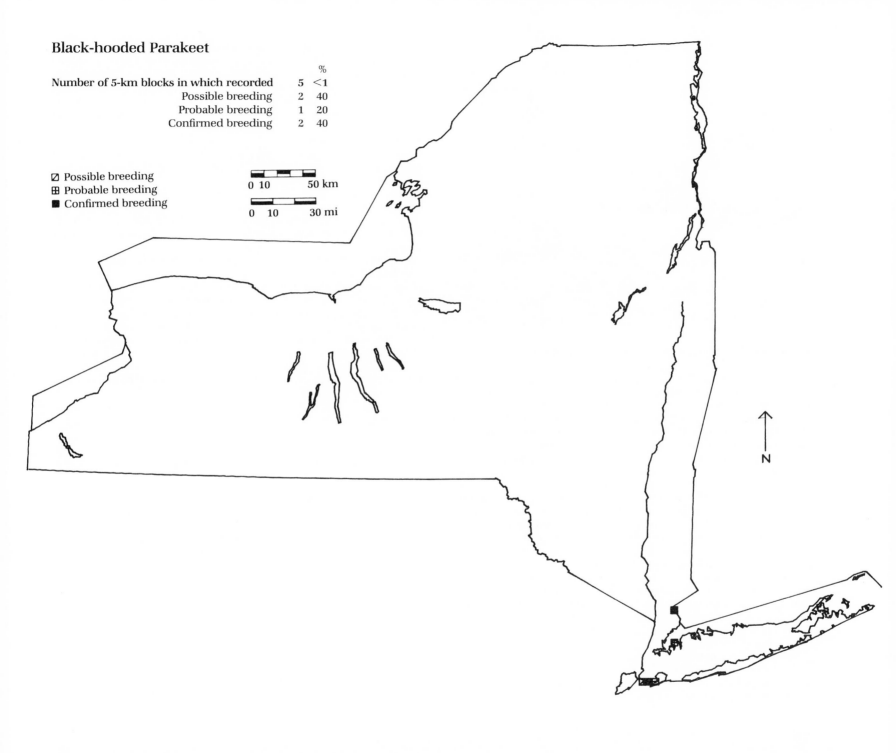

Black-hooded Parakeet

Number of 5-km blocks in which recorded	5	% <1
Possible breeding	2	40
Probable breeding	1	20
Confirmed breeding	2	40

◩ Possible breeding
⊞ Probable breeding
■ Confirmed breeding

0 10   50 km

0 10   30 mi

N

Black-billed Cuckoo *Coccyzus erythropthalmus*

The Black-billed Cuckoo is a fairly common nesting species in New York except in heavily forested areas and near large urban centers. It is a more northern species than its yellow-billed relative (Godfrey 1966; Peterson 1980) and is more widely distributed in the state. Robbins et al. (1986) found that BBS data for the period 1965–79 indicated a significant increase in numbers of this species in the eastern United States.

The Black-billed Cuckoo's movement into the state probably followed the course of human occupation, taking advantage of the edge habitat produced by the removal of the primeval forest. By the 1900s it was a fairly common summer resident of New York and was generally distributed except in the colder portions of the state (Eaton 1914). Both Eaton and Bull (1974) believed that throughout much of the southern half of the state, populations of the two species of cuckoo appeared to be about equal. Farther north, however, the more numerous Black-billed Cuckoo was found in the highlands where the Yellow-billed Cuckoo was absent. Eaton called it slightly more common in western New York than the Yellow-billed Cuckoo, and Saunders (1942) found it to be a common summer bird in the mountains of Allegany State Park, where it was more abundant than the Yellow-billed Cuckoo except in the valleys. According to Bull (1974) it was the only cuckoo breeding on the barrier beaches of the south shore of Long Island.

The distribution of the Black-billed Cuckoo compared with that of the Yellow-billed Cuckoo reflects not only their continental distributional differences but also altitudinal differences. The Black-billed Cuckoo was found more frequently in northern New York south to the latitude of the Great Lakes Plain and the Mohawk Valley. In the lower Hudson Valley and Coastal Lowlands, where both latitude and altitude exert their effects, the distribution of the two species appears similar. The black-billed species was usually found at higher elevations along the Appalachian Plateau, the Tug Hill Transition, the Adirondack Foothills, and at lower elevations in the Catskill Peaks. Both species tend to be distributed along river valleys and extending into the highlands, particularly in the Adirondacks and Central Appalachians.

The Black-billed Cuckoo is very secretive during nesting and is almost always noted first by its call, which can be difficult to distinguish from that of the Yellow-billed Cuckoo. Atlas workers were urged to attempt to see, as well as to hear, the two species to avoid making an incorrect identification. Probably because the Black-billed Cuckoo is more abundant, its nests were found twice as often as those of the Yellow-billed Cuckoo. Most "confirmed" records, however, were of adults with food for young.

The Black-billed Cuckoo seems to prefer more wooded areas than does the Yellow-billed Cuckoo (Harrison 1975). Its nesting habitat includes brushy pastures, shrubby hedgerows at edges of fields, and dry, open upland woods and groves. Within its habitat it requires low, dense, shrubby vegetation (DeGraaf et al. 1980). Nests are built in thick bushes, tangles of vines, and sometimes in small hemlocks or other evergreens. Saunders (1936) estimated 7 pairs occupied the aspen-cherry habitat at Quaker Run Valley in Allegany State Park. The nest is a small platform of twigs, loosely woven and lined with catkins, cottony fibers, dry leaves, or pine needles. It is more substantially lined than that of the Yellow-billed Cuckoo (Harrison 1975).

Stephen W. Eaton

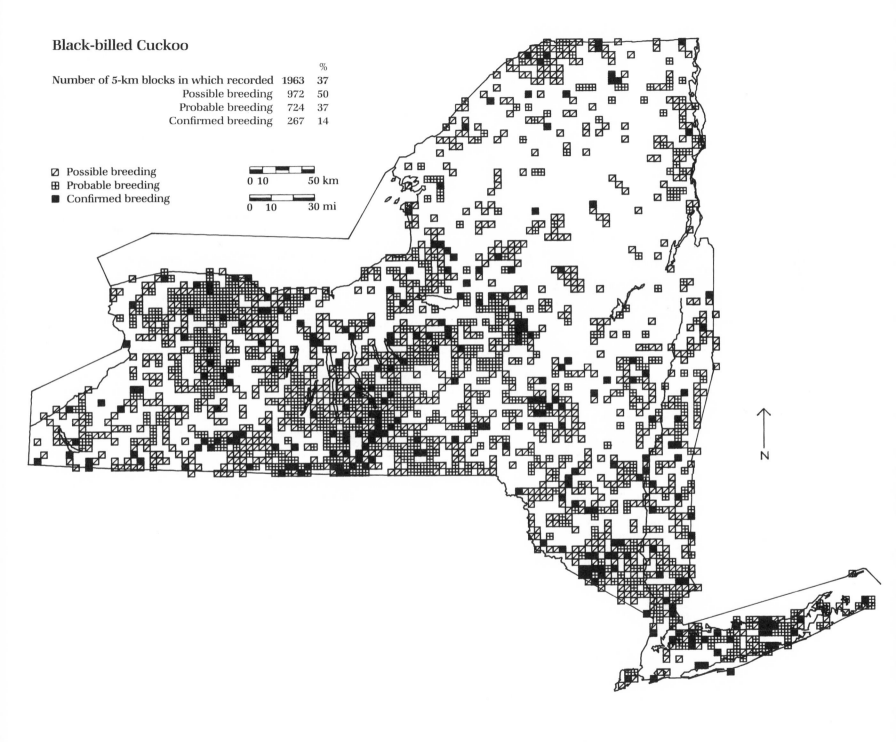

Black-billed Cuckoo

		%
Number of 5-km blocks in which recorded	1963	37
Possible breeding	972	50
Probable breeding	724	37
Confirmed breeding	267	14

☑ Possible breeding
⊞ Probable breeding
■ Confirmed breeding

0 10   50 km

0 10   30 mi

N

Yellow-billed Cuckoo *Coccyzus americanus*

The more southern of the two cuckoo species that inhabit New York, the Yellow-billed Cuckoo is a rather uncommon to fairly common breeding species. In North America it was estimated to be almost ten times more abundant than the Black-billed Cuckoo (Van Velzen and Robbins 1969). In New York, however, it is less abundant, being found in a much more restricted area of the state than the Black-billed Cuckoo. The population in the eastern United States, including New York, appears to have increased recently. Robbins et al. (1986) said BBS data showed that there had been a significant increase in Yellow-billed Cuckoo numbers in eastern and central parts of the country during the period 1965–79.

Eaton (1914) called the Yellow-billed Cuckoo a fairly common summer resident of the Hudson Valley and Coastal Lowlands and in the rest of the state at lower elevations, being present only in the valleys of the Central Adirondacks and Catskill Peaks. In the Adirondacks, this species was very scarce. One summer report from Essex County was of an adult that flew into a plate glass window of the Lake Placid Club (Benton 1951b). Saunders (1942) said it was fairly common in Allegany State Park, particularly in the Allegheny River Valley. Bull (1974) called this cuckoo a local breeder at low elevations, rare northward, and absent in the mountains, which describes its distribution today as well.

Atlas observers found the Yellow-billed Cuckoo best represented in areas of the state that were below 305 m (1000 ft) in elevation. The distribution pattern is fairly local in areas where it occurs, including along the Great Lakes Plain, particularly Orleans, Monroe, and Genesee counties, and also in Livingston County along the Genesee River Valley.

Farther east it was found in various areas in the Finger Lakes Highlands and central Appalachian Plateau, generally along the valleys of the Susquehanna River. It was also found in southeastern New York, including Queens, Bronx, and Richmond counties, and on the Coastal Lowlands away from the barrier beaches, as Bull (1974) also described. It usually avoids heavily forested and large urban areas.

The Yellow-billed Cuckoo is often found in oak–northern hardwood forests particularly along the Appalachian Plateau and southeastern New York, where there are recurring high levels of gypsy moth infestation. Bull (1974) pointed out that this species was subject to marked fluctuations, and Tate (1981) stated that throughout most of its range it is cyclical and responsive to caterpillar population outbreaks.

This cuckoo, like its New York congener, was more often heard than seen, and its secretiveness in general, and relatively short incubation and nestling periods (Hamilton and Hamilton 1965; Potter 1980) made it difficult to "confirm." Most of the "confirmed" records were of fledglings.

The Yellow-billed Cuckoo nests in open second-growth woods, abandoned fields with scattered bushes and small trees, and overgrown or abandoned fruit orchards (Bull 1974). It is typically found in successional shrubland and along stream banks with dense thickets. Its habitat must contain low, dense, shrubby vegetation (DeGraaf et al. 1980). Saunders (1936) found it in cherry-aspen and oak-hickory sprout in Allegany State Park.

Bent (1940) said that unlike the cuckoo found in Europe, which always lays its eggs in nests of other species, both North American species usually build their own nests and rear their own young, although they are very poor nest builders. Bendire (1895) speculated that there was some latent trace of parasitism in these cuckoos. The Yellow-billed Cuckoo not only places its eggs in the nest of the Black-billed Cuckoo, but also in the nests of Wood Thrush, American Robin, Gray Catbird, Cedar Waxwing, Mourning Dove, Northern Cardinal, and Red-winged Blackbird (Bendire 1895; Harrison 1975). Nolan and Thompson (1975) theorized that the parasitism that does occur may be the result of an insect outbreak that triggers egg production. The female is ready to lay but without a nest, leaving it no alternative but to deposit its eggs in the nest of another bird.

The nest is a shallow, frail platform, composed of small rootlets, sticks, or twigs, and thinly lined with bits of moss, grass, pine needles, or catkins (Harrison 1975). The pair bond is maintained during incubation; one cuckoo brings pine needles, grass, and small sticks to its incubating partner before relieving it at the nest (Eaton 1979).

Stephen W. Eaton

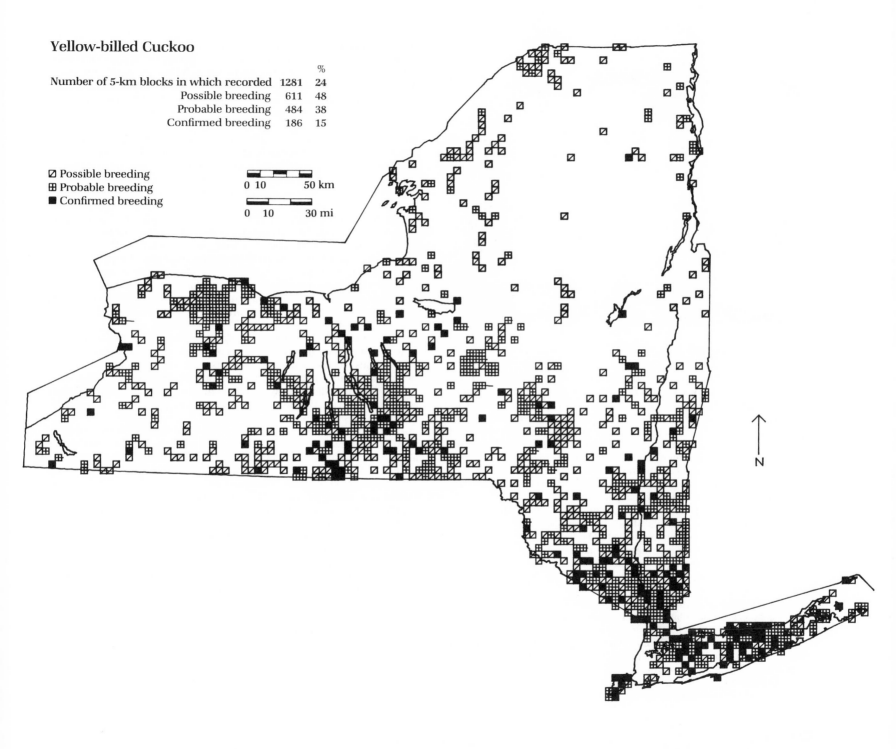

Yellow-billed Cuckoo

		%
Number of 5-km blocks in which recorded	1281	24
Possible breeding	611	48
Probable breeding	484	38
Confirmed breeding	186	15

◨ Possible breeding
⊞ Probable breeding
■ Confirmed breeding

0 10   50 km

0 10   30 mi

N

Common Barn-Owl *Tyto alba*

The Common Barn-Owl, one of our most distinctive owls with its monkeylike face, is a very local and uncommon breeder throughout the state but is absent from the Adirondacks. Eaton (1914) stated that this nocturnal species, formerly called barn owl, "has been regarded as a rare bird in this state but, . . . probably is more common than the paucity of records would lead us to suppose." The same can be said of this species today. Since Eaton's time increased numbers of observers throughout the state have greatly expanded our knowledge of the Common Barn-Owl's abundance here, but it remains a very difficult species to locate, and many individuals have certainly escaped detection.

There is evidence that this species has declined somewhat in the past century in agricultural regions of northern and western New York. A number of the early breeding records cited by Eaton (1914) were reported from Wayne and Monroe counties along Lake Ontario, where this owl was reported in only one Atlas block as "possible." Bull (1974) also listed breeding records for 13 locations in the lowlands along Lake Erie and the Niagara River, where only three Atlas blocks had "possible" records. And, in northern Clinton County along the Canadian border, where there were no Atlas records, a farmer remembered regularly seeing "monkey-faced" owls around local farm buildings when he was growing up (Peterson, pers. comm.). It appears that Common Barn-Owl populations in New York may be following a trend similar to that recently documented in Ohio (Colvin 1985). There, the Common Barn-Owl increased with the spread of agriculture and clearing of forests in the 1800s and then began to decrease after the 1950s as farms were abandoned, commercial fertilizers replaced the need for meadows as part of crop rotations, edge areas were eliminated, and farming shifted from grains to soybean and corn.

In New York the Common Barn-Owl exhibits an extremely spotty distribution for a species seemingly well adapted for coexistence with humans. This is probably due to its overall rarity, the difficulty in locating this quiet nocturnal owl, and the fact that New York is at the northern limits of the species' breeding range, although it has bred in Quebec and breeds in southern Ontario (Ontario Breeding Bird Atlas, preliminary map). Atlas data show the Common Barn-Owl to be well represented only on Long and Staten islands, the southern portion of the central Hudson Valley, and in agricultural lands of northwestern Livingston County in western New York. The species is almost completely absent from the St. Lawrence Plains and all forested highlands in the state such as the Adirondacks and Catskills. The Vermont Atlas "confirmed" breeding in four locations in the central portions of the Lake Champlain valley across the lake from New York (Laughlin 1985), but searches on the New York side turned up only one "possible" and one "probable" breeding record.

This species is extremely difficult to locate because of its nocturnal habits and because it does not give a regular territorial call, as do other owl species. Many individuals and nestings certainly escaped the efforts of Atlasers. Of 40 "confirmed" records, 25 were of nests and 15 were of dependent fledged young.

The Common Barn-Owl is found almost exclusively in the presence of humans, nesting in agricultural areas, small towns, and cities (Bull 1974). This species requires open areas such as cropland or marsh to hunt over, never hunting in woods. It will settle in virtually any area where there are populations of rodents on which to feed, as noted by Bull (1974). "They are prevalent in the vicinity of garbage dumps where they are attracted by the numerous rodents."

The Common Barn-Owl has been reported nesting in every month of the year in New York (Bull 1974). This species usually nests in old or abandoned buildings, hollow trees, holes in banks or cliffs, or in nest boxes built for them (Terres 1980). Of 40 New York nests cited by Bull (1974), 3 were in hollow trees and the rest were in man-made structures. The most commonly used site was abandoned or unused silos. Other nest sites listed by Bull were barns, water storage towers, church steeples, bridge abutments, the exterior window ledge of a lighthouse, and in the cable housing of a canal lock. Contrary to this, Colvin (1986) found in New Jersey that tree cavities were the most frequently used nest site and speculated that it is simply easier to locate the barn-owl in man-made structures. In southwestern New Jersey Common Barn-Owl nest boxes have been readily used by the barn-owls, but populations in those areas do not seem to have increased because of this nest box program (Colvin 1986). The same nest site is often used for many years, though often not by the same owls (Colvin 1985). The nest itself is usually placed in a dark corner in the chosen site (Cruickshank 1942).

Steven C. Sibley

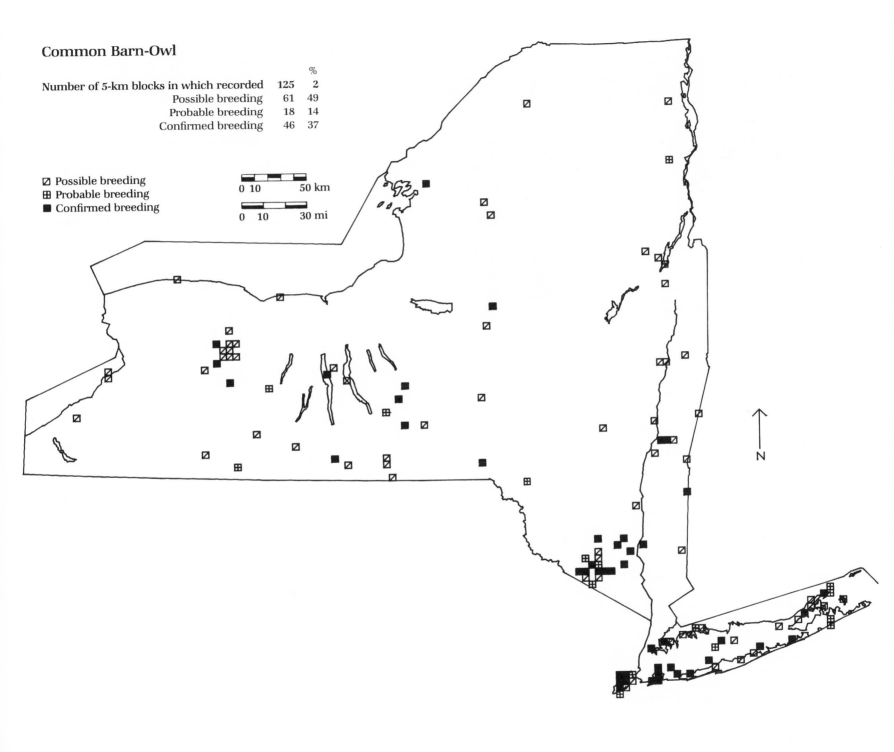

Common Barn-Owl

		%
Number of 5-km blocks in which recorded	125	2
Possible breeding	61	49
Probable breeding	18	14
Confirmed breeding	46	37

◨ Possible breeding
⊞ Probable breeding
■ Confirmed breeding

0 10   50 km

0 10   30 mi

N

Eastern Screech-Owl *Otus asio*

A very nocturnal North American owl (VanCamp and Henny 1975), the Eastern Screech-Owl was considered by DeGraaf et al. (1980) to be uncommon to rare in the Northeast. In New York, however, it appears to be fairly common except in the northern part of the state. Both Eaton (1914) and Bull (1974) indicated that it was widely distributed throughout New York except in the north, where it was mostly absent. Eaton considered it the commonest owl.

There is some evidence the Eastern Screech-Owl has declined in this century in New York and elsewhere. The reversion of farmland to forest in New York may have had a negative impact on the species, since this owl is found in smaller woodlots and open areas (DeGraaf et al. 1980). Van Camp and Henny (1975) found that cutting of forests seemed to have had a positive effect on its abundance in northern Ohio. In a discussion of population trends for the period 1935–72, Temple and Temple (1976) indicated that this owl had decreased significantly in the Cayuga Lake basin in central New York. Fruit orchards there had been sprayed with pesticides, and Temple and Temple suggested pesticide contamination of its prey might have affected this population. Beardslee and Mitchell (1965) indicated that it was less common than in former times in western New York. In New York City the Eastern Screech-Owl formerly bred in Central and Prospect parks (Bull 1964); however, Davis reported that it was no longer found there (Tate 1981). The Eastern Screech-Owl was added to the *American Birds'* Blue List in 1981 (Tate 1981).

The density of the Eastern Screech-Owl may be influenced locally in part by the abundance of other owls in an area. Trautman (1940), writing about central Ohio, believed that the Eastern Screech-Owl increased when the Barred and Great Horned Owl numbers decreased.

The Atlas map shows a distribution pattern that is partially a reflection of coverage. The Eastern Screech-Owl is rarely observed during daylight hours, when most Atlas work was done. Craighead and Craighead (1956) called it the most difficult owl to census because it roosts in tree hollows during the day and leaves little evidence of its presence, although it does respond to tape recordings of its call, even during daylight hours. The concentrations of Eastern Screech-Owl records on the map are at areas where intensive evening owl searches were made using tape-recorded calls. It would seem that the Eastern Screech-Owl should have been recorded in many more Atlas blocks in southern New York, except for the higher elevations of the Allegany Hills and Catskills, where the species may have always been absent, and in large tracts of contiguous forest. The Eastern Screech-Owl was not found around New York City or the heavily developed west end of Long Island; however, it was found around other urban areas, as indicated in heavily populated Westchester County. Eaton (1914) found it nesting within the boundaries of such cities as Rochester, Monroe County, and Buffalo, Erie County, and it still nests in these cities today (Andrle, Spahn, pers. comms.). The absence of Eastern Screech-Owl records is noticeable in areas where block busting during June and July was the main source of coverage. Little time was available to search for this species, although it is responsive to taped recordings of its call after the young have fledged (Martin pers. comm.).

Those Atlas observers who spent many evenings early in the breeding season searching for this owl were usually successful. It was fairly easy to predict areas where playing tape-recorded Eastern Screech-Owl calls would elicit responses from one or more individuals; particularly wooded streams or small swampy areas, cemeteries, and apple orchards. Records were also obtained from local residents, especially farmers.

Within its nesting habitat the Eastern Screech-Owl prefers to feed in open areas such as fields and meadows, and along wooded edges of fields and streams (DeGraaf et al. 1980). The nest is placed in natural cavities and also in holes made by the Northern Flicker or Pileated Woodpecker. Often it is found in abandoned or seldom-used buildings or in nesting boxes. The entrance hole must be about 7.6 cm (3 in) in diameter. Thomas et al. (1979) said that the minimum nest height was 4.6 m (15 ft) and that the minimum dbh of nest trees was about 30.5 cm (12 in). No nest is built and no material is placed in the cavity. Eggs are laid in rotted chips, leaves, or rubble already there (Harrison 1975).

Stephen W. Eaton

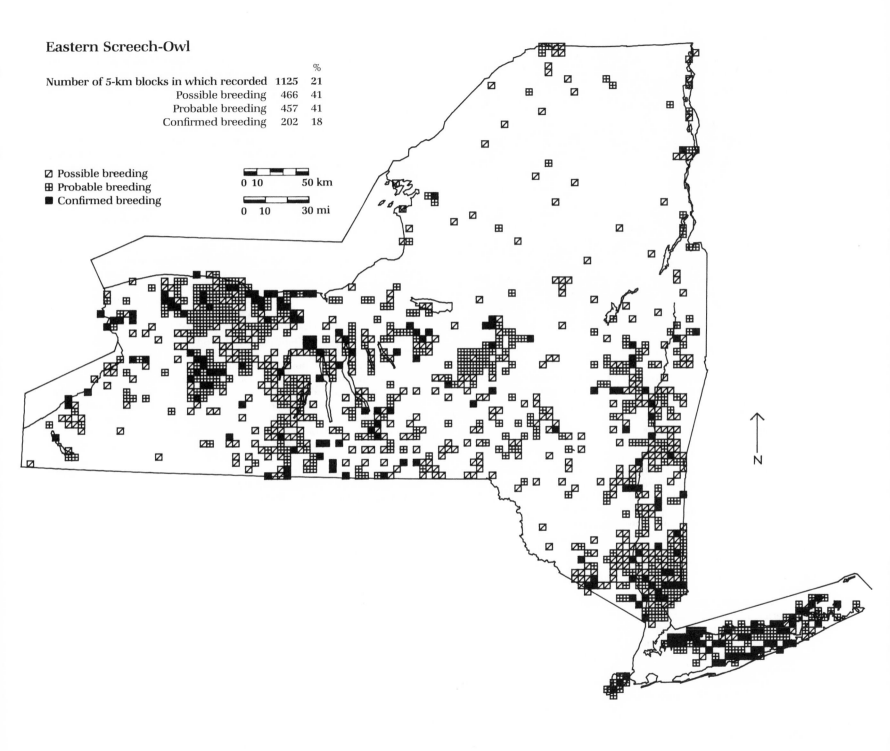

Eastern Screech-Owl

		%
Number of 5-km blocks in which recorded	1125	21
Possible breeding	466	41
Probable breeding	457	41
Confirmed breeding	202	18

☑ Possible breeding
⊞ Probable breeding
■ Confirmed breeding

0 10   50 km

0 10   30 mi

N

Great Horned Owl *Bubo virginianus*

This muscular nocturnal predator of medium-sized birds and mammals is a fairly common permanent resident in New York, especially in productive bottomland woods where its prey, the eastern cottontail, striped skunk, and muskrat, abound.

Although the Great Horned Owl is a fairly common species today, in the early 1900s Eaton (1914) called it uncommon except in wooded areas. He said that the lack of hollow trees suitably situated for nesting sites and the almost absolute certainty of the destruction of broods reared in open nests anywhere except in the wildest districts had been important factors in the gradual decline of the species. This was a period when farmers and sportsmen considered the Great Horned Owl fair game. Bull (1974) called it an uncommon to fairly common resident. Killing of this owl has lessened thanks to enforcement of protective legislation and to public education, and this species seems to have increased. DeGraaf et al. (1980) commented that the Great Horned Owl may have become more tolerant of human activity and occasionally is observed at parks in cities and towns.

Atlas observers found this owl along the Great Lakes Plain, in the northern Cattaraugus and Finger Lakes highlands, the Central Appalachians, the Hudson Valley and adjacent highlands, and the western Mohawk Valley. It was surprisingly well distributed on northern and eastern Long Island. On the western Appalachian Plateau and in the Adirondack and Catskill areas it was recorded less frequently. In these areas it appears to be associated with the great river systems and their tributaries, such as the Allegheny, Chemung, and Susquehanna rivers,

and the various smaller systems that flow from the Central Adirondacks. In the Allegany Hills it is restricted mainly to the Allegheny River Valley (Eaton 1981). Where soils are unproductive in adjacent highlands, such as the upper Genesee River Valley and the Neversink Highlands, few large mammals occur to serve as prey for this owl. It still occurs on Staten Island but was not recorded from the other boroughs of New York City.

Because special efforts were required to locate nocturnal species such as the Great Horned Owl, it is under-represented on the Atlas map. But since it was often possible to observe it during the day, the Great Horned Owl is better represented than other New York owl species. Moreover, because the Great Horned Owl is familiar to many, records were often obtained from farmers and local residents. This owl was relatively easy to "confirm" because of its long period of parental care and its vocal and visual conspicuousness. Often the mobbing calls of the American Crow or Blue Jay made an observer aware of the owl's presence.

The Great-Horned Owl is one of the first species to nest in New York each year. Its nesting habitat is usually a bottomland, mature forest adjacent to a good feeding area. It tends to occupy the same territories year after year, sometimes even the same nest (Hagar 1957). Almost any type of forest will do, but it prefers more extensively wooded areas to small woodlots (DeGraaf et al. 1980). Hagar (1957) found that wooded tracts of more than 8 ha (20 a) were used most commonly in southeastern Madison County, the smallest used being 6 ha (15 a). In a study of Great Horned Owl territories near Ithaca, Tompkins County, Baumgartner (1939) found them to be about a quarter of a mile wide and about a mile long. In optimum range, such as continuous woodland along a river valley, the territorial aggressiveness of the male probably determines nesting density; densities average from 1 to 3 pairs per 2.6 sq km (1 sq mi). There normally are no other species of large owls nesting or feeding extensively within the Great Horned Owl's territory (Baumgartner 1939).

The nest is usually an old tree nest of a large bird such as the Red-tailed Hawk, Osprey, Bald Eagle, Great Blue Heron, or American Crow. Hagar (1957) found that even nests newly built by hawks were at times taken over by this owl. Sometimes it uses an old squirrel's nest, a natural tree cavity, a rock ledge (Bull 1974), or a cave. This owl has been known to nest on the ground in logs or among rocks. Very little, if any, material is added to the nest, but down and contour feathers accumulate (Harrison 1975).

The Great Horned Owl has been persecuted for more than two centuries but still survives and probably has made a comeback as the public attitude towards hawks and owls has changed.

Stephen W. Eaton

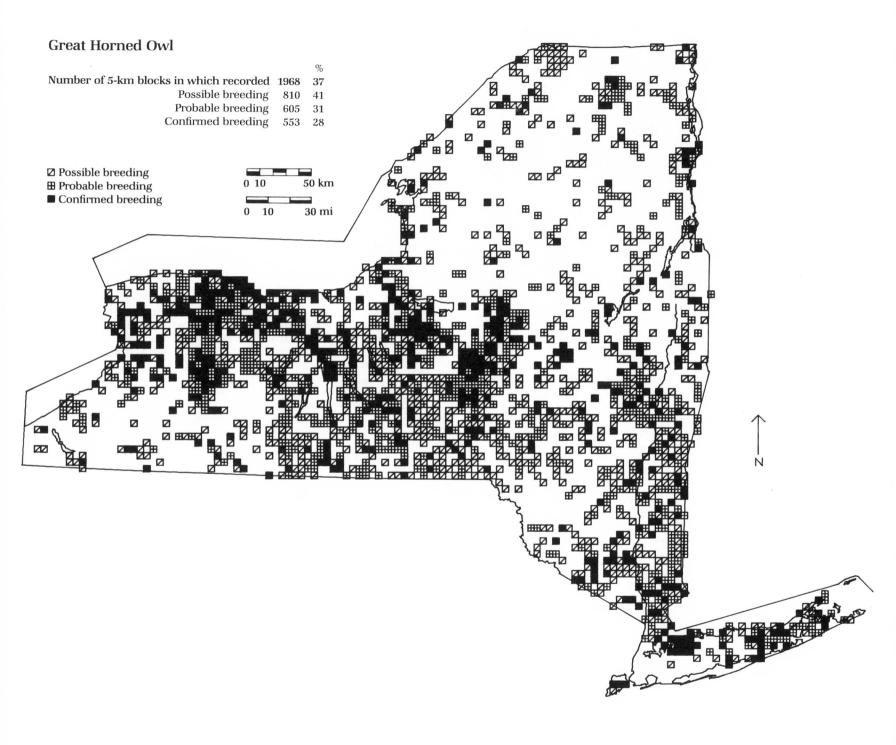

Great Horned Owl

Number of 5-km blocks in which recorded		%
Number of 5-km blocks in which recorded	1968	37
Possible breeding	810	41
Probable breeding	605	31
Confirmed breeding	553	28

◩ Possible breeding
⊞ Probable breeding
■ Confirmed breeding

0 10   50 km

0 10   30 mi

N

Barred Owl *Strix varia*

The Barred Owl is only accidental on Long Island but is widespread over other parts of the state, where it varies from being rare to common, depending upon the type and amount of forest cover available.

Over most of New York the range of the Barred Owl appears unchanged since Eaton (1914) declared, "It is undoubtedly the commonest owl in the Adirondacks, and is still common in all the more wooded districts of the State." Nevertheless, deforestation in the 19th century and subsequent reforestation during the 20th have undoubtedly had a considerable impact. On the Coastal Lowlands this has always been a rare bird, Bull (1964) noting: "One of the mysteries of bird distribution is that of the Barred Owl on Long Island," adding that the species had been found at only five localities: Jericho, Nassau County; and in Suffolk County at Heckscher State Park, East Patchogue, Sag Harbor, and Shelter Island. No actual nesting has been recorded, however, in more than 40 years, and Atlas observers found the Barred Owl only in Suffolk County near Setauket and on Fishers Island. Between 1905 and 1908 Chapin and Cleaves located 11 nests on Staten Island (Bull 1974), where Atlas searches failed to turn up even a single calling Barred Owl. In the highly urban areas of the Coastal Lowlands, this owl is disappearing.

The Barred Owl is most densely distributed in the Adirondack High Peaks, Central Adirondacks, Western and Eastern Adirondack Foothills, and Lake Champlain Valley, with a peripheral concentration on the Tug Hill Plateau. Other areas also show contiguous populations, including the Cattaraugus Highlands, Central Appalachians, Catskill Peaks, surrounding highlands and hills, and the Hudson Valley and neighboring high areas. This owl is apparently scarce in the Drumlins north of the Finger Lakes, the Eastern Ontario Plains, St. Lawrence and Malone plains, and the Champlain Transition south of Quebec. Although many gaps in the range exist because of the difficulties in locating a largely nocturnal species in remote areas with no resident observers, a special effort to encourage owlers by articles in several publications paid obvious dividends. The Atlas map, somewhat surprisingly, resembles the ranges of Northern Goshawk or Broad-winged Hawk more closely than it does those of the diurnal raptors of wet woodlands, the Cooper's and Red-shouldered hawks. Perhaps the size or contiguity of the forest tract is as important as any other habitat considerations; the Barred Owl is generally most concentrated in areas of the state with heavy forest cover.

Unlike the Great Horned Owl, which shows toleration for fragmented forests and woodlots, the Barred Owl seems to prefer larger unbroken woodlands, either coniferous or deciduous, and often mixed, with sufficient old growth to provide suitable nesting trees (Bent 1938). Since this owl also shows an affinity for low, wet woods, northern white cedar often served the author as a useful indicator tree for owling, yet the Barred Owl is frequently heard or seen on forested ridges. Eaton (1914) noted that this owl "breeds wherever it finds swampy woods or forests of sufficient extent to secure it protection from its one great enemy, civilized man." Shooting, leghold traps, and motor vehicles take a toll, but the reforestation of the state should be having a beneficial effect. The forests of New York are now at their most extensive since before 1825 (Considine 1984; Stevens 1985).

The Barred Owl nests in tree cavities, old hawk or squirrel nests, and even barns. Snags used as nest trees were found most commonly in northern hardwood, lowland hardwood, or oak forests (Evans and Conner 1979). Hole nests are usually in deciduous trees, although conifers are sometimes chosen; the owl's choice seems about evenly divided between living and dead trees (Bull 1974). The nest hole size is about 20 cm (7.9 in) (Evans and Conner 1979). Nests in New York have been found in American chestnut, oak, birch, and basswood, at heights from 6 to 12 m (20–40 ft) above the ground (Bull 1974). When the nest is in a natural cavity, "a number of fluffy feathers or bits of long, soft, gray down are scattered about, clinging to the trees or underbrush in the vicinity of the nest, or seen waving in the breeze on the nest itself; these are very helpful in locating the nest" (Bent 1938). Stick nests are usually those built originally by Cooper's Hawks or Red-shouldered Hawks (Bent 1938), and the Barred Owl's breeding habitat, if not its range, is similar to that of these hawks (Bull 1974). Atlas observers located nests with young in 15 blocks, nests with eggs in nine, and occupied nests in four, plus one used nest.

John M. C. Peterson

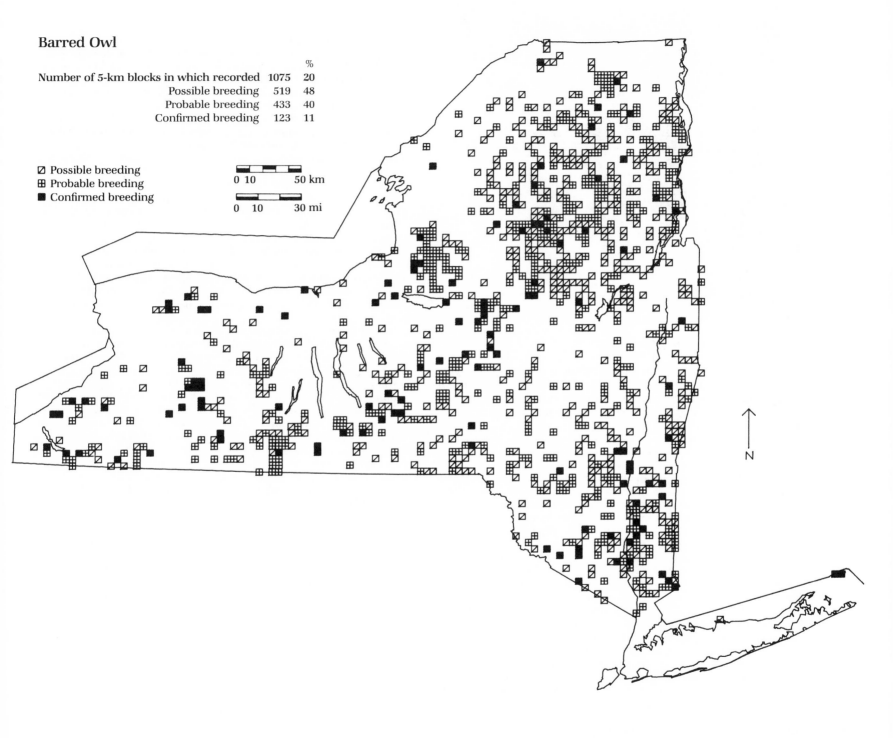

Barred Owl

		%
Number of 5-km blocks in which recorded	1075	20
Possible breeding	519	48
Probable breeding	433	40
Confirmed breeding	123	11

◩ Possible breeding
⊞ Probable breeding
■ Confirmed breeding

0 10   50 km

0 10   30 mi

N

Long-eared Owl *Asio otus*

Largely nocturnal, frequently silent, and capable of assuming cryptic poses in its thickly forested habitat, the Long-eared Owl is very uncommon in New York, although it may be more widespread than the Atlas map suggests. It was recorded in only 81 blocks and "confirmed" in just 15.

The Atlas map portrays a range that differs in several respects from Bull's (1974). His map showed 54 known breeding localities in New York and suggested two populations: one in western New York, extending across the Appalachian Plateau and Great Lakes Plain; another on the Coastal Lowlands extending northward up the Hudson Valley to the Taconic Foothills. Because there were no records from the Adirondacks and Catskills, the Long-eared Owl was indicated in Bull by a question mark over northern New York; but this lack of records was thought to be "due, in part, to lack of observers in those areas" (Bull 1974). Earlier in this century Eaton (1914) made much the same assessment, finding it "not very uncommon about dense wooded swamps and hillsides in most parts of the State, but . . . apparently uncommon in the Adirondack forests."

Atlas observers found Long-eared Owls in widely scattered locations in forested areas in upstate New York; a seemingly detached population was noted on the Coastal Lowlands. The Long-eared Owl was not recorded in the southeastern part of the state. There were no records from the Catskills and except for the Helderberg Highland, none in the surrounding foothills, where it might be expected to occur, nor was it located east of the lower Hudson Valley. No records were obtained in the

Allegany Hills. Although Saunders (1942) knew of only one record there (July 1930), he believed the owl was more prevalent. Whether the Atlas records portray a range change is difficult to determine, but it seems unlikely. Because the Long-eared Owl is difficult to find, it was undoubtedly under-recorded in many regions. Also, in those ecozones where it was found, this owl probably went undetected in many blocks. The reforestation of New York during this century and the establishment of countless conifer plantations should provide excellent nesting habitat for this owl. Despite the intensity of Atlas efforts, an even more focused search appears necessary in order to better understand the Long-eared Owl's breeding status in New York.

Locating the Long-eared Owl is made difficult by its confusing vocalizations. Competitive calling during courtship consists of a carefully spaced series of short *whoo-whoo-whoo* calls. Sometimes single low hoots are given, at other times as many as 20 in a series. Noncompetitive calling may include a variety of *whoo*s, *wuh*s, and chickenlike clucks. Other vocalizations include catlike mewing, foxlike cries, grouselike whistles, puppylike barks, and Screech-Owl-like whistles, as well as various moans, shrieks, and twitters (Bent 1938; Armstrong 1958). With such a rich and varied repertoire, the Long-eared Owl proved difficult for many fieldworkers to identify, even those familiar with the basic courtship calls.

During the breeding season this medium-sized owl occupies dense woods, usually thick conifers or mixed forest, often near water (Bent 1938; Armstrong 1958; AOU 1983). Although Armstrong (1958) believed that the Long-eared Owl hunted more in woodlands than in open areas, Marti (1976) and others stated that it hunts almost exclusively in open areas. Around lakes and ponds in the Adirondacks, where Atlas observers heard these owls calling, openings were generally small or lacking. This suggests that owls in this area probably are woodland hunters. An Atlas observer in Albany County found a nest with young in a mixed coniferous and deciduous wet woodlot. The approximate size of the woodlot and an adjacent northern hardwood area was 4–4.7 ha (10–12 a), with open fields surrounding the woods (Gawalt, pers. comm.).

Rather than build its own nest, the Long-eared Owl uses old nests of other birds, most frequently those of crows and hawks, although Audubon found a nest in Pennsylvania that the owls had evidently made themselves. Used nests are slightly repaired and then lined with dried grasses, leaves, and feathers (Bent 1938). Of the New York nests that Bull (1974) described, 6 were in pines, 3 in hemlocks, 2 each in cedars and maples, and 1 each in apple and sassafras trees, with 1 in a " 'natural hollow' " in a maple. Heights ranged from 3.7 to 13.7 m (12–45 ft). The Albany County nest with 4 young was located in a white pine tree (Gawalt, pers. comm.).

John M. C. Peterson

208

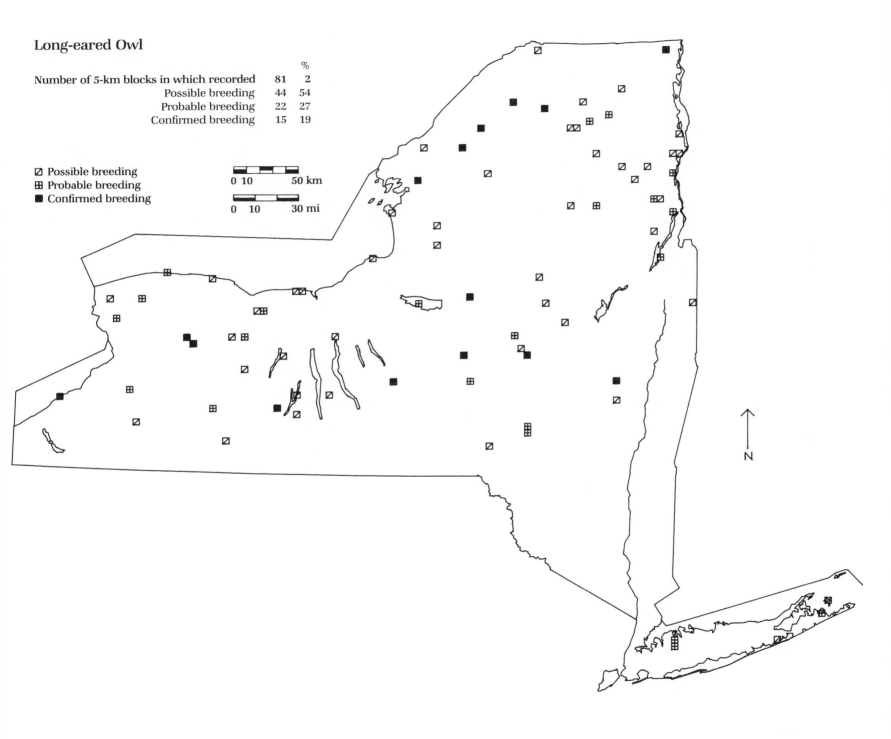

Long-eared Owl

Number of 5-km blocks in which recorded	81	% 2
Possible breeding	44	54
Probable breeding	22	27
Confirmed breeding	15	19

☑ Possible breeding
⊞ Probable breeding
■ Confirmed breeding

0 10   50 km

0 10   30 mi

N

Short-eared Owl *Asio flammeus*

The Short-eared Owl quarters the marshes, dunes, and fields with easy, buoyant flight hunting its favorite food, the meadow vole. It is a rare breeder in New York and opportunistic in breeding, needing an abundant supply of voles and open habitat for nesting.

According to Tate (1981), it appears to have declined in New York, Ohio, Indiana, and Ontario, and has been on the *American Birds'* Blue List since 1976 (Tate 1986). It is listed as a Species of Special Concern in New York. Clark (1975) stated that "ecological changes due to agriculture, urban development, and succession" had made former breeding grounds no longer suitable. In addition to habitat loss, Fimreite et al. (1970) suggested that in Canadian agricultural areas this owl may be a victim of contamination from heavy use of pesticides and herbicides. In New York primary Short-eared Owl habitat is found in agricultural areas, including the Great Lakes Plain, where toxic chemicals commonly are used.

Eaton (1914) said that as a resident species this owl was widely distributed, and its presence during the breeding season apparently was dependent upon the availability of nest sites and freedom from persecution. In the late 1800s and early 1900s nesting was reported from Suffolk County; near Buffalo; Canandaigua, Ontario County; and near Brockport and Chili, Monroe County (Eaton 1914). It was reported to be quite common at marshes of the Seneca River above and below Montezuma NWR, and in the wetlands near the eastern end of Lake Ontario. Its numbers apparently declined, as Bull (1974) called the Short-eared Owl scarce during the breeding season, indicating only two main historic areas of concentration on Long Island and in western New York. On Long Island he reported ten breeding localities from Suffolk County and five from western Long Island.

The Atlas map shows only one "confirmed" record in upstate New York near Potsdam, St. Lawrence County, and only four on Long Island, with no records east of Great South Bay where breeding had formerly been documented. One of the "confirmed" records was at Floyd Bennett Field in Queens County (Spencer 1981). The few scattered records from western and central New York are observations of individual birds. Several observations were reported in 1981, and a pair was observed in 1983 in the Livingston County farmland of the Genesee River Valley. In Jefferson County 4 owls were observed at Ashland WMA and two were seen at the Watertown Airport in mid-June (Chamberlaine 1980). The two "probable" records in Franklin County were from the Brandon Burn, a large, previously forested area that burned in the early 1900s and grew back mainly to grasses and low shrubs (Peterson, pers. comm.). There are a few records along the Lake Champlain Valley where summer records had not previously been documented.

According to Clark (1975), the Short-eared Owl is opportunistic in its choice of breeding sites, and its breeding range changes depending on the food supply. In a study at this species' breeding ground in Manitoba, he found that nesting was sporadic, with fewer breeding pairs when its primary prey, the meadow vole, was scarce. In addition, Clark described several behaviors of the Short-eared Owl that are characteristic of an irruptive population, including "nomadic movement, specialized feeding habits, plasticity with regard to time and locale of breeding, and flexible fecundity." It should not be surprising, then, that the distribution of this owl has changed in New York and that it appears in different areas from year to year.

This owl is not difficult to locate during the nesting season because it is active during the day and has an elaborate courtship display that can go on for days (Clark 1975). Atlas observers were advised against trying to locate the nest of this and other ground-nesting raptors.

The Short-eared Owl is a bird of open country. In much of its range it seems to prefer grassland habitat, but it also nests in stubble and in rush (Clark 1975). In New York its breeding habitat varies with the physiographic region. In the Coastal Lowlands it nests on bare sand with a scattering of beach grass. On the Great Lakes Plain it has been found nesting on dry, flattened cat-tail in a wetland; in a field of goldenrod, spiraea, and grass; in an uncultivated field among alfalfa stubble; and in a fallow field of wheat stubble (Bull 1974).

Within its nesting habitat it selects drier sites for its nest (Clark 1975). The nest is generally in a slight depression on the ground and sparsely lined with grasses, weed stalks, stubble, and feathers. Sometimes it is just on flattened vegetation at a chosen spot, and rarely it is found in an excavated burrow. The same site may be used in successive years (Harrison 1975).

Stephen W. Eaton

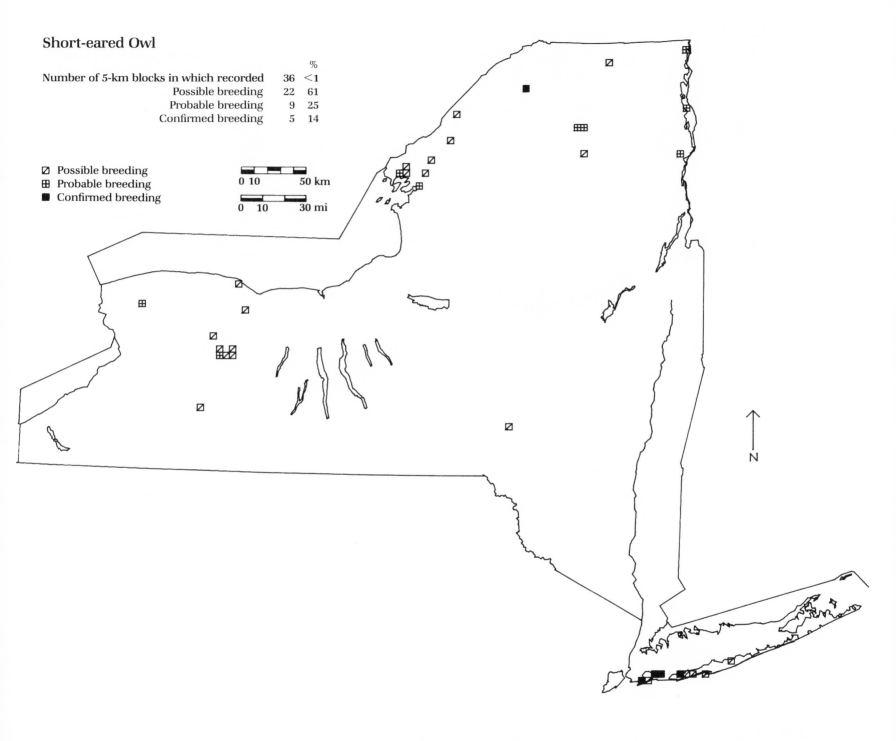

Short-eared Owl

		%
Number of 5-km blocks in which recorded	36	<1
Possible breeding	22	61
Probable breeding	9	25
Confirmed breeding	5	14

⊠ Possible breeding
⊞ Probable breeding
■ Confirmed breeding

0 10   50 km

0 10   30 mi

N

Northern Saw-whet Owl *Aegolius acadicus*

The smallest of the owls that breed in New York, the Northern Saw-whet Owl is fairly common in the Adirondack region but decidedly rare elsewhere in the state. It is so nocturnal that nesting individuals are extremely difficult to locate, which should be considered when assessing its rarity.

Eaton (1914) and Bull (1974) regarded the species as rare, but both indicated that it was probably often overlooked. Beehler (1978) in a work about the Adirondack Park, the stronghold of the species, called it "uncommon." Griscom (1923), however, had earlier stated dogmatically, "Its supposed rarity may confidently be stated as a myth." Gordon (1959) also pointed out that although this was the least known resident owl, it was probably more common than realized and found in every corner of the state. Earlier, Bent (1938) also followed this same line of thinking when he said this owl was "probably much commoner than it [was] generally supposed to be" in North America.

This presumption may be the result of the significant number of owls observed during late fall and winter migratory movements of this species. For example, on 29 October 1967, Ward and others walked through a couple of miles of pine groves between Oak Beach and Cedar Beach in Suffolk County and found the astounding total of 26 birds (Bull 1974). Cohen (1966) mist netted 31 individuals between 30 September and 30 November, all caught at night in his backyard banding station in Atlantic Beach, Nassau County. The origin of the birds found along the barrier beaches in the fall and winter is unknown.

There are huge areas of the state where there is no evidence of breeding. Breeding is concentrated in the heavily forested areas of the Adirondacks with few scattered records elsewhere. The scarcity of records in the Catskills is somewhat surprising and may just represent a lack of evening observation there. Breeding was documented in the Catskill foothills in 1986 after the Atlas in an area where birds had been heard calling the previous winter. On Long Island breeding was "confirmed" in Lattingtown, Nassau County, on the north shore, only the fourth documented record on Long Island (Spencer 1981). The results of the Atlas study do not represent any change from previous knowledge, but owls were under-recorded in general by Atlas observers. The Northern Saw-whet Owl calls infrequently while incubating, the time when most observers were in the field. In addition, many observers may have overlooked the call of this species or are not familiar with it.

The Northern Saw-whet Owl is found in forested wetlands and conifer groves, but it also occurs in deciduous woodlands. "Mature forests with living and dead, mixed trees are preferred nesting habitats" (DeGraaf et al. 1980). Atlas observers gave various descriptions of habitat. On Long Island a pair nested in the same woodlot as an Eastern Screech-Owl. The woods were deciduous containing mainly ash, locust, and Norway maple, with spicebush and arrowwood making up the understory. In the Adirondacks it was found in large alder shrub swamps, northern white cedar swamps, along remote pond shores, and also in mature conifers bordering a lake (Carroll and Peterson 1983). One of the few Catskill records was of an owl calling from a tamarack swamp adjacent to a large red pine plantation (J. Carroll, pers. comm.). In a study of the home range of the Northern Saw-whet Owl, Forbes and Warner (1974) found that this owl makes use of edges most often and seems to prefer wooded lowland habitat.

The Northern Saw-whet Owl is a hole nester, using mostly deserted woodpecker holes and sometimes natural cavities. It brings no material into the nest, using whatever is in the hole from the previous occupant, wood chips if it was a woodpecker or plant material if it was a mammal (Harrison 1978). Evans and Conner (1979) found the minimum dbh of the nest tree to be 30 cm (11.8 in) and the nest hole 6.1 cm (2.4 in) in diameter. Although the owl will sometimes nest in boxes meant for Wood Ducks or Eastern Screech-Owls (Bull 1974), these are almost always in heavily wooded country. A most unusual box nesting took place on Long Island (Schaeffer 1968) adjacent to a salt marsh, on the edge of an irrigation ditch in totally open country.

Emanuel Levine

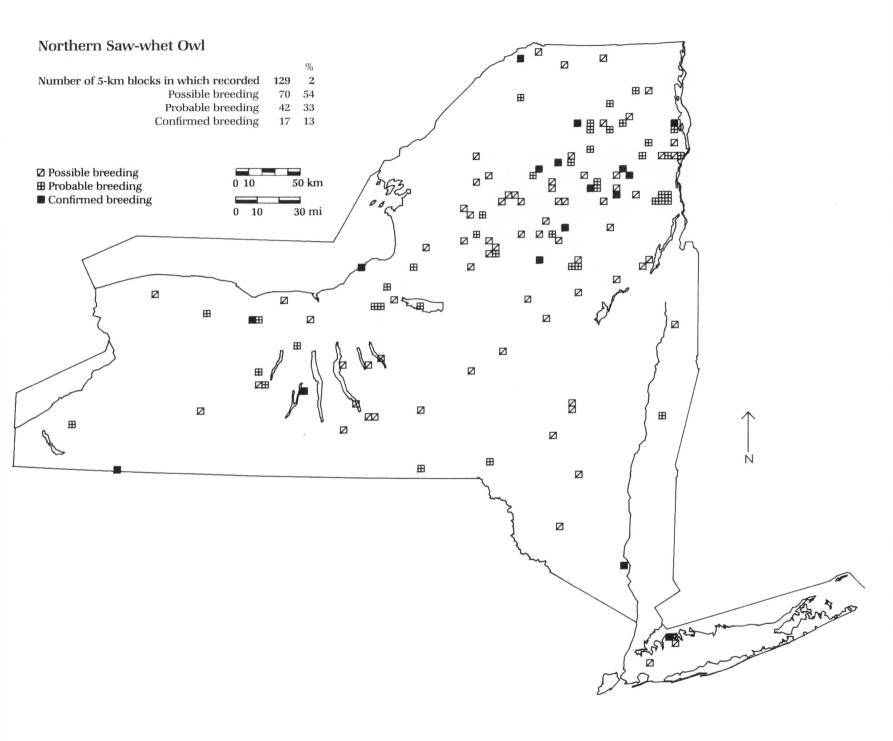

Northern Saw-whet Owl

		%
Number of 5-km blocks in which recorded	129	2
Possible breeding	70	54
Probable breeding	42	33
Confirmed breeding	17	13

◩ Possible breeding
⊞ Probable breeding
■ Confirmed breeding

0 10   50 km

0 10   30 mi

N

©KLA-C 1986

Common Nighthawk *Chordeiles minor*

The crepuscular Common Nighthawk appears to be in trouble, declining throughout most of New York; it deserves close attention over the coming decade. An uncommon, extremely local breeder throughout the state, it is listed as a Species of Special Concern by the NYSDEC and is on the 1986 *American Birds'* Blue List (Tate 1986).

The Common Nighthawk originally nested only on barren ground in open areas. The shift from natural to man-made sites occurred in the mid-1800s when mansard and flat gravel roofs were first introduced. Nesting on these roofs was first recorded in Philadelphia in 1869, and by 1880 roof nesting was common throughout the Northeast (Gross 1940). The reasons for this shift are not clear but may have been due, in part, to a reduction in its natural nesting habitat by the intensive use of land for agriculture in upstate New York from the late 1800s to early 1900s and the subsequent reforestation in many areas, with active prevention of forest fires.

As it invaded these new sites, its numbers seemed to increase (Gross 1940). Eaton (1914) said that "in New York it is found in every county of the state as a summer resident, but is somewhat local in its breeding," and Griscom (1923) described it as a common summer resident on Long Island. Declines in numbers, however, were noted around New York City and Westchester County, where the nighthawk disappeared as a breeder in Prospect Park by 1920 and from Central Park by 1930 (Carleton 1958). According to Cruickshank (1942), "It no longer nests in New York City, has virtually disappeared from the pastures of Westchester County, and very few towns in this county can now boast of breeding pairs."

Although at one time "hundreds" of Common Nighthawks "could be seen nightly circling the floodlight-lit upper stories of the Westchester Lighting Company building" (Reilly, pers. comm.), nighthawks in Westchester County are now in severe trouble and even roof-nesting birds in White Plains and Larchmont have disappeared within the past decade (Weissman, pers. comm.). In Rochester numbers are way down (Spahn, pers. comm.), and in nearby Oswego and northern Cayuga counties the species also seems to be declining (Smith and Ryan 1978). The Common Nighthawk is not declining in all areas, however, as populations in the Adirondacks seem to be at least equal to if not larger than in historic times (Peterson, pers. comm.). Although BBS data for the period 1965–79 do not show any change in nighthawk populations in the state, it should be noted that this species would be poorly sampled on BBS routes (Robbins et al. 1986).

During the Atlas survey the Common Nighthawk was located throughout the state, including in New York City, but only rarely away from cities and towns. Because it is usually active for only a few hours around dawn and dusk each day, making it more difficult for observers to locate, the Common Nighthawk probably occurs in more blocks than Atlas records indicate. It was even more difficult to "confirm" breeding (only 45 blocks), with "confirmed" records evenly divided between nests with young, nests with eggs, and recently fledged young.

Rooftops in cities, towns, and villages are the most common nesting sites in the state, but the Common Nighthawk still nests regularly in some open areas, like the Brandon Burn, a large tract of land in Franklin County that was burned in a forest fire in 1908. The fire was so hot that even the topsoil was scorched, and now only lichens, grasses, and low shrubs, mainly blueberries, are found there, with a few scattered clumps of aspen trees and black spruce clones. The area also is heavily browsed by a large deer herd (Barnett, pers. comm.). Natural breeding sites on eastern Long Island included "sandy openings in mixed pine-scrub-oak barrens, on bare ground in pastures and fields, on sand dunes, on gravel beaches, and on flat rocks and logs in the open" (Bull 1974).

Terres (1980) stated that the eggs are "laid on a spot chosen by the female on open barren rocks or bare gravelly soil of pastures, beaches . . . ; old fields, also on ground in vineyards, corn and potato fields, and even gardens; sometimes atop stump or fence rail up to 8 ft. above ground." The nest consists merely of the spot where the eggs are laid, as no vegetation or other material is removed from or added to the site to make a nest (Gross 1940). Gross described a nighthawk that nested in an American Robin's old nest in Maine for five consecutive years.

Steven C. Sibley

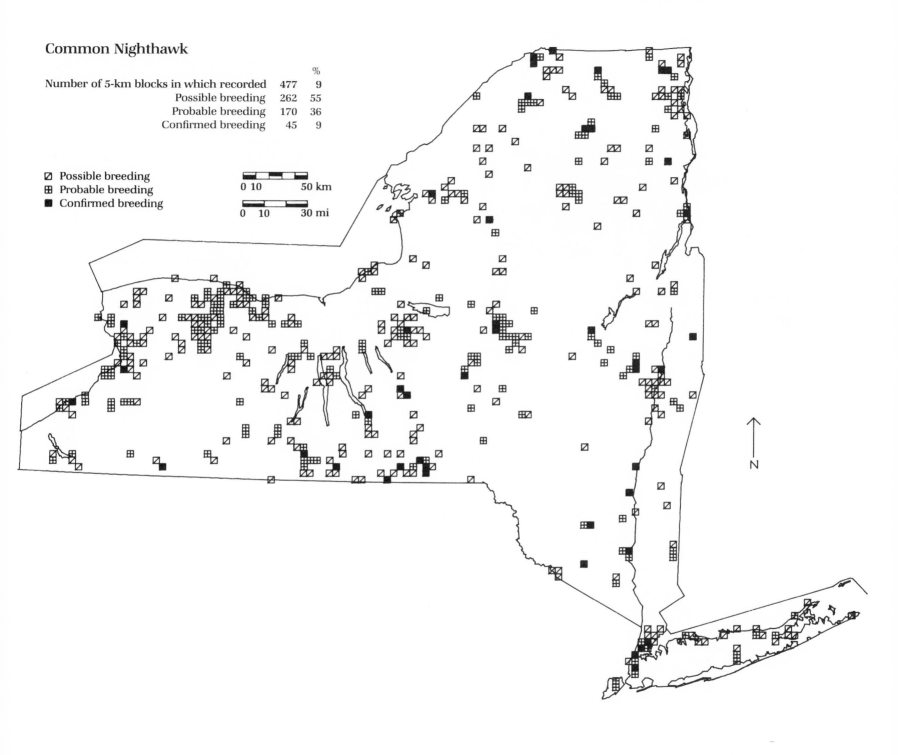

Common Nighthawk

		%
Number of 5-km blocks in which recorded	477	9
Possible breeding	262	55
Probable breeding	170	36
Confirmed breeding	45	9

▧ Possible breeding
⊞ Probable breeding
■ Confirmed breeding

0 10   50 km

0   10   30 mi

N

©KLA-C 1987

Chuck-will's-widow *Caprimulgus carolinensis*

Until recently the Chuck-will's-widow, a large, crepuscular, and nocturnal goatsucker, bred only in the oak and pine woodlands from Florida to southern Maryland (Davis 1972). Like a number of other austral species, it has been extending its range northward for the past several decades.

Appearing in New Jersey in 1922, it became an established breeder there by the mid-1960s. In New York the first specimen was taken in 1933 at Riverhead (Suffolk County), on Long Island (Bull 1964). With the exception of 2 individuals observed in Monroe County in 1959 and 1961 (Miller 1959, 1961), there were no more New York reports until 1969. Since then it has been seen and heard almost yearly in Suffolk County, Long Island, and occasionally on Staten Island and in Central Park in New York City. The state's first nest was found in June 1975 at Oak Beach on Long Island (Davis 1975).

From 1977 to 1981 the Chuck-will's-widow was heard at Oak Beach, and breeding was "confirmed" there by Atlas observers in 1981. A transition area of maritime dune and shrubland between Oak and Cedar beaches was systematically crisscrossed until the well-camouflaged adults flew about 1.8 m (6 ft) away and then gaped at the intruders with partly outspread wings. Two 3–4-day-old chicks were found crouched on the carpet of pine needles and leaves from where the adult birds had been flushed (Greenlaw, pers. comm.). Additional Atlas records of this species came from similar areas along Long Island's south shore and in the pine barrens of central Nassau County.

This species is difficult to locate during the day and, except on moonlit nights, calls only from sunset to darkness. It sings most frequently when the moon is more than half-full (Cooper 1981). Unless it is calling, a special effort has to be made to locate this species, and some may have been missed by Atlas observers.

In southern states the usual nesting habitat of this species is in mixed oak and pine woods with little, if any, undergrowth (Sprunt 1940). At Oak Beach the nest was placed in a thin ground cover of low poison-ivy seedlings in the shade of a Japanese black pine. The surrounding habitat where the Chuck-will's-widows foraged consisted of bayberry, and shrubby thickets of greenbrier and poison ivy, in which were openings covered with beach heather (Greenlaw, pers. comm.). On Staten Island this species was found in an upland habitat of semi-open scrub oak near a small stream running through dry, clayey soil (H. Fischer, v.r.). During daylight hours the Chuck-will's-widow rests on the ground in dense undergrowth or perches on a low limb or log. Once having chosen a roosting spot, it tends to return to the same place repeatedly (Sprunt 1940). The eggs are laid on the ground within a few feet of the same spot year after year (Sprunt 1940).

It may be just as well that the Chuck-will's-widow is an uncommon bird. The experience of A. Sprunt, Jr., is not one that most would appreciate. One night a chuck outside his bedroom window called 834 times without stopping (Sprunt 1940).

Gordon M. Meade

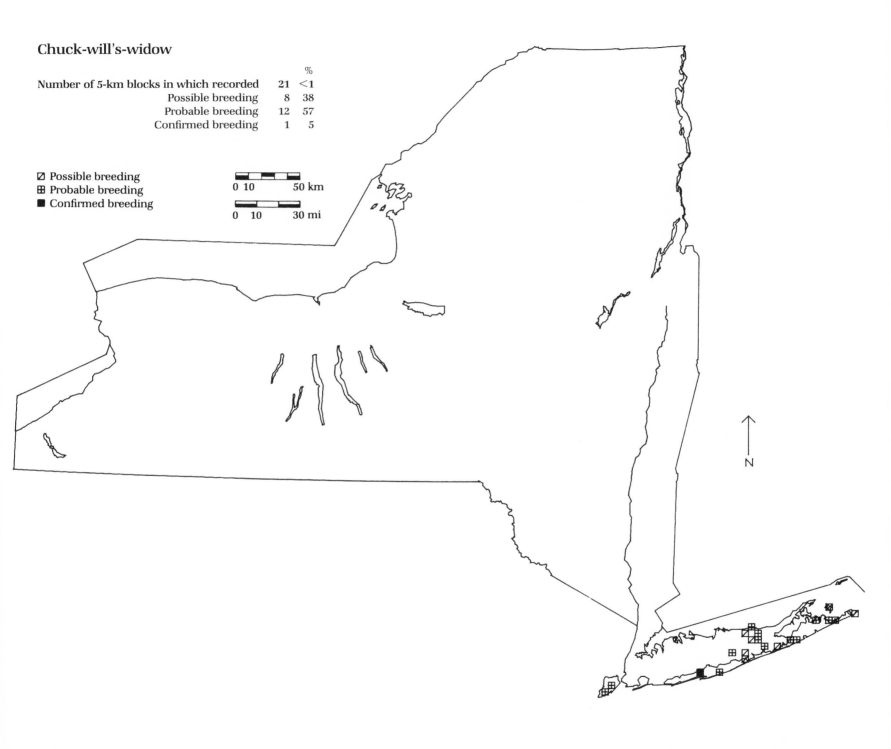

Chuck-will's-widow

		%
Number of 5-km blocks in which recorded	21	<1
Possible breeding	8	38
Probable breeding	12	57
Confirmed breeding	1	5

☑ Possible breeding
⊞ Probable breeding
■ Confirmed breeding

0 10   50 km

0   10   30 mi

N

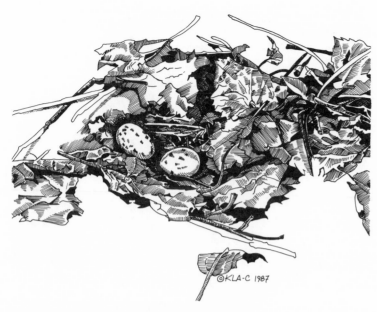

Whip-poor-will *Caprimulgus vociferus*

Although the Whip-poor-will is rather uncommon in the state and only rarely seen, most people would immediately recognize the call of this vociferous night bird. There is some evidence that the Whip-poor-will has declined in the state since the early 1900s. Stoner (1932), speaking of the Oneida Lake region, said "Local residents are unanimous in their statements that the bird is much less common here now than it was a few years ago." Stoner's comments are typical of those made in almost every 20th-century publication on New York birds.

The reasons for the apparent decline of this species are very poorly known. One hypothesis is that the decline was due to the decrease of large Saturnid moths, a major part of its diet (Robbins et al. 1986). The moths are believed to have declined initially because of the cutting of the state's forests in the 1800s, and then more severely in the early 1900s with the spread of industrial pollution and use of pesticides. The Whip-poor-will is now increasing in numbers and expanding its range in the south-eastern United States, possibly because of the reforestation of cleared lands (Cooper 1981). New York has been undergoing reforestation since the 1930s, but there is no indication the species is increasing. Although nocturnal species are poorly sampled, BBS data from 1965 to 1979 showed no increase or decrease in the state (Robbins et al. 1986). The Whip-poor-will is probably most abundant on Long Island. For example, at least 100 calling birds were counted in Connetquot River State Park, Suffolk County, on 8 June 1974 (Bull 1976).

Today the Whip-poor-will's distribution seems to be much more restricted than in the early 1900s, with almost a complete absence of Atlas records from the western and central portions of the state except for a population south of the Finger Lakes, which is confined to northern hardwood forests in river valleys below 305 m (1000 ft). Atlas data show the species to be widespread in the northern hardwood forests of the Hudson Valley and in a ring of low-elevation forested areas surrounding the Adirondacks, mainly white pine– and oak–northern hardwoods, or aspen–gray birch–paper birch forests. It is entirely absent from large areas of the Adirondacks and Catskills which are heavily forested and above 305 m (1000 ft), and there are just a few records from the Central Tug Hill, the Tug Hill Transition, and near Oneida Lake. The Whip-poor-will was rarely found in heavily populated areas, although it continues to occur on Staten Island and throughout Suffolk County. The discovery of the river valley distribution pattern south of the Finger Lakes was previously unknown, as was the scarcity of this species from the Central Tug Hill. The uncertainty over the status of this species is exemplified along the Lake Ontario shore in Orleans and Monroe counties, where the Whip-poor-will was found at several locations during the mid-1970s (Kibbe, pers. comm.) but was not located during the Atlas. It is unclear whether it was simply missed or no longer present.

The Whip-poor-will is a nocturnal bird, usually most vocal just after dark and just before dawn, and well camouflaged when roosting during the day. This makes it difficult to locate and exceptionally difficult to "confirm." For this reason, it is undoubtedly under-recorded on the Atlas map, and only about 3% of the Atlas records are "confirmed."

This species shows a preference for dry woodland such as the pitch pine–scrub oak barrens on Long Island and deciduous woods inland (Bull 1974). Saunders (1942) and others have also noted a preference in upstate New York for the drier oak-hickory forests over beech-maple woods. However, along the Black Creek marshes in St. Lawrence County, numerous Whip-poor-wills were found in wet woods at the edge of the marshes (Kibbe, pers. comm.).

This species most often occurs in woods with scattered or adjacent fields or open areas rather than in unbroken tracts of forest (Eaton 1914). These open areas serve as foraging sites for this bird, which hunts visually and therefore is most active on bright moonlit nights. Its breeding is synchronized with the lunar cycle. Raising of the young, a time when food demands are high, occurs during periods around full moon (Mills 1986).

The Whip-poor-will lays its eggs on the ground and does not build a nest. The nest site is usually in dry, open woodland without dense underbrush (Tyler 1940a). Eggs are placed on a bed of leaves, often just under the overhanging branches of a small bush, but never tucked in against the base of a tree or shrub (Tyler 1940a). On Long Island, Raynor (1941) found 3 nests in extensive second-growth oak woodland with some pitch pine.

Steven C. Sibley

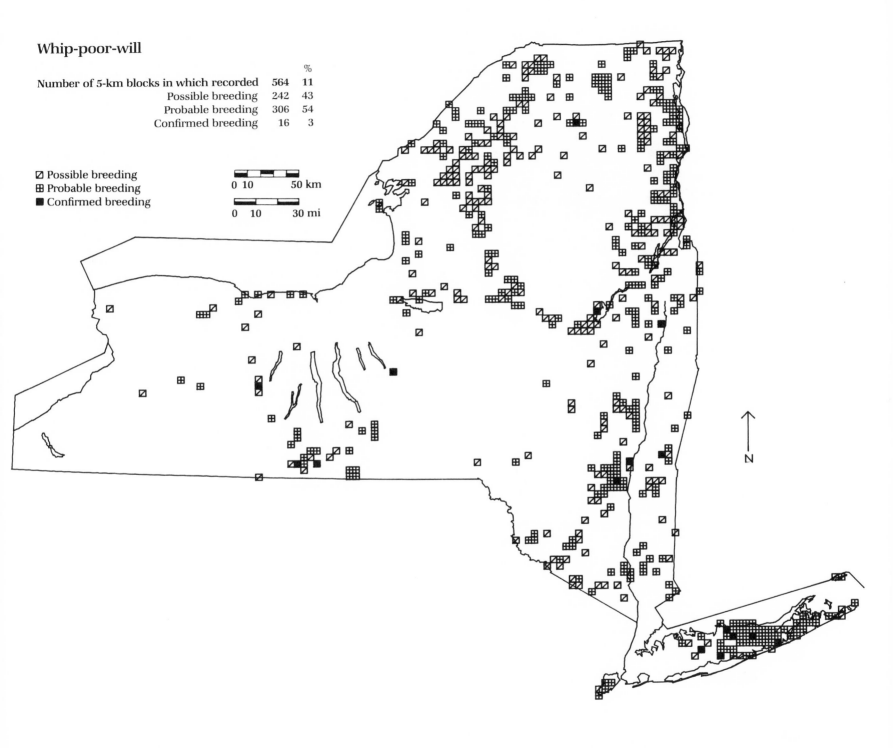

Whip-poor-will

		%
Number of 5-km blocks in which recorded	564	11
Possible breeding	242	43
Probable breeding	306	54
Confirmed breeding	16	3

◩ Possible breeding
⊞ Probable breeding
■ Confirmed breeding

0 10   50 km

0 10   30 mi

N

Chimney Swift *Chaetura pelagica*

The tiny Chimney Swift, whose flight silhouette is affectionately referred to by birders as a cigar with wings, is a fixture in the summer skies over most cities, towns, and villages in the state. It is a common breeder throughout New York. Although the size of its population does not appear to have changed in the past century, this species probably increased greatly during the initial settling of New York in the late 1700s (Tyler 1940b). By adapting from nesting in tree cavities and on cave walls to chimneys and building walls, the Chimney Swift shifted its distribution away from forests. In the urban sprawl of New York City this species has noticeably decreased since the 1900s (Siebenheller 1981). Although it did show a significant decrease in New York State, it increased in the eastern United States as a whole on BBS routes from 1965 to 1979 (Robbins et al. 1986).

Eaton (1914) stated that the Chimney Swift bred in every county of the state, and the same holds true today. In large urban areas this species is somewhat confined to large city parks and other open or wooded areas. Although it is difficult to see from the Atlas map, nearly all Chimney Swift sightings correspond to the location of a city, town, or small village. Most gaps in the distribution seem the result of a lack of suitable nest sites in areas of extensive agriculture or a lack of suitable feeding areas. The Atlas revealed that the Chimney Swift is still present throughout New York City. It is much more widely distributed in the Adirondacks than expected and much less widely distributed than expected in other parts of the state. This species may feed up to a few miles from the nest site (Norse

and Kibbe 1985), so many "possible" records may represent birds that were actually nesting in an adjacent block.

Because of its active aerial pursuit of insects and constant chatter in flight, the Chimney Swift is easily located and observed. It is, however, rather difficult to "confirm," and 51% of all Atlas records were only sightings of the birds. Breeding was "confirmed" in only 16% of the blocks, most often by observers seeing adults entering or leaving a nest site.

The Chimney Swift does all of its feeding on the wing, so it is hard to define its habitat preferences; however, it is often seen near water, and Atlas records correspond well with the distribution of river systems in the state.

Before western European settlers arrived, this species probably nested exclusively on cave walls and inside hollow trees (Bull 1974). Palmer (1949) wrote that the Chimney Swift nested in chimneys in Maine as early as 1671. By the early 1800s in eastern Pennsylvania, breeding was entirely limited to chimneys; no natural nest sites were known (Wilson 1812). Since 1900 there have been only a handful of records in New York of the Chimney Swift nesting in hollow trees (Bull 1974). Most Chimney Swifts now nest either in chimneys or on the inner walls of buildings.

Fischer (1958), in his landmark study of this species in New York, determined that the two main nest site requirements were darkness and shelter from the weather. The nest is constructed of small twigs, which are attached to the wall and to each other with the swift's gluelike saliva, forming a semicircular bowl in which the eggs are laid. The nests are usually too weak to be reused in subsequent years (Dexter 1962). However, an unusual circular cup nest built on a ledge in an air shaft in a building in Ohio was reused when 10 years old (Dexter 1981a).

The Chimney Swift gathers the twigs for its nest by using its feet to snap off the tips of dead branches while in flight. Nest construction takes about 18 days. The first eggs are actually laid when the nest is only half completed; both birds continue to add twigs and saliva until just before the eggs hatch (Fischer 1958). Occasionally, from 1 to 3 unmated Chimney Swifts will help a mated pair raise its young (Dexter 1981b).

Steven C. Sibley

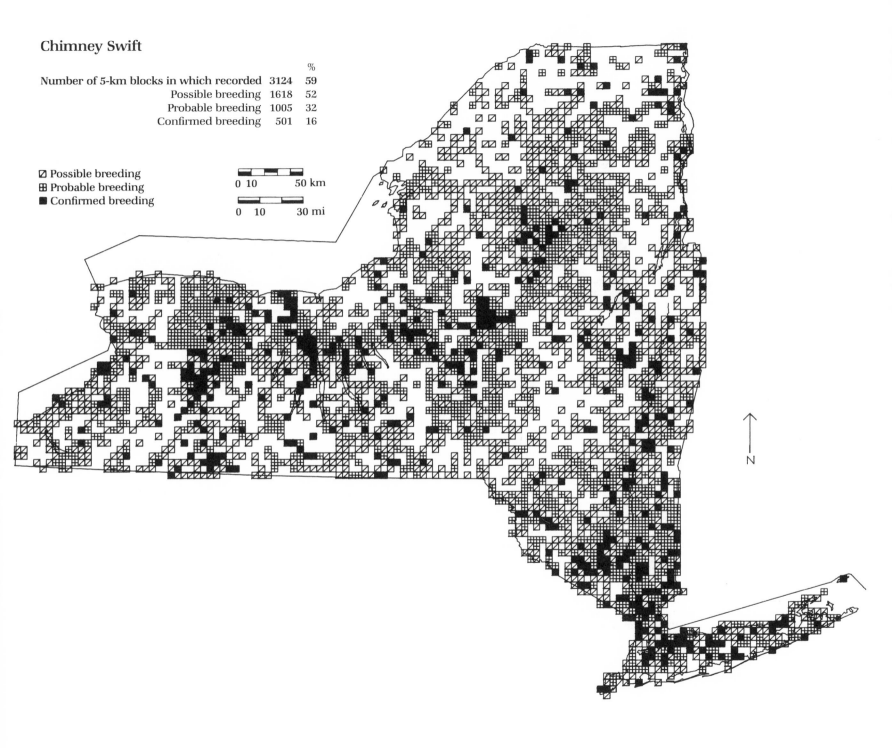

Chimney Swift

Number of 5-km blocks in which recorded	3124	% 59
Possible breeding	1618	52
Probable breeding	1005	32
Confirmed breeding	501	16

◨ Possible breeding
⊞ Probable breeding
■ Confirmed breeding

0 10   50 km

0 10   30 mi

N

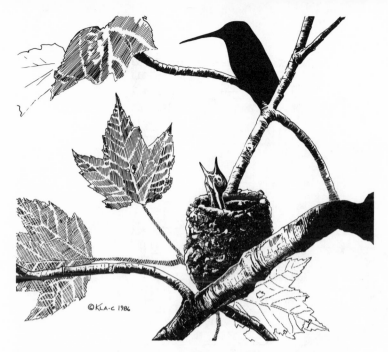

©KLA-C 1986

Ruby-throated Hummingbird *Archilochus colubris*

The diminutive Ruby-throated Hummingbird, one of the most well-known and popular birds, is a common breeder in almost every portion of the state. There appears to have been virtually no change in this species' status in New York over the past century other than its extirpation from New York City because of destruction of all suitable habitat.

The species has been absent from New York City for some time, as Cruickshank (1942) stated that it does not nest within the city. This hummingbird did nest on Staten Island until 1955, however (Siebenheller 1981). The Atlas map indicates its absence in the New York City area and in the southwestern portion of Long Island. As with many other woodland species, in Westchester County the Ruby-throated Hummingbird occurs almost exclusively in the northern half of the county because of a lack of suitable habitat in the southern half (Weissman, pers. comm.). The Ruby-throated Hummingbird is also absent from other urban areas and the less forested areas of the state like the Eastern Ontario and St. Lawrence plains and Mohawk Valley.

Because of its small size, rapid flight, and quiet nature, the Ruby-throated Hummingbird is a difficult species to "confirm" as a breeder. Almost half of all "confirmed" records were of recently fledged young. Despite the difficulty, nests were located in 39% of all blocks in which the species was "confirmed." Seeing a hummingbird is much easier than locating a nest, and Atlas coverage was probably quite good for this species, as observers concentrated their efforts on flowering gardens, clumps of lilac, old apple orchards, and bare snags along streams and roadsides. The male hummingbird often performs its courtship display flights in these open settings.

This species occurs in a wide variety of habitats but is rarely found in cities or other densely populated areas. Bull (1974) stated that the Ruby-throated Hummingbird "prefers rural areas . . . such as gardens, orchards, roadside thickets and trees, and forest clearings often near streams." In the Oneida Lake region Stoner (1932) found that the hummingbird occurred in "cutover areas, about low bushes in open places in the woods and about cottages and farmyards." In the Adirondacks, Eaton (1914) saw the species in small clearings and along steams up to 1067 m (3500 ft) elevation on Mount Marcy, Essex County. Robbins (1984) found that the Ruby-throated Hummingbird was generally more abundant in woodlands larger than 7 ha (17.3 a).

The Ruby-throated Hummingbird feeds at a variety of sources, from native and cultivated flowers (Austin 1975) to hummingbird feeders and even sapsucker drilling holes, where it feeds on the sap and small insects attracted to the sap (Miller and Nero 1983). In New York the hummingbird is often found at sapsucker drilling sites where yellow birch thrives (Reilly, pers. comm.). Today, many people attract these sprightly birds to their yards with sugar-water hummingbird feeders or by planting gardens of showy, nectar-producing flowers such as cardinal flower, bee-balm, columbine, and trumpet creeper (Spofford 1985). Although the species shows no preference for flowers of any color, most hummingbird flowers are bright red or orange (Miller and Miller 1971).

Nests of the Ruby-throated Hummingbird are almost always placed on the edge of a clearing, such as along a road or stream, on a thin, down-sloping branch, usually with other branches providing shelter overhead (Saunders 1936; Terres 1980). Typical is the nest found by one Atlas observer along the Vlomankill, Albany County. The nest was located approximately 3 m (10 ft) above the ground on a downward-hanging hemlock limb over the bank of the stream (J. Carroll, pers. comm.). Of 11 nests located by Saunders (1936) in Allegany State Park, 9 were over water and all were on thin down-sloping branches.

The nest itself is camouflaged, sturdy, and very small; it is always placed on the upper surface of a limb or in the fork of a limb out away from the trunk. It is attached to the branch with silk from spiders' webs gathered by the female; the male rarely helps in nest construction (Johnsgard 1983). Typically, the Ruby-throated Hummingbird's nest is "compactly built of soft down from ferns, milkweed, fireweed, thistles, young oak leaves fastened in place by hummer with her bill" (Terres 1980). It is decorated with pieces of lichen or moss. The nest is lined with plant down (Johnsgard 1983). A Ruby-throated Hummingbird nest found during the Atlas survey in a pine in Quogue, Suffolk County, was built almost entirely of roofing insulation (Salzman, pers. comm.).

Steven C. Sibley

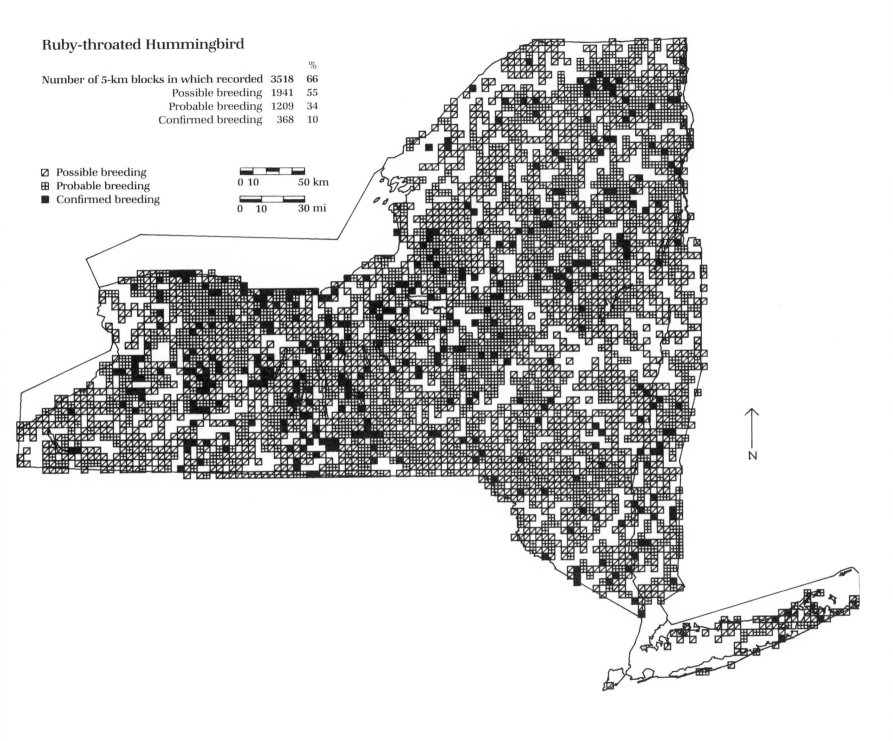

Ruby-throated Hummingbird

		%
Number of 5-km blocks in which recorded	3518	66
Possible breeding	1941	55
Probable breeding	1209	34
Confirmed breeding	368	10

☑ Possible breeding
⊞ Probable breeding
■ Confirmed breeding

0 10   50 km

0 10   30 mi

N

© KLA-C. 1987

Belted Kingfisher *Ceryle alcyon*

"Kingfisher of kingfishers" is the rough translation of the Belted King-fisher's scientific name (Forbush 1929), and indeed this common bird is an expert at catching fish, the main component of its diet. Eaton (1914) said, "The Belted Kingfisher is a common summer resident and breeds in every county in New York State." Seventy years later this bird is still common throughout the state, and Atlas data show no change in its distribution. Although Judd (1907) thought that the Belted Kingfisher was more common along the inland ponds and streams in Albany County and along the Hudson River than in other places in the state, no current evidence supports this observation. The species' familiar rattle may be heard wherever there is a lake, pond, wooded creek, river, bay, or estuary. The number of individuals within a given breeding area is, however, limited, for this species is highly territorial and requires a large area along a waterway for foraging (Cruickshank 1942).

Found in every ecozone in the state, the Belted Kingfisher is equally at home in such different areas as the Adirondack High Peaks, the Great Lakes Plain, and the Coastal Lowlands. Within each ecozone, however, it is usually confined to those waterways having nearby banks that can be excavated for nest sites. Atlas observers had no trouble finding the Belted Kingfisher and located active nests in 424 blocks and used nests in 225. Observers described various methods of "confirming" this species. All high banks were checked for nest holes having the two conspicuous grooves at the bottom where the kingfisher drags its feet as it enters the hole. Sometimes fish were spotted in the bill of a kingfisher as it flew to its nest. In addition, because the Belted Kingfisher and Northern Rough-

winged and Bank swallows often use the same nesting bank, a search made when these swallows were observed often resulted in "confirmed" records of swallows and of kingfishers (Carroll and Peterson 1982). The likelihood of encountering this species at sand and gravel operations was also high.

According to Bent (1940), swift, rocky mountain streams and par-ticularly trout brooks are this species' preferred habitat. But DeGraaf et al. (1980) defined its habitat as "ponds, lakes, rivers, and streams that con-tain fish." Indicating that turbid water hampers fishing, they noted that fewer kingfishers are found at lakes with large waves than at small clear bodies of water. In coastal areas this species is found along tidal estuaries and brackish intertidal shores (Cruickshank 1942).

The Belted Kingfisher usually nests near a body of water if a suitable bank in which to excavate its nesting burrow is available. It will, however, select a site at some distance from water, such as a railroad cut, hillside, cliff, or quarry, if no suitable bank is present (Bent 1940). Nests located as much as a mile away from water have been recorded (Cornwell 1963). Although Bent (1940) also noted several instances of nests in tree cavities, none have been documented in New York.

Both the male and female dig the nesting burrow, using their bills as probes and their feet as shovels. They dig a roughly circular tunnel about 8.9–10 cm (3.5–4 in) wide and 7.6–8.9 cm (3–3.5 in) high that slopes upward from the entrance and may be as short as 0.9 m (3 ft) or as long as 4.6 m (15 ft). The egg chamber at the end of the tunnel widens out. Evidently there is no lining material brought in by the nesting pair, but the nest hole is often cluttered with fish scales, fish bones, and crusta-cean parts probably from pellets ejected by young or the female during a previous nesting (Bent 1940).

Apparently, the Belted Kingfisher did not suffer the reproductive set-backs from DDT which other fish-eating species such as the Osprey did. Smith (pers. comm.) suggested that this might have been because the kingfisher feeds on small fish and lower forms of aquatic life, such as crayfish, insects, and mollusks, which would not have had DDT con-centrations comparable to those of the larger prey fish consumed by the Osprey.

Emanuel Levine

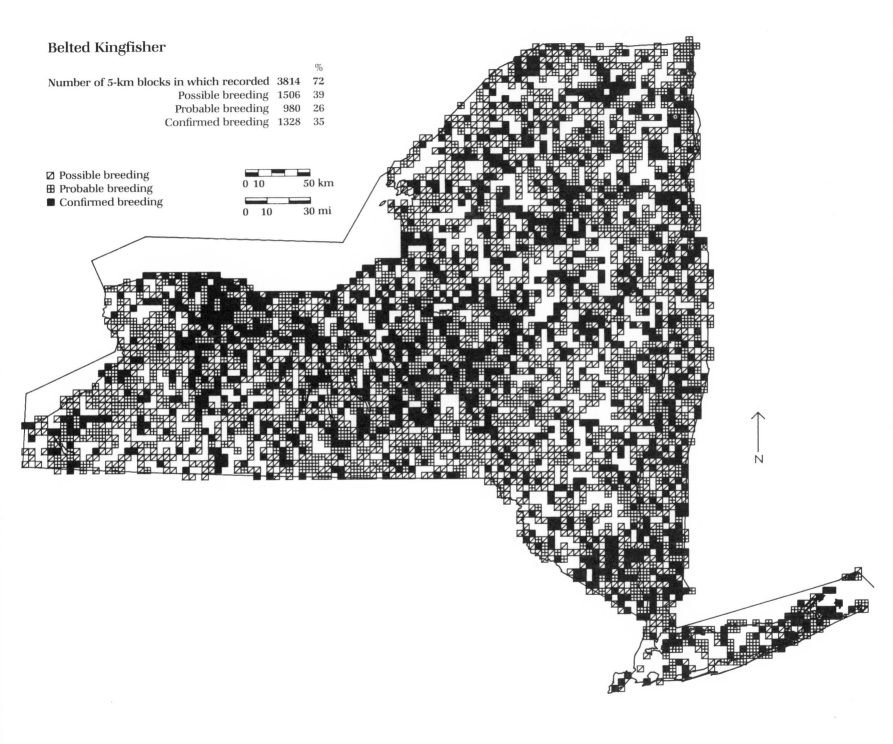

Belted Kingfisher

		%
Number of 5-km blocks in which recorded	3814	72
Possible breeding	1506	39
Probable breeding	980	26
Confirmed breeding	1328	35

◩ Possible breeding
⊞ Probable breeding
■ Confirmed breeding

0 10   50 km

0   10   30 mi

N

©KA-C 1987

Red-headed Woodpecker *Melanerpes erythrocephalus*

The Red-headed Woodpecker is one of the familiar, easily recognized birds of towns and countryside throughout much of eastern North America. In New York this species is a locally uncommon to fairly common breeding bird in the lowlands, occurring primarily below 305 m (1000 ft). In the low elevations of the southeastern part of the state, however, it is now rare.

Already in 1844 De Kay referred to this species as being less common in the Atlantic coastal areas than formerly. Eaton (1914), Bull (1974), and others reported its continuing decline, both as a breeder and a migrant, especially in southeastern New York. Many explanations are offered for this decline. When the European Starling became established, it competed with the woodpecker for nest cavities. Apparently it can oust the woodpecker from an occupied hole (Harrison 1975). Along country roads cars have taken a heavy toll on this fly-catching woodpecker (Eaton 1914; Bull 1974). Various introduced diseases have decimated some of this woodpecker's preferred trees (e.g., American chestnut and American beech). In addition, the food preferences of this species have led to conflict with humans. Eaton (1914) reported that it was the third most complained about bird species because of its assault on small cultivated fruits, including cherries, some species of apples and many berries, and also corn in milk.

Recently, however, there have been reports of small increases from southeastern New York (DiCostanzo 1983a, 1984a, 1984b, 1984c, 1985; Treacy 1985) and of a range expansion into higher elevations (Gretch et al. 1982; Mack 1982; J. Peterson 1983, 1984a). Robbins et al. (1986) found that

for the period 1965–79 numbers of Red-headed Woodpeckers significantly increased on BBS routes in the Northeast.

The center of the Red-headed Woodpecker population in New York is in the Great Lakes Plain, the adjacent Cattaraugus and Finger Lakes highlands, the St. Lawrence Plains, and the Mohawk Valley, where its preferred habitats—bottomland swamps and open woodlands—occur in abundance. This distribution generally cuts off sharply at the edge of the Appalachian Plateau, where elevations increase, except in an area southeast of Oneida Lake. Elsewhere in the state almost all of the "confirmed" breeding records are from river valleys, especially in areas north and west of the central Hudson Valley. There is only one "confirmed" record from the Adirondacks, in the Western Adirondack Foothills near Bay Pond in Franklin County.

The Red-headed Woodpecker was an easy species for Atlas observers to detect. It is noisy and aggressive when defending its breeding and food-storage territories against other avian invaders. Its striking red, black, and white color pattern was readily identified as it made fly-catching sorties from exposed perches. Atlas observers "confirmed" this species in 37% of the blocks, an average rate of confirmations for woodpeckers. The most commonly used codes were ON (adults entering or leaving a nest site), FY (adults carrying food or feeding young), and FL (recently fledged young).

The habitat preferences of the Red-headed Woodpecker are united by the requirement of open woods with dead trees. This opening can be formed by old burns, the thinning of a dying forest, tracts of trees killed by periodic or beaver-produced high water in a bottomland, or clearing or thinning the forest for residential areas, farms, or wooded parklands (Eaton 1914).

The nest of the Red-headed Woodpecker is located in a hole in a dead, barkless tree, limb, or stub, utility pole, or fence post 1.5–24.4 m (5–80 ft) above the ground and surrounded by open space (DeGraaf et al. 1980). Both sexes help excavate an opening about 4.6 cm (1.8 in) in diameter and dig out a cavity 20.3–61 cm (8–24 in) deep (Harrison 1975). No nesting material is added. The species' affinity for utility poles as nest sites has led to some conflict with humans. The introduction of creosote-treated poles is reported to have made the poles attractive to the Red-headed Woodpecker, but creosote is lethal to both the eggs and the young (Skutch 1985).

Robert G. Spahn

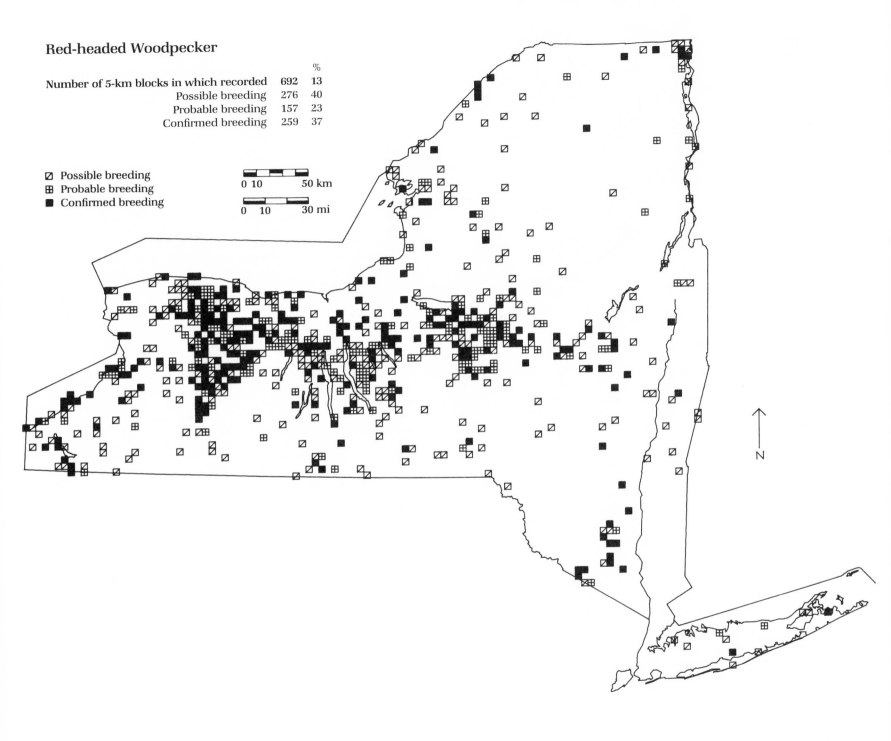

Red-headed Woodpecker

Number of 5-km blocks in which recorded 692   13 %
          Possible breeding   276   40
          Probable breeding   157   23
          Confirmed breeding   259   37

◪ Possible breeding
⊞ Probable breeding
■ Confirmed breeding

0 10   50 km
0 10   30 mi

N

©KLA-C 1986

Red-bellied Woodpecker *Melanerpes carolinus*

The Red-bellied Woodpecker resides in two distinct areas of New York: in the southeastern counties and along the north shore of Long Island, and in the Genesee River and Finger Lakes areas. Wherever it occurs it is now a fairly common, conspicuous, and colorful bird of woodlands, countryside, and suburban yards. It is seen increasingly at backyard feeders.

These two disjunct populations are shown clearly on the Atlas map. The wide separation between the two groups presents an intriguing question. How did these populations reach quite separate locations? The Long Island–southeastern group occupies the north shore of Long Island and extends up the Hudson River Valley on both sides of the river as far north as Columbia and Greene counties. A review of the regional reports in *The Kingbird* from the mid-1950s through 1985 shows that the range and numbers of this population expanded steadily.

The Genesee Valley–Finger Lakes group lies within a wide area that begins at the southeast corner of Lake Ontario, sweeps east around Oneida Lake, down below the Finger Lakes, then west of the Genesee Valley, and back to Lake Ontario at the northwest corner of Niagara County. *The Kingbird* regional reports depict an increasing population within this area, with remarks such as "now recorded in every month," "now too common to be mentioned," and "continues its long-term climb."

Between the two groups there is a wide gap, with only a few scattered breeding reports to the southeast toward the Susquehanna River water-

shed. To the southwest there are a few reports along Lake Erie and the Great Lakes Plain.

The southeastern group seems to have entered New York from northern New Jersey, where it appeared about 1955. By the late 1960s and early 1970s breeding was reported on Long Island, and its spread after that was rapid up the Hudson Valley (Bull 1964, 1974).

The origin of the Genesee Valley–Finger Lakes group is more obscure. It is clear that the Red-bellied Woodpecker has been in the area for many decades because Eaton (1914) knew of breeding records in Erie and Yates counties in the 1890s and 1910s. He considered it local and uncommon in western New York. Seeber (1963) was unable to explain its genesis, which he felt was too discontinuous from birds in Ohio to have come from there. Bull (1974) suggested it came up the Mississippi and Ohio River valleys into Illinois, Indiana, Ohio, and southeastern Michigan, and then through the Niagara Frontier corridor between Lakes Erie and Ontario into New York. Breeding has been documented in the Niagara Frontier region and also along the northern shores of Lake Erie and Lake Ontario in Canada (Ontario Breeding Bird Atlas, preliminary map). Also, this woodpecker might have come from the south following the Susquehanna and Genesee river valleys up to the Great Lakes Plain. Bull (1974) pointed out, however, that the species is a rare breeder immediately to the south in northern Pennsylvania.

This species began to increase in the 1950s, an increase that accelerated in the 1960s. This may have been due to two factors. In the 1960s residential bird feeding began to be more popular, giving the Red-bellied Woodpecker an increased food supply; second, it was during the 1950s and 1960s that elms were dying from Dutch elm disease, particularly in river swamps. These dying trees attracted wood borers and other insects, providing additional food and nesting sites. The resulting growth in population provided pressure for range expansion (Bull 1974).

In northern climates like New York's, the species' favorite breeding habitats are flooded wooded swamps, openings in mature oak forest, Appalachian oak–hickory woods along roads and near pastures, forested stream bottoms, and beech-maple mesic forest. Where large trees are near residences in towns and villages, the bird frequently spills over to become a feeder haunter but still retains its characteristic wariness (Bull 1974).

When selecting a nest tree, this bird prefers northern hardwoods, lowland hardwoods, oak, and pine (Evans and Conner 1979). Once the site is selected, both sexes begin excavating. It takes the pair about 7–10 days to dig a cavity 30.5 cm (12 in) deep through a 4.6-cm (1.8-in) entrance hole 1.5–21.3 m (5–70 ft) above the ground. The birds may use a cavity from a former year after making some repairs (Bent 1939), and they may usurp nesting cavities of other species.

Gordon M. Meade

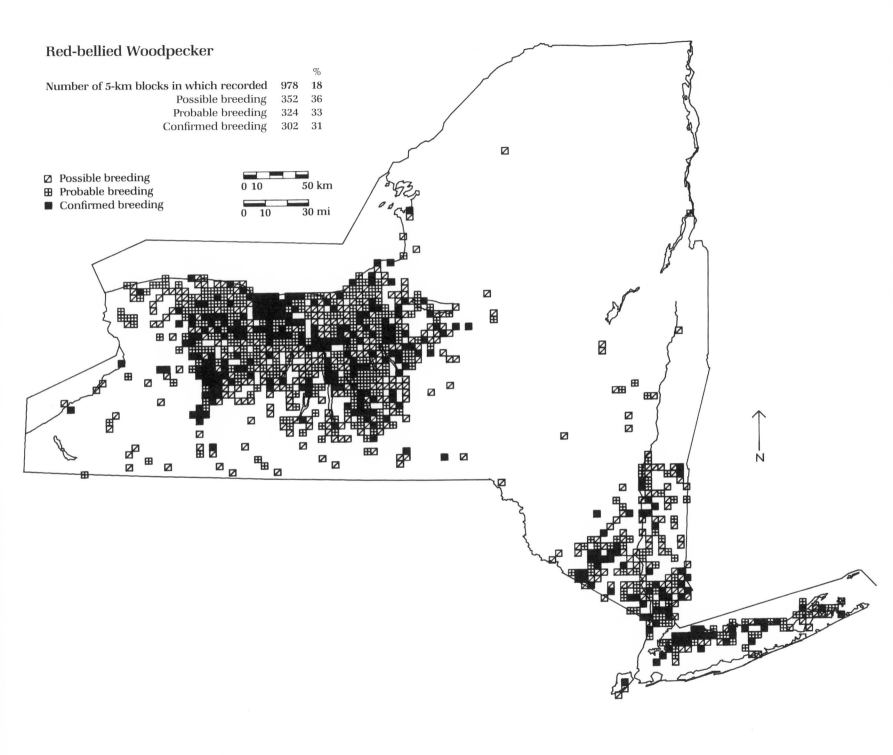

Red-bellied Woodpecker

		%
Number of 5-km blocks in which recorded	978	18
Possible breeding	352	36
Probable breeding	324	33
Confirmed breeding	302	31

◫ Possible breeding
⊞ Probable breeding
■ Confirmed breeding

0 10   50 km

0 10   30 mi

N

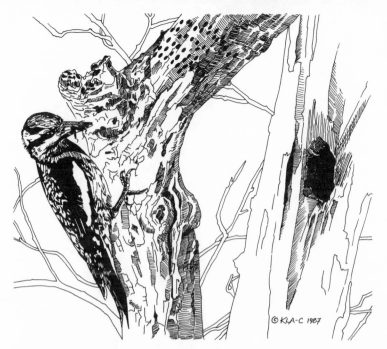

© KLA-C 1987

Yellow-bellied Sapsucker *Sphyrapicus varius*

A handsome woodpecker, the Yellow-bellied Sapsucker is a common breeder at higher elevations in much of the state but uncommon at lower elevations. Its distribution appears to have changed in the last decade. It is now almost totally absent from the Great Lakes Plain, where it formerly bred in Niagara, Orleans, and Wayne counties (Bull 1974), but has expanded into other areas.

Both Eaton (1914) and Bull (1974) reported the Yellow-bellied Sapsucker to be a species of the Adirondack and Catskill mountains. Bull also indicated that it bred at higher elevations in central New York and in the western part of the state, where according to Beardslee and Mitchell (1965) it was uncommon. Saunders (1942) included this species in his list for the Allegany State Park and called it an uncommon summer resident. This species appeared to be very rare in some areas where it is now established. In 1948 a nest was found in Cayuga County, the first documented nesting there since 1880 (Benton 1950). The Atlas shows several blocks in that county where breeding was "confirmed." The spread of this species into lower elevations of the state was hardly noticed until Atlas fieldwork began. The only mention in *The Kingbird* was made by Pitzrick (1977), who reported that adult sapsuckers were feeding young near Conewango Creek, Cattaraugus County, at an elevation of about 392 m (1285 ft), "lowest to date in the county."

The Atlas map illustrates a population spreading out from its former breeding areas in the Catskills and Adirondacks; it also shows a strong presence all along the Appalachian Plateau. Most records are at elevations above 152 m (500 ft), although there are numerous areas where it occurs at lower elevations. For example, on the St. Lawrence Plain breeding was documented at elevations under 152 m (500 ft) that were described by Walker (pers. comm.) as "wooded tongues of the Adirondacks." It does not now breed and has not been known to breed south of the Catskills, Neversink Highlands, and Mongaup Hills.

Since the sapsucker is anything but secretive in the breeding season, this change probably represents a true expansion and is not just a reflection of the intensity of Atlas fieldwork. McIlroy and Lehman (pers. comms.) indicated that new habitat for this species was provided when the state acquired much hilltop land that has been managed for forest growth, and also when abandoned subsistence farmland reverted to woodland.

The Yellow-bellied Sapsucker was easily located by Atlas observers, particularly after the eggs hatched. Some observers termed this woodpecker's nest tree a "talking tree," referring to the noise of the food-begging young in the tree (Carroll and Peterson 1984). Nests with young were found in 178 blocks, 25 more than the much more common Hairy Woodpecker.

Breeding takes place in either deciduous or mixed deciduous and evergreen forests but within the forests and not in clearings (Bull 1974). The nest hole is freshly excavated each year, mostly by the male (Kilham 1977). The nest tree may be dead or living but with decaying heartwood. Kilham (1971a) found that in New Hampshire the nest tree most frequently used by the sapsucker was mature aspen infected with the fungus conks of the false tinder. This fungus causes decay in the heartwood but does not affect the sapwood, which provides a strong shell to protect the nest cavity. The diameter of the hole is only 3.3–3.8 cm (1.3–1.5 in), which hardly seems big enough for the bird (Harrison 1978). Kilham (1977) observed a female that had difficulty entering the nest through the nest hole and even more difficulty getting out. The smooth-sided excavation is 10.2–12.7 cm (4–5 in) across and 15.2–35.6 cm (6–14 in) deep, with no lining in the bottom other than a few wood chips; it may be anywhere from 1.5 to 18.3 m (5–60 ft) above the ground (Harrison 1978). The young are fed both sap and insects that are attracted to the sap.

On its breeding ground the Yellow-bellied Sapsucker may inflict lasting damage to trees in the area of the nest. In addition to drinking the sap, it also eats the cambium layer of the tree through which it drills. Sometimes smaller trees are completely girdled and die (Tyler 1939). Birds of 35 species have been recorded at sapsucker pits, either drinking sap or catching insects attracted to the sugar-rich exudate (Skutch 1985).

Emanuel Levine

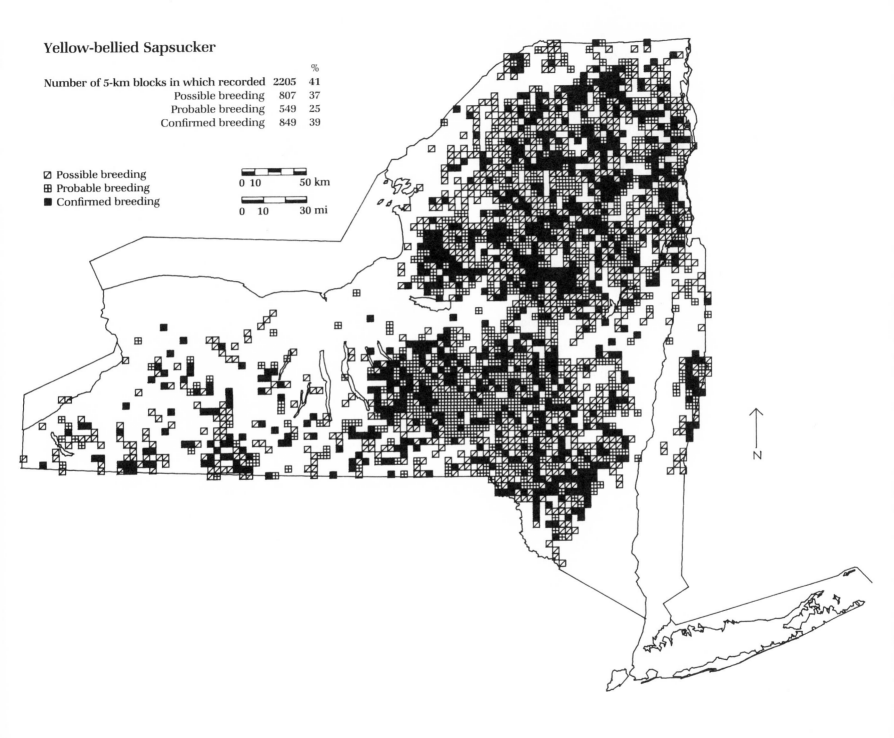

Yellow-bellied Sapsucker

		%
Number of 5-km blocks in which recorded	2205	41
Possible breeding	807	37
Probable breeding	549	25
Confirmed breeding	849	39

◨ Possible breeding
⊞ Probable breeding
■ Confirmed breeding

0 10   50 km

0 10   30 mi

N

©KLA-C 1987

Downy Woodpecker *Picoides pubescens*

The Downy Woodpecker is a widely distributed breeder and is fairly common to common throughout most of New York. In fact, this smallest of North American woodpeckers is fairly common throughout most of this continent and is absent only from the tundra/alpine regions and the southwestern deserts.

The range of the Downy Woodpecker has remained nearly constant; however, its abundance probably increased during this century. In New York much farmland was abandoned during the last 100 years. The regrowth of forests throughout much of New York, as documented for Tompkins County by Caslick (1975), has increased nesting habitat for this bird. The amount of suitable habitat and therefore the density of this bird's population is probably greater now than at any other time during this century.

This woodpecker can be expected to be found breeding throughout all deciduous forests of New York and was recorded during Atlas coverage in areas at a variety of elevations and with considerably varying forest cover. Eaton (1914) and Bull (1974) also reported the woodpecker as very widely distributed in New York. Areas with fewer records of breeding are the western end of the Appalachian Plateau, and northern New York, especially the Adirondacks. Except for the mountain spruce–fir forests of the Adirondacks and areas with little forest cover, the Downy Woodpecker could be expected to occur in most blocks. Its absence in some areas may be correlated to lack of intense Atlas coverage.

Locating the Downy Woodpecker was most easily done either just before nesting, when the birds noisily enforced territory boundaries, or after nesting, as they moved about with young. Atlas observers were able to "confirm" the Downy Woodpecker more often than the Hairy Woodpecker, probably because it is found more often around populated areas. Over two-thirds of the "confirmed" records were of fledglings, and one-third were of nests.

The habitat of the Downy Woodpecker is diverse, as indicated by its nesting in many forest types throughout North America. It requires only a moderate area of some sort of deciduous vegetation, including pine-oak (Conner 1981) and hemlock–northern hardwood (Kisiel 1972) forests. However, pure pine plantations (Conner 1981) or mountain spruce–fir forests (Kisiel 1972) are marginal habitats. It nests in abandoned orchards, young and mature forests, and along wooded stream banks in fields. The Downy Woodpecker is quite tame and often nests close to human activity. Flexibility in behavior and habitat selection undoubtedly have contributed to the wide distribution and moderate abundance of this bird.

The Downy Woodpecker forages on smaller diameter stems and higher in the trees than the Hairy and Pileated woodpeckers and the Yellow-bellied Sapsucker. It taps at the surface more often but drills into the subcambium and excavates less often than any of the others (Lawrence 1967; Kisiel 1972; Conner 1981).

Jackson (1970) observed that male Downy Woodpeckers forage on smaller branches and on smaller live trees than do the females. If food supply is limited within a breeding territory, this difference between sexes may increase the amount of food a pair can bring to their young by reducing competition.

Some studies have suggested that the Downy Woodpecker learns to adapt its feeding to changing conditions so that rates of obtaining food are maximum (Lima 1984). Confer and Paicos (1985) analyzed feeding on insects in goldenrod galls. Woodpeckers frequently used the fly larva exit tunnel in some galls to extract it. If they failed to find an exit tunnel the gall was likely to be abandoned, without the woodpeckers' spending time to drill a new hole.

Nest building is shared by both sexes (Harrison 1975) and may serve to form or reinforce the pair bond. Usually it is the female that selects and initiates excavation of the nest cavity (Lawrence 1967). Nests are located from 1.5 to 15.2 m (5–50 ft) above ground. Usually the nests are drilled in dead wood. The entrance hole has a diameter of about 3.2 cm (1.25 in) and the cavity is 20.3–30.5 cm (8–12 in) deep. Both sexes share in the incubation and feeding of young.

John Confer

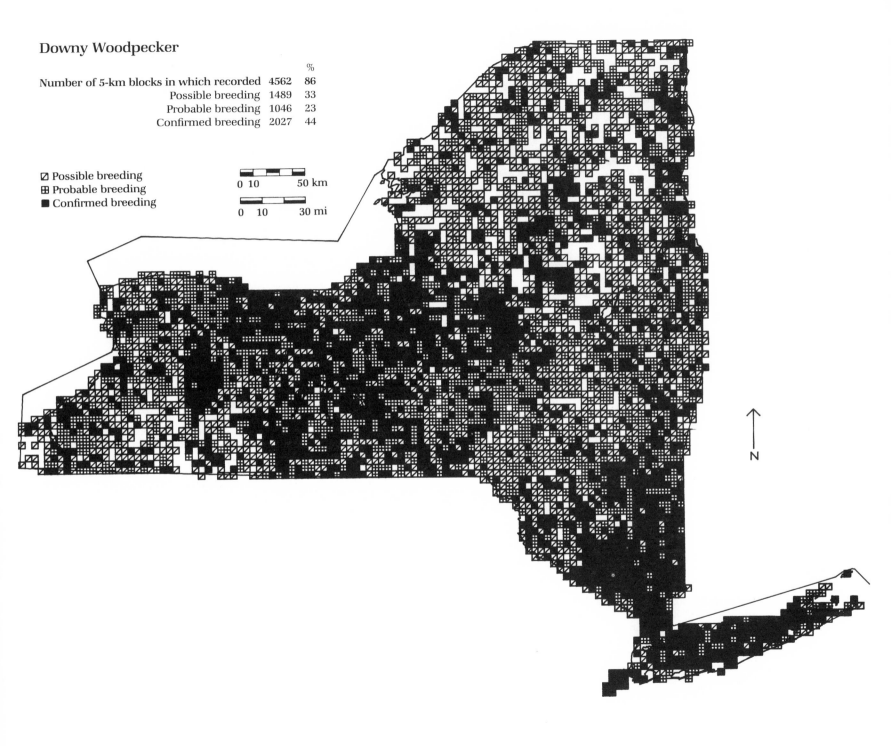

Downy Woodpecker

		%
Number of 5-km blocks in which recorded	4562	86
Possible breeding	1489	33
Probable breeding	1046	23
Confirmed breeding	2027	44

☑ Possible breeding
⊞ Probable breeding
■ Confirmed breeding

0 10   50 km

0 10   30 mi

N

©KLA-C 1987

Hairy Woodpecker *Picoides villosus*

The Hairy Woodpecker, nearly identical to but larger than the closely related Downy Woodpecker, is a fairly common breeder in wooded areas throughout the state. A bird of extensive woodlands, it would be expected that this species probably decreased statewide during the widespread forest clearing of the 1800s, and then slowly increased during the 1900s as farms were abandoned and allowed to grow back to woods. However, the literature (Eaton 1914; Bull 1974) provides no evidence of such fluctuations. This species has apparently remained fairly common throughout the past century or more.

The Hairy Woodpecker is found in all regions except the New York City area, the outer south shore of western Long Island, and the heavily agricultural sections of the Eastern Ontario and St. Lawrence plains, where few extensive woodlands exist. In western New York the scattered distribution may be a factor of coverage. Around New York City the Hairy Woodpecker is almost completely absent from Bronx, New York, Kings, and Queens counties.

This species tends to be very quiet and inconspicuous during May and June, when the nest cavity is being excavated and eggs laid and incubated (Kilham 1983). During that period it is hard to locate and even more difficult to "confirm" breeding. Despite this, the Hairy Woodpecker was "confirmed" in one-third of all the Atlas blocks in which it was reported, with 42% of the "confirmed" records of recently fledged young. Young birds accompany their parents for several weeks after fledging and often show up at bird feeders at this time. Also, well-grown young make a great deal of noise in the nest cavity begging for food, and nests are fairly easy to locate at this stage, as is the conspicuous behavior of adults carrying beakfuls of food to the young.

The Hairy Woodpecker is more a bird of extensive woodlands than is the Downy Woodpecker, but it is still found in nearly all forest habitats in the state, including suburban neighborhoods, woodlots, and the edges of clear-cuts and fields. Conner and Adkisson (1977) found that the Hairy Woodpecker accepted a wide range of habitats, including clear-cuts with standing dead trees. This adaptability may have been what allowed this species to remain fairly common throughout the forest clearing of the 1800s and early 1900s.

Bull (1974) described the Hairy Woodpecker's preferred nesting habitats as "extensive tracts of forest with plenty of large trees, dead stubs, and fallen logs. In such places it is found in mountains, river bottoms, and wooded swamps." Saunders (1942) considered it a bird of the extensive beech-maple mesic and Appalachian oak–hickory forests of Allegany State Park, as opposed to the second-growth forests, woodlots, and open areas occupied by the Downy Woodpecker. However, habitat used by the Hairy Woodpecker is more like that used by the Downy Woodpecker than that used by the Northern Flicker, Pileated Woodpecker, or Red-headed Woodpecker (Conner and Adkisson 1977). Extensive woodlands are the only habitat in which this species is usually more common than the Downy Woodpecker.

The Hairy Woodpecker nests in tree cavities that it usually excavates in a dead tree or dead limb of a living tree infected with fungal heart rot (Conner et al. 1976). The nest hole is located from 1 to 18.3 m (3–60 ft) above ground (Eaton 1914; Bull 1974) and is circular or slightly elliptical (higher than wide). Of 26 nests in Pennsylvania, the average height above ground was 9.1 m (30 ft). The Hairy Woodpecker shows no preference for nesting in any one tree species (Bent 1939).

The interior dimensions of the nest were described by Eaton (1914). The nest opening is "about 2 inches in diameter, leading backward 2 or 3 inches through the solid wood, then downward for 8 to 16 inches, where the cavity is enlarged and a few chips left on the bottom as a bed for the eggs." No nesting material is used or brought to the nest cavity (Bent 1939).

The male plays the principal role in selecting the nest site and does the majority of the excavation, although both sexes are involved (Lawrence 1967). Excavation of the nest takes from 1 to 3 weeks, depending upon the hardness of the wood (Bent 1939).

Steven C. Sibley

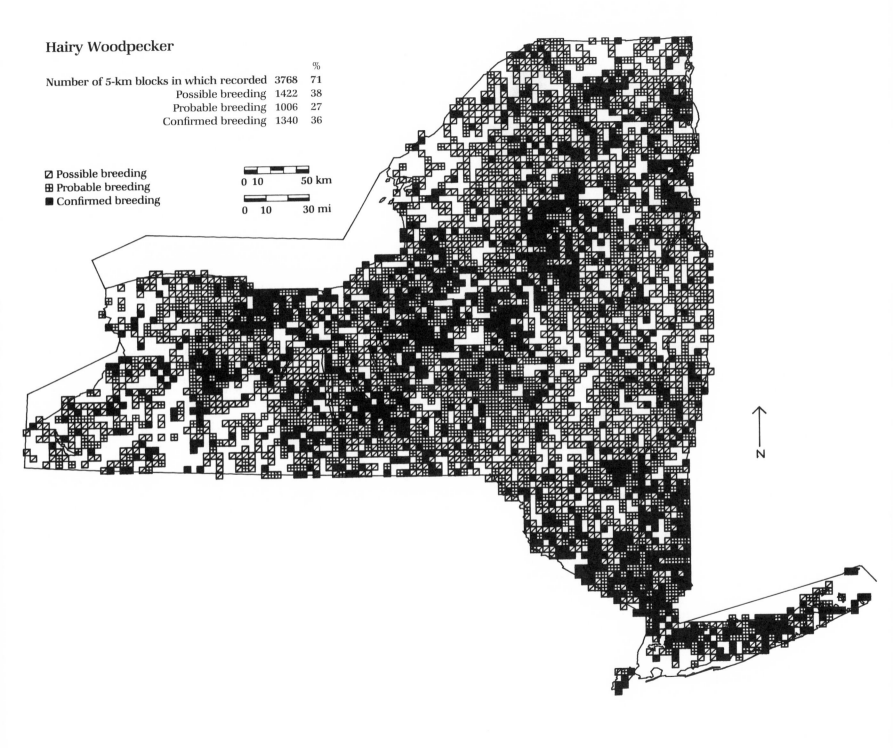

Hairy Woodpecker

Number of 5-km blocks in which recorded		%
Number of 5-km blocks in which recorded	3768	71
Possible breeding	1422	38
Probable breeding	1006	27
Confirmed breeding	1340	36

◩ Possible breeding
⊞ Probable breeding
■ Confirmed breeding

0 10   50 km

0 10   30 mi

N

Three-toed Woodpecker *Picoides tridactylus*

The Three-toed Woodpecker is a rare breeder in conifer forests of the Adirondacks, and certainly the least abundant woodpecker nesting in New York.

Between 1874 and 1877 Roosevelt and Minot (1929) found the "Banded-backed Woodpecker," as they called it, much less common than the Black-backed Woodpecker in Franklin County. About the same time Merriam (1879) wrote: "So far as I am aware this rare woodpecker is only found along the eastern border of Lewis County, in the Adirondack region, where it is a resident species; and even here it is much less common than its congener, the Black-backed Woodpecker." In 1883 he found "numerous" nests, most bordering the flooded inlet of Seventh Lake, Hamilton County. Eaton's party found it breeding in the Adirondack High Peaks during 1905, from 610 m (2000 ft) at the Marcy Swamp to 1219 m (4000 ft) on the slopes of Mount Marcy, Essex County (Eaton 1914). He judged it to be "quite uniformly distributed, but . . . less common than the Black-backed Woodpecker, evidently about one-half as common as that species," although in an earlier appraisal of the same trip he found it "nearly as common as the Black-backed species" (Eaton 1910). More recently Bull (1974) noted that the Three-toed Woodpecker had been found breeding in just half as many localities (9) as the Black-backed Woodpecker (18 known locations). Both Saunders (1929a) and Bull (1974) indicated that the Three-toed Woodpecker had become scarcer.

Atlasers found the species in 22 blocks, and in many of those blocks Three-toed Woodpeckers were observed in several locations. Within New York the Three-toed Woodpecker is restricted to the Adirondack High Peaks, Central Adirondacks, Sable Highlands, Western Adirondack Foothills, and Central Tug Hill. Although certainly rare, the Three-toed Woodpecker is probably under-recorded, since only cursory coverage of prime habitat was possible in many remote wilderness areas. Moreover, the bird is rather quiet, especially during incubation, and taps softly when feeding.

Throughout its holarctic breeding range this species exhibits a variety of plumages (Scott 1983; Bull 1974). The western North American subspecies is more white-backed or barred, while the *bacatus* subspecies in the Adirondacks can be quite black on the back, showing only partial barring or a few flecks of white at times, as LaFrance (1973) described. It is possible that at least some dark-backed Three-toed Woodpeckers were incorrectly recorded as Black-backed Woodpeckers by Atlasers.

Within the mountain spruce–fir and spruce–fir–northern hardwood forest communities, the Three-toed Woodpecker prefers areas where dead timber remains after fires or logging. It is also commonly found in swamps and bogs with scattered standing trees (DeGraaf et al. 1980). Whether near black spruce bogs or among the mountain spruce–fir forest of the Adirondack High Peaks, the Three-toed and Black-backed Woodpeckers occupy the same habitat. Bull (1974) suggested that interspecific competition may be reduced by different foraging heights, or that the smaller, lighter bill of the Three-toed Woodpecker may be adapted "for probing less deeply for its food and for digging into softer or rotten wood." Carleton (1980) noted that the Three-toed Woodpecker "prefers smaller live and dying balsams and spruce trees in thick stands on mountain slopes or in high altitude swamps" of Essex County, which suggests that a recent die-off of red spruce may be exploited by this woodpecker and its black-backed congener, should bark beetles and borers attack these dying stands.

The first New York nest, and apparently first North American nest, of the Three-toed Woodpecker was found by C. L. Bagg and C. Hart Merriam on 4 June 1878 (Merriam 1878b). The location was apparently just north of McKeever in the Town of Webb, Herkimer County. The nest hole was 2.4 m (8 ft) high in a spruce, the opening 3.8 cm (1.5 in) in diameter, and the cavity 25.4 cm (10 in) deep. Evans and Conner (1979) found that this woodpecker uses both hard and soft snags. They found that nest tree heights averaged about 9 m (30 ft) with an optimum dbh of 35 cm (14 in).

John M. C. Peterson

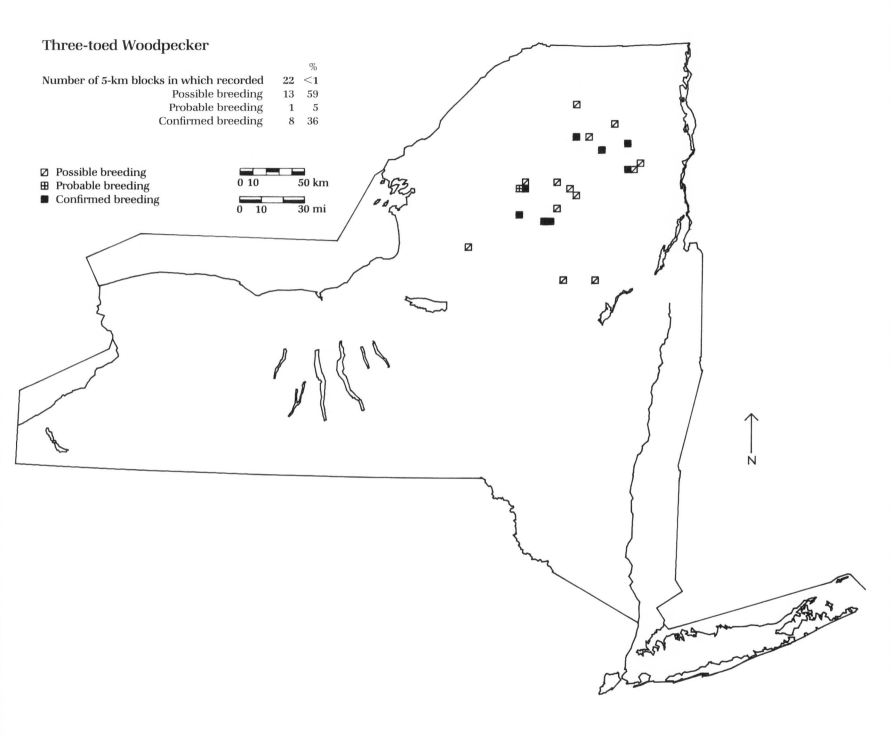

Three-toed Woodpecker

		%
Number of 5-km blocks in which recorded	22	<1
Possible breeding	13	59
Probable breeding	1	5
Confirmed breeding	8	36

◩ Possible breeding
⊞ Probable breeding
■ Confirmed breeding

0 10   50 km

0 10   30 mi

N

Black-backed Woodpecker *Picoides arcticus*

Widely distributed over the Adirondacks, where it was recorded in 114 Atlas blocks, the Black-backed Woodpecker is a fairly common nesting species of balsam fir and spruce forests. Within portions of its range where hardwoods dominate and suitable conifers exist only in localized pockets, this woodpecker is rather uncommon.

No historical change in distribution or abundance of the Black-backed Woodpecker in New York has been documented, although the effects of lumbering, fires, insect infestations, tree diseases, artificial impoundments, and beaver activity must have had considerable local impact. Roosevelt and Minot (1929) described the species as "Common; second in numbers to the Hairy only" in the 1870s. Following the turn of the century Eaton (1914) continued to find it "fairly common in all portions of the spruce and balsam belt of the Adirondacks, there ranking next to the Sapsucker and Hairy woodpecker in abundance and probably much more plentiful than the American three-toed woodpecker and the Downy woodpecker." Achilles (1906) also called it "fairly common" in denser spruce and balsam above 914 m (3000 ft). More recent authors have described the Black-backed Woodpecker as "uncommon" (Bull 1974; Beehler 1978; Carleton 1980). Although descriptive terms differ, there is no evidence that the population has actually changed.

Local increases in the Black-backed Woodpecker population may have occurred, however, in response to massive infestations of the wood-borer and bark-borer that follow in the wake of forest maladies such as spruce budworm infestations, or forest fires, flooding, and other disasters that kill conifers. Fresh food sources may now be increasing for this opportunist as high-altitude conifers die off.

In the Northeast the Black-backed Woodpecker nests in Maine, New Hampshire, and Vermont, as well as northern New York (AOU 1983). As the Atlas map indicates, its range in New York is confined to the Central Adirondacks, Adirondack High Peaks, Sable Highlands, and Western Adirondack Foothills, with two records from the Central Tug Hill. As with many boreal species, the Black-backed Woodpecker occupies a habitat that is often remote. Because of its often local occurrence, especially in boggy pockets or on slopes of remote peaks, this can be a difficult bird to find under any circumstances. The habits of this woodpecker made the Atlas survey more difficult. Observers usually were not surveying early in the season when drumming was most prevalent. By the time Atlasers started fieldwork, many pairs were already sharing incubation duties. Even when feeding, this woodpecker is known as a soft tapper, remaining motionless for long periods, camouflaged against a dark conifer trunk. The map is undoubtedly incomplete, though generally correct in outline. The possibility is also strong that dark-backed Three-toed Woodpeckers could at times have been mistaken for this species.

The Black-backed Woodpecker is found in both mountain spruce–fir and spruce–fir–northern hardwood forests where blowdowns or disease provide its rather specialized diet of wood-boring larvae and bark-borers that flourish only in recently killed timber (Van Tyne 1926; AOU 1983). As Yunick (1985) observed, "These birds are not harbingers of ecological tranquility." Most of the Atlas observations were near wet areas, either marshy or boggy places near a beaver pond, lake, stream, or river or in swampy openings. Typical feeding is done rapidly but not rhythmically; it leaves a random pattern of irregularly spaced holes, the size and depth dependent upon the location of the wood-borers (Chapman, pers. comm.).

The first Black-backed Woodpecker nest in New York appears to have been found near Inlet, Hamilton County, on 27 May 1883 by C. Hart Merriam. Construction of a dam at Sixth Lake had raised the water of Seventh Lake, and the nest containing 1 fresh egg was in a dead spruce left standing in 1.8 m (6 ft) of water, the opening 1.5 m (5 ft) above the surface (Bendire 1895; Bent 1939). Atlas observers recorded Black-backed Woodpecker nests with young in ten blocks, occupied nests in an additional four, and some blocks had multiple nesting records. A nest hole in Newcomb was drilled 6.1 m (20 ft) high in a sound, treated power pole; 2 other nests were in live balsam trees in wooded campsites, the holes surrounded by debarked areas. Other nests were in a cedar near a boggy lake shore (A. and W. Chapman, pers. comm.) and in a dying spruce in a closely packed spruce stand (D. Niven, v.r.), and several were in old, dead conifer stubs. Bull (1974) also mentioned nests in tamarack and white birch, noting heights from 1.2 to 12.2 m (4–40 ft) above the ground.

John M. C. Peterson

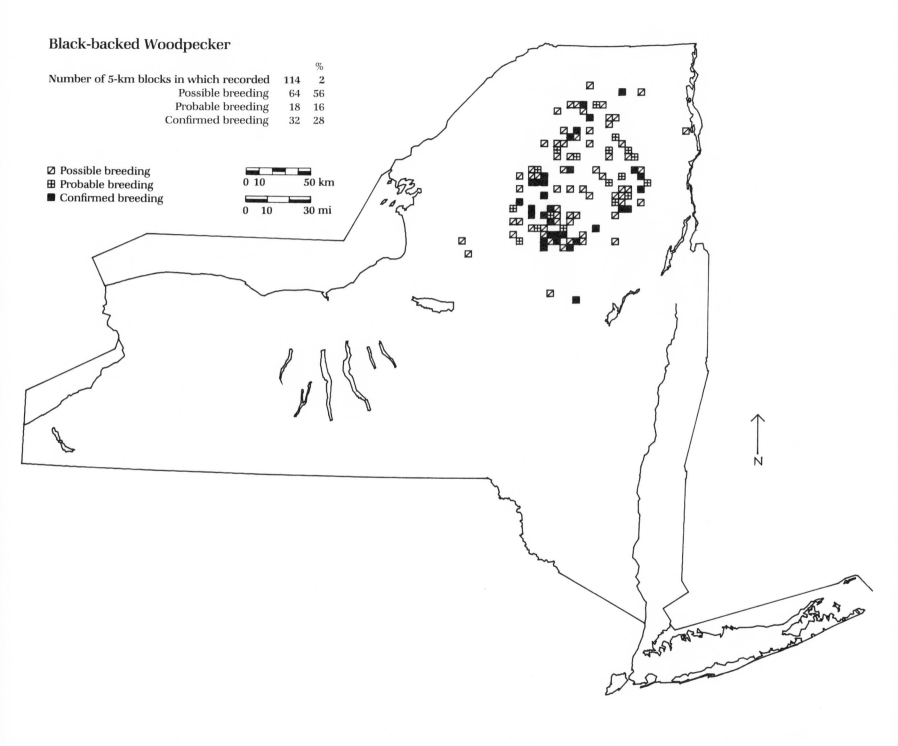

Black-backed Woodpecker

		%
Number of 5-km blocks in which recorded	114	2
Possible breeding	64	56
Probable breeding	18	16
Confirmed breeding	32	28

☒ Possible breeding
⊞ Probable breeding
■ Confirmed breeding

0 10   50 km

0 10   30 mi

N

©KLA-C 1986

Northern Flicker *Colaptes auratus*

Known until recently as the yellow-shafted flicker, the Northern Flicker has also been called yellowhammer and golden-winged woodpecker for the bright yellow color of the shafts of its wing and tail feathers. This large woodpecker, a common breeder throughout the state, is probably the most conspicuous of the nine species of woodpeckers known to breed in New York. Eaton (1914) described it as a common summer resident and "one of our dominant species." However, although still common and widespread, the Northern Flicker has been declining in numbers in recent years. Significant declines were noted in 20 states and provinces, including New York, on BBS routes from 1965 to 1979. This decline is considered largely due to the European Starling, which competes with the Northern Flicker for nest holes (Robbins et al. 1986). In New Hampshire starlings took over newly excavated flicker holes so regularly that a behavioral study of the Northern Flicker was moved to a more wooded site where no starlings were present (Kilham 1983).

The Northern Flicker is still one of the most widespread breeding species in New York. Only 13 other species were recorded in more blocks during the Atlas survey. The largest gaps in its distribution, at the northern and extreme western portions of the state, are forested areas above 610 m (2000 ft) in elevation. In the Adirondacks these gaps correspond with spruce–fir–northern hardwood forest.

This species was easily located by Atlas observers because it prefers open areas and gives its loud, distinctive calls frequently throughout the spring and early summer. Breeding was also rather easily "confirmed" because of the bird's abundance and conspicuousness. Two codes, ON

(adults seen entering or leaving a nest site) and FL (fledglings), accounted for 67% (1544) of all "confirmed" records.

There is almost no habitat in the state where the Northern Flicker is not found. It nests everywhere from wilderness woodland to city parks and woodlots in agricultural areas (Eaton 1914) but is most abundant in woodlands and wood edges (Stauffer and Best 1980) with open areas for foraging (DeGraaf et al. 1980).

The Northern Flicker typically nests in a tree cavity that it excavates, although it will often use a nest box or an old cavity (Kilham 1983). It has also been known to excavate holes in hay stacks, telephone poles, and buildings, and to use old Belted Kingfisher burrows (Bent 1939) or even underground hollows (Dennis 1969). One pair nested on the ground in open sand dunes on Fishers Island, Suffolk County, in June 1916 (Pearson 1916).

New cavities, usually excavated each year, take 5–19 days to complete (Lawrence 1967). The male selects the site and performs most of the excavation (Kilham 1983). The flicker shows little preference for any tree species for nesting and usually selects dead trees or limbs but is more likely than other woodpeckers to nest in live trees (Stauffer and Best 1980). It generally locates the nest anywhere from 1.2 to 12.2 m (4–40 ft) from the ground (Bull 1974), but if taller trees are available, it may choose a location as high as 18.3 m (60 ft) (Kilham 1983). Most nests are located near the top of broken-off stubs of dead trees (Lawrence 1967) in forest edges or groves bordering fields (DeGraaf et al. 1980). The Northern Flicker, like other woodpeckers, invariably nests near the top of the trunk, where there is less rain runoff into the cavity, and if the trunk is not exactly vertical, on the side angled downward, where rain is much less likely to fall into the nest hole (Conner 1975).

Nest cavities range from 17.8 to 91.4 cm (7–36 in) deep; the opening, from 5.1 to 10.2 cm (2–4 in) in diameter (Dennis 1969; Bent 1939). The cavity is widened at the bottom, where it has an average diameter of 19.5 cm (7.67 in) (Burns 1900). No material is added to the nest to form a lining, but enough wood chips are left in the cavity to form a soft bed for the eggs (Bent 1939).

Steven C. Sibley

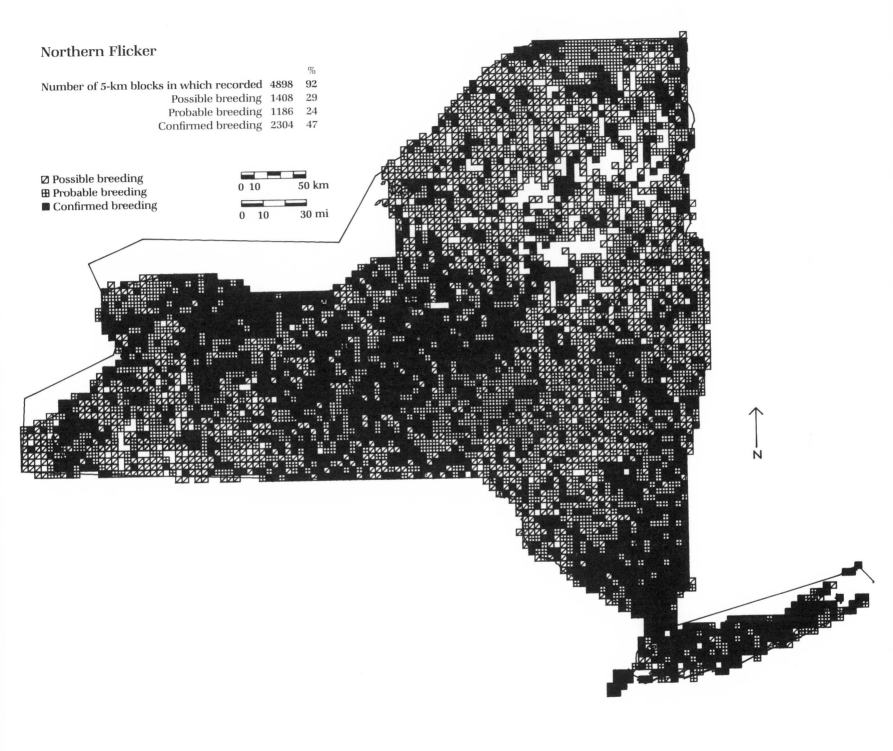

Northern Flicker

Number of 5-km blocks in which recorded	4898	%
		92
Possible breeding	1408	29
Probable breeding	1186	24
Confirmed breeding	2304	47

◩ Possible breeding
⊞ Probable breeding
■ Confirmed breeding

0 10   50 km

0 10   30 mi

N

©KLA-C 1986

Pileated Woodpecker *Dryocopus pileatus*

Notable for its large size and striking plumage, the Pileated Woodpecker commonly reveals its presence by the large rectangular feeding holes it excavates in tree trunks. Because it is adept at keeping out of sight and is silent much of the time, even experienced observers are thrilled to encounter one. This woodpecker varies from being very uncommon and local to fairly common in New York. The Pileated Woodpecker probably declined drastically following settlement in the Northeast. Land clearing and logging with accompanying fires removed its habitat, and shooting also was quite common. It has recovered considerably in this century with forest regrowth, protective laws, and also because it has adapted to nonwilderness areas (Christy 1939; Short 1982).

De Kay (1844) described this woodpecker as abundant in uncleared forests but almost unknown in the coastal areas of the state. Eaton (1914) stated that the Pileated Woodpecker was mainly confined to the Adirondacks and Catskills and was not common there. He thought it would never return to most other areas of the state, although he suggested it was beginning to reestablish itself in some localities. Pearson (1917) and others in the early 1900s believed it unable to adapt to new conditions and thought it might not survive the loss of virgin forests, but by the 1920s a rebound was noticeable, and Bull (1974) called its recovery since Eaton's time phenomenal.

Atlas observers found the Pileated Woodpecker in every county from Westchester northward, but it did not breed on the Coastal Lowlands. It was frequent in, but not limited to, heavily forested and mountainous regions of the state. Although there are noticeably fewer records from

such low-elevation areas as the Great Lakes and St. Lawrence plains and Mohawk Valley, there were many records from the Hudson and Lake Champlain valleys. Even highly agricultural, suburban, and some urban areas were home for a few pairs in localities with suitable habitat. This widespread distribution generally resembles recent findings in Vermont (Norse 1985a).

As the Pileated Woodpecker increased, it may have lost some of its former wariness (Hoyt 1957), but Atlas observers heard its calls or drumming more often than they actually saw the bird. It was a rather difficult species to "confirm," with extra time and perseverance usually required to locate the nest site. But during routine observations, young recently out of the nest were occasionally noted, sometimes at the edge of a clearing or in semi-open areas. Probably because of its territory size and type, this woodpecker was "confirmed" much less frequently than any other woodpecker species.

The Pileated Woodpecker inhabits mixed coniferous-deciduous forests, deciduous forests, second-growth woods that have some large trees, and even suburban parks (Short 1982). Various forest communities are suitable. In eastern New York the author observed them in hemlock–northern hardwood, spruce–fir–northern hardwood, pine–northern hardwood, and floodplain forest. It favors valleys, especially near water. In Tompkins County near Ithaca, Hoyt (1957) found few if any nests far from water. She mentioned nest trees in swampy areas and others near a stream or lake. Conner et al. (1975) found no Pileated Woodpecker nest trees over 150 m (492 ft) from water, and most less than 50 m (164 ft). Most nest cavities are excavated in large dead trees, often tall stubs, in dead parts of living trees, and rarely in live wood. The nest tree is typically in shade in mature, dense forest. Christy (1939) listed records of one nest each in hemlock and pitch pine, and 30 in trunks of various deciduous trees.

The nest hole is excavated by the pair, the male doing most of the work (Hoyt 1957). Christy (1939) stated that the entrance commonly, though not always, faces east or south, but Hoyt did not find any preference. Conner (1975) found that Pileated Woodpecker nest hole entrances face slightly downward; the slope of the trunk appeared to be the most important factor in nest orientation. He suggested that openings facing even slightly downward tend to protect the nest from rain. New nests are dug each year, and occasionally a hole is started and then abandoned and a new nest dug.

Paul F. Connor

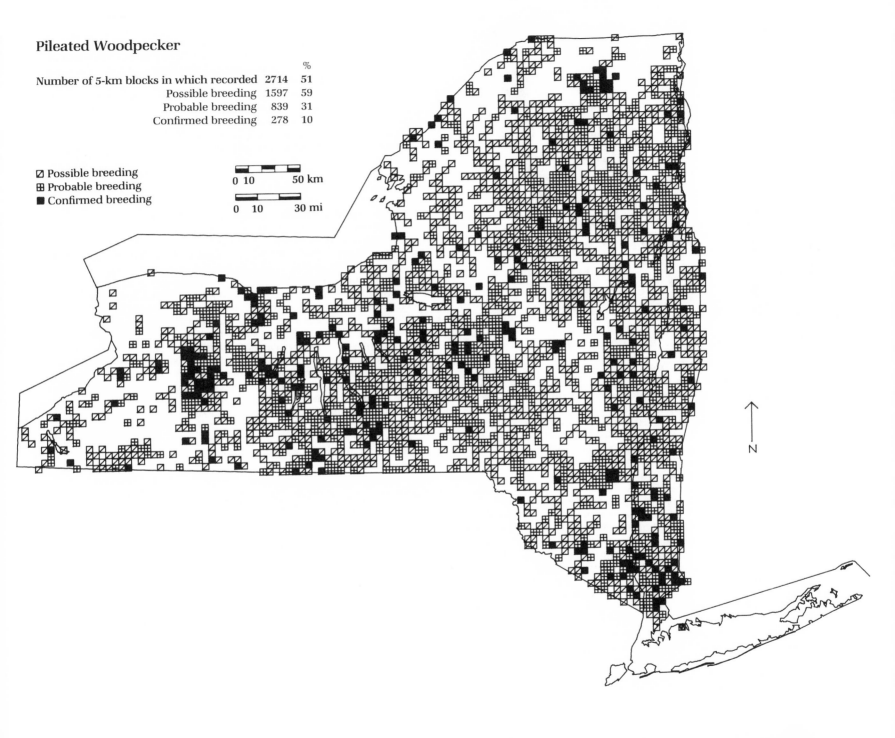

Pileated Woodpecker

Number of 5-km blocks in which recorded	2714	% 51
Possible breeding	1597	59
Probable breeding	839	31
Confirmed breeding	278	10

◨ Possible breeding
⊞ Probable breeding
■ Confirmed breeding

0 10   50 km

0 10   30 mi

Olive-sided Flycatcher *Contopus borealis*

The distinctive "quick-three-beers" whistle of the Olive-sided Flycatcher is a familiar summer sound over the Adirondack region, where the species is fairly common, although its distribution is often thin or quite local. In the Catskill Peaks and in the Central Tug Hill this is a rather uncommon flycatcher, also local, but it is rare elsewhere in the state.

Except for a small, isolated population found by Atlas observers in the Rensselaer Hills, no significant changes have been recorded over the past century. Roosevelt and Minot (1929) thought the bird was not uncommon around the St. Regis lakes, Franklin County, between 1874 and 1877. To Merriam (1881), the Olive-sided Flycatcher was a common summer resident of the Adirondacks, breeding about the middle of June. Eaton (1914) declared it a fairly common summer resident of the Adirondacks and Catskills, "breeding from an altitude of 1500 feet to the highest portions of the mountains," and mentioned records of breeding from Madison County and not far from Albany. Saunders (1929a) saw a number of Olive-sided Flycatchers in Essex County in 1925 but failed to find any in 1926. During the same period he found them to be quite common in neighboring Franklin County. The Atlas map is strikingly similar to that presented by Bull (1974), except for additional records in Fulton and Rensselaer counties.

Given the difficulties of access to Olive-sided Flycatcher habitat, the map is quite good, although certainly incomplete. Observers found that the Olive-sided Flycatcher favors small boggy ponds, swampy ends of lakes, marshy streams, wet backwaters of rivers, quaking bogs, and old beaver meadows. Most areas had dead standing trees, which it uses as

singing and feeding perches, and were usually bordered by forest. Some birds were located by following a remote brook to its source at a tiny mountaintop pond, where the only Olive-sided Flycatcher in the block would be found. The surrounding forest was usually coniferous or mixed, with black spruce frequently mentioned in northern areas, red spruce farther south, as well as balsam fir, tamarack, and hemlock. Bordering shrubs were commonly speckled alder in the north, rhododendron and mountain laurel in the south. Several Catskill observers noted an Adirondack flavor provided by sphagnum mosses and pitcherplants.

In Rensselaer County the species was found mainly in bogs at higher elevations and where there are extensive swamps with dead trees. According to Paul Connor (pers. comm.), "This area apparently presents marginal habitat on the fringe of its breeding range."

Elevations where the Olive-sided Flycatcher was found were lower than Eaton (1914) mentioned, usually between 168 m (550 ft) and 842 m (2764 ft). Early in this century he found that the burned lands of the Adirondacks, slashings, and the borders of flowed lands were major habitats. Such areas of human disturbance are now far less frequent, and many of today's observers found that beavers—largely extirpated in the last century—were responsible for a large proportion of Olive-sided Flycatcher breeding sites. The cool, deep ravines of the Catskills described by Hough (Bull 1974) seem much less favored than the habitats already described, with most Catskill records above 457 m (1500 ft).

After wintering in South America, the Olive-sided Flycatcher arrives on the northern breeding grounds during May and into June (Bull 1974). The nest is a loosely formed cup of twigs and grasses, sometimes lichens of the genus *Usnea;* it is lined with finer plant materials and hair (Harrison 1978). Adirondack nests were built on an outer branch from 7.6 to 13.7 m (25–45 ft) high in balsam fir or spruce. Although easily identified by song (approximately 70% of Atlas records were of singing males), the Olive-sided Flycatcher was difficult to "confirm"—only 6 nests were found. I found it became very difficult to find as singing began to slow during late June.

John M. C. Peterson

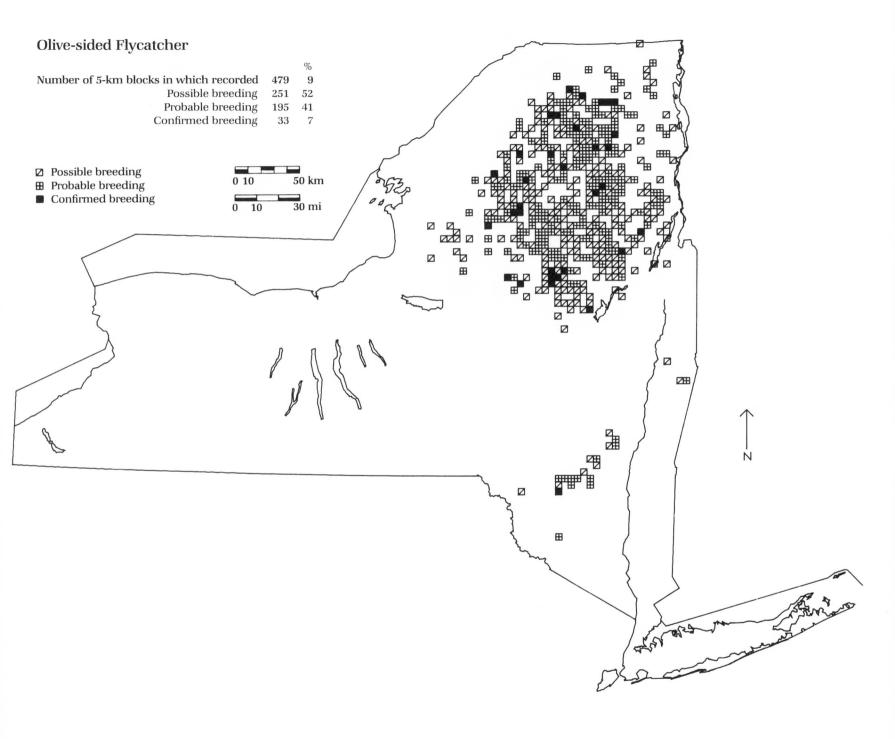

Olive-sided Flycatcher

		%
Number of 5-km blocks in which recorded	479	9
Possible breeding	251	52
Probable breeding	195	41
Confirmed breeding	33	7

◩ Possible breeding
⊞ Probable breeding
■ Confirmed breeding

0 10   50 km

0 10   30 mi

N

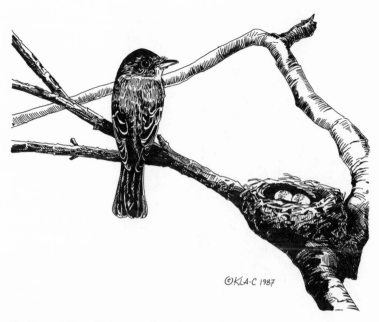

©KLA-C 1987

Eastern Wood-Pewee *Contopus virens*

The Eastern Wood-Pewee, the most well-known of the woodland fly-catchers, is a common breeder throughout New York. Its abundance as a breeding species in New York in the past century appears unchanged, although its numbers almost certainly decreased during the heavy lumbering and agricultural period of the 1800s and also in the New York City region, where Cruickshank (1942) said that it was "slowly decreasing in the ever expanding metropolitan area." Eaton (1914) stated that "in New York it is universally distributed as a summer resident and breeds commonly in every county in the State." This species showed a very slight but significant decline on USFWS BBS routes in North America from 1965 to 1979 and was still considered common and widespread (Robbins et al. 1986).

The Atlas map shows the Eastern Wood-Pewee throughout New York except in the immediate vicinity of New York City, where there is no suitable woodland habitat. It is sparsely distributed in high elevations of the Adirondacks characterized by large unbroken tracts of spruce-fir forests, in particular, and in heavily agricultural areas of western New York where only scattered woodlots exist.

A persistent caller at all hours of the day throughout the summer, this species was easily located by Atlas observers. Although the Eastern Wood-Pewee's *pee-ah-wee* call is distinctive, observers in the Adirondacks had to be careful not to confuse it with the *chur-wee* call of the Yellow-bellied Flycatcher. In more than 31% of all Atlas blocks this species was reported as a "probable" breeder when a male was heard singing on territory. Despite the ease with which this species can be located, its

secrecy around its nest makes breeding difficult to "confirm." The Eastern Wood-Pewee was "confirmed" in only 15% of all the blocks in which it was reported. Most "confirmed" records were of adults seen carrying food to their young.

An "edge" species found mainly at forest margins and openings (Hespenheide 1971), the Eastern Wood-Pewee is common in fragmented and open forest tracts. It adapts well to human alteration of the state's woodlands and is equally abundant in forests of all sizes as long as there are openings (Robbins 1984). Hespenheide (1971) found that when this flycatcher occurred in areas where the forest vegetation appeared uniform and without openings, the canopy layer was incomplete. The Eastern Wood-Pewee also nests in fruit orchards and in large shade trees in city parks and along village streets (Bull 1974). In Allegany State Park it is a common bird of mature beech-maple, oak-hickory, and river-valley forests (Saunders 1942).

The nest of this flycatcher is a thick-walled cup almost invariably placed on the upper surface of a horizontal limb or in the fork of a small limb far out from the trunk. Nests are built anywhere from 1.8 to 16.8 m (6–55 ft) above ground (Eaton 1914; Ellison 1985c), although usually between 6.1 and 9.1 m (20–30 ft). The nest is built from a wide variety of materials. It "is constructed of small twigs, rootlets and grass stalks neatly matted together, and over the outside a coating of greenish and grayish lichens is invariably affixed" (Eaton 1914). Terres (1980) stated that the nest is "built of weed stems, plant fibers, spider cocoons, string, lined with wool, horsehair, bits of thread, grasses" and is extremely well camouflaged with a covering of lichens. Occasionally "the center of the nest is so loosely constructed that when it is placed in the horizontal fork one may see through it from the ground" (Eaton 1914).

Steven C. Sibley

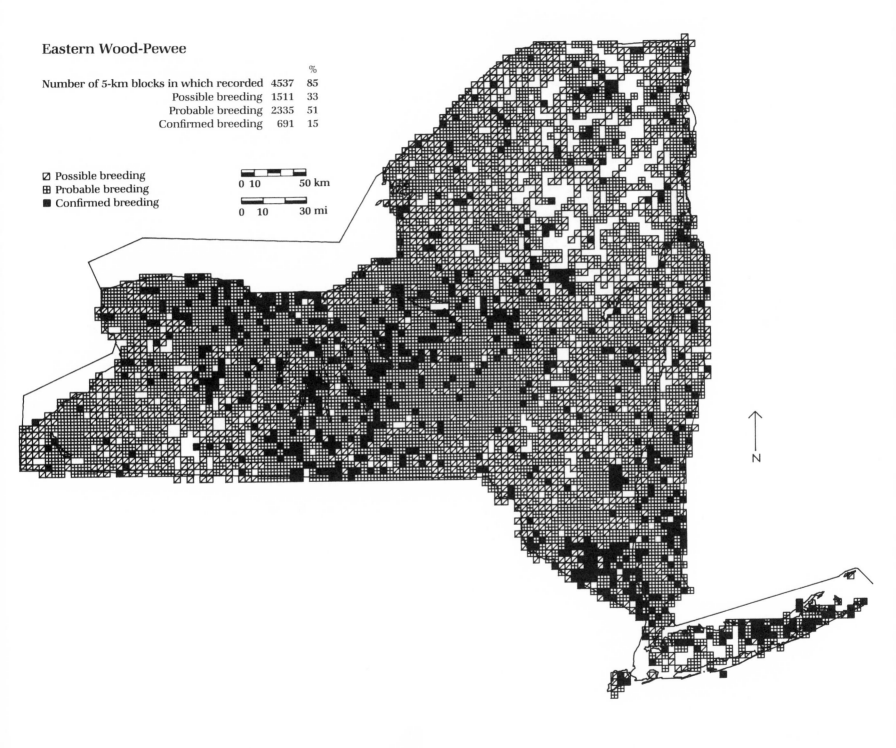

Eastern Wood-Pewee

		%
Number of 5-km blocks in which recorded	4537	85
Possible breeding	1511	33
Probable breeding	2335	51
Confirmed breeding	691	15

☑ Possible breeding
⊞ Probable breeding
■ Confirmed breeding

0 10   50 km

0 10   30 mi

N

Yellow-bellied Flycatcher *Empidonax flaviventris*

Within the shaded, damp spruce forests of the Adirondacks the Yellow-bellied Flycatcher is a fairly common bird. On the higher slopes of the Catskills it is uncommon, and on the Tug Hill Plateau and elsewhere in the state the species is locally rare. Within its range the Yellow-bellied Flycatcher seems more abundant than the more obvious and certainly better recorded Olive-sided Flycatcher.

Theodore Roosevelt and his friend, H. D. Minot, felt that the Yellow-bellied Flycatcher was rather rare in Franklin County during the years 1874–77 (Roosevelt and Minot 1929). (The future president was at that time between the ages of 15 and 18.) During the same era Merriam (1881) also described it as a rather rare summer resident in the Adirondack region. Eaton's party found it at a number of locations in the High Peaks of Essex County during 1905, including Boreas Pond, Elk Lake, Indian Head, Mount Marcy, and Mount Skylight (Eaton 1914). Saunders (1929a) found it along the trail to Algonquin Peak between 762 m (2500 ft) and 1219 m (4000 ft), and on the trail to Avalanche Pass. He judged this the least common of the three small flycatchers of the region (Yellow-bellied, Alder, and Least flycatchers), finding it in second-growth birch and cherry of old burns that were then so common.

Both Eaton (1910) and Bull (1974) published range maps; the Atlas distribution more closely resembles that shown by the earlier author. Eaton's Adirondack and Catskill ranges are strikingly similar to those found by Atlas observers, and he noted a small discrete population on the east side of Lake George where there was one Atlas record and on the Tug Hill Plateau. Shading on Bull's map showed breeding closer to Lake Champlain and Quebec than Atlas observers found. Evidence of breeding was recorded in Madison County, as Eaton's map suggested.

Extralimital summer records of Yellow-bellied Flycatchers, like that in Madison County, must be treated with caution since this is often one of the last birds to arrive on the northern breeding grounds. The Yellow-bellied Flycather is rare before late May, sometimes still arriving during June. The period of song and nesting is fairly short, with birds mostly silent by late July and beginning to leave by early to mid-August.

Since this species is similar in plumage to other flycatchers of the genus *Empidonax*, most were located by experienced observers familiar with their habitat and voice. Field workers compared the listless *per-wee* or *chu-wee* song to the *pee-a-wee* song of an Eastern Wood-Pewee, and found the *killik* call difficult to distinguish from the somewhat snappier *che-bek* of the Least Flycatcher. The number of Yellow-bellied Flycatcher records rose as observers learned to pay careful attention to pewees and leasts when in boreal forest. This species is difficult to "confirm"; most Atlas "confirmed" records were of adults carrying food to young or of recently fledged young. Given the inaccessibility of their habitat and the vocal similarities to other flycatchers, it seems certain that some were missed.

The Yellow-bellied Flycatcher prefers younger stands of cool spruce or hemlock–northern hardwood forest, dense and shady enough to provide a thick, mossy mat that covers the ground, rocks, fallen trees, and old stumps. In New York such forest is found in large tracts only in the Adirondack High Peaks, Central Adirondacks, Western and Eastern Adirondack Foothills, and among the Catskill Peaks. It may inhabit relatively low black spruce areas of the Adirondacks at 549–610 m (1800–2000 ft), mixed forest with hemlocks on middle slopes (also about 600 m) in the Catskills (J. Carroll, v.r.), or high-elevation red spruce on the Adirondack High Peaks (pers. obs.). In the Central Adirondacks this flycatcher is sometimes found near cliffs or large boulders just below mountain summits where the rocks and shade may provide cooler temperatures (McKinney, pers. comm.). The continuing die-off of red spruces, attributed to acid precipitation, may have had an impact on this species in the Adirondack High Peaks.

The Yellow-bellied Flycatcher is a ground-nesting bird that places its nest in mossy hollows or cavities near hummocks, roots of blowdowns, or low rocks (Bendire 1895). Only a single nest was found during Atlas fieldwork. In addition, Walter Chapman (pers. comm.) saw a bird carrying grass on 23 June 1983, downstream from the dam at Beaver Flow, Essex County. This was the only Atlas record of nest building. The nest is a cup of mosses, grasses, stems, and rootlets, lined with fine plant materials, and well hidden (Bendire 1895). A nest found by Major Bendire in Herkimer County "was one of the neatest and most cunningly hidden pieces of bird architecture I have ever seen."

John M. C. Peterson

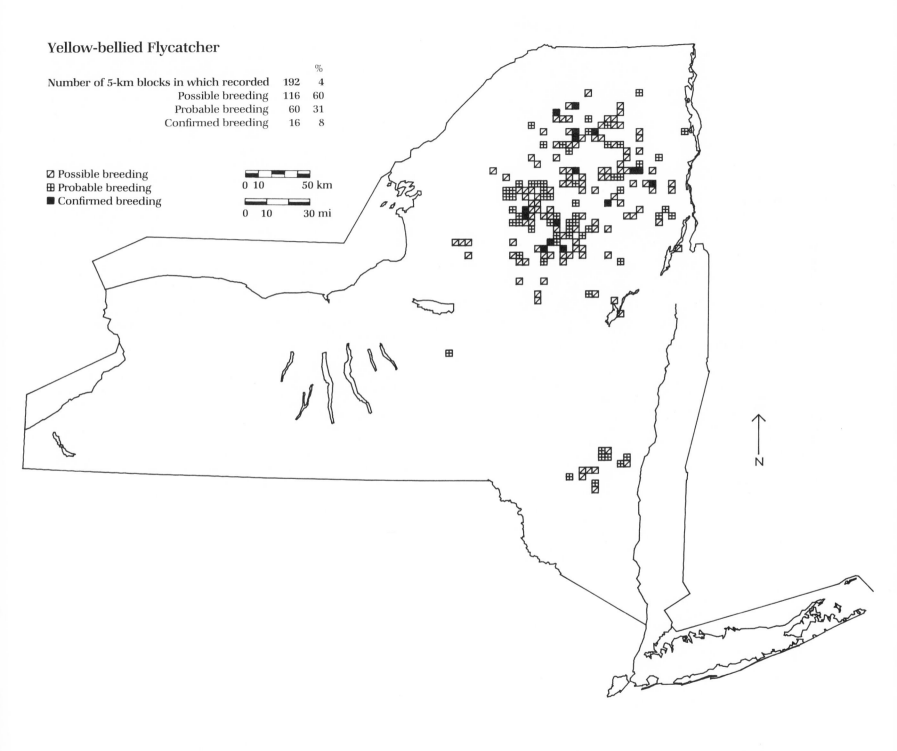

Yellow-bellied Flycatcher

		%
Number of 5-km blocks in which recorded	192	4
Possible breeding	116	60
Probable breeding	60	31
Confirmed breeding	16	8

☑ Possible breeding
⊞ Probable breeding
■ Confirmed breeding

0 10   50 km

0 10   30 mi

N

©KLA-C 1987

Acadian Flycatcher *Empidonax virescens*

The most southern member of the difficult to identify *Empidonax* flycatchers, the Acadian Flycatcher has made a definite comeback in New York. Rather uncommon, it is found mainly in the western part of the state, the lower Hudson Valley, and eastern Coastal Lowlands.

In the early 1900s it was common in the lower Hudson Valley as far north as the southern edge of the highlands, fairly common in the western portion of Long Island, but local and uncommon in Suffolk County. In the upper Hudson Valley and in western New York it was considered irregular (Eaton 1914). Bull (1974) delineated two disjunct historical populations, one in the Erie-Ontario Plain and Finger Lakes Highlands, and a second in the lower Hudson Valley and Coastal Lowlands, but indicated that no known breeding had taken place in the state since an abortive attempt in 1957. In extreme western New York no proven breeding had occurred in more than 60 years. Robbins et al. (1986) said the Acadian Flycatcher declined slightly during the first half of the period 1965 to 1979 but increased slightly after about 1973 on the Appalachian Plateau.

Atlas workers found that this species has become reestablished, apparently moving back into the state along the river valleys. A population was found in Allegany State Park, which is bordered by the Allegheny River. Baird (1986b), in a study of the birds of Quaker Run Valley in the park, estimated that 54 pairs bred there, where it had been absent from 1921 to 1937 (Saunders 1942). It was also found in Chautauqua County, where several small ravines cut from the Appalachian Plateau onto the Lake Erie Plain. The source of these invading Acadian Flycatchers appears to be

Ohio and western Pennsylvania. The recovery of the forest in this area from devastation at the turn of the century by the chemical wood industry, particularly along many small streams, has furnished habitat for this species, which nests mainly in shady hemlock ravines.

Atlas observers also found the Acadian Flycatcher along the Great Lakes Plain. These may have spilled over from a southern Ontario population established along the northern shore of Lake Erie (Godfrey 1966; Bull 1974). The source of this species in the central Appalachian Plateau is not so readily apparent, but it may have pushed up the Susquehanna River Valley into the Finger Lakes region. In southeastern New York it also seems to have returned along stream valleys. Eaton (1914) called the Acadian Flycatcher fairly common on western Long Island, but Griscom (1923) said it had disappeared by 1923. Bull (1974) said it was last known to nest on Long Island in 1926 at Orient, Suffolk County. But it has returned; it was found by Atlas workers along the north and south shores and especially in the eastern part of Suffolk County, near Montauk, Sag Harbor, and along the Peconic River.

The habitat of the Acadian Flycatcher varies from hemlock–northern hardwood glades in Chautauqua and Cattaraugus counties and in southeastern counties (B. Weissman, v.r.) to beech-maple mesic hardwood forest on the Great Lakes Plain and in Nassau County (R. Cioffi, v.r.). Most Atlas observers reported the species associated with swiftly moving streams and rocky ravines. Todd (1940) said that it liked the cool, damp forest of the bottomlands, where there is little undergrowth, and often followed the ravines of streams to higher ground, but that it avoided second-growth, orchard, and shade trees, which were favored by some of its congeners. Its special habitat requirements are mature, dense woodlands with tall trees, closed canopy, and open spaces in the understory for feeding (DeGraaf et al. 1980).

Todd (1940) called the nest a rather untidy structure of the vireo type, but much shallower. Harrison (1975) said it is placed on a lower branch of a large tree, far out from the trunk and usually shaded by leafy branches 2.4–6.1 m (8–20 ft) from the ground. It is built by the female of fine dry plant stems, plant fibers, tendrils, and catkins. There is a slight lining of grass stems, fine rootlets, plant down, and spider webs. Usually, long streamers of grass or other material hang below the nest, giving it a misleading, trashy appearance from below. Despite its appearance, it is strongly constructed (Christy 1942).

Stephen W. Eaton

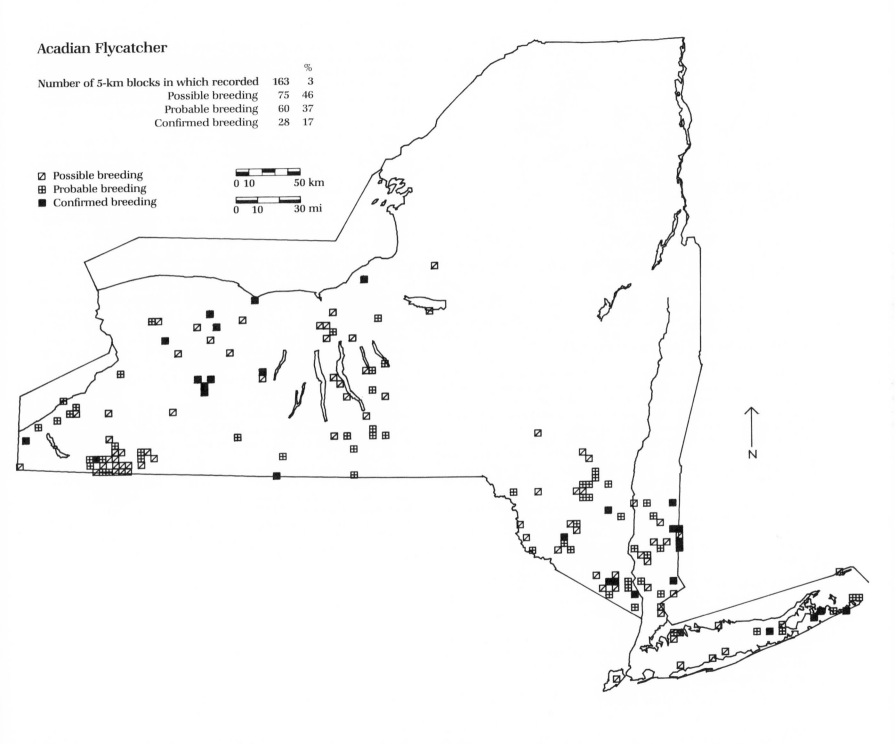

Acadian Flycatcher

		%
Number of 5-km blocks in which recorded	163	3
Possible breeding	75	46
Probable breeding	60	37
Confirmed breeding	28	17

◪ Possible breeding
⊞ Probable breeding
■ Confirmed breeding

0 10   50 km

0 10   30 mi

N

©KLA-C 1987

Alder Flycatcher *Empidonax alnorum*

The Alder Flycatcher, an unassuming bird best distinguished from others of the genus *Empidonax* by its voice, is distributed widely but sparsely in wet thickets throughout most of New York. It is more common in northern parts of the state and virtually absent from the Coastal Lowlands. Its breeding history is somewhat difficult to determine because it was considered conspecific with the nearly identical Willow Flycatcher until 1973 (AOU 1983). Many early writers failed to distinguish between the two song types, which were thought to represent regional differences in voice or possibly different subspecies rather than separate species.

Eaton (1914), for example, stated that the Alder Flycatcher was a summer resident in most of the state, fairly common in the Catskills and Adirondacks and in the colder swamps of central and western New York. However, because he called members of both species Alder Flycatchers, he failed to recognize the Alder Flycatcher's absence from Long Island and the Willow Flycatcher's absence from much of northern New York. Cruickshank (1942) reported that the Alder Flycatcher was a fairly common though local resident in Rockland County. He probably was referring to the Willow Flycatcher, as the Alder Flycatcher is nearly absent from that area today.

Both species apparently were uncommon in New York in the 1800s. Eaton (1914) noted that Traill's flycatcher (the original name for both species) was unknown to De Kay in 1850. He also stated that the species had evidently extended its range in parts of the state, for despite thorough searching by egg collectors, no nests and eggs were found between 1860 and 1885 in many portions of western New York where it was known to breed in the early 1900s. Probably he was referring to the Willow Flycatcher, which was originally considered the western subspecies and was more common in areas south and west of New York such as the open country of Ohio. Recently, however, the Willow Flycatcher has spread dramatically north and east; it apparently is replacing the Alder Flycatcher in some areas (Stein 1963).

Nevertheless, as the Atlas map indicates, the Alder Flycatcher is still the dominant species in northern portions of New York; it also is found, though sparsely, throughout most other regions. It seems to avoid river valleys and, except for one "possible" and three "probable" records, is absent from the Coastal Lowlands, perhaps because of competition from the Willow Flycatcher, which prefers lowland areas.

These two flycatchers presented problems for Atlasers. They can be reliably distinguished only by voice, yet they sing only sporadically, usually just in the spring. Furthermore, the songs are quite similar, and even well-trained birders sometimes cannot tell them apart. Habitat also is an unreliable indicator. The Alder Flycatcher is generally more common at high elevations where there are streams and lakes in wooded areas, and the Willow Flycatcher is more abundant at lower elevations, in open, brushy areas with grassland habitats. Nevertheless, the range and habitats of the two species overlap, and both sometimes can be heard singing at the same time in the same place. Also, both are quiet and elusive, staying well hidden in thickets, appearing but occasionally to pursue flying insects or to perch near the top of a bush. Nests are also difficult to find, which may explain the high ratio of "probable" to "confirmed" breeding records.

The Alder Flycatcher usually breeds in dense, low, damp thickets of alder, willow, buttonbush, elderberry, and red osier dogwood along banks of small streams, shores of ponds, and borders of marshes (Terres 1980). It also inhabits boggy, shrubby, boreal coniferous forests (Aldrich 1953).

The nest is built from 38.1 to 83.8 cm (15–33 in) above the ground on a horizontal branch or the fork of a small shrub. The brown untidy structure resembles that of a Song Sparrow or Indigo Bunting and is loosely built of coarse grasses, cat-tail leaves, mosses, pine needles, rootlets, bark, and twigs; it is lined with fine grass. Numerous streamers hang down from the rim or the bottom to well below the bulk of the nest. The Alder Flycatcher uses much less cottony material in its nest than does the Willow Flycatcher (Stein 1958; Terres 1980).

Richard E. Bonney, Jr.
Judith L. Burrill

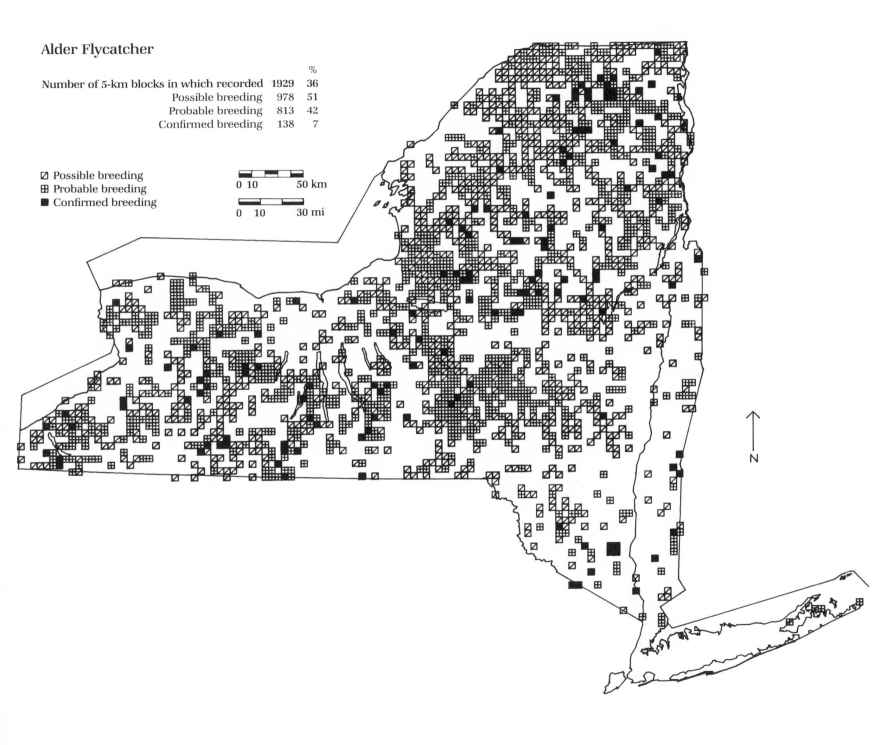

Alder Flycatcher

Number of 5-km blocks in which recorded		%
Number of 5-km blocks in which recorded	1929	36
Possible breeding	978	51
Probable breeding	813	42
Confirmed breeding	138	7

☑ Possible breeding
⊞ Probable breeding
■ Confirmed breeding

0 10   50 km

0 10   30 mi

N

© KLA-C 1987

Willow Flycatcher *Empidonax traillii*

The Willow Flycatcher is an inconspicuous bird that looks nearly identical to the Alder Flycatcher. It lives in slightly more open country than the alder but is best distinguished by its sneezy *fitz-bew* song. The Willow Flycatcher is fairly common throughout much of the southern and lower elevations of New York. Its history in the state is unclear because it was considered conspecific with the Alder Flycatcher until 1973 (AOU 1983). Most early writers did not distinguish between the two species, which together were known as Traill's flycatcher (see further discussion under Alder Flycatcher).

Originally, the Willow Flycatcher was considered the western subspecies of Traill's flycatcher. It was common in the Ohio River Valley and the Mississippi River drainage, and it apparently overlapped little with the Alder Flycatcher. Recently, however, it has spread north and east and in some areas may be replacing the Alder Flycatcher (Stein 1963).

Stoner (1932) stated that reports indicated that the Traill's flycatcher was becoming more widely distributed in the state. He was probably referring to the Willow Flycatcher. More definitive were Arthur A. Allen's observations in Ithaca, Tompkins County, which he reported to Stein (1958): only *fee-bee-os* (Alder Flycatcher) were present before 1940, but the area was infiltrated by *fitz-bew*s (Willow Flycatcher) shortly afterward, and after 1950 the Alder Flycatcher was displaced by the Willow Flycatcher in most areas. It is possible that the Willow Flycatcher is continuing to expand its range even farther east. Only since the 1960s has the species begun to spread into adjacent Vermont (Norse 1985b).

One cautionary note: Eaton's (1914) map of the known breeding range of Traill's flycatcher in 1907 omits a large portion of southern New York, where both Willow and Alder flycatchers breed presently, a fact which suggests that both species may have increased their New York ranges during this century.

Atlasing shows the Willow Flycatcher to be absent from or sparse in most high elevations in New York, including the Adirondacks, Catskills, and southern Appalachian Plateau. It is most widely distributed in the Great Lakes Plains, the northern sections of the Appalachian Plateau, the Mohawk Valley, and the lower Hudson Valley, areas of low elevation and abundant suitable habitat.

The Willow Flycatcher is found in dry upland pastures thickly overgrown with shrubs, along streams and lake edges in grassland areas with shrubs, and in swampy thickets, especially of willow and buttonbush (AOU 1983, Norse 1985b). In general, the Willow Flycatcher is found in drier locations than the Alder Flycatcher (Norse 1985b).

Very few "confirmed" records were obtained. Finding nests in the shrubby habitat is very difficult. Also, because the Willow Flycatcher ceases its singing for a period in early summer, some birds could have been missed in blocks that were surveyed only during that time. (See further discussion of difficulties of atlasing under Alder Flycatcher.)

The nest is usually built from 0.5 to 2.7 m (1.5–9 ft) above ground in the upright fork of a shrub but sometimes on a horizontal limb, often in a willow. The gray structure resembles the nest of a Yellow Warbler; made of shredded milkweed bark, tufts of cat-tail, grasses, silky material of aspen and willow, and cottony material, it is often melded together by wind and rain. It is lined with fine grasses and sometimes with feathers. Nests built in upright crotches often have silky streamers hanging from the nest bottom (Stein 1958; Terres 1980).

Richard E. Bonney, Jr.
Judith L. Burrill

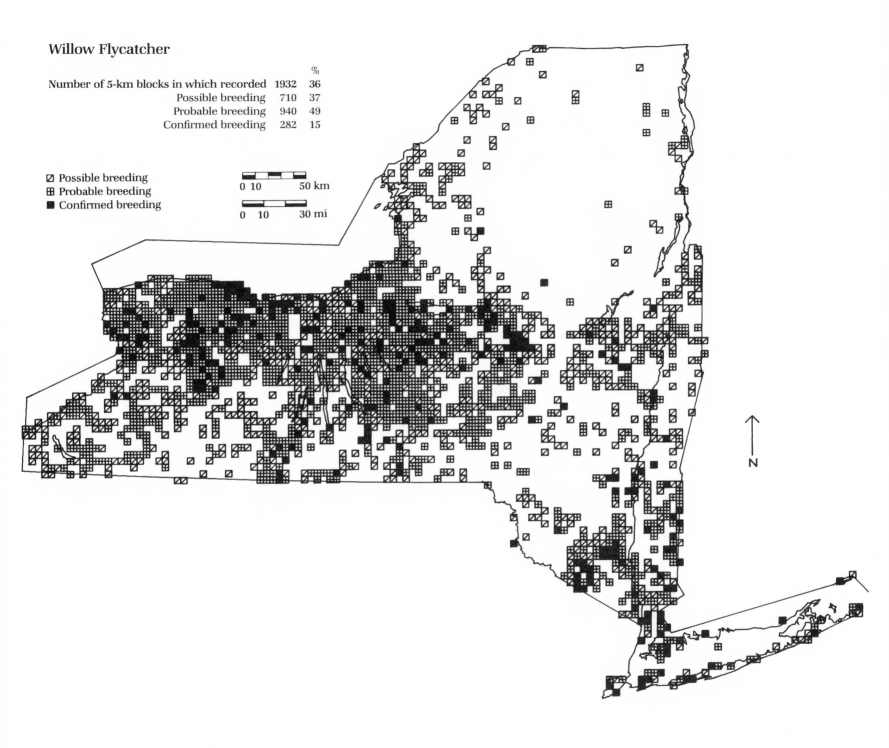

Willow Flycatcher

		%
Number of 5-km blocks in which recorded	1932	36
Possible breeding	710	37
Probable breeding	940	49
Confirmed breeding	282	15

◪ Possible breeding
⊞ Probable breeding
■ Confirmed breeding

0 10   50 km

0 10   30 mi

N

©KLA-C 1987

Least Flycatcher *Empidonax minimus*

The Least Flycatcher is one of the smallest yet noisiest members of its genus, persistently uttering a brisk, emphatic *che-bec* song in the breeding season. It is common to abundant over most of the state except on Long Island and around New York City, near the southern limit of its range on the coast, where it is very rare and local. In 1982, partly based on reports from the Northeast, this species was placed on *American Birds'* Special Concern list (Tate and Tate 1982), then later was listed as a Species of Local Concern only (Tate 1986). Numbers have been reported down in southeastern mainland New York.

Eaton (1914) stated that this species was common throughout the state; however, his statement apparently did not apply to Long Island. Griscom (1923) stated that as a summer resident the Least Flycatcher occurred there only rarely and locally on the north shore; Bull (1974) listed 13 known Long Island breeding sites, 12 being scattered along the north shore from Queens County east to Fishers Island. Griscom (1923) noted that in the New York City region it was steadily decreasing with the loss of rural areas and no longer nested regularly anywhere in Bronx County and vicinity. On Staten Island active nests were found by Chapin in 1905 and 1908, but it has probably not nested there since about 1950 (Siebenheller 1981).

Beehler (1978) called it an abundant breeder in the Adirondack Park and remarked that all early records indicated this species was common there. From 1966 to 1979 the Adirondacks (and New Hampshire) maintained the densest populations of this species on the USFWS BBS (Robbins et al. 1986). On the Tug Hill Plateau a road count of bird species was conducted by Peakall and Rusk (1963), who found it to be the commonest flycatcher of taller deciduous woods. A similar count conducted in 1970 indicated that its numbers had considerably increased (Rusk and Scheider 1970). Recent reports for other regions usually suggested a stable or even increasing population.

However, according to observers, several southeastern counties above New York City recently showed a decline, a phenomenon also reported in New Jersey. Treacy (1982b, 1983b) reported that in 1982 New York Atlas observers found Least Flycatcher numbers low and reported many more Acadian Flycatchers than usual. In 1983 numbers of Least Flycatchers were reported down, notably in Rockland and Dutchess counties.

Atlas results reveal a distribution apparently much like the less-detailed historical one. The absence of the Least Flycatcher in and near New York City reflects continued urbanization there, and also it has nearly disappeared from the north shore of Long Island. The only suggestion of a concentration in the Coastal Lowlands is in central Suffolk County, including south shore localities. The otherwise widespread distribution attests to a wide habitat tolerance, although habitat is limited in highly agricultural regions of New York.

A relatively bold and active *Empidonax*, the Least Flycatcher is easily located by its frequent singing, which continues into early July. Also, the conspicuous territorial and courtship activities of the males are frequently noted early in the season. This species can be rather quiet during the late nesting period, especially when caring for the young, which probably accounts for the somewhat low number of "confirmed" records.

The Least Flycatcher is a breeding bird of deciduous and mixed forests, but it prefers semi-open areas: forest edge, open woodland, stream and pond borders, and also orchards and parks. In the northern Allegany Hills, Saunders (1936) found it in successional northern hardwood forests, primarily aspen-cherry, a diversified habitat along stream valleys, and in small unsprayed orchards, but not in beech-maple mesophytic or Appalachian oak-hickory forests. In the Adirondacks it was common at lower elevations and in aspen growth on old burns (Saunders 1929a). Breckenridge (1956) observed that nesting Least Flycatchers in a forest with scattered small openings require a certain degree of openness among the tree limbs intersecting the zone of flycatcher use beneath the canopy. Johnston (1971) found this species numerous in a parklike modified forest and nearly absent from surrounding closed-canopy forests.

The small nest is placed at a crotch or on a horizontal branch and is often 0.6–7.6 m (2–25 ft) above ground in almost any kind of tree, or even shrub. Resembling the nest of the Yellow Warbler or American Redstart, it is composed of shredded bark, grasses, and other fibers, and is lined mainly with plant down and fine grasses (Bent 1942; Harrison 1975).

Paul F. Connor

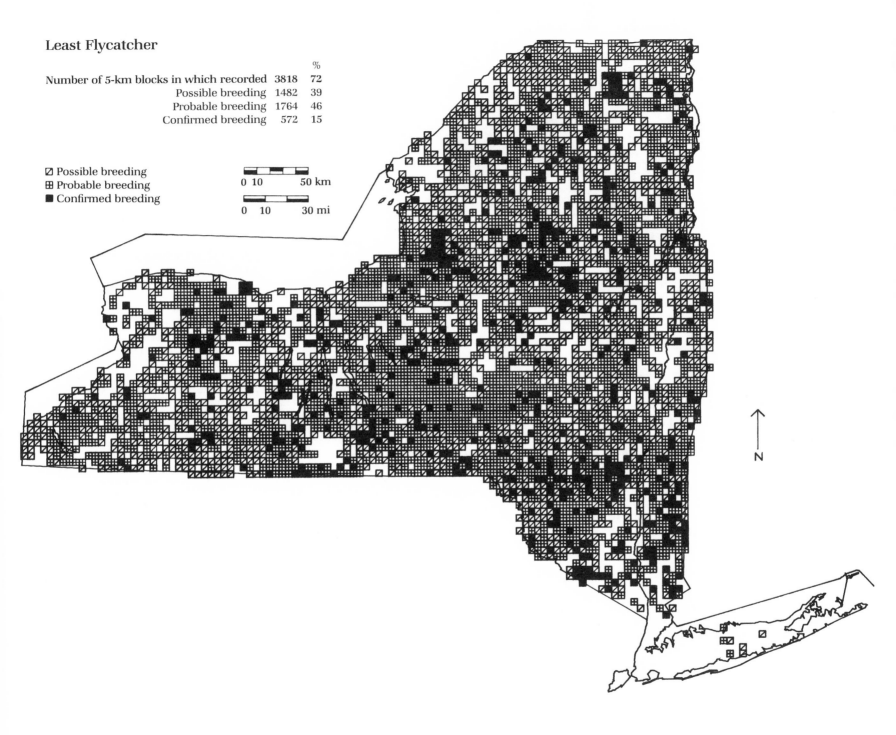

Least Flycatcher

Number of 5-km blocks in which recorded	3818	%
		72
Possible breeding	1482	39
Probable breeding	1764	46
Confirmed breeding	572	15

◫ Possible breeding
⊞ Probable breeding
■ Confirmed breeding

0 10   50 km

0 10   30 mi

N

Eastern Phoebe *Sayornis phoebe*

The Eastern Phoebe, a familiar flycatcher of summer camps, woodland glades, and mossy cliffs, is a common bird in New York except in the mountains and large urbanized areas. Eaton (1914) stated that it was probably the most common member of the flycatcher family, being a summer resident throughout the state except in the mountain spruce–fir forests of the Catskills and Adirondacks. Bull (1974) called it a widespread breeder inland but local in the Coastal Lowlands. Beardslee and Mitchell (1965), writing of the birds of the Niagara Frontier region, stated that formerly it was considered abundant but that by the mid-1960s it could barely qualify as a common summer resident. They noted that where insecticides were sprayed the bird was absent or its numbers low. Robbins et al. (1986) determined that Eastern Phoebe numbers remained stable from 1965 until the severe winter of 1976–77, when a period of continous subfreezing temperatures reduced the population in the northern part of its winter range. Thereafter a gradual recovery began.

Atlas workers found the Eastern Phoebe generally distributed, with only a few areas where it was not reported. In the Central Adirondacks, Adirondack High Peaks, and Central Tug Hill, it was generally absent except along the many streams draining from the highlands. It remains a local breeder in the Coastal Lowlands.

The Eastern Phoebe was an easy bird to "confirm" because it nests in man-made structures and is vocally and visually conspicuous. Active nests were found in 1409 blocks, more blocks than for any other species except American Robin and Barn Swallow. Used nests were found in 426 blocks; only used nests of the Northern Oriole were reported more often.

Like many other passerine birds that venture north early, the Eastern Phoebe is double-brooded, giving Atlas observers more opportunity to "confirm" breeding. Although some have suggested that the species may have 3 broods, Clark and Eaton (1977) found that there is not enough time after the second clutch for another 45 days of parental care before the intervention of molt and the regression of reproductive organs.

Although the Eastern Phoebe may be found in any of the forest types, it usually builds its nest near water. It forages over streams and ponds for the insect life emerging from their surfaces and, in addition, uses mud to construct its nest. It breeds where there are suitable nest sites, where water is available, and where there is some sort of shelter from wind and storm (Eaton 1914). Weeks (1984) found that the species required wooded areas for foraging within approximately 50 m (164 ft) of the nest site. It has become increasingly dependent upon humans for nest sites, such as open porches, bridges, open barns, and outhouses, but nests can occasionally still be found on sheltered cliffs (DeGraaf et al. 1980). This must have been its natural nest site in pre-Columbian times.

The nest is usually placed on a shelflike projection such as a window, rafter of a farm building, bridge girder or trestle, or it may be plastered to a rocky ledge, or to concrete or wooden walls (Harrison 1975). The nest, built by the female over a period of 3–13 days, is typically less than 4.6 m (15 ft) from the ground (DeGraaf et al. 1980). It is a large nest, considering the size of the bird, and is well constructed of weeds, coarse grasses, fibers, and mud, then covered with mosses. The lining is of fine grasses and hair (Harrison 1975).

The Eastern Phoebe is "one of the very commonest foster parents of the young Cowbird" (Friedmann 1929), but this varies greatly in New York. A population in Allegany State Park, Cattaraugus County, was parasitized only 1.2% of the time and another population in Cattaraugus County only 5.7% (Clark and Eaton 1977). Jones (1975) found a population in Seneca County to be parasitized 33% of the time. He calculated that it took the loss of 27 Eastern Phoebes to produce a total of 15 Brown-headed Cowbirds. According to Weeks (1984), Eastern Phoebes that nested on bridges experienced less parasitism than those that nested at other sites.

Stephen W. Eaton

Eastern Phoebe

Number of 5-km blocks in which recorded 4283   80   %
           Possible breeding   768   18
           Probable breeding   771   18
        Confirmed breeding 2744   64

☑ Possible breeding
⊞ Probable breeding
■ Confirmed breeding

0 10   50 km
0 10   30 mi

N

Great Crested Flycatcher *Myiarchus crinitus*

The large and colorful Great Crested Flycatcher is a common and widely distributed breeder. Breeding has been "confirmed" in all but three counties of the state: Kings, New York, and Bronx. Concrete and buildings replace earth and trees in those counties.

Eaton (1914) stated that in the Adirondacks and Catskills the Great Crested Flycatcher was found in the "valleys almost to the heart of those regions" and that it was fairly common in upland areas above 305 m (1000 ft). Beehler (1978) described the breeding status of the Great Crested Flycatcher in Adirondack Park as "fairly common" along the outskirts of the park where elevations were below 610 m (2000 ft). He pointed out that the Great Crested Flycatcher in the Adirondacks had probably benefited greatly from the replacement of much conifer growth with deciduous forest. Beardslee and Mitchell (1965) called it a "fairly common summer resident" in western New York. Bull (1974) merely called it a "widespread breeder in the lowlands."

That elevation has been one of the limiting factors in the breeding distribution of this species is borne out by atlasing. The map shows a widespread distribution with fewer records in the higher elevations of the Catskills and Adirondacks. Because this is a bird of deciduous and mixed woods, it is not found in the spruce-fir forest of the Central Adirondacks.

Although the Great Crested Flycatcher is very vocal in the early part of the breeding season, it becomes quiet and more difficult to find as the season progresses. Where block busting was conducted only in the later part of the breeding season, this flycatcher was probably missed. "Con-

firmed" records were fairly difficult to obtain. Nests were located in 228 blocks. Territorial behavior was observed very often for this species, but the greatest number of records were of calling birds.

This species prefers deciduous or mixed woods, and although it inhabits heavily forested areas, it is also seen along edges and in openings, including swamps, fruit orchards, and among scattered trees in cropland (DeGraaf et al. 1980). Bent (1942) stated that the Great Crested Flycatcher, originally a bird of forest interiors, had moved into more open areas and forest edges. Hespenheide (1971) found this flycatcher to exhibit little preference for either densely wooded or open areas; it occupies a wide range of habitats.

Like all *Myiarchus* flycatchers, the Great Crested Flycatcher nests in cavities, preferring natural cavities in middle-aged to mature trees. It uses larger woodpecker holes and birdhouses less frequently (DeGraaf et al. 1980). Nests have been recorded in hollow logs, stovepipes, rain gutters, and tin cans. The majority of sites are below 6.1 m (20 ft). When occupying a large natural cavity, this flycatcher fills the cavity with a foundation of trash and leaves before building the cup. The nest cup is a soft lining of feathers and fur, pine needles, rootlets, plant fibers, and similar material (Bent 1942).

Legend has it that all Great Crested Flycatcher nests contain shed snakeskin. Although some of the nests do contain snakeskins, many contain such shiny and similar material as cellophane, wax paper, or onionskin to which the bird is attracted during nest building (Bent 1942).

Emanuel Levine

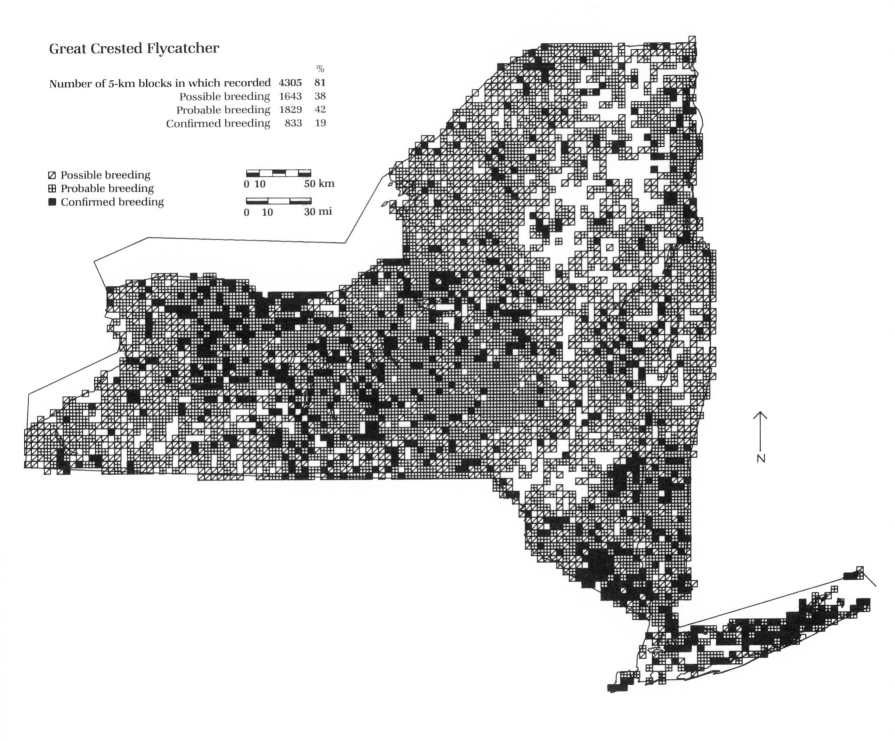

Great Crested Flycatcher

		%
Number of 5-km blocks in which recorded	4305	81
Possible breeding	1643	38
Probable breeding	1829	42
Confirmed breeding	833	19

☑ Possible breeding
⊞ Probable breeding
■ Confirmed breeding

0 10   50 km

0 10   30 mi

N

©KLA-C 1986

Eastern Kingbird *Tyrannus tyrannus*

The Eastern Kingbird is a conspicuous flycatcher and a very common breeder throughout the state. It has been a common breeder for over a century, with little apparent change in numbers (Eaton 1914; Bull 1974). Saunders (1942) speculated, however, that the Eastern Kingbird was much less common before the extensive clearing of forests during the 19th century. Openings along rivers, streams, lakes, and naturally open meadows and marshes were probably the original habitats of this species. The Eastern Kingbird has proven very adaptable and can be found today in almost all open habitats, often in close association with humans.

Atlas records indicate the Eastern Kingbird is found virtually everywhere in the state except for the New York City area and the heavily forested regions of the Adirondacks, the Tug Hill Plateau, and the Allegany Hills, where open areas are not found. Atlas coverage for this species was excellent. The Eastern Kingbird is extremely territorial and is often observed tormenting a passing crow or kestrel. Most of the "probable" records were of birds exhibiting territorial behavior. The kingbird is often perched on roadside wires, its characteristic silhouette unmistakable. The species is also easily "confirmed," as it builds a large bulky nest, often in a conspicuous location, and its fledglings are very noisy when hungry. Most "confirmed" Atlas records were of adults carrying food, feeding young, or of recently fledged young.

In New York the Eastern Kingbird "breeds in generally open country, especially in cultivated areas—farms, orchards, along rural roadsides—and also along lake and river shores, and in open woodlands, swamp edges and clearings" (Bull 1974). In the Oneida Lake basin Stoner (1932)

described typical habitats as "low willow and alder flats along the lake, in and about old orchards where it frequently nests, along roadsides supporting low bushes and trees, and in low meadows bearing sparse and intermittent growths of scrub willows or other shrubs or small trees." A recent comparative study of Eastern Kingbirds nesting in lakeshore and in upland habitats in eastern Ontario found no major differences in biology or behavior between the sites (Blancher and Robertson 1985). However, because of moderated temperatures along the lake shore, pairs there bred earlier. Lakeshore pairs foraged more efficiently because of the availability of both over-land and over-water habitats, and the increased foraging efficiency resulted in more rapid growth of nestlings.

As would be expected for a common species found in a wide variety of habitats, the Eastern Kingbird builds its nest in a bewildering array of locations. Some of these are summarized by Terres (1980), who stated that this species "usually builds well out on horizontal limb of isolated tree or orchard tree or in low shrubs growing along water's edge . . . or on stumps or snags above water far from shore of lake or pond, in hollows of trees near water, in open on fence posts, and in suburbs may nest in rain gutter of house." Other curious locations include one on the winch of a moored houseboat (Bull 1974), in vines on a stone wall (Eaton 1914), and in electric street lights, a rain gauge, on telephone poles, and in old oriole nests (Bent 1942). Nests are placed anywhere between 0.6 and 18.3 m (2–60 ft) above ground, although usually 3–6.1 m (10–20 ft) up (Bent 1942).

Both sexes help build the nest, which is rough and unkempt on the outside and rather large and bulky for the size of the bird. "Its exterior is constructed of small twigs and dry weed stems, mixed with cottonwood down, pieces of twine, and a little hair. The inner cup is lined with fine dry grass, a few rootlets, and a small quantity of horsehair." Other materials such as cloth, feathers, and sheep wool are used, if available in the vicinity of the nest (Bent 1942).

Steven C. Sibley

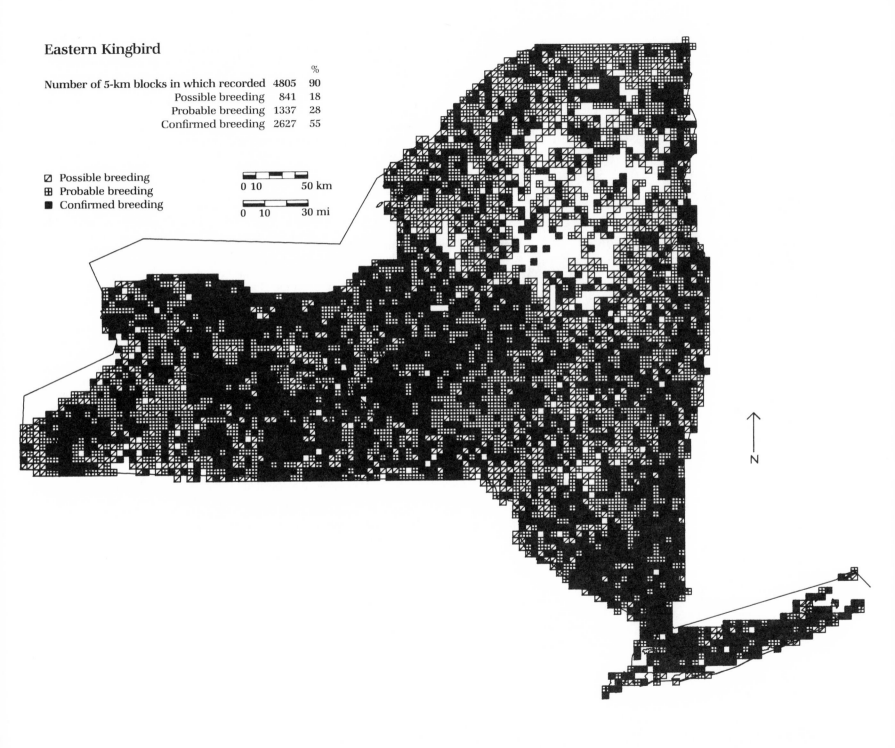

Eastern Kingbird

		%
Number of 5-km blocks in which recorded	4805	90
Possible breeding	841	18
Probable breeding	1337	28
Confirmed breeding	2627	55

☒ Possible breeding
⊞ Probable breeding
■ Confirmed breeding

0 10   50 km

0 10   30 mi

N

Horned Lark *Eremophila alpestris*

A hardy species of fields and prairies, the Horned Lark is a fairly common breeder locally in western and central New York and on Long Island, and uncommon and very local throughout the rest of the state. The Horned Lark did not always nest in New York, since it was not until the mid-19th century that statewide clearing of forests created suitable habitat. By about 1880 the total number of farms and acreage in farms reached its peak, with about 75% of the state's land area in agriculture (Richmond and Nicholson 1985). Hurley and Franks (1976) stated, "The fields produced by agriculture created vast artificial 'barrens' that were good nesting habitat for the Horned Lark, formerly restricted to early nesting when the prairie grasses were still short." Hurley and Franks decribed an expansion eastward across New York from Ontario and Michigan and into New England.

First recorded breeding in New York in 1875 at Buffalo, Horned Larks were also found breeding in Monroe and Lewis counties in 1876, in Queens County on western Long Island by 1879, in Albany County in 1881, Essex County by 1900, and in Clinton County near the Canadian border by 1905 (Bull 1974). After the report of young Horned Larks in Queens County at Long Island City in 1879, the next confirmed breeding on Long Island was not until 50 years later in 1936 on a golf course in Idlewild (Cruickshank 1942). After the 1936 nesting this species rapidly expanded its range on Long Island and was known to be breeding at more than ten localities, from Montauk Point on the eastern tip to Jamaica Bay by 1942 (Cruickshank 1942).

Eaton (1914) stated that "this species . . . has now found conditions favorable to its habitation in the eastern states and has gradually been spreading year after year till now we must call it one of the common birds of the open field." There is some evidence that the Horned Lark is now decreasing in numbers throughout the state (Robbins et al. 1986) as agricultural practices change and farms are abandoned, allowing land to grow back to forest. Recent increases in the amount of cropland and decreases in pastureland (Richmond and Nicholson 1985) must also have a negative impact on this species. Scheider (1975), speaking of the Horned Lark in central New York, stated that it "continues to grow ever more scarce—if corn becomes the local cropland king, it will disappear altogether as a breeder here." In recent years corn and alfalfa have replaced grains as the major crops in the state (Richmond and Nicholson 1985), usually to the detriment of this species. However, in some areas, such as around Rochester, it can often be seen in poor sections of corn fields where the corn is stunted or does not grow at all (Spahn, pers. comm.).

The Horned Lark occurs only in open areas with bare ground or short grass. In New York this species inhabits "airports, golf courses, fields, pastures, sandy beaches and dunes, and barren wastes" (Bull 1974).

Distribution of breeding Horned Larks roughly corresponds to the distribution of open areas throughout the state, with most breeding in lower-elevation agricultural areas of the Erie-Ontario Plain in western New York and the Finger Lakes, and on Long Island in dune areas along the coast. The Horned Lark is also evenly distributed in the Eastern Ontario Plains, St. Lawrence Plains, and the Champlain Transition.

This species is very difficult to locate in the extensive open areas it inhabits. The Horned Lark is also often missed because it begins to breed in March, and the first brood is often fledged before May. It was definitely overlooked in many blocks where it certainly occurs, both in western and central New York and in the Eastern Ontario and St. Lawrence plains. The Horned Lark is most easily located by diligent scanning of suitable fields and by listening for the male's tinkling aerial courtship song. Breeding was also difficult to "confirm," with most records (55%) of adults feeding recently fledged dependent young.

In New York nests have been found "in stubble fields of wheat and rye, in newly planted corn, among young tomato, potato, and strawberry plants, on overturned sod clumps, bare ground, gravel road shoulders, sand dunes with beach grass, and on short-grass strips of airports and golf links" (Bull 1974).

The nest consists of a shallow depression in the ground dug out by the female with her beak and feet (Bent 1942), although the nest is sometimes placed in a natural depression such as in a cow footprint (Eaton 1914). The nest is usually constructed next to or partly under a tuft of grass, rocks, or dirt clod with clods, pebbles, or similar items placed around the edges (Bent 1942). A cup of stems, grasses, and leaves is built in the nest depression, with a finer inner lining of feathers, plant down, and hairs (Terres 1980).

Steven C. Sibley

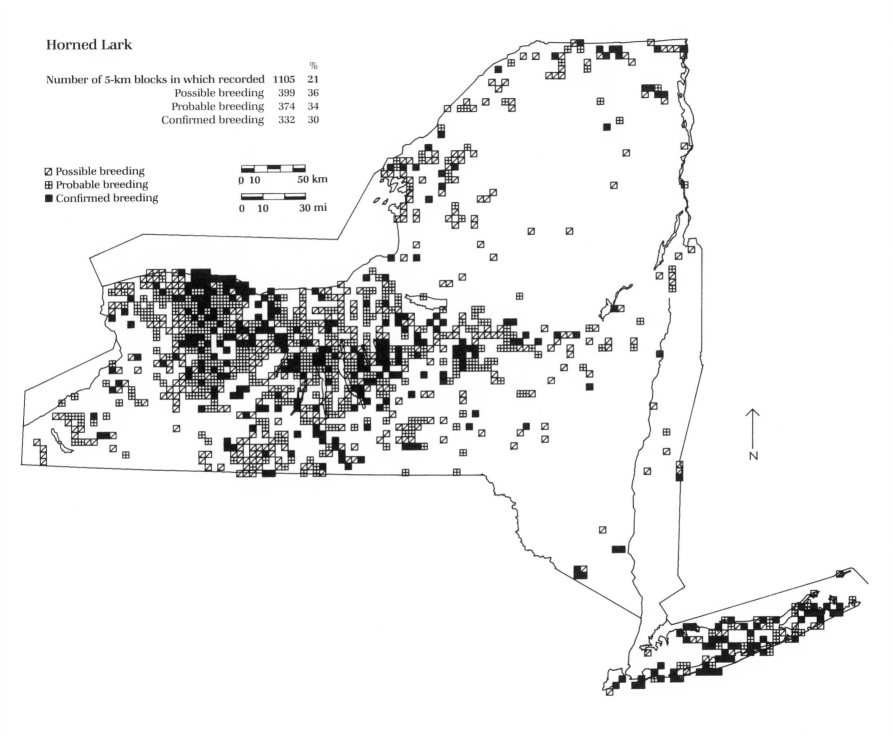

Horned Lark

Number of 5-km blocks in which recorded   1105 21   %

		%
Possible breeding	399	36
Probable breeding	374	34
Confirmed breeding	332	30

◩ Possible breeding
⊞ Probable breeding
■ Confirmed breeding

0 10   50 km

0 10   30 mi

N

Purple Martin *Progne subis*

The Purple Martin, a popular yard bird, is locally common at lower elevations in many parts of the state. Before New York was settled in the 1700s, this species was probably uncommon and extremely local, although native Indians thought highly of the birds and had for years put up hollow gourds and calabashes in which they could nest (Wilson and Bonaparte 1870).

With the settling of New York and clearing of the land in the 1800s this species increased. However, when the European Starling and House Sparrow were introduced in the late 1800s, they began to usurp martin nesting sites, and the Purple Martin declined from 30 to 50% statewide between 1860 and 1910 (Eaton 1914). Eaton stated that if sparrows were not kept out of martin houses, the species would "be extirpated within a generation in most localities." The increased popularity of martin houses in the past 50 years and the determination of martin house owners to exclude the Tree Swallow, European Starling, and House Sparrow have stabilized martin populations, but its continued existence depends largely upon efforts of martin house owners.

The Purple Martin breeding range in New York is confined to low elevations of the state, as indicated on the Atlas map; it is almost completely absent from the Adirondacks, Catskills, and Tug Hill Plateau, which are above 305 m (1000 ft). The Purple Martin is found mainly along the shores of Lakes Ontario, Erie, and Champlain, Oneida Lake, and the Finger Lakes, as well as major rivers and the coast. The distribution of the Purple Martin in neighboring Ontario and Pennsylvania has a similar pattern (Ontario and Pennsylvania Breeding Bird atlases, preliminary maps). This pattern is determined mainly by nest site and prey availability, because there are generally more flying insects in wet than in dry areas.

Since the Purple Martin is found very locally and often around vacation cottages, as well as in parks in towns and cities, it is easy to locate, and Atlas coverage was probably extremely good. The exclusive use of houses, gourds, and other types of nest boxes put out for these birds in the open also made it easy to "confirm" breeding, as 64% of all Atlas reports were in this category. Half of the "confirmed" records were of birds observed entering or leaving nest compartments.

The habitat of the Purple Martin must be near water and have an available nest site—in New York that is usually a house mounted on a pole. In addition to rivers and lake shores, this species occurs in meadows next to ponds and near coastal marshes (Bent 1942). The Purple Martin is dependent upon an abundance of flying insects for its food, a dependency that makes this species susceptible to cold, wet weather, which keeps insects down. In the spring of 1966 an especially severe period of such weather in western New York caused the starvation deaths of many martins, reducing populations in that area from 75 to 80% (Benton and Tucker 1968).

Before there were humans in North America, the Purple Martin nested in natural cavities in trees and cliffs (Bent 1942). Nesting by this species in such cavities has been rare in the East in the past century but is still regular in the western United States (Armistead 1983). The Purple Martin formerly often nested in the crevices and cornices of buildings in Rochester (Monroe County), Syracuse (Onondaga County), Utica (Oneida County), and other cities in the state, but it was forced out by the more aggressive House Sparrow and European Starling (Bul! 1974). However, there were reports of the Purple Martin nesting in building cornices in Buffalo as recently as the early 1970s (Bull 1974).

Today, almost all the state's nests are in specially designed martin houses, some of which are quite elaborate and contain as many as 200 compartments. The recommended dimensions for a Purple Martin nest compartment are 20.3 × 20.3 cm (8 × 8 in) with an entrance 5.1 cm (2 in) in diameter and 3.8 cm (1.5 in) above the floor. The meager nest is built by both sexes and is constructed of twigs, feathers, mud, rags, paper, string, straw, shreds of bark, grass, or leaves, or some combination of these (Bent 1942). Occasionally, a small rim of mud up to 6.4 cm (2.5 in) high is built between the nest and the entrance hole, perhaps to keep the eggs from rolling out (Bent 1942). Males arrive at nesting areas sometimes weeks before the females and select one nest compartment to defend (Allen and Nice 1952). Allen and Nice discovered that when the female arrives, she too selects a nest compartment, thereby acquiring a mate, rather than selecting a mate first. European Starlings and House Sparrows can be discouraged by waiting to put up martin houses until the first male martin "scouts" appear in early spring, and by cleaning out all nests and material put in nest compartments by unwanted species.

Steven C. Sibley

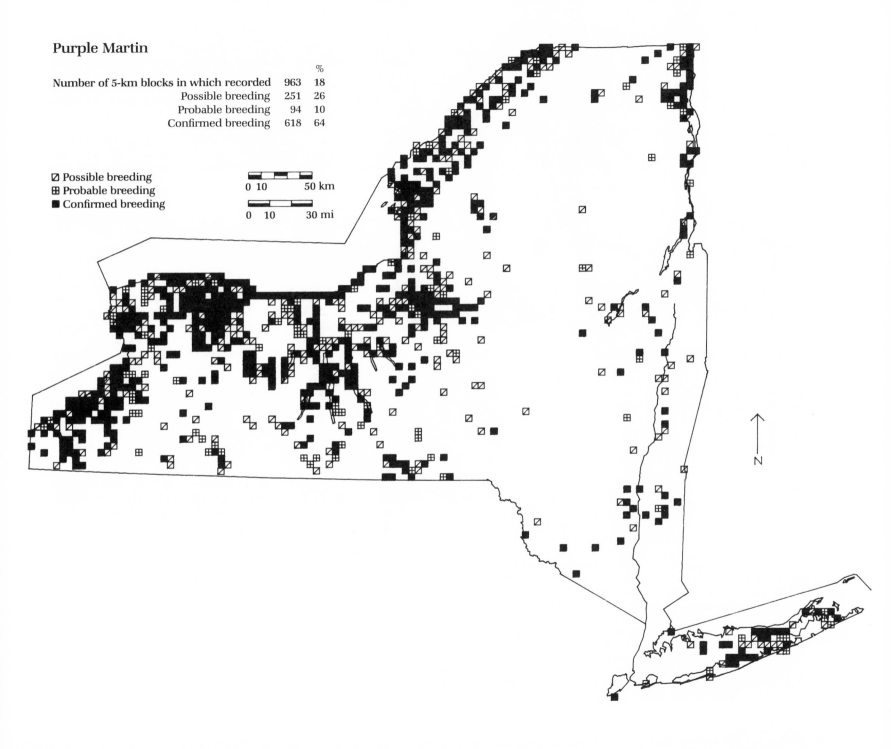

Purple Martin

		%
Number of 5-km blocks in which recorded	963	18
Possible breeding	251	26
Probable breeding	94	10
Confirmed breeding	618	64

☑ Possible breeding
⊞ Probable breeding
■ Confirmed breeding

0 10   50 km

0 10   30 mi

N

©KLA-C 1986

Tree Swallow *Tachycineta bicolor*

The Tree Swallow is a common breeder throughout most of the state. Of New York's six breeding swallow species, none is more widely distributed. Although this species has been common in upstate New York at least since the early 1800s, it did not occur in many areas on Long Island until the early 1900s. At that time, in southeastern New York, there was only one known breeding locality in all of Dutchess County, with only a few pairs observed (Griscom 1933), and only occasional breeding records in the New York City region and the western third of Long Island (Cruickshank 1942). On Staten Island there was only a possible 1877 breeding record, until the first substantiated nesting in 1945. However, by 1950 the species was a regular breeder (Siebenheller 1981). A similar pattern occurred over the rest of Long Island during this same period, and the Tree Swallow is now a common breeder, especially on the eastern half.

This species is susceptible to short-term population declines caused by extremes of weather. Severe weather during the winter of 1939 and 1940 killed thousands of Tree Swallows overwintering in the southern United States, and the following summer breeding populations throughout the Northeast were far below normal (Cruickshank 1942). Colder-than-normal springs during the Atlas survey are believed to have reduced Tree Swallow populations on Long Island (Salzman, pers. comm.).

The Tree Swallow is absent as a breeder only from the innermost areas of New York City, including all of Bronx, New York, and Kings counties. Although it now occurs across most of Long Island, it is still found more often in upstate New York. In addition to breeding throughout most of the state, the Tree Swallow occurs at a wide range of elevations, being found up to 914 m (3000 ft) in the Adirondacks (Peterson, pers. comm.). In Erie and Niagara counties the Tree Swallow is believed to be absent from some blocks because of the spraying of insecticides and a lack of nest cavities, both natural and man-made (Andrle, pers. comm.) Nest boxes put up largely for the Eastern Bluebird but used by this species probably have increased its abundance and made available many open fields with farm ponds that were not previously used because of a lack of nest sites.

This swallow was reported from 87% of all Atlas blocks covered in the state and "confirmed" in 66% of the blocks in which it was reported. Because it spends most of its day flying over open areas capturing insects, this species is easy to locate and was probably not missed even by blockbusting efforts. The Tree Swallow is also easy to "confirm," as it nests in the open and can readily be seen entering its nest hole throughout the breeding season. Of all "confirmed" records, 32% were of adults seen entering nest holes, 23% were of nests with young, and 23% were of recently fledged dependent young.

The Tree Swallow occurs in open areas, always in the vicinity of water, where there are large populations of flying insects on which to feed. River valleys, lakes, marshes, flooded swamps, and beaver ponds with many dead and hollow trees standing in the water or along the shore are its preferred habitats, but it is quite common around fields and meadows if suitable nesting sites are available and there is some open water nearby (DeGraaf et al. 1980).

When not using nest boxes, this species nests in natural cavities or woodpecker holes in trees, or rarely in cliff crevices (Peterson, pers. comm.) or in roots of upturned trees (Bull 1974). Formerly, it nested in gourds put out by native Indians for the Purple Martin (Tyler 1942). It now takes readily to nest boxes and houses put out for the martin and Eastern Bluebird and is capable of driving out both of these species and, occasionally, the House Sparrow. Nests in natural tree cavities are located from 1.2 to 9.1 m (4–30 ft) above the ground or water (Cruickshank 1942). The territory of this swallow is restricted to the immediate vicinity of the nest site (Kuerzi 1941).

The nest is built by the female with little or no help from the male and is constructed of dry grass, straw, and, rarely, pine needles. These materials are hollowed out to make a nest cup and lined with feathers that the bird gathers nearby; it shows a marked preference for white feathers (Tyler 1942).

At most nests individuals other than the breeding pair visit almost daily, sometimes in groups as large as 20. These birds do not help raise the young or affect nesting success in any way (Lombardo 1986) but seem only to be searching for potential future nest sites (hatching-year birds) or breeding opportunities (sexually mature birds) (Lombardo 1987).

Steven C. Sibley

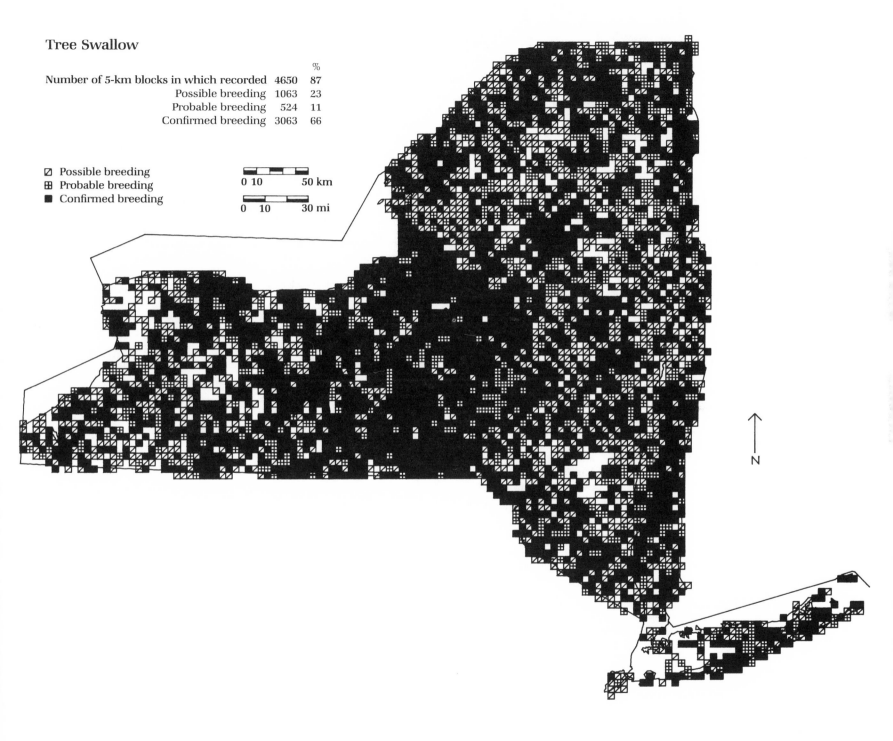

Tree Swallow

Number of 5-km blocks in which recorded	4650	% 87
Possible breeding	1063	23
Probable breeding	524	11
Confirmed breeding	3063	66

◫ Possible breeding
⊞ Probable breeding
■ Confirmed breeding

0 10   50 km

0   10   30 mi

N

wing as other swallows and can be confused with the Bank Swallow. Some observers may have missed this species. Most of the "confirmed" records were of birds seen entering or leaving the nest cavity or hole.

Unlike other species that have recently expanded their ranges, like the Northern Cardinal and Tufted Titmouse, the Northern Rough-winged Swallow is migratory and its range expansion has, therefore, not been influenced by mild winters or bird feeding. However, humans may have played a major role by providing nest sites such as concrete or rock bridges, dams, and sand mines (Cruickshank 1942).

The Northern Rough-winged Swallow nests in an open area that has an available nest site and is near water, most often near a stream (DeGraaf et al. 1980). The nest is placed in a hole in a bank, in a natural rock fissure, in a masonry crevice, or in crannies under bridges and other structures (Harrison 1978). It will also use an abandoned Bank Swallow or Belted Kingfisher hole (DeGraaf et al. 1980). Lunk (1962) in a definitive study offered no positive evidence that it either digs its own burrow or enlarges an existing burrow. Unlike most swallows, it nests alone or with only a few others of its species, but it can be found sharing a bank with a Bank Swallow colony (DeGraaf et al. 1980). The nest is built of grasses, weeds, and other loose plant material; the cup is lined with finer grass fibers and rootlets (Harrison 1978).

Emanuel Levine

Northern Rough-winged Swallow
Stelgidopteryx serripennis

The Northern Rough-winged Swallow, so named because of the serration along the outer primary feather of each wing (Lunk 1962), is a locally fairly common breeder near waterways throughout most of the state. It was evidently not observed by naturalists in New York until the last third of the 19th century (Eaton 1914). On Long Island, Griscom (1923) called this swallow "very rare and irregular," and Cruickshank (1942) said, "Long Island is by far the poorest major locality for the Rough-winged Swallow." According to Bull (1974) this species was not even listed until the year 1870, but by the mid-1970s it was breeding in most of the river valleys and lake shores throughout New York, though not in the Catskills and Adirondacks.

New York is now near the northern limit of this swallow's breeding range; the actual northern edge in this part of the continent is southern Ontario (Ontario Breeding Bird Atlas, preliminary map). Its distribution in New York is widespread but mainly tied to the riverine, lacustrine, and coastal estuarine communities. Although reported frequently in such low elevations as the areas along Lakes Erie and Ontario and coastal Long Island, it was found even in the Adirondack and Catskill regions. In New York and elsewhere it occurs above 610 m (2000 ft) and has been recorded in California as high as 1981 m (6500 ft) (Dingle 1942). Like the Tree Swallow and Purple Martin, the Northern Rough-winged Swallow is an obligate cavity-nesting swallow whose distribution pattern is mainly determined by available nest sites (Muldal et al. 1985).

The Northern Rough-winged Swallow is not as easily identified on the

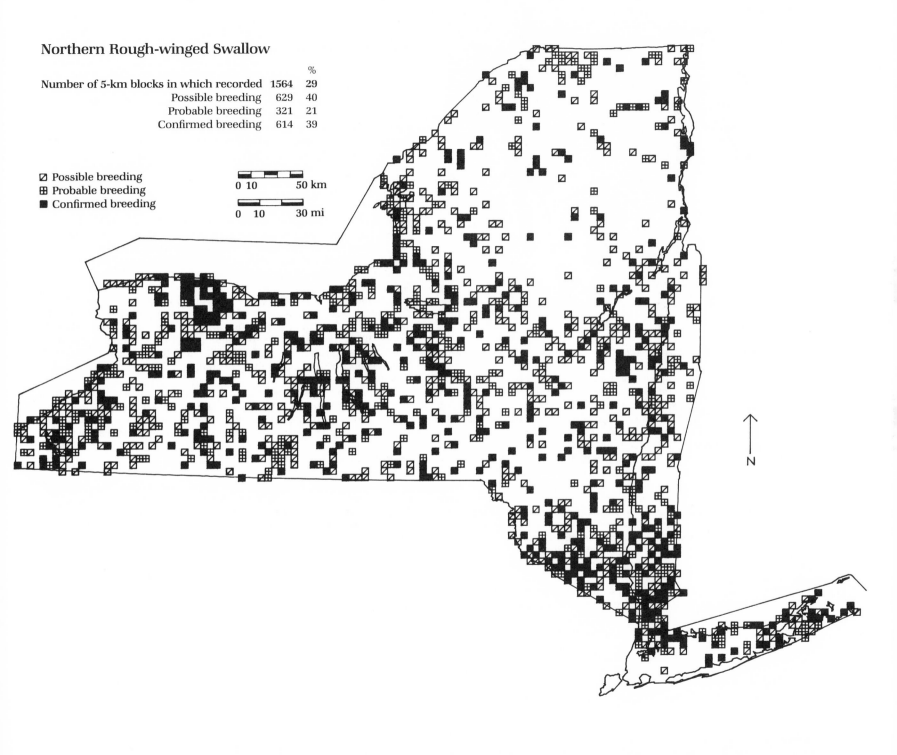

Northern Rough-winged Swallow

		%
Number of 5-km blocks in which recorded	1564	29
Possible breeding	629	40
Probable breeding	321	21
Confirmed breeding	614	39

☑ Possible breeding
⊞ Probable breeding
■ Confirmed breeding

0 10   50 km

0 10   30 mi

N

©KLA-C 1987

Bank Swallow *Riparia riparia*

The gregarious Bank Swallow is an unusual bird that digs its nest hole into the ground. Since it prefers to nest near water, it is restricted largely to soft banks of ponds, lakes, and streams or nearby quarries. Where such habitat occurs, the bird is common, even abundant, but because the specialized habitat is irregularly distributed, the Bank Swallow is a widespread but local breeder throughout most of New York.

Eaton (1914) stated that the species was generally distributed throughout the state and was very abundant in some localities where sand banks were plentiful, such as the shore of Lake Ontario, the Genesee River Valley, the Hudson Valley, and on Long Island. There were large portions of New York where this swallow was absent, however, such as on well-drained uplands where sandbanks were sparse and soils hard.

The Bank Swallow may have changed its distribution on the Coastal Lowlands in recent years. Griscom (1923) stated that in the New York City area the bird was found only along the outer beaches of Long Island's south shore and was local or absent elsewhere, and Cruickshank (1942) said that only a few pairs nested on the north shore in Suffolk County and that most of the nesting was concentrated on Long Island's extreme eastern tip. There were few Atlas records from the south shore, but several on the north shore and quite a few throughout the eastern third of the island.

Farther inland, Griscom (1923) noted that the Bank Swallow was once reported as a common summer resident at Ossining, Westchester County, but that there were no recent records, a finding confirmed by Atlasing. And Cruickshank (1942) said that "astonishingly few" pairs were nesting in Rockland and Westchester counties, where Atlasers found them nesting in several blocks.

These changes in distribution are not surprising, considering that favored nesting areas—sand and gravel pits—are quite transient. From year to year they appear and disappear and change in size because of human activities and erosion. This affects not only the distribution but also the size of nesting colonies (Stoner 1932).

Atlasing showed that the Bank Swallow is distributed sparsely but fairly evenly throughout the state. It is found in suitable habitat in every ecozone, and its distribution corresponds closely to that of New York's major lakeshore and river systems, probably because these areas have very sandy soils. The Bank Swallow was fairly easily found by Atlasers who used topographic maps to locate quarries or sand banks, where an active Bank Swallow colony could often be found.

The Bank Swallow usually nests near water, in the steep banks of streams and creeks where the soil may be sand, clay, or gravel. It also nests in the banks of lakes, bays, or oceans, and occasionally is found some distance from water in the steep sides of sand and gravel pits, highway and railroad embankments, and even in piles of sawdust or coal dust (Stoner 1932; Bull 1974; Terres 1980) or dry well walls (Harrison 1975). Kiviat et al. (1985) found a colony nesting in an eroding bank of sandy dredge spoil along the Hudson River in Columbia County. According to Stoner (1932), the bird prefers to nest in the vicinity of human habitation, or in sand and gravel pits being worked, rather than in abandoned ones. This is probably because mining operations keep banks freshly exposed and clean (Freer 1979).

A nesting colony is a busy place, often with hundreds of birds darting in and out of nest holes, wheeling in all directions and emitting a continuous buzzing twitter. Colony size varies but can be in the hundreds. Freer (1979) reported on two colonies; one with 197 active nests and another with 218 nests. The burrows are placed near the top of the vertical side of a bank and dug straight into the bank or slightly upward (Terres 1980). Both the male and female participate in digging, using their bills and feet (Stoner 1936). The entrance holes are shaped like a flattened ellipse about 5.1–6.4 cm (2–2.5 in) wide and 3.8 cm (1.5 in) high. The depth of the burrow appears to be dependent on the soil type, being deeper when soil is fine and loamy and shorter when it is gravelly (Stoner 1932), but is usually from 45.7 to 91.4 cm (18–36 in) deep (Eaton 1914).

The floor of the burrow is flat and the ceiling is arched. The end of the burrow is enlarged to hold a nest composed of straw, grass, and feathers (Eaton 1914; Harrison 1975). The feathers, ordinarily white and from domestic fowl, are placed in the nest to form a lining generally completed before the young are hatched (Stoner and Stoner 1952). The nests also may contain rootlets, pine needles, bits of wool, and horsehair (Terres 1980).

Judith L. Burrill
Richard E. Bonney, Jr.

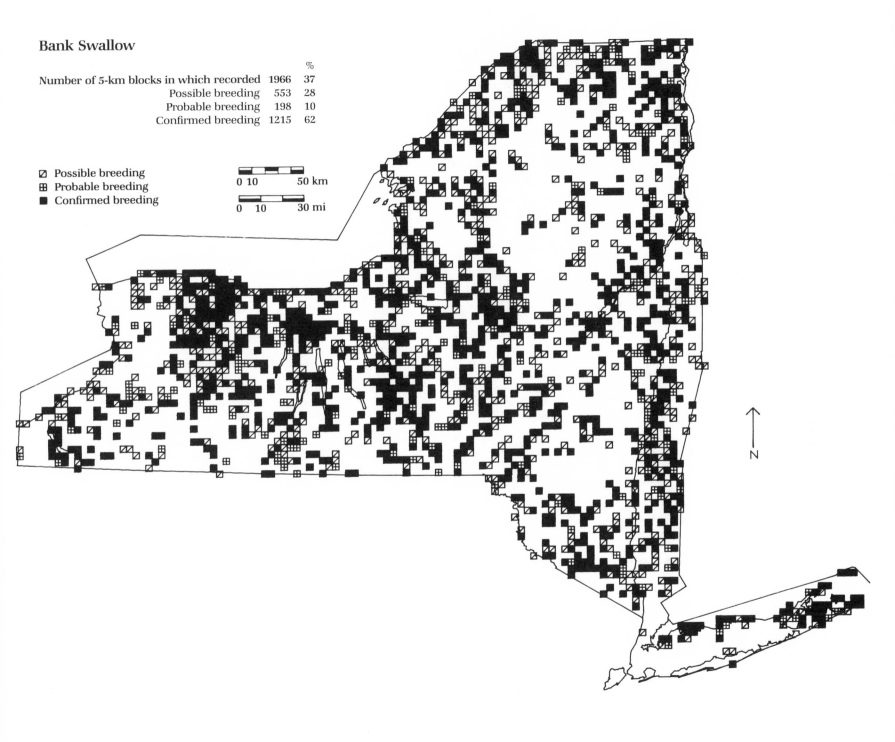

Bank Swallow

Number of 5-km blocks in which recorded	1966	% 37
Possible breeding	553	28
Probable breeding	198	10
Confirmed breeding	1215	62

◩ Possible breeding
⊞ Probable breeding
■ Confirmed breeding

0 10   50 km

0 10   30 mi

N

Cliff Swallow *Hirundo pyrrhonota*

The colonial-nesting Cliff Swallow is fairly common in some areas of the state. Although colonial nesting is rather rare in passerine birds, it is highly developed in the Cliff Swallow (Emlen 1954). Colonies found by Speier (1965) in the Olean area of Cattaraugus County had 31, 16, 9, and 4 nests.

De Witt Clinton first reported the Cliff Swallow in New York in 1817 at Whitehall, near the southern end of the Lake Champlain canal (Baird et al. 1905). By the early 1900s it was known locally throughout the state and was most common in the Catskills and Adirondacks (Eaton 1914). Eaton believed, however, that it had almost entirely disappeared from many sections in which it had been common earlier. Large barn colonies began to fail in the early to mid-1900s, partly because of the invasion of the House Sparrow. This sparrow often usurps the nest of the Cliff Swallow and may even repel it from surrounding nest sites (Weeks 1984). In addition, the surfaces of painted barns were too smooth for the Cliff Swallow's nest to adhere (Gross 1942). By the early 1970s few barn colonies remained. The number of Cliff Swallows has subsequently increased as the construction of bridges across major highways and large dams has provided new nesting sites (Weeks 1984; Robbins et al. 1986).

Atlas workers found the Cliff Swallow in four areas: the Catskills and adjacent ecozones; the Tug Hill area and Oswego Lowlands; the St. Lawrence Plains and Transition; and the Lake Champlain Valley and Transition. Concentrations were also found along the Allegheny River and its tributaries in southwestern New York; the headwaters of Tonawanda Creek and the Genesee River, the Chemung and Susquehanna rivers, the Delaware River and its tributaries in central New York; and the headwaters of the Mohawk and Black rivers and tributaries of the St. Lawrence and the Chazy, Saranac, and Ausable rivers in northern New York. Its last recorded breeding on Long Island was in 1924 (Bull 1964). This may be a result of the lack of suitable nest sites and the right kind of mud and of too much sand for nest building (Forbush 1929).

In the East the Cliff Swallow has depended mainly on barns for nest sites, whereas in the West it often uses natural nest sites (Weeks 1984). Eaton (1914) stated that it almost always nested in communities under the eaves of barns, although it occasionally plastered its nest under the protective ledges of mountain cliffs. In addition to an appropriate nest substrate, the nesting habitat must contain both open areas with shallow water for insect foraging, such as marshes, estuaries, and bottomland old fields, and a source of mud for nest building (Emlen 1954). The Cliff Swallow requires an unobstructed flight path to, under, and through its nesting area. If the area near the nest site becomes overgrown with vegetation, the bird's use of the area will lessen (Weeks 1984).

The nest is made of pellets of mud or clay, roofed over, and shaped like a retort. A narrow entrance leads into an enlarged chamber. The typical nest contains 900 to 1200 mud pellets and takes from 1 to 2 weeks to complete (Harrison 1975). The inside chamber is lined with a few dried grass stems, feathers, and other similar materials. Forbush (1929) stated that the Cliff Swallow, unlike the Barn Swallow, makes "bricks" without straw; therefore, it must have clay or clayey mud for building material. Kilgore and Knudsen (1977) found that the Cliff Swallow selected mud with a higher sand, lower silt content and with less organic material than that chosen by the Barn Swallow. They speculated that such mud is more pliable and easier to mold. If there is too much sand in the mud, the nest will fall when it dries (Forbush 1929). An unusual nest site was that of a single Cliff Swallow in the middle of an active Bank Swallow colony in a sandpit at Powder Mill Park, Monroe County (Davis 1971).

To encourage nesting on barns that are painted and perhaps too smooth to allow mud to bind well, one can place a two-by-four 5 inches below the eave, flat against the wall. This will give the purchase necessary to hold the nest (Forbush 1929). As many as 2000 nests were reported by Buss (1942) on a Wisconsin barn where nesting was thus encouraged.

Stephen W. Eaton

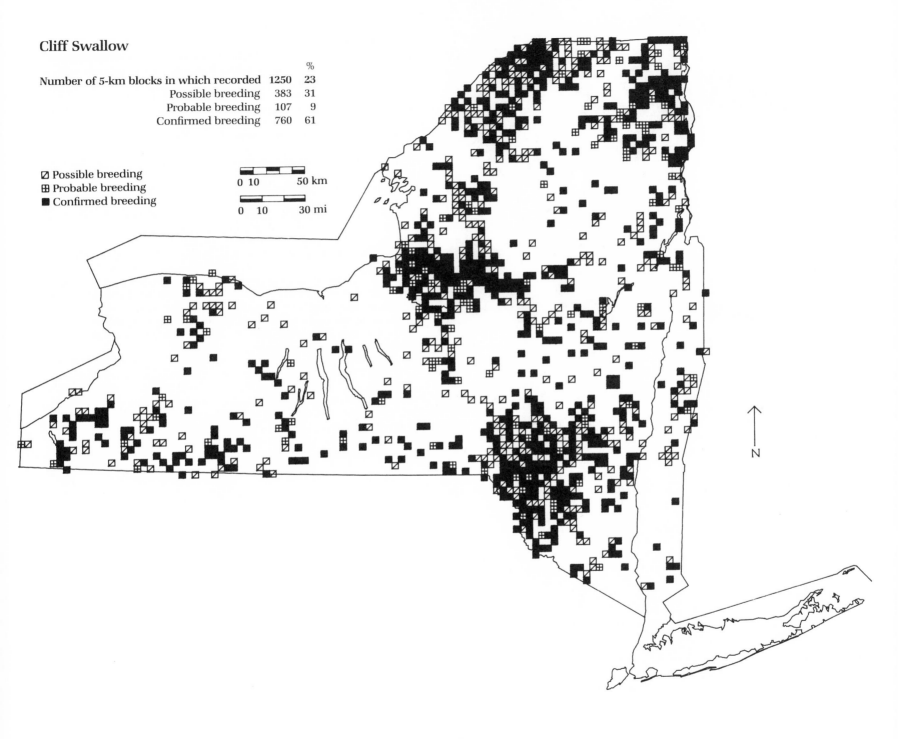

Cliff Swallow

Number of 5-km blocks in which recorded		%
Number of 5-km blocks in which recorded	1250	23
Possible breeding	383	31
Probable breeding	107	9
Confirmed breeding	760	61

☑ Possible breeding
⊞ Probable breeding
■ Confirmed breeding

0 10   50 km

0 10   30 mi

N

©KLA-C 1986

Barn Swallow *Hirundo rustica*

The Barn Swallow is a very common species found in open country. Even as early as the mid-1800s De Kay (1844) called it "one of our most common visitors." Both Eaton (1914) and Bull (1974) considered it a widespread breeding species.

This swallow was probably not so common in pre-Columbian times. Baird et al. (1905) stated that this species originally nested in caves, overhanging rocky cliffs, and similar localities. As the country was settled, the species gradually left these natural sites for farm buildings, and its numbers greatly increased. Its acclimation to human habitation, at least in the East, seems to have preceded that of the Cliff Swallow by perhaps a century (De Kay 1844; Baird et al. 1905).

Atlas observers found the Barn Swallow in most of New York except for some urban areas, and parts of the Central Adirondacks, Adirondack High Peaks, and Central Tug Hill. This swallow was recorded in more blocks than all but 10 other species. Its distribution depends mainly on the availability of nest sites near open foraging areas and of mud for nest construction (Samuel 1971). The Barn Swallow was one of the most frequently "confirmed" species because it nests around human habitation and fledglings are easily observed perched on telephone or power lines. More Barn Swallow nests were found than nests of any other species.

The Barn Swallow can be found in both rural and suburban areas. It nests either alone or in aggregations of from 2 to over 40 pairs. Colony size appears to be correlated with the size of the nest site or number of entryways into it (Snapp 1976). Nest sites include open barns, sheds, protected areas on buildings and bridges, as well as natural sites such as cliffs and shallow caves (Bent 1942) with unobstructed entryways (Snapp 1976; Weeks 1984). Most nests in New York are in man-made sites, but a nest was found on a shale cliff overlooking Lake Ontario in Oswego County (Bull 1974). Shields (1984), studying a Barn Swallow population at Cranberry Lake, St. Lawrence County, found that most breeding adults remained faithful to previously used colonies, as well as to clusters within colonies, nests, and mates when the latter were alive and available.

The Barn Swallow's use of straw to help bind clay during nest construction may account for its distribution, which is widespread compared to that of the Cliff Swallow. The mud nest is usually on vertical floor joists or horizontal crossbeams in barns and similar structures. Nests are seldom placed less than 3 m (10 ft) apart on the same beam (Snapp 1976). The masonry is started with a mud disk attached to a vertical surface around which the cuplike structure is developed. Both sexes work together and take 6–12 days to complete the nest (Bent 1942; Harrison 1975). The mud is always carried in the swallow's mouth and throat, and is pressed out through the partly opened mandibles as a semiliquid mass. The globules of mud, initially small and later large, are laid down in tiers. The mud cups are lined with rootlets, dried grass, and feathers. Nests in narrow cavities or holes with supporting sides and floor do not need the mud foundation and are simply made of grasses or feathers (Bent 1942).

Stephen W. Eaton

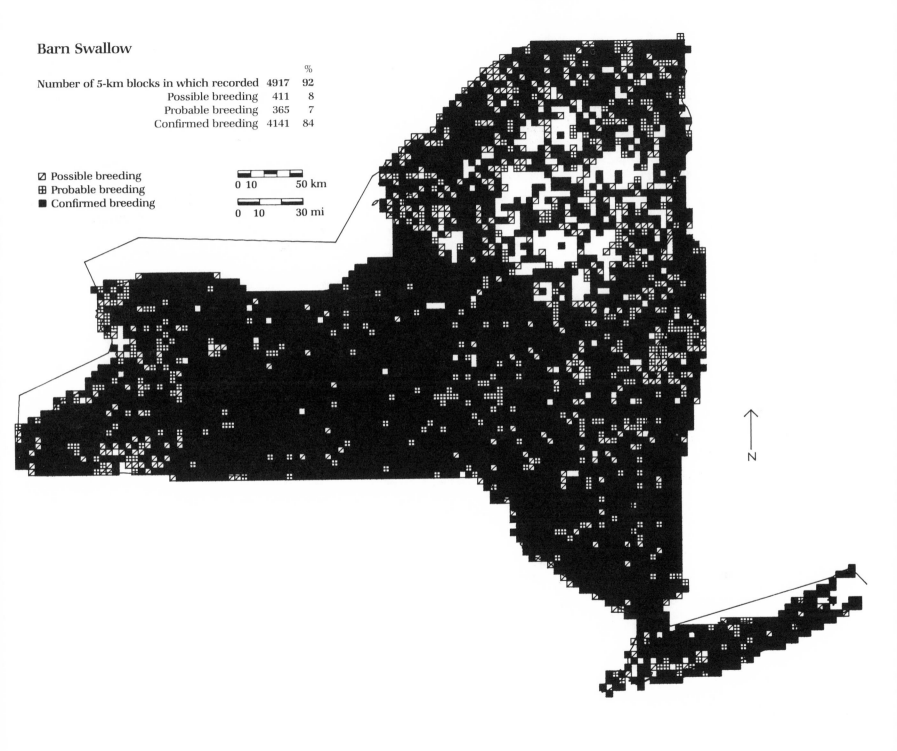

Barn Swallow

Number of 5-km blocks in which recorded 4917 92
Possible breeding 411 8
Probable breeding 365 7
Confirmed breeding 4141 84

☒ Possible breeding
⊞ Probable breeding
■ Confirmed breeding

0 10 50 km

0 10 30 mi

N

J Page
©1987

Gray Jay *Perisoreus canadensis*

The Gray Jay is a fairly common resident restricted to spruce–fir–northern hardwood forests within portions of the Central Adirondacks and Western Adirondack Foothills. Within the Adirondack High Peaks of Essex County this northern jay is rare.

During his 1840 Adirondack visit De Kay (1844) saw Gray Jays "at the sources of the Saranac in June, where they appeared to be numerous." Around the St. Regis lakes, Franklin County, between 1874 and 1877 Roosevelt and Minot (1929) found them "locally common in the thicker woods." Eaton (1914) felt that the Gray Jay was a fairly common resident in denser portions of the Adirondacks, both in black spruce–tamarack swamps and on the wooded mountain slopes. A map of breeding distribution was presented by Bull (1974), which showed records from 18 localities in six counties, including Silver Lake, Clinton County, a county from which there were no Atlas records. The Gray Jay was once reported to be a fairly common resident of Clinton County by George Shattuck (Eaton 1910) but today is known only as an accidental winter visitor (Warren 1979).

Atlas observers located the Gray Jay in 95 blocks in the Adirondacks, and in many blocks there were multiple sightings, or sightings over several years within the same block. This jay undoubtedly breeds in many other locales from which no records were obtained. The Atlas map suggests several disjunct populations, which is probably not entirely accurate, although the High Peaks birds may be disconnected from the mostly contiguous main range.

The Gray Jay is a bird of medium to mature boreal spruce forest (Erskine 1977), especially black spruce in New York. On higher ground the Gray Jay occurs in white and red spruce, but it is most often observed in lower-elevation black spruce. Where balsam fir predominates, the Gray Jay is usually replaced by the Blue Jay (Erskine 1977). I have observed the Gray Jay in stands of varying age and density, from broken bog forest to dense tracts opened only by blowdowns. The logging of spruce, first for lumber, then for pulpwood, has removed much of this habitat over the past century or more. Large areas of southern Franklin County that may once have had the Gray Jay are now covered with middle-aged red and Scotch pine, planted in the wake of railroad fires at the turn of the century (Donaldson 1921). Donaldson stated that plantings were set out in many parts of Essex and Franklin counties following the other great fires of 1903: 520,000 transplants in 1905, 548,000 in 1906, and 150,000 in 1907, or over a million trees, mostly pines or Norway spruce, planted in just three years. Perhaps this explains, in part, the separation of the High Peaks population. The reasons for the scarcity of Atlas records from the High Peaks (only two blocks) is unclear. Reports from Atlas fieldwork and previous *Kingbird* records do not approach Eaton's (1914) description of finding the birds "at intervals in all the forests from the Ausable lakes to Skylight camp on the slopes of Mt. Marcy."

The nest is usually placed against the trunk of a small spruce, on a horizontal branch or in a crotch. It is a cup of various twigs, *Usnea* lichen, bark, and grasses (Bent 1946); the thick and soft walls are lined with lichen, bark fibers, feathers, moss, hair, and fur. Caterpillar cocoons are often added to the nest (Wilmore 1977). This insulated lining is essential for a bird that nests in February or early March (Eaton 1914).

Before the Atlas study there was no published account of a nest, either with eggs or young, being found in New York (Bull 1974). The first nest in the state, a used nest, was reported by Atlas observers. Found 26 July 1983 at Long Pond outlet, St. Lawrence County, it was built next to the trunk of a 3 m (10 ft) young black spruce in thick woods of moderate sized trees. It was placed about 1.2 m (4 ft) above the sphagnum substrate, and the nest materials included grouse feathers and fur (perhaps varying hare) (NYSDEC files).

John M. C. Peterson

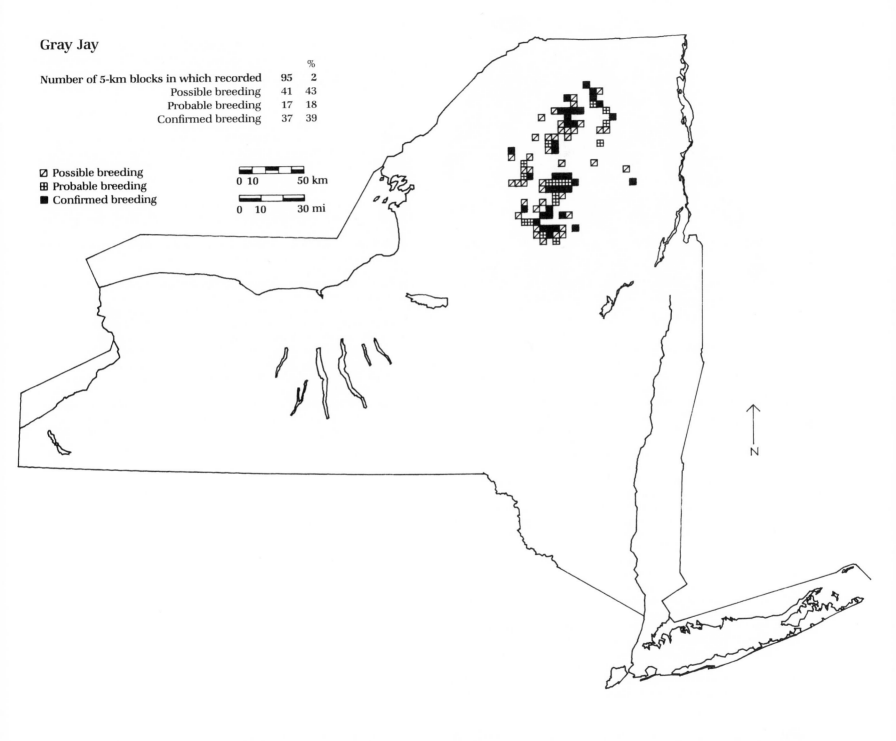

Gray Jay

Number of 5-km blocks in which recorded	95	% 2
Possible breeding	41	43
Probable breeding	17	18
Confirmed breeding	37	39

◪ Possible breeding
⊞ Probable breeding
■ Confirmed breeding

0 10   50 km

0 10   30 mi

N

Blue Jay *Cyanocitta cristata*

The Blue Jay, one of New York's most abundant and well-known birds, is a very common breeder almost everywhere in the state. Until the early 1900s, however, this species was a forest bird and not commonly found in the state's cities and towns (Eaton 1914). Eaton described it as common "only in the less settled districts," saying that it preferred evergreen or mixed woods. In the agricultural areas of western New York, he said it was "confined mostly to the larger forests, swamps and ravines." These statements are no longer true. The Blue Jay has gradually moved into cities and towns, becoming as abundant in suburbs and urban parks as it is in extensive forests in upstate New York.

In New York City, Carleton (1958) listed no breeding records for this species in Prospect Park and listed it as only an occasional breeder in Central Park. The Blue Jay now breeds every year in both locations and at other sites within the city (Salzman, pers. comm.). Only the concrete alleys of New York City seem completely without this species. Griscom (1923) and Cruickshank (1942) considered it absent along the outer beaches of Long Island, on Gardiners Island, and on portions of Orient Point, Suffolk County. All of these areas now have breeding populations. In 1921 it was considered unusual to see a Blue Jay within the village on Staten Island, Richmond County, an area where this species is now common and breeds every year (Siebenheller 1981).

As indicated on the Atlas map, the Blue Jay is a widespread breeder in the state. Of the 230 species "confirmed" as breeding in New York during the Atlas, only 3 were reported from more blocks. Although quiet and secretive early in the breeding season, the Blue Jay is easily seen and relatively easy to "confirm" once the young have hatched or left the nest. The Blue Jay was "confirmed" in 50% of the blocks in which it was reported. Most "confirmed" records (51%) were of recently fledged young following their parents or of adults seen carrying food to their young (30%).

Today the Blue Jay breeds in a wide range of habitats, including city parks and residential neighborhoods as well as extensive forests and woodland swamps. Although in forested areas this species seems to prefer oak woodlands, it can be found in all types and sizes of woods (Bull 1974). The variety of sites in which the Blue Jay nests has increased with its invasion of populated areas over the past 50 years. Away from human habitation, it typically nests in the crotch of a tree trunk, the fork of a branch near the trunk, or, occasionally, near the tip of a horizontal branch at heights of from 1.5 to 15.2 m (5–50 ft) above the ground, although usually below 6.1 m (20 ft) (Tyler 1946a). In populated areas the species often nests in ornamental shrubs next to houses and only a few feet off the ground. Some very unusual locations include a metal cross-beam of a street lamp, exposed tree roots on the side of a 1.8 m-high (6 ft) dirt bank next to a road, and the eaves of buildings (Tyler 1946a), as well as a fire escape in Brooklyn Heights (Salzman, pers. comm.), and an old American Robin's nest on a rafter inside a shed (Rosche 1966).

Both adults construct the nest, gathering small branches and twigs by breaking them off trees and shrubs with their bills (Terres 1980). A false nest or base made of twigs is constructed on the chosen site before the actual nest is built (Hardy 1961). The bulky nest is then constructed using materials such as twigs, bark, moss, lichens, paper, rags, string, wool, leaves, plant fibers, dry grasses, and strips of bark. These materials are occasionally held together with mud. Fine rootlets or, in populated areas, cloth or paper, are used to line the cup of the nest (Tyler 1946a).

Steven C. Sibley

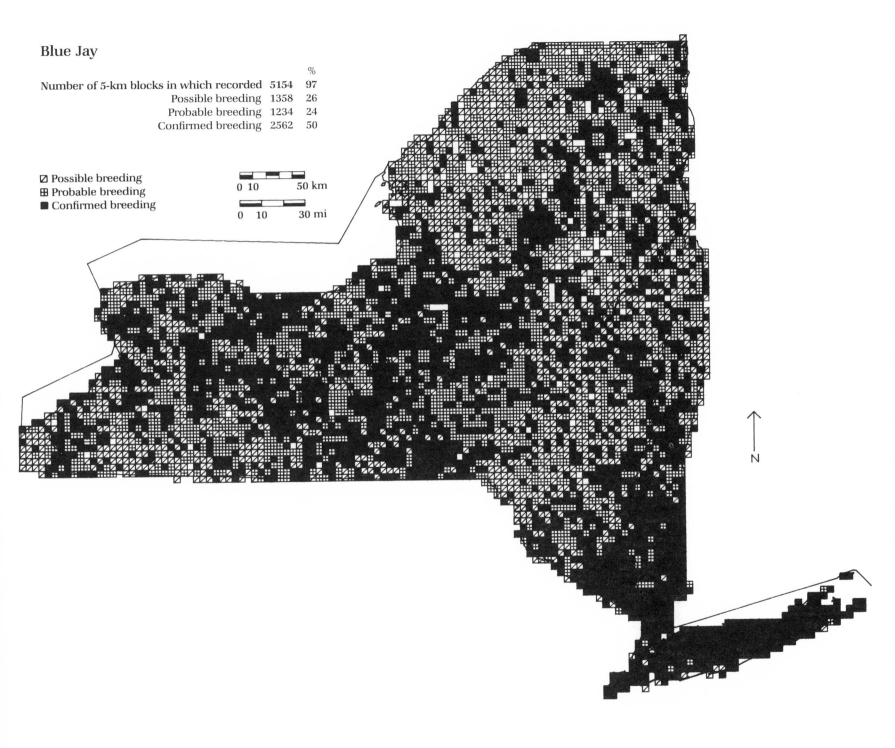

Blue Jay

Number of 5-km blocks in which recorded	5154	% 97
Possible breeding	1358	26
Probable breeding	1234	24
Confirmed breeding	2562	50

☑ Possible breeding
⊞ Probable breeding
■ Confirmed breeding

0 10   50 km

0 10   30 mi

N

American Crow *Corvus brachyrhynchos*

A noisy and gregarious open-country bird, the American Crow is also known as common crow (AOU 1983). Because of its ability to decimate agricultural crops, it has been widely persecuted for many years. Attempts to control the crow have been largely futile, however, and it remains a very common breeding species in suitable habitats throughout much of New York. Eaton (1914) stated that it bred in every county, entering the forested regions of the Catskills and Adirondacks along the cleared land and river valleys. He also noted that the American Crow was found in the High Peaks region of the Adirondacks, but was replaced by the Common Raven in the wildest portions of the western Adirondacks. Bull (1974) stated that the American Crow is a widespread breeder throughout the state.

The American Crow is probably more common today than it was in the early 1800s, when much of the state was covered with virgin forest. According to Eaton (1914), when De Witt Clinton visited Seneca Lake near Geneva, Ontario County, in 1810, he was told that there were numerous ravens in the area, but no crows. By 1910, however, after much of the land had been cleared, the raven had disappeared from Ontario County, and the crow was abundant. Today, much of New York is reverting to heavily forested land, and the raven is again increasing. Therefore, it is possible that the number of American Crows will decrease somewhat from the high levels of the earlier years of this century, although Robbins et al. (1986) discovered that the American Crow increased slightly on BBS routes in New York from 1965 to 1979.

Examination of the Atlas map indicates that the American Crow is found throughout most of the state. It is most often "confirmed" in the western and southeastern portions of the state—in the Great Lakes Plain, the Appalachian Plateau, the St. Lawrence Plains, the lower Hudson Valley, and the Coastal Lowlands. These regions were well covered by Atlasers and contain a large amount of agricultural and urban land. The American Crow is notably absent only from parts of the Adirondacks and Catskills.

During much of the year the American Crow is highly conspicuous because of its large size, loud voice, and gregarious habits. During the breeding season, however, it is quiet and secretive, and its nest can be difficult to find. Therefore, the species may breed even more widely than indicated by the map.

The American Crow is most numerous in open areas, such as agricultural land, urban areas, and tidal flats, which are adjacent to open forest and woodland having cover suitable for nesting and roosting (Bull 1974; AOU 1983). Because human activity provides and maintains suitable habitat, this species has thrived and undoubtedly will continue to prosper (Bull 1974).

The American Crow usually nests in woodlands but sometimes chooses a tall, isolated tree or occasionally even a small tree or shrub in a hedgerow. Depending on the location, the nest may be anywhere from 1.8 to 30.5 m (6–100 ft) above the ground, although 5.5–18.3 m (18–60 ft) is the common range (Terres 1980). The nest is generally built in a crotch of a tree or in the fork of a sturdy branch close to the trunk. The bird seems to prefer coniferous trees, but it commonly uses deciduous trees in regions where conifers are scarce (Bent 1946).

Both sexes may construct the nest, which is a bulky, well-built structure with a large foundation of materials such as sticks, twigs, or cornstalks, and a well-formed central cup lined with bark, plant fibers, grasses, mosses, leaves, roots, cow hairs, twine, or rags. The nest is so deeply hollowed that usually only the tip of an incubating bird's tail can be seen projecting over the edge (Eaton 1914).

Richard E. Bonney, Jr.

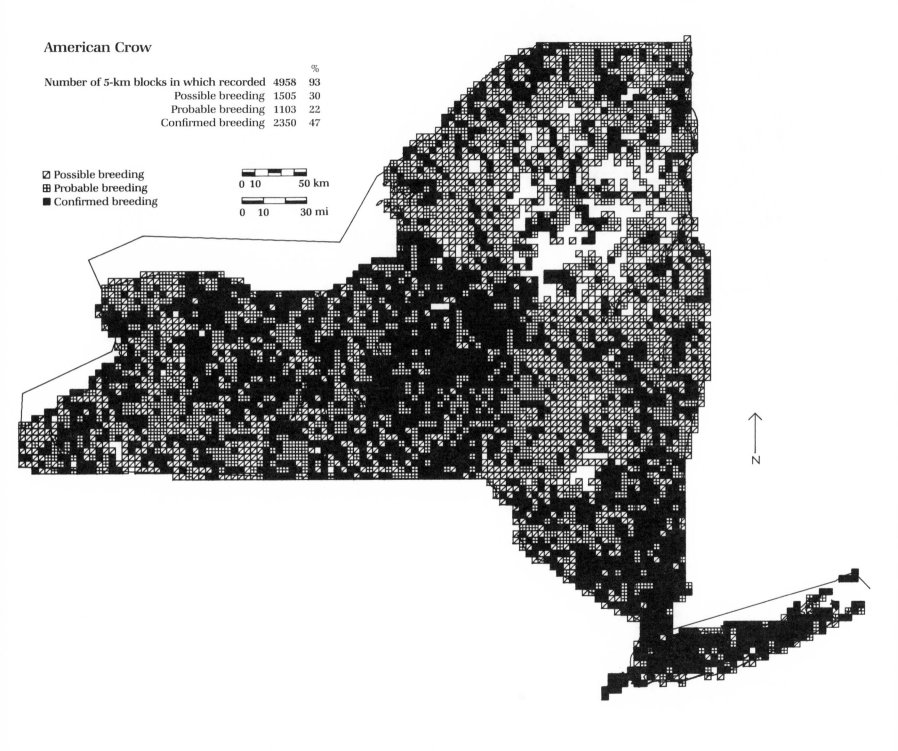

American Crow

Number of 5-km blocks in which recorded	4958	% 93
Possible breeding	1505	30
Probable breeding	1103	22
Confirmed breeding	2350	47

◨ Possible breeding
⊞ Probable breeding
■ Confirmed breeding

0 10   50 km

0 10   30 mi

N

Fish Crow *Corvus ossifragus*

A conspicuous denizen of the coasts and tidal rivers, the Fish Crow is an uncommon to locally common breeder throughout most of its restricted range in the Coastal Lowlands and southern Hudson Valley. On Long Island, Fish Crows are few and generally breed only in isolated pairs away from American Crows (Salzman, pers. comm.). This crow is rare away from southeastern New York, where its presence outside its normal habitat is somewhat mysterious, although DeGraaf et al. (1980) described breeding in "fertile farmland well inland from the coast (100+ miles, 160 km) in Pennsylvania and Maryland."

It appears that the Fish Crow has increased its range in recent years. Consider first the Coastal Lowlands population. In the early 1900s the Fish Crow was apparently confined to the western portion of Long Island (Eaton 1914). Breeding occurred at scattered locations along the south shore from Brooklyn, Queens County, to Montauk, Suffolk County, at the tip (Cruickshank 1942). Atlasing, however, found the species breeding throughout all of Long Island, except for small portions of the interior.

Consider also the Hudson Valley population. Eaton (1914) stated that the Fish Crow was confined to the lower Hudson Valley. It was reported as far north as West Point, Orange County, and occasionally to Poughkeepsie, Dutchess County, but very rarely farther north. Bull (1974) stated that the species bred as far north as Poughkeepsie on the east bank of the Hudson and as far as Esopus, Ulster County—about 16 km (10 mi) farther north—on the west bank. Atlasing, however, found evidence of breeding as far north as East Greenbush, Rensselaer County, "confirming" it at Castleton-on-the-Hudson, nearly 48.3 km (30 mi) north of Esopus.

Finally, consider the breeding records in Tompkins and Broome counties. Bull (1974) stated that the Fish Crow was unknown outside of southeastern New York, but atlasing "confirmed" breeding in one location in Ithaca, Tompkins County, south of Cayuga Lake, and "possible" breeding was recorded at another Ithaca location and also in Vestal, Broome County, near the Susquehanna River, where Fish Crows have been reported for several years but no nesting has been documented (J. Shepherd, v.r.). The Ithaca Fish Crows were first discovered on 16 March 1974 (Comar 1974). The first nest was found in 1979 (Benning 1979), and another in 1981 with three young (C. and L. Leopold, v.r.).

It seems that much of the range expansion has occurred within recent years. Poole (1964) noted that the species had been increasing and expanding its range northward in Pennsylvania. Northward expansion along the Hudson may be limited, however; the end of the tidal Hudson, the Troy Dam, Rensselaer and Albany counties, is not far from the northernmost Atlas records.

The Fish Crow is usually found in the immediate vicinity of the seashore, along beaches, bays, inlets, lagoons, and swamps, or inland along major watercourses (AOU 1983), especially tidal rivers (Eaton 1914). It prefers to nest in coniferous trees, with cedar, pine, and spruce being favored in drier localities. In low, wet woodlands, oak, ash, maple, and black gum are used (Bull 1974). In Ithaca the nest was found in a developed suburban area dotted with mature white pines and Norway spruces (C. and L. Leopold, v.r.).

The Fish Crow usually breeds in small colonies of 2 or 3 pairs with nests in separate trees, usually from 4.6 to 12.2 m (15–40 ft) from the ground but occasionally much higher (Bent 1946; Bull 1974; Terres 1980). The nest closely resembles that of the American Crow, being made of sticks, twigs, bark, and grasses, and having an inner cup lined with strips of grapevine bark, fine grasses, and pine needles (Eaton 1914).

Richard E. Bonney, Jr.

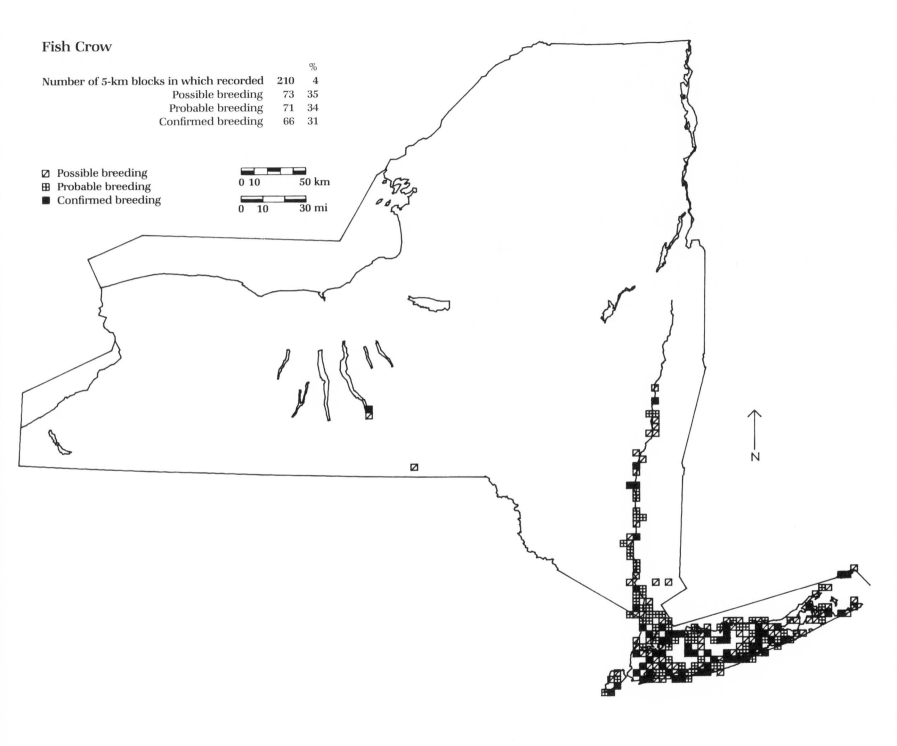

Fish Crow

		%
Number of 5-km blocks in which recorded	210	4
Possible breeding	73	35
Probable breeding	71	34
Confirmed breeding	66	31

◪ Possible breeding
⊞ Probable breeding
■ Confirmed breeding

0 10   50 km

0 10   30 mi

N

Common Raven *Corvus corax*

Following a recent dramatic recovery, the Common Raven is again a fairly common breeding bird throughout most of the Adirondacks but is uncommon and local in other parts of New York. The Common Raven is listed as a Species of Special Concern by the NYSDEC.

Recent excavations at a rich fossil site in Genesee County indicated that the Common Raven was an inhabitant of New York as long as 9500 years ago (Steadman, pers. comm.). As recently as 1810 great numbers of Common Ravens were seen along Seneca Lake, where a century later it was unknown (Eaton 1914). By the mid-1800s the numbers of Common Ravens had decreased as the species retreated into the more remote regions of the state. By the 20th century the Common Raven had clearly declined in the Adirondacks. Eaton (1910, 1914) failed to find the species in Essex County in 1905 and was told that it was found there only in late fall or winter.

A review of New York records for the period 1914–50 indicated that there were only ten reports of the Common Raven in the Adirondacks. On 27 May 1968 Carleton (pers. comm.) heard a Common Raven while climbing Knob Lock Mountain, Essex County, and the next day Greenleaf Chase found a pair and a nest being built at Chapel Pond 6 miles away. On 1 June, Carleton reported on the cliff site above the pond: "Three young standing on ledge in or next to nest, changing positions" (Delafield 1968). The reestablishment of the species began in earnest in the 1970s. Of 600 reports of the Common Raven in New York from the turn of the century to 1979, 87% were from the period 1970–79 (Bishop 1980).

The decline of the Common Raven outside the Adirondacks in New York was probably due to disturbance and destruction of nesting territories, persecution by humans, and a decrease in the food supply that resulted from loss of forest habitat to agriculture and human settlement. Within the Adirondacks loss of forest habitat from logging and fires and the disturbance of nesting territories that resulted was probably the major factor (Bishop 1980). Additional stresses may have been imposed by the loss of moose and decline of the white-tailed deer in the late 1800s, concurrent with declines among predators, like the wolf and lynx, upon which the raven, as a scavenger, depended for winter food (Bishop 1980; Knight and Call 1980).

The reforestation of New York, a change in attitudes toward wildlife, and an increase in the raven's food supply from road kills, coyote predation, and an increasing deer population are thought to be responsible for the recovery of this corvid (Bishop 1980). There are several factors, however, that may still have an adverse impact on the Common Raven, including the planned closing of landfills that are foraging areas for this species (Knight and Call 1980) and rock climbing, which causes disturbance at the nest (Marquiss et al. 1978). Nevertheless, the Common Raven now appears to have a secure hold in the state, and further range expansion can be expected as forest cover increases.

In an intensive search for breeding Common Ravens from 1977 to 1979, Bishop (1980) found only 14 nests, all within the Adirondack Park. The Atlas map shows "confirmed" breeding in 44 blocks in the Adirondacks. Breeding was also "confirmed" in the Catskill Peaks and Lake Champlain Valley. Other observations were in the Taconic Highlands, Tug Hill Transition, and on the Appalachian Plateau, from areas of heavy forest cover. Vermont may have been the source of ravens seen in New York's Rensselaer and Washington counties (Oatman 1985). Records from the Central Appalachians represent a continuum of the western Pennsylvania Appalachian population (AOU 1983; Pennsylvania Breeding Bird Atlas, preliminary map). Although the Common Raven is a large and vocal bird, coverage of remote areas was sometimes cursory, and ravens undoubtedly occur in many more blocks than indicated on the map. Most Atlas records were obtained by the relatively few observers familiar with vocalizations, field marks, and habits of this large corvid.

The Common Raven occupies a variety of habitats in wooded parts of the state, from boreal forests to hardwood ridges, "but most frequently in mountainous or hilly areas, especially in vicinity of cliffs, a preferred nesting site" (AOU 1983). The nest is a large, thick bowl built of sticks and lined with a layer of soil at the bottom and a layer of thick hair, sometimes mixed with bark, grass, moss, and other insulating materials (Knight and Call 1980). All New York nests, except those on a mine structure and an open pit mine in Essex County, have been on rock cliffs; however, the raven is strongly suspected of tree nesting in the Adirondacks and elsewhere.

John M. C. Peterson

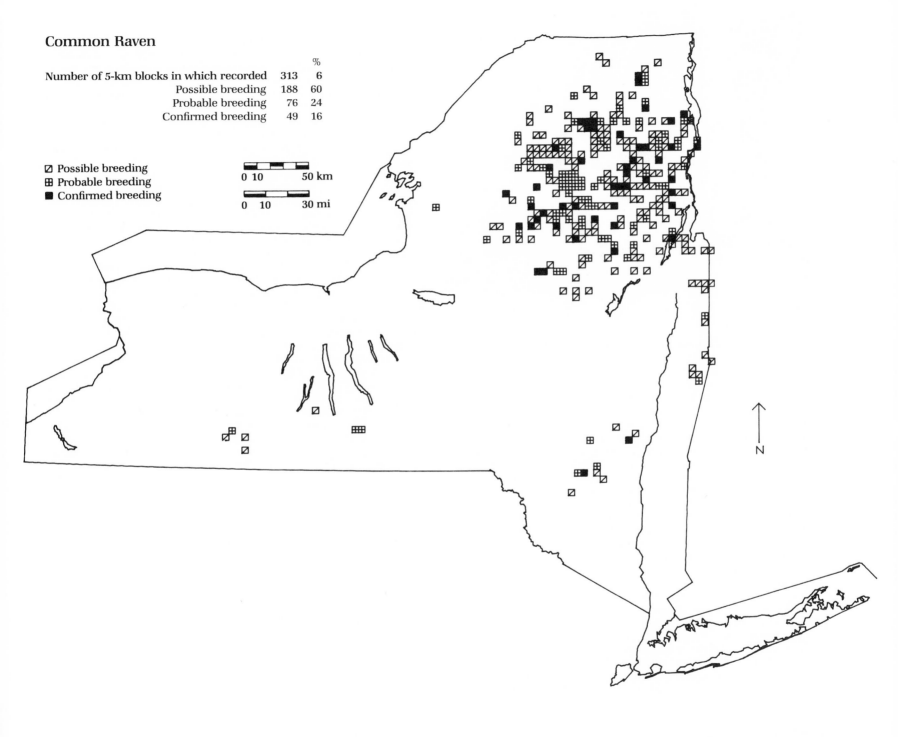

Common Raven

Number of 5-km blocks in which recorded	313	%
		6
Possible breeding	188	60
Probable breeding	76	24
Confirmed breeding	49	16

◨ Possible breeding
⊞ Probable breeding
■ Confirmed breeding

0 10   50 km

0 10   30 mi

N

Black-capped Chickadee *Parus atricapillus*

Named for its call, *chick-a-dee-dee*, and its black cap, the Black-capped Chickadee is one of the best known birds in the Northeast and is a common breeding species in New York. Statewide BBS data for the period 1965–79 show a significant increase in this species' abundance (Robbins et al. 1986).

Urbanization of the New York City region in the late 1800s and early 1900s caused the chickadee population to decline, but in the 1940s and 1950s a resurgence apparently occurred (Siebenheller 1981). Cruickshank (1942) stated that the Black-capped Chickadee was "gradually disappearing as a summer resident all around the metropolis." From the 1920s to the 1950s in Central and Prospect parks it was only a regular fall transient and rare at all other times. No breeding was documented until 1950 in Prospect Park and until 1954 in Central Park (Carleton 1958). Siebenheller (1981) stated that this species has increased on Staten Island since the 1940s, when it was a rare nester. Although the reasons for this recent trend are unclear, it does correspond to an increase in backyard bird feeding.

The distribution shown by the Atlas map does not represent any change from what was previously known. The Black-capped Chickadee breeds throughout the state and is generally absent only from New York City and portions of Long Island's south shore. Saunders (1929a) indicated that this species was found at elevations as high as 1219 m (4000 ft). In New York State it is most abundant in the Adirondacks (Robbins et al. 1986). Although this chickadee is rather silent in late May and early June during the peak of nesting activity, its soft *fee-bee* song in early spring

and the noisy chatter and feeding of young make it an easy species to locate. Only four species were recorded in more Atlas blocks, and few had a higher rate of "confirmed" records. Nests were not found as often as recently fledged young or adults feeding young.

The Black-capped Chickadee is a bird of the woods, preferring mixed deciduous-coniferous forest (DeGraaf et al. 1980). However, it also can be quite common as a breeding species in more open habitats such as orchards, suburban neighborhoods, and city parks, as well as in hedgerows and woodlots with mean widths of as little as 15 m (49.2 ft) (Stauffer and Best 1980). Throughout the state it also breeds in coniferous woodlands ranging from conifer plantations to the pitch pine–scrub oak barrens of the Coastal Lowlands and the spruce-fir forest of the Adirondack High Peaks. The Black-capped Chickadee often nests in comparatively open areas and feeds and rests in deeper woods. Nesting territories in a variety of habitats in New York averaged 5.3 ha (13.2 a) (Odum 1941a).

Nests are almost invariably placed in small cavities in soft dead trees or limbs within 3 m (10 ft) of the ground (Stauffer and Best 1980). However, the Black-capped Chickadee also readily uses nest boxes and occasionally nests as high as 15.2 m (50 ft) (Tyler 1946b). The chickadee often excavates the nest cavity itself by pecking and tearing away rotting wood (Odum 1941b). Unusual nest sites include the open tops of tree stumps and fence posts, a cavity I found in the top of a wooden guard rail in the middle of a picnic area in Ithaca, Tompkins County, and an open nest "cupped in a knot-hole on an upper rail of a split-rail fence, roofed over only by thick multiflora rose" in Fayetteville, Onondaga County (Scheider 1958). The size of the entrance to the nest cavity varies greatly; it is rarely round (Odum 1941b). The cavity itself ranges from 12.7 to 20.3 cm (5–8 in) deep (Eaton 1914; Odum 1941b), with the inside diameter of the bottom of the cavity measuring 6.4–7.1 cm (2.5–2.8 in) (Odum 1941b). A small cup of materials such as plant fiber, hair, fur, moss, feathers, and insect cocoons is constructed in the bottom of the cavity (Tyler 1946b). Both sexes participate in excavating the nest cavity, but only the female brings material to line the nest (Odum 1941b).

Steven C. Sibley

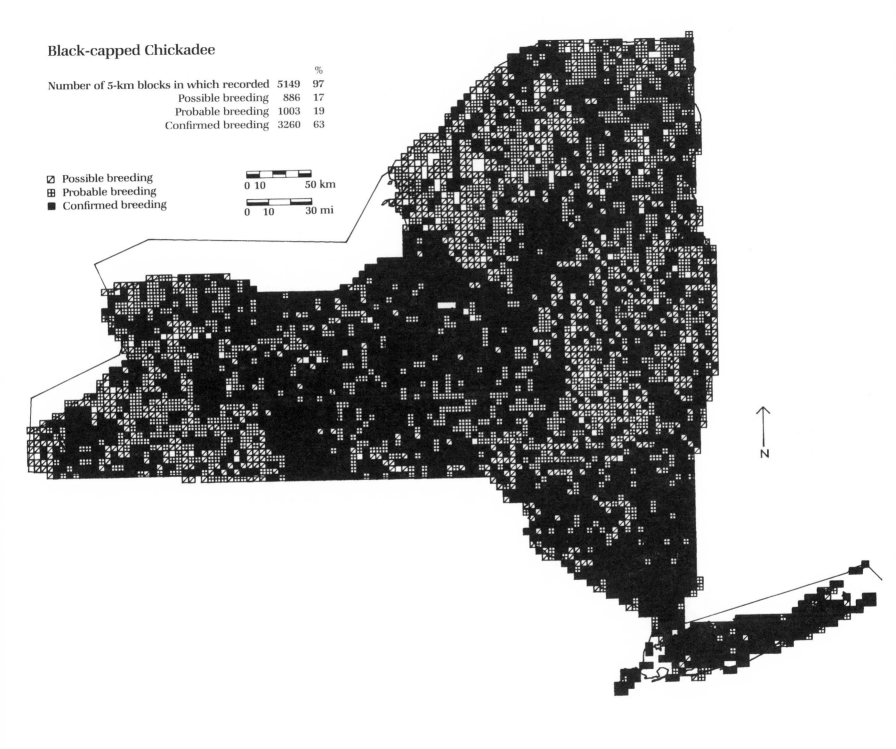

Black-capped Chickadee

		%
Number of 5-km blocks in which recorded	5149	97
Possible breeding	886	17
Probable breeding	1003	19
Confirmed breeding	3260	63

◪ Possible breeding
⊞ Probable breeding
■ Confirmed breeding

0 10   50 km

0 10   30 mi

N

Boreal Chickadee *Parus hudsonicus*

Although absent over much of the state, the Boreal Chickadee is fairly common throughout the spruce–fir forests of the Adirondacks, where its wheezy *chick-sa-day-day* is a familiar sound, from lowland bogs to the High Peaks. Early observers from Roosevelt onward seldom mentioned numbers of Boreal Chickadees encountered, but by the late 1960s bands of up to 25 birds were being regularly reported from the Adirondacks between late July and late March (Yunick 1984). However, as Yunick indicated, there has been an increase in observer activity as well.

Theodore Roosevelt found what was then known as the Hudsonian chickadee in small flocks at Bay Pond, Franklin County, at an elevation of 485 m (1590 ft) in early August 1874 or 1875 (Roosevelt and Minot 1929). During early 1882 Headley found it as abundant as the Black-capped Chickadee near Big Moose Lake, Herkimer County. In April of the same year Merriam collected four specimens, and Ralph and Bagg found birds breeding the following year in Hamilton and Herkimer counties. Ralph and Bagg reported an Oneida County record in 1886, and at the close of the 19th century eggs were collected at Jock Pond, Herkimer County, in July 1898 (Eaton 1914).

During 1905 Eaton found what by then was called the Acadian chickadee at several locations in the High Peaks of Essex County (Eaton 1914). Saunders (1929a) observed the Boreal Chickadee commonly in the High Peaks from about 1219 m (4000 ft) to timberline, noting: "In the high mountains it is not much in evidence in early summer, but later it is quite common on the McIntyre and Marcy trails, and is almost certain to be found by those who look for it." Like the Black-capped Chickadee, the Boreal Chickadee is relatively secretive during the peak of breeding activity.

Bull (1974) presented a map of the 20 known breeding localities, half of them in Essex County. Atlas observers recorded the Boreal Chickadee in 123 blocks, a sixfold increase over the accumulated records of a century, and with fewer than a third in Essex County. The Atlas range, however, is nearly the same as that delineated by Bull, especially in the obvious gap along the Essex-Hamilton-Warren county boundaries. Much of this area appears to have been heavily logged for pulpwood at frequent intervals in the past. The native spruce–balsam fir is replaced by a successional forest of aspen, cherry, maple, and other broad-leaved trees, mixed with some white pine and hemlock, making the area unsuitable for this chickadee. The "possible" Atlas record in a conifer stand at the edge of a beaver meadow near Lyonsdale, Lewis County (M. Milligan, v.r.), was not unexpected in light of the 1886 Oneida County record at nearby Remsen (Eaton 1914).

Although recorded in 123 blocks, the Boreal Chickadee was undoubtedly missed in some. The map, however, is generally accurate, not only suggesting areas of concentration but also in showing certain gaps within the range where this species could not be found and habitat was no longer appropriate. Most "confirmed" records came from family groups that included recently fledged young, easily distinguished from the dingy, soiled adults by their clean white cheeks and fresh juvenile plumage, prior to the parents' post-breeding molt.

The favored habitat of the Boreal Chickadee is mixed spruce and balsam fir, sometimes interspersed with other species such as yellow birch. Tamaracks may be present in areas of the northern Adirondacks, especially in Essex and Franklin counties, but are much less common in much of Hamilton and Herkimer counties and unusual on the upper slopes of the High Peaks. Tamaracks, then, do not seem an essential component. Some two-thirds of the Adirondack Park is private land where logging continues to remove tracts of black spruce. Even on the secure state lands within the High Peaks red spruce are dying from the assault of destructive pest organisms (Birmingham, pers. comm.) and from other causes, one of which may be acid rain (Tripp 1983).

The Boreal Chickadee excavates a cavity or nests in a woodpecker hole in a dead tree or stub from 1.8 to 4.6 m (6–15 ft) above the ground. It seems to show a preference for yellow birch in the Adirondacks, although it has employed sugar maple and spruce saplings (Bull 1974). McLaren (1975) noted that "softness of the heartwood rather than a preference for a specific species of tree [seemed] to determine the nest site choice." Atlas observers located just 2 nests with young and none with eggs. Breeding was "confirmed" from about 457 m (1500 ft) near Bay Pond Bog to about 1128 m (3700 ft) near Indian Falls.

John M. C. Peterson

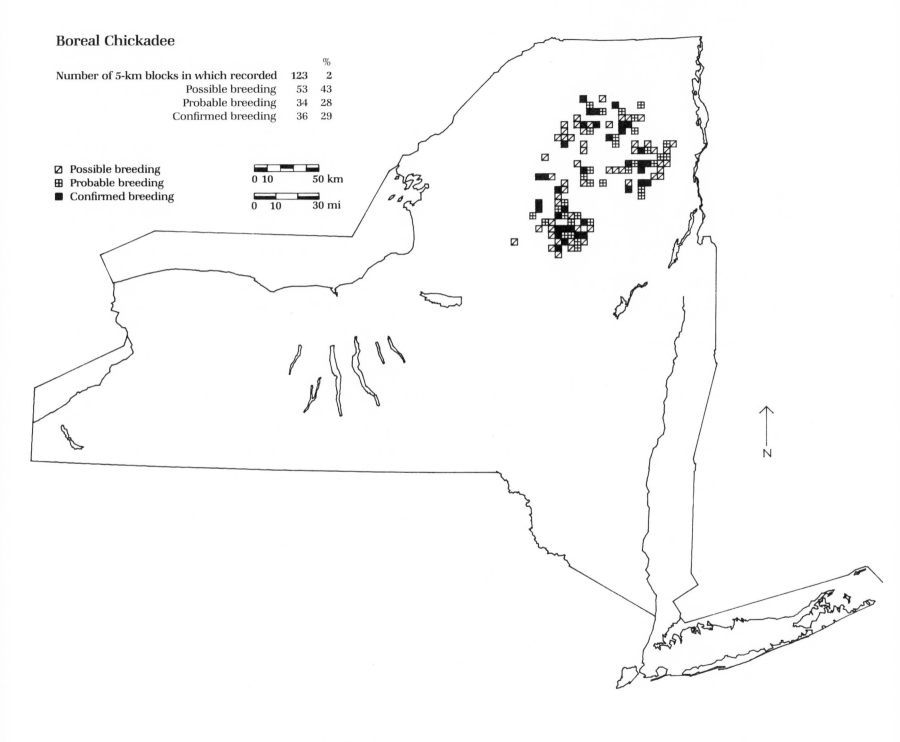

Boreal Chickadee

Number of 5-km blocks in which recorded	123	% 2
Possible breeding	53	43
Probable breeding	34	28
Confirmed breeding	36	29

◩ Possible breeding
⊞ Probable breeding
■ Confirmed breeding

0 10   50 km

0 10   30 mi

N

CJ Page © 1987

Tufted Titmouse *Parus bicolor*

This active, spritely, but somber-colored bird is one of several species that have infiltrated New York from the south in the past few decades, achieving widespread colonization. Although in the early 1900s Eaton (1914) knew of only one definite in-state breeding record, on Staten Island, today the Tufted Titmouse is a common species in many parts of the state.

Bull (1974) said that before 1950 the Tufted Titmouse was rare and local north and east of New Jersey, especially east of the Hudson River. It was extremely rare on Long Island. In 1954 there appeared to be an explosive population increase and expansion from northern New Jersey up to Rockland County, New York, and into Connecticut (Bull 1974). By 1971 it had spread east to western Suffolk County and locally up the Hudson Valley as far as Schenectady.

In western New York another influx was occurring. In the fifth edition of the AOU Check-list (AOU 1957) the species was said to be resident in extreme southern New York (Chautauqua and Chemung counties) but moving northward. An analysis of the number of Tufted Titmice seen per party on CBCs in the 1950s indicated a steady increase in the number of counts on which they were recorded, as well as the number of birds on the counts (Audubon Field Notes 1950, 1954, 1958). The pattern of increase seemed to indicate that the major invasion route was up the Susquehanna River–Finger Lakes corridor, with some spread from northwestern Pennsylvania up the Allegheny River into Chautauqua and Cattaraugus counties (Eaton 1959). By 1974 breeding was occurring in a broad band from the southern state line up to Lake Ontario, just to the east of Oneida Lake and west to the south shore of Lake Erie southwest of Buffalo (Bull 1974).

The Atlas map shows a more extensive distribution, with evidence of breeding to the Canadian border in Clinton County and a few records around the east end of Lake Ontario in Jefferson County. Comparison of the Atlas map with Bull's (1974) distribution map for this species makes it clear that the two populations are moving toward each other. There is still, however, a sparsely inhabited zone extending from Delaware County, through the Catskills, Herkimer County, the Tug Hill Plateau, and the Adirondacks. Much of this territory lies at or above 305 m (1000 ft) elevation. The Tufted Titmouse also occurs sparsely in western New York.

Why has this species had such a burgeoning range growth in the 1960s, 1970s, and 1980s? Eaton (1959) suggested these possibilities: an amelioration of the climate and the maturation of forest growing up on abandoned farmland. Boyd (1962) wrote about the increase of the Tufted Titmouse in western Massachusetts and gave data that seemed to support the climate-warming theory. Range extension began there in 1940, and by 1954 it was seen annually in Springfield, Massachusetts, in increasing numbers. In the period 1900 to 1909 the mean average temperature at Springfield was 8.1°C (46.7°F). For the period 1950 to 1959 it had risen to 9.3°C (48.8°F). Climatic research in England found that the global mean temperature showed little change in the 19th century, marked warming to 1940, steady conditions to the mid-1970s, and then a rapid warming to 1984 (Jones et al. 1986).

Because of its fairly constant activity, its easily heard voice, and its combative conduct, the Tufted Titmouse is readily observed and was not often missed in its territory by Atlas observers.

This industrious forager flits about searching and probing for food in its favored habitat of deciduous woodland, suburbs with mature shade trees, moist bottomlands, and swamps, adding residential feeding stations in the colder months (Bent 1946; DeGraaf et al. 1980).

When nesting time arrives, it seeks a cavity in live or dead wood, such as an old woodpecker or squirrel hole, a lightning-strike split, a hole in a fence post, a hollow metal pole, or even a man-made box (Bent 1946). Rarely, the nest hole will be excavated by the bird itself in a dead stub (Forbush 1929). The nest site can be from 0.9 m to 27.4 m (3–90 ft) high. Sometimes the same nest cavity is used over again for several years. The nest builder uses bark strips, grass, rags, wool, cotton, string, leaves, and often snakeskin. Leaves are most often used when they are wet and can be poked into a ball and crammed into the cavity (Bent 1946).

Gordon M. Meade

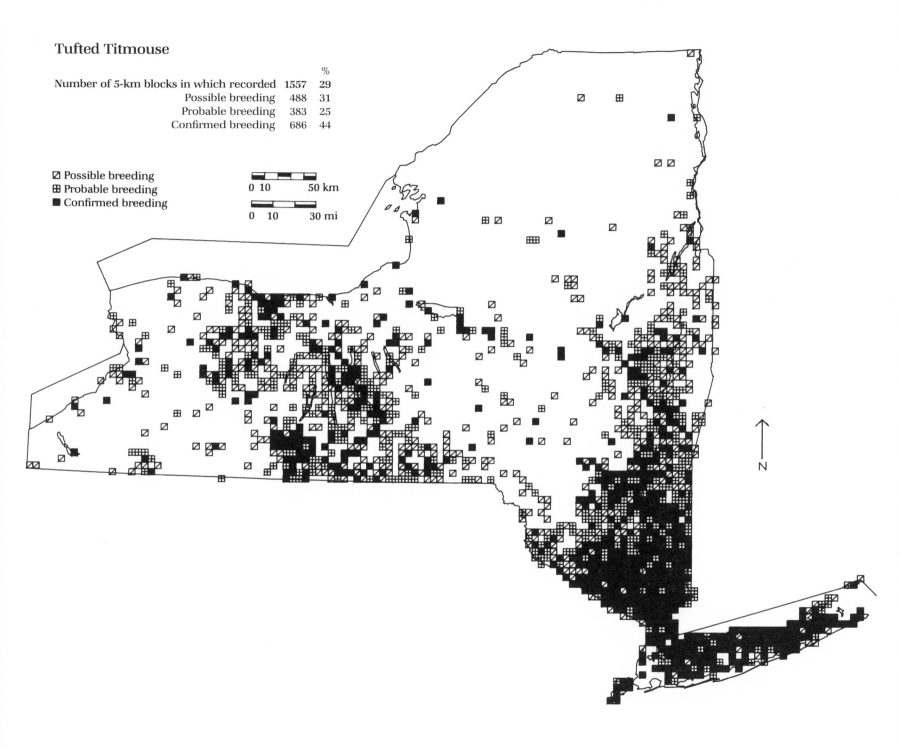

Tufted Titmouse

		%
Number of 5-km blocks in which recorded	1557	29
Possible breeding	488	31
Probable breeding	383	25
Confirmed breeding	686	44

☑ Possible breeding
⊞ Probable breeding
■ Confirmed breeding

0 10   50 km

0 10   30 mi

N

C J Page ©1987

Red-breasted Nuthatch *Sitta canadensis*

The Red-breasted Nuthatch breeds over much of New York but is not as generally distributed as its larger white-breasted relative. Within the conifer forests of the Adirondacks this small nuthatch is a common breeder, its nasal *ank-ank-ank* a familiar accompaniment to its active movement over the trunks of spruces or branches of balsam firs. Outside the Adirondacks, where abundance is largely determined by the availability of conifer growth, birds may be rare to fairly common.

A map showing the Red-breasted Nuthatch's "known breeding range, 1906," was among those presented by Eaton (1910). That map showed a range confined to the Central Adirondacks, the Eastern and Western Adirondack Foothills, Central Tug Hill, and Catskill Peaks, plus extralimital nesting in Erie, Madison, and Yates counties. A more recent map of breeding distribution was prepared by Bull (1974), who noted that there had been considerable range extension to lower elevations since Eaton's time. Besides showing a slight expansion in the Adirondacks to the Canadian border and eastward to Lake Champlain, the map showed breeding Red-breasted Nuthatches in nine counties around the Catskills. Bull also mapped 40 localities outside the mountains, including seven sites on the Great Lakes Plain. The known southeastern breeding limits in 1974 were in the northern Shawangunk Mountains, Ulster County.

The Atlas map illustrates a continued 20th-century range expansion all across the state and southeast, to include the Coastal Lowlands. Several new pocket populations of this species have been found on the Appalachian Plateau and in the Rensselaer Hills, Taconic Mountains, and Manhattan Hills. The most notable additions, however, have been on Long Island and Staten Island. These small nuthatches have now penetrated every physiographic region of the state except the other boroughs of New York City, the Hudson Highlands, and Triassic Lowlands.

Sedentary much of the time, this little bird periodically deserts its breeding grounds to take part in southward flights, returning in spring to find new areas to settle (Tyler 1948a; Bull 1974). With the gradual reforestation of much of the state during this century, the Red-breasted Nuthatch has had new opportunities placed in its path, especially as the conifer plantations established in the 1930s have matured.

Although a quite small denizen of dark forests, the Red-breasted Nuthatch tends to emit nasal *ank* notes or strings of *enks* when disturbed, making it easy to locate. Though certainly missed in some blocks, it was recorded in 1690, and we now have a good picture of its range. Most "confirmed" records came from family groups with recently fledged young (162 blocks) or adults feeding young (133 blocks). The ON (adults seen entering or leaving an occupied nest) code was employed in 34 blocks where the contents of active nests could not be determined. Atlas observers also noted nests with young in 21 blocks, with eggs in 4.

Favored habitat is coniferous, often spruce, or mixed woods with a strong conifer flavor. Density is usually greatest in balsam fir, hemlock, pine, or spruce stands of medium maturity. A resident pair may share the same area continuously, remaining mated for more than one season (Tyler 1948a).

The Red-breasted Nuthatch begins excavation of a nest cavity in a dead snag or rotted stub in spring, although it uses nest boxes or old woodpecker holes on occasion. Evans and Conner (1979) found that in northeastern forests the optimum dbh of snags used as nest sites by this nuthatch was 30.5 cm (12 in), and the entrance hole of the cavity was found to be about 3 cm (1.2 in). A curious feature of the nest hole is the pitch that both adults bring from balsam firs, spruces, or pines and smear around the opening. In one case, a female Red-breasted Nuthatch was found dead, stuck to the pitch. Speculation about the reasons for this adaptation includes Kilham's (1972) suggestion that the sticky pitch can prevent predators from entering the nest. The height of nests varies considerably, from as low as 1.8 m (6 ft) to as high as 15.2 m (50 ft). The cavity is lined with a cup of grasses, rootlets, and hair or fur (Tyler 1948a; Bull 1974).

John M. C. Peterson

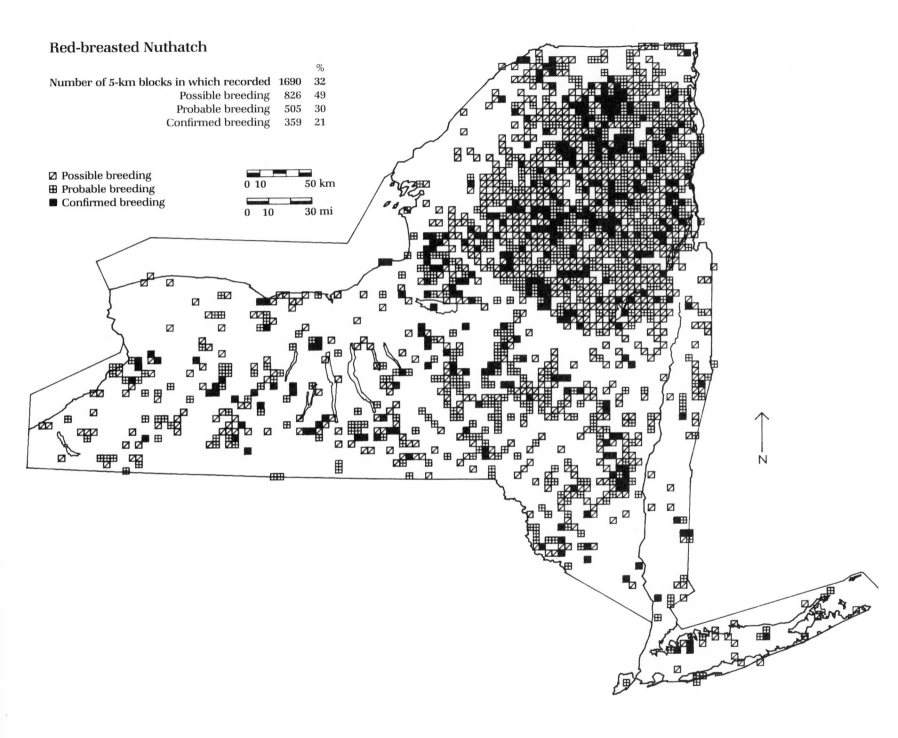

Red-breasted Nuthatch

		%
Number of 5-km blocks in which recorded	1690	32
Possible breeding	826	49
Probable breeding	505	30
Confirmed breeding	359	21

◩ Possible breeding
⊞ Probable breeding
■ Confirmed breeding

0 10   50 km

0 10   30 mi

N

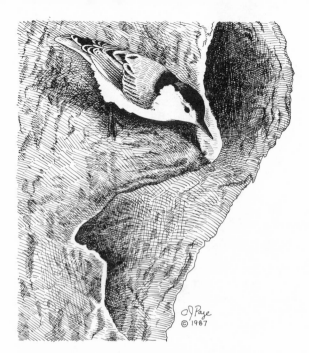

White-breasted Nuthatch *Sitta carolinensis*

A bark-gleaning bird of deciduous woodlands, the White-breasted Nuthatch is a fairly common breeder virtually throughout the state. It is often seen clinging upside down to tree trunks of all sizes. Eaton (1914) stated that this nuthatch was generally distributed throughout the state except for the balsam fir belt of the Catskills and Adirondacks. He added that it was uncommon in Keene Valley, Essex County, and around Old Forge, Herkimer County, although a few invaded the cleared land and valleys of the Adirondacks up to the edge of the coniferous woods where it was replaced by the Red-breasted Nuthatch. Eaton further noted that Bicknell had found the White-breasted Nuthatch uncommon in the valleys of the Catskills, but that it was present in most other low-elevation parts of the state, and Saunders (1942) noted that it is a common nester in Allegany State Park. Bull (1974) reported that the species was a widespread but local breeder, rare on the coastal plains and in the higher mountains.

The White-breasted Nuthatch may have expanded its distribution somewhat in recent years. Griscom (1923) noted that the species was common around New York City in winter but uncommon in summer. Cruickshank (1942) confirmed this 20 years later, adding that the bird was only a casual visitor along the coast and on the eastern tips of Long Island. Atlasing, however, found the species in many blocks in the New York City region and throughout most of Long Island. Also, Judd (1907) stated that the species was rare in Albany County in summer, whereas Atlasing found evidence of breeding in 80% of the blocks in the county.

On the other hand, Robbins et al. (1986) discovered a significant decrease in numbers of the White-breasted Nuthatch on BBS routes from 1965 to 1979.

The Atlas map clearly indicates the absence of the White-breasted Nuthatch from portions of the mountain spruce–fir forest of the Adirondacks, as well as parts of the spruce–fir–northern hardwood forests and successional northern hardwoods of northern and western New York. The absence of records may be partly due to less intense block coverage in those areas. The distribution of "confirmed" breeding records of this nuthatch bears a striking resemblance to the map of Atlas coverage showing blocks having 75 or more species. The White-breasted Nuthatch is very quiet during the early part of the breeding season and may have been missed where coverage was less intense.

The White-breasted Nuthatch is found primarily in open deciduous woodlands but also inhabits more dense deciduous and mixed deciduous-coniferous forests, orchards, groves, isolated shade trees, swamps, or urban and suburban areas having a quantity of large trees with suitable nest sites (Eaton 1914; Saunders 1942; Stoner and Stoner 1952; Bull 1974; AOU 1983). Ellison (1985d) suggested that in Vermont the presence of trees such as beech, oak, and hickory may be important in determining the species' abundance, as the White-breasted Nuthatch is scarce where these trees are few.

The White-breasted Nuthatch is an early breeder, with nest-building activity most evident in April (Ellison 1985d). It builds in a cavity of a deciduous tree, generally from 1.5 to 15.2 m (5–50 ft) from the ground, with the entrance often through a rotted-out knothole, which the birds may enlarge slightly (Eaton 1914; Bull 1974). It seems to prefer holes at least twice as wide as its body (Kilham 1971b), and sometimes makes use of an abandoned nesting hole of the Downy Woodpecker, as well as artificial nest boxes. The nest itself is composed of leaves, mosses, soft grass, shreds of inner bark, rabbit fur, wool, cow hair, and feathers (Eaton 1914; Terres 1980).

Richard E. Bonney, Jr.

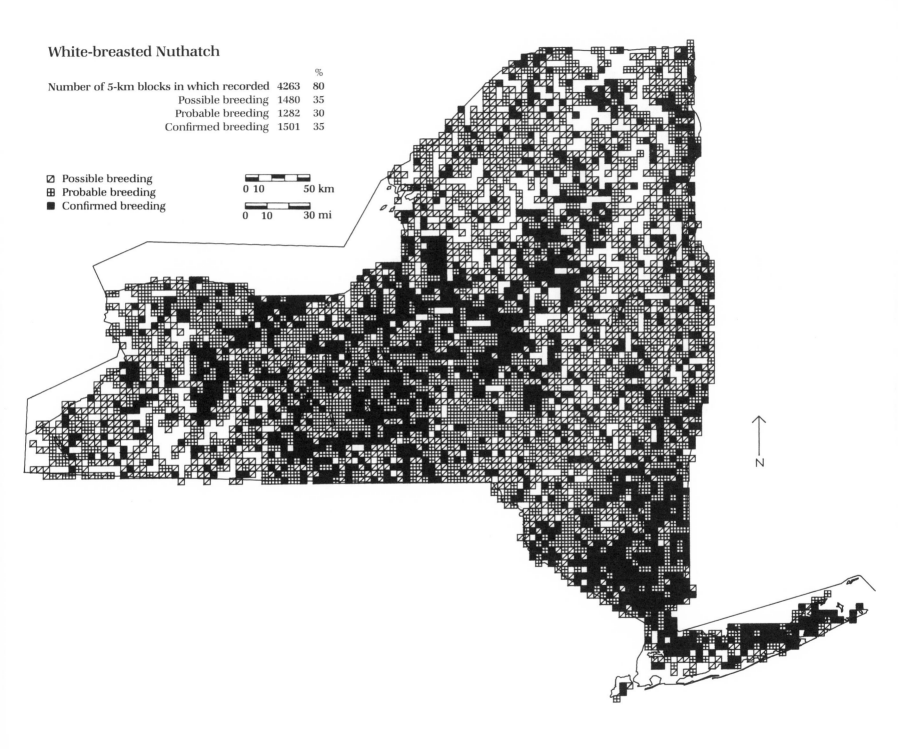

White-breasted Nuthatch

		%
Number of 5-km blocks in which recorded	4263	80
Possible breeding	1480	35
Probable breeding	1282	30
Confirmed breeding	1501	35

Possible breeding
Probable breeding
Confirmed breeding

0 10 50 km

0 10 30 mi

N

C J Page
© 1987

Brown Creeper *Certhia americana*

The unobtrusive Brown Creeper, which seemingly spends its life in near-sighted searching for food on tree trunks, is now a fairly common breeder in many parts of the state. It is, however, rather uncommon and decidedly local in other parts.

There has been a steady push southward of the species' breeding range on the East Coast in the 1960s, 1970s, and 1980s. In the 1957 edition of the AOU Check-list, Bronx County was the southern limit. However, the 1983 check-list extends the coastal southern limit to the lowlands of Virginia, Maryland, and Delaware (AOU 1957, 1983). This expansion must have been due, in part, to the reforestation in the eastern United States and New York, which has undoubtedly provided additional habitat for this species. In addition, Davis (1978) noted a southern range expansion in Michigan, which he attributed to the dying of American elms from Dutch elm disease. The loose bark from the dead trees provided nest sites and insects provided a good food supply. Because the bark of the American elm falls off completely in a short time, Davis felt that the expansion would be temporary. An increase of the Brown Creeper in central New York was attributed by Rusk and Scheider (1966) and Scheider (1971b) to the vast numbers of trees that died from Dutch elm disease there.

Eaton (1914) spoke definitively of populations of this species only in the Adirondacks and the Catskills, advising that the Brown Creeper was "an abundant summer resident" in those regions, but relegated it in the rest of the state to merely a "local summer resident." Beardslee and Mitchell (1965) called this species "rare" in western New York. Griscom (1923) listed no breeding records on Long Island or in Westchester County, and

Cruickshank (1942) showed only one breeding record in that area in Bronx County in 1926. The first confirmed breeding record on Long Island came in 1947, followed by additional records in 1950 and 1963 (Bull 1964). Atlasing subsequently "confirmed" breeding in 11 blocks in Westchester and 15 on Long Island.

Breeding was documented by the Atlas at higher elevations, but there are blocks throughout the state where the Brown Creeper breeds below 152 m (500 ft), and on Long Island it is found virtually at sea level. The Brown Creeper was found mainly in areas with 50% or greater forest cover. Areas of concentration recorded by Atlas work are the Central Adirondacks, the Tug Hill Transition, the western Mohawk Valley the eastern Appalachian Plateau, including the Catskill region, the lower Hudson Valley, Finger Lakes Highlands, the eastern Cattaraugus Highlands, and Allegany Hills. East of the Hudson River it was found in the Rensselaer Hills, southern Taconic Highlands, and adjacent Hudson Highlands and Manhattan Hills. Breeding was "confirmed" in every county in the state except Niagara and the five counties that make up New York City.

Despite that fact that the song of the Brown Creeper is not as well known as the bird itself, Atlas observers had little trouble locating this species. Males, both when establishing territories and maintaining territories, often answer other male songs in adjacent territories (Davis 1978). Nests were found in 70 blocks by those observers who patiently followed the Brown Creeper after noting it carrying food. Sometimes the food was for fledglings, but often the foodbearer disappeared under a piece of loose bark. Finding fledglings or simply observing the bird with food was certainly easier, however, than finding the nest.

Although the bird shows a decided preference for swampy woodlands or upland areas near streams (Davis 1978), nesting is recorded as well in dry uplands, in both coniferous and deciduous forests (Tyler 1948b). An additional habitat requirement may be openings in the forest canopy at the nest site to allow light penetration (Davis 1978). There must be dead or dying trees in the area, with loose shingles of bark adhering to them. The nest is almost invariably placed behind one of those shingles of bark, usually at heights not exceeding 6.1 m (20 ft) (Tyler 1948b). According to Davis (1978) the bark in the area of the nest is free from dirt and rotting material. Occasionally a nest has been found in a knothole or woodpecker hole (Tyler 1948b).

The nest is made of twigs, bark, and bark scale bound together and to the rough bark surface by spider egg cases and insect cocoons. The nest is not attached to the tree trunk. The nest cup is centrally situated upon the base structure and is made of fine bark and wood fibers. The female does most of the nest building, although the male has been observed bringing materials to the nest site (Davis 1978).

Emanuel Levine

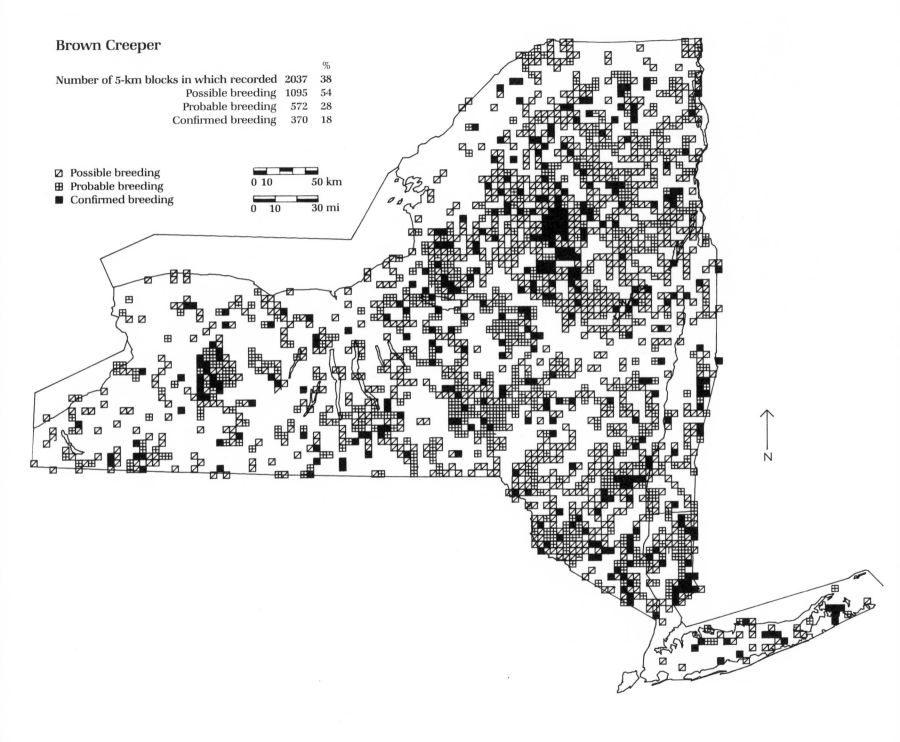

Brown Creeper

		%
Number of 5-km blocks in which recorded	2037	38
Possible breeding	1095	54
Probable breeding	572	28
Confirmed breeding	370	18

▨ Possible breeding
⊞ Probable breeding
■ Confirmed breeding

0 10   50 km

0 10   30 mi

N

CJ Page © 1987

Carolina Wren *Thryothorus ludovicianus*

If birds had mottos or slogans, the Carolina Wren's might well be "If at first you don't succeed, try, try again." Repeatedly this species appears in an area in New York for a season, or a year or two, and then disappears, only to reappear later. It is persistent and fairly hardy, and feeding stations will pull it through some winters, but during hard winters when there are ice storms, long periods of sub-zero cold, and heavy snow, its populations can be decimated. With a succession of mild winters it usually returns and breeds again (Bull 1974). During the period 1966–79 BBS data documented the effects of a variety of winter weather. The population was level from 1966 to 1970 during a period of normally cold winters, then increased dramatically during several mild winters until 1977, when it dropped to near its 1966 level following two unusually severe winters (Robbins et al. 1986).

In upstate and interior New York during the first four decades of the 20th century the Carolina Wren was considered by ornithologists to be a rare straggler (Short 1896; Eaton 1914). Eaton knew of only seven observations and two breeding records. Horsey (1938) said it was rare in the Rochester area, Monroe County, and Beardslee and Mitchell (1965) also said it was rare in the Niagara Frontier region. A change in its status was noted in the Niagara Frontier region by a gradual increase after 1953 despite the tolls of severe winters (Beardslee and Mitchell 1965).

Bull (1974) stated that the Carolina Wren was locally a fairly common resident on Long Island and in the lower Hudson Valley, uncommon at lower elevations elsewhere, and virtually absent in northern New York and in the mountains. Beginning in the 1950s it had shown a marked increase upstate, especially in the mid-Hudson and eastern Mohawk valleys; along the Great Lakes Plain, near Syracuse, Onondaga County, and Rochester and in the western Finger Lakes; and in the lower Genesee River Valley and other river valleys of the Appalachian Plateau. Beddall (1963) believed that the Carolina Wren, like several other southern species, although to a lesser degree, has been undergoing a range expansion, possibly under the advantageously long period of above-average temperatures that began in the 1930s. Concurrently the expansion was tempered by the species' ambivalence to the presence of man; some individuals accommodate well to man, others do not.

The breeding range today, as shown by the Atlas, has not changed much since Bull's (1974) assessment. It is still found mainly on Long Island except for the densely populated southwestern end, in the lower Hudson Valley, along the Mongaup River in Sullivan County, and Esopus Creek in Ulster County. It was recorded in the Susquehanna River Valley and watershed, up through the Finger Lakes corridors to Lake Ontario, and along the southern shore of Lake Erie. The northernmost "confirmed" record was along Lake Champlain in Essex County. Obviously, the distribution of this species follows closely the river systems of the state.

The Carolina Wren, though uncommon, is not a bird that would be easily missed by Atlas observers. It is rarely still for a moment. It will disappear suddenly into one hole and come out at another while nervously jerking its body and tail. If approached, it will pop into some dense cover while emitting repeated chatter (Bent 1948).

Away from human habitation, the Carolina Wren breeds in woodlands, thickets, brushy hollows, swamps, and along stream beds. It is typically found near water in dense tops of fallen trees, brush heaps, and rocky places. The nest is placed in a hole in a tree, an open tree crotch, a hole in a bank, or among tangled roots. This wren has also become a bird of farmyards and gardens, where it will nest in almost any nook or cranny with enough room. Sometimes the sites are bizarre, like the one built in a tool box of a tractor that was in daily use. The nest material can be anything soft and pliable: grasses, weed stalks, strips of inner bark, leaves, mosses, rootlets, and feathers, and even a cast-off snakeskin. The lining may be fine grass, rootlets, hair, feathers, or moss (Bent 1948). Sometimes the nest is a matted ball of sticks with a side entrance and central cavity built in a low shrub or in grasses. (DeGraaf et al. 1980).

Gordon M. Meade

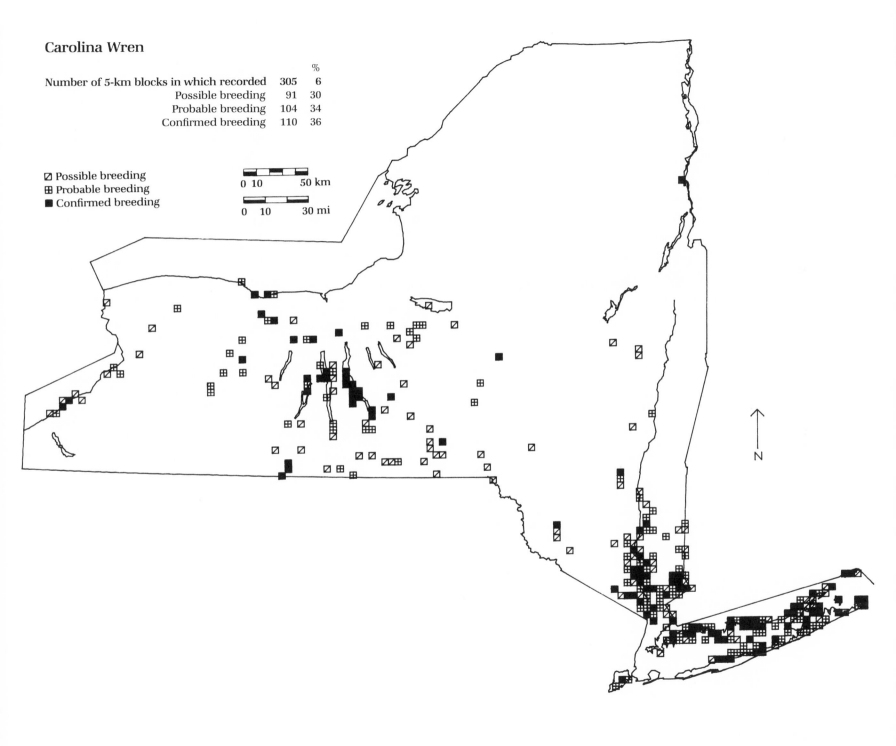

Carolina Wren

Number of 5-km blocks in which recorded	305	% 6
Possible breeding	91	30
Probable breeding	104	34
Confirmed breeding	110	36

◨ Possible breeding
⊞ Probable breeding
■ Confirmed breeding

0 10   50 km

0 10   30 mi

N

House Wren *Troglodytes aedon*

A perky, thicket-loving bird, the House Wren is a common breeder throughout all of New York except for portions of the Adirondack region. Both Eaton (1914) and Bull (1974) considered it a common summer resident except in the higher mountains.

The House Wren apparently had a period of decline and recovery earlier in this century in New York, similar to that reported in New England by Ellison (1985e). Eaton (1914) stated that the species exhibited periods of scarcity in central and western New York between 1887 and 1914. By 1914, however, the House Wren was again well established throughout that area. Eaton (1914) believed that the temporary disappearance was related to wintering-ground calamities, but he also noted that the House Sparrow had an unfavorable influence on the wren population by usurping nest cavities before the birds returned in the spring.

Griscom (1923) added further evidence that the House Wren had declined. He stated that the species was a generally common summer resident throughout the New York City region but was relatively uncommon near the sea and in the suburbs, where it had decreased markedly in the late 1800s and early 1900s owing to competition from the House Sparrow and the European Starling. He also noted that the House Wren was a common summer resident in northern Westchester County but had steadily decreased in the suburban districts; in Central Park it bred at least as late as 1908, but by 1923 had become an uncommon transient, and it also had become an uncommon and decreasing summer resident in the Bronx region. On Long Island the House Wren was still a common summer resident except at Orient (Suffolk County) at the eastern tip and at Mastic (Suffolk County) and Long Beach (Nassau County) on the south shore.

Twenty years later the situation seemed much the same in the New York City region. Cruickshank (1942) stated that the House Wren apparently could not survive in densely settled areas and had decreased steadily around the metropolis since the beginning of the century. He further noted that on Long Island the species had decreased as a summer resident to the east in Suffolk County. Nevertheless, the House Wren seems to have recovered in recent years. Robbins et al. (1986) noted that the House Wren had increased on BBS routes in New York from 1965 through 1979.

Atlasing showed the bird to be well distributed throughout most of the state, including the outer beaches and peninsulas of Long Island. The species is noticeably absent only from parts of the Adirondack region including the Adirondack High Peaks, Central Adirondacks, and Eastern and Western Adirondack Foothills. These are areas where the House Wren never has been common because it rarely breeds at elevations above 458 m (1502 ft), and never over 641 m (2102 ft) (Ellison 1985e). Also, these areas are probably too heavily forested and have early breeding season temperatures that are too cold for the House Wren to lay eggs (Kendeigh 1963).

The House Wren was easily located by Atlas observers, as it is frequently found around human habitation and its song and scolding call are easily recognized. Most of the "confirmed" records are of active nests or of adults seen entering or leaving nest holes, mainly nest boxes.

The House Wren breeds primarily in thickets, shrubbery, and brushy areas in partly open habitats, and is fond of gardens, orchards, and estates, especially those that are neglected and unkempt. It also breeds in farmlands, woodlands, and swamps that contain an abundance of dead trees, stumps, brush piles, and fallen trees (Eaton 1914; Bull 1974; AOU 1983).

The House Wren nests in a huge variety of sites, stuffing its messy nest into an astonishing assortment of crannies including cavities in trees and buildings, upturned tree roots, stone walls, tin cans, mailboxes, iron pipes, pockets of clothing left hanging outdoors, old shoes, nail kegs, fence posts, and the abandoned nests of other birds (Stoner and Stoner 1952; Bull 1974). The House Wren also nests in artificial nest boxes (Terres 1980). Nests are generally located below 3 m (10 ft) and are bulky structures composed of sticks, twigs, and grass stalks, generally decorated with spider webs and lined with soft materials such as grass, feathers, hair, wool, spider cocoons, and catkins (Eaton 1914; Gross 1948a; Stoner and Stoner 1952). Nest building could not be used by Atlas observers as evidence of breeding for this species because, whether mated or not, it builds a large number of nests within its territory, most of which are never used.

Richard E. Bonney, Jr.

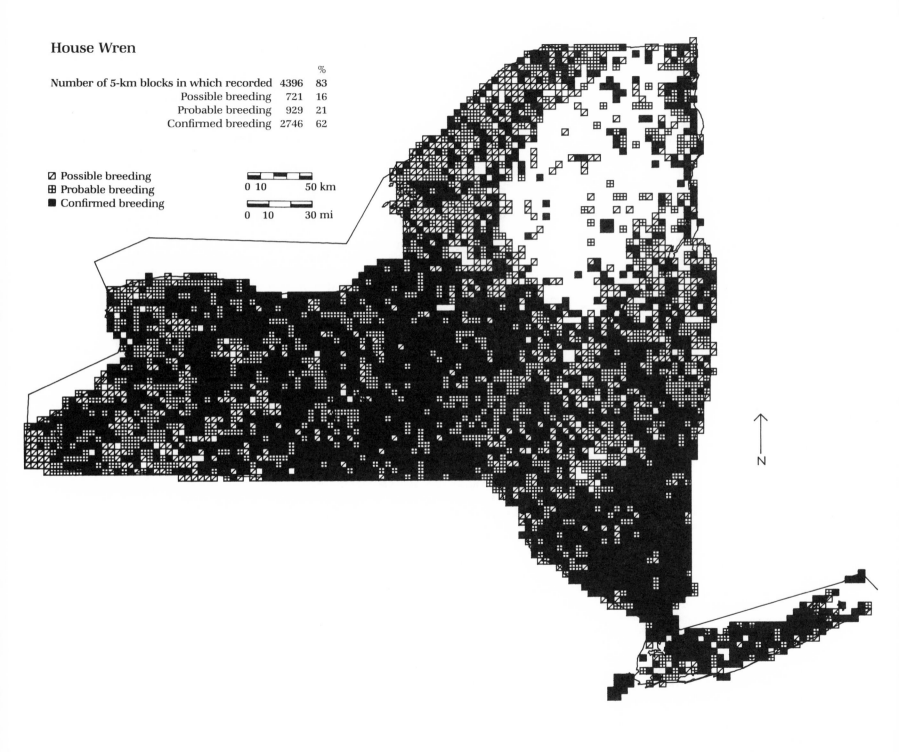

House Wren

		%
Number of 5-km blocks in which recorded	4396	83
Possible breeding	721	16
Probable breeding	929	21
Confirmed breeding	2746	62

☑ Possible breeding
⊞ Probable breeding
■ Confirmed breeding

0 10   50 km

0 10   30 mi

N

Winter Wren *Troglodytes troglodytes*

The Winter Wren, a diminutive songster of dark, dank forests at high elevations, is a common breeder in the Catskills, Adirondacks, and Tug Hill regions of New York. It also breeds in several other scattered locations, including cool, swampy areas at low elevations. Eaton (1914) stated that the bird was an abundant summer resident of the Adirondacks and higher Catskills, where "its tinkling, rippling melody" could be heard in the spruce-fir forests. He also said that some bred in various localities of central New York. Bull (1974) called it a common breeder in the mountains, local elsewhere at high elevations, and rare to the southeast.

The Winter Wren may have increased its New York range somewhat during this century. Eaton (1981) stated that as a breeding species it was increasing with the recovery of the forest, and that records in Cattaraugus County showed considerable range extension since 1909. Also, Pink and Waterman (1967) stated that no nest ever had been found in Dutchess County, whereas atlasing had one "confirmed" and five "probable" records there. And Rosche (1967) said that the bird was very rare in Wyoming County, whereas atlasing turned up several records there. However, according to Robbins et al. (1986), the Winter Wren population was very seriously affected by the severe winters of 1976–77 and 1977–78, reversing an increase documented by BBS data from the Northeast during the previous decade.

Atlasing found that the Winter Wren bred throughout most of the Adirondacks: in the Central Adirondacks, Adirondack High Peaks, Sable Highlands, Eastern and Western Adirondack Foothills, Central Tug Hill, and Tug Hill Transition. These findings coincide with those of Bull (1974), who stated that this wren was quite numerous in both the Adirondack and Tug Hill regions. The species is also found throughout much of the Catskill region, especially in the Catskill Peaks and Delaware Hills. Additionally there are records clustered in the Allegany Hills, Cattaraugus Highlands, and Rensselaer Hills, and reforestation areas in Cortland and Chenango counties also provide good habitat. The Winter Wren was not found in the Coastal Lowlands.

It is unlikely that many Winter Wrens were overlooked. The bird's long and distinctive territorial song made "probable" breeding easy to record. In a study of 2 singing males in New York, Kroodsma (1980) found the song to be variations on two basic song types repeated up to 40 times. The Winter Wren was hard to "confirm"—only 11 nests were found.

The Winter Wren prefers the deep mountain forest, hiding in brush heaps, rock-piled ravines, woodland thickets, along banks of rushing mountain streams, and on moss-covered logs of the humid slopes (Eaton 1914; Cruickshank 1942). Bull (1974) stated that it occupied two distinct nesting habitats. First, it was found in montane forests consisting primarily of spruce and balsam fir. Second, it was found less commonly in more lowland areas in "cold" bogs and swamps, especially hemlock–northern white cedar bogs upstate, and even more rarely in hemlock-rhododendron swamps in the lower Catskills and to the southeast of them. Except in swamp and ravine locations, Atlas records generally came from elevations above 305 m (1000 ft).

The Winter Wren usually nests from 0.2 to 0.8 m (9–30 in) from the ground, in a cavity in the upturned roots of a fallen tree, or in a hole in a stump, hollow tree, log, or stream bank (Eaton 1914; Bent 1948; Bull 1974). The bulky nest, quite large for the size of the bird, is composed of fine twigs, coarse mosses, and lichens, reinforced with a few fine twigs of spruce and fir. The interior is well lined with mosses, grasses, mammal fur, and the soft feathers of various birds. The Winter Wren sometimes builds extra nests that may be decoys. These are thought to be built by the male and usually are not lined (Eaton 1914; Bent 1948). Nest building, therefore, could not be considered evidence of breeding for Atlas observers.

Richard E. Bonney, Jr.

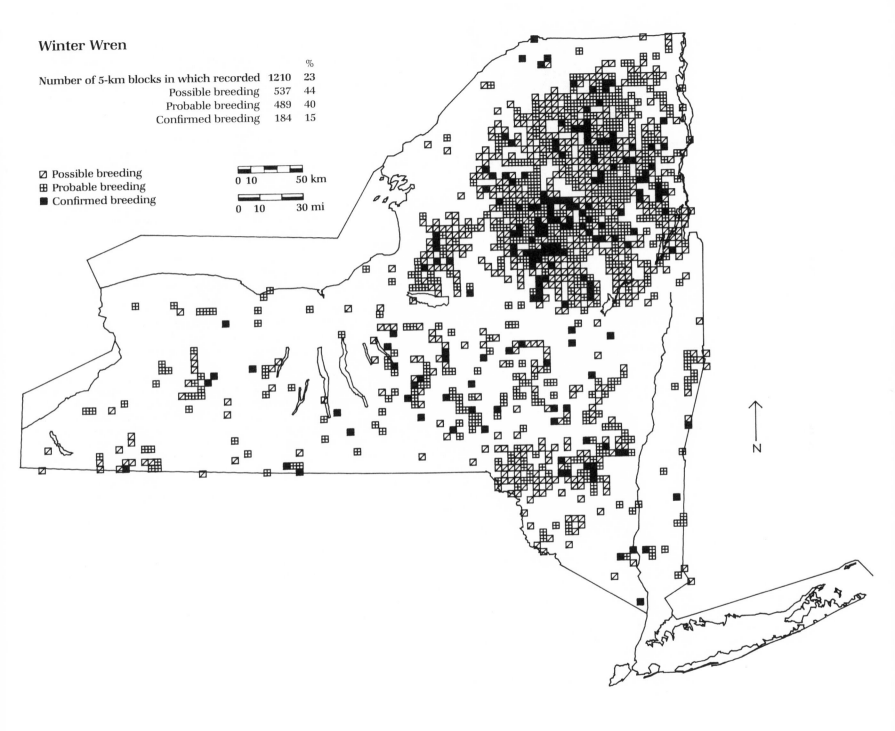

Winter Wren

Number of 5-km blocks in which recorded 1210 23 %
Possible breeding 537 44
Probable breeding 489 40
Confirmed breeding 184 15

◩ Possible breeding
⊞ Probable breeding
■ Confirmed breeding

0 10 50 km

0 10 30 mi

N

CJ Page
© 1987

Sedge Wren *Cistothorus platensis*

The tiny, elusive Sedge Wren is very rare to uncommon and local in New York. It is especially scarce in the eastern and southern portions and is listed in New York as a Species of Special Concern. Also, it is on *American Birds'* Special Concern List, owing to a widely reported decline in much of its breeding range (Tate and Tate 1982; Tate 1986).

Historical records indicated that the Sedge Wren never was common or well known in the state, but it has certainly declined in this century. De Kay (1844) stated that the species was not numerous. Eaton (1914) termed it local and uncommon throughout, except for a few colonies in the lower Hudson Valley and in parts of central and western New York. Bull (1964) stated that this species had decreased during the previous 15 years in the New York City region; Beardslee and Mitchell (1965) reported that it bred in western New York but that since 1952 it had decreased somewhat. Bull (1974) termed this wren rare to uncommon, absent in the higher mountains, and least rare in southeastern and parts of central and western New York, but that it bred on Long Island only before 1962.

The reasons for this wren's decline are not clear, but habitat loss seems important locally. Bull (1964, 1974) stated that a number of Long Island sites were obliterated by building developments or through the burning of brackish coastal meadows, and lower Hudson Valley localities were also destroyed. Cruickshank (1942) mentioned habitat alteration as causing its recent disappearance from New York City. The author noted this species at Oakwood, Staten Island, from 5 June to 1 August 1943, and found 3 or more singing males and an empty lined nest near the ground on 2 July; but this was the last known Staten Island nest, according to

Siebenheller (1981). Gretch (1984) noted that habitat in Clinton County near farmland could be temporary, after finding that a 1982 nest site was grazed by cows later that summer and a 1981 area plowed and planted. Sedge Wren habitat is also lost when meadows are replaced through plant succession. According to Griscom and Snyder (1955), the breeding population in Massachusetts crashed badly after the severe winter of 1940 and has not recovered.

Atlas surveying indicated an apparent absence of the Sedge Wren from all of southeastern New York, except for a singing, nest-building male observed in Westchester County (T. Burke, v.r.). Upstate records are scattered or sporadic, with extensive areas apparently unoccupied, including most of the eastern and southwestern counties. The Sedge Wren was recorded most frequently on the Great Lakes and St. Lawrence plains. It was usually found at elevations under 152 m (500 ft) but also at elevations from about 305 to 610 m (1000–2000 ft) at some upland localities.

The Sedge Wren is difficult to find or observe, for it runs about mouse-like near the ground in dense sedge or grass and, if flushed, quickly drops back down. Local abundance may fluctuate, and in a given year it may or may not occur in a seemingly favorable location; it also may desert a nesting locality if the ground becomes too dry or too wet (Cruickshank 1942). Atlas observers in St. Lawrence County noted that this wren bred late, with territories established in July; Kibbe (1985c) found the same for Vermont. Thus, as well as being scarce, this little wren may also be overlooked.

Typical habitat in the Northeast is moist sedge meadows, commonly with grasses and scattered shrubs. Terres (1941) found a pair in an area with heath shrubs and densely growing sedges, at 457 m (1500 ft) in the western Adirondacks. In coastal brackish tidal marshes the habitat in New York was salt-meadow grass interspersed with herbs and low shrubs (Bull 1974). In Clinton County, Gretch (1984) found the Sedge Wren in patches of bulrush in a transition zone between a deeper streamside marsh and a bordering pasture; somewhat similar habitat was reported in St. Lawrence County (M. Milligan, v.r.). Several Atlas observers mentioned that the Marsh Wren was found breeding in marshy or wetter areas near or bordering the Sedge Wren breeding sites.

The nest resembles that of the Marsh Wren but is nearly globular, not as oblong, and built close to the ground or shallow water, rarely higher than 0.6 m (2 ft). It is well hidden, woven into growing vegetation, about 10.2 cm (4 in) high or a little larger, and is composed of dried or green sedges, and lined with plant down, fur, feathers; the inconspicuous side entrance is about 2.5 cm (1 in) wide (Walkinshaw 1935). Males, some of which are polygynous (Crawford 1977), build numerous unlined dummy nests; the brood nest is one selected and lined by the female.

Paul F. Connor

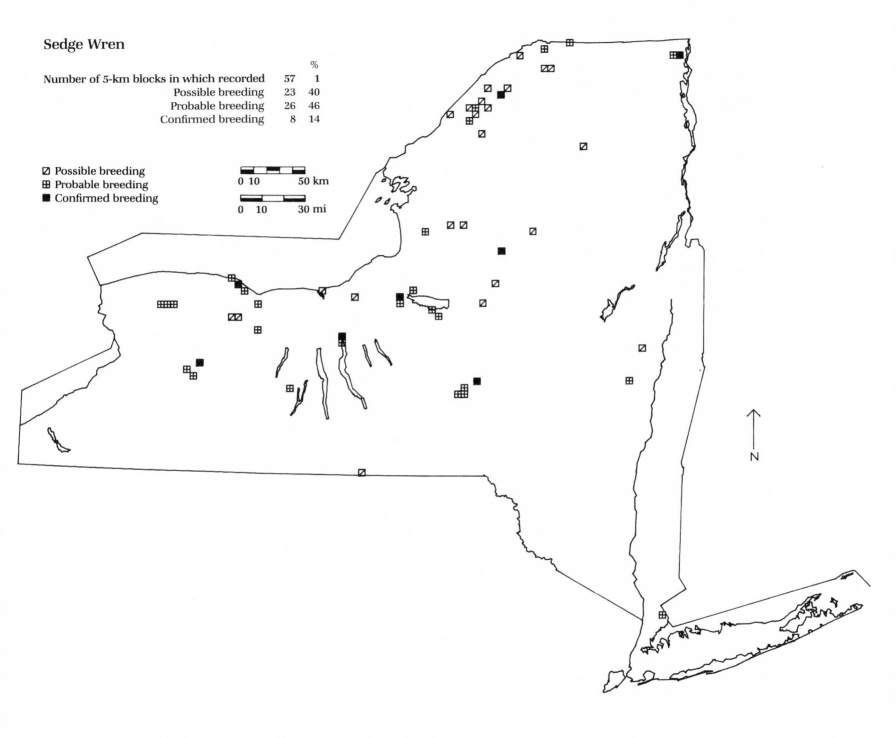

Sedge Wren

Number of 5-km blocks in which recorded	57	% 1
Possible breeding	23	40
Probable breeding	26	46
Confirmed breeding	8	14

☑ Possible breeding
⊞ Probable breeding
■ Confirmed breeding

0 10   50 km

0 10   30 mi

N

Marsh Wren *Cistothorus palustris*

The animated Marsh Wren, with its reedy, bubbling song, is one of the most characteristic songbirds of New York's larger marshes, where it is fairly common to abundant. No widespread change in abundance has been documented for any region of the state. However, many local populations have been, and continue to be, adversely affected by draining and filling of wetlands.

The Marsh Wren was recorded by Eaton (1910) as a summer resident on 20 lists published by regional authorities from 1844 to 1907. Eaton (1914) stated that this wren was common in the extensive marshes of the coast, the Hudson River, Lakes Erie and Ontario, and the Finger Lakes, and was very abundant in marshes along the Seneca River, Tonawanda Swamp (Oak Orchard WMA and Iroquois NWR), and the Niagara River. It had also been found along the Appalachian Plateau. Although Eaton made no mention of breeding in northern New York, Bull (1974) indicated that it was found there but was less numerous and rare at higher elevations. Beehler (1978) cited just three nesting season reports from the Adirondacks, and later Warren (1979) reported three breeding localities for Clinton County.

Habitat destruction has affected local breeding populations throughout New York. Cruickshank (1942) stated that in the New York City region the Marsh Wren was locally common in reed and cat-tail beds but indicated that these were being destroyed and this wren becoming rarer. On Staten Island the Marsh Wren formerly nested abundantly in extensive salt marshes but has declined since landfilling began about 1950 (Siebenheller 1981, pers. obs.). In Hudson River marshes in Rensselaer County I noted a decline by 1979 of the Marsh Wren, Least Bittern, Virginia Rail, and Common Moorhen due to drainage and habitat deterioration. In Cattaraugus County the Marsh Wren has greatly decreased since the 1960s (Eaton 1981).

Atlas surveying indicates that the Marsh Wren is well distributed on the coast and Hudson River, on the Great Lakes and St. Lawrence plains, in central New York, and near both ends of Lake Champlain, but that it is local elsewhere. Although the survey revealed previously unknown nesting localities in northern New York, a loss is apparent, particularly in the Finger Lakes region, when the Atlas map is compared with Eaton (1914). Very few or none were reported at the outlets of Canandaigua and Seneca lakes and the heads of Seneca and Cayuga lakes; also, results indicate few locations along the Seneca and Niagara rivers, but it was recorded in many blocks at Tonawanda and Oak Orchard WMAs and Iroquois NWR. This species was found at only a few locations in the Mohawk Valley and Catskill areas, and along stretches of the Appalachian Plateau.

The Marsh Wren is a conspicuous bird in the breeding season, and its nest is easily found. The male displays by singing in flight as it flutters above the marsh; it also sings much of the time, often day and night, throughout the breeding season (Kroodsma and Verner 1978). The Marsh Wren normally tends to remain on territory for the season (in contrast to the mobility of the Sedge Wren), and bird observers are frequently attracted to the marsh habitat. Thus, in Atlas work, the Marsh Wren was readily located, with nests found quite often.

In New York prime habitats are shallow marshes of cat-tail–bulrush inland, tidal cat-tail marshes along the Hudson River, and salt and brackish tidal marshes on the coast. Narrow-leaf cat-tail is much preferred to common cat-tail (Welter 1935), and Saunders (1942) believed the Marsh Wren inhabited the latter only where mixed with narrow-leaf cat-tail, common reed, or bulrush. On the coast, grasses (common reed, cordgrass, etc.) and shrubs, especially saltmarsh-elder, are characteristic. Because their territories are small, these wrens are numerous in preferred habitats where overall avifaunal diversity is low (Kroodsma and Verner 1978).

The large domed nests are lashed to cat-tails or other plants and are usually 0.3–0.9 m (1–3 ft) above water, or higher when built later in the season in growing vegetation or in shrubs (Bent 1948; Harrison 1984). Males, some of which are polygynous, build unlined dummy nests that are used as courting sites (Verner 1963). The female builds the brood nest or may line a dummy nest. She weaves the outer walls of long pliable water-soaked cat-tail or other leaves, with an opening on one side. Thereafter, working from the inside, the bird places sedge and grass leaves to reinforce the outer wall; then it adds a compact insulating layer of cat-tail down, feathers, rootlets, or other material. The innermost lining consists of feathers and shredded leaves. A sill, an inward projection of the inner lining, forms a floor to the opening (Welter 1935).

Paul F. Connor

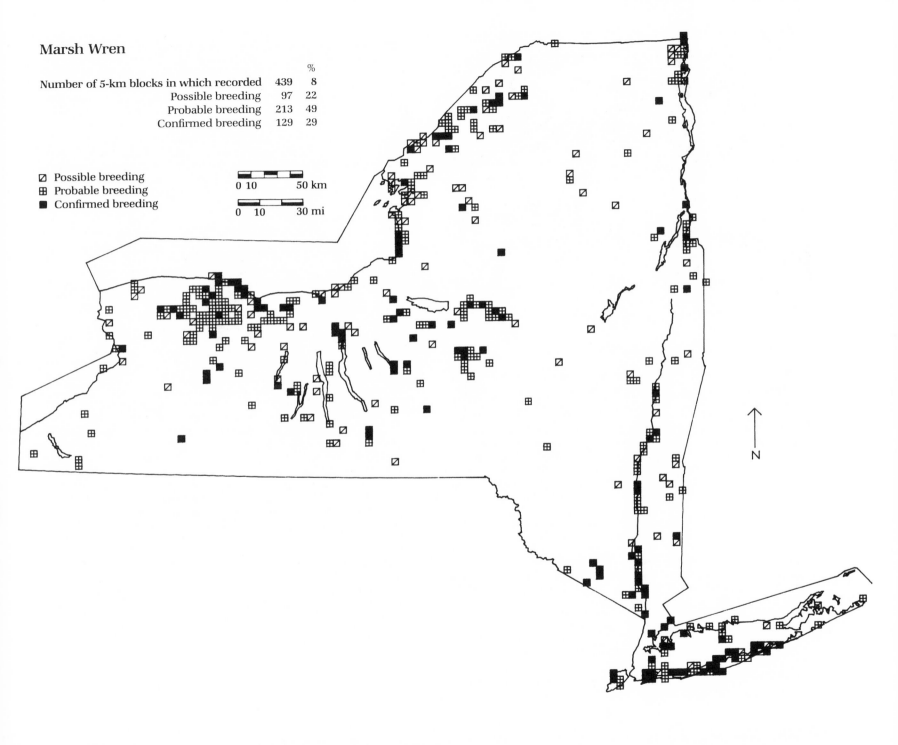

Marsh Wren

Number of 5-km blocks in which recorded	439	% 8
Possible breeding	97	22
Probable breeding	213	49
Confirmed breeding	129	29

◨ Possible breeding
⊞ Probable breeding
■ Confirmed breeding

0 10   50 km

0 10   30 mi

N

Golden-crowned Kinglet *Regulus satrapa*

A diminutive bird, the Golden-crowned Kinglet is a fairly common to common local breeding species in suitable habitats. Eaton (1914) stated that it was one of the most common summer birds of the higher Adirondacks. Saunders (1929a) also considered this kinglet common there and even found it in second-growth spruce where trees were only 3–4.6 m (10–15 ft) high. In the late 19th and early 20th centuries the Golden-crowned Kinglet was occasionally found breeding in isolated places at the southern perimeter of the Adirondacks, and there were a few scattered breeding records elsewhere in the state (Bull 1974). Otherwise, it was known to breed only in the native spruce of the higher Catskills and Adirondacks.

During Atlas surveying this kinglet was found both in native spruce forests and in planted spruce and mixed spruce-pine stands. In addition to the Adirondacks and Catskills it was also recorded in native spruce in the Tug Hill Plateau and sparingly in the Taconic Highlands. During the 1930s conifers, including spruces, were planted widely in reforestation areas in many parts of the state, usually in hilly terrain. Private individuals planted conifers on a smaller scale in other places as well. Such stands are largely absent from the Erie-Ontario Plain, Eastern Ontario Plains, St. Lawrence Plains, and the Mohawk and Hudson valleys. The species was not discovered breeding at all in the Coastal Lowlands. As the spruce plantations matured, the Golden-crowned Kinglet moved into many of them (Andrle 1971a). This expansion has resulted in a disjunct distribution statewide because such reforestation areas and private spruce stands are irregularly distributed. In addition, the stands in the state areas are often varying distances apart and separated by, or interspersed with, pines and native deciduous or mixed forest. An examination of the Atlas map illustrates the wide extent of breeding that has been documented outside the native spruce areas.

Most plantations containing breeding kinglets lie from about 366 to 732 m (1200–2400 ft) elevation in the state's hilly stream-dissected uplands. In native spruce the Golden-crowned Kinglet is found from about 600 to 1198 m (1970–3930 ft) in the Adirondacks (Peterson, pers. comm.). Weissman (pers. comm.) mentioned a nest in a plantation at less than 122 m (400 ft) in Westchester County. The plantations may be many hectares in extent or as small as 1 ha (2.5 a) or less and still harbor a breeding pair or two of kinglets. It has also been found in spruce rows near houses (Lehman, pers. comm.). Elevation does not seem to be a factor affecting its presence in these tracts.

The Golden-crowned Kinglet also breeds in pine, mixed pine and spruce, balsam fir and hemlock groves, and in bogs with conifers. Occasionally it nests in scattered evergreen trees or groves in parks or cemeteries. Habitats in which it has been found breeding are bogs, northern white cedar swamp, black spruce–tamarack swamp, hemlock-hardwoods swamp, hemlock–northern hardwood forest, spruce flats, spruce–fir–northern hardwood forest, mountain spruce–fir forest, and conifer (spruce or mixed spruce-pine) plantations.

Atlas coordinators remarked on the difficulty of hearing and finding the kinglets in dense dark plantations, so no doubt some have been overlooked. They also mentioned that it may be missed because of its early nesting. Weissman (pers. comm.) reported that some apparently suitable spruce groves were unoccupied. Some plantations have probably not been surveyed because they are distant from roads.

Lumbering and thinning in state reforestation and private spruce stands are reducing kinglet populations and breeding range locally. The maturation of small planted spruces is having the opposite effect. Spruce die-off in the Adirondacks may also be a factor diminishing kinglet populations (Peterson, pers. comm.).

The female Golden-crowned Kinglet builds the nest, usually from about 1.8 to 18.2 m (6–60 ft) above ground, and most often near the higher height. It is suspended from and woven into stems, normally of pendant twigs below a horizontal limb, or in a twig fork under foliage often far out on a branch. It is a deep, thick cup constructed of green mosses, lichens with some plant fibers, dead grass, and sometimes pine needles. It is lined with delicate strips of inner bark, fine rootlets, fibers, hair, and often feathers, usually of grouse but also those of Hermit Thrush and Ovenbird (Bent 1949), with quills pointing downward and tips arching over the two layers of usually 8 to 9 eggs. Foliage over the top is sometimes very dense, forming a canopy that shields it from the sun and partially from rain.

Robert F. Andrle

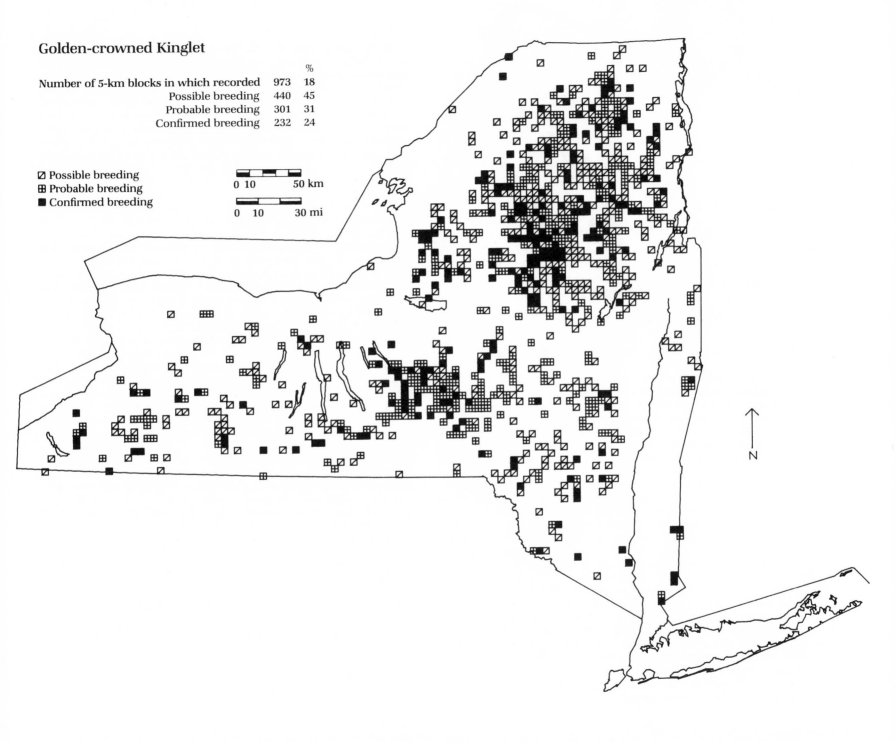

Golden-crowned Kinglet

Number of 5-km blocks in which recorded	973	% 18
Possible breeding	440	45
Probable breeding	301	31
Confirmed breeding	232	24

☑ Possible breeding
⊞ Probable breeding
■ Confirmed breeding

0 10   50 km

0 10   30 mi

N

Ruby-crowned Kinglet *Regulus calendula*

The Ruby-crowned Kinglet is fairly common throughout the Adirondacks, mainly in coniferous forests where spruce is a major component. It is absent from most of the rest of the state, with confirmed breeding documented only in Allegany State Park in 1977.

This kinglet, like the Golden-crowned Kinglet, has enjoyed remarkable success in New York during the 20th century. Both in abundance and distribution, this tiny bird has experienced an explosion. The reasons for this spread are not clearly understood. A century before the Atlas survey Merriam (1881) knew the bird only as a migrant in the Adirondack region. The first summer sighting in the state was made on 19 July 1905, at 1219 m (4000 ft) on Mount Marcy, Essex County, by Eaton (1910, 1914). A Ruby-crowned Kinglet was evidently carrying food to its young, but as Eaton "had no means of securing the specimen or of definitely verifying the observation," he believed there was still no "definite breeding record for the state." A male in full song was found on Whiteface Mountain, Essex County, at about 1189 m (3900 ft) on 16 June 1922 (Kittredge 1925).

Today it seems remarkable that the great student of bird song, Aretas Saunders, failed to encounter a single Ruby-crowned Kinglet during his two summers in the Adirondacks. He had even written on variations in Ruby-crowned Kinglet song (Saunders 1919) before visiting Essex and Franklin counties, and could hardly have missed the loud, rising triplets from this tiny songster. Saunders knew of Eaton's record and also climbed Marcy (although he apparently did not descend to the site of the old Skylight camp where the 1905 sighting was made), but he never heard or saw this species (Saunders 1929a).

The first recognized breeding record for the state was of a Ruby-crowned Kinglet nest with young at Bay Pond, Franklin County, in June 1942 (Parkes 1952b). The nest was collected and deposited in the Cornell University collection (Bull 1974). Bull provided a map showing 16 known summer localities in Essex, Franklin, Hamilton, and Herkimer counties, and presciently noted: "Further field work should produce more breeding records—especially in St. Lawrence and Hamilton counties; there has not even been a summer report in the former county to my knowledge." Another nest was collected at Heart Lake, Essex County, in 1978 (Nickerson, pers. comm.). Nesting has also been documented outside the Adirondacks. On 2 July 1977 a Ruby-crowned Kinglet nest with young was found on an island in Red House Lake in Allegany State Park. It was 12.6 m (41.3 ft) above the ground in a Norway spruce. An adult kinglet was seen nearby with 5 fledged young on 24 July (Andrle 1978).

Difficulties in obtaining thorough coverage of many Adirondack blocks within the expected range of the Ruby-crowned Kinglet meant that the species was under-recorded, as the gaps in the map suggest. The main range is confined to the Adirondack High Peaks, Central Adirondacks, and Western Adirondack Foothills and Transition. Yet even within this area there are many blocks where coverage was good which had no records of Ruby-crowned Kinglet, because the habitat was generally unsuitable. As long ago as 1923 a pair of Ruby-crowned Kinglets summered at Rochester, Monroe County, but they were thought to be "left over" from spring migration (Dye 1924). Atlas observers found birds in a mature conifer stand in Letchworth State Park, Wyoming County; in a conifer stand on the shore of Lake Erie, Erie County; and elsewhere there were scattered records from the Central Tug Hill and Transition, where it was found in spruce-bordered bogs or swampy areas, in the Champlain Transition (at Churubusco Bog, a boreal pocket just south of the Quebec border). These records usually represent individuals or pairs observed on several occasions during the summer.

Typical habitat is medium to mature spruce (black, white, or red), sometimes with balsam fir mixed, and yet there is a great deal of variation from boreal bog forest to mixed woodlands with a strong spruce component. The Ruby-crowned Kinglet is virtually absent from the hemlocks so favored by the Golden-crowned Kinglet. Although 19th-century logging and forest fires at the turn of the century destroyed or disrupted much spruce habitat, there is no evidence that the Ruby-crowned Kinglet was ever common before those disruptions. At higher elevations the present die-back of spruce could have a local impact but remains undocumented.

The nest is a surprisingly thick globular cup, compactly built of mosses and woven to the branch of a spruce (Nickerson, Peterson, pers. obs.). Only the fourth record of a New York nest, containing eggs, was located during the Atlas at North Meadow, Essex County, on 21 June 1981.

John M. C. Peterson

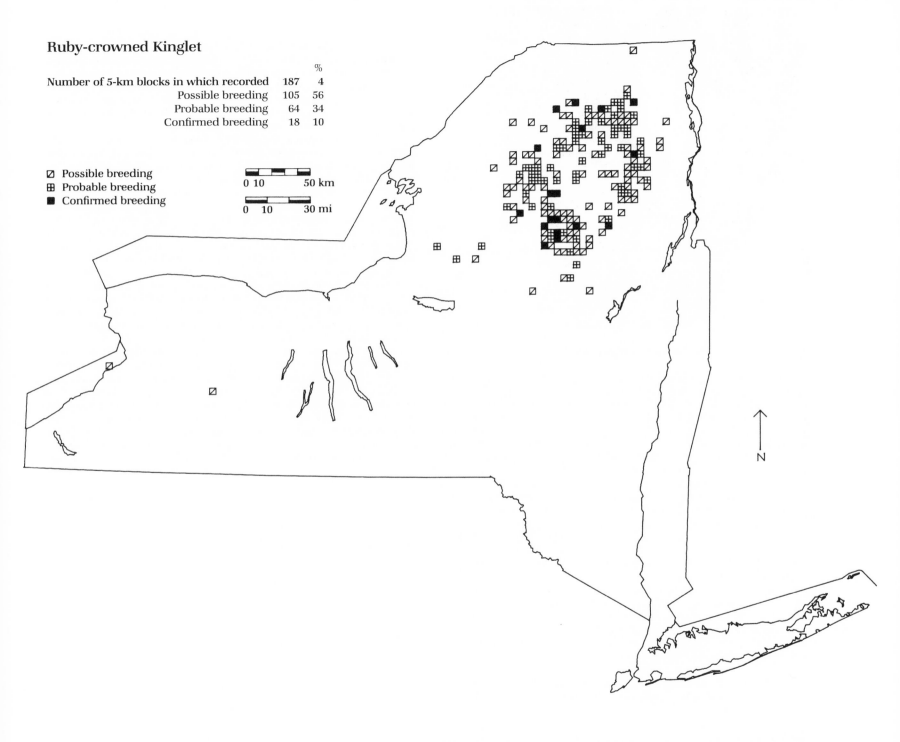

Ruby-crowned Kinglet

		%
Number of 5-km blocks in which recorded	187	4
Possible breeding	105	56
Probable breeding	64	34
Confirmed breeding	18	10

◩ Possible breeding
⊞ Probable breeding
■ Confirmed breeding

0 10   50 km

0 10   30 mi

N

Blue-gray Gnatcatcher *Polioptila caerulea*

Finding the sprightly Blue-gray Gnatcatcher, a fairly common breeding species in New York, is one of the delights of spring birding.

The Blue-gray Gnatcatcher has been recorded during the summer in the state at least since 1844 (De Kay 1844). Eaton (1914) stated that it was a regular but rather rare visitor on Long Island and in central New York from 1849 to 1908. In 1890 it successfully raised young at Coldwater, near Rochester, Monroe County (Short 1896). In the spring of 1920 a pair unsuccessfully attempted to nest in Rochester (Horsey 1938). Seasonal reports from the early 1950s to 1973 reveal that during this time the species steadily increased in number and expanded its range. Phrases such as "phenomenal increase," "notable penetration," "unprecedented flight," "spreading into new sectors," and "first nesting of the state" appear repeatedly in regional reports in *The Kingbird*. Bull (1974) considered the species quite prevalent in western, central, and southeastern counties but found relatively few records from the Mohawk Valley, Long Island, and the New York City area.

The Atlas map shows that the Blue-gray Gnatcatcher now inhabits areas mainly at elevations of 152 m (500 ft) or lower, along the main valleys of the Hudson, Susquehanna, and Genesee rivers. Concentrations are still found in central and southeastern New York, but the species has moved up the Hudson Valley, branching into the Mohawk and Lake Champlain valleys. It is now found at several places on the north shore and throughout the eastern half of Long Island.

This gnatcatcher's distribution is probably well represented on the map because its behavior almost guarantees that it will be seen or heard wherever it is present during the breeding season. It constantly utters its distinctive, penetrating *spee* call note. While caring for young in the nest, it seems oblivious of humans, flying directly to the nest with food. Also, the growing young become very noisy, allowing even the inexperienced to find the nest easily (Weston 1949). As a result "confirmed" breeding was most frequently recorded as NY (nest with young), NE (nest and eggs), FY (feeding young), and FL (fledglings).

The Blue-gray Gnatcatcher favors habitat that includes densely foliaged trees along watercourses and timbered swamps, and it often nests in tops of the tallest trees. Beautifully crafted nests are usually placed on a horizontal limb of 2.5–5.1 cm (1–2 in) diameter or in a 90° fork from 1.5 to 22.9 m (5–75 ft) above the ground in a variety of tree species. The cup-shaped nest is compactly built of soft plant fibers bound together with insects' silk and spider webs. The plant fibers may be sycamore fuzz, leaf down, dandelion, or thistle down. Fibrous materials such as fine bark strips, grasses, tendrils, feathers, horsehair, or even the wool of the stems of cinnamon ferns form the nest lining (Weston 1949).

The elasticity of the tiny, superbly woven nest is a result of the thick, resilient foundation that resists the stresses of weather, buffeting, and family struggles. An adult bird must crowd into the nest's narrow dimensions of 3.2 × 3.2 cm (1.25 × 1.25 in) inside, which explains why the bird can assume no other position than with its bill and tail pointing upward (Weston 1949).

Possibly unique among American birds is the Blue-gray Gnatcatcher's behavior of sometimes tearing up a completed or partly built nest and using the torn materials to build a new nest a short distance away from the first (Weston 1949). Hargrave (1933) ascribed such behavior to a change of conditions which made the site of the bird's first nest unsuitable or undesirable. Root (1969) felt that the reason might be the need to conserve time and energy in locating material for a new nest. Watching a Blue-gray Gnatcatcher pair building a nest offers an education in perseverance and ingenuity.

Gordon M. Meade

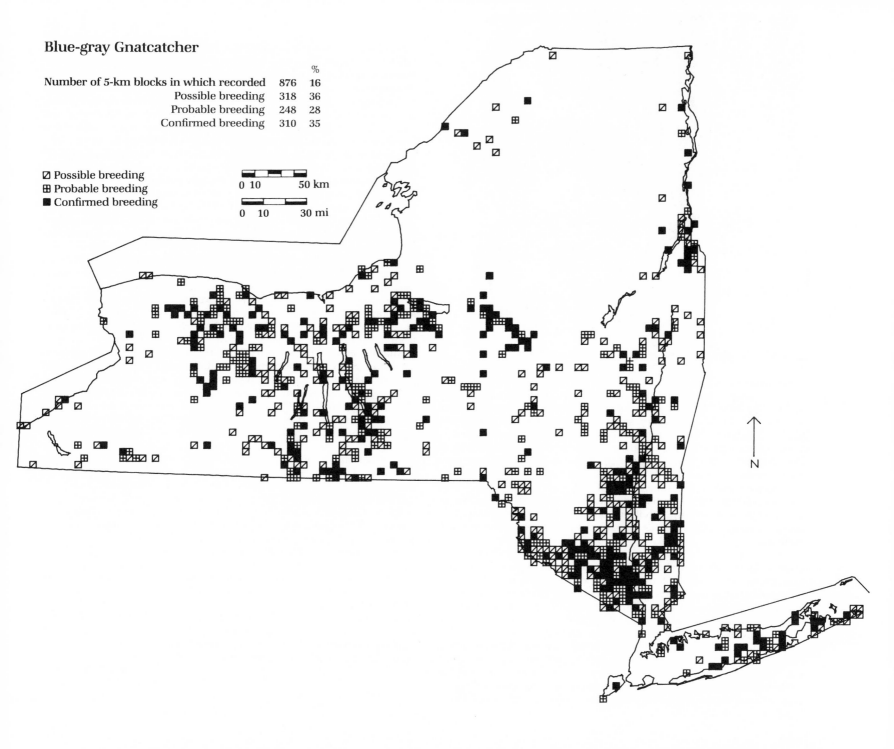

Blue-gray Gnatcatcher

		%
Number of 5-km blocks in which recorded	876	16
Possible breeding	318	36
Probable breeding	248	28
Confirmed breeding	310	35

☑ Possible breeding
⊞ Probable breeding
■ Confirmed breeding

0 10   50 km

0 10   30 mi

N

Eastern Bluebird *Sialia sialis*

New York's state bird, the Eastern Bluebird is widely distributed throughout the state, with greatest concentrations in the farm country of midwestern New York. In most locales this open-field species is rather uncommon. The earliest nesting songbird and harbinger of spring, the bluebird has a breeding history greatly affected by human settlement.

The Eastern Bluebird was probably uncommon throughout most of New York when the first European settlers arrived and the land was thickly forested wilderness. As settlers cleared land for farms, however, the bluebird and several other birds of the open country apparently increased. By the turn of the century when about 75% of New York was in farmland (Richmond and Nicholson 1985), the bluebird was a familiar species; Eaton (1914) said it was a common summer resident of all parts of the state.

Other writers confirmed its abundance. Judd (1907) said of Albany County that the Eastern Bluebird was "as numerous now as when our grandfathers were boys." Stoner (1932) said it was a common summer resident in the Oneida Lake terrain, and Saunders (1942) wrote that it was common in the valleys of Allegany State Park and the surrounding region, found wherever there was open country.

Nevertheless, in the early 1900s the bluebird began to decline in some areas, including the suburbs of New York City (Griscom 1923) and areas around Allegany State Park (Saunders 1942). By 1960 the Eastern Bluebird was decreasing greatly in most suburban areas and was even declining around farms and orchards (Bull 1974). BBS data for the period 1965–79 showed a significant decline for this species in the eastern United States (Robbins et al. 1986).

In addition to the effects of harsh weather (Robbins et al. 1986), three other factors seemed to have been responsible. First, much of New York farmland was reverting to forest, so the bluebird's favored habitat was disappearing. Second, there was a decrease in available nest sites. Trees and wooden fenceposts with nest cavities were removed as small farms gave way to large operations with fewer small pastures and orchards and more large fields with metal fenceposts. Third, increasing numbers of the introduced European Starling and House Sparrow began to usurp bluebird nest holes. In addition, the bluebird may have been highly susceptible to pesticides sprayed on farm fields (Zeleny 1976).

Today, however, the Eastern Bluebird has begun to increase again in some areas, partly as a result of "bluebird trails" where scores of nest boxes have been erected in open habitat by bluebird lovers. In areas where bluebird numbers are dense, new boxes are often occupied within hours, suggesting that there are more birds than nest sites (Chadwick, pers. comm.). Competition for nest boxes with other bird species continues to be a problem. Of 8572 boxes placed for use by the Eastern Bluebird, 75% were occupied by other species (Chadwick 1987a). More boxes are put up every year, however, so the bluebird population may continue to grow in suitable habitat. Reports received from The Nest Box Network, a National Audubon Society project to encourage the placement of nest boxes throughout New York, revealed that 7068 bluebirds fledged in 1986 from 2138 nest boxes (Chadwick 1987a).

Atlasing showed the birds to be distributed fairly widely throughout most ecozones except for parts of the Adirondacks, Central Tug Hill, and northern Tug Hill Transition, heavily forested areas where there is little open habitat. It is also notably absent along much of the Great Lakes and St. Lawrence plains and Coastal Lowlands. Nesting Eastern Bluebirds are very conspicuous and nest boxes were sought out by Atlasers, which probably explains the relatively high percentage of "confirmed" breeding records.

The Eastern Bluebird is largely a rural bird of open country, found in cropland, pastures, fruit orchards, vineyards, gardens, parks, stumpy lots, roadsides, the edges of open woodlands, and by wetlands. In the Adirondacks the species inhabits fens and bogs, and in developed areas is found frequently on golf courses (Eaton 1914; Bent 1949; Bull 1974; Terres 1980). Open areas interspersed with scattered trees, posts, or poles for perching are preferred (Pinkowski 1977).

The bluebird nests in cavities, which it never excavates. Instead, it uses holes in poles, posts, stumps, trees, or hollow limbs created by woodpeckers or natural decay, and nest boxes located near open areas (Judd 1907; Saunders 1942). The nest site is usually from 1.5 to 5.1 m (5–17 ft) above the ground but has been found at 10.7 m (35 ft). The bluebird prefers nest boxes between 0.9 and 1.5 m (3–5 ft) high (Zeleny 1976). The nest is a loosely built cup of grasses and weed stalks lined with finer grasses (Harrison 1975). The lining may also include pine needles, fine twigs, hair, and feathers (Terres 1980).

Judith L. Burrill
Richard E. Bonney, Jr.

316

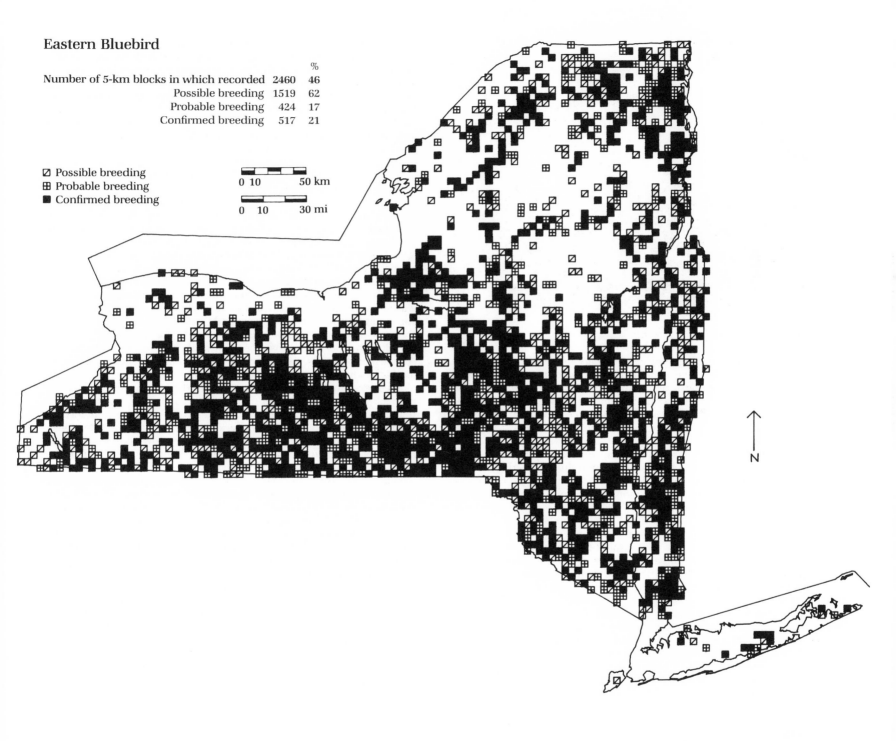

Eastern Bluebird

		%
Number of 5-km blocks in which recorded	2460	46
Possible breeding	1519	62
Probable breeding	424	17
Confirmed breeding	517	21

◩ Possible breeding
⊞ Probable breeding
■ Confirmed breeding

0 10   50 km

0 10   30 mi

N

Veery *Catharus fuscescens*

The unusual, gurgling song of the Veery, a secretive but common thrush, is heard in damp woodlands throughout most of New York. Eaton (1914) wrote that it was a common summer resident in a large portion of the state but rather uncommon on Long Island. He also noted that it was practically absent from the mountain spruce–fir forests of the Catskills and Adirondacks, where it was replaced by the Swainson's and Gray-cheeked Thrushes.

The Veery apparently has always been considered common in suitable habitat. This species was the most common thrush breeding in Albany County (Judd 1907; Stoner and Stoner 1952) and in the Oneida Lake region, where it was found in moist woodlands all around the lake (Stoner 1932). Pink and Waterman (1967) said it was very common in Dutchess County, and Beardslee and Mitchell (1965) said it was the most common member of its genus in the Niagara Frontier region. The only different assessments were in Wyoming County, where Rosche (1967) said it was "fairly common," and in Allegany State Park, where Saunders (1942) found the Veery to be the least common of the four thrushes, which is not surprising since the bird is generally not found at higher elevations.

In the New York City region, particularly on Long Island, the Veery has apparently increased in numbers and expanded its range in recent years. Griscom (1923) noted that it was a common summer resident of the richer woodlands in the New York City region, especially in northern Westchester County, but that it decreased to the south and was mainly a transient on Long Island, although some occasionally nested on the north shore. Twenty years later Cruickshank (1942) noted that a few

Veerys bred on the south shore between Idlewild (now John F. Kennedy International Airport), Queens County, and Woodmere, Nassau County, and that some had been heard singing through June in ideal habitat in southern Nassau County and probably bred there as well. Raynor (1959) documented the expansion of the Veery into eastern Long Island, and Bull (1976) said that more than 100 Veerys had been singing in Connetquot River State Park in Suffolk County, central Long Island, on 14 June 1975. Atlasing found the Veery breeding in scattered blocks all along the Coastal Lowlands.

Throughout other portions of the state Veery populations are likely to have increased as land cleared in the 1800s reverted to forest. However, recent studies (e.g., Wilcove 1985) have shown that ground-nesting birds are particularly susceptible to predation in fragmented forests, so Veery populations may be declining around urban and suburban areas. For example, the Veery was an abundant nesting species on Staten Island until about 1950 but declined very quickly thereafter and has not been found nesting there since 1955 (Siebenheller 1981).

Atlasing found the Veery throughout virtually all of New York, in every ecozone. It is absent from part of the Adirondacks, where elevations are too high for the bird to breed. In northeastern montane forests the Veery was found to breed up to about 762 m (2500 ft) (Noon and Able 1976). In some parts of the Great Lakes Plain and the Mohawk and Hudson valleys there are areas where suitable forest cover for this species is lacking. In addition, the species may have been missed where coverage was limited. "Confirmed" records were relatively easy to obtain because adults with food for young are conspicuous.

The Veery inhabits forests having thick undergrowth and is most common in low, wet forests, either deciduous or coniferous, especially those that are flooded early in the season and have a dense ground cover of shrubs, moss-covered logs, ferns, and other herbs (Eaton 1914; Dilger 1956; Bull 1974). The Veery also can be found in willow thickets and dense patches of birch and blueberry (Stoner 1932), on hillsides where there are springs (Saunders 1942), and hemlock ravines and brushy burnt-over woods (Terres 1980). In examining the differences in Veery and Wood Thrush habitat in Connecticut, Bertin (1977) found that, although both thrushes were found in moist areas, the Veery was found more often in thickets and early successional woodland. In more mature forests the Veery occupied north-facing slopes or wet depressions where the microclimate was cooler. Presumably, this is an adaptation of a northern species to the southern portion of its range.

The Veery usually nests on the ground, at the base of a shrub, or among weeds, roots, and fallen branches or tree sprouts. It also may nest in low shrubs or in saplings as much as 1.8 m (6 ft) off the ground (Bull 1974; Terres 1980). The bulky nest is built of leaves, bark strips, weed stalks, and grasses and is lined with rootlets or fine grasses (Terres 1980). Between the lining and the outside of the nest is usually a layer of well-rotted wood or mud (Eaton 1914). If the ground is wet, the Veery often lays a thick foundation of dead leaves, thereby constructing a nest large for the size of the bird (Tyler 1949).

Richard E. Bonney, Jr.

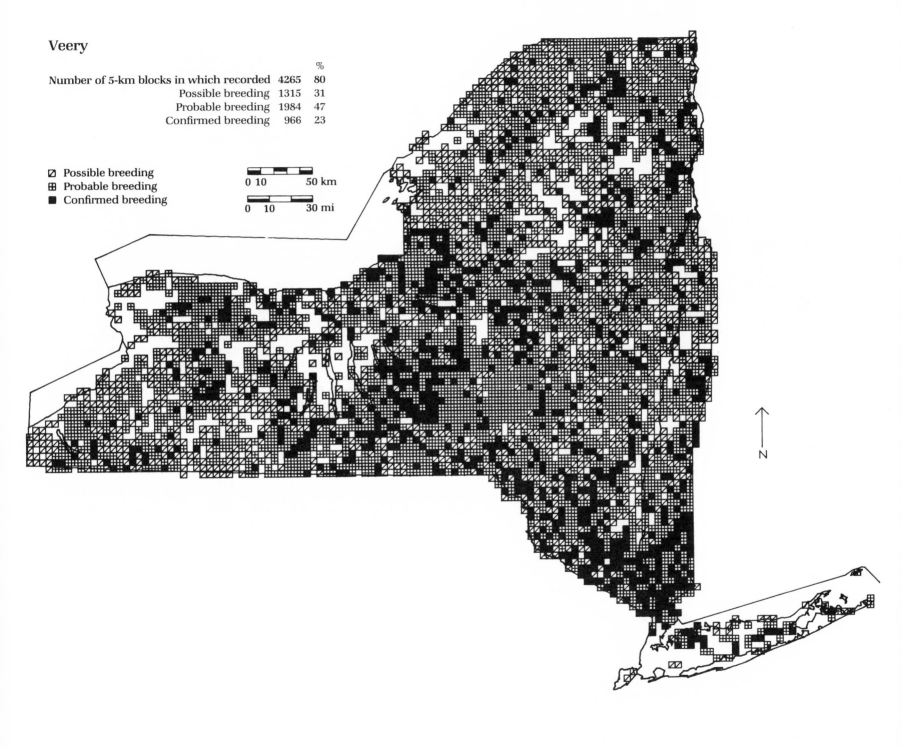

Veery

Number of 5-km blocks in which recorded   4265   80 %
                          Possible breeding   1315   31
                          Probable breeding   1984   47
                         Confirmed breeding   966   23

☒ Possible breeding
⊞ Probable breeding
■ Confirmed breeding

0 10   50 km
0 10   30 mi

N

Gray-cheeked Thrush *Catharus minimus*

In the conifer forests at high elevations in the Adirondacks and Catskills, the rather reclusive Gray-cheeked Thrush is a fairly common breeder. Because of its affinity for the dense and often stunted stands of spruce and balsam fir on the cool upper slopes, this thrush is quite local.

A century ago little was known of this species as a breeding bird in New York. Merriam (1881) noted, "I have taken it as late as June 1, but am not sure that it breeds" (in the Adirondacks). That same year, however, in the Catskills, Eugene P. Bicknell discovered a new breeding thrush on Slide Mountain, Ulster County (Wallace 1949). Smaller than other races of Gray-cheeked Thrush, Bicknell's Thrush is a form or subspecies restricted to the mountains of Maine, Massachusetts, New Hampshire, New York, Vermont, and portions of the Maritimes and Quebec. This appears to be a relict population, its mountaintop breeding grounds almost islands south of the contiguous main range of the species. Merriam (1884) then reported: "In my cabinet is a specimen of this recently described Thrush which I shot in Lewis County, near the western border of the Adirondacks, May 24, 1878," or over three years before Bicknell's discovery.

Credit for recording the first confirmed breeding of this thrush in the Adirondack High Peaks of Essex County belongs to Eaton's party, which found it feeding its young 12 July 1905, on the southwest side of Mount Marcy, at an elevation of 1325 m (4348 ft). Saunders (1929a) saw an adult "evidently worried about its young, at about 3,800 feet, along the Marcy Trail, July 25, 1926." Weyl (1927a) encountered a pair on the Marcy Trail that same summer, "at an elevation of about 4,000 feet."

No changes in historical distribution or abundance in this century and

earlier are known. The Atlas map closely resembles that presented by Bull (1974), showing the northern and southern populations within the state. The Adirondack birds are found in mountainous areas of Clinton (Lyon Mountain), Essex, Franklin, and Hamilton counties, as Bull also showed. The Gray-cheeked Thrush in the Catskills is found on the upper slopes in Greene and Ulster counties, its range quite similar to the specific localities mapped and listed by Bull. The record of "possible" nesting on Mount Pisgah in Delaware County is, however, a new location.

The map provides an incomplete view of the breeding range, although the general outline is correct. Many higher peaks where the Gray-cheeked Thrush undoubtedly occurs were never visited. Many peaks, including a number that were visited, have no trails to the summit, and all present an arduous climb. The Gray-cheeked Thrush sings its ethereal, spiral song for shorter periods than other thrushes, mostly at false dawn (including flight songs before sunrise) and again in the evening gloaming, just as darkness descends. On cloudy days the song period is longer (Dilger, pers. comm.). Since camping is now prohibited above 1219 m (4000 ft), observers could not be present when singing was most active. The song itself is not familiar to many people, and the variety of call notes is known to only a few. Gray-cheeked Thrush was "confirmed" in just eight blocks: six FY (adults with food for young) and two FL (fledglings).

In the Adirondacks this thrush occupies the boreal zone of red spruce and balsam fir up through the treeline, where black spruce, lacking competition, replaces the red up to the open rock face of the summit. In a study of thrushes on Mount Mansfield in Vermont, Noon and Able (1978) found that the Gray-cheeked Thrush occupied the stunted spruce-fir forest at elevations from about 1100 to 1300 m (3600–4265 ft), an area that excluded other thrush species. In the Catskills red spruce and balsam fir typify the highest elevations, where the Gray-cheeked Thrush is found. The lowest elevations at which it has been found appear to be 838 m (2750 ft) at Lake Colden in the Adirondacks (Carleton 1980) and 1006 m (3300 ft) in the notch between Slide and Cornell mountains in the Catskills (Hough 1964). From Eaton (1914) onward many have heard its song from the 1629-m (5344-ft) summit of Mount Marcy or atop 1281 m (4204 ft) Slide Mountain.

Bull (1974) stated, "I have seen only two reports of New York nests containing eggs: one nest with three eggs, the other with four," and no nests—either with eggs or young—were found by Atlas workers. The nest is usually placed low in a conifer, sometimes in a birch, from 0.9 to 3.7 m (3–12 ft) above the ground, commonly lower near the summit. It is constructed of fresh moss and small twigs and lined with duff and rootlets (Wallace 1949).

John M. C. Peterson

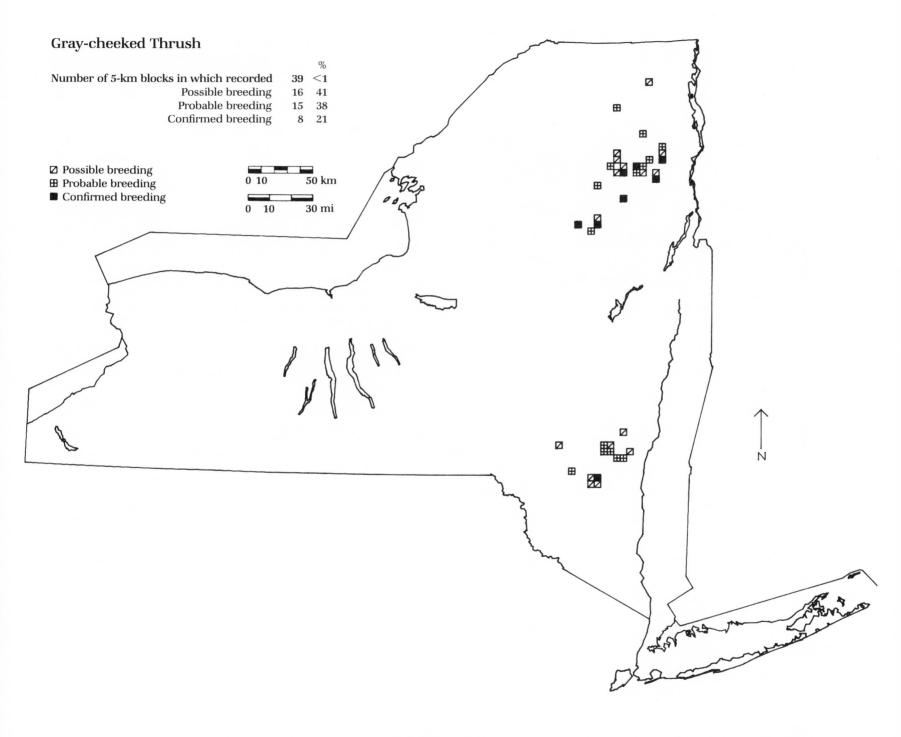

Gray-cheeked Thrush

		%
Number of 5-km blocks in which recorded	39	<1
Possible breeding	16	41
Probable breeding	15	38
Confirmed breeding	8	21

☒ Possible breeding
⊞ Probable breeding
■ Confirmed breeding

0 10   50 km

0 10   30 mi

N

C.J.Page
© 1987

Swainson's Thrush *Catharus ustulatus*

Formerly known as the olive-backed thrush, the Swainson's Thrush is rather uncommon to common at moderately high elevations across the state. Beehler (1978) considered the Swainson's Thrush "fairly common in appropriate habitat" within the Adirondack Park. In Essex County, Carleton (1980) described the species as a "common summer resident from about 2,000' to 3,000', rarely lower."

There appear to have been few changes in the abundance of Swainson's Thrush since Roosevelt and Minot (1929) judged it "the commonest thrush" just above 488 m (1600 ft) near the St. Regis lakes, Franklin County, over a century ago. Eaton (1914) recognized Swainson's Thrush as "a common summer resident both in the Catskills and in the Adirondacks," also indicating that it bred sparingly on the highest mountains near the Pennsylvania border. He dismissed reports from "the colder gullies and swamps of western New York," thinking it "likely that Veery eggs had been misidentified." Bull (1974) also rejected these records from the Great Lakes Plain, noting just 11 breeding localities outside the Catskill and Adirondack–Tug Hill districts: seven sites around the Allegany Hills, two in the Eastern Adirondack Foothills, one in the Taconics, and one in northwestern Chenango County near Pharsalia. His suspicion that Swainson's Thrush would be found breeding west of the Catskills and along the southern tier section of the Central Appalachians has not been documented by the Atlas, except for a population in Chenango County.

The main North American range of the Swainson's Thrush extends north to the treeline, with isolated southern populations in New York,

Pennsylvania, and West Virginia. The distribution of this species within New York is better understood since completion of Atlas fieldwork. The Adirondack breeding range extends to both the Eastern and Western Adirondack Foothills. Elsewhere around the state more restricted populations are found on the Central Tug Hill, the Rensselaer Hills, Taconic Mountains and Foothills, Helderberg Highlands, Schoharie Hills, Catskill Peaks, Delaware Hills, Allegany Hills, and Cattaraugus Highlands, and on parts of the Appalachian Plateau (most notably in northwestern Chenango County).

Most Swainson's Thrushes were located by their upward sliding songs or *whit* calls and were probably missed in a number of blocks—especially in the Adirondack heartland—by observers unfamiliar with these vocalizations or unable to provide coverage at dawn and dusk when thrushes are most active. Although there is some altitudinal overlap in the thrushes, Swainson's Thrush is generally found at higher elevations than the Hermit Thrush, and lower than the Gray-cheeked Thrush. This distribution is dramatically illustrated by the Atlas map, with Swainson's Thrush generally located from about 549 m (1800 ft) up to somewhere near 1219 m (4000 ft) in the High Peaks of Essex County. In a distributional analysis of thrushes on Mount Mansfield in Vermont, Noon and Able (1978) found similar elevational differences. The Hermit Thrush bred from lower elevations up to about 1000 m (3281 ft). The Swainson's Thrush was found from just over 700 m (2297 ft) to about 1100 m (3609 ft), overlapping the range of the Hermit Thrush at lower elevations. However, little overlap was observed between the ranges of the Gray-cheeked Thrush, which was found from about 1100 (3609) to 1300 m (4265 ft), and the Swainson's Thrush.

Although tolerant of mixed and sometimes deciduous forests, the Swainson's Thrush favors conifer habitats. At lower elevations within its range, it may occupy hemlock–northern hardwood forest (Bull 1974). The Adirondack population favors spruce-fir forest, usually black spruce in lower areas, red spruce on the slopes of the High Peaks. White spruce may be intermixed on well-drained sites, and there may be an admixture of birches and other deciduous trees. In disturbed forests, or on bogs, the Hermit Thrush often replaces the Swainson's Thrush (pers. obs.). In Chenango County this thrush was found in the large reforestation areas (mostly Norway spruce and white spruce) according to Lehman (pers. comm.). At relatively high elevations in Cattaraugus County a zone of hemlock provided nesting habitat (Dilger, pers. comm.).

The nest is generally placed low in a balsam fir, eastern hemlock, small spruce, or other tree, and is usually well concealed (Bull 1974). The layered nest cup is made of coarse plant materials and mud, lined with fine grasses and plant fibers (Bent 1949). During the six years of Atlas fieldwork, nests were recorded in only 15 blocks: 10 with eggs, 3 with young, and 2 used nests.

John M. C. Peterson

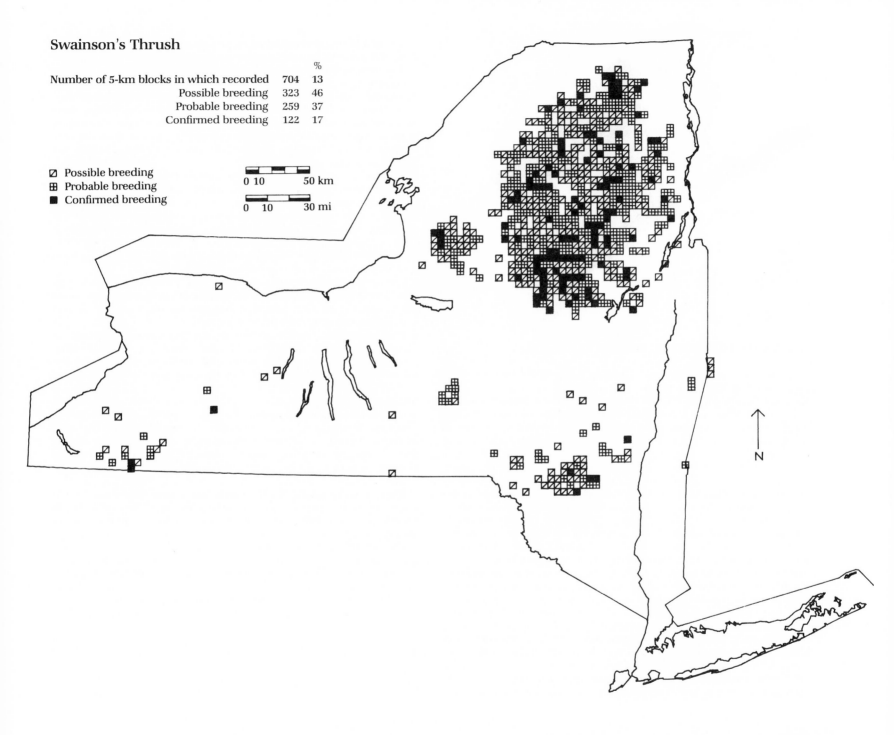

Swainson's Thrush

		%
Number of 5-km blocks in which recorded	704	13
Possible breeding	323	46
Probable breeding	259	37
Confirmed breeding	122	17

◨ Possible breeding
⊞ Probable breeding
■ Confirmed breeding

0 10   50 km

0 10   30 mi

N

Hermit Thrush *Catharus guttatus*

The flutelike voice of the Hermit Thrush is one of the most beautiful songs heard in New York forests. It is a fairly common species with a somewhat unusual distribution. Primarily a high-elevation bird, it inhabits the Adirondack, Catskill, and Allegany regions, as well as a few other locations, usually above 305 m (1000 ft). It also is found, however, in the Coastal Lowlands and in other lowland areas with sandy soil and scrub vegetation.

Eaton (1914) said that the species was a summer resident of the mountainous areas of the Catskills and Adirondacks and of the cooler portions of the Allegany Plateau near the Pennsylvania border, as well as in the colder swamps, gullies, and hill slopes of eastern, central, and western New York. Bull (1974) agreed, stating that this thrush was especially numerous in the mountains but was fairly widely distributed, from the cool, moist, wooded slopes of the Adirondack and Catskill mountains to the hot, dry, sandy pitch pine–scrub oak barrens of Long Island.

Stoner (1932) found the Hermit Thrush to be a fairly common summer resident in the Oneida Lake region, breeding in the more heavily wooded sections on the north side of the lake. According to Saunders (1942), it was the most common breeding thrush in the forests of Allegany State Park. Judd (1907) documented breeding in Albany County, presumably in the Helderberg Highlands and Albany pine bush. Griscom (1923) and Cruickshank (1942) both noted that the Hermit Thrush was locally common on Long Island, mainly in the pine barrens. However, it recently may have expanded on Long Island: Bull (1974) noted that, besides the pine barrens, the bird was found only along the south shore of Suffolk County,

whereas Atlasing found it breeding at scattered locations all along the Coastal Lowlands, especially on the north shore.

The Hermit Thrush was documented by Atlas observers in all areas previously described. It was found in all ecozones throughout both the Adirondack and Catskill regions. It also was found breeding in scattered locations along the Appalachian Plateau, especially in the Allegany Hills and southern Madison and Chenango counties where huge reforestation areas of mixed conifer species are present. It was found in the pine barrens of Long Island, the Albany pine bush, and the sand drumlins of Fort Drum and Rome sand plains.

Because this distribution is for the most part related to altitude, with the birds being recorded most often at high elevations, the breeding of the Hermit Thrush on Long Island is, as Griscom (1923) put it, "inexplicable," especially considering that the species is very rare in the higher hills of nearby New Jersey. Besides elevation, this species' distribution is probably affected by temperature and precipitation. It is found mainly in areas where mean July temperatures are under 21°C (70°F) and annual precipitation is above 88.9 cm (35 in).

The Hermit Thrush is difficult to "confirm." However, it was relatively easy to record by song and by the alarm call that it gives when an observer enters its territory. The majority of "confirmed" records were of nests with young and of recently fledged young.

The Hermit Thrush prefers cool, damp forests, especially coniferous or mixed (Eaton 1914; Dilger 1956), but also nests in a variety of other locations where undergrowth is thick. In the Catskills it nests on the lower boulder-strewn slopes and in the deep wooded areas of cool ravines or depressions (Hough 1964). In Wyoming County it is found most frequently on hemlock knolls, as well as in black spruce and tamarack bogs (Rosche 1967). In the Oneida Lake region it prefers mixed evergreen-deciduous forest types (Bull 1974), and in Allegany State Park it occurs in all types of forest except mature maple-beech (Saunders 1942). In contrast, on Long Island it breeds in the pine barrens where the ground is covered with little but bearberry and pine-barren sandwort (Griscom 1923).

Dilger (1956) found that the Hermit Thrush seems to prefer edges—old burns, along streams and bogs, powerline cuts—and Noon et al. (1979) encountered the bird more often in disturbed habitat and in pioneer stands than in mature forests.

The Hermit Thrush nearly always nests on the ground, usually in open spaces in dense woodlands (Stoner 1932). The nest is made of dried grasses, leaves, bark fibers, or mosses and is lined with fine grasses or pine needles, plant fibers, or rootlets (Gross 1949). Beneath the lining usually is a cup of mud or damp rotten wood (Eaton 1914).

Richard E. Bonney, Jr.

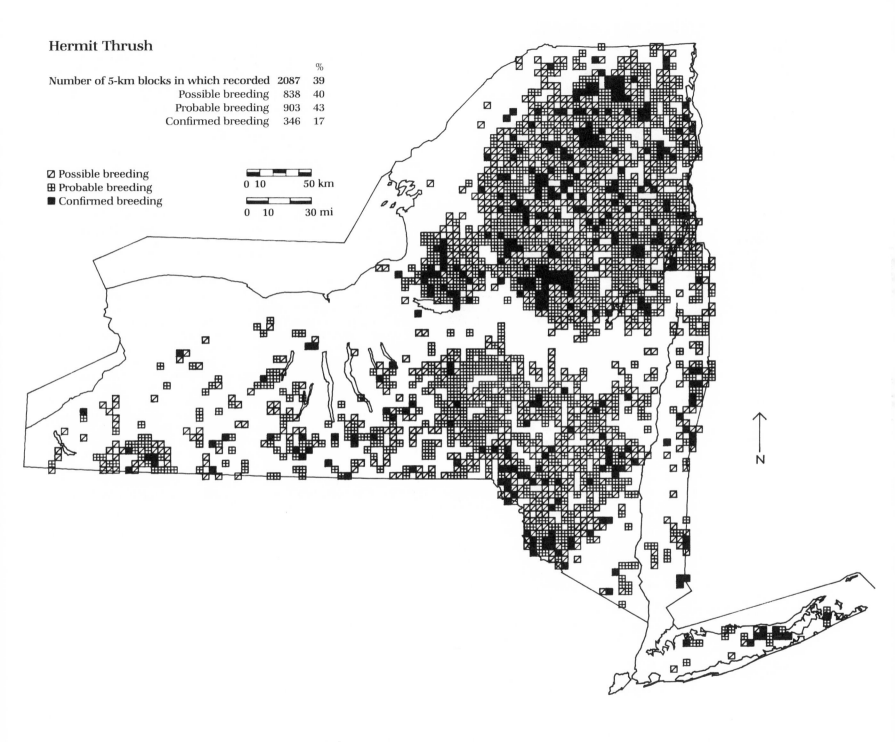

Hermit Thrush

Number of 5-km blocks in which recorded	2087	% 39
Possible breeding	838	40
Probable breeding	903	43
Confirmed breeding	346	17

◩ Possible breeding
⊞ Probable breeding
■ Confirmed breeding

0 10   50 km

0 10   30 mi

N

Wood Thrush *Hylocichla mustelina*

The Wood Thrush, a common species in New York, exudes peacefulness as it sings its beautiful, liquid song at dusk. Hearing its ethereal melody in the darkening forest is one of the joys of early summer throughout most of New York. Both Eaton (1914) and Bull (1974) considered the species common in southeastern, central, and western New York, uncommon to rare in more northern areas, and absent from the higher mountains.

The thrush's range has changed rather dramatically in the past century; only since 1890 has it become a common species in the northeastern United States (Weaver 1949). According to Bull (1974), it has been expanding its range northward in the Adirondacks, penetrating deeper into wilder deciduous forests. It also may be expanding its distribution on the Coastal Lowlands. Cruickshank (1942) stated that it was known only as a rare transient visitor in the pine barrens and on the coastal strip of Long Island, but Atlas observers found the bird breeding in both these areas. At the same time, the species may be decreasing in heavily populated areas; Cruickshank (1942) reported a gradual decrease around the outskirts of New York City.

Eaton (1914) noted that in certain years the species seemed to diminish in woodlands where once it had been abundant; he suggested that competition with the Veery might be a factor. However, in a study of thrush species conducted in the mountainous areas of New York and Vermont, Noon and Able (1978) found that although the ranges of the Wood Thrush and Veery overlapped along an elevational gradient, they are separated by the preference of the Veery for a greater shrub density

within its breeding habitat. Robbins et al. (1986) found that the numbers of Wood Thrushes increased on BBS routes in New York from 1965 to 1979.

The Atlas map indicates that the Wood Thrush is found throughout most of New York. It is widespread in every ecozone, being noticeably absent only from portions of the Central Adirondacks, Adirondack High Peaks, and Western Adirondack Foothills, and from large urban areas. The numerous records from the southeastern and midwestern regions of the state in part reflect the affinity of the species for nesting in wooded suburban areas, which are common in those regions.

The ratio of "confirmed" to "probable" records is surprisingly low, considering that the nest can be quite readily found. "Probable" records were fairly easy to obtain, as the bird frequently utters a loud alarm note when an observer enters its territory, and its song is well known.

The Wood Thrush prefers cool, moist, open woodlands, mixed or deciduous, with a rich undergrowth of shrubs and small trees, particularly if they are adjacent to bogs and small streams (Stoner 1932; Cruickshank 1942). It is most abundant in edge situations associated with northern hardwood forests (Dilger 1956). It also is found in towns and villages having numerous shade trees with patches of dense shrubbery (Weaver 1949; Bull 1974). Although not generally found on high mountains, the Wood Thrush was found as high as about 750 m (2461 ft) on Mount Mansfield in Vermont (Noon and Able 1978), and occupied habitats there which were described as having a little coniferous understory with a high degree of canopy cover and height. Apparently, the Wood Thrush requires trees at least 12 m (39.4 ft) high in its territory, possibly for song perches (Bertin 1977).

The nest is usually built from 1.8 to 3.7 m (6–12 ft) above the ground but may be as high as 15.2 m (50 ft). It may be placed in a cluster of shrubbery, in a crotch of a sapling or tree, or saddled on a limb, sometimes cemented to the branch with mud (Weaver 1949; Terres 1980). The structure resembles that of the American Robin: a firm, compact cup of dead leaves, weed stalks, mosses, grasses, and sometimes twigs reinforced with mud or occasionally rotten wood or leaf mold. The inner lining consists of fine dark rootlets and a few grass leaves (Eaton 1914; Saunders 1942; Weaver 1949).

Richard E. Bonney, Jr.
Judith L. Burrill

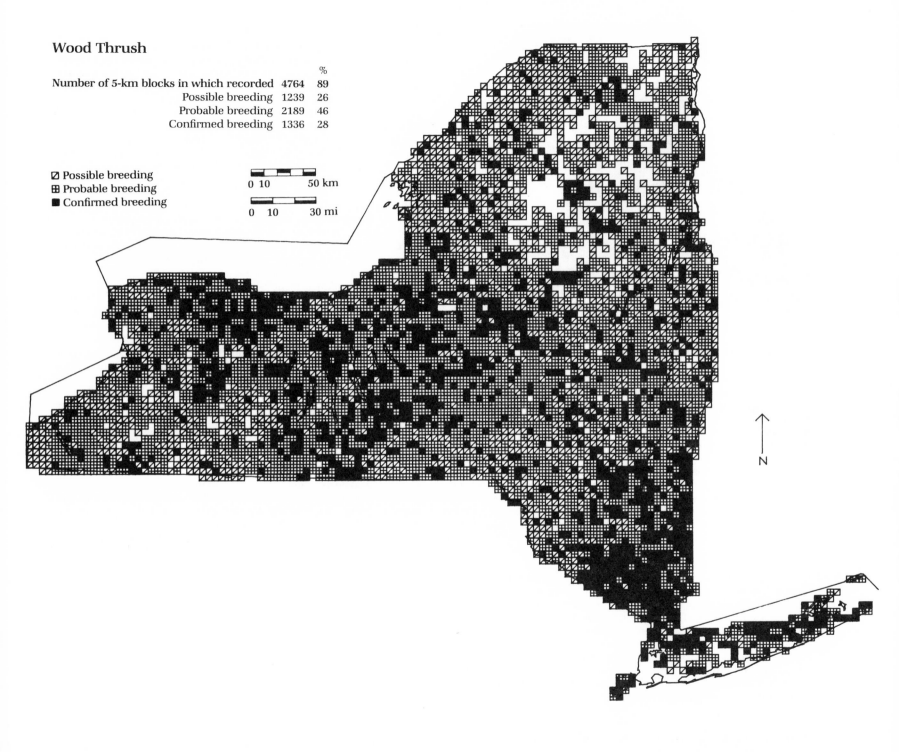

Wood Thrush

Number of 5-km blocks in which recorded		%
Number of 5-km blocks in which recorded	4764	89
Possible breeding	1239	26
Probable breeding	2189	46
Confirmed breeding	1336	28

☑ Possible breeding
⊞ Probable breeding
■ Confirmed breeding

0 10   50 km

0 10   30 mi

N

American Robin *Turdus migratorius*

The American Robin is perhaps the most common and widespread breeding bird species in New York. Both Eaton (1914) and Bull (1974) considered it to be an abundant summer resident throughout the state. The American Robin was found in more Atlas blocks than all species but the Song Sparrow.

This thrush probably has always been numerous. Its original preference for forest habitats would have made it well adapted for breeding in New York's presettlement wilderness. Although the American Robin has remained abundant by shifting its habitat preference, in recent years its numbers may have decreased slightly because of the use of pesticides. Bull (1974) noted that in some New York suburban communities excessive spraying with pesticides had greatly decreased the local population, and mortalities of robins attributed to diazinon and organophosphates, such as chlorpyrifos, continue to be reported, particularly where these pesticides are applied to lawns (Stone and Gradoni 1985, 1987).

The Atlas map indicates that the American Robin is found in every ecozone. Its absence is most notable in parts of the Adirondack High Peaks, Central Adirondacks, Western Adirondack Foothills, and Western Adirondack Transition. This may reflect a lack of intensive coverage in these areas rather than the absence of robins. Forest robins seem to be secretive and sometimes are hard to find (Peterson, pers. comm.).

The American Robin may have expanded its range on the Coastal Lowlands during this century. Griscom (1923) reported that formerly it was only a rare spring and regular fall transient on Long Island, whereas a pair or two had bred there from 1918 to 1923. Also, Cruickshank (1942) observed that the species could not breed on the great coastal marshes and that it seldom bred on the outer strip of Long Island. However, during the term of the Atlas project breeding was "confirmed" in almost all Long Island blocks.

The American Robin is among the easiest species to "confirm" because its nest is large and conspicuous, and because adults carrying nest materials, food, or fecal sacs are easily sighted (Kibbe 1985d). In addition, the breeding season in New York is long, with nesting beginning in late April and ending in July (Howard 1967). The American Robin was "confirmed" in more Atlas blocks than any other species.

Originally a forest thrush, this extremely adaptable bird now nests commonly near human habitations. In fact, some writers (Eaton 1914; Stoner 1932; Cruickshank 1942) have noted that the bird prefers such nesting sites and is found most commonly in parks, suburban yards, gardens, and country estates. However, the American Robin is still frequently found in forests—deciduous, coniferous, lowland, and montane—as well as woodlands, forest borders, scrublands, thickets, orchards, and cultivated lands (Terres 1980; AOU 1983). Undoubtedly the species' adaptability is responsible for its abundance.

The American Robin uses a tremendous variety of nest sites, placing its nest on virtually any broad surface that will support it and has shelter from above (Terres 1980). Nest sites include the ground, window sills, ledges under the eaves of houses, porches, sheds, and barns, in vines alongside buildings, on top of stumps, and in crotches of nearly all types of trees and bushes, usually from 1.2 to 7.6 m (4–25 ft) but sometimes as high as 24.4 m (80 ft) from the ground (Eaton 1914; Bull 1974). Howell (1942) found that 50% of the nests in Ithaca, Tompkins County, were within 3 m (10 ft) of the ground. He also observed that evergreen trees were often used for early season nests and deciduous trees for later ones.

Klimstra and Stieglitz (1957) found that males never participated in nest building, although they perched nearby. The nest is made of mud held together with grass stems and weed stalks, and lined with fine grass blades (Eaton 1914; Stoner and Stoner 1952). Heavy rains can wreak havoc with the mud nests, which may explain why the American Robin chooses sheltered locations (Saunders 1942). Robins sometimes nest very close together; Stoner (1932) writes of a cottage porch on which six pairs built nests and fledged young simultaneously.

Richard E. Bonney, Jr.

American Robin

		%
Number of 5-km blocks in which recorded	5189	97
Possible breeding	283	5
Probable breeding	261	5
Confirmed breeding	4645	90

☒ Possible breeding
⊞ Probable breeding
■ Confirmed breeding

0 10   50 km

0 10   30 mi

N

Gray Catbird *Dumetella carolinensis*

The Gray Catbird, also known as common or northern catbird (AOU 1983), is an alert thicket-loving bird named for its mewing call. It is found very commonly throughout all of New York except for a few portions of the Adirondacks and outlying areas.

Apparently the Gray Catbird has been common in New York at least since the early 1900s. Eaton (1914) noted that it was one of the dominant species throughout the settled portions of the state, and that it even penetrated the spruce-fir forests of the Catskills and Adirondacks along the clearings and river valleys up to an elevation of 610 m (2000 ft). Griscom (1923) said that it was an abundant summer resident throughout the New York City region, and it was found to be a common bird of the Oneida Lake region (Stoner 1932), in Allegany State Park (Saunders 1942), in Dutchess County (Pink and Waterman 1967), and in the Niagara Frontier region (Beardslee and Mitchell 1965). Bull (1974) noted that it was a widespread breeder, although rare to absent in the mountains at high elevations.

One hint that the numbers of Gray Catbird may have increased in recent years comes from Cruickshank (1942). He stated that the species had just recently become a fairly common summer resident on some sections of the outer strip of Long Island, a region where atlasing found the bird to be distributed widely. Also, Robbins et al. (1986) showed a general increase of the Gray Catbird from 1965 to 1979 on BBS routes in the East, including central and western New York. The species is likely to remain abundant even as shrubby habitat reverts to forest throughout much of the state, for it readily nests in shrubbery in suburban and other heavily populated areas.

Atlasing found the Gray Catbird to be one of the most widely distributed birds in the state. Its absence is noted only in the Central Adirondacks, Adirondack High Peaks, and northern part of the Western Adirondack Foothills, probably because these areas are at high elevations and have deep forests with little shrubby habitat. The Gray Catbird was fairly easy to "confirm" because the adults are noisy and aggressive at the nest, especially when young are present.

The Gray Catbird can be found wherever low, dense thickets of shrubs, vines, and briars occur, from suburban yards and city parks to roadsides, farmlands, forest edges, streamsides, swamp borders, and even coastal sand dunes (Gross 1948b; Bull 1974; DeGraaf et al. 1980; Terres 1980). It avoids unbroken woodland and is seldom found in coniferous habitat.

The Gray Catbird's nest is usually hidden from 0.9 to 3 m (3–10 ft) off the ground, although nests have been recorded on the ground and as high as 18.3 m (60 ft). It is found in swamp rose, hawthorn, and other trees affording thorny protection, as well as vine tangles, garden shrubs, willows, and other bushes and small trees, even evergreens (Gross 1948b). Most nest building is done by the female (Stokes 1979). The structure is rather bulky and composed of long sticks, twigs, and straw, with dead, often rotted or skeletonized, leaves. A carefully woven inner lining is usually made of reddish brown rootlets but sometimes of pine needles, fine shreds of bark, and horsehair (Eaton 1914; Gross 1948b; Terres 1980).

Some unusual nest records exist. One Gray Catbird nest was found in the cavity of a dead apple tree (Gross 1948b). Near Albany a pair of Gray Catbirds and a pair of American Robins both helped to construct a nest that was begun by the catbirds. In all, 7 eggs were laid, 3 by the catbird and 4 by the robin, and both species shared the tasks of incubating and feeding the young: 2 robins and 1 catbird hatched; all fledged (Benton 1961a).

Judith L. Burrill
Richard E. Bonney, Jr.

Gray Catbird

Number of 5-km blocks in which recorded	4920	% 92
Possible breeding	3004	61
Probable breeding	1139	23
Confirmed breeding	777	16

◩ Possible breeding
⊞ Probable breeding
■ Confirmed breeding

0 10   50 km

0 10   30 mi

N

Northern Mockingbird *Mimus polyglottos*

Mark Catesby, the English botanist, first described the Northern Mockingbird, "mock-bird," in his pioneering work, *The Natural History of Carolina, Florida and the Bahama Islands*, in 1731 (Allen 1951). Presumably, the species did not then breed in New York, but today it is a common breeding species in cities, towns, and around farms in rural areas of southeastern and central New York, and it is fairly common where it occurs elsewhere. In recent years BBS data indicated that the Northern Mockingbird was decreasing in parts of its range. However, in New York it increased significantly (Robbins et al. 1986).

De Kay (1844) considered the Northern Mockingbird rare in New York, having seen it in Queens County and having collected one in Rockland County. Eaton (1914) cited numerous reports in southeastern New York and on Long Island from the 1870s to the 1890s. By the first decade of the 20th century it was reported in the interior of the state as well. Both Eaton and this author observed that, although the Northern Mockingbird seemed to arrive in early winter, it seldom was observed after January. During the 1920s and 1940s occasional sightings were reported throughout the state. An isolated nesting occurred in 1925 in Erie County (Smith 1927), but others were not found until 1960 at Syracuse, Onondaga County, and Branchport, Yates County (Burtt 1960). Between 1951 and 1970 a widespread increase occurred around the state; breeding was documented at over 100 locations (Bull 1974). This expansion is not readily explained, but Beddall (1963) suggested several possible reasons: climatic warming favoring northward movement; the increase of residential areas providing additional suitable habitat for the species; and the

bird's innate exploratory drive, which operates under advantageous conditions.

The Atlas map shows a limited distribution for the Northern Mockingbird, although it will no doubt continue to expand into available habitat. Like other southern invading species, it has spread from Long Island and southeastern New York up the Hudson Valley into the Lake Champlain Valley. In central New York it apparently has moved through the Susquehanna River Valley and Finger Lakes region and appears to be spreading east and west along the Great Lakes Plain. Thus far, it is not found at higher elevations.

In the Northeast the Northern Mockingbird is partial to thickets, porch vines, house shrubbery, and yard plantings near human habitation. It also can be found in open woodland edges, pastures, woodlots, and prairie-like stretches with occasional bushes or small trees (Sprunt 1948). In some areas it prefers to nest in multiflora rose tangles (DeGraaf et al. 1975). The nest site must have available a food source of edible fruits, high perches for singing, and dense, low shrubbery for cover (DeGraaf et al. 1980).

When nest-building time comes, both adults gather twigs, grass, and rootlets to build a rather bulky but sturdy nest (Sprunt 1948). The nest is built with a strong outer layer of loosely laid twigs with a compact inner layer of plant, hair, and other available materials. Laskey (1962) found that the nest lining was always made of brown rootlets. It is placed from 0.5 to 3 m (1.5–10 ft) above the ground, most often from 0.9 to 3 m (3–10 ft) in a shrub or vine, but sometimes in trees (DeGraaf et al. 1980). DeGraaf et al. (1975) found that nests in residential areas were placed higher than nests in rural areas, presumably to more readily avoid predation, and Laskey (1962) found that late nests were placed higher than early nests.

Gordon M. Meade

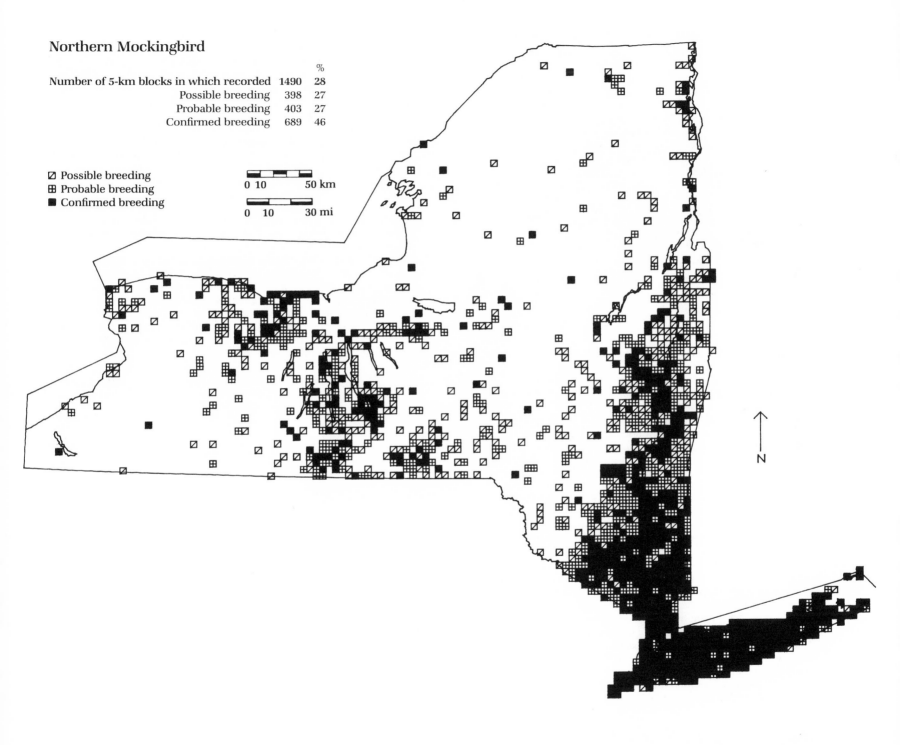

Northern Mockingbird

		%
Number of 5-km blocks in which recorded	1490	28
Possible breeding	398	27
Probable breeding	403	27
Confirmed breeding	689	46

◩ Possible breeding
⊞ Probable breeding
■ Confirmed breeding

0 10   50 km

0 10   30 mi

N

©KLA-C 1987

Brown Thrasher *Toxostoma rufum*

The Brown Thrasher is named for its habit of thrashing noisily about in thick undergrowth. Yet, for all its scuffling, it is usually a reclusive bird, easier to hear than to see.

The Brown Thrasher is found in all the lower elevation regions of New York, but it is nowhere very common. Eaton (1914) said that its range very closely matches the range of the Gray Catbird but that the Brown Thrasher was less common than its relative, especially in colder parts of the state. Bull (1974) considered the Brown Thrasher a widespread breeder but less numerous upstate, being rare to absent in the mountains and in the northeastern part of the state. Even in the Allegany Hills in southwestern New York, Saunders (1942) said that it was not found at high elevations, although it was common in the river valleys. Altogether, the evidence suggests that the species is limited more by altitude than by latitude.

The numbers of thrashers may have declined in recent years. Apparently the bird was never very common upstate (Judd 1907; Stoner 1932; Beardslee and Mitchell 1965), nevertheless, Robbins et al. (1986) found that its numbers had decreased on BBS routes conducted in New York from 1965 to 1979. Brown Thrasher numbers probably have been affected by changing land use in New York. Like other thicket birds, it may have increased as forested lands were cleared for agriculture late in the 19th century. Rosche (1967) noted that the species had become more abundant in Wyoming County during the late 1950s and early 1960s and said that, as abandoned farmland grew into successional shrubland, this was one of the first birds to appear. But as cleared land reverts to forest, the species probably will decrease in many areas, although additional habitat created from shrubs and gardens around human development may offset overall declines. Note that the Atlas map shows a widespread distribution in the heavily populated southeastern portion of the state, including Long Island, where Bull (1974) said it was particularly numerous.

Atlasing shows the Brown Thrasher to be widespread in New York except in portions of the Allegany Hills, Catskill Peaks and adjacent ecozones, the Adirondacks, and the Tug Hill Plateau. Presumably these areas are too high and too heavily forested to support the species.

Dry, open country is the usual haunt of the Brown Thrasher. It inhabits thickets and vine tangles along fencerows and forest edges, shrubby undergrowth, bushy hillsides, and scrubby fields (Stoner 1932; Saunders 1942; Bull 1974). It prefers thorny thickets and uses tall saplings as singing perches (Rosche 1967); often the birds are found in suburban yards and estates (Cruickshank 1942).

The Brown Thrasher usually nests within 2.4 m (8 ft) of the ground and occasionally directly upon it (Bull 1974). The nest is built in dense thorny bushes, small trees, or vine-covered tangles, with viburnum and hawthorn being preferred. The bulky, basketlike structure is made of twigs, dead leaves, coarse weed stalks, grasses, and thin bark strips, sometimes with a lining of softer material such as rootlets or grasses (Stoner 1932; Stoner and Stoner 1952; Terres 1980). Occasionally the thrasher will lay its eggs in the nests of other birds (Bent 1948).

Richard E. Bonney, Jr.
Judith L. Burrill

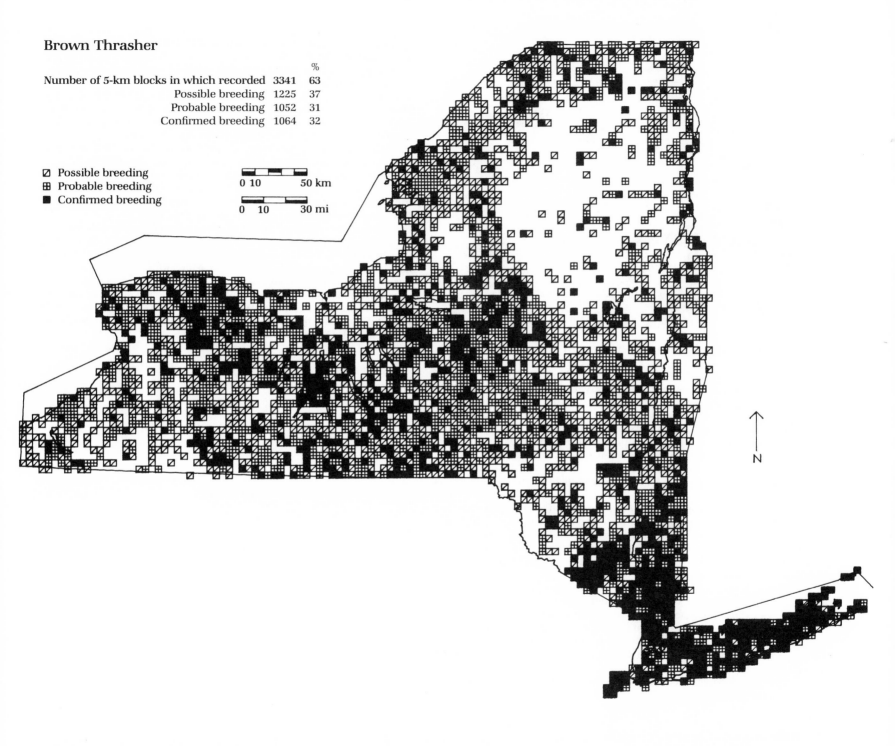

Brown Thrasher

		%
Number of 5-km blocks in which recorded	3341	63
Possible breeding	1225	37
Probable breeding	1052	31
Confirmed breeding	1064	32

◩ Possible breeding
⊞ Probable breeding
■ Confirmed breeding

0 10   50 km

0   10   30 mi

N

Cedar Waxwing *Bombycilla cedrorum*

Named for the red, waxy tips found on the flight feathers of most adults, the elegantly plumaged Cedar Waxwing is best known for its berry eating. A common breeder throughout most of the state, this species was more abundant on BBS routes in the Adirondacks than anywhere else in North America (Robbins et al. 1986). Its status in New York has changed somewhat since 1900. It has disappeared from portions of the New York City region as a result of urbanization (Griscom 1923) and was last confirmed breeding on Staten Island in 1967 (Siebenheller 1981). During the past two decades, however, it has increased slightly in the state (Robbins et al. 1986).

Except for the densely populated region in and around New York City and Nassau County on Long Island, the Cedar Waxwing breeds throughout the state. It is found at most elevations both near the coast and in upstate New York in a wide variety of upland and lowland habitats, but especially near water. Only 11 species were recorded in more Atlas blocks than the Cedar Waxwing.

This bird was easily located because of its characteristic high-pitched call (inaudible to those with high-frequency hearing loss) and its preference for exposed perches near the tops of trees and shrubs. Most of the gaps in the Atlas map in upstate New York are probably the result of insufficient coverage rather than the absence of the waxwing. It was recorded in 92% of all Atlas blocks and was "confirmed" in 31% of the blocks in which it was recorded.

The Cedar Waxwing prefers habitats that contain open areas with trees and shrubs or edges in more rural areas (Cruickshank 1942), usually near water and fruiting trees or bushes (Lea 1942). This species avoids dense forests (DeGraaf et al. 1980).

Two aspects of the Cedar Waxwing's breeding behavior are noteworthy. One is its habit of nesting semicolonially; in one case observers found 4.4 active nests per 0.4 ha (1 a); in another, 20 nests per 0.9 ha (2.3 a) (Rothstein 1971). Few North American passerine birds, aside from swallows, nest colonially. Semicolonial nesting of this species in New York has been described in Essex (Saunders 1929a), Suffolk (Bull 1974), and Tompkins counties (Keller, pers. comm.). Even during the breeding season it travels in flocks around the nesting area (Lea 1942). In addition, it nests later in the season than all passerines other than American Goldfinch (see Appendix C). Although present in New York throughout the spring, the Cedar Waxwing does not usually begin breeding until early June and has been known to nest as late as October (Bull 1974).

Nests are placed in trees and small shrubs of many varieties from 1.2 to 7.6 m (4–25 ft), rarely to 15.2 m (50 ft), above the ground (Bull 1974), usually in a crotch next to the trunk (DeGraaf et al. 1980). The rather large and bulky nest resembles that of the Eastern Kingbird and has a well-formed inner cup. Materials used include grasses, leaves, strips of bark, mosses, rootlets, plant fibers, and man-made materials such as twine and cloth (Tyler 1950). The Cedar Waxwing regularly pirates nest materials from old and active nests of other bird species, in addition to gathering its own materials (Tyler 1950, pers. obs.).

Steven C. Sibley

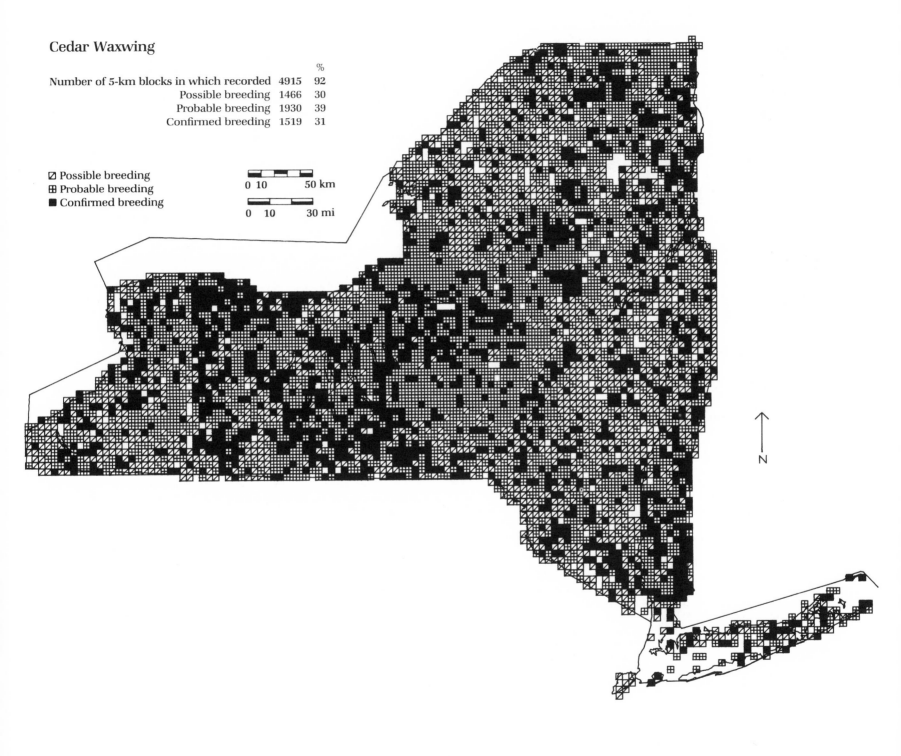

Cedar Waxwing

		%
Number of 5-km blocks in which recorded	4915	92
Possible breeding	1466	30
Probable breeding	1930	39
Confirmed breeding	1519	31

☑ Possible breeding
⊞ Probable breeding
■ Confirmed breeding

0 10   50 km

0 10   30 mi

N

Loggerhead Shrike *Lanius ludovicianus*

The Loggerhead Shrike is perhaps the most seriously declining species in New York, one of only seven birds listed as Endangered. Although formerly it was a fairly common breeding species in western and central New York (Eaton 1914), it is now a very rare and local breeder and is even rare as a migrant. This decline, which has occurred in the entire Northeast and nearly all of the continent, is documented in the *American Birds'* Blue List for 1971–1986 (Tate 1986). The only concentration of breeding of the Loggerhead Shrike in the northeastern part of the continent now appears to be in Ontario, where Atlas observers recorded it in 148 squares (Ontario Breeding Bird Atlas, preliminary map).

Breeding was first documented near Buffalo in 1869, nine years after breeding occurred in Hamilton, Ontario, suggesting movement into the state through the Niagara Frontier region, though it seems obvious that some of De Kay's (1844) northern butcherbirds were really this species. In the early 1900s Loggerhead Shrike numbers had apparently increased, particularly in western New York, as land was cleared for agriculture (Eaton 1914). The peak of abundance apparently occurred in the waning decades of the 19th century, substantiated by the records of oologists and summarized in Bull (1974). Bull also noted a progressive decline through the 1930s and 1940s that has continued in recent years, the reasons for which are not clear. Certainly changes in land use may have played a part, as some abandoned farmland has become heavily overgrown. Also, apple orchards, a former primary breeding habitat, have been heavily sprayed. Pesticide contamination has been cited by many as a potential factor in this decline (Cadman 1985; Novak 1986a; Robbins et

al. 1986). The role of automobile road kills, as suggested by Eaton (1953), may also be significant. Within a few days after fledging, 4 of 7 young produced from 3 nests in Orleans County were found dead on the road in an area with very low traffic density (Davids, Symonds, pers. comms.). During a 1986 study of nesting Loggerhead Shrikes, fledged young were observed to sit on road surfaces for periods as long as five minutes (Novak, pers. comm.). The Loggerhead Shrike has a reported average clutch size of 4–6 (Harrison 1975). Of 11 nestings reported with sufficient details during the Atlas period, one fledged 1, seven fledged 2, two fledged 3, and only one hatched 4 or 5 young, all of which disappeared within five days in this last case.

Comparison of the Atlas map with the historical map of breeding distribution in Bull (1974) reveals only three Atlas records (all in Orleans County) in two of the areas that historically had the heaviest concentrations of shrikes. These are west of Monroe County (where even the few sites on the Atlas map have been deserted since 1983) and east of Oneida Lake, Oneida County (where there are no Atlas records). It should be noted that the three "confirmed" Atlas records in Orleans County represent one block with a nest with eggs and two blocks with fledged young. The last stronghold of this species in New York is still along the St. Lawrence Plains, as was reported in Bull (1974). Only two other areas had "confirmed" breeding: the Eastern Ontario Plains and Lake Champlain Valley. The other scattered sites are nearly all observations of single birds. The search in 1986 turned up only 2 nests in Franklin County (Novak 1986a), with one additional nesting reported from St. Lawrence County (Crowell and Smith 1986).

The Loggerhead Shrike prefers open fields and scrubby clearings with thickets and hedgerows having hawthorn and apple among the tree species. Its habitat must contain trees or shrubs with thorns or a multitude of small crotches, as this bird impales or wedges its larger prey in order to tear loose the flesh (Pearson 1936). All of the recent nestings in the St. Lawrence Plains have been in old or active pastures with scattered hawthorn (Novak 1986a; Walker, pers. comm.), similar to the primary breeding habitat reported in nearby Ontario (Cadman 1985). This species sometimes sits on a high exposed perch, then drops and with a low undulating flight moves to its next perch, sweeping upward to land. However, it often perches low to the ground on small shrubs or trees and can be very elusive. One of the anomalies of this species' decline is that there appears to be an abundance of suitable habitat, unoccupied or from which the birds are even now disappearing.

In keeping with its habitat preference, the Loggerhead Shrike builds its nest in the center of a dense shrub or tree, often a hawthorn or apple, 1.5–4.6 m (5–15 ft) from the ground. The nest is a bulky structure made of sticks, coarse grass, bark strips, and leaves; it is lined with fine grasses, softer bark, and patches of hair or feathers (Eaton 1914).

Robert Spahn

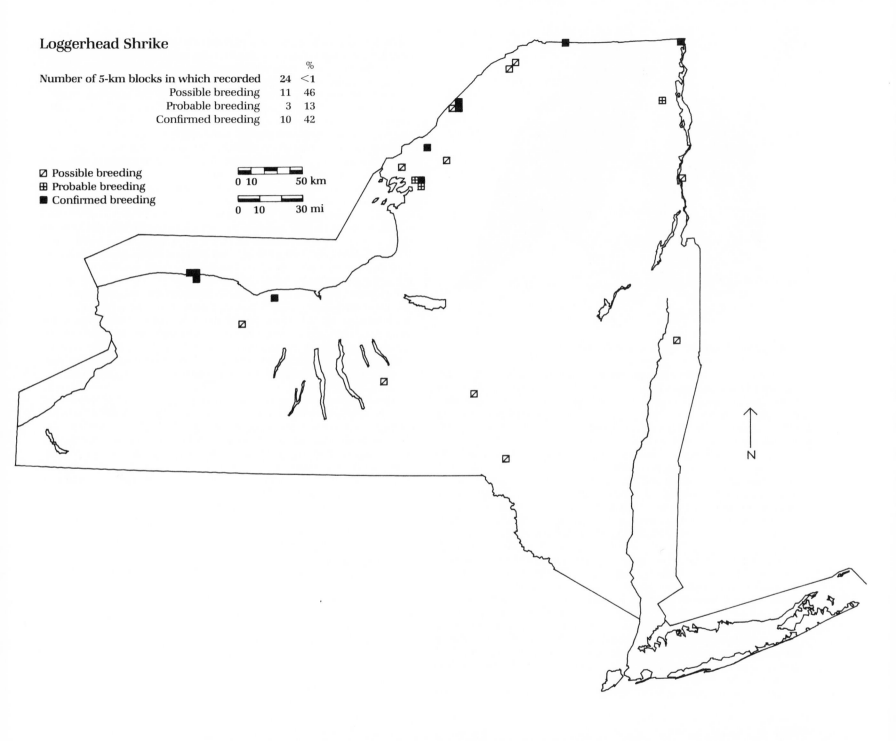

Loggerhead Shrike

		%
Number of 5-km blocks in which recorded	24	<1
Possible breeding	11	46
Probable breeding	3	13
Confirmed breeding	10	42

◨ Possible breeding
⊞ Probable breeding
◼ Confirmed breeding

0 10   50 km

0 10   30 mi

N

European Starling *Sturnus vulgaris*

The European Starling, also known as "Common Starling" (AOU 1983), has been a U.S. resident for only one century. Introduced from Europe in 1890, this species has proved so adaptable that now it is abundant throughout almost all of New York except for forested portions (Bull 1974). As parts of New York revert to forest, it is possible that European Starling populations may decline slightly.

In the late 1800s several unsuccessful attempts were made to introduce the bird into the United States. Then, in April 1890, 60 individuals were released in Central Park, New York City, and the species became permanently established (Eaton 1914). For a few years the starling was confined to metropolitan New York. By the turn of the century, however, it was spreading rapidly, appearing in the lower Hudson Valley in 1905, off extreme eastern Long Island in 1908, in Albany, Albany County, in 1911, in Buffalo, Erie County, in 1914, in Ithaca, Tompkins County, in 1916, and at Lake Champlain in 1917. By 1922 the European Starling was firmly established everywhere except in the mountains and was one of the most common species in the state. In 1960, 300,000 starlings were present in one roost in Syracuse, Onondaga County; 20,000 passed the hawk lookout at Derby Hill, Oswego County, on 16 April (Bull 1974).

According to Miller (1967) the success of this species can be attributed to several factors: "Its omnivorous feeding habits, the availability of nesting sites, the presence of a minimum of natural competitors, predators, and parasites, and its extremely high breeding potential (two broods of 3–6 young a year)."

The European Starling is found virtually everywhere in New York except for portions of the Central Adirondacks, Adirondack High Peaks, and Eastern and Western Adirondack Foothills. It also is missing from part of the Tug Hill Plateau and a few blocks in the Allegany Hills, Catskill Peaks, Delaware Hills, and Neversink Highlands. Heavy forest cover probably makes these areas unsuitable for this species. The European Starling is not only easy to find, it is easy to "confirm." Over 88% of the Atlas records are "confirmed," with observations of nests, adults carrying food and feeding young, and fledglings most often recorded.

When first established in New York, this species was principally a bird of disturbance communities, preferring parks, gardens, lawns, cultivated fields, pastures, and garbage dumps (Cruickshank 1942). It proved to be very adaptable, however, and today is found in all types of open and semi-open habitat, including lightly wooded areas, although it still is most abundant near cities and towns.

This hole-nesting species uses a variety of nest sites. It seems to prefer tree cavities or old woodpecker holes, especially those of the Northern Flicker. It also nests in birdhouses, hollows under the eaves of buildings, air-conditioning vents, deserted holes of the Bank Swallow, and many other locations (Bent 1950; Bull 1974). The nest itself is bulky and loosely constructed of grass or straw, sometimes mixed with twigs, rootlets, corn husks, cloth, paper, strings, or green leaves; it is lined with fine grasses and feathers. The size of the nest depends on the size of the cavity, but the inner cup is usually about 7.6 cm (3 in) in diameter. The nest site may be anywhere from 0.6 to 18.3 m (2–60 ft) from the ground, but is usually between 3 and 7.6 m (10–25 ft). It may be used for several successive years and can become quite foul smelling, causing a public nuisance (Bent 1950).

The European Starling has proved to be a detriment to other hole-nesting species, such as the Eastern Bluebird, Great Crested Flycatcher, Northern Flicker, and Red-headed Woodpecker, largely because the starling establishes territories and claims nest holes before these species can do so. In addition, the starling is very aggressive and can drive some species from their nests.

Richard E. Bonney, Jr.

European Starling

		%
Number of 5-km blocks in which recorded	4622	87
Possible breeding	378	8
Probable breeding	162	4
Confirmed breeding	4082	88

☑ Possible breeding
⊞ Probable breeding
■ Confirmed breeding

0 10   50 km

0   10   30 mi

N

©KLA-C 1987

White-eyed Vireo *Vireo griseus*

Between 1844 and 1974 the only region of New York that consistently maintained a breeding population of White-eyed Vireos was the lower Hudson Valley and Long Island, where it was considered common by De Kay (1844), Eaton (1914), and Bull (1974). Today it is still found mainly in southeastern New York but with a somewhat expanded range.

Robbins et al. (1986), reporting the results of the first 15 years of the BBS, showed that this species experienced a significant increase in the East, including New York, with an especially strong increase in southern New England, where it appears to be pushing northeastward.

A review of regional reports in *The Kingbird* from 1950 to 1985 indicated that in some areas this species has been observed with increasing frequency and in greater numbers. This has occurred in Chautauqua and Erie counties, where Beardslee and Mitchell (1965) noted a slight influx from 1954 to 1956; from Rochester south to Letchworth State Park in Wyoming County, where there were numerous reports; and in Region 3, where there were 21 reports from 1954 to 1985. Around Syracuse, Onondaga County, it has been seen almost every year since 1960. It was surprising to find few Atlas records from these areas.

Bull's (1974) assertion that the White-eyed Vireo was at the northern limit of its range around the New York City area is no longer true. Westchester County is now part of its range, with breeding "confirmed" also in Putnam County. The beginning of a movement up the Hudson Valley is suggested by Atlas records from Dutchess, Columbia, Rensselaer, Warren, and Hamilton counties. Elsewhere the records are quite scattered, with "possible" breeding in Allegany and Tompkins counties, and "probable" breeding documented in Cattaraugus, Monroe, and Franklin counties.

The presence of this vireo, whose striking song differs markedly from that of other vireos, probably did not go undetected by Atlas observers. Three to nine distinct, short, phonetic notes are delivered emphatically, separated by brief pauses. Most songs include a loud, accented *whee* or *wheeyo* (Bent 1950). However, a silent White-eyed Vireo is difficult to find because of its skulking, will-o'-the-wisp behavior in the densely tangled vegetation it prefers. Its nest is also difficult to find (Atlas observers recorded only 4), but if the nest is found and approached, this vireo becomes one of the most fearless birds, a sputtering defender of its "castle" (Bent 1950).

It seems to favor two types of habitat: low, swampy thickets on the banks of small streams, or old fields often on hillsides which tend to be rather dry but perhaps with a stream or common reed wetland nearby. Old fields that have grown up to shrubs and small trees are preferred, with such plants as wild cherry and plum, witch-hazel, sumac, dogwood, alder, sassafras, and blackberry. A ground cover of greenbrier, grasses, and ground pine are also usually present (Bent 1950; B. Butler et al., v.r.). Atlasers found that drier, somewhat higher second-growth habitat was preferred; reports from swampy or moist areas were fewer.

In its hard-to-penetrate habitat the White-eyed Vireo conceals the nest so that it is found only after an exhausting search, or just fortuitously. It is located 0.3 to 2.4 m (1–8 ft) off the ground, rarely higher. Even when seen alone as a museum specimen, the nest would probably be recognized as a vireo's nest, though which species would be hard to identify. The nest is narrow, shaped like an inverted cone, quite pointed at the bottom, and beautifully woven of strips of bark and grasses mixed with soft plant fibers and down bound together with spider silk. On the outside it is decorated with lichens, green mosses, bits of wasp's nest, and scraps of paper (Bent 1950). Alexander Wilson, pioneer ornithologist, said that the bird was called "the Politician," apparently in sly allusion to its habit of decorating the nest with the commonest material available—newspaper (De Kay 1844).

Gordon M. Meade

342

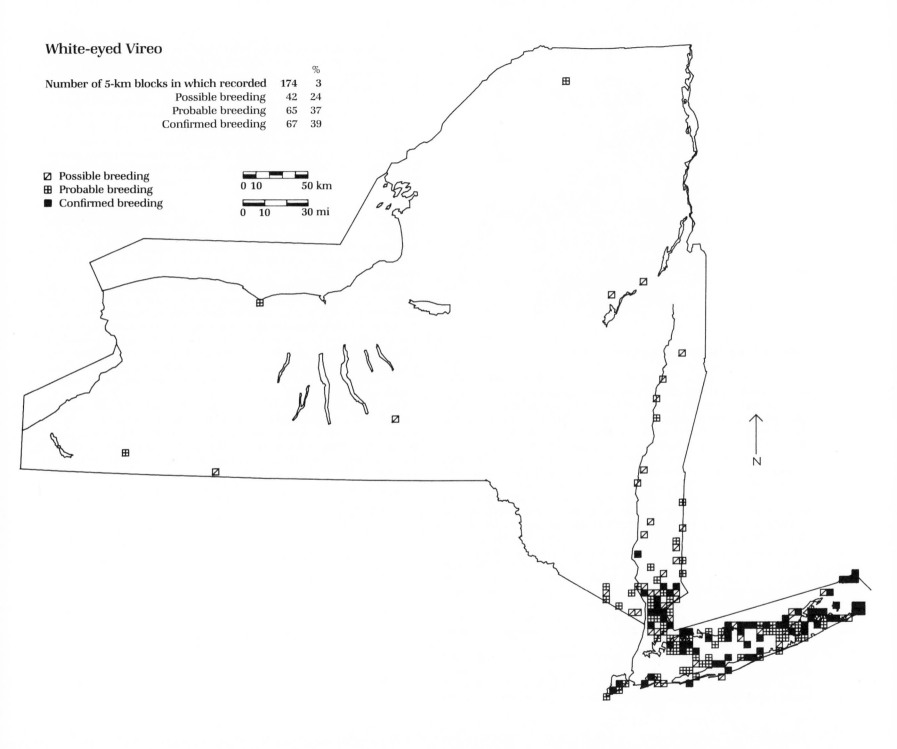

White-eyed Vireo

		%
Number of 5-km blocks in which recorded	174	3
Possible breeding	42	24
Probable breeding	65	37
Confirmed breeding	67	39

◫ Possible breeding
⊞ Probable breeding
■ Confirmed breeding

0 10   50 km

0 10   30 mi

N

©KLA-C 1986

Solitary Vireo *Vireo solitarius*

The Solitary Vireo is the first of its genus to return in the spring and the last to depart in the fall, and in migration it usually occurs in small numbers. However, in its breeding habitat its territory is large, and only a solitary bird is usually heard singing. This handsome vireo is a common breeder in the Adirondacks, Tug Hill Plateau, and Catskill Peaks. Elsewhere in the state it is a fairly common summer resident at higher elevations and uncommon at lower elevations. BBS data for 1965–79 document an increase in this vireo's population in the Northeast (Robbins et al. 1986).

Eaton (1914) described the Solitary (then called the blue-headed) Vireo as a summer resident of the Adirondacks and higher reaches of the Catskills. He specifically pointed out that the species did not nest in western New York and mentioned just two isolated records in Ithaca, Tompkins County, in 1893 and in 1913. Saunders (1923), however, said it was well distributed although uncommon in the Allegany Hills, and Beardslee and Mitchell (1965) called it "fairly common in the higher parts of Allegany Park." Bull (1974) indicated that the breeding range also extended into the Appalachian Plateau. He pointed out that although singing males were present in Orange, Rockland, and Putnam counties, there had been no positive proof of breeding in the southeastern portion of the state.

The Atlas shows breeding "confirmed" in these three counties, and outlines what appears to be continued range expansion within the state, which undoubtedly is partly due to reforestation of agricultural lands at many higher elevations. In addition, this vireo exploits many habitat types, including damp hemlock ravines at lower elevations. Although most Atlas records were above 305 m (1000 ft), it appears that the vireo's nesting preference may be a factor of temperature rather than elevation. Most Atlas observers were familiar with the song of the Solitary Vireo, although in areas where hardwoods predominate, it might have been confused with that of the Red-eyed Vireo. This vireo does not sing as frequently as other species (Sabo 1980), but when it does sing, it is fairly tame and easy to locate; however, its presence was difficult to "confirm."

The Solitary Vireo is the only vireo species in New York which has a habitat tolerance for coniferous woods, although within the conifer stand its territory is usually associated with a deciduous component (Bent 1950). It may be found in northern hardwood forests, mixed woods, and also coniferous woods. It prefers woodlands with "openings in the canopy and a dense understory" (DeGraaf et al. 1980). In a study in the White Mountains of New Hampshire, Sabo (1980) found that it is important for the Solitary Vireo to have a stream within its territory. The bird's territory is large—5 ha (12.4a)—and sometimes follows along streams for as much as 300 m (984 ft).

In the Allegany Hills the Solitary Vireo was found in beech-maple mesic, cherry-aspen, and sometimes in Appalachian oak-hickory forests (Saunders 1923). It is also found in the hemlock–northern hardwood forests in the Catskills and surrounding high elevations. In many other areas, however, it is found in red pine plantations and sometimes in spruces (Lehman, J. Carroll, pers. comms.). In the Adirondacks it is largely a species of mixed coniferous and deciduous woods, tending more toward coniferous and being more common on cool, north slopes (Peterson, pers. comm.).

The nest is rarely placed at heights over 6.1 m (20 ft) and sometimes as low as 0.9 m (3 ft). It is a neat cup; like all vireo nests, it is suspended by its upper edges from a forked branch and bound to the branch with plant fibers, hairs, and spider webs (Harrison 1978). The nest is very difficult to find; however, once found, an incubating bird may be quite fearless about allowing itself to be lifted off the nest (Bent 1950).

Emanuel Levine

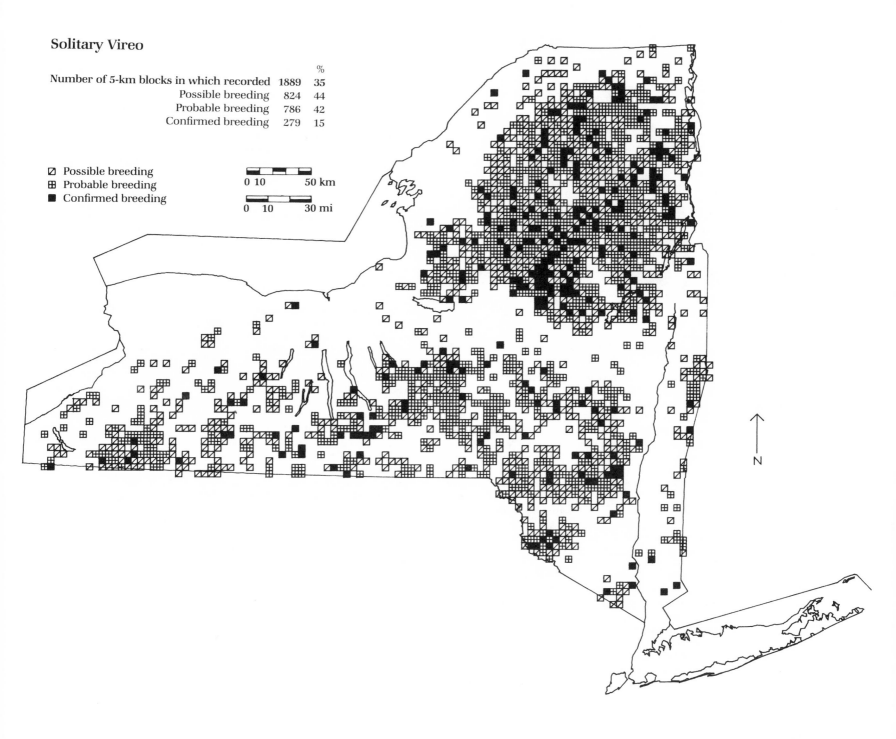

Solitary Vireo

Number of 5-km blocks in which recorded	1889	35 %
Possible breeding	824	44
Probable breeding	786	42
Confirmed breeding	279	15

☑ Possible breeding
⊞ Probable breeding
■ Confirmed breeding

0 10   50 km

0 10   30 mi

N

©KLA-C 1987

Yellow-throated Vireo *Vireo flavifrons*

Remarkably, De Kay (1844) considered the White-eyed Vireo to be more common than the Yellow-throated Vireo in New York. Where the Yellow-throated Vireo occurred with the Red-eyed and Warbling Vireos, Eaton (1914) believed it to be generally less common and more restricted in its distribution than either of the other two vireos. Bull (1974) thought that the Yellow-throated Vireo was more common in upstate New York than in the southeastern part of the state, where he considered it "decidedly rare to uncommon and very local in distribution." Temple and Temple (1976) reported a significant 20-year decline of this species in the Cayuga Lake basin, beginning in 1955. They speculated that the bird's reported preference for nesting in elms, most of which have died from Dutch elm disease, may have contributed to the decline. Since 1966 Yellow-throated Vireo populations have shown an increasing trend in New York (Robbins et al. 1986), although the species still appears to be rather uncommon in most of the state.

Eaton (1914) said, "I have found that in some localities where it was common years ago it has practically disappeared and made its appearance in other localities where it was formerly unknown." This may be due to its feeding habits. Because the Yellow-throated Vireo feeds extensively on Lepidoptera larvae that are characterized by periodic population eruptions (McAtee 1926), its populations may fluctuate locally in response to food availability. This bird's food habits also make it potentially vulnerable to both chemical and bacteriological control measures that focus upon gypsy moths but that employ control agents that nonselectively destroy all Lepidoptera larvae, thereby reducing available food for the vireo.

Atlas data show that the Yellow-throated Vireo is distributed principally in a band running from northwest to southeast mostly east of the Genesee River and south of the Mohawk River. It occurs at elevations ranging from near sea level on the north shore of Long Island to about 610 m (2000 ft), although it appears to be most prevalent at elevations below 305 m (1000 ft). The pattern of distribution of the Yellow-throated Vireo in New York probably relates to its requirement for forested areas and wooded streamsides (Bull 1974), since the band in which it occurs includes areas of more than 50% forest cover and a relatively greater density of rivers and streams than elsewhere in the state. Probst (1979) included the Yellow-throated Vireo among those species of birds that favor open, mature stands of trees, with large, spreading tree crowns. He observed that such conditions are most common in floodplain forests but that they also could be found in other forests where incomplete timber harvesting had occurred or where there had been heavy natural mortality of trees.

The Yellow-throated Vireo occurs mainly in open woods and swamps but also can be found in fruit orchards, shade trees around houses and along village streets, and in wooded areas along streams (Bull 1974). Forest fragmentation, particularly in heavily developed urban areas, appears to have considerable negative effect on the Yellow-throated Vireo. In studies of songbird populations of Wisconsin woodlots, Ambuel and Temple (1982) found significant declines in abundance of Yellow-throated Vireo. Whitcomb et al. (1981) reported that the Yellow-throated Vireo was more susceptible than other neotropical migrants to forest fragmentation, including increased vulnerability to Brown-headed Cowbird brood parasitism. Eaton (1914) noted the local disappearance of nesting Yellow-throated Vireos in years following known instances of brood parasitism by cowbirds. Cowbirds occur more frequently within woodlots smaller than 40.5 ha (100 a) than in larger woodlots (derived from Brittingham and Temple 1983); the incidence of brood parasitism by this species is thus higher in smaller forest fragments. In view of the susceptibility of the Yellow-throated Vireo to forest fragmentation and associated brood parasitism by cowbirds, the distribution and abundance of this species should be monitored carefully throughout the state.

The nest of the Yellow-throated Vireo typically is placed in a deciduous tree at a height of 6.1–18.3 m (20–60 ft), although nests have been described as low as 0.9 m (3 ft) above the ground. Conifers rarely are used for nesting. The nest is typical of other birds in the same family—a compact, tightly woven, open cup suspended from the horizontal fork of a branched twig. It is attached to the fork with spider webs and is almost completely covered with lichens held in place by strands of spider web and neatly lined with fine grasses (Bent 1950).

Charles R. Smith

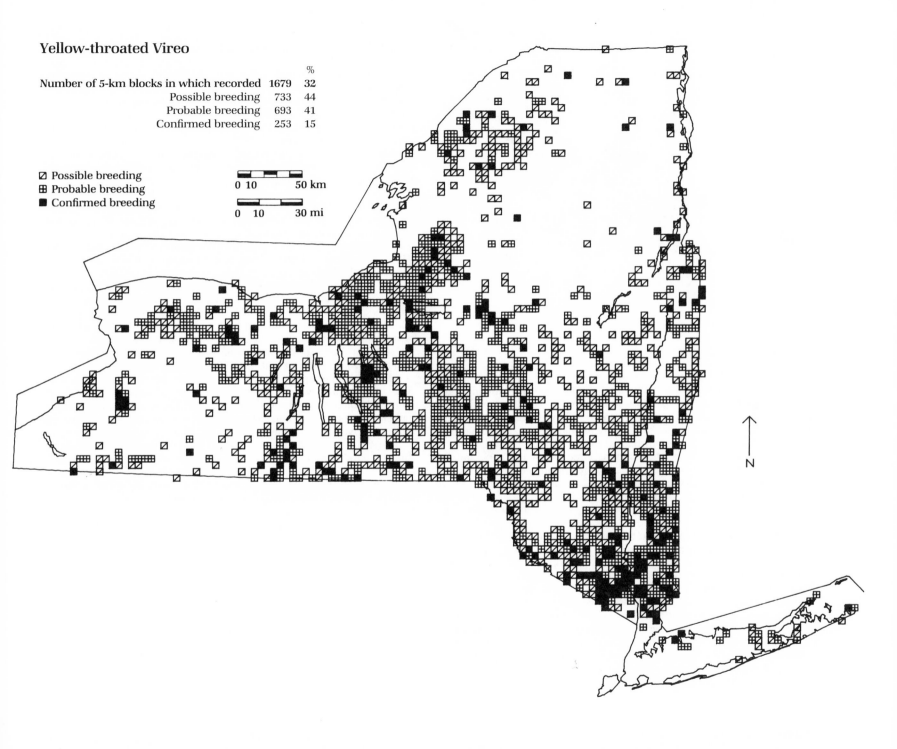

Yellow-throated Vireo

		%
Number of 5-km blocks in which recorded	1679	32
Possible breeding	733	44
Probable breeding	693	41
Confirmed breeding	253	15

☑ Possible breeding
⊞ Probable breeding
■ Confirmed breeding

0 10   50 km

0 10   30 mi

N

©KLA-C 1987

Warbling Vireo *Vireo gilvus*

The most striking feature of the otherwise nondescript Warbling Vireo is its loud, finchlike warble. The bird usually remains hidden in the treetops; with a less conspicuous song it would be far less noticeable, even though the species is fairly common at low elevations throughout most of the state, being found regularly in shade trees and orchards in most towns, villages, and suburban areas.

Eaton (1914) stated that the Warbling Vireo was not as generally distributed as the Red-eyed Vireo but that it undoubtedly bred in every county except those in the interior of the Catskills and Adirondacks. Bull (1974) said that it was a fairly widespread but local breeder, rare on the coastal plain and absent in the higher mountains but sometimes present at their edges.

Cruickshank (1942) discussed the history of the Warbling Vireo in the New York City area. Formerly a common summer resident over most of the region, it decreased early this century, dropping to a low in 1922. Griscom (1923) had already noted the disappearance of the bird from Central Park and its virtual disappearance from the Bronx, where Cruickshank (1942) said that only 4 pairs were still breeding. Both writers mentioned the bird's rarity on Long Island, where only a few scattered pairs were known to nest, primarily along the north shore. Since then there has been a slow but steady increase in most rural areas but a continuous decline in the more densely populated suburbs. Bull (1974) suggested that suburban declines may have resulted from the spraying of diseased elms, presumably because the birds die from pesticides sprayed on their favored trees. Such declines probably have occurred over much of the state.

Beardslee and Mitchell (1965) reported that its numbers apparently had diminished considerably, since it no longer could be classed as a breeding species throughout the Niagara Frontier region. Robbins et al. (1986), however, found a significant increase in the Warbling Vireo on BBS routes from 1965 to 1979 in the East especially from Pennsylvania north to Ontario and Massachusetts.

Atlasing showed that the Warbling Vireo is fairly well distributed at low elevations throughout most of the state. It is now largely absent from the Adirondack High Peaks, the Central Adirondacks, the Eastern and Western Adirondack Foothills, and the Central Tug Hill. It also is absent from much of the Catskill Peaks, the Delaware Hills, the Allegany Hills, several other areas along the Appalachian Plateau, and most of the Coastal Lowlands. Clearly, its distribution is related to elevation, although its limited distribution on Long Island must involve some other factor. Bull (1974) noted that breeding locations in the central Catskills and Adirondacks are all at low altitudes.

The Warbling Vireo favors open woods and isolated trees of open country such as elms and silver maples in villages, city parks, golf courses, and farmyards. It is also frequently found in elms, willows, alders, locusts, silver maples, and sycamores along streams, lakes, and swamps (Cruickshank 1942; Bull 1974; Terres 1980). Saunders (1942) theorized that the Warbling Vireo originally was restricted to trees along stream borders but that with the coming of civilization, it adopted the similar habitat created by shade trees along city streets. This is consistent with James (1976), who found that choice of habitat was based on the general habitat configuration rather than on the composition of the tree species, although elms and silver maples are commonly found along city streets, as well as along stream banks.

The Warbling Vireo suspends its basketlike nest from the forked branch of a shade tree at a height of 4.6–12.2 m (15–40 ft) from the ground. The nest of fine grassy bark strips, plant down, and spider webs resembles that of the Red-eyed Vireo but is smaller and more compact and not as well ornamented on the outside (Eaton 1914).

Richard E. Bonney, Jr.

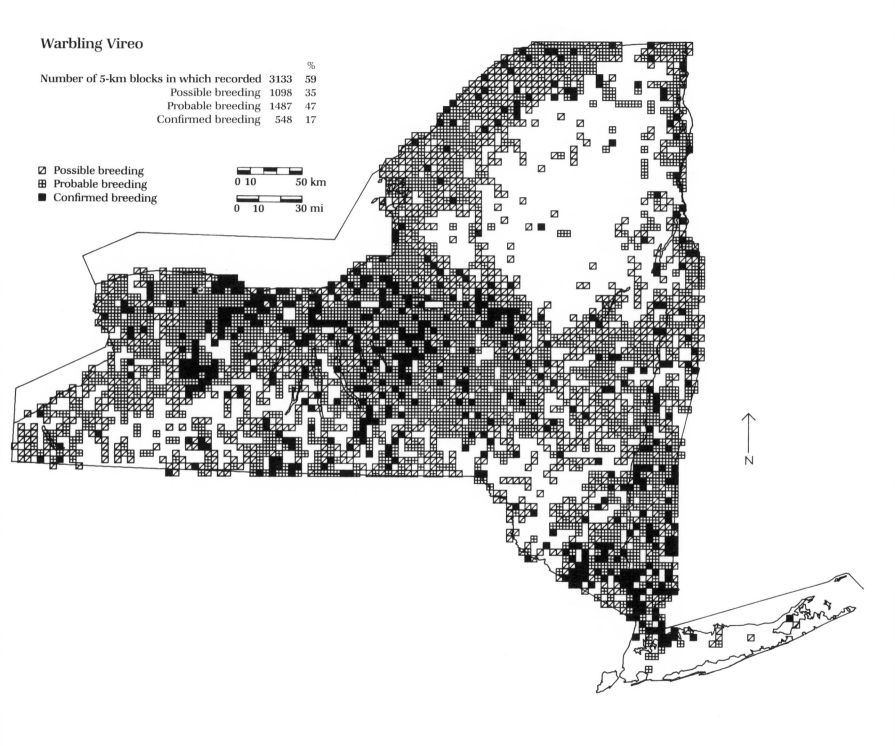

Warbling Vireo

		%
Number of 5-km blocks in which recorded	3133	59
Possible breeding	1098	35
Probable breeding	1487	47
Confirmed breeding	548	17

◨ Possible breeding
⊞ Probable breeding
■ Confirmed breeding

0 10   50 km

0 10   30 mi

N

©KLA-C 1987

Philadelphia Vireo *Vireo philadelphicus*

A rather nondescript bird of second-growth hardwoods, the Philadelphia Vireo is uncommon in its restricted breeding grounds within the woodlands of the Adirondacks and Tug Hill Plateau.

In April 1903, two years before an unsuccessful search for the vireo by Eaton (1914), a farmer near Lake Placid, Essex County, lost control of a grass fire that smoldered in the forest duff until 3 June and then sprang to life. This crown fire burned 8 miles in just two and a half hours and destroyed Adirondak Loj on the Heart Lake Road (Donaldson 1921). Along this same road, on 12 July 1926, Aretas Saunders found a singing male Philadelphia Vireo, the first summer record for the state. The vireo was in aspen groves along North Meadow Brook at 564 m (1850 ft), and Saunders (1929a) noticed: "While charred stumps and blackened stubs still stand in spots, the area is now covered with a growth of aspen and fire cherry. Here and there young maples, spruces, and balsams are coming in, and the locality is no longer unsightly, but indeed attractive." Geoffrey Carleton searched this same area 7 July 1932, finding a singing male that engaged in chases with two other Philadelphia Vireos, and on a return visit 6 July 1933 Carleton also found a singing male. "The birds were in second growth about 20 or 25 feet high, mainly maple, poplar and cherry, mixed with open brambly patches—land once burnt over" (Carleton 1935).

The conflagration that created this habitat was but one of thousands of forest fires that swept the Adirondacks for 6 weeks during the dry summer of 1903, creating habitat for the Philadelphia Vireo. Records of this vireo were reported from many of the burned-over areas as the forest began to grow back. Although Beehler (1978) reported "numerous summer sightings from the central Adirondacks," there appear to be only about a dozen documented New York records prior to 1980, from just three counties: Essex, Franklin, and Hamilton.

The sixth edition of the AOU Check-list (AOU 1983) does not include New York within the recognized distribution of the Philadelphia Vireo, although the fifth edition (AOU 1957) mentioned that the species had been "recorded in the breeding season" in northeastern New York (the Adirondacks). New York is at the southern edge of the breeding range, and Atlas observers encountered the Philadelphia Vireo in 46 blocks at scattered northern locales throughout the Adirondack High Peaks, Central Adirondacks, Eastern Adirondack Foothills, Sable Highlands, and Western Adirondack Foothills, as well as the Central Tug Hill. The Atlas map may only begin to sketch the outlines of the main range; this vireo is certainly not common, but as Saunders (1929a) noted, "Its plain coloring and the similarity between its song and that of the Red-eye may . . . cause observers to overlook it."

The habitat of the Philadelphia Vireo was described by Barlow and Rice (1977) as "deciduous growth dominated by aspen or yellow birch with understory saplings of the same species or of alder or nearly pure stands of alder." Because of the great Adirondack forest fires of the late 19th and early 20th centuries, as well as the clear-cutting by loggers during the same era, there is certainly a great deal of open deciduous and mixed second-growth habitat where the Philadelphia Vireo has yet to be found. The gradual abandonment of farms throughout New York during this century is also beginning to provide pioneering stands of aspen, birch, alder, and other species. The Philadelphia Vireo is generally found from about 457 to 792 m (1500–2600 ft) in New York, and there are an increasing number of areas outside the Adirondacks suitable for this species, Tug Hill being an example. Meanwhile, the habitat created by the 1903 fires, where the earliest discoveries were made, is maturing. Most areas where Philadelphia Vireos were first found are "forever wild" Adirondack Forest Preserve and, as they slowly develop toward climax forest, will no longer be suitable for this vireo.

Nest site selection and nest building are done by the female; the male, however, shares in incubation (Barlow and Rice 1977). The nest near Marcy Dam, Essex County, in 1963 was 5.5 m (18 ft) high in a sugar maple (Sheffield and Sheffield 1963). The 1975 nest bordering Marcy Brook, Essex County, was also in a sugar maple but placed 15.2 m (50 ft) above the ground, the nest tree growing on a steep slope. This nest was a typical vireo cup woven to the fork of small branches, the exterior mainly yellow birch strips, with some poplar down and shreds of bark (pers. obs.). The contents of this nest, if any, could not be seen and there is no positive record of a nest with eggs for the state. No nests were found by Atlas observers, although food-carrying or feeding young were recorded in ten blocks, and recently fledged young were noted once.

John M. C. Peterson

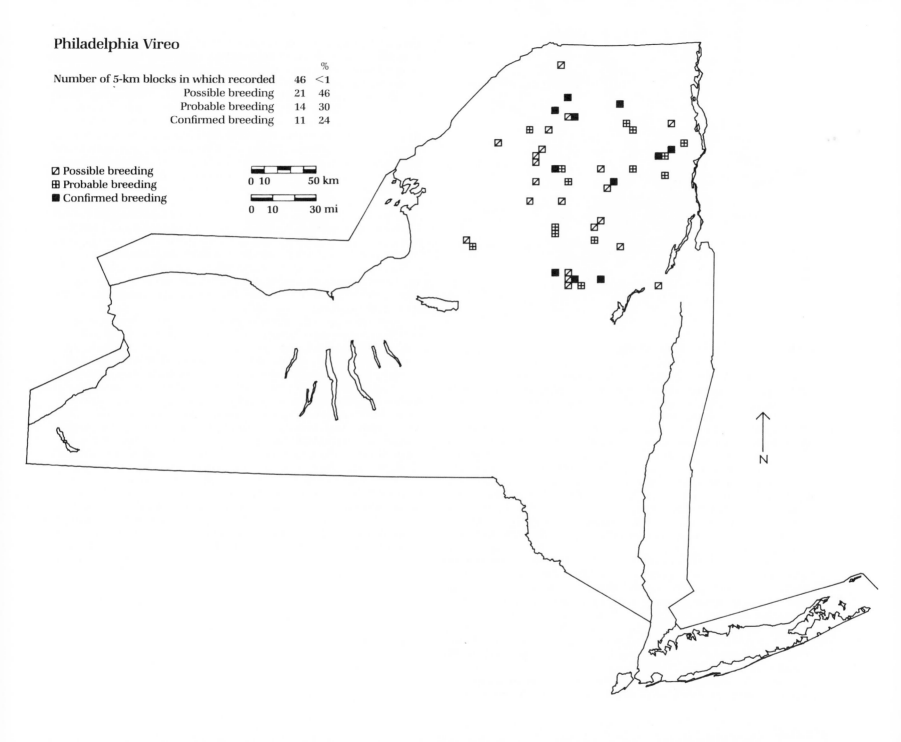

Philadelphia Vireo

Number of 5-km blocks in which recorded	46	<1
Possible breeding	21	46
Probable breeding	14	30
Confirmed breeding	11	24

◪ Possible breeding
⊞ Probable breeding
■ Confirmed breeding

0 10   50 km

0 10   30 mi

N

Red-eyed Vireo *Vireo olivaceus*

The Red-eyed Vireo is a dull bird both in appearance and song. It is nevertheless quite conspicuous, being widespread and noisy; it breeds commonly in groves and woodlands throughout much of New York, where it sings tirelessly from the treetops all summer long.

Eaton (1914) said that the Red-eyed Vireo was distributed uniformly throughout the state, being found in every wooded area in southern and western New York. He added that he had found it breeding in the Adirondacks as high as the summits of the Bartlett Ridge 1164 m (3820 ft) and Mount Colvin 1237 m (4057 ft) in the Adirondack High Peaks, and that it seemed to be nearly as common in the north woods as in the groves of western New York. Bull (1974) stated that this species was a widespread breeder, the most numerous and widespread member of its family.

There is some indication that the Red-eyed Vireo may have decreased somewhat during the last century in the New York City and Long Island areas. Cruickshank (1942) indicated that it was steadily decreasing in all heavily settled areas around the city, and Siebenheller (1981) stated that this vireo was once a very common breeder on Staten Island but was presently decreasing in numbers. These decreases were, in part, due to destruction of woodland habitat in these urban and suburban areas; however, the Red-eyed Vireo is also heavily parasitized by the Brown-headed Cowbird (Bull 1974), which has recently increased in numbers.

BBS data for the period 1965–79 documented an increase of about 2% annually for this species in the eastern part of its range, and it was considered to be the most common woodland species in the eastern deciduous forest (Robbins et al. 1986). Local populations, however, can be affected by variation in certain prey species. In a study in New Hampshire, Holmes et al. (1979) found that Red-eyed Vireos failed to produce young in a year following a severe decline in defoliating caterpillars.

Atlasing recorded the Red-eyed Vireo "probable" or "confirmed" throughout virtually all of New York, in all ecozones. The male Red-eyed Vireo is very obvious, as it spends a great deal of time in territorial defense, especially during the incubation period (Holmes et al. 1979), and the song is easily identified. A high number of "confirmed" records were reported. The noisy food-begging calls of the young are frequently heard and nests are easy to find; more than 330 were located. The Red-eyed Vireo was eighth on the list of most often reported Atlas species. In intensely surveyed areas "confirmed" records are numerous, suggesting that nesting could have occurred in most blocks.

The Red-eyed Vireo breeds in open deciduous forest, second-growth woodland, and large shade trees in city parks and suburban areas (Bull 1974; AOU 1983). Woodlands with an undergrowth of slender saplings from 1.8 to 4.6 m (6–15 ft) high are favored, as the bird shows a preference for low, shaded undergrowth (Stoner and Stoner 1952). It is also found in wooded clearings, borders of burns, along brooks or bog openings in forests, and in residential areas wherever there are orchards, parks, gardens, or tree-lined streets (Terres 1980).

Stoner (1932) noted that the species was found in nearly every woodlot visited in the Oneida Lake region, except some of the deepest and most dense hemlock tracts on the north side of the lake. However, he did find it in some dense mixed woods and alder thickets. In Allegany State Park, Saunders (1942) found the Red-eyed Vireo to be the most common species of the mountain forest, being present at all elevations and in all forest types—Appalachian oak-hickory, cherry-aspen, even the higher beech-maple mesic, where hemlock was absent and few birds occurred. Probably the abundance of young woodlands throughout much of the state is at least partly responsible for the species' widespread distribution.

The Red-eyed Vireo weaves dainty, thin-walled, basket-shaped nests of plant fibers and strips of bark, rootlets, and pine needles, smoothly lined with finer strips and more pine needles. The firm structure is suspended from a fork near the end of a branch in a shrub or the low limb of a tree, usually from 0.6 to 3 m (2–10 ft) but up to about 18.3 m (60 ft) from the ground. The outside is ornamented with spider webs and other whitish materials such as light-colored pieces of birch bark or lichens (Eaton 1914; Stoner 1932; Bull 1974; Terres 1980).

Richard E. Bonney, Jr.

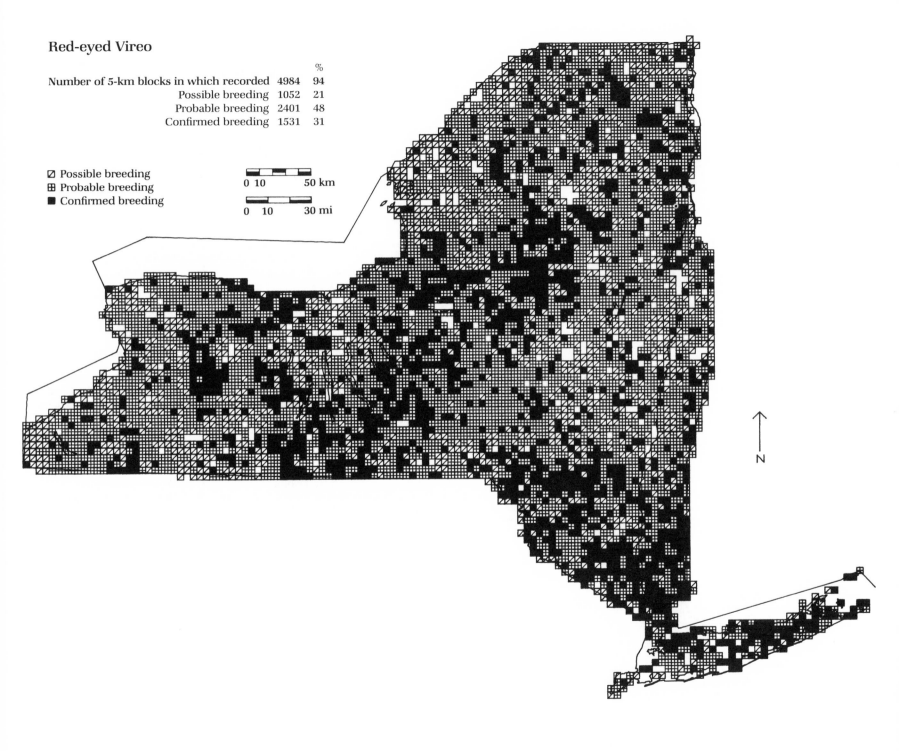

Red-eyed Vireo

		%
Number of 5-km blocks in which recorded	4984	94
Possible breeding	1052	21
Probable breeding	2401	48
Confirmed breeding	1531	31

◨ Possible breeding
⊞ Probable breeding
■ Confirmed breeding

0 10   50 km

0 10   30 mi

N

Blue-winged Warbler *Vermivora pinus*

For more than a century the Blue-winged Warbler has been expanding from its original range west of the Appalachians into much of the northeastern United States (Gill 1980). This expansion has brought it into contact with the Golden-winged Warbler. Two hybrid forms, the Brewster's and Lawrence's, occur in areas of sympatry, or range overlap. Interactions between the two species and their hybrids are described in the next two accounts.

The Blue-winged Warbler expanded into the New York City region in the late 1800s, probably from southeastern Pennsylvania (Eaton 1914; Short 1962; Bull 1974; Gill 1980). By 1907 this warbler was common on the western end of Long Island, and by 1930 it was well established up the Hudson Valley to Dutchess County. The Atlas documents the continued northward range expansion of this species to its present status as a common breeder as far as the southeastern Mohawk Valley.

Range maps by Short (1962) and Bull (1974) show a discontinuous distribution of the Blue-winged Warbler in New York. Apparently the Blue-winged Warbler expanded into upstate New York from northwestern Pennsylvania, moving first into the western Appalachian Plateau and the western Great Lakes Plain, and later moving both east and north. Despite nesting reports as far back as 1867 in Chemung County and 1889 in Niagara County (Benton, pers. comm.), the Blue-winged Warbler did not become well established upstate until between 1938 and 1960 (Scheider 1959; Benton 1960; Short 1962; Beardslee and Mitchell 1965). Once the warbler arrived in an area, very rapid increases in its abundance were noted. For example, the first observation for Tompkins County was in 1941; by the mid-1950s the Blue-winged Warbler was more numerous than the Golden-winged Warbler, and by the mid-1970s it exceeded the latter by about 8:1 (Confer and Knapp 1979, 1981). Since the mid-1950s the bird has been common in Tompkins County and now is nearly as abundant as the Yellow Warbler or Common Yellowthroat (pers. obs.). The Atlas survey shows that the formerly separate Hudson Valley and upstate New York populations are now continuous.

Breeding was "confirmed" for the Blue-winged Warbler throughout most of the western and southern regions of New York. There are few records, however, in the higher elevations between the Delaware and Hudson river valleys, and it is virtually absent from northeastern New York. Yet, the range of the Blue-winged Warbler has been expanding for a century and further range expansion is anticipated. The temporal pattern of northward expansion in New York complements its range expansion throughout eastern United States as summarized by Gill (1980) and as recently documented for Vermont (Clark 1985).

The Blue-winged Warbler nests in a broad range of secondary succession growth (Confer and Knapp 1981), but despite its species name, *pinus*, it does not utilize conifers. Territories are occasionally located in swampy areas, although dry areas are more frequently used. In Tompkins County these warblers nest in lightly grazed pastureland with few large trees, as well as in areas with up to 90% tree cover (Confer and Knapp 1981).

The expansion of the Blue-winged Warbler in the northeastern states appears to be correlated with historical patterns of land use (Gill 1980). The Blue-winged Warbler expanded into various regions of the Northeast only after farmland was abandoned and secondary succession created appropriate habitat. Climatic warming during this century may have facilitated this expansion (Bull 1974).

Nests are built on or close to the ground and are shaped like a deep, narrow cone. Large, coarse material is used on the outside, while fine shreds of grapevine and sometimes horsehair or split grass stems form the inside. Several descriptions emphasize that the inner lining is not woven in a circular fashion, but instead is laid across the nest in a crisscross pattern (Bent 1953; Harrison 1975). Many older publications contain numerous nest descriptions, but birds may have recently evolved a superior ability to hide their nests because nests are now very difficult to find (pers. obs.).

John Confer

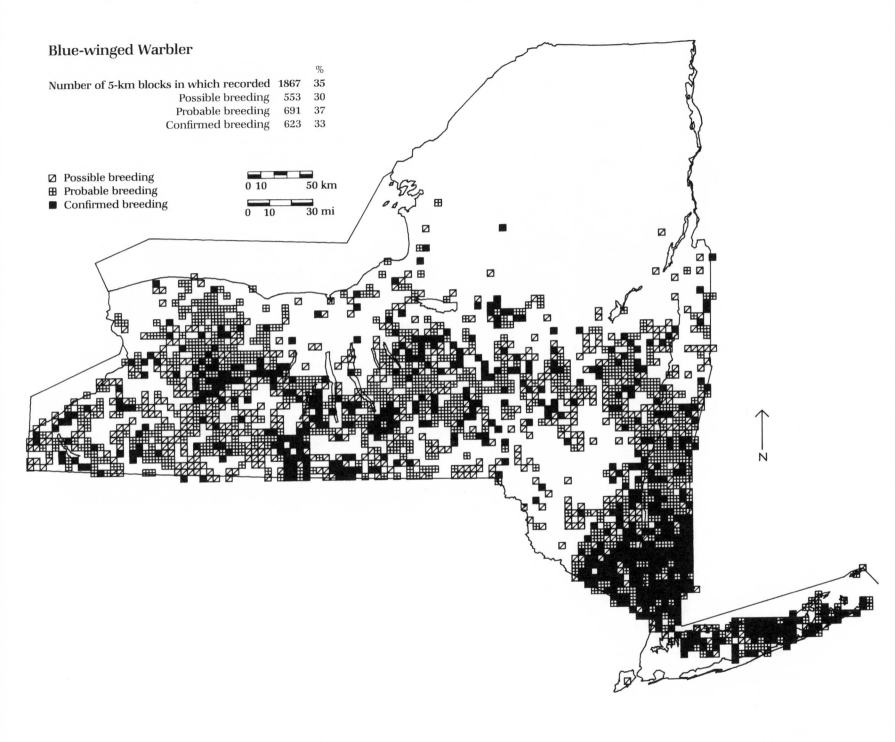

Blue-winged Warbler

		%
Number of 5-km blocks in which recorded	1867	35
Possible breeding	553	30
Probable breeding	691	37
Confirmed breeding	623	33

◨ Possible breeding
⊞ Probable breeding
■ Confirmed breeding

0 10   50 km

0 10   30 mi

N

Golden-winged Warbler *Vermivora chrysoptera*

For more than a century the Golden-winged Warbler has expanded its range in much of northeastern United States. This expansion began earlier than that of the Blue-winged Warbler and has extended to higher altitudes and latitudes (Gill 1980). The northern limits continue to expand, but the golden-winged is losing much of its southern range. Within this shifting distribution, the abundance of the Golden-winged Warbler can change rapidly. For example, in Broome and Tompkins counties the bird has changed from common in the 1940s to rare in the 1980s (Confer and Knapp 1981). At the same time it is increasing in abundance at the new, northern limits of its range. But even in the center of its current distribution the golden-winged is highly localized and not common throughout a wide area.

Early ornithologists knew little of this uncommon bird (Tyler 1953). The earliest nesting for New York probably occurred in the southern part of the state in the late 1800s. Benton (1960) noted the first documented arrival in upstate New York to be 1867, with nesting in 1896. By 1914 the golden-winged was known to breed in the highlands east of the Hudson River from the coast northward into Rensselaer County, and in scattered locations in the Finger Lakes Highlands, Schoharie Hills, and central and southeastern portions of the Great Lakes Plain (Eaton 1914; Bull 1974). By 1972 this bird was widespread and fairly common throughout central and western New York (Scheider 1959; Benton 1960) and had nested as far north as Saratoga County (Bull 1974).

The Atlas documents a continuation of this range expansion, part of a widespread northward movement (Gill 1980). Nesting now occurs in the Mohawk Valley and in the lower elevations completely surrounding the Central Adirondacks. The clumped distribution pattern illustrated by the Atlas map may reflect this shifting population. Atlas observers familiar with the song who were in the field in May and early June, or in late June or early July to see the young shortly after leaving the nest, had no difficulty locating this species. The nest, however, was extremely difficult to find. Only 7 nests of this species were reported.

The Golden-winged Warbler nests in areas with scattered patches of grass, thick brush, and a few trees (Eaton 1914; Tyler 1953; Scheider 1959; Ficken and Ficken 1968; Confer and Knapp 1981). Some accounts indicate a preference for wet or swampy areas. Confer and Knapp (1981) suggested, however, that wetness itself was not a requisite, but that cut-over swamps and wet fields frequently provide the thick brush and scattered small trees that are preferred. This kind of habitat is mainly produced by succession following abandonment of farmland. The chronology of expansion of the golden-winged into various regions of New York, and indeed throughout New England, corresponds with the time when various regions had large amounts of land in early, secondary successional stages (Confer and Knapp 1981).

Eaton (1914) and Confer and Knapp (1981) indicated that appropriate habitat would occur only briefly during succession. Many regions of New York that once provided suitable nesting habitat no longer have large tracts of recently abandoned land. It is possible that the golden-winged population will decrease as its nesting habitat becomes rarer (Confer and Knapp 1981). Gill (1980, 1985) speculated that the golden-winged will continue to decline, perhaps even to extinction, as a result of interactions with the Blue-winged Warbler. It is very noteworthy that the Atlas documents nesting of the Golden-winged Warbler in southern New York in areas that have been co-occupied with the Blue-winged Warbler for this entire century, apparently contradicting the pessimistic predictions of both studies. It appears that stable populations of the Golden-winged Warbler may be able to co-occur with the Blue-winged Warbler.

Considering that the Golden-winged and Blue-winged Warbler readily hybridize, accounts of the nest structure are surprisingly different (Eaton 1914; Tyler 1953). Both species build their nests at the base of a cluster of plant stems, but in the following respects the nests differ. The golden-winged nest is frequently on the ground; the blue-winged nest is attached to plants a few centimeters above the ground. The golden-winged nest is cup-shaped but not especially deep or narrow; the blue-winged is several centimeters deep. The outside of the golden-winged nest is disorderly with the petiole of leaves protruding outward; the blue-winged nest is tightly woven. The lining of the nest of the golden-winged is not especially fine and is often dark red; the blue-winged nest lining is fine.

John Confer

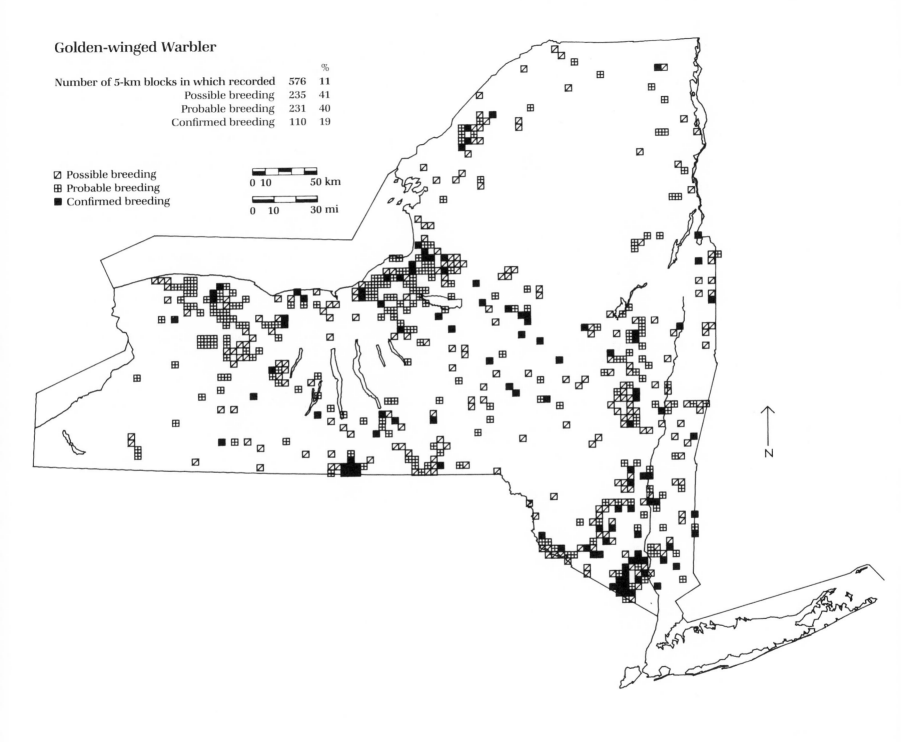

Golden-winged Warbler

Number of 5-km blocks in which recorded	576	11 %
Possible breeding	235	41
Probable breeding	231	40
Confirmed breeding	110	19

◫ Possible breeding
⊞ Probable breeding
■ Confirmed breeding

0 10   50 km

0 10   30 mi

N

© DAS 1987

Blue-winged × Golden-winged Warbler Hybrids
Vermivora pinus × Vermivora chrysoptera

The so-called Brewster's Warbler and Lawrence's Warbler are the result of hybridization between the Blue-winged Warbler and Golden-winged Warbler. Historical records document that the Blue-winged and Golden-winged warblers were allopatric in colonial times. Now that they have become sympatric, their fertile hybrids show that they are very closely related. Hybrids did not occur until northeastward range expansion by the former species brought it into contact with the latter. The hybrids were formally described in 1874, although New York specimens for the Brewster's and Lawrence's warblers were first obtained in 1832 and 1879, respectively (Parkes 1951). Faxon (1911) provided convincing evidence of the hybrid nature of these forms.

Hybrids are found wherever the Blue-winged and Golden-winged warblers are sympatric, as shown on the Atlas maps. Hybridization is frequent in areas where both are common. Short (1963) analyzed over 1000 specimens from the major museums of the northeastern United States and concluded that between 42 and 89% were hybrids, with the estimate varying according to the criteria for the pure form. Most of the specimens, thought to be hybrids when examined in the hand, would not appear to be hybrids when observed with binoculars in the field.

The Golden-winged Warbler has declined in abundance and even disappeared from portions of its range. In part, this is due to the loss of suitable habitat (see Golden-winged Warbler account), but hybridization might contribute to this decline. Although both Bull (1974) and the Atlas document the coexistence of the Golden-winged Warbler with the Blue-winged Warbler in the southern Hudson Valley for nearly a century, it is much more common for expansion of the Blue-winged Warbler into the range of the Golden-winged Warbler to be followed by the disappearance of the latter within 20–50 years (Gill 1980, 1985; Confer and Knapp 1981). Ficken and Ficken (1967) suggested that hybrids have a reduced ability to obtain mates, although their observations about this have not been fully accepted (Ficken and Ficken 1969; Short 1969; Gill and Murray 1972). Even if hybridization is disadvantageous, in many circumstances this would be equally disadvantageous for both species and would not explain the relative decline of the Golden-winged Warbler. Hybridization might be more deleterious to the Golden-winged Warbler if the Blue-winged Warbler is increasing more rapidly than the Golden-winged Warbler or if the Golden-winged Warbler is more likely to mate with hybrids. The two questions of a possible disadvantage for hybrids, and a relatively greater disadvantage for the Golden-winged Warbler than for the Blue-winged Warbler, require further work.

The pattern of inheritance of color among these birds has been analyzed (Parkes 1951; Short 1963; Gill 1980). The Brewster's Warbler is the most common hybrid form. All crosses between a pure Blue-winged Warbler and a pure Golden-winged Warbler produce a Brewster's Warbler. Thus, it seems that: the black eye stripe is dominant to the black eye and throat patch; the white breast and gray back are dominant to the yellow breast and olive-green back; the wing patch is dominant to the two wing bars; and the yellow color of the wing patch or bars is dominant to white. As a first approximation, the Brewster's Warbler can be thought of as having at least one dominant gene for each color characteristic. The Lawrence's Warbler seems to be homozygous recessive for all color characteristics, which accounts for its rareness compared with the Brewster's Warbler.

A full description of the genetic basis for these color patterns is more complex. All combinations of color pattern and intermediate hues of gray to yellow-green and white to yellow occur (Short 1963; Gill 1980). These

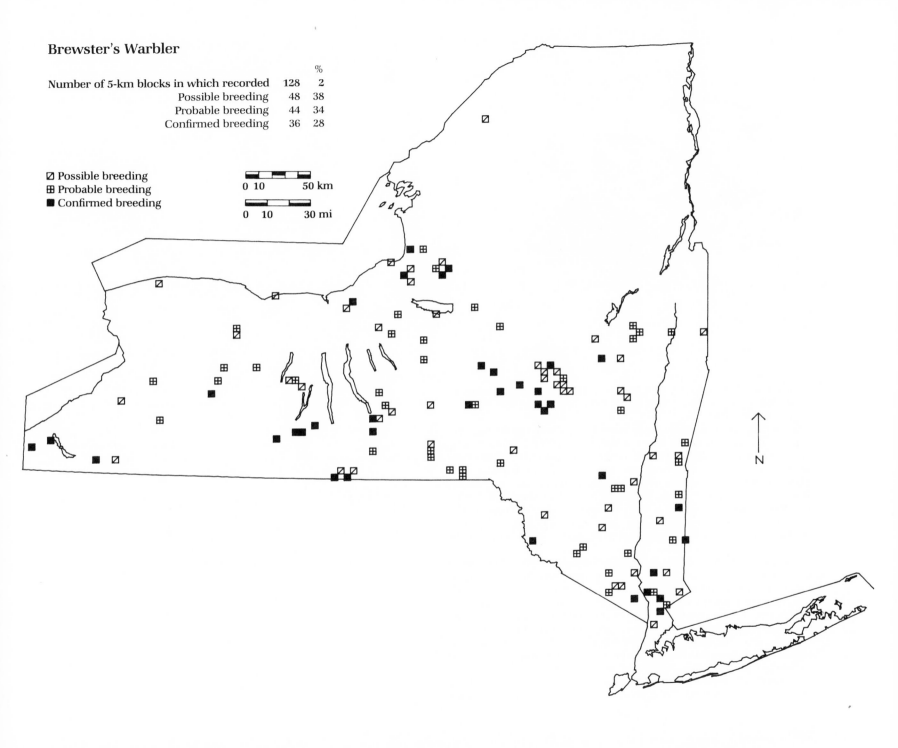

Brewster's Warbler

		%
Number of 5-km blocks in which recorded	128	2
Possible breeding	48	38
Probable breeding	44	34
Confirmed breeding	36	28

◪ Possible breeding
⊞ Probable breeding
■ Confirmed breeding

0 10   50 km

0 10   30 mi

N

intermediate forms suggest that factors such as multiple genes, incomplete dominance, or modifying genes influence the expression of the basic color pattern.

In atlasing for these hybrids observers found that some hybrid forms did not fit the criteria for either a Brewster's Warbler or Lawrence's Warbler and recorded the hybrid it most closely resembled. Additionally, hybrids may sing the song of either parent or even alternate songs, and observers were asked to make a visual identification for each record. The habitat and nest construction of crosses of the pure species, or various combinations of the hybrids with a pure species, does not differ noticeably from that for matings within each species. Whatever difference might exist, it has not been sufficient to be noted in the extensive, recent literature about this complex.

It is intriguing to consider population changes of these two warbler species and their hybrids, and possible causes over the last 150 years: range expansion to the north for both species; range contraction in the south for the Golden-winged Warbler with the unusual circumstance of prolonged co-occurrence in the lower Hudson Valley; uncertain consequences of hybridization; changes in human land practices. These and other factors may determine the ultimate balance between these two closely related species. Previous studies, this Atlas, and future studies will help document the outcome of this interesting example of ongoing evolution.

John L. Confer

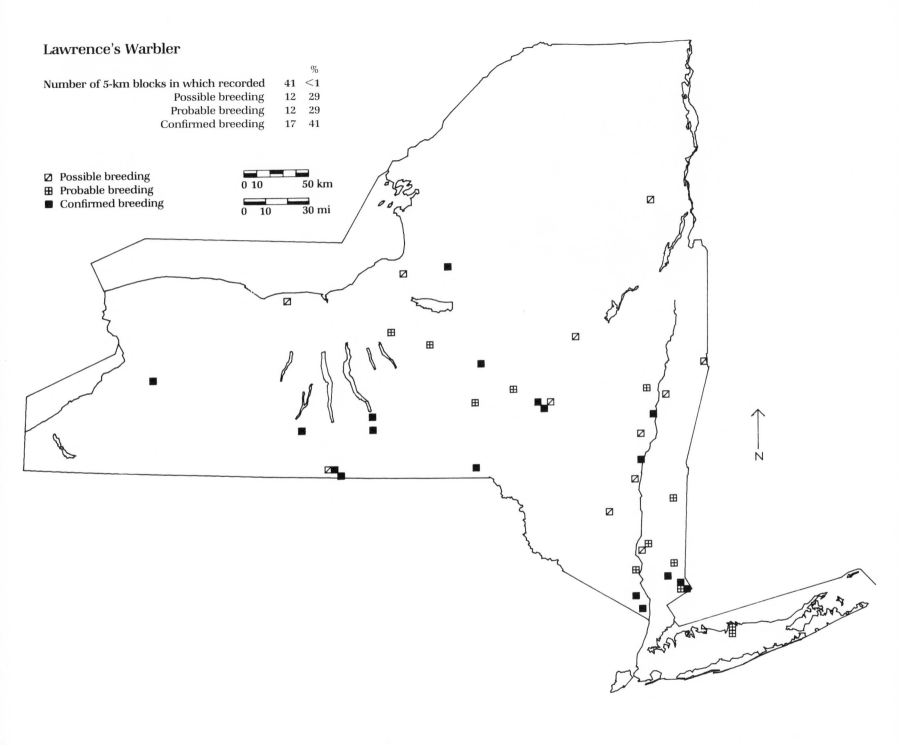

Lawrence's Warbler

		%
Number of 5-km blocks in which recorded	41	<1
Possible breeding	12	29
Probable breeding	12	29
Confirmed breeding	17	41

☑ Possible breeding
⊞ Probable breeding
■ Confirmed breeding

0 10   50 km

0 10   30 mi

N

©DAS 1987

Tennessee Warbler *Vermivora peregrina*

Although experienced Atlas observers, once they became attuned to the song of the Tennessee Warbler, sometimes encountered several singing males in a single day, records were obtained in only 27 Atlas blocks. Even within the Adirondacks this warbler appears rather uncommon. No documented nest with eggs or young has ever been found in New York.

Over a century ago Merriam (1878a) wrote of the Tennessee Warbler in Lewis County, "I have taken two of this species (May 19 and 23, 1877), the only record of its appearance in this locality." Roosevelt and Minot (1929) simply included the species on their Franklin County list compiled in 1877 and 1878 as: "? Tennessee Warbler." A few years later Merriam (1881) wrote of this bird in the Adirondack region: "Breeds. Not rare in suitable localities. Generally prefers hardwood areas." Eaton's party found 17 warbler species in Essex County during the summer of 1905, but not the Tennessee (Eaton 1910), and he later concluded, "The full text of Doctor Merriam's notes . . . does not indicate that the nest of the Tennessee Warbler was found in Lewis County, or that it was ever seen there later than May 29, which may well be a migration date" (Eaton 1914). Nevertheless, Eaton thought it probable that the Tennessee Warbler might be found breeding in "the North Woods" but thought the earlier evidence insufficient to include it as a summer resident of the state. More solid evidence was finally obtained. Saunders (1929a) had several encounters in Essex County in 1925 along the Heart Lake Road near Lake Placid and above Indian Falls on the trail to Mount Marcy. The following summer he found an adult male feeding two fledglings near the Heart Lake Road, for the only solid confirmation prior to the Atlas. Without elaboration, Saunders noted, "The Tennessee Warbler has increased in recent years, and now it may be not uncommon in many places in the Adirondacks."

For a bird that favors young second growth, conditions must have been perfect by the 1920s. When the forest preserve was created in 1885, vast tracts of the Adirondacks were waste lands, left barren of trees by lumbering. Forest fires in the late 1800s and early 1900s burned thousands of acres of state and private land (Donaldson 1921), and the succession of young aspen and conifers that sprang up was ideal for the Tennessee Warbler. Yet only about ten summer sightings were added over the next half century, most in deciduous growth at moderate to high elevations (Bull 1974; Carleton 1980; McKinney, pers. comm.). The map of distribution published by Bull (1974) showed breeding and summer records from just five locales, but reports were increasing in the decade prior to the Atlas.

The Tennessee Warbler was recorded in 27 blocks in the Central Adirondacks, the Adirondack High Peaks, and the Western and Eastern Adirondack Foothills during the course of the Atlas fieldwork, but the species was probably under-recorded, since most Atlas records were obtained by just a few observers. The difficulty observers had hearing and recognizing its song may explain the paucity of records. The song is two-parted, similar to that of the Nashville Warbler, but observers who learned to discriminate between the two felt that "it always sounds as if the Tennessee has problems getting started" (D. Niven, v.r.). The failure of some birders to find this warbler involves more than song recognition. The pitch of the song ranges from 6600 to 9150 cycles per second, compared with about 4000 cycles per second for the average passerine (Brand 1938). The sound, as loud as it is, may simply be beyond the hearing range of some birders.

The Tennessee Warbler's habitat is often second growth, with vegetation elements having a strong boreal affinity. The rather misleading name was provided by Alexander Wilson, who discovered the bird in migration along the Cumberland River early in the 19th century (Bent 1953). On the Adirondack breeding grounds this warbler favors somewhat wet areas of young deciduous growth, often aspen, with a mixture of balsam fir, spruce, or tamarack, northern shrubs, and often a ground cover of sphagnum mosses. The areas are relatively open, with edge frequently provided by roads or power lines, but even within the hamlet of Blue Mountain Lake, Hamilton County, where I found a male singing behind a power substation on a residential street, the plant associations were largely boreal.

The Tennessee Warbler usually builds its nest into a sphagnum mound or other concealed area. The nest is constructed of thin, light-colored grasses, lined with finer grasses or hair. Nest locations vary from the shaded borders of boreal acid bogs with black spruce to dry, upland sites that lack the sphagnum moss ground cover of the typical bog (Bent 1953).

John M. C. Peterson

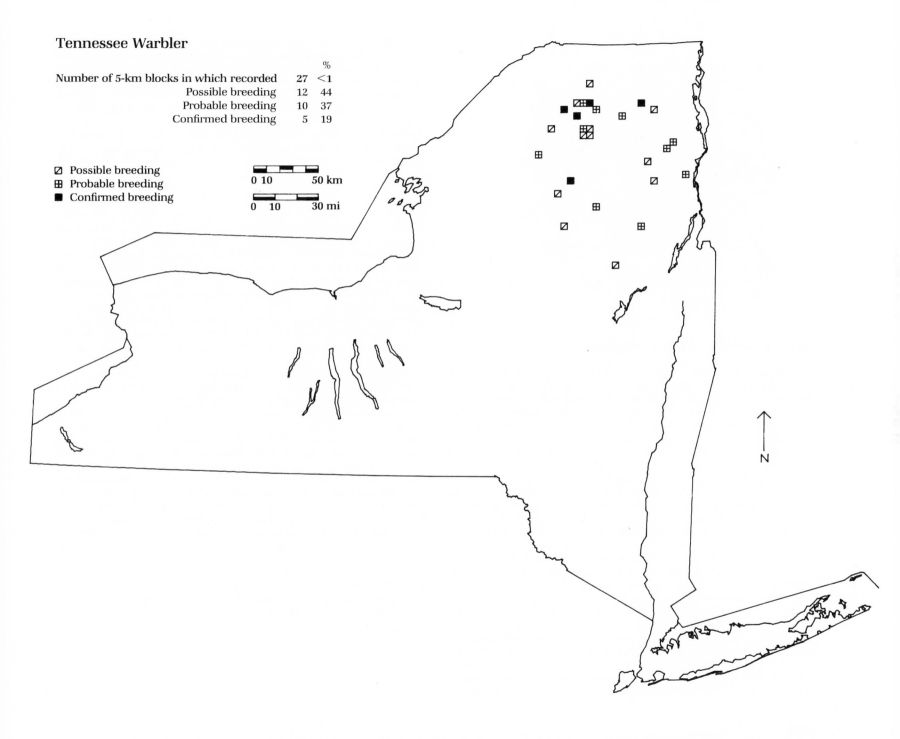

Tennessee Warbler

		%
Number of 5-km blocks in which recorded	27	<1
Possible breeding	12	44
Probable breeding	10	37
Confirmed breeding	5	19

☒ Possible breeding
⊞ Probable breeding
■ Confirmed breeding

0 10   50 km

0 10   30 mi

N

© DAS 1987

Nashville Warbler *Vermivora ruficapilla*

Generations of ornithologists have commented upon the changes in abundance and range of the Nashville Warbler, a small, ground-nesting bird, ever since Alexander Wilson encountered the first specimen near Nashville, Tennessee, in 1811 (Bent 1953). In New York the Nashville Warbler is a rather uncommon to common breeder over much of the state, found mostly at higher elevations.

Little was known of the Nashville Warbler a century and a half ago, when it was regarded as a southern bird (Bent 1953). Numbers apparently increased during the mid-19th century but seemed to decline again in the early years of the 20th (Bent 1953). In New York, Eaton (1914) knew the Nashville Warbler as a locally common breeder in northern portions of the state, much as it is today. At that time the species was also found locally in eastern New York from Albany County to Orange County. Nests had been found in the counties of Monroe, Onondaga, and Tompkins, as well, yet Eaton commented, "I never met with a breeding pair in any of the western counties of the State" and noted that in his experience the Nashville Warbler was "extremely rare as a summer resident" throughout central and western New York at the turn of the century.

Bull (1974) agreed that the Nashville Warbler was "common enough in the Adirondacks and Catskills." He urged that "the species should be sought during the nesting season in Putnam, Orange, Rockland, and northern Westchester counties," and he noted three records from Dutchess County, plus single records from Orange and Westchester counties. Although Atlas records were obtained from Orange County, no breeding Nashville Warblers were recorded anywhere in Dutchess, Putnam, Rock-

land, or Westchester counties between 1980 and 1985. Of western New York Bull (1974) stated: "The Nashville Warbler is apparently absent as a breeder from the lowlands of the lake plain and some distance back from the Great Lakes as well." He had heard of no nesting records from Niagara or Wayne counties; however, Atlas records were obtained from the latter.

Thus, although the evidence is sketchy, the range of the Nashville Warbler seems little changed in eastern New York but may well have expanded in western portions during the past century. The reversion of farmland and old fields, especially in the hill country along the Appalachian Plateau, to successional forest has provided suitable, though temporary, additional habitat for this species.

The Nashville Warbler occupies both dry and wet habitats, and is generally distributed through deciduous, mixed hardwood-coniferous, and boreal coniferous forest types (Temple et al. 1979). Dry tracts can include successional shrubland of alder, willow, and birch thickets, old burns (Saunders 1929a), cut-over areas, or successional old fields reverting either to aspen-cherry thickets or stands of gray birch (Bull 1964, 1974). In the Helderberg Highlands the Nashville Warbler has been associated with the late seral stage leading to a mature beech-maple-hemlock forest, specifically with scattered stands of tall trees and shrubs. In various stands of mixed forest in the north central or northeastern United States, population densities of this warbler were greatest in an area 73% spruce-fir and 27% aspen–white birch, and the Nashville Warbler is likely to become more abundant with an increasing abundance of pine (Temple et al. 1979).

The wet terrain chosen by this species is typically the edges of tamarack swamps (Saunders 1929a) or spruce-sphagnum bogs (Bull 1974). In lowland conifer forests the Nashville Warbler may benefit from the increase in edge created by logging, provided that large blocks of forest remain uncut, both for dense cover and as a seed source to regenerate cut-over areas (Dawson 1979). At higher elevations in the Adirondack High Peaks it may frequent small forest openings, the steep slopes providing drainage, and the shade and dampness of small spruce flats offering a thick ground cover of moss (pers. obs.). In the alpine krummholz it may occupy stands of dwarfed birch almost up to timberline (Saunders 1929a). Although it seems unlikely that wet habitats have increased, except perhaps where beavers have flooded areas of black spruce–tamarack, the abandonment of many farms and the effects of logging operations continue to provide increased dry, brushy areas marked by dense stands of young alder, aspen, birch, cherry, and willow.

The nest is a small compact cup of rootlets and fibers, lined with hair and set on the ground. Sometimes it is placed in a hollow in moss; at other times, built against the base of shrubs and partially concealed by surrounding vegetation (Harrison 1978).

John M. C. Peterson

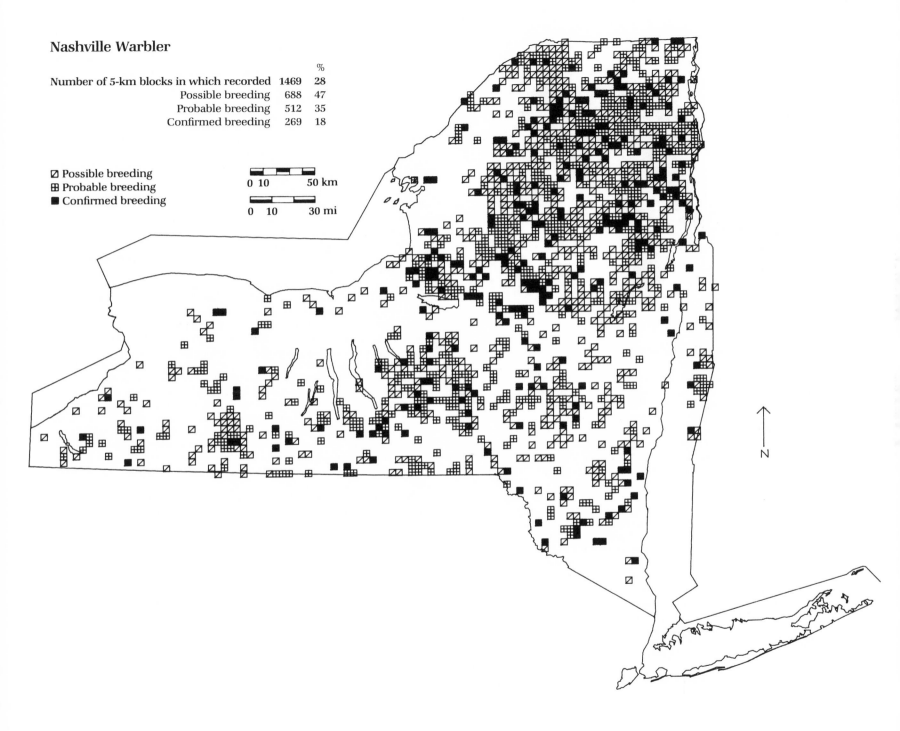

Nashville Warbler

		%
Number of 5-km blocks in which recorded	1469	28
Possible breeding	688	47
Probable breeding	512	35
Confirmed breeding	269	18

◩ Possible breeding
⊞ Probable breeding
■ Confirmed breeding

0 10   50 km

0 10   30 mi

N

© DAS 1987

Northern Parula *Parula americana*

Early in this century the Northern Parula bred locally throughout New York, especially where old man's beard or *Usnea* lichen was abundant. Eaton (1914) thought, "It is probably commoner as a breeding species in the swamps of Long Island and in the Catskill and Adirondack districts than in other portions of the State, although I have noticed a few pairs nesting in the gullies of the Finger Lakes region and in various scattered peat swamps of western New York." Today, however, only within the Adirondacks is this warbler still fairly common. Elsewhere, it is rare, or at best uncommon, having suffered a considerable decline since 1900.

When this species was known as the blue yellow-backed warbler, Roosevelt and Minot (1929) described it as "very common" in Franklin County between 1874 and 1877; Merriam (1881) considered it a "tolerably common summer resident" of the Adirondacks. During the same era the Northern Parula was a common breeder on the Coastal Lowlands, especially eastern Long Island. Between 1879 and 1882 two oologists collected 40 sets of eggs on Shelter Island, Suffolk County. One collector alone secured 24 sets, "all in beard moss clumps in wet situations" (Bull 1974).

During his 1905 visit Eaton found the Northern Parula to be fairly common in Essex County, especially in the Marcy Swamp and swamps near Boreas Ponds and Elk Lake. He cautioned, "It is by no means generally distributed in the North Woods, but almost entirely confined to swamps . . . practically confined to the localities where usnea moss is fairly abundant." In the ravines of Canandaigua, Seneca, and Cayuga lakes he found it nesting on the damp south side of gullies where

hemlocks were abundant, but where there was almost no *Usnea* (Eaton 1914).

On Long Island by 1924 there were only 6 pairs found on Gardiners Island, and 20 pairs on Shelter Island the following year (Bull 1964). By the early 1930s the Northern Parula had disappeared from both islands, and the last confirmed nesting on the Coastal Lowlands was in 1951. Bull (1964) also noted, "It is unreported as a breeding species in Westchester, Putnam, Rockland, and Orange counties." A decade later he concluded, "With the possible exception of the Adirondacks region, where Parula Warblers are still fairly common—although local—breeders, the nesting population of the state has crashed badly in recent years and has even disappeared from Long Island" (Bull 1974). He suspected the almost total absence of the Northern Parula from the Catskills was "more imaginary than real," but except for three "possible" Atlas records in the Mongaup Hills and Neversink Highlands, the absence appears real.

The Northern Parula nests in open hardwood or coniferous forests, woodlands, and swamps (AOU 1983) but displays a close affinity for trees festooned with the hanging *Usnea* lichen. By the turn of the century *Usnea* was disappearing and so, too, the Northern Parula. "Why did the Usnea lichen disappear?" asked Bull (1964). He could only conclude, "Whether it was due to possible climatic change, disease, or other cause, is uncertain," later adding "drainage" as another possibility (Bull 1974). Yet, as early as Eaton's time, botanists understood that although long-lived and insensitive to drought or temperature extremes, "lichens appear to be very easily affected by the presence in the air of noxious substances such as are found in large cities or manufacturing towns. In such districts lichen vegetation is entirely or almost entirely absent" (Crombie and Blackman 1911). Pendant lichens are often the first to go, and the last record of *Usnea* in Manhattan was in 1900, and in Ithaca, Tompkins County, at about the same time (Harris, pers. comm.). The *Usnea* lichen, considered extremely sensitive to sulfur dioxide, has been declining over much of its range during this century, although there is some evidence to suggest that, with the passage of clean-air laws, it may be returning (Harris, pers. comm.; Ketchledge, pers. comm.). Yet, even within the Adirondack High Peaks and Central Adirondacks, the author and others have noted that *Usnea* appears less abundant today than it did little more than a decade ago (Chase, Ketchledge, pers. comms.), possibly as a consequence of acid precipitation.

In New York the nest of the Northern Parula is usually, but not invariably, hidden in hanging clumps of *Usnea*, with a side entrance above the hollowed-out cup, much like the woven basket of a Northern Oriole (Bent 1953). A few nests in the state have been built in clumps of drift grass or leaf skeletons and pine needles (Bull 1974). Only 5 nests were located during the Atlas survey.

John M. C. Peterson

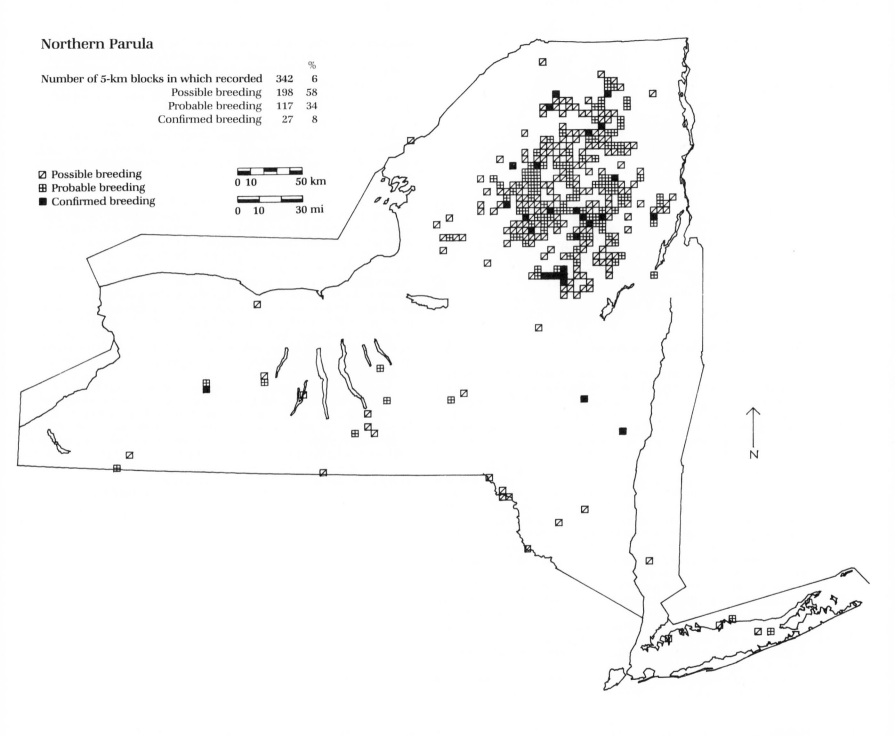

Northern Parula

		%
Number of 5-km blocks in which recorded	342	6
Possible breeding	198	58
Probable breeding	117	34
Confirmed breeding	27	8

◨ Possible breeding
⊞ Probable breeding
■ Confirmed breeding

0 10   50 km

0 10   30 mi

N

Yellow Warbler *Dendroica petechia*

This canary-colored bird of open country thickets is perhaps the best known warbler in New York. It is distributed widely throughout the state, being very common in all regions except the Adirondacks. Eaton (1914) stated that it was a common summer resident throughout New York except in the mountain spruce–fir forests of the Adirondacks and Catskills, which it penetrated as far as clearings and river valleys extended. Bull (1974) said that it was a widespread breeder.

Robbins et al. (1986) found that the Yellow Warbler increased on BBS routes in the Northeast, including New York, from 1965 to 1979. However, the warbler appears to be decreasing near population centers. For example, Griscom (1923) wrote that the species was a common summer resident in all rural sections of the New York City area but had decreased greatly in the immediate vicinity of the city. Cruickshank (1942) said that the Yellow Warbler had decreased slightly in the expanding New York City suburbs. Siebenheller (1981) noted that it had diminished greatly in the last 30 years on Staten Island.

Population decreases could be related to clearing of suitable habitats for buildings, or to spraying of suburban areas with pesticides. Declines also could be due to cowbird parasitism, as the Yellow Warbler is the most heavily victimized bird in New York (Bull 1974). This species may decrease further as more of the state's open lands revert to forest.

Atlasing showed the Yellow Warbler to be very widely distributed in New York. It was a "probable" or "confirmed" breeder in nearly every block except in the Central Adirondacks, Adirondack High Peaks, and Western Adirondack Foothills, regions probably too heavily forested for this species, or that may be at elevations too high. Although the Yellow Warbler is distributed throughout most of the Catskill and Allegany regions, Saunders (1942) noted that in Allegany State Park the species was found only in valleys where willow thickets occurred along streams or rivers. This suggests that elevation, as well as forest cover, may be a factor determining distribution.

The Yellow Warbler is extremely conspicuous and easy to "confirm." It was recorded in more blocks than any other warbler except the Common Yellowthroat. One cautionary note: the song, the most commonly used criterion for recording "probable" breeding, can be confused with that of the American Redstart or Chestnut-sided Warbler, even by experienced observers.

Nests were found in 524 blocks, and fledglings and adults carrying food or feeding young were observed in about 2000 blocks. Distraction displays were recorded in 64 blocks. One such display was described by an observer: When she closely approached a tree where a Yellow Warbler was singing, the bird appeared to fall from the tree and float to the ground. Then it fluttered along the ground away from the tree, apparently grievously injured (J. Carroll, pers. comm.).

The Yellow Warbler commonly inhabits gardens, shrubs, vines, fence-rows, and shade trees of towns, suburban areas, parks, and orchards (Eaton 1914; Stoner 1932). Away from civilization it is fond of low brushy growth in wet areas, especially willow thickets along the edges of swamps, streams, and lakes (Stoner 1932; Cruickshank 1942). The Yellow Warbler also is found in dense thickets of poison ivy and bayberry on coastal sand dunes (Bull 1974).

The female Yellow Warbler does most of the nest construction. The cup-shaped structure is usually placed within a few meters of the ground in a shrub or bush but may be as high as 12.2 m (40 ft) in a sapling or small tree. The nest is large, compact, and firmly woven into the crotch of the supporting branch. Building materials include plant fibers, especially milkweed bark; fine grasses; pieces of bark; and down from willows, poplars, ferns, and other plants. The lining is made of plant down and fine grasses, hair, or feathers (Eaton 1914; Stoner 1932).

Richard E. Bonney, Jr.

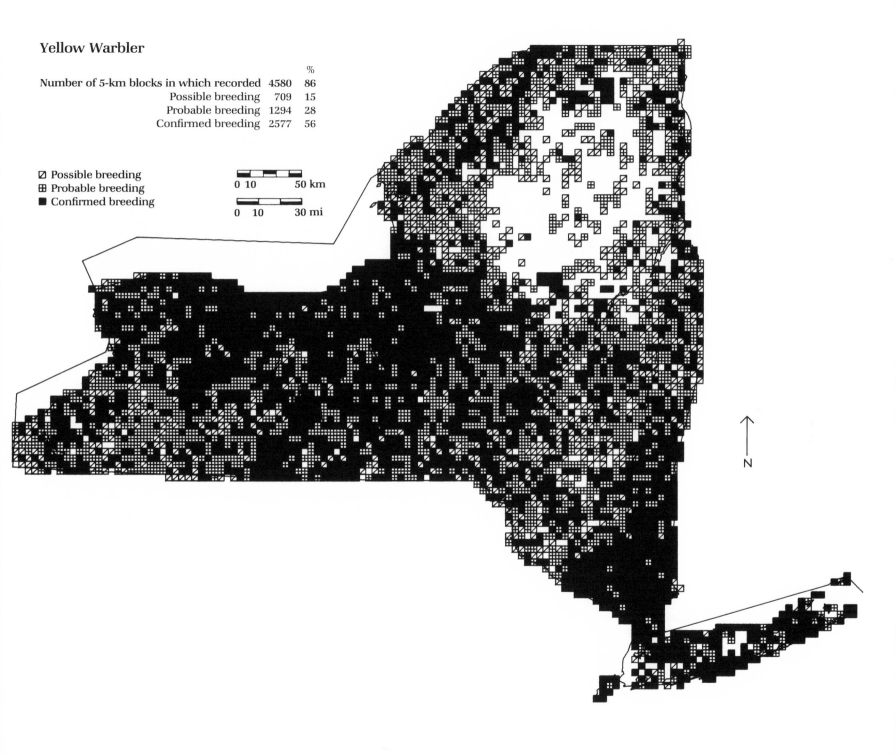

Yellow Warbler

		%
Number of 5-km blocks in which recorded	4580	86
Possible breeding	709	15
Probable breeding	1294	28
Confirmed breeding	2577	56

◩ Possible breeding
⊞ Probable breeding
■ Confirmed breeding

0 10   50 km

0 10   30 mi

N

© DAS 1987

Chestnut-sided Warbler *Dendroica pensylvanica*

Audubon said that he searched the borders of Lakes Ontario, Erie, and Michigan without finding the Chestnut-sided Warbler (Audubon and Chevalier 1840–1844), and Atlas data show that would be mostly true today as well. It is a common warbler that was found throughout much of New York except for the agricultural areas along the Great Lakes Plain and other heavily farmed locations. It was also absent from urban, heavily forested, and high-elevation areas. USFWS BBS data for 1966–74 ranked the Chestnut-sided Warbler fourth in frequency of occurrence among the warblers that nest in the state (Bystrak 1975).

At one time this warbler was rare in New York (De Kay 1844), but it probably started to increase in the late 1800s as the forests of the state were cut. (Lumbering got an early start in New York, reaching a peak about 1850, when New York was first among the states in lumber production, contributing one-fifth of the national total [Stout 1958]). By the early 1900s, as succession advanced, the Chestnut-sided Warbler was found in all sections of New York, although uncommonly on Long Island and only locally in the southeastern portion. It was especially common in the bushy pastures of eastern, central, and western New York and the outskirts of the Adirondacks (Eaton 1914). In discussing the birds of the Mount Marcy region, Essex County, Eaton (1910) said that the Chestnut-sided Warbler was only slightly less common and less generally distributed than the Black-throated Green, Black-throated Blue, Yellow-rumped, and Magnolia warblers, and the Ovenbird. More recently Bull (1974) indicated that it was a widespread breeder, rare and local only on the Coastal Lowlands.

The greatest change in the distribution of this species since the early 1900s appears to be its expansion as a breeding bird into southeastern New York and on Long Island, where it is absent only from the developed urban areas. Spencer (1981), however, said it was declining locally in central Suffolk County because of habitat changes. The species does not seem to be confined to any physiographic region but can be found in most sections where there is early second-growth type habitat. Where good soils encourage agriculture, as along the Great Lakes Plain, the Chestnut-sided Warbler is rare or absent. It is also absent from the highest peaks of the Adirondacks and Catskills, but elsewhere in these mountainous areas it is widely distributed in edge and open habitats.

The Chestnut-sided Warbler was reported by Atlas observers most frequently singing on territory and feeding young. It was easily observed, and its song is distinctive early in the season and early in the morning. Later in the season and in the mid-day heat, this species, the American Redstart, and Yellow Warbler can become nearly impossible to separate by song (Ficken and Ficken 1962).

The Chestnut-sided Warbler is a bird of open areas (Kendeigh 1945). Its preferred habitat in western New York and most of the state is along the edges of woods, in bushy pastures, and along neglected roadsides, usually in wilder and more deserted situations than where the Yellow Warbler is found (Eaton 1914). In the Mount Marcy region of the Adirondacks it was found in slashings up to an altitude of 914 m (3000 ft) (Eaton 1914). In the St. Regis lakes area of Franklin County, Saunders (1929a) found it inhabiting low second-growth hardwoods, open aspen-cherry growth of old burns, and roadside shrub. In Allegany State Park he found it in cherry-aspen and second-growth Appalachian oak-hickory forests (Saunders 1942). Except when it sings perched high in a tree, this species spends most of its time in brushy habitat within 3 m (10 ft) of the ground where it nests, forages, and also on occasion sings (Kendeigh 1945).

The nest has been found in briars and bushes (Eaton 1914), in a sugar maple sapling, on a spruce bough, in a choke-cherry tree, and in a clump of ostrich fern, all from 0.5 to 0.6 m (1.6–2 ft) from the ground (Eaton 1981). Saunders (1938) said that of 10 nests found in Allegany State Park, 3 were in blackberry bushes, 3 were in young sugar maples, 3 in hornbeams, and 1 in a young beech. They ranged from 0.4 to 1.2 m (1.3–4 ft) from the ground.

The nest is a loosely woven, thin-walled cup made of coarse or fine strips of inner bark, shredded weed stems, grasses, and plant down. It is lined with fine grasses and sometimes hair, and built entirely by the female in about 5 days (Harrison 1975).

Stephen W. Eaton

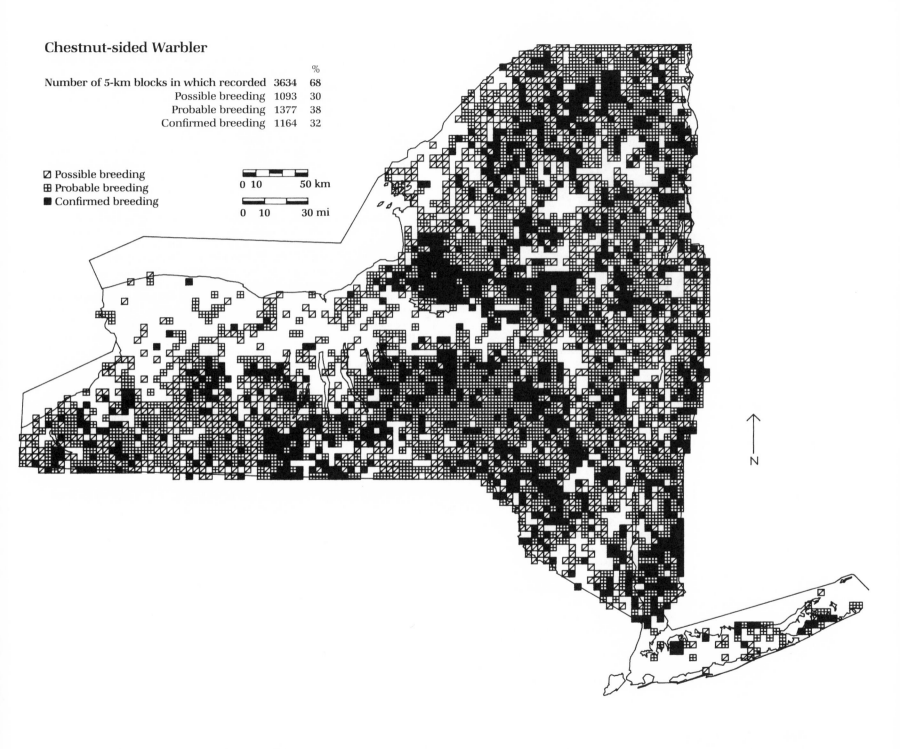

Chestnut-sided Warbler

		%
Number of 5-km blocks in which recorded	3634	68
Possible breeding	1093	30
Probable breeding	1377	38
Confirmed breeding	1164	32

◪ Possible breeding
⊞ Probable breeding
■ Confirmed breeding

0 10   50 km

0 10   30 mi

N

© DAS 1987

Magnolia Warbler *Dendroica magnolia*

Wilson (1811) called the Magnolia Warbler the black and yellow warbler, an appropriate name that has since been discarded. Its striking plumage makes it a favorite warbler among birders in the state, where it is a fairly common breeder at elevations above 305 m (1000 ft). BBS data for the period 1965–79 showed that its numbers apparently increased somewhat, particularly in the Adirondacks (Robbins et al. 1986).

Eaton (1914) referred to this species as a common summer resident of the Catskills and Adirondacks. In central, western, and southwestern New York he stated that it bred sparingly in swamps, gullies, and cooler hillsides. His map of its distribution (Eaton 1910) differs only slightly from the Atlas map. Since Eaton's time many small farms, particularly on the Appalachian Plateau, have been abandoned, much of the land returning to forest, which provides habitat for this warbler. In addition many hilltop farms were planted with Norway spruce and red pine during the 1930s, and it began nesting there. Bird lists from some of the counties that experienced this phenomenon—Wyoming (Rosche 1967), Cattaraugus (Andrle 1971b), and Onondaga (Bull 1974)—indicated that the Magnolia Warbler bred there in the 1960s and 1970s.

The Atlas map shows that this warbler is found at various high-elevation forests. In the northern part of its range it is found primarily within the spruce–fir–northern hardwood forest of the Adirondacks, the hemlock–northern hardwood forest of the Tug Hill Transition, and the conifer plantations of the Central Tug Hill. Forests in the Adirondacks and Tug Hill cover about 85% of the land area (Stout 1958). Along the Appalachian Plateau this species was found in most of the state forests containing predominately Norway spruce and red pine. Elsewhere it occurs in hemlock–northern hardwood forests. In most areas it was not found to breed below 152 m (500 ft). The small populations that once inhabited the gullies of Keuka, Owasco, and Canandaigua lakes in the Finger Lakes region (Eaton 1914; Benton 1949) appear to be gone or at least were undetected by Atlas observers, but the species was found in hemlock ravines elsewhere, including on the Great Lakes Plain.

The Magnolia Warbler, a rather secretive bird, is easily identified by its song, particularly when habitat is used as a point of reference. Its nest is difficult to find because it is placed in dense cover and thick undergrowth. Fledged young spend much time in this type of cover and are therefore difficult to locate during their long period of dependency on adults.

The Magnolia Warbler is found mainly in areas of spruce or hemlock at forest edges or clearings. Usually, young conifers are well represented in its breeding habitat (DeGraaf et al. 1980). On the Appalachian Plateau it often nests in second-growth hemlock bordering an opening in the woods (Bent 1953). In the Helderberg Highlands, Kendeigh (1945) found this warbler almost exclusively in hemlock and other conifers, both in the forest interior as well as along the edge. In the Adirondacks it appears to be more common in second-growth coniferous and mixed coniferous forests rather than in mature forests (Saunders 1929a). Saunders noted that it was quite common near the edges of tamarack and balsam fir swamps and on old burns, where young spruce were beginning to come in. Twice he found it nesting in black spruce. He also stated that it did not occur above 1219 m (4000 ft), which the lack of Atlas records at these elevations corroborates.

The nest is well hidden in thick growth near the top of the tree or out on a horizontal limb. Loosely made of fine grass stems, small twigs, and weed stalks, it is lined with black rootlets and moss stems. The use of coal-black rootlets is characteristic of the species and distinguishes it from similar nests of other species (Harrison 1975). As Bent (1953) noted, these rootlets form a background against which the eggs are shown in striking contrast.

Stephen W. Eaton

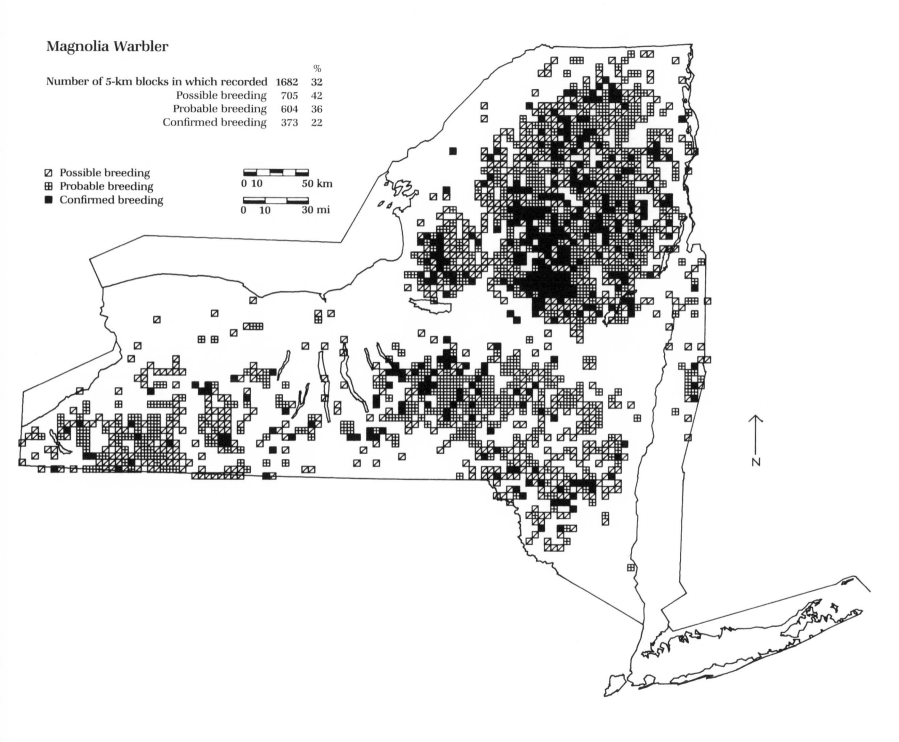

Magnolia Warbler

		%
Number of 5-km blocks in which recorded	1682	32
Possible breeding	705	42
Probable breeding	604	36
Confirmed breeding	373	22

◪ Possible breeding
⊞ Probable breeding
■ Confirmed breeding

0 10   50 km

0 10   30 mi

N

©DAS 1987

Cape May Warbler *Dendroica tigrina*

The Cape May Warbler is a rare breeder in the spruce forests of the Adirondacks, also accidental in spruce plantations and older reforestations well outside the Adirondack Park.

Even as a migrant this warbler was very rare in New York State before 1900 (Bull 1964), but during this century it has become fairly numerous during spring and fall passage. Eaton (1914) reported that "we have been unable as yet to record it positively as a breeding species in the Adirondack district, although the author searched for it diligently during the spring and summer of 1905, and various bird students who are perfectly familiar with the species have looked for it in the same region without success." Finally, on 4 July 1947, an adult female was discovered at North Meadow, near the Heart Lake Road, Essex County, at an elevation of 579 m (1900 ft): "The bird moved actively around, holding food in the bill, and was found to be giving it to two young standing in branches of spruce trees" (Carleton et al. 1948). Carleton and others did find the Cape May Warbler still frequenting North Meadow during the Atlas survey. A second confirmation came on 23 June 1962, when Rusk and Scheider saw a female feeding recently fledged young at Madawaska, Franklin County (Bull 1974), where the elevation is about 488 m (1600 ft).

Although Atlas observers were unable to "confirm" breeding, they came close. On 24 July 1985 a female Cape May Warbler and 2 young were found along Potter Brook, St. Lawrence County, at an approximate elevation of 472 m (1550 ft). Because the fledglings were capable of sustained flight, this observation shows only as a frustrating "possible" record. The observers were sure, however, that these birds had bred somewhere in the Adirondacks (Leukering, pers. comm.). Other "possible" and "probable" Adirondack records came from the High Peaks of Essex County, the Sable Highlands, on the 75,000-acre Boreal Heritage Preserve being formed by the Nature Conservancy in western Franklin County, and at scattered Hamilton County locations in the Central Adirondacks. Birds in the Adirondacks were found from elevations of 366 m (1200 ft) along the East Branch of the Sacandaga River, Hamilton and Warren counties, to almost 914 m (3000 ft) along the shores of Avalanche Lake, Essex County.

The Cape May Warbler, though rare even in the Adirondacks, may have been under-recorded during the Atlas project. Habitat, habits, and song combine to provide difficulties, even for experienced observers. The bird tends to keep to the tops of trees, and as Brewster (1938) noted, "It has a habit of singing on the extreme pinnacle of some enormous fir or spruce, where it will often remain perfectly motionless for ten or fifteen minutes at a time; on such occasions the bird is extremely hard to find." Whether the forest is dense or open, the foliage itself is always dark and dense, even in full sun. The bird's song is weak, and the pitch is so high that many observers are unable to hear it.

In my experience the Cape May Warbler is most abundant in stands of medium-aged spruce, about 25–75 years old, usually with some regeneration of younger balsam fir. It favors areas hit by spruce budworm infestations (Erskine 1977), which may account for its scarcity in otherwise suitable spruce–balsam fir habitat. Many of the million acres of the Adirondacks that burned between 1880 and 1913 may now be good Cape May Warbler habitat. Saunders (1929a) described the Heart Lake Road as he saw it earlier in this century: "At one time this road was bordered by primeval forest, much of which was cut long ago; and this was followed in 1903 by a devastating forest fire. While charred stumps and blackened stubs still stand in spots, the area is now covered with a growth of aspen and fire cherry. Here and there young maples, spruces and balsams are coming in." Two decades later the Cape May Warbler was nesting here (Carleton et al. 1948). Yet even as areas reach the proper successional stage, other spruce stands are logged or begin to experience die-off, while older, untouched stands reach the climax stage of maturity less favored by these warblers.

There have been only two records, from as many counties, of female Cape May Warblers feeding fledglings, but no nest with eggs or young has ever been found in New York. The Cape May Warbler nests much as it forages and sings, high in spruces, within a meter or so of the very top, perhaps 9.1–18.3 m (30–60 ft) from the ground. Resembling a ball of sphagnum and placed near the trunk, the bulky nest is woven of grasses and twigs, with a thick, felted lining of fur and feathers (Bent 1953).

John M. C. Peterson

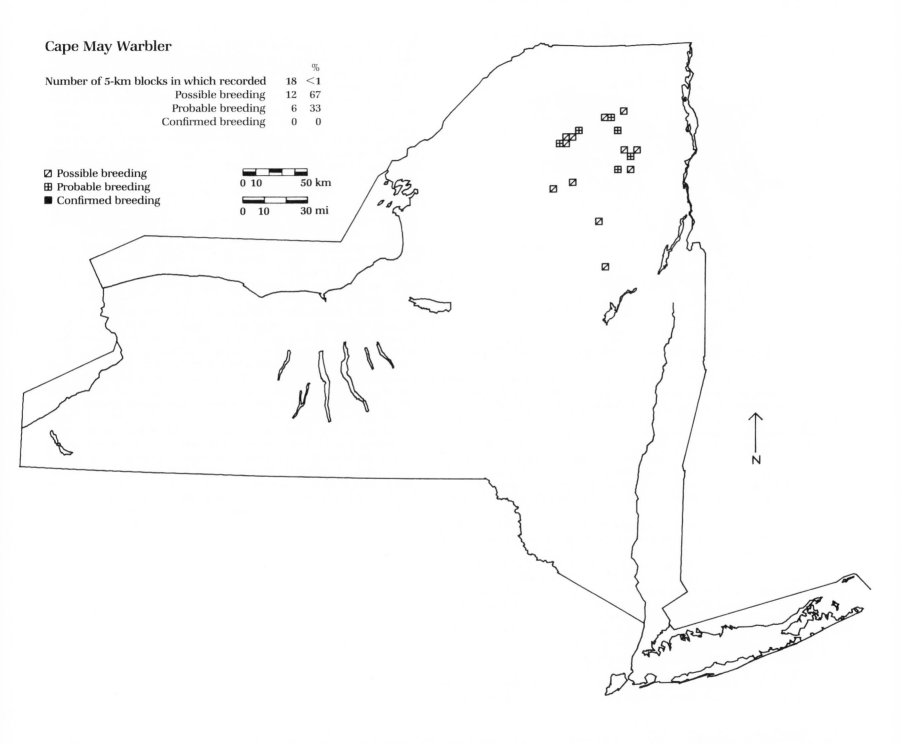

Cape May Warbler

		%
Number of 5-km blocks in which recorded	18	<1
Possible breeding	12	67
Probable breeding	6	33
Confirmed breeding	0	0

◩ Possible breeding
⊞ Probable breeding
■ Confirmed breeding

0 10   50 km

0 10   30 mi

N

© DAS 1987

Black-throated Blue Warbler *Dendroica caerulescens*

The Black-throated Blue Warbler, found in deciduous and mixed woodland undergrowth, is common at high elevations throughout much of New York, especially in the Adirondacks and Catskills and along the eastern Appalachian Plateau. According to Eaton (1914) the abundance of this warbler changes markedly from year to year.

Both Eaton (1914) and Bull (1974) considered it a widespread breeder in these higher locations. Eaton (1914) also indicated that it bred in nearly every ravine in the Finger Lakes region, and Bull (1974) noted that it nested at lower elevations in other portions of the state, although it was rare or absent in the low-lying agricultural regions of the western Great Lakes Plain.

The Atlas map shows the bird's affinity for mountainous country. It is found mainly above 305 m (1000 ft) and is widely distributed throughout all of the Adirondack ecozones, Central Tug Hill, and the Tug Hill Transition. It also is well distributed throughout the greater Catskills area and at higher elevations in the northern hardwood forests along the eastern Appalachian Plateau. In addition it is found in the Rensselaer Hills and Taconic Mountains in eastern New York and in the Allegany Hills in western New York, where Saunders (1942) considered it a common bird of the mountain forest. It is found infrequently in other parts of western New York, where Beardslee and Mitchell (1965) indicated that it bred, but sparingly. The bird is virtually absent from low elevations, including the major river valleys and the Coastal Lowlands.

The majority of Atlas records were of singing birds. This is not surprising because the male Black-throated Blue Warbler continues to sing even after the young have hatched (Black 1975), giving Atlas observers the opportunity to record singing well into the breeding season. Also, breeding is fairly easy to "confirm" because adults are exceptionally tame when feeding young (Terres 1980).

Although this warbler is a mountain dweller, it is found more commonly in deciduous and mixed woodlands than in spruce-fir forests (Eaton 1914). It seems to prefer wooded hillsides and frequents the low bushes and smaller trees that form the forest undergrowth (Cruickshank 1942). In a study of the Black-throated Blue Warbler in the Hubbard Brook Experimental Forest in New Hampshire, Black (1975) found that the bird had a patchy distribution in the forest understory: it nested and foraged mainly where the foliage was denser. Bull (1974) noted that the bird sometimes nested in second-growth forest with clearings, and that nests were found in mountain laurel, hemlock, spruce, and maple. In the southern part of the state it appears to avoid the dry oak and oak–northern hardwood forests, probably because the understory in these forests is sparse. In Allegany State Park, Saunders (1942) said that this warbler lived in undergrowth in beech-maple mesic forests, both young and mature, and also in Appalachian oak-hickory forests, but not in any other successional forests.

The Black-throated Blue Warbler can place its nest as low as 10.2 cm (24 in) from the ground but may nest as high as 6 m (20 ft) (Bull 1974) in a bush or low sapling in dense forest underbrush, usually near clearings, old roads, or trails. However, it usually nests from 0.3 to 1.2 m (1–4 ft) (Harrison 1975). The low placement of the nest can have an adverse affect on the species' productivity. In New Hampshire, Black (1975) found that breeding success varied from year to year but was usually lower than 40%, a result of heavy predation by ground-active predators such as the chipmunk.

The neat, compact, thick-walled nest is made of bark shreds, grass blades, leaves, and beech buds with a lining of fine rootlet fibers, hair, and mosses. Most nests are decorated with decayed spongy pieces of light-colored wood, fastened to the outside with strands of spider webs. The bird prefers to nest in American beech or sugar maple (Eaton 1914; Saunders 1942; Terres 1980). The female does most of the nest construction with occasional help from the male (Bent 1953).

Richard E. Bonney, Jr.

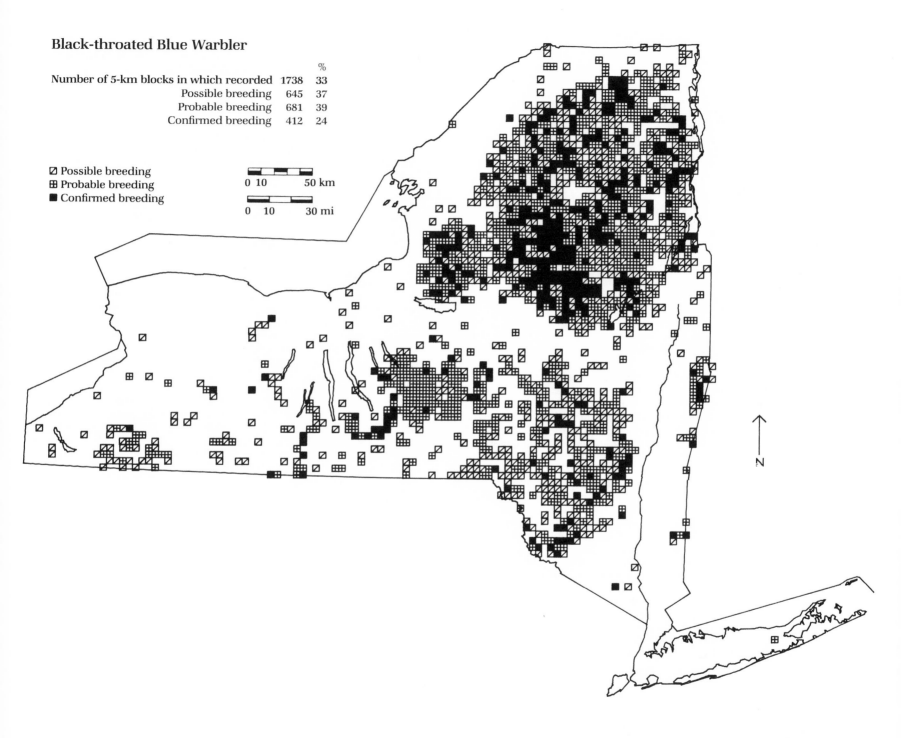

Black-throated Blue Warbler

Number of 5-km blocks in which recorded	1738	% 33
Possible breeding	645	37
Probable breeding	681	39
Confirmed breeding	412	24

☑ Possible breeding
⊞ Probable breeding
■ Confirmed breeding

0 10   50 km

0 10   30 mi

Yellow-rumped Warbler *Dendroica coronata*

Nesting from near sea level on Long Island and along the Lake Champlain Valley to the stunted spruce just below the rocky dome of Mount Marcy, Essex County, the familiar Yellow-rumped Warbler with its bright yellow rump patch has one of the widest altitudinal ranges of any New York bird. This is a common warbler over much of northern New York, where it extends even into lower elevations, and uncommon to fairly common at higher elevations elsewhere in the state. It is locally rare at low elevations.

Early in this century Eaton (1914) surmised: "In New York the breeding range is apparently confined to the spruce belt of the Catskills and Adirondacks. Its breeding at Utica [Oneida County] and Buffalo [Erie County] which has been reported has never been confirmed by later observations although it may occasionally, like other species, breed casually in various parts of the State." In 1922 a nest with eggs was discovered in the Lake Champlain Valley near Plattsburgh, Clinton County (Bull 1974). Then, beginning in the early 1950s, the Yellow-rumped Warbler began to turn up away from the Adirondacks and Catskills; Bull (1974) noted: "This species, like the Golden-crowned Kinglet, was discovered nesting in spruce and red pine plantings, although not as widespread as that bird. Nests were found to a lesser extent in white pine and hemlock." The map of breeding distribution presented by Bull showed 25 known localities in 14 counties outside the Adirondack and Catskill montane regions. The southernmost known records were in the Shawangunk Hills, Ulster County, near Lake Awasting and Mohonk Lake during 1952.

The breeding range of the Yellow-rumped Warbler has expanded greatly in the state during this century. Atlas fieldwork found it breeding locally as far south as Dutchess County in the Taconic Highlands and even in Suffolk County on the Coastal Lowlands. This warbler appears now to be more widespread than the Golden-crowned Kinglet. Breeding is concentrated in the Central Adirondacks, Adirondack High Peaks, Sable Highlands, and Western Adirondack Foothills, outward to the Eastern and Western Adirondack, Champlain, and St. Lawrence transitions, and eastward to the Lake Champlain Valley. To the west, a peripheral population on the Central Tug Hill and Tug Hill Transition is barely separated from this main range by the Black River Valley. Elsewhere the Catskill population is now blurred, with the Yellow-rumped Warbler presently nesting over a much-expanded range south of the Great Lakes Plain and Mohawk Valley. This population extends across the Appalachian Plateau to the Catskill Peaks, Delaware Hills, Helderberg Highlands, Mongaup Hills, Neversink Highlands, and Schoharie Hills, with a detached range across the Hudson Valley in the highlands of the Taconic Mountains and Rensselaer Hills.

The Atlas map is incomplete, given the usual difficulties of observers' access to remote areas, and there is also the possibility that some records are of late migrants found singing in suitable breeding habitat. Nevertheless, the Atlas range follows altitudinal boundaries and ecozones so closely that, in spite of gaps, the map appears to be quite accurate.

This warbler occupies a wide range of forested habitats but, according to my observations, prefers coniferous, particularly mountain spruce–fir forest, as well as the gnarled alpine krummholz (forest of stunted trees) that creeps toward and clings to the alpine summits of the Adirondack High Peaks. It is also found near the edges of young coniferous woods and in mixed woods (DeGraaf et al. 1980). Both the gradual regrowth of young forest and planting of conifers during this century have benefited the Yellow-rumped Warbler. Even on plains, lowlands, and valleys where distribution is thin, there were enough Atlas records to suggest a continued range expansion. This warbler has one of the most varied of feeding habits, which might explain, at least in part, its success as a breeding species (MacArthur 1958).

Atlas observers located 25 nests. Bull (1974) reported that the height of 10 New York nests ranged from 1.2 to 3.0 m (4–10 ft), all in evergreens. MacArthur (1958) found the median nest height to be from 4.6 to 6.1 m (15–20 ft). The cup of twigs, bark strips, moss, and other materials is lined with hair and feathers and is placed on a branch, away from the trunk (Harrison 1978).

John M. C. Peterson

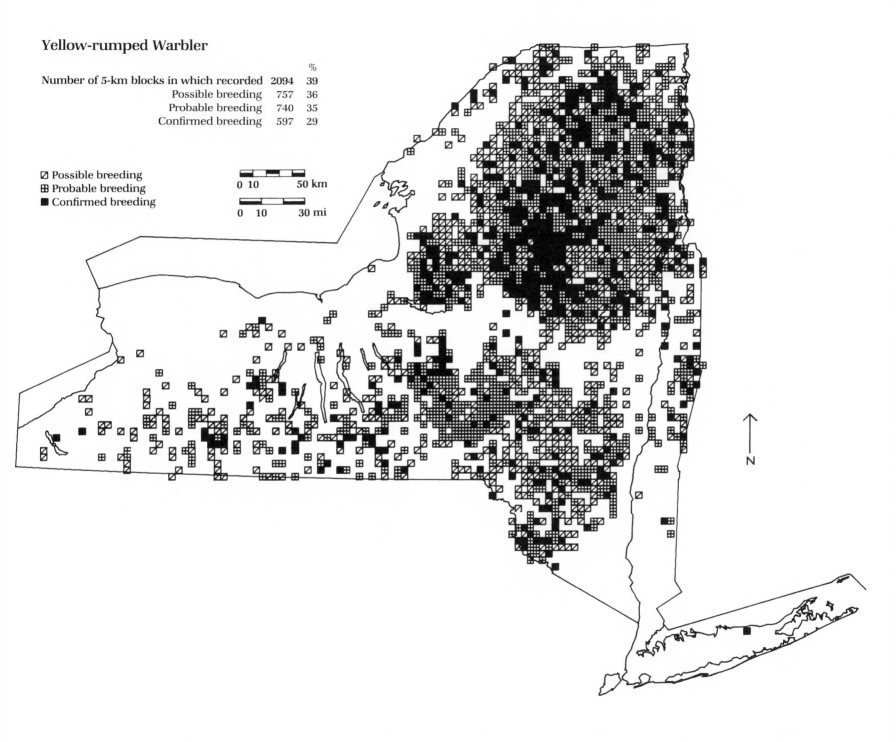

Yellow-rumped Warbler

		%
Number of 5-km blocks in which recorded	2094	39
Possible breeding	757	36
Probable breeding	740	35
Confirmed breeding	597	29

☑ Possible breeding
⊞ Probable breeding
■ Confirmed breeding

0 10   50 km

0 10   30 mi

N

Black-throated Green Warbler *Dendroica virens*

A woodland species, the Black-throated Green Warbler is fairly common in the coniferous and mixed coniferous and deciduous forests of New York. Its status appears to have changed little in the past one hundred years, although with the planting of conifer forests in the state, additional habitat became available, and its range must have expanded.

Early in this century it was an abundant breeding species in the Adirondacks and Catskills and was found quite commonly in the mixed and coniferous forests in all parts of the state (Eaton 1914). Eaton indicated that it was especially common "in the hemlock woodlands near the Pennsylvania border, in the wooded gullies of the central lakes region, in all the cooler swamps of central and western New York, and locally in southeastern New York, even on Long Island." Bull (1974) called it common and widely distributed, except in the New York City area. On Long Island, he said it was found mainly on the north shore and less frequently in central Long Island. In 1939 8 pairs were breeding within a 3-mile area on the Orient peninsula on the eastern tip of the island (Bull 1974). On the south shore it was found in Suffolk County only at Sayville, Mastic, and Georgica (Bull 1974). Elliott (1958) very early pointed out that land development had probably eliminated the Black-throated Green Warbler from the West Hills area, where it had nested for 15 years.

The most obvious change in its status is on Long Island, where there were very few Atlas records, none "confirmed." Long Island observers consider this warbler a rare breeding species (J. McNeil, pers. comm.). Its breeding status may have been overstated in the past. The species is still quite generally distributed along the Appalachian Plateau, although

there must have been a great decrease in its population in these areas when the forests were clear-cut to support the chemical wood industry, especially in Delaware and Cattaraugus counties (Stout 1958).

The Black-throated Green Warbler is slightly more austral than the Magnolia Warbler but more boreal than the Chestnut-sided Warbler. Its distribution is quite similar to that of the Magnolia and Blackburnian Warblers but extends farther down the Hudson. These species seem to overlap vertically in the same forests, the Magnolia Warbler being found nearest the ground, the Black-throated Green Warbler in the central part of the canopy, and the Blackburnian Warbler in the pointed tops of mature conifers (Kendeigh 1945).

Like most warblers, this species was most often "confirmed" by Atlasers who watched adults feeding dependent young or observed young fledglings. It was often heard singing on territory. The map, including the "possible" and "probable" records, probably accurately represents its total breeding range. Almost all the warblers offer great challenge to the nest finder, and as a group they are very difficult to "confirm."

The habitat of this species on the Appalachian Plateau is most often mature or second-growth hemlock–northern hardwood forest (Saunders 1942; Kendeigh 1945) and conifer plantations (Rosche 1967; Bull 1974). The conifers planted in the 1930s and later were mostly Norway and white spruce, red and white pine, and European larch. Klingensmith (1970) found 21 pairs of Black-throated Green Warblers in a 15.4-ha (38-a) plantation of red and white pines in Allegany County. Andrle (1971b) found 6 pairs in a 60.7-ha (150-a) plantation of spruce and pine in Cattaraugus County. On Long Island it formerly nested in pitch pine (Bull 1974), and in western Suffolk County in oak-hickory woodland (Bull 1964). In the Adirondacks, Saunders (1929a) said that it occurred in all forests except perhaps those above 1219 m (4000 ft). Exposure to harsh conditions at high elevations reduces the stature of the forest (Able and Noon 1976).

On the Appalachian Plateau the Black-throated Green Warbler usually nests 3–12.2 m (10–40 ft) from the ground on a lateral branch of a hemlock (Kendeigh 1945; Bull 1974) but sometimes also in beech, maple, and yellow birch (Saunders 1942a). Palmer (1949) said that in Maine (which is probably similar to our Adirondack region), of 15 nests, 5 were in hemlocks, 4 in firs, 2 in spruces, 2 in birches, 1 in cedar, and 1 in a grapevine on the side of a building.

The nest is a compactly built structure with a deep cup saddled to a branch or fork of a conifer in thick foliage. It is built of fine bark, twigs, mosses, grasses, lichens, and spider webs. The lining is made of thick felted hair, fur, fine stems, rootlets, or feathers. Where white birch is common, this warbler weaves strips of this bark into the outside body of the nest. Sometimes this makes a nest easier to find (Harrison 1975).

Stephen W. Eaton

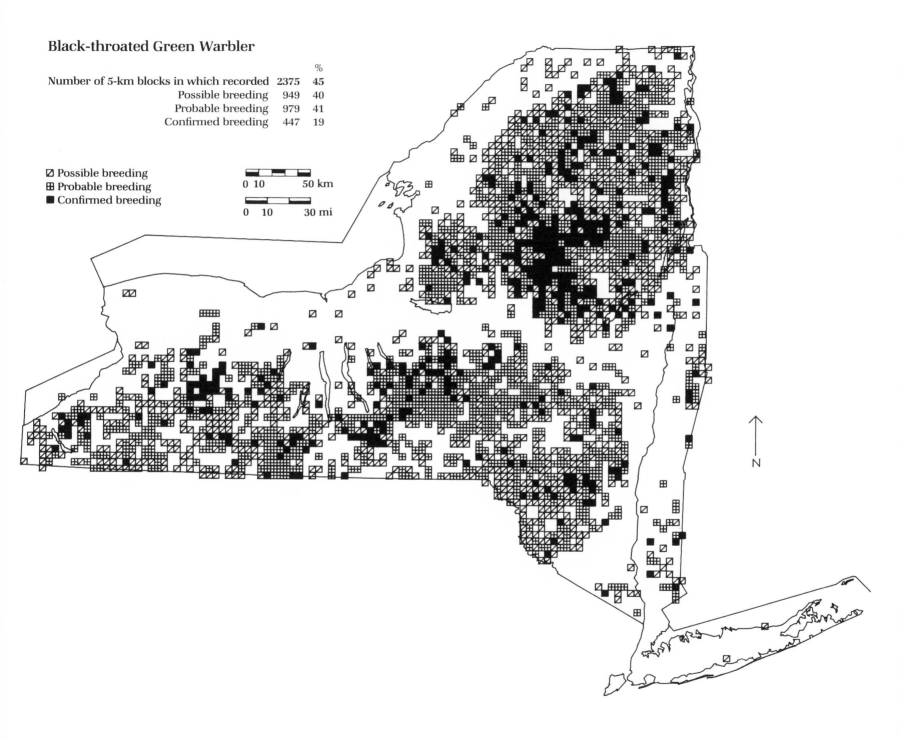

Black-throated Green Warbler

		%
Number of 5-km blocks in which recorded	2375	45
Possible breeding	949	40
Probable breeding	979	41
Confirmed breeding	447	19

◨ Possible breeding
⊞ Probable breeding
■ Confirmed breeding

0 10   50 km

0 10   30 mi

N

© DAS 1987

Blackburnian Warbler *Dendroica fusca*

The Blackburnian Warbler was named after an English woman, Mrs. Blackburn, a "patron to ornithology." This black bird with the flaming throat is one of the most breathtaking "jewels of the bird world" (Bent 1953). It is a locally common breeding bird found mainly at high elevations in the state. The status of this species has apparently not changed over the course of this century; both Eaton (1914) and Bull (1974) use the word "common" to describe breeding populations.

The Atlas map shows this species to be concentrated in the Adirondacks, Catskills, Tug Hill Plateau, and Allegany Hills, all of which are heavily forested and include elevations over 305 m (1000 ft). In addition, it is found in the Appalachian Plateau, particularly its eastern section, and in the Rensselaer Hills, as well as locally elsewhere. This warbler is not confined to high elevations, however. For example, on the St. Lawrence Plains and Transition, it occurs at elevations below 152 m (500 ft) alongside the Yellow-bellied Sapsucker and Solitary Vireo. These areas are low-altitude extensions of the Adirondacks. Although Bull identified the southeastern nesting limits in the state (1974) as Putnam, Westchester, and Orange counties, no nesting was discovered in either Putnam or Westchester counties during the Atlas survey. The distribution of this species has changed somewhat from that described by Eaton in the early part of this century (1914). No longer concentrated only in the mountains, it has expanded its range, particularly along the eastern Appalachian Plateau, where Atlas observers found it primarily in reforested areas.

The Blackburnian Warbler's high-pitched song is difficult to recognize and, for some birders, even to hear. Once spotted, however, the bird cannot be confused with other warbler species. It is well represented on the Atlas map and was "confirmed" in more than 20% of the blocks in which it was reported, mainly by observations of adults carrying food for young.

The Blackburnian Warbler is found most often in deep coniferous forests and swampy woods at higher elevations (Bent 1953). *Usnea*-draped black spruce woods are its preferred habitat (DeGraaf et al. 1980). Over the past century this ecological niche has probably been one of the least disturbed. The greatest threat to this species is not on its breeding ground but on its wintering ground, where forest habitat is being destroyed at a rapid pace. It should be noted, however, that *Usnea* lichens are fast disappearing throughout the state (Bull 1964), and what effect this has had or will have on this warbler is unknown.

At lower elevations this species nests between 152 and 305 m (500–1000 ft) in mixed and second-growth woods that contain considerable hemlock and spruce, but especially hemlock (Eaton 1914). DeGraaf et al. (1980) indicated that it is associated with very tall hemlock trees and also noted that it spends most of its life foraging in the tops of tall trees. According to Kendeigh (1945), it sings and feeds at heights of 10.7–22.9 m (35–75 ft). Adaptability, the mark of a successful species, seems to characterize the hemlock warbler, as this species was once called (Bent 1953).

The nest is usually in a conifer, anywhere from 1.5 to 24.4 m (5–80 ft) high but most often between 6.1 and 15.2 m (20–50 ft) high, well concealed by foliage or *Usnea* lichens (Bent 1953). A compact cup of twigs, lichens, and rootlets, it is lined with plant fibers (Harrison 1978) and may be placed either near the trunk or farther out toward the tips of the branches (MacArthur 1958).

Emanuel Levine

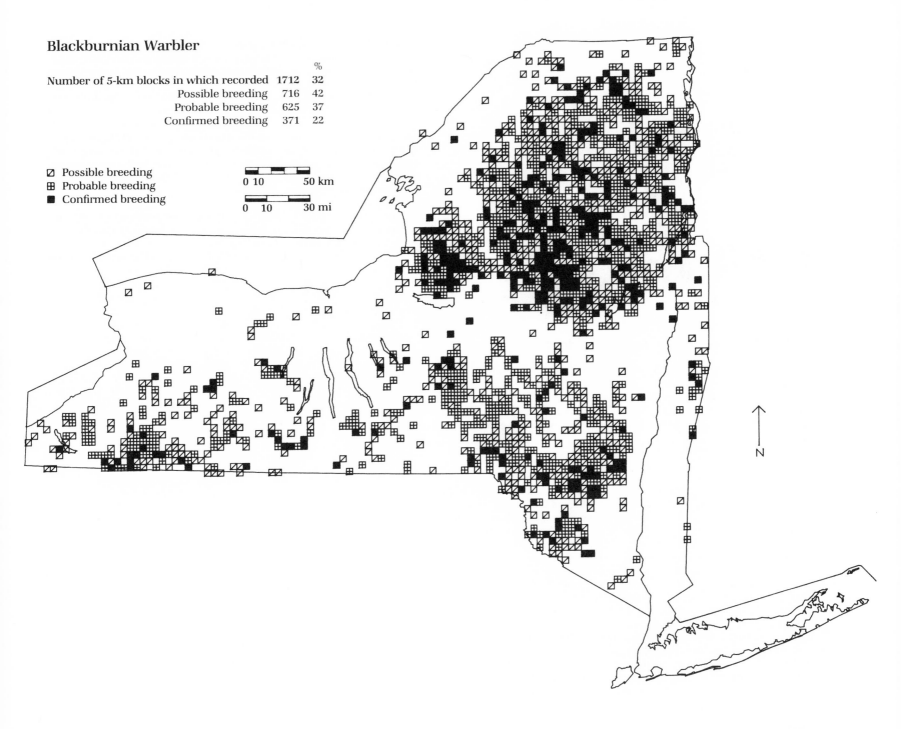

Blackburnian Warbler

		%
Number of 5-km blocks in which recorded	1712	32
Possible breeding	716	42
Probable breeding	625	37
Confirmed breeding	371	22

☑ Possible breeding
⊞ Probable breeding
■ Confirmed breeding

0 10   50 km

0 10   30 mi

N

Yellow-throated Warbler *Dendroica dominica*

The song of the Yellow-throated Warbler heard in June 1984 was barely audible above the rushing waters of Catskill Creek in Greene County. The warbler foraged in the uppermost limbs of a huge sycamore tree, pausing now and then to sing. Surprised Atlas observers watched a pair building a nest and subsequently feeding young, the first nesting record for the state.

In July of that same year, in Allegany State Park, Cattaraugus County, other Atlas observers watched a pair of Yellow-throated Warblers make trips back and forth to their nest in a red pine. Two nests in one year represent quite a coincidence, especially since the locations of both nests are considerably north of the known edge of the species' range.

The Yellow-throated Warbler is considered a southeastern bird. Eaton (1914) indicated that it wandered into New York on occasion and called it the "rarest of our accidental visitants of the warbler family"; Bull (1974) agreed. In Eaton's time the species' eastern range extended only from central Delaware and southern Maryland to Florida. Recently, that northern edge has moved to central New Jersey, central Pennsylvania, and central Ohio (AOU 1983). Atlas observers in Pennsylvania and Ohio have documented breeding records in the northern areas of those states as well (Bart 1985; Pennsylvania Breeding Bird Atlas, preliminary map). This species appears to be expanding its range northward. Smith (1978) documented the expansion in West Virginia: "As the bird has moved north in West Virginia it has begun appearing in numbers in sycamore woods where it seems to have largely been absent in the past and has become more numerous in pine forests where it has bred for many years."

The habitat of this species varies, depending on its regional location. Everywhere, however, it prefers tall trees—sycamore, oak, pine, or cypress (Mengel 1964). The Catskill Creek birds were located in riverbottom habitat at an elevation of approximately 55 m (180 ft). Along the creek were large sycamores with scattered cottonwood, hemlock, and a few white pine. The Cattaraugus birds were located in a fairly open stand of planted pine (red, white, and Scotch) on the northwestern and western shores of Science Lake at an elevation of about 567 m (1860 ft). The habitat surrounding that nest site consisted of hemlock–northern hardwood forest. Dominant tree species included sugar maple, black cherry, American beech, yellow birch, and hemlock (Baird 1984). Both nests were located high in pine trees, 13.7 m (45 ft) in Greene County and 19.8 m (65 ft) in Cattaraugus County.

The Yellow-throated Warblers at Catskill Creek had white lores, indicating they were probably of the *albilora* subspecies (D. Gagne, v.r.). It is likely the Allegany birds were of that race as well (Baird 1984).

Janet R. Carroll

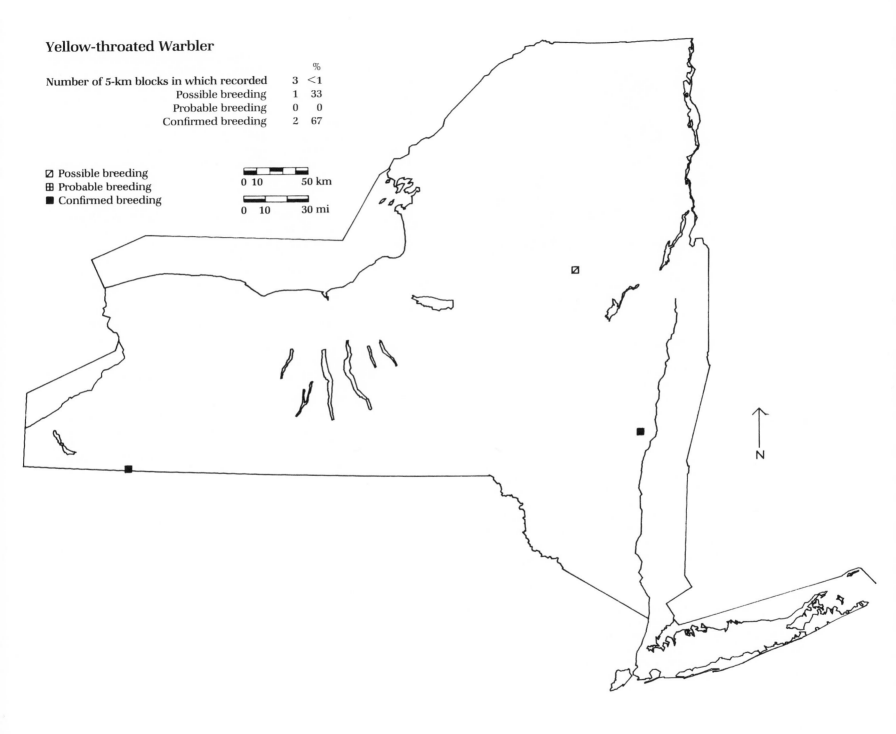

Yellow-throated Warbler

		%
Number of 5-km blocks in which recorded	3	<1
Possible breeding	1	33
Probable breeding	0	0
Confirmed breeding	2	67

◩ Possible breeding
⊞ Probable breeding
■ Confirmed breeding

0 10   50 km

0 10   30 mi

N

©DAS 1987

Pine Warbler *Dendroica pinus*

Aptly named, the Pine Warbler is locally common in the remaining pitch pine–scrub oak barrens of the Coastal Lowlands "but has decreased in recent years" (Bull 1974). Over most of upstate New York this warbler is a rare to fairly common breeder where various pine species grow. It does not appear to be abundant in any of the northeastern forests (Capen 1979). Robbins et al. (1986) found that during 1965–79, Pine Warbler numbers fluctuated: increasing from 1973 to 1975, declining following the hard winter of 1976–77, and then rapidly recovering.

Eaton (1914) observed, "It is a summer resident throughout the southeastern, eastern, western and central portions of the state, especially on Long Island, but has not been noticed in the Canadian areas of the Catskills and Adirondacks." Theodore Roosevelt, for example, recorded breeding at Oyster Bay, Nassau County (Eaton 1914), but failed to find the Pine Warbler in Franklin County (Roosevelt and Minot 1929). Similarly, Bull (1974) stated that this warbler was "absent in the mountains and at higher elevations generally," although he noted breeding at Elizabethtown, Essex County, and Warrensburg, Warren County. He mapped four disjunct populations: Long Island, the Hudson–Lake Champlain valleys, the Great Lakes–St. Lawrence plains, and the Finger Lakes region. Of the Long Island population, Bull commented that the Pine Warbler was "generally distributed in the eastern third, but fast dwindling in western Suffolk Co."

The Atlas map differs from these earlier descriptions of the Pine Warbler's breeding range. Records were obtained in the Adirondack High Peaks, Central Adirondacks, and Sable Highlands, as well as in the surrounding foothill and transition ecozones. There were also records from the Catskill Peaks, the Delaware, Mongaup, and Schoharie hills, and the Neversink Highlands. Although the Long Island population is similar to that in Bull's description, the range extends somewhat farther west into Nassau County. Most records in the state were scattered and local, but concentrations away from the Coastal Lowlands were found along the Black River Valley, Western Adirondack Foothills, St. Lawrence Transition, and Lake Champlain Valley.

The Pine Warbler was a difficult species for many observers to locate largely because they were unfamiliar with its haunts and songs, and the Atlas map is probably somewhat incomplete. The bird is rather nondescript, resembling many confusing fall warblers, or immatures, even in breeding plumage. Most confusing, however, are the songs, actually rapid trills of two types. One song is similar to the trill of a Worm-eating Warbler or Chipping Sparrow, although lower pitched and more musical. The other song type is similar to the song of the Yellow-rumped Warbler (Borror and Gunn 1985). Even some experienced observers confessed difficulty in distinguishing among these trills, and they believed that the Pine Warbler may have been overlooked or its songs mistaken for those of another species.

The Pine Warbler is one of only two species, the other being the endangered Kirtland's Warbler, that is found exclusively in pine forests in the northeastern states. Unlike the Kirtland's Warbler, which occurs in the seedling, sapling stage of forest growth, the Pine Warbler is found only in mature forests (Capen 1979). As previously mentioned, the habitat may include almost any species of pine, as long as the trees are generally well spaced. This warbler commonly uses pitch pines on the Coastal Lowlands but favors white or red pines in many upstate locales. Undoubtedly, the cutting and burning of pines on Long Island has had a negative effect; Bull (1964) noted, "This species has decreased markedly within the past 10 or 15 years." Elsewhere, however, the planting of pines and the reforestation of New York has probably had a positive impact, as has the logging of spruce and subsequent deliberate regeneration of pine within the Adirondacks.

The Pine Warbler is an early nester; its eggs are found in New York as early as 4 May (Bull 1974). Most "confirmed" records came from sightings of parents feeding young or recently fledged young; only 3 nests with eggs and none with young were found. One New York nest was described in Eaton (1914) as being "composed of strands of grapevine bark, fine rootlets and horsehairs, decorated with bits of brown and white spider cells and lined with a compact mass of animal hair and fluffy feathers with a thick ring of woolly material around the rim."

John M. C. Peterson

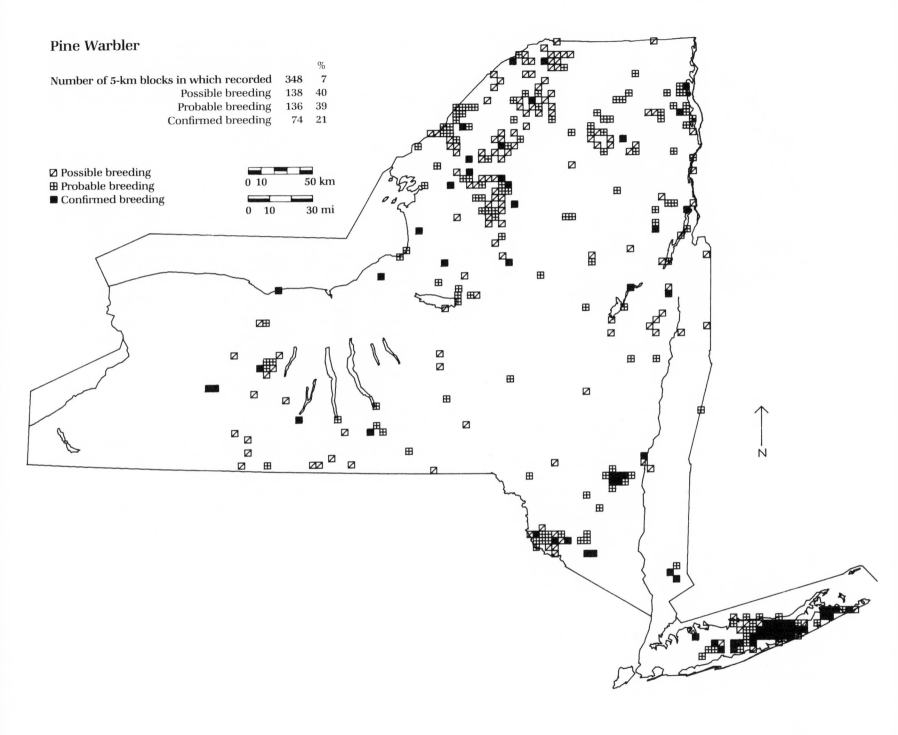

Pine Warbler

		%
Number of 5-km blocks in which recorded	348	7
Possible breeding	138	40
Probable breeding	136	39
Confirmed breeding	74	21

☑ Possible breeding
⊞ Probable breeding
■ Confirmed breeding

0 10   50 km

0 10   30 mi

N

©DAS 1987

Prairie Warbler *Dendroica discolor*

The Prairie Warbler is a southern species, and New York is near the northern limit of its range, which helps to explain why it is largely confined to southern portions of the state. It is fairly common in the Coastal Lowlands and the Hudson Valley, and it is spreading more recently westward into the southern Appalachian Plateau. New York was the only state in which the Prairie Warbler increased on BBS routes between 1965 and 1979 (Robbins et al. 1986).

The Prairie Warbler has increased its range in New York during the past half-century. Even in the southern Hudson Valley it seems to be more widely distributed. Although Griscom (1923) said that it was rare away from Long Island, Cruickshank (1942) noted that it was increasing in northern Westchester and Rockland counties. Pink and Waterman (1967) said that the bird was a fairly common breeder in Dutchess County, where breeding had been unknown before 1933, and atlasing found the species well distributed throughout the Hudson Valley all the way north to Albany. Also, Eaton (1914) stated that the species was virtually unknown as a breeder in central and western parts of the state, but now there are a number of breeding locations in south central New York, particularly in the southern Appalachian Plateau.

Bull (1974) discussed the species' arrival in the state. Apparently it did not become an established breeder in most upstate regions until after 1943. In fact, except on Long Island and the lower Hudson Valley, the breeding distribution of the Prairie Warbler remains irregular, perhaps because it is still becoming established.

The expansions in range, which are occurring not only in New York but also in Canada and New England, are probably related to the proliferation of numerous scrubby fields resulting from abandonment of small farms. As these revert to forest, Prairie Warbler habitat will once again diminish, so it is unlikely that the bird will ever become abundant in upstate New York. Many writers (e.g., Bent 1953; Terres 1980) have noted the species' erratic and spotty distribution, even in areas where it is fairly common.

It is interesting to consider the near absence of the Prairie Warbler in the northern part of the state despite its presence as a breeder to the north in Ontario (Ontario Breeding Bird Atlas, preliminary map). Because of this irregular distribution, Bull (1974) suggested that the species was invading New York not from the south but from the west by way of Michigan, probably via the Mississippi Valley. A western invasion might explain the "hole" in the species' southern distribution, that is, the greater Catskill area. Perhaps the bird has not yet arrived there, or it might be absent because of the high elevation or a lack of shrubland.

Atlasing confirmed Bull's (1976) note that the species is still expanding its New York range. Besides the numerous "confirmed" records in the southern Appalachian Plateau outside of the bird's original New York range, breeding has been documented at several other locations to the north, including "confirmed" records in Schenectady, Saratoga, and Montgomery counties. Although some of these locations may have been overlooked before the Atlas project, their surprising number suggests that the species is still invading New York and will perhaps become widespread even into the Great Lakes Plain.

The Prairie Warbler is not very conspicuous in its shrubby habitat and was identified most often by its ascending, buzzy notes. Few nests were found, and most "confirmed" Atlas records were of adults carrying food or feeding young.

The Prairie Warbler is poorly named, for it prefers not grasslands but open, scrubby fields. On Long Island the species is particularly numerous in pitch pine–scrub oak barrens kept open by fire, and inland it prefers dry hillsides partially covered with a brushy growth of bushes and saplings, especially red cedars (Eaton 1914; Bull 1974). The northernmost Atlas records came from Essex County, where a pair was observed in a dry juniper-covered hillside; Jefferson County at Limerick Cedars, a calcareous pavement barrens with mixed shrubs and open areas; and Lewis County, from a field overgrown with hawthorn, blackberry, and other shrubs.

The Prairie Warbler nests from 0.3 to 3 m (1–10 ft) above the ground, usually in barberry or low hickory, dogwood, pitch pine, or cedar. The nest is attached to a branch fork and is a compact and firmly woven, thick-walled structure of plant down, bark shreds, straw, and dry leaves; it is bound with spider silk and lined with hair and feathers (Bent 1953; Terres 1980).

Richard E. Bonney, Jr.

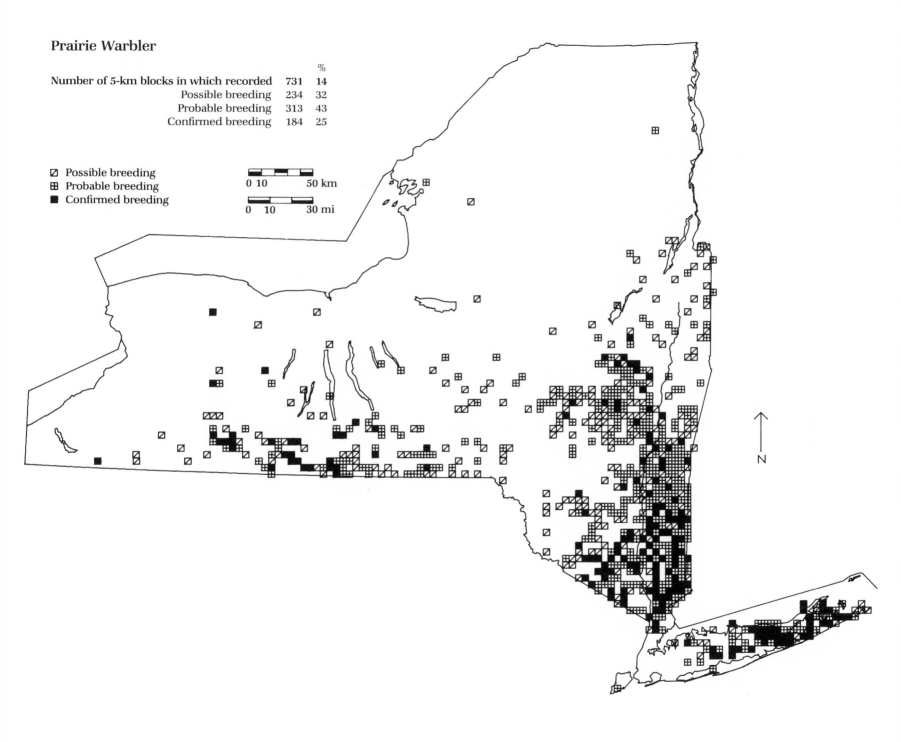

Prairie Warbler

		%
Number of 5-km blocks in which recorded	731	14
Possible breeding	234	32
Probable breeding	313	43
Confirmed breeding	184	25

◨ Possible breeding
⊞ Probable breeding
■ Confirmed breeding

0 10   50 km

0 10   30 mi

N

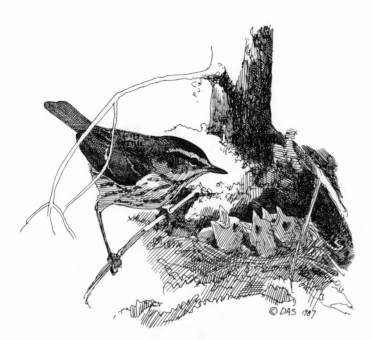

© DAS 1987

Palm Warbler *Dendroica palmarum*

The twitch-tail, as the Palm Warbler was once called, is a species most New York birders check off only during migration. Breeding was "confirmed," however, in 1984, a first state record. Adirondack Atlas workers found a nest with eggs of the eastern subspecies, *D. p. hypochrysea*, at Bay Pond Bog, Franklin County. In July 1986 a pair of Palm Warblers was observed at nearby Spring Pond Bog (Peterson, pers. comm.).

In the sixth edition of the AOU Check-list (AOU 1983) the breeding range was described as mainly to the north of New York in Canada, extending only into Maine in the United States. A downy fledgling Palm Warbler had been observed in 1980 in Floating Island bog in northeastern New Hampshire, and in 1985 Atlas observers found evidence of breeding, though not "confirmed," in the same area (Sutcliffe, pers. comm.). The record in New York could represent birds that stopped short of their northern breeding grounds, yet it is possible the Palm Warbler may have nested previously in Adirondack bogs, where habitat is similar to that in the more northern parts of its range but where few observers go during the spring or summer.

Knight (1908) considered the Palm Warbler to be one of the most common breeding warblers in northern and eastern Maine. In the large bogs where the Palm Warbler nests, it forms loose colonies. Knight speculated that the number of individuals on Bangor Bog, located just outside Bangor, Maine, reached into the hundreds.

The Palm Warbler commonly breeds in boreal bogs, less often in various aged spruce stands, and is one of the few species of North American birds restricted to the boreal conifer forest during the breeding season (Erskine 1977). Bangor Bog, as described by Knight (1908), and Bay Pond Bog, Franklin County, in the Adirondacks are similar. Both bogs have open expanses of peatland, poor fen in the case of Bay Pond Bog. Dense alder thickets form an edge between the open bog and the black spruce–tamarack swamp. Bangor Bog is considerably larger, covering several square miles, but Bay Pond Bog is only about 145.7 ha (360 a), just over one-half square mile (Knight 1908; J. Peterson 1984a). There are no bogs in the Adirondacks close to the size of Bangor Bog and few other bogs of comparable size to Bay Pond Bog. Peterson (1984a) suggested bog size might be the major limiting factor for the Palm Warbler in New York.

At both Bay Pond and Spring Pond bogs, the Palm Warbler was observed in the black spruce–tamarack swamp among the scattered trees, rather than near the open mats or shrub swamps (Peterson, pers. comm.). Palmer (1949) quoted A. E. Brower, who stated that "the birds greatly prefer the small irregular patches of bog along the edges of open bog." The Adirondack nest was found on a sphagnum hummock near the bog edge built on the mat at the base of a clump of spruces. The nest was being built on 6 July and egg laying had begun by 8 July (J. Peterson 1984a). Typically, this species is a very early breeder, beginning nest building usually in mid-May. Newly hatched young have been recorded as early as 30 May (Palmer 1949). The Adirondack nest was "made of grasses woven in a small cup around the woody stems of Labrador Tea and lined with a few feathers" (J. Peterson 1984a).

Janet R. Carroll

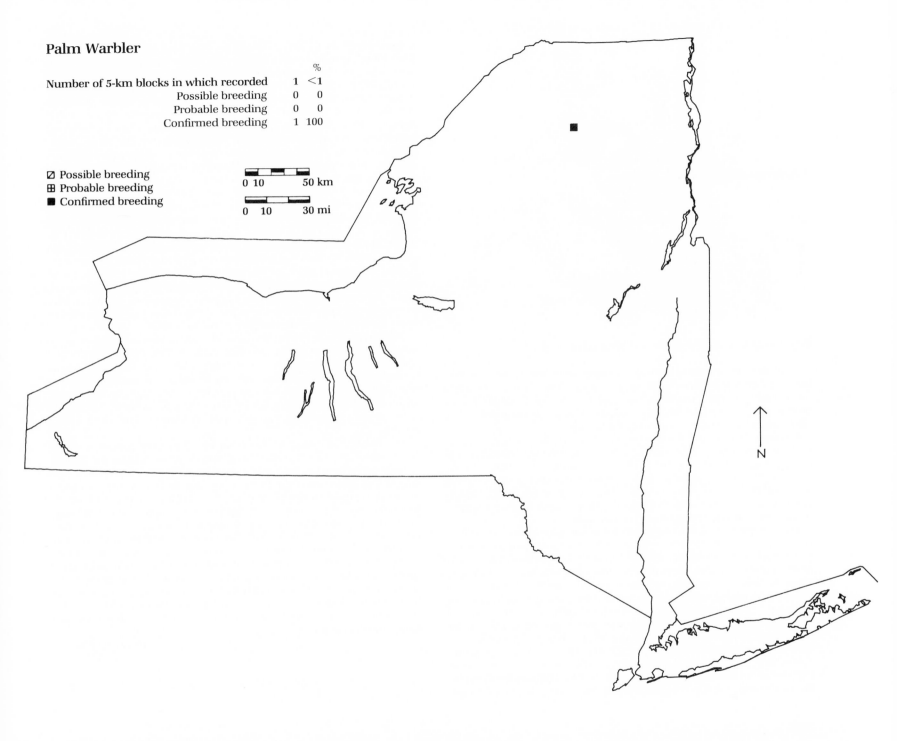

Palm Warbler

Bay-breasted Warbler *Dendroica castanea*

Unproven as a nesting bird in the state until the 1920s, the Bay-breasted Warbler remains something of an enigma, rather difficult to locate and rarely encountered. New York is at the southern edge of its breeding range, and within the state this warbler is confined to the Adirondacks, where it is an uncommon breeder.

Merriam (1881) found the Bay-breasted Warbler to be rather rare, occurring in the Adirondacks during spring migration but "not known to breed." Eaton (1914) reported, "Although I searched for this warbler through the highest portion of the Adirondacks during the breeding season of 1905, neither I nor any one of my five assistants could find any evidence of its residence in that region." Their search of Essex County's mountains may have been somewhat too high.

The summers of 1924 through 1927 were watershed years for the Bay-breasted Warbler in the Adirondacks, with 13 records from three counties. All sightings were made at moderate elevations between 497 m (1631 ft) and 853 m (2800 ft), on dates ranging from 26 June to 23 July. The first of these records was obtained at Pillsbury Lake, Hamilton County, on 14 July 1924, where Charles Johnson (1927) "found the species in full song and shot a specimen in order to verify the identification." On 22 July of the next year Saunders saw an adult male at the Lake Placid Club for the first Essex County record. The next year on 8 July 1926, he saw another male at Lake Clear Junction for the first Franklin County record (Saunders 1929a). A few days later, on 11 July, a singing male was seen along Deer Brook on the slope of Snow Mountain, Essex County (Weyl 1927a). Before the end of the month, the first confirmed record of breeding was obtained in Essex County: "On July 23, 1926, a female of this species was found feeding fully fledged young, near North Hudson, N.Y. One of the birds was collected and the identification is positive" (Weber 1927). Weyl and Livingston added more Essex County records during 1927 from the Ausable River, Elk Lake, Giant Mountain trail, Upper Ausable Lake, Wilmington Notch, and made an important find along the road at St. Huberts: "At this spot (elevation, 1200 ft [366 m]) a nest with four fledglings was discovered in a hemlock by Mr. Livingston on June 29" (Weyl 1927b).

The two "confirmed" Atlas records came from Hamilton and Herkimer counties, and were of adults observed carrying food for young. Bull (1974) suggested that the Bay-breasted Warbler seemed to be replaced at higher elevations by the Blackpoll Warbler, "but more field work remains to be done to determine this." Morse (1979) also suggested that the two species may exclude each other in most places, noting that in the many bird censuses in northern coniferous forests, they have been recorded together at the same site only in the province of New Brunswick. Blackpoll Warblers were both visiting a nest site with food and building a nest near the Bay-breasted Warbler at West Lake (G. Lee, v.r.). Other Adirondack and Tug Hill Plateau Atlas records came from Essex, Franklin, Lewis, Oneida, and St. Lawrence counties, but the 11 blocks in Hamilton were more than in any other county. Atlas workers found this warbler as high as Flowed Land, Essex County, at an elevation of 841 m (2760 ft).

Scarce to begin with and then hard to find amid dense spruces, the Bay-breasted Warbler is equally difficult to hear. The song is high pitched and weak, and the ending has no strength at all; it just fades and dies (F. Scheider, v.r.). Observers with good hearing may fail to notice the song or may confuse it with the similar short version of the song of the Black-and-white Warbler. For many birders, however, the pitch is simply too high to hear. The Atlas map gives a good general indication of the range of this species.

The Bay-breasted Warbler has been found to favor conifer or sometimes mixed growth, often near water. Spruce is consistently mentioned, usually native species, but birds in Herkimer County frequented a Norway spruce plantation near a large open bog (W. Henrickson, v.r.) and a 30- to 40-year-old spruce plantation (M. Milligan, v.r.). Other trees may include balsam fir, hemlock, pine, birch, willows, and shrubs. Sites were along rivers, open water courses, sluggish streams, and beaver ponds. Like the Cape May Warbler, the Bay-breasted Warbler is often more numerous in forests with spruce budworm infestation (MacArthur 1958).

The nest, loosely constructed of grasses and twigs with a lining of rootlets and hairs, is generally placed on the horizontal branch of a spruce or other conifer (Bent 1953). Bull (1974) reported that no New York nest with eggs was known, and none was found during the Atlas period. Only 4 nests with young have been found in the state, but none was added by Atlas observers.

John M. C. Peterson

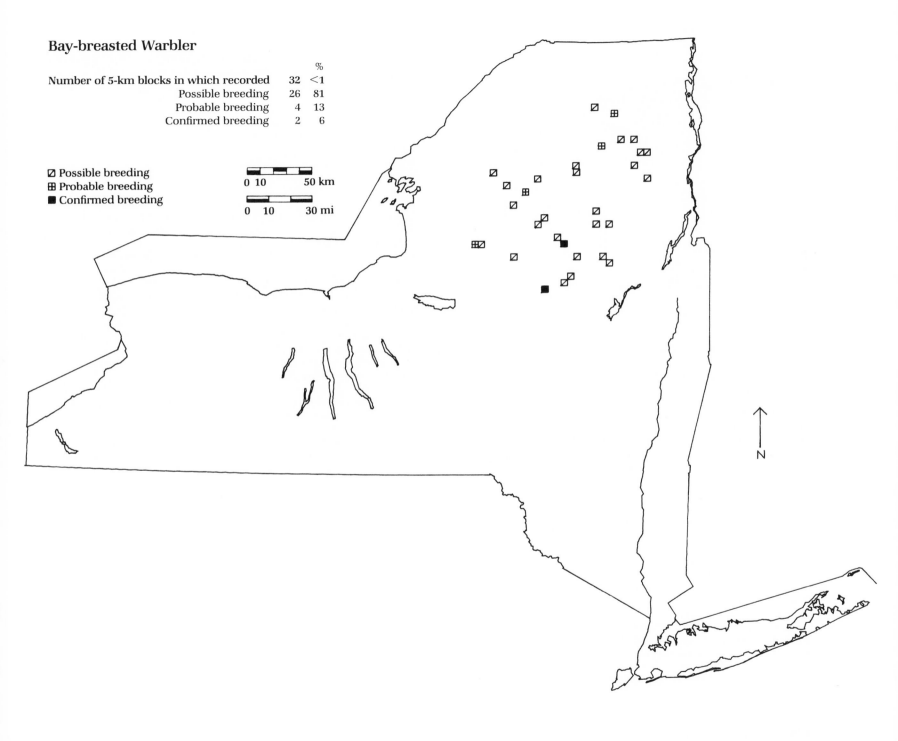

Bay-breasted Warbler

		%
Number of 5-km blocks in which recorded	32	<1
Possible breeding	26	81
Probable breeding	4	13
Confirmed breeding	2	6

☒ Possible breeding
⊞ Probable breeding
■ Confirmed breeding

0 10   50 km

0 10   30 mi

N

©DAS 1987

Blackpoll Warbler *Dendroica striata*

The Blackpoll Warbler is found in the high-elevation spruce-fir forests of the Adirondacks and Catskills where the species is common. It is fairly common in similar habitat at lower elevations, but more local in distribution. The Blackpoll Warbler is probably most abundant within the Adirondack High Peaks of Essex County, where a hiker may pass from one territory to another and hear its high-pitched song rising and falling almost without interruption. Eaton's party found it "breeding quite commonly" in the High Peaks in the summer of 1905 (Eaton 1914).

Atlas fieldwork has helped clarify the range of the Blackpoll Warbler, especially at lower elevations. The Adirondack population extends westward into the Western Adirondack Foothills of Franklin, Hamilton, Herkimer, Lewis, and St. Lawrence counties much farther than the map in Bull (1974) indicated. Silloway (1923) found the Blackpoll Warbler near Cranberry Lake, St. Lawrence County, and Saunders (1929a) found one near Lake Clear, Franklin County; both areas are outside the shaded range Bull (1974) suggested. The elevation of Cranberry Lake is 453 m (1485 ft); Lake Clear is 494 m (1621 ft). The Blackpoll Warbler was found in many blocks in the Adirondacks well below the previously accepted altitudinal boundaries. Essex County observers found it as low as 305 m (1000 ft) near Johnson Pond, 524 m (1720 ft) near the Boreas River, and 610 m (2000 ft) at Preston Ponds. Around Sagamore Lake, Hamilton County, which is also at about 610 m (2000 ft), as many as 8 singing males were found in a day, in habitat containing spruce and that containing maple. This area had been heavily logged, creating low growth, but the Blackpoll Warblers were in tall trees, on sloping and well-drained terrain (F. LaFrance, v.r.).

Within the Perkins Clearing area of Hamilton County it occupied a mixed transition zone of northern hardwoods and spruce near 762 m (2500 ft), as well as more typical conifer swamps bordering bogs at about the same elevation. Within the Catskills the Blackpoll Warbler was found from about 792 m (2600 ft) on the slopes of Graham Mountain in Ulster County to 1114 m (3655 ft) on the top of High Peak mountain in Greene County.

The Blackpoll Warbler is one of the relatively few species of birds which breeds within the boreal forest and its bordering transition zone (Erskine 1977). In New York its range seems determined by shallow soils that are generally moist and poor in nutrients, rather than by temperature, although climate may play a role at higher elevations (DiNunzio 1984). Where northern hardwoods give way to birch and mountain maple, then to cold, wet forests of red spruce and balsam fir, the Blackpoll Warbler becomes increasingly abundant. On mountains that rise above 1219 m (4000 ft), its range may continue upward into the subalpine zone of stunted balsam fir and black spruce which extends almost to the summit on southeastern exposures, away from the prevailing winds (pers. obs.). This boreal zone occupies an altitudinal layer that is much wider in the Adirondacks than in the Catskills (Stout 1958). Although Atlas observers succeeded in finding the Blackpoll Warbler over a much wider breeding range than was previously known, the inaccessibility of some boreal areas undoubtedly makes the map incomplete. Furthermore, the song—an extremely high-pitched series of short notes, loudest in the middle and weak at the start and finish—may be out of the hearing ranges of some birders. Most Atlas "confirmed" records came from adults carrying food to young, although distraction displays were reported in four blocks. Newly fledged young were reported from only a single block. Given that egg dates as late as 10 July have been reported in New York, more fledglings could probably have been observed if Atlas fieldwork had extended later in the season.

The somewhat bulky nest is usually placed against the trunk of a spruce 0.3–3 m (1–10 ft) above the ground, and is built of twigs and spruce sprigs, bark, grasses, mosses, and lichens. A liberal lining of feathers is characteristic, the kind varying with availability (Gross 1953a). Only 1 nest with eggs was found during the Atlas project, but Bull (1974) reported 10 nests in New York, half containing 4 eggs, the other half 5.

John M. C. Peterson

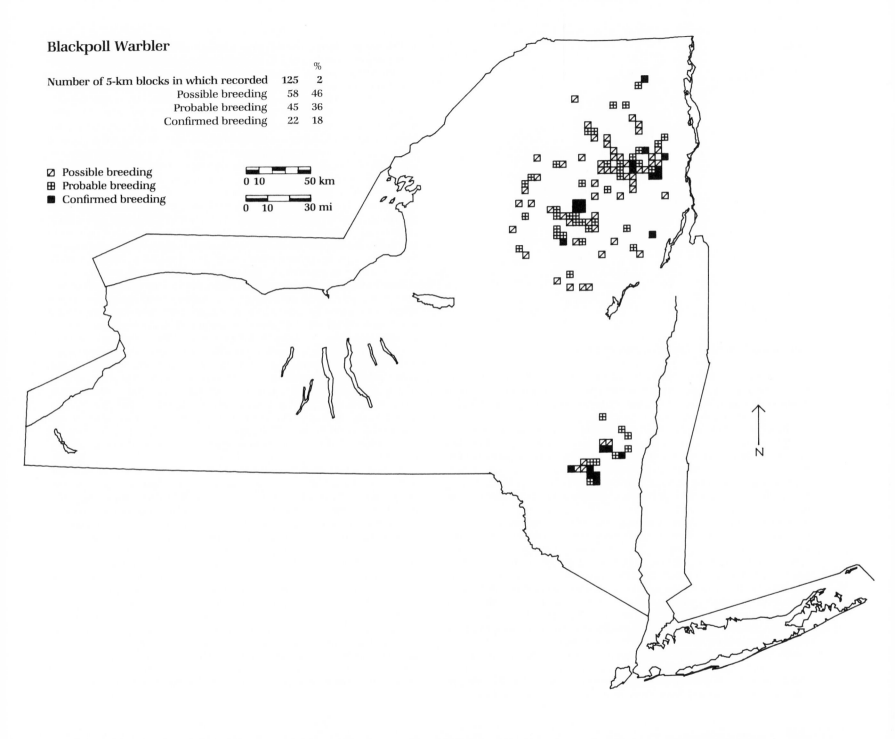

Blackpoll Warbler

		%
Number of 5-km blocks in which recorded	125	2
Possible breeding	58	46
Probable breeding	45	36
Confirmed breeding	22	18

▨ Possible breeding
⊞ Probable breeding
■ Confirmed breeding

0 10   50 km

0 10   30 mi

N

© DAS 1987

Cerulean Warbler *Dendroica cerulea*

One of the smallest of its family, the Cerulean Warbler frequents the high canopy of tall deciduous forest trees. In New York it is rare in many counties, but locally it is fairly common. It has been slowly expanding its range in the state for decades.

Eaton (1910, 1914) documented the Cerulean Warbler as a locally common breeding species in central and western New York near Lake Ontario as early as the 1870s and 1880s. He stated that it had evidently invaded the state from the Mississippi River Valley and was very rare in eastern New York. Later, in 1922, the Cerulean Warbler was found nesting in Dutchess County, the first breeding record for eastern New York (Griscom 1923).

Bull (1974) mapped two major and one minor breeding areas of the Cerulean Warbler: A, the Great Lakes Plain west of Wayne County and B, the Finger Lakes Highlands northeast to Oneida Lake and vicinity; also area C, the isolated population in Dutchess County, on the east side of the Hudson River, since 1922 or earlier. Although A and B undoubtedly arrived from the west through Ontario between Lakes Erie and Ontario, Bull (1974) thought C probably came from the west also, since the gradually spreading southeastern population did not reach northern New Jersey to the south before the early 1950s, and there were no Cerulean Warblers then to the southwest, in Pennsylvania.

Atlas surveying revealed a complex range expansion in this species. Observations in western New York indicated that it has spread southward along the Genesee River and apparently increased in the Allegany Hills near the Allegheny River, with other new localities found eastward to Chemung County. The Allegany Hills population may have originated to the south, since the Cerulean Warbler is now found in western Pennsylvania (Pennsylvania Breeding Bird Atlas, preliminary map). Elsewhere, new breeding localities were found in the upper Mohawk Valley, but much of this agricultural region lacks sufficient habitat. Northward, this species seems to be spreading in the Indian River Lakes area, and sightings to the east near rivers in St. Lawrence County suggest further expansion; the probable source is in nearby Ontario (Ontario Breeding Bird Atlas, preliminary map). Although the Hudson River population seems to be stabilized, Atlas records indicate widespread, if localized, expansion in the hill country west of the river: a thriving population in the Hudson Highlands and new breeding records in nearby ecozones, including "confirmed" breeding in 1984 at 536 m (1760 ft) elevation near High Point mountain, Ulster County (F. Murphy, v.r.). Further expansion seems likely if the trend continues: in the Appalachian Plateau, in areas peripheral to the Adirondacks, and in mostly unoccupied eastern and northeastern New York. A single singing male near Lake Champlain, 11 June 1983 (J. Mason, v.r.), is augmented by the presence of a small isolated breeding colony in Vermont near the lake (Ellison 1985f). Few Cerulean Warblers were found on Long Island except, remarkably, a small disjunct breeding colony in eastern Suffolk County (Salzman 1983). Some historical sites were unreported on the survey, including the Niagara River area in western New York, localities that apparently experienced loss of habitat.

This warbler is among the most difficult to observe, ranging higher overhead than most warblers, and often obscured from view by treetop leaves and branches. Several Atlas observers reported "pishing" singing males down to close viewing range. Also, adults visit lower, more visible levels when feeding young. Fortunately, the Cerulean Warbler is a persistent singer all day, and it continues to sing in the rain, when it may descend to lower branches. Singing has been reported in some areas as less frequent in June than in May, and may cease by the end of June (Deed 1984).

In most of New York this warbler nests in wooded swamps, in deciduous forest in stream bottoms, and along lake and river shores with numerous tall trees (Bull 1974). On wooded Schodack and Houghtaling islands in the upper Hudson River in Columbia and Rensselaer counties, I have observed the Cerulean Warbler in May and June, chiefly in groups of tall cottonwoods. Another major habitat is drier open, upland forest, mostly oak, or oak and maple, as reported by Atlas observers in hilly Orange, Sullivan, and Ulster counties. Eaton (1981) listed oak woods as habitat in Cattaraugus County, besides silver maples along the Allegheny River. The colony in eastern Suffolk County was found in black locust trees bordered by deep woods and fields (Salzman 1983).

The nest is placed well out from the trunk, often above an open area, and is saddled at the fork of a horizontal branch. The cup is shallow and made of fine strips of bark, grasses, and weed stems, and it is lined with fine reddish fibers consisting of moss stems, rootlets, or strips of bark; externally it is grayish, firmly bound together with spider silk and decorated with whitish lichens (Chapman 1907; Eaton 1914; Bent 1953).

Paul F. Connor

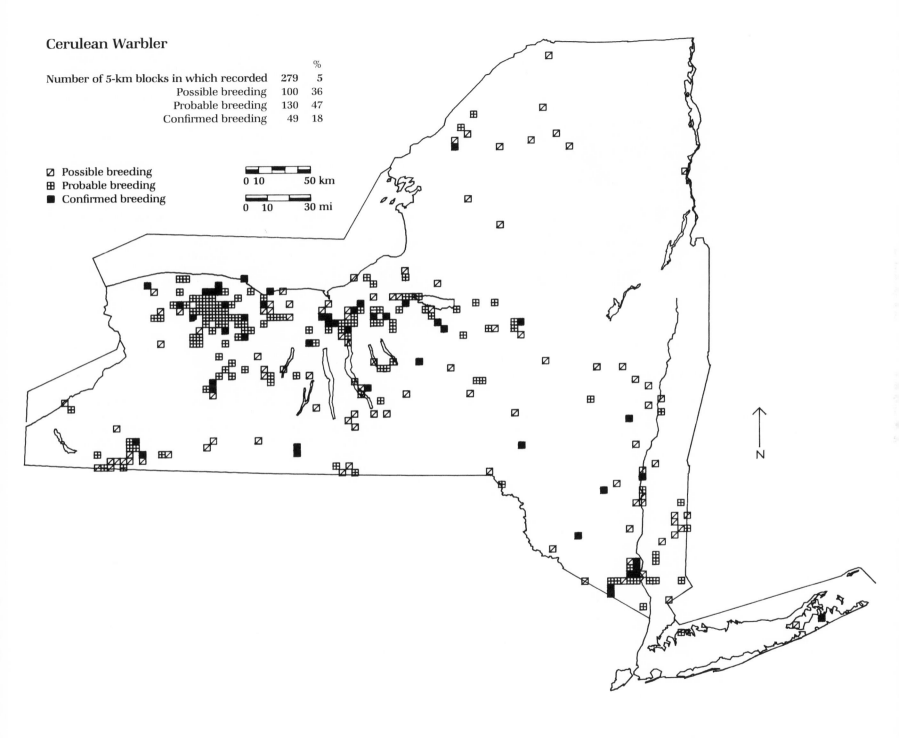

Cerulean Warbler

Number of 5-km blocks in which recorded		%
Number of 5-km blocks in which recorded	279	5
Possible breeding	100	36
Probable breeding	130	47
Confirmed breeding	49	18

◩ Possible breeding
⊞ Probable breeding
■ Confirmed breeding

0 10   50 km

0 10   30 mi

N

©DAS 1987

Black-and-white Warbler *Mniotilta varia*

Many species of warblers breed in New York, but the Black-and-white Warbler is the only one that can be found in both the high elevations of the Adirondacks and the low elevations of the Hudson Valley. Structurally, the Black-and-white Warbler is much like the nuthatch: it has an unusually long hind toe and claw for a warbler, a shortened tarsus, and a lengthened forelimb (Eaton et al. 1963). This adaptation allows it to forage for lepidopterid and other insect larvae on the larger limbs and trunks of trees. This bark-gleaning foraging method and its ability to catch food on the wing (Harrison 1984) allow the Black-and-white Warbler to exploit many food sources and may help explain why its range of distribution in the state and in eastern North America is broader than the ranges of most warblers. It is a fairly common breeder over much of New York, except in the western part of the state.

The Black-and-white Warbler was considered a local or fairly common summer breeding species on Long Island, in the southern Hudson Valley, and in the lowlands of western New York to the edges of the Catskills and Adirondacks. However, within the cooler portions of the state, such as the Appalachian Plateau, Catskills, Tug Hill Plateau, Taconic Highlands, and Adirondacks, Eaton (1914) found this species to be a common breeder, even at timberline on Mount Marcy in Essex County. In western New York, however, it has been rarely reported as a breeding species (Eaton 1914; Beardslee and Mitchell 1965).

The Atlas map shows a breeding distribution pattern that appears to have changed little in the last century. This species is concentrated in the eastern part of the state; it is found quite regularly on the eastern half of the Appalachian Plateau, and less regularly on the western half, where it occurs mainly on those steep, rocky slopes having a southern exposure, especially in the Allegany Hills. Eaton (1981) found it inhabiting the oak woods along the slopes above the Allegheny River in Cattaraugus County. Atlas observers found it along the slopes adjacent to the Susquehanna River and its tributaries, where oak is also the main tree species. The Black-and-white Warbler was found sparingly in the cropland and dairy-farming country along the Great Lakes Plain, the Mohawk Valley, and the Eastern Ontario Plains and Black River Valley where the elm–red maple–northern hardwood forests have been fragmented. It is scarce in the northern hardwood forests of western New York.

The Black-and-white Warbler usually can be located when singing males arrive on their breeding ground early in the season, but when nesting is well along, it becomes very quiet and may be overlooked (Harrison 1984). Where block busting was done late in the breeding season, it may have been missed. Although distraction displays were used quite often to "confirm" ground-nesting warblers, only 16 such displays were reported for the Black-and-white Warbler. The female will often sit tight and not leave the nest until nearly stepped upon (pers. obs.). However, when she is disturbed from the nest, she "trails painfully over the dead leaves in an effort to lead the intruder away" (Chapman 1907).

This warbler prefers to breed in "mature or second-growth deciduous or mixed woodland" (DeGraaf et al. 1980). It is found in alpine krummholz in the Adirondacks (Eaton 1914) and in deciduous woodlands on the Coastal Lowlands (Bull 1964). The nest is placed on the ground under leaf litter near the base of a tree or stump, near a rock, or under a log or fallen tree branch; it is occasionally built on a short stump (Harrison 1975). When the Black-and-white Warbler approaches the nest, it uses a neighboring tree trunk as surface, then sneaks along the leaf litter to its nest (pers. obs.). The nest is usually hidden from above by leaves, much like the Ovenbird's nest. It is built of dry, skeletonized leaves, inlaid with grasses, weed fibers, inner bark strips, rootlets, and sometimes hair (Bent 1953).

Stephen W. Eaton

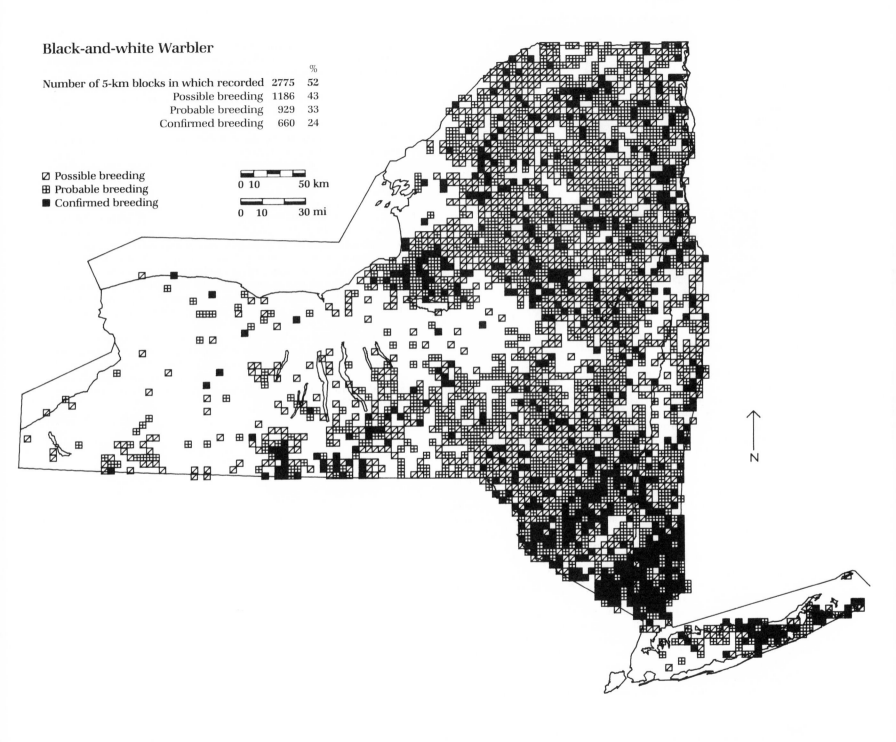

Black-and-white Warbler

		%
Number of 5-km blocks in which recorded	2775	52
Possible breeding	1186	43
Probable breeding	929	33
Confirmed breeding	660	24

◪ Possible breeding
⊞ Probable breeding
■ Confirmed breeding

0 10   50 km

0 10   30 mi

N

©DAS 1987

American Redstart *Setophaga ruticilla*

The American Redstart, a butterfly among the warblers, is a common breeding bird throughout the state except in areas of intensive agriculture and where urban sprawl has eliminated its habitat. BBS data collected in New York during the period 1966–74 ranked the American Redstart fifth among New York warbler species (Bystrak 1975). Robbins et al. (1986) reported that BBS data indicated an increase for this warbler, particularly in New York. They stated that one of the highest counts came from the Adirondacks.

Early information on this species can be found under the *Empidonax* flycatchers, the group in which it was placed by Audubon and Chevalier (1840–44) and De Kay (1844). In the early 1900s it was a common summer resident in all portions of the state (Eaton 1914). It was found nesting in Central Park in New York City, and in the deciduous forests of eastern, central, and western New York. Mention was also made of it breeding in slashes left by the McIntyre Iron Company on the slopes of Mount Skylight and Mount Marcy in Essex County. Bull (1974) called it a widespread breeding species except for on the Coastal Lowlands, where it was considered rare and local.

The Atlas map indicates that the American Redstart is distributed widely throughout New York with localized gaps occurring mainly in areas where cropland and orchards replace deciduous forest such as near the Finger Lakes, and also in developed areas such as New York City and western Long Island. Its distribution appears to have changed little during the past 80 years.

This strongly territorial warbler is both vocally and visually conspic-uous throughout the summer (Ficken 1962). During the warmer parts of the day and late in the breeding season its song has a less emphatic ending and is difficult to distinguish from the songs of both the Yellow and the Chestnut-sided warblers. The American Redstart was more difficult to "confirm" than other common warblers such as the Yellow Warbler and Common Yellowthroat, both of which nest closer to the ground and are therefore more conspicuous at the nest than the redstart. Hickey (1940) found that many unmated American Redstarts wander during the breeding season. Some of the "possible" records of this species were probably of these wandering males and do not reflect actual breeding.

The American Redstart inhabits many forest types, especially second-growth forest (DeGraaf et al. 1980). On Long Island it was found in red maple–hardwood swamps and in upland deciduous woods (Bull 1964). In the Adirondacks it was recorded in American beech–spruce–hemlock associations in Hamilton County (Noon et al. 1982a), black spruce–American beech–maple in Essex County (Noon et al. 1982b), and white ash–basswood in St. Lawrence County (Maxwell and Smith 1980). Elsewhere it was found in shrub swamps and forested uplands and wetlands, mainly deciduous. An adaptable species, it breeds in a variety of habitats (Benson 1939). It uses different foraging methods in response to habitat differences (Sherry 1979), which may help explain its extensive geographical range in New York and in North America.

The nest is placed in a fork of a tree branch or shrub 1.2–9.1 m (4–30 ft) from the ground. It is a compactly woven cup of plant down, bark fibers, small rootlets, and grass stems, and lined with fine grasses, weed stems, hair, and sometimes feathers. On the outside it is decorated with lichens, birch bark, bud scales, and plant down, and bound with spiders' silk (Harrison 1975).

Stephen W. Eaton

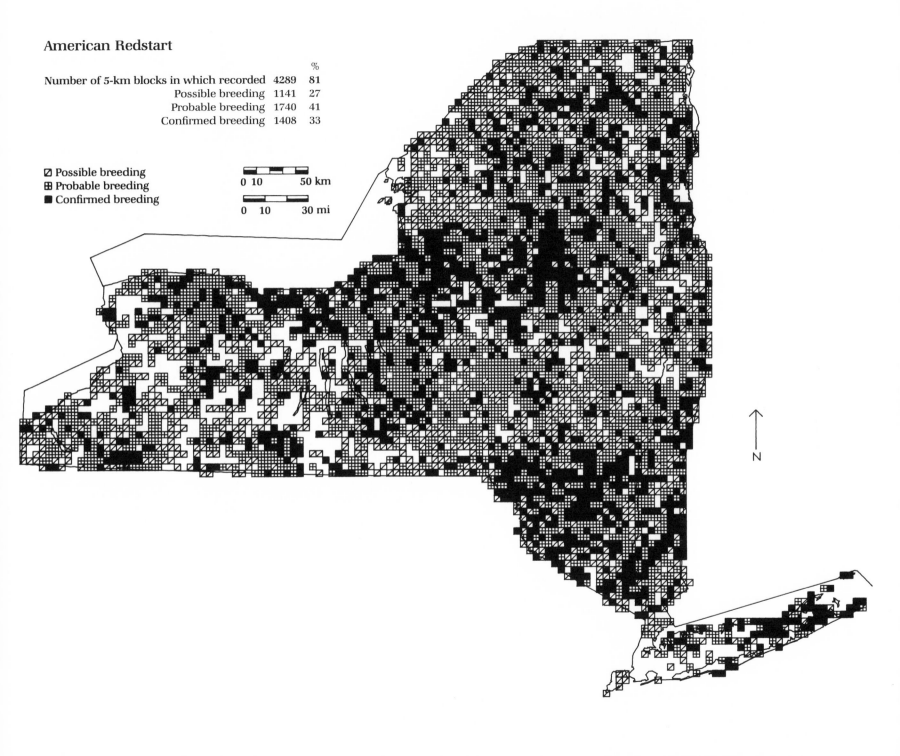

American Redstart

Number of 5-km blocks in which recorded 4289   81 %
              Possible breeding   1141   27
              Probable breeding   1740   41
             Confirmed breeding   1408   33

◩ Possible breeding
⊞ Probable breeding
■ Confirmed breeding

0   10   50 km

0   10   30 mi

N

© 1987
DAS

Prothonotary Warbler *Protonotaria citrea*

The Prothonotary Warbler formerly and perhaps more appropriately called the golden swamp warbler, is a rare, locally breeding bird in New York. Eaton (1914) referred to it as an accidental visitant to New York, and he thought that it probably did not breed within the state's limits, being mainly an inhabitant of southern swamp forests. The first record of attempted breeding in the state was in 1910 when a male was seen building a nest near Ithaca in the marsh at the head of Cayuga Lake, Tompkins County (Allen 1911). The first successful New York nest was found in 1931 at Oak Orchard Swamp (Beardslee and Mitchell 1965), where a colony still flourishes (Eaton 1985). Confirmed as breeding at Montezuma NWR in 1948 (Parkes 1952a), it has continued to nest there for almost 40 years (Benning 1983).

The Prothonotary Warbler also bred in the flooded woodlands of Oneida Lake in Onondaga County until 1968 (DeBenedictis 1982). Stoner (1932) saw a male and a female in early June 1928 near the mouth of Chittenango Creek, which enters Oneida Lake from the south. The first documented nesting in that area, however, was at Short Point in 1944 (Bull 1974). In 1957 the species was reported at another site 2 miles west of Short Point at Muskrat Bay (Scheider 1957). These three sites were along about 6 miles of the southwestern shoreline. Sporadic nesting has occurred elsewhere in the state, including near Ashville, Chautauqua County, in June 1952 and at Riverside Marsh, south of Frewsburg, Chautauqua County, along the Conewango Creek Valley, in 1963 (Beardslee and Mitchell 1965).

In the 1970s male Prothonotary Warblers were heard singing during June on Long Island. However, nesting was not recorded there until 1979, at Nissequogue River State Park in northwestern Suffolk County (Wheat 1979). In 1983 another breeding site was located at Belmont Lake State Park in southwestern Suffolk County (DiCostanzo 1983b).

The Prothonotary Warbler, like the Acadian Flycatcher, Tufted Titmouse, and Kentucky Warbler, is a southern species which has been gradually moving northward. As is typical of a bird at the edge of its range, it is found nesting in an area one year and then not again. Atlas observers "confirmed" this warbler at the historical sites at Oak Orchard WMA and Montezuma, and also at a new location along Delta Lake north of Rome in the drainage of the Mohawk River in Oneida County. The "probable" record in Albany County was at Black Creek Marsh, where a pair was observed many times one year during the breeding season but not "confirmed" (J. Carroll, pers. comm.). Atlas observers "confirmed" breeding at the two Long Island sites mentioned above, as well as at another location near Wyandanch, Suffolk County, in 1983. The upstate New York birds could have come from an Ontario population located along Lake Erie (Ontario Breeding Bird Atlas, preliminary map), or they could have come up the Susquehanna River corridor. The Long Island colonies must be an extension of successful populations in New Jersey (Bull 1964).

This species is generally found in wooded swamps, along edges of streams or ponds, or in flooded riverbottom that is shaded by hardwoods such as oak, elm, maple, and ash (DeGraaf et al. 1980). At Oak Orchard Swamp it nests in an area of flooded willow and buttonbush (Beardslee 1932; Beardslee and Mitchell 1965); at Montezuma it is found in red maple–hardwood swamp (Parkes 1952a); and on Long Island it nests in flooded forest of red maple, black oak, and black gum, with an understory of white alder, *Clethra alnifolia* (Wheat 1979).

The Prothonotary and Lucy's warblers are the only North American warblers that nest in cavities, including birdhouses. The male Prothonotary Warbler sometimes begins building the nest before the female arrives. It selects the territory and nest site, but the female does most of the building (Harrison 1975). In one case a male built a nest in a mailbox, but it was not used (Beardslee and Mitchell 1965). A nest located in Long Island was in a cavity in the stump of a red maple, 1 m (3 ft) from the ground. The nest itself consisted of grass, strips of purplish bark, short pieces of very thin twigs, and small dry pieces of sphagnum moss (Wheat 1979). Although the nest is completed in 6–10 days, the eggs are not laid until several days later (Harrison 1975).

Stephen W. Eaton

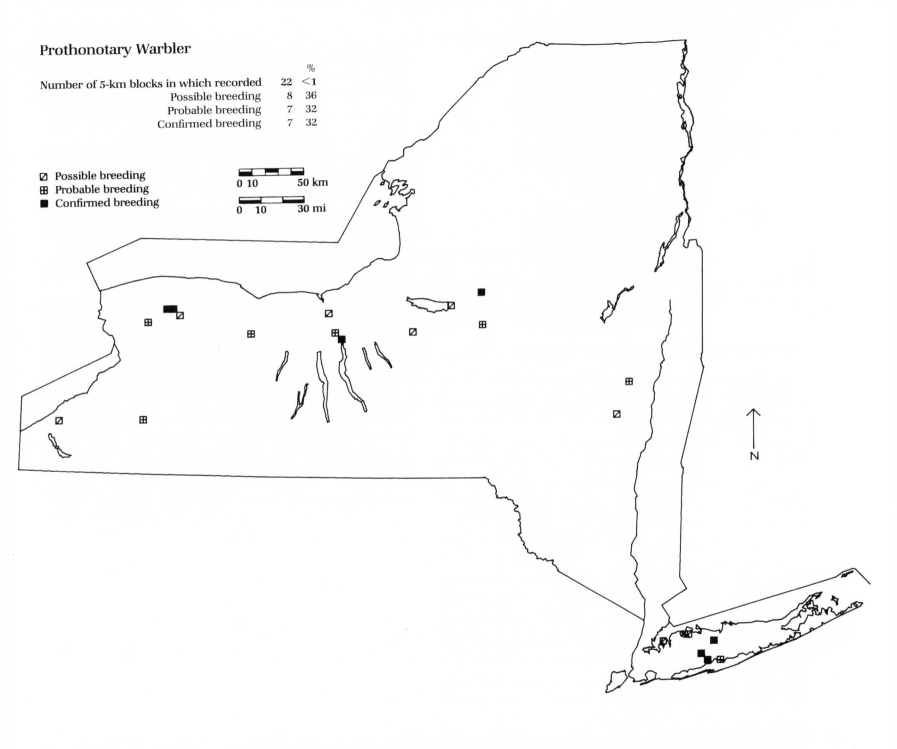

Prothonotary Warbler

Number of 5-km blocks in which recorded		%
	22	<1
Possible breeding	8	36
Probable breeding	7	32
Confirmed breeding	7	32

◩ Possible breeding
⊞ Probable breeding
■ Confirmed breeding

0 10   50 km

0 10   30 mi

N

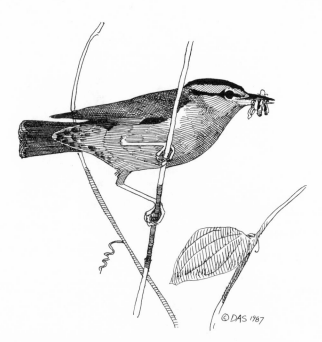

© DAS 1987

Worm-eating Warbler *Helmitheros vermivorus*

A reclusive bird, the Worm-eating Warbler seldom eats worms. It is an unusual warbler, being primarily a ground forager and a ground nester. It is locally common and found primarily in the lower Hudson Valley, with a few scattered breeding records on Long Island and along the eastern Appalachian Plateau.

Eaton (1914) and Griscom (1923) said that the species was confined almost entirely to the lower Hudson Valley, where it was common in northern Westchester County, and a few places near New York City and on the north shore of Long Island. Eaton also noted a few reports from western New York, but no nests had been found. In addition, Bull (1974) indicated that it was a locally common breeder in the Susquehanna-Chemung River watershed from Binghamton, Broome County, to Elmira, Chemung County. He said it was a rare and local breeder in the upper Hudson Valley north to Albany County. Bull noted only 28 breeding records in the state.

Atlasing indicated a distribution more widespread than that suggested by these writers. The species is distributed widely in the lower Hudson Valley—throughout Westchester, Putnam, and Rockland counties—and fairly widely up the Hudson Valley to east central Greene County, although not into Albany County. A first "confirmed" record for Columbia County was documented during Atlas fieldwork; observers spotted a pair with recently fledged young in a wooded area near the Taconic State Park. Breeding was "confirmed" in 97 blocks. Along the Appalachian Plateau, breeding was "confirmed" in Delaware and Chenango counties where Bull's (1974) map showed no records, and other scattered records

in several more new areas, including Schuyler, Tompkins, Cortland, Madison, and Schoharie counties. Virtually all breeding locations are along or near river systems. Statewide only 6 active nests were found, most "confirmed" records being of adults carrying food or feeding young. The "probable" records are mostly of singing males. The Worm-eating Warbler could have been overlooked by some observers since it is wary and difficult to see. It is also quiet, and when it does sing, it sounds very much like a Chipping Sparrow.

What appears to be a change in the distribution of this species is at least in part a reflection of the intense observations of the Atlas project. Griscom (1923) noted that the increased observations during the last decade had shown the species to be far less rare than previously supposed. Robbins et al. (1986) did discover, however, that the Worm-eating Warbler increased overall on BBS routes throughout the entire eastern United States from 1965 to 1979.

Where this bird is found, it may be quite numerous. Bull (1974) noted that 10 pairs bred on a 16.2-ha (40-a) tract in Grassy Sprain, Westchester County, in 1941, and 12 pairs along the Chemung River near Elmira in 1943.

A ground dweller, the Worm-eating Warbler prefers dense undergrowth on hillsides of deciduous woods, often in ravines. It also inhabits swampy thickets and wet areas (Eaton 1914; Bull 1974). It can be seen walking slowly and deliberately among dry leaves. Typical habitat included a location in Greene County, where observers watched a pair feeding young. The birds were observed in the understory of a small tract of state forest at an elevation of 107–183 m (350–600 ft). The primarily deciduous woods had heavy undergrowth and a small stream running through a gorge at one corner (P. Hunt, v.r.).

The Worm-eating Warbler nests on the ground, usually on the steep side of a wooded hillside or ravine, or near the edge of a swamp or stream. The nest is well hidden in a drift of dry leaves, under a bush or against a sapling, by the side of a mossy log, or at the edge of a stone amid herbaceous growth. The structure is composed of leaves, grasses, bark strips, and rootlets, usually lined with fine grass and hair or with the stems of maple seeds or the stalks of hair moss (Eaton 1914; Terres 1980).

Richard E. Bonney, Jr.

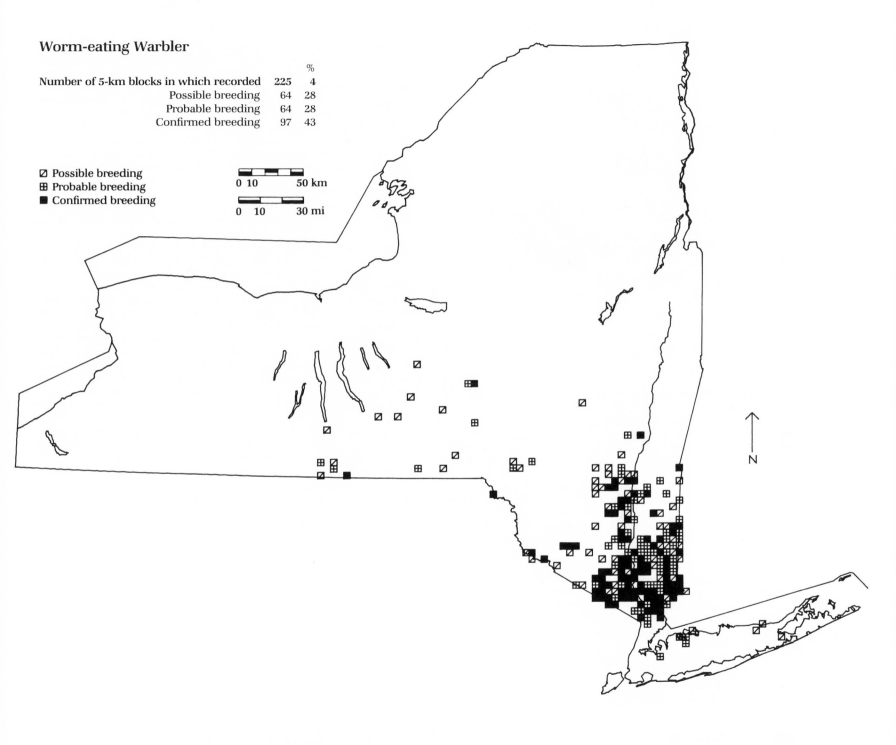

Worm-eating Warbler

Number of 5-km blocks in which recorded	225	% 4
Possible breeding	64	28
Probable breeding	64	28
Confirmed breeding	97	43

☑ Possible breeding
⊞ Probable breeding
■ Confirmed breeding

0 10   50 km

0 10   30 mi

N

© DAS 1987

Ovenbird *Seiurus aurocapillus*

The familiar "teacher-bird," the Ovenbird is one of the most common warblers inhabiting deciduous woodlands. Even though it is one of the most frequent hosts of the Brown-headed Cowbird, it persists in good numbers wherever there are large woodlots of at least 14.2 ha (35 a) or large blocks of forest (MacClintock et al. 1977). According to Robbins et al. (1986), the Ovenbird appears to be increasing in the eastern part of its range. It was ranked third among the 31 species of warblers recorded on 112 BBS routes in New York for the period 1966–74 (Bystrak 1975).

De Kay (1844) referred to this species as a common bird in the summer, and Eaton (1914) said it was uniformly distributed in all woodlands from the slopes of Mount Marcy, Essex County, to the groves of Staten Island, Richmond County. He said it was fully as dominant a species as the Yellow Warbler, though it required woodland. Bull (1974) said it was distributed statewide and was one of the most common members of the family. Its status today appears nearly unchanged. Given the human impact on the forests of New York since the 1700s, this statement is certainly a tribute to the Ovenbird's adaptability.

As Atlas data indicate, the Ovenbird has been displaced by agriculture along the Great Lakes Plain, Black River Valley, and Mohawk Valley, all areas with productive soil. In Nassau and Suffolk counties Atlas workers found it still breeding except in the more developed areas. It was gone from Bronx, New York, Kings, and Queens counties as a breeding species, but was still found on Staten Island.

The Ovenbird is probably one of the easiest warblers to "confirm" and, as Harrison (1975) described, he has never set out to find its nest without accidentally finding it by flushing the tight-sitting female from the nest. The long period of dependency by the fledglings also helps in "confirming" this species, which can be readily found by its loud, nearly always recognizable song and call note.

This species is usually found in mature forests, either deciduous or mixed. According to DeGraaf et al. (1980), it requires "open forest with little underbrush and an abundance of fallen leaves, logs, and rocks." Temple et al. (1979) found that Ovenbird numbers increase as the density of trees increases. It can be found in the following deciduous forests in New York: beech-maple mesic forest in Cattaraugus County (Saunders 1938), red maple–gray birch, St. Lawrence County (Van Riet 1983), and maple-oak in Ulster County (Stapleton 1980). It is also found in mixed forests: pitch pine–oak, Suffolk County (Clinton et al. 1980) and beech-spruce-hemlock, Hamilton County (Noon et al. 1982a). Saunders (1942) reported that it was absent from the successional hardwood forests such as cherry-aspen, but although the Ovenbird is considered a species of mature forests, the dbh of the largest trees in the beech-maple mesic forest of Allegany State Park at that time, where the Ovenbird was one of the most abundant species, ranged up to 30.5 cm (12 in) (Saunders 1936).

In the Adirondacks, Saunders (1929a) said it occurred in mixed forest but not in pure conifer forests. He also found it in small numbers in the denser aspen growths on old burns, but only where the growth had reached a considerable height and formed a dense stand. Eaton (1914) noted it commonly as high as 1067 m (3500 ft) on several Adirondack mountains. He said it occurred where growth was predominantly spruce and pine with only a few deciduous trees intermingled.

In a study done in Allegany State Park from 1983 to 1985, Baird (1986b) repeated Saunders' (1936) work of 1930 and 1931. The Ovenbird in Saunders' study ranked third in abundance of all pairs of birds in the valley (1196 pairs), and in Baird's study it had slipped to a rank of sixth (506 pairs) behind the Red-eyed Vireo, the American Redstart, Black-throated Green and Blackburnian warblers, and the Dark-eyed Junco.

The nest is built on the ground, in a depression in dead leaves with an arched top of dead leaves, forming a structure not unlike an old dutch oven, with the opening at the ground level to 2.5 cm (1 in) above. Invisible from above, it is made of grasses, plant fibers, weed stems, leaves, rootlets, mosses, and bark. It is lined with fine rootlets, fibers, and hair. The female constructs the nest alone in about 5 days (Harrison 1975).

Stephen W. Eaton

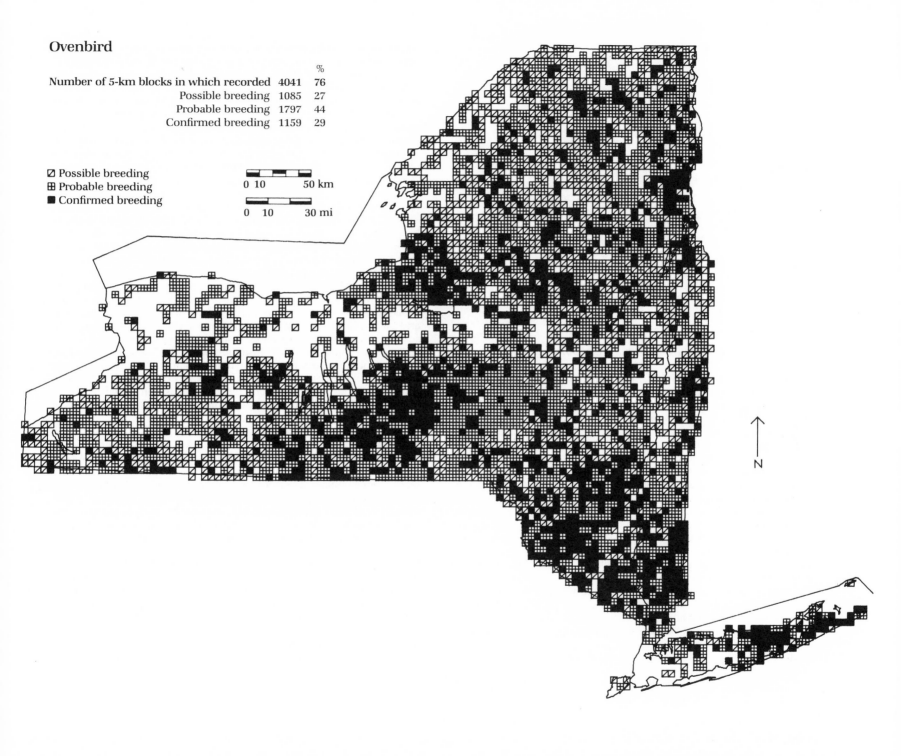

Ovenbird

		%
Number of 5-km blocks in which recorded	4041	76
Possible breeding	1085	27
Probable breeding	1797	44
Confirmed breeding	1159	29

◪ Possible breeding
⊞ Probable breeding
■ Confirmed breeding

0 10   50 km

0 10   30 mi

N

Northern Waterthrush *Seiurus noveboracensis*

A warbler of wooded swamps, the Northern Waterthrush is a fairly common breeder throughout the state south to Rockland County. Its name is translated from Latinized Greek as the New York tail waver (Forbush 1929). It is found wherever forested wetlands of sufficient size occur, which in New York seems to be over 8 ha (19.8 a) (Eaton 1957).

This species has been observed throughout the state for many years. Although in the 1870s Roosevelt and Minot (1929) called this waterthrush a rare summer resident around St. Regis lakes, Franklin County, Eaton (1914) said it was a common summer resident of the Adirondacks and Catskills and in many swamps in central and western New York. Eaton's (1910) distribution map of this species showed it as breeding throughout the Adirondacks, the Tug Hill Plateau, and Catskills, with scattered areas of occurrence across the Appalachian Plateau and Great Lakes Plain. In the Mount Marcy area, Essex County, he referred to the Northern Waterthrush as common in swamps up to 838 m (2750 ft). Bull (1974) mentioned several low-lying areas along the Great Lakes Plain where the Northern Waterthrush bred: Oak Orchard and Bergen swamps, Orleans and Genesee counties; Montezuma NWR; and Snake Creek, Oswego County, and Stony Point and Henderson Pond, Jefferson County. He also documented breeding in Putnam, Westchester, and Orange counties in southeastern New York.

Much habitat loss occurred from wetland draining in New York after World War II, adversely affecting the Northern Waterthrush. For example, Potter Swamp, Ontario and Yates counties, and the Urbana Swamp, Steuben County, which formerly possessed good Northern Waterthrush populations (Eaton 1914), are now mainly devoted to truck farming.

The Northern Waterthrush was found in 1134 Atlas blocks across the state, going unreported only in Niagara County and the New York City area, and Nassau and Suffolk counties on Long Island. The spotty distribution of this waterthrush is determined by the presence of wooded swamps or bogs in an area. In the Adirondacks it appears to follow the streams flowing in all directions from the central highlands. The Atlas map reflects fairly accurately the fact that forested wetlands are scarce in western New York (Beardslee and Mitchell 1965).

It seems quite probable that this species has been under-recorded in some areas. The Northern Waterthrush was not reported from Montezuma NWR and vicinity, where an abundance of red maple–hardwood swamps occur. Although it was reported from many blocks on the Great Lakes Plain, there are many more forested wetlands in this area where breeding should occur. The Northern Waterthrush stops regular singing near the middle of June (Eaton 1957), which may explain why its presence could have been missed by Atlas personnel working then. A secretive species, this waterthrush seeks cover immediately when disturbed. Distraction displays, observed in 26 blocks, were described by one observer as follows: The female "arched her back and lowered her head and tail forming a very symmetrical convex curve with her body. She scampered mouse-like for short distances along a horizontal branch about a foot above the grassy hummocks. . . . She lifted her wings high over her shoulders and fluttered them in a manner that gave the illusion that she could not fly. Her show of helplessness continued as she rolled sideways off the branch . . ." (Cook 1983). Its nest is well hidden, and Atlas observers were able to find only 30, usually "confirming" the species by observing adults with food.

The Northern Waterthrush in New York is essentially a bird of red maple–hardwood swamps on the Great Lakes Plain and of hemlock–northern hardwood swamps in the valleys and uplands of the Appalachian Plateau. In the Adirondacks and the Tug Hill Plateau it is found in black spruce–tamarack and spruce–balsam fir swamps (LaFrance 1975), in northern white cedar swamps, and in alder thicket swamps (Palmer 1949). Its territory ranges from 0.8 to 1.5 ha (2–3.2 a), and the territories are adjacent to one another in one section of a swamp, and are not distributed equally throughout the whole area (Eaton 1957).

The nest is frequently placed on the vertical face of a downed tree in the wooded swamp, where deeper water occurs below and the roots of the tree have been dislodged. A few nests have been found on the sides of clumps of cinnamon fern (Eaton 1957). The nest is built by the female and made mostly of mosses and lined with moss sporophytes, deer hair, small rootlets, and sedge leaves (Eaton 1957). Two nests examined critically had a decided entranceway of leaves resembling the nest of the Louisiana Waterthrush (Eaton 1957).

Stephen W. Eaton

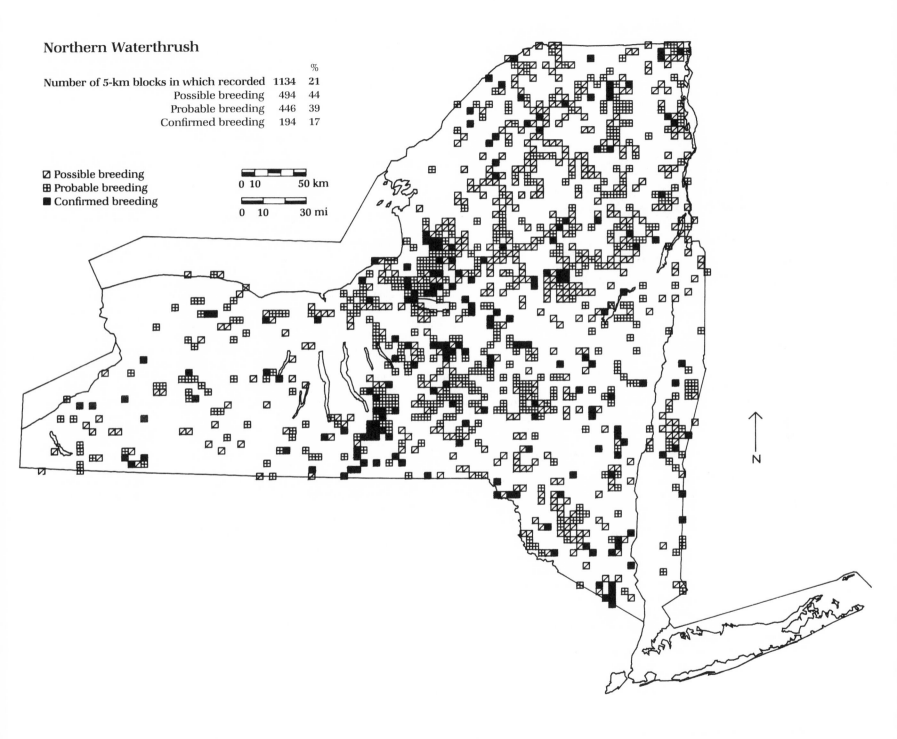

Northern Waterthrush

		%
Number of 5-km blocks in which recorded	1134	21
Possible breeding	494	44
Probable breeding	446	39
Confirmed breeding	194	17

◩ Possible breeding
⊞ Probable breeding
■ Confirmed breeding

0 10   50 km

0 10   30 mi

N

©DAS 1987

Louisiana Waterthrush *Seiurus motacilla*

A large, riparian warbler, the Louisiana Waterthrush is a fairly common to common species found along wooded, swift-flowing streams. Eaton (1914) referred to it as a common summer resident in the lower Hudson Valley, and local north to Lake George. In central and western New York he considered it a typical bird of the ravines, particularly in the area of the Finger Lakes and Genesee Valley. Bull (1974) documented some range extension along the Lake Champlain Valley to Port Henry, Essex County, and almost to Boonville, Oneida County, in the Black River Valley. Saunders (1942) considered this waterthrush only a fall migrant in Allegany State Park, and Beardslee and Mitchell (1965) called it an uncommon summer resident in the Niagara Frontier region. Eaton (1981) found it nesting along most of the wooded streams of southern Cattaraugus County along the drainage basin of the Allegheny River. It appears to have responded to the regrowth of the forest and clearing of the streams.

Atlas workers have increased the knowledge of its range. Although still essentially a bird of the southeastern part of the state, it is now known to be well distributed across the Appalachian Plateau and the Mohawk Valley; it has even invaded the Central Adirondacks, occurring along the south and north branches of the Moose River in Herkimer and Hamilton counties and on the Tug Hill Plateau. This warbler is found as far up the Lake Champlain Valley as Plattsburgh, Clinton County, and still breeds on Long Island in Nassau and Suffolk counties, north of the terminal moraine.

Its distribution distinctly follows the major river systems of the state. It does not occur on the Great Lakes Plain, on most of the Eastern Ontario and St. Lawrence plains, in the Adirondack High Peaks, and in the Sable Highlands. As its common name suggests, it is a southern species nearly reaching its northern limits here.

The Louisiana Waterthrush is relatively easy to "confirm," if enough time is spent early in the season walking the streams along which it breeds. It establishes territories along medium- to high-gradient streams, which usually flow through several forest types. One pair will occupy approximately 400 m (1312 ft) along a stream, or about 2.5 pairs per kilometer (0.6 mi), if forest cover is continuous (Eaton 1958).

Between 1840 and 1920 in the Allegheny River Valley it must have been greatly reduced during the extensive lumbering of the region (Eaton 1981). Logs were floated to market in the larger streams, and many small lumber mills emptied sawdust into the streams. Removal of the canopy of trees from the streams' banks allowed sunlight to penetrate, which increased water temperatures. This affected the aquatic insect life upon which the Louisiana Waterthrush fed. It had still not recovered when Saunders (1942) studied the birds of Allegany State Park from 1921 to 1940. It is quite possible that during the 18th century the species was common along the undisturbed river systems of the Appalachian Plateau, the Catskills, and lower elevations of the Adirondacks, and that this waterthrush is now returning to its original range. Its early history is unfortunately unclear. De Kay (1844) only reported on the Northern Waterthrush, which he apparently confused with the Louisiana Waterthrush, saying, "It is partial to the neighborhood of brooks . . . and . . . inhabits Louisiana."

Both the male and female construct the nest, although the female takes a leading role (Eaton 1958). They place it on the ground in a depression they line with old leaves and near the top of a stream cutbank. A pathway of leaves often leads to the well-concealed nest. Plant stems, pine needles, or small hemlock twigs form the outside of the nest bowl. Inner linings are made of small rootlets, small plant stems, horsehair, grass culms, or moss sporophytes. Nests are built in 2–3 days; the lining is added over 2–3 more days until the first egg is laid. Renesting occurs if the first is destroyed, but, like most warblers, the bird is single-brooded if the first nest is successful.

In the Finger Lakes region the nests are often parasitized by the Brown-headed Cowbird, which usually removes 1–2 eggs. As many as 3 waterthrushes and 2 cowbirds have been successfully fledged from 1 nest (Eaton 1958).

Stephen W. Eaton

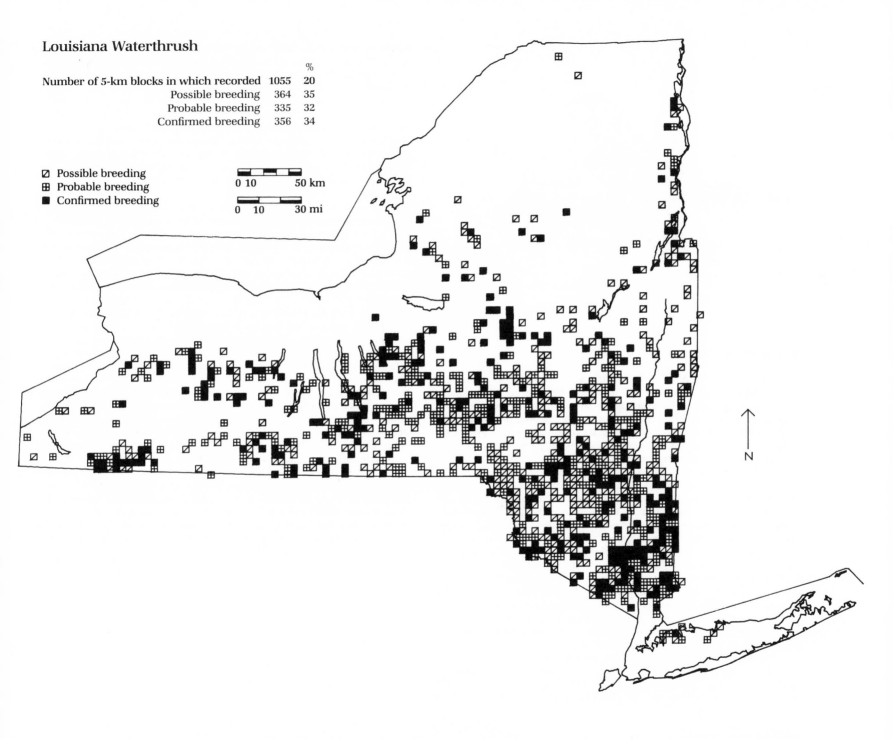

Louisiana Waterthrush

		%
Number of 5-km blocks in which recorded	1055	20
Possible breeding	364	35
Probable breeding	335	32
Confirmed breeding	356	34

◩ Possible breeding
⊞ Probable breeding
■ Confirmed breeding

0 10   50 km

0 10   30 mi

N

Kentucky Warbler *Oporornis formosus*

An elegant southern warbler with black "sideburns," the Kentucky Warbler is a very uncommon and local breeder in New York, the northern edge of its range. Its numbers have increased in the state since the 1950s, but it is not as common as it was a century ago. The Kentucky Warbler is one of many southern birds noted for its northward range expansion since the 1960s. The fact that this warbler is only now reoccupying a breeding range that it held over 100 years ago is often overlooked.

The peak abundance of this species in the state may have been during the 1870s, when Fisher (1878) found 16 individuals and 4 nests between 21 May and 5 July 1875, near Ossining, Westchester County. Since 1900 the greatest number of Kentucky Warblers reported in the literature occurred in 1928, when 6 breeding pairs were found between Elmsford and Grassy Sprain Reservoir and west to Irvington, all in Westchester County (Bull 1964). Eaton (1914) described the Kentucky Warbler as a rare summer resident on Long Island and a common breeder in the lower Hudson Valley. He also listed two proven breeding records away from those areas, both in Cortland County, in 1903 and 1906. As early as the late 1880s, however, Chapman (1889) noted that it was disappearing from traditional nesting areas on the New Jersey side of the Palisades; it continued to breed occasionally along the Palisades until 1942 (Cruickshank 1942) but not afterward (Bull 1964).

In 1973 a pair nested in Huntington, Suffolk County, for the first confirmed nesting for Long Island (Ewert 1974) and the first breeding record in the state since 1942. Since 1973 it has been found with increasing regularity on Long Island and at traditional sites in the lower Hudson Valley, such as Grassy Sprain Reservoir, where it disappeared as a breeding species in 1939 (Bull 1964) but was recorded breeding during the Atlas in 1981 and 1984. The reasons for this species' disappearance from New York during the mid-1900s and its subsequent return are not understood.

The Kentucky Warbler's distribution is concentrated in the lower Hudson Valley, with scattered records in the Coastal Lowlands and smaller incursions in western and central New York up the Allegheny, Delaware, and Susquehanna river valleys from Pennsylvania. The records in central and western New York were of territorial singing males and may have represented lone males, a common situation for a species at the northern edge of its range. Breeding has been recorded only twice in the past in western and central New York (Bull 1974).

A secretive bird, the Kentucky Warbler spends most of its time on or near the ground in the dense understory of forests. However, this species was not easily missed by Atlas observers as the male is a very persistent singer and the song is loud and ringing, suggestive of the song of the Carolina Wren. Although the Kentucky Warbler may possibly be underrecorded because of its rarity, the Atlas map almost certainly gives an accurate picture of its current distribution, as observers are always alert for rare birds.

In New York the characteristic habitat listed by Atlas observers was forest with a dense understory of shrubs and vines in or adjacent to either a stream, swamp, or reservoir. In addition, in almost every case the site was a beech-maple mesic forest, and many of the sites were on hillsides, some with rocky outcrops and small ravines. The most atypical site was also the westernmost, a deciduous woodland recently ravaged by a tornado and subsequent logging activity so that it consisted of new herbaceous growth amidst bulldozed piles of trees and brush, with only scattered standing timber. Even this site, however, was on a slight slope 91 m (300 ft) above a stream (W. Dister, v.r.). In Rockland and Westchester counties two blocks where Atlas observers found the Kentucky Warbler were subsequently bulldozed for development.

The Kentucky Warbler nests on or near the ground, usually placing the nest in a dense tangle of vines or in the base of a shrub or root of a tree (Eaton 1914). The nest is rather bulky and is constructed mostly of leaves. Occasionally the birds will build a large base of leaves up to 10.2 cm (4 in) above the ground into which the nest is built (Bent 1953). They also use twigs and rootlets, and they line the interior with finer, usually dark rootlets and cow or horse hair, if available (Bent 1953).

Steven C. Sibley

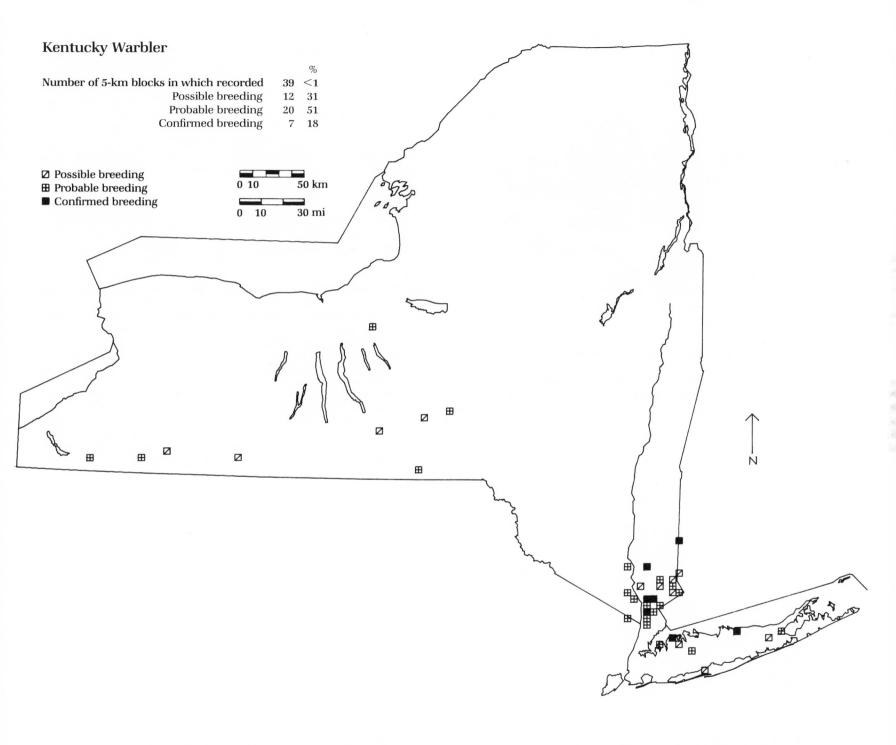

Kentucky Warbler

		%
Number of 5-km blocks in which recorded	39	<1
Possible breeding	12	31
Probable breeding	20	51
Confirmed breeding	7	18

☑ Possible breeding
⊞ Probable breeding
■ Confirmed breeding

0 10   50 km

0 10   30 mi

N

©DAS 1987

Mourning Warbler *Oporornis philadelphia*

"The singular appearance of the head, neck, and breast, suggested the name," wrote Alexander Wilson in his original description of the Mourning Warbler, first published in 1810. The specific name, *philadelphia*, was based on the collection of a specimen found "on the border of a marsh, within a few miles of Philadelphia" (Wilson and Bonaparte 1876). The Mourning Warbler rarely was encountered by early ornithologists. Wilson himself, in spite of his extensive travels, saw and collected only the one specimen, and Audubon saw very few Mourning Warblers during his wanderings (Griscom and Sprunt 1979). Although relatively widespread in New York, the Mourning Warbler remains rather uncommon, although it can be locally abundant in optimal habitat. Robbins et al. (1986) reported that North American populations of the Mourning Warbler have remained essentially stable since 1966.

De Kay (1844) considered the Mourning Warbler a rare and little-known species in New York and made no mention of evidence of its nesting in the state. Reed and Wright (1909) reported no nests of the Mourning Warbler from the Cayuga Lake basin of the Finger Lakes region. Little can be deduced from the reports of these pioneering ornithologists except that the Mourning Warbler was rarely encountered. Whether it was truly rare in colonial times, or simply obscure and secretive, is difficult to determine. It is possible that the Mourning Warbler, along with the Chestnut-sided Warbler (also rarely encountered by Wilson and Audubon), is more common today than in Wilson's time, a beneficiary of the widespread clearing of forests between 1800 and 1850 and subse-

quent regrowth to successional shrubland (Bull 1974; Griscom and Sprunt 1979).

Although Atlas data show a breeding range for the Mourning Warbler that has spread outward somewhat from its centers of concentration since Eaton's (1910) time, this could be the result of the increased coverage of the Atlas project and not of a change in the species' distribution. It shows concentrations in the Adirondacks, the Appalachian Plateau east of the Finger Lakes Highlands, the Allegany Hills and Cattaraugus Highlands, and the Great Lakes Plain from the Central Tug Hill westward, at elevations between 30 and 914 m (100–3000 ft). And as predicted by Bull (1974), increased fieldwork revealed previously unreported populations from the Rensselaer Hills, contiguous with nesting populations in western Massachusetts (DeGraaf and Rudis 1986).

In New York the Mourning Warbler occupies two distinct habitat types (Eaton 1914; Bull 1974). It can be found in upland sites of brushy second growth with dense tangles of blackberry and raspberry, which pioneer following fires or clear-cutting of timber, and in wet bottomland woods among dense growths of ferns, skunk-cabbage, and marsh marigold. It is largely absent from oak and oak–northern hardwood communities, which are typical of drier, upland sites throughout the state. On the Connecticut Hill WMA, Tompkins and Schuyler counties, Keller (1980) found densities of the Mourning Warbler to be 4–13 pairs per 40.5 ha (100 a) on areas that had been clear-cut 4 and 5 years previously. No Mourning Warblers were found on areas 1–3 years after clear-cutting. The Mourning Warbler obviously benefits from clear-cutting or the devastation of forest fires.

The nest is built on or near the ground, usually at the base of a clump of ferns or flowering plants, making it especially vulnerable to predators like the raccoon and striped skunk (Wilcove 1985). The bulky flattened nest is constructed of dead weeds and grasses; it is lined with fine, dead grass and usually with an inner cup of fine strips of black inner bark or black rootlets. Eaton (1914) observed that the lining of the nest typically was of some kind of black material.

Charles R. Smith

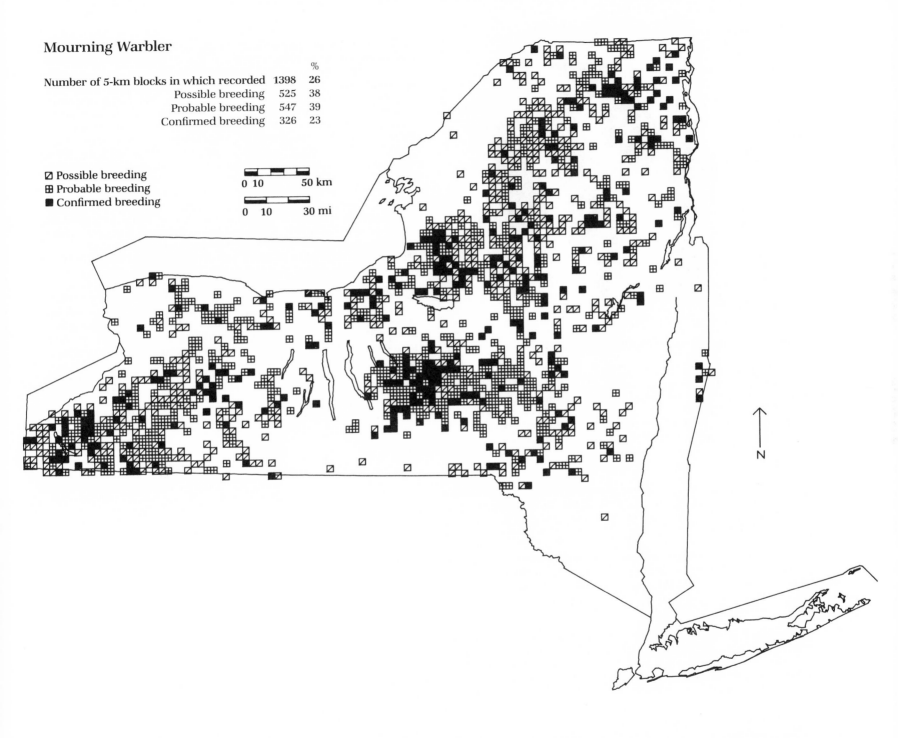

Mourning Warbler

Number of 5-km blocks in which recorded		%
Number of 5-km blocks in which recorded	1398	26
Possible breeding	525	38
Probable breeding	547	39
Confirmed breeding	326	23

◨ Possible breeding
⊞ Probable breeding
■ Confirmed breeding

0 10　　　50 km

0 10　　　30 mi

N

©DAS 1987

Common Yellowthroat *Geothlypis trichas*

A well-known warbler with the domino mask in the adult male, the Common Yellowthroat is a widespread and abundant breeding bird in New York. In fact, only the Song Sparrow and American Robin are recorded in more Atlas blocks. It is found in a variety of habitats and at a wide range of altitudes from the bayberry-covered dunes of outer Long Island beaches to above 914 m (3000 ft) in the Adirondacks.

There is no hard evidence that the Common Yellowthroat is more or less abundant or widespread today than in the past. Virtually every author that mentioned distribution, e.g., Chapman (1906) and Eaton (1914), called the species widespread, common, or abundant. Subjective analysis suggests, however, that the species may have undergone a period of substantial increase when forest clearing and rural road opening by colonists increased brushy thickets and woodland edge, providing expanded opportunities for breeding. More recent abandonment of farmland and its growth to successional forest should have the opposite effect. Furthermore, increased urbanization, suburban sprawl, highway building, and marsh drainage should also have a strongly negative impact on breeding opportunities. However, Robbins et al. (1986) documented an increase for this species in New York during the period 1965–79.

The Atlas map clearly shows the widespread distribution of this species. Blank blocks on the map may be artifacts of coverage rather than evidence of the species' absence. There may be some blocks, however, where unbroken forest affords no breeding habitat for it.

Since the Common Yellowthroat has a loud and persistent call and song, and it readily responds to "pishing," its presence in breeding season is difficult to miss. This was also one of the easier species to "confirm" with observations of adults feeding young.

The habitat of the Common Yellowthroat includes shrubs, thickets, or tangles in moist, marshy, or streamside areas, and along overgrown roadsides or other edge habitat. Stewart (1953) cited dense herbaceous vegetation and small woody plants in damp or wet situations. It is also found in old fields with clumps of shrubbery, in cat-tails, or even on dry hillsides (Gross 1953b). Many authors cite the extreme variability of the bird's favored habitats, but in New York it appears to prefer to breed in or near moist or wet locations. Kendeigh (1945) noted that in the Helderberg Highlands this warbler appeared to require a dense growth of low vegetation that is most commonly found in wet areas.

This species normally nests within a few inches of the ground (Gross 1953b), rarely to 1.5 m (5 ft) (Chapman 1907), often backed against a shrub stem. The nest, constructed by the female, is a bulky cup, occasionally partially roofed over. It is loosely woven of mixed vegetation such as grass and weed stems, leaves, and bark strips, and it is lined with finer materials. It is attached to grass tussocks, reeds, greenbrier, or shrubs and is amazingly well concealed, as has been attested to by Atlas records. Only 8% of the "confirmed" records represent nests. Nests found by Kendeigh (1945) were at the edge of a pond in a clump of lilies and on a patch of fern at the edge of a hemlock-beech forest. The species is the seventh most frequent host to the eggs of the Brown-headed Cowbird (Friedmann 1929).

Robert Arbib

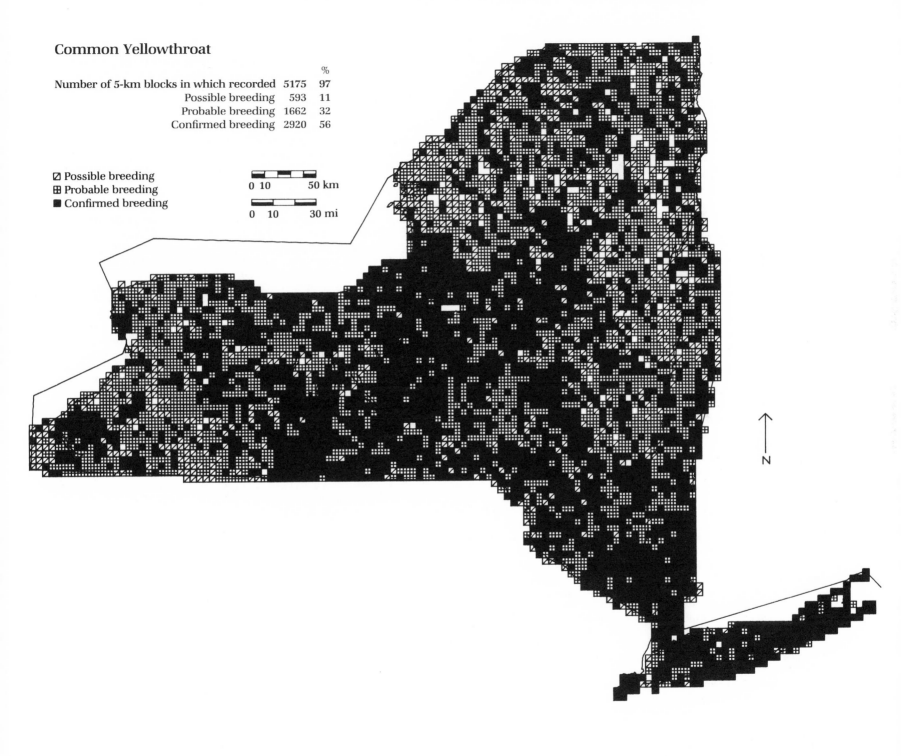

Common Yellowthroat

Number of 5-km blocks in which recorded 5175 97
Possible breeding 593 11
Probable breeding 1662 32
Confirmed breeding 2920 56

%

☒ Possible breeding
⊞ Probable breeding
■ Confirmed breeding

0 10 50 km

0 10 30 mi

N

Hooded Warbler *Wilsonia citrina*

The Hooded Warbler, though locally distributed in New York, is usually fairly common in areas where it occurs. For example, Baird (1983) found that out of 14 warbler species that inhabited Quaker Run Valley in Allegany State Park, the Hooded Warbler was the sixth most abundant. He estimated that 290 pairs nested there in 1983, compared with Saunders's (1936) estimate of 23 breeding pairs in 1930 and 1931.

De Kay (1844) considered the Hooded Warbler to be rare in the state, but by the early 19th century it had apparently become established. Eaton (1914) reported it to be abundant near Highland Falls, Orange County, and in Palenville, Greene County. In the interior he called its distribution local, but large breeding colonies were noted on the Appalachian Plateau and Great Lakes Plain. Bull (1974) stated that it was a "locally common breeder in the lower Hudson Valley." He considered the species rare, especially in the western part of the Finger Lakes region, and rare and local in southwestern New York; but it was apparently absent in many of the southern counties along the Pennsylvania border and on Long Island.

The Atlas map shows the Hooded Warbler absent or scarce in some areas where it bred formerly, but it has expanded into others. In southeastern New York the Hooded Warbler was reported mainly in the Hudson Highlands and Manhattan Hills. It was also reported in isolated blocks in Ulster and Sullivan counties. Although Eaton (1914) reported the Hooded Warbler in Greene County, no Atlas records were from there. Only a few Atlas records were from Wayne, Monroe, and Orleans counties, where Eaton (1910) showed it to be fairly generally distributed, and

no Atlas records were from Madison County. Breeding sites in Onondaga County shown in Bull (1974) have largely disappeared. The species was found, however, in the northern hardwood forest of the Tug Hill Plateau and in the eastern Drumlins and Oswego Lowlands. The Hooded Warbler still occupied areas south and east of Buffalo referred to by Bull (1974), but the Atlas documented three new areas of concentration: in Wyoming County, especially in Letchworth State Park; in southern Cattaraugus County; and in Chautauqua County. The size of the Cattaraugus populations may be of relatively recent origin, as Saunders (1942) said that it was uncommon in Allegany State Park, but according to Beardslee and Mitchell (1965) it was the most common breeding warbler in the Big Basin area of the park. It avoids the large block of oak–northern hardwood forest south of the Finger Lakes.

The habitat of the Hooded Warbler is mature forest of more than 14.2 ha (35 a) which contains a well-developed understory (MacClintock et al. 1977). In Westchester County it was found on wooded slopes containing an understory of mountain laurel (Bull 1964). Silloway (1920) found a nest at Palisades Interstate Park in a shallow ravine where scattered blackberry canes grew. Scheider (1959) said that in three very disjunct areas of central New York the Hooded Warbler was found in well-drained open maple woods containing a thick but not impenetrable undergrowth of sapling maples. The density of this species appears to depend somewhat upon forest type. Baird (1986a) described its density as 0.7 pairs per 40.5 ha (100 a) in the mature northern hardwood forest. In the Appalachian oak-hickory forest he found the density had increased to 5 pairs per 40.5 ha (100 a). In Maryland MacClintock et al. (1977) found an isolated block or island of mature forest less than 14.2 ha (35 a) which lacked Hooded Warblers, but a similar control area within a large 160-h (400-a) tract had 10 pairs per 40.5 ha (100 a).

The nest is built from near the ground to 1.2 m (4 ft) above (Bull 1974) in sugar maple saplings, mountain laurel, or herbaceous plants such as blackberry canes. It is neat and compactly built with a well-woven rim. Leaves, grasses, bark strips, plant down, and weeds make up its main body, and it is lined with black rootlets, soft grasses, plant fibers, and, in one case, porcupine fur (Harrison 1975; Eaton 1981).

Eaton et al. (1963) showed that this species had the largest eye-ring diameter of 32 species of warblers examined. The resultant increased light-gathering capacity might be an adaptation for life in the dense understory under a closed canopy.

Stephen W. Eaton

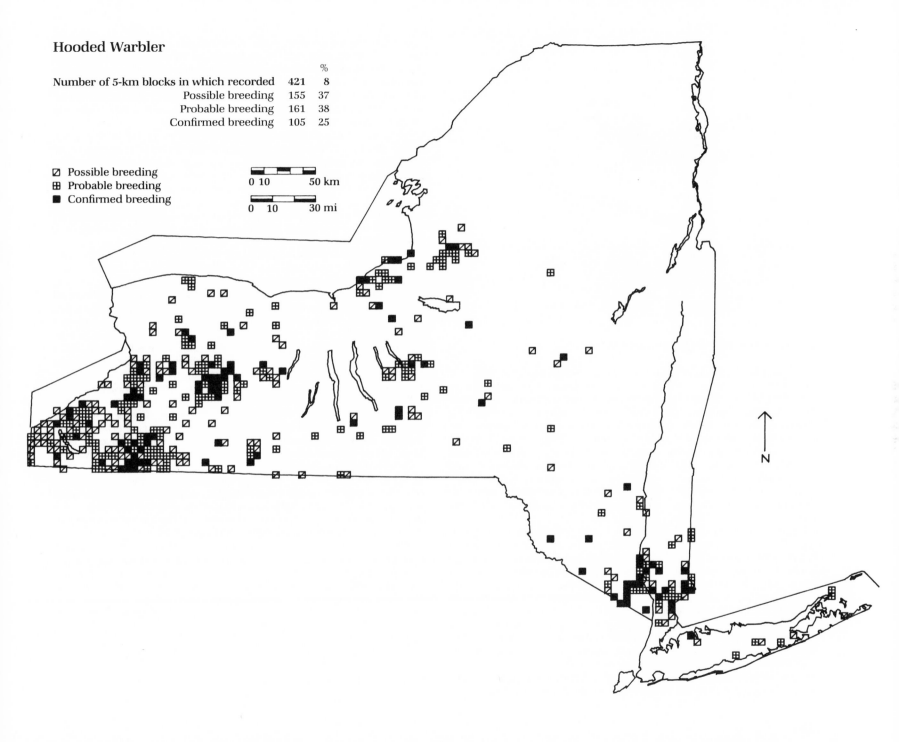

Hooded Warbler

		%
Number of 5-km blocks in which recorded	421	8
Possible breeding	155	37
Probable breeding	161	38
Confirmed breeding	105	25

☑ Possible breeding
⊞ Probable breeding
■ Confirmed breeding

0 10   50 km

0 10   30 mi

N

Wilson's Warbler *Wilsonia pusilla*

Wilson's Warbler was not confirmed as a nesting species in New York until 1978, just 2 years before Atlas fieldwork began. Even after intense Atlas efforts, only three satisfactory records were obtained. Its distribution, however, suggests that Wilson's Warbler may be found nesting anywhere within the Adirondacks where suitable habitat occurs. The three Atlas records were distributed among as many counties: Essex, Franklin, and Hamilton. The breeding range of the Wilson's Warbler extends from Alaska to Newfoundland and into the western United States, but in the Northeast it is found only as far south as coastal Maine, New Hampshire, and northern Vermont. The Adirondacks and New England are at the periphery of the range of this boreal warbler (AOU 1983).

As recently as 1974 Bull stated that "no summering—let alone breeding—has been substantiated in New York to date," adding, "reports in midsummer from the Adirondacks are unconfirmed." Many years before a Maine observer had noted of Wilson's Warbler "that it has not been more often discovered during the nesting season is on account of the favorite habitats being rarely visited by ornithologists. . . . There seems to be no doubt at all that a person acquainted with the habits of the present species and the Yellow Palm Warbler as well, can go into territory . . . and find both species in many localities where other observers have failed to see them, provided that suitable tracts of spruce and hackmatack [balsam poplar] bog exist in the region" (Knight 1908).

On 30 June 1978 Dan Nickerson located and photographed a male and female Wilson's Warbler in North Meadow, near the Heart Lake Road, Essex County. This was the same area studied by Saunders in the summers of 1925 and 1926, and where he had confirmed breeding of the first Tennessee Warbler for the state in 1926. This same area was visited by Carleton in 1947, providing the first state nesting record for Cape May Warbler. Nickerson returned the following day, 1 July, and found the nest and 3 eggs of a Wilson's Warbler; this was the first, and thus far only, nest of this species found in New York (Nickerson 1978).

Not surprisingly, perhaps, the first Atlas encounter also took place at North Meadow, where a pair was seen 7 June 1980. As Nickerson (1978) noted, this area was formerly used as a pasture, which may account for the relatively open spaces between clumps of 4.6–9.1 m (15–30 ft) spruce, tamarack, and balsam fir, these clearings often filled with a dominant growth of meadow-sweet or Spirea that grows to a height of 0.3 to 1.5 m (1–5 ft). Much of the ground is covered with moss and lichen, with willow and alder providing a transition between the conifers and an adjoining beaver-flooded marsh.

A single male was observed 22 June 1984 in a small area of black spruce and marsh just off the access road to the Bartlett Carry Club, Franklin County (T. Dudones, v.r.). A few weeks later, on 10 July, another male was sighted near a beaver meadow on the southwest side of the Wilson Pond trail off Route 28, Hamilton County, beyond Grassy Pond. Appearing in response to vocal imitations of a Barred Owl, the bird was quite agitated as it perched on side branches of a spruce. The area is a grassy, sedgy beaver meadow bordering a stream, surrounded by a thick growth of shrubs and saplings, and is enclosed in unbroken northern hardwood forest with pockets of spruce. The bird later appeared in response to the taped song of Wilson's Warbler (S. Laughlin, v.r.). Note that beavers had an influence at two of these sites.

Habitat of the Wilson's Warbler during the breeding season is described as "shrubby and brushy areas (especially near water), bogs, and thickets in riparian woodland, in boreal and montane regions" (AOU 1983). Certainly there is an abundance of such habitat within the Adirondacks, yet most is apparently unoccupied by this warbler. The nest found at North Meadow in 1978 was typical, located beneath a hummock of moss: "The nest was made mostly of grass and held three eggs. Each egg was creamy pink in color with reddish-brown speckles. These were densely concentrated toward the blunt end of the egg and more lightly dispersed over the rest of the egg" (Nickerson 1978).

The song, a one- to four-parted trill with a chattery or staccato quality, is extremely variable. On 5 July 1982 in excellent boreal bog-type habitat near Sabattis, Hamilton County, the taped song of a Wilson's Warbler elicited an identical response; when the singer was located, the author and Nickerson identified it as a male Nashville Warbler.

John M. C. Peterson

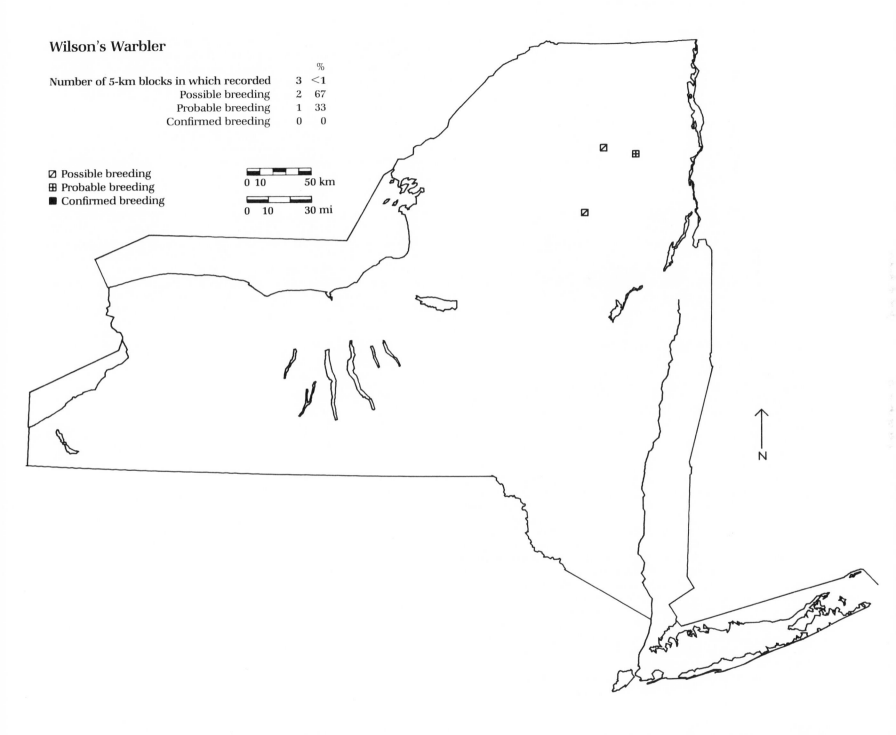

Wilson's Warbler

		%
Number of 5-km blocks in which recorded	3	<1
Possible breeding	2	67
Probable breeding	1	33
Confirmed breeding	0	0

◰ Possible breeding
⊞ Probable breeding
◼ Confirmed breeding

0 10   50 km

0 10   30 mi

N

Canada Warbler *Wilsonia canadensis*

The Canada Warbler is a shy, retiring bird that often remains hidden in thickets, occasionally making itself visible when engaged in bouts of animated fly catching. It is fairly common in cool, humid woodlands throughout high elevations in New York. It is also found, uncommonly, in wet areas at low elevations, including a few places in the Coastal Lowlands. Bull (1974) said that of the northern warblers, this species was the most widely distributed in the state.

The Canada Warbler may be more widely distributed in New York than it was earlier in this century. In adjacent Vermont, Ellison (1985g) suggested that the species may have increased since that time as forest cover increased. In New York, however, evidence for population growth is scanty. Stoner (1932) stated that the species was more common in the Oneida Lake region than it had been 25 years earlier. Bull (1974, 1976) noted only two breeding locations on Long Island (one certain, one possible), whereas the Atlas project found five (four "possible" and one "probable"). Atlasing also found more breeding locations in Westchester and Rockland counties than were known to Bull (1974). Thus, it is possible that the species may be extending its range southward.

As the Atlas map shows, the Canada Warbler is most widely distributed throughout the Adirondacks, Tug Hill Plateau, the Catskill region, and the eastern Appalachian Plateau. This distribution is generally correlated with elevation but is complicated by the species' tolerance of cool, wet areas at lower elevations as well. The Atlas project documented breeding over a greater area than was previously known.

Because the Canada Warbler was difficult for Atlasers to detect, it may be under-recorded. Not only does the bird stay well hidden in the shrubby undergrowth, but its song is not familiar to many birders. The nest is difficult to locate, and only 8 were found. On the other hand, the Canada Warbler does become very agitated when an observer enters its territory, making loud chipping noises as it flies from shrub to shrub.

The Canada Warbler breeds in varied habitats. It is almost always found in dense undergrowth, tangled thickets, or vines. The undergrowth may be in open woodland, either deciduous or mixed; forest edges; stream banks; and bogs or swamps (Eaton 1914; Stoner 1932; Bull 1974). The common denominator seems to be cool temperatures and high humidity, although there are exceptions. For example, a pair was observed in shrubs in a dry, sandy area in Saratoga County (J. Carroll, pers. comm.).

The Canada Warbler usually builds its well-concealed nest on or near the ground in moist thickets or dense fern growths. The bulky and rather formless structure is constructed of dead leaves, weed stalks, grasses, moss, and bark strips, and is lined with rootlets or hairs (Eaton 1914; Stoner 1932). Some nests may be placed as high as 0.6 m (2 ft) among bushes, on mossy hummocks, at the base of moss-covered logs or rocks, in earthen banks, or in cavities of stumps (Bent 1953; Bull 1974).

Richard E. Bonney, Jr.

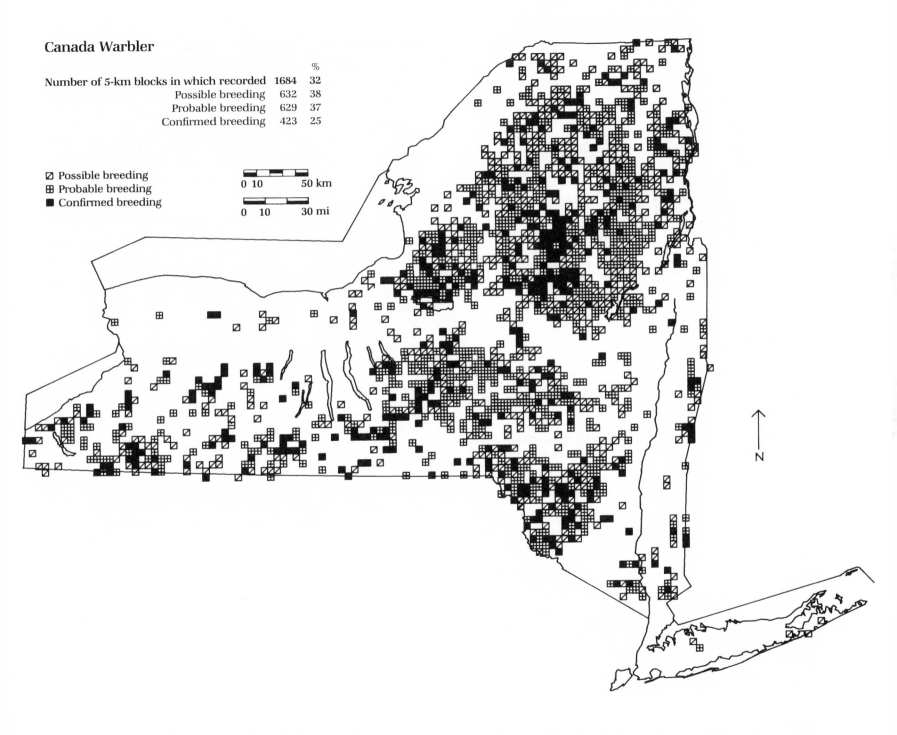

Canada Warbler

Number of 5-km blocks in which recorded 1684   %
                                              1684 32
                  Possible breeding 632 38
                  Probable breeding 629 37
             Confirmed breeding 423 25

▧ Possible breeding
⊞ Probable breeding
■ Confirmed breeding

0 10   50 km

0 10   30 mi

N

©1987 DAS

Yellow-breasted Chat *Icteria virens*

A large warbler, the Yellow-breasted Chat has a song that Harrison (1984) described as "the alarm call of a wren; a series of nasal quacks; a wolf whistle; a fog horn; and a chuckling, high-pitched laugh." This song is one New York birders seek out, since the Yellow-breasted Chat is an uncommon and local breeding species found sometimes in loose colonies or as single pairs (Benton 1960; Mudge 1962). BBS data for the period 1965–1979 documented a "sharp and steady decrease" of the species in the eastern portion of its range (Robbins et al. 1986). Because of an apparent decline, it was on the *American Birds'* Blue List for 6 years, 1976–1981, but was taken off because the decline appeared to be quite localized (Tate 1981).

The Yellow-breasted Chat may have decreased in south central and southeastern New York since the early 1900s. Eaton (1914) called it a common summer resident in the lower Hudson Valley and on Long Island. Bull (1964) reported 22 pairs in ten localities in 1940 on Long Island, and 30 pairs in 11 localities in Westchester and Rockland counties; over 12 pairs were reported on Fishers Island in 1950. Atlas observers "confirmed" the Yellow-breasted Chat in only five blocks on Long Island, including one on Fishers Island. It was also "confirmed" in one block on Staten Island and in two blocks in Westchester County but none in Rockland, where it historically bred. There were, however, several breeding records from Orange County. To the north of these two counties along the Hudson Valley, it is almost entirely absent.

Eaton (1914) found the Yellow-breasted Chat extremely local in other sections of southern, central, and western New York where it occurred in the Delaware, Susquehanna, and Chemung river valleys and near the southern ends of the Finger Lakes. Its distribution apparently has changed little since then. Benton (1960) said that in central New York it was a regular summer resident, locally common at the south end of the Cayuga Lake basin (also indicated by Eaton's 1910 map) but not at the north end. Scheider (1959) documented nesting in the Camillus Valley, Onondaga County, in 1957 and 1958. Atlas observers documented breeding in central New York in the Susquehanna and Chemung river valleys, along the Genesee River, and in several blocks around the Finger Lakes, although none of the Finger Lake records were "confirmed." It was also found along the Great Lakes Plain. In western New York the chat was described as a rare breeding species, with only one known breeding record from 1890 (Beardslee and Mitchell 1965). Atlas observers "confirmed" breeding in three widely scattered locations in the Niagara Frontier region, but generally, the species was scarce.

The northernmost breeding records known by Bull (1974) were from Oneida and Washington counties. Evidence of breeding was found but not "confirmed" by Atlas observers as far north as Saratoga County. The breeding distribution of this species is restricted to certain brushy hillsides or brushy tangles in the lowlands (Benton 1951a; Mudge 1962). This southern warbler is found at elevations mainly below 305 m (1000 ft); however, Baird (pers. comm.) found a nest in the Allegheny River Valley at 411 m (1350 ft).

The Yellow-breasted Chat is essentially a southern species; as one would expect, it entered New York along the Coastal Lowlands, the Hudson Valley, and the river systems of the Appalachian Plateau. It scattered from these points of entry into the valleys of the Finger Lakes and Great Lakes Plain. This warbler may also have come into New York from Ontario, where Atlas observers found this species along Lake Erie and the Niagara peninsula (Ontario Breeding Bird Atlas, preliminary map).

The habitat where Mudge (1962) found 10 pairs on Long Island was described as dense brushy tangles, in areas where the predominant plants were greenbrier, smooth sumac, poison ivy, bayberry, viburnum, bush honeysuckle, highbush blueberry, American bittersweet, wild grape, red cedar, dogwood, and sassafras, with an occasional apple tree. On a hillside overlooking Cayuga Lake, where Benton (1951a) found 4 singing males, the habitat was a slope covered with scattered clumps of northern blackberry, red cedar, hawthorn, and roses. The area was pastured, but the thorny growth maintained the habitat for brush-loving birds.

The nest is placed 0.3–1.8 m (1–6 ft) above the ground in a bush, briar tangle, vines, or a low tree. It is bulky and made of leaves, vines, weed stems, and grasses; it is lined with fine grasses or plant stems (Harrison 1975).

Stephen W. Eaton

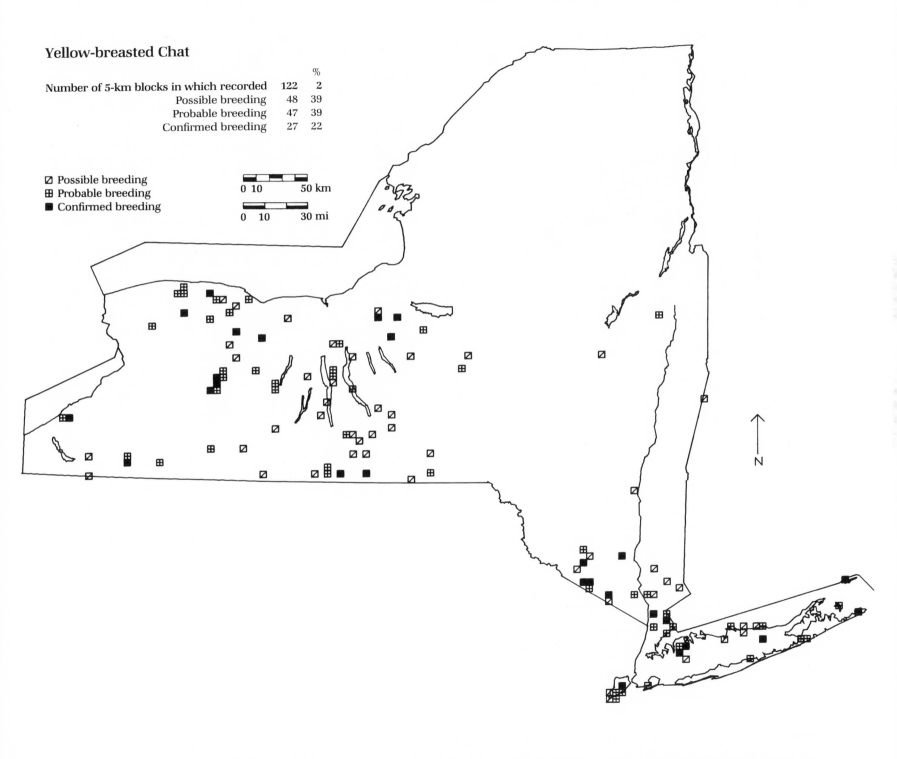

Yellow-breasted Chat

		%
Number of 5-km blocks in which recorded	122	2
Possible breeding	48	39
Probable breeding	47	39
Confirmed breeding	27	22

☑ Possible breeding
⊞ Probable breeding
■ Confirmed breeding

0 10   50 km

0 10   30 mi

N

Scarlet Tanager *Piranga olivacea*

The black-winged red-bird of De Kay (1844), the Scarlet Tanager, is the state's only regularly occurring representative of a tropical group (subfamily Thraupinae) that includes over 230 species (Perrins and Middleton 1985), a number nearly equivalent to all the known breeding species of New York. It is a common breeding species in New York forests. The highest densities reported in the Northeast on BBS routes for the period 1965–1979 included the Adirondacks and the Appalachian Plateau (Robbins et al. 1986).

Little real evidence exists to assess the historical status of this species. Eaton (1914) considered the Scarlet Tanager a fairly common summer resident of the forests of the state but less common in the more cultivated areas of southern, central, and western New York. Reed and Wright (1909) reported that its abundance had increased in the Cayuga Lake basin since 1899. The increase coincides with the increase of forest habitat, a result of widespread reforestation in New York that began at the end of the 19th century.

Atlas data show that the Scarlet Tanager is widespread, occurring either as an actual or potential breeding species in most survey blocks. It seems more sparsely distributed in areas of the state where forest cover is less than 50%, such as the agricultural areas along the Erie-Ontario and Eastern Ontario plains. It is also absent from large urban areas and tracts of mountain spruce–fir forests of the Adirondacks. It occupies a wide range of elevations, from sea level on Long Island to nearly 1524 m (5000 ft) in the Adirondacks.

The Scarlet Tanager occurs in most of the forest types represented in the state, with a distinct preference for woodlands of 20 ha (49.4 a) or larger in size (Robbins 1984). Bull (1974) noted that it preferred oak forest in southern New York. Since the Scarlet Tanager is found in mature deciduous forest, it is potentially sensitive to forest fragmentation. Whitcomb et al. (1981) showed that it is more susceptible than average to stresses associated with forest fragmentation, and Ambuel and Temple (1982) reported a significant declining trend for this species in fragmented Wisconsin forests. As an open-cup nester, the Scarlet Tanager is vulnerable to brood parasitism by the Brown-headed Cowbird. This vulnerability contributes to the tanager's sensitivity to forest fragmentation (Brittingham and Temple 1983).

Increases and declines of this species related to the availability of prey have also been documented. Holmes and Sherry (1986) considered the Scarlet Tanager a foliage forager that preys extensively on large flying insects, often by hawking them from the air, rather as a flycatcher does. They reported instances of increased abundance of the Scarlet Tanager in New Hampshire coincident with major irruptions of Lepidoptera caterpillars. The Scarlet Tanager can also show a marked population decline in response to adverse weather early in the breeding season. Following a period of rain and cold during late May 1974, many birds were found dead along highways in New Hampshire and Vermont, and breeding Scarlet Tanagers in the Hubbard Brook Experimental Forest in New Hampshire declined by 33% during 1974 and 67% in 1975 and 1976, compared with the average population numbers the preceding 4 years (Zumeta and Holmes 1978). The authors concluded that the depression of insect activity by the cold, wet weather, leading to reduced availability of food, was a significant factor contributing to the subsequent local decline of the Scarlet Tanager. Tanager numbers in the Hubbard Brook Experimental Forest did not return to their pre-1974 levels until 1983 (Holmes and Sherry 1986).

In a detailed account of the life history of the Scarlet Tanager, Prescott (1965) described a typical nest: it is placed in a cluster of leaves (or with leaves shading the nest), usually midway out on a nearly horizontal branch, with an unobstructed view to the ground below, and with open flyways from adjacent trees to the nest. The nest usually is from 3.7 to 9.1 m (12–30 ft) above the ground in a deciduous tree, although hemlock and tamarack occasionally are used. It is a fragile structure, loosely constructed of fine twigs and lined with finer twigs and rootlets, and often the eggs can be seen through the nest material and counted by an observer on the ground (Eaton 1914).

Charles R. Smith

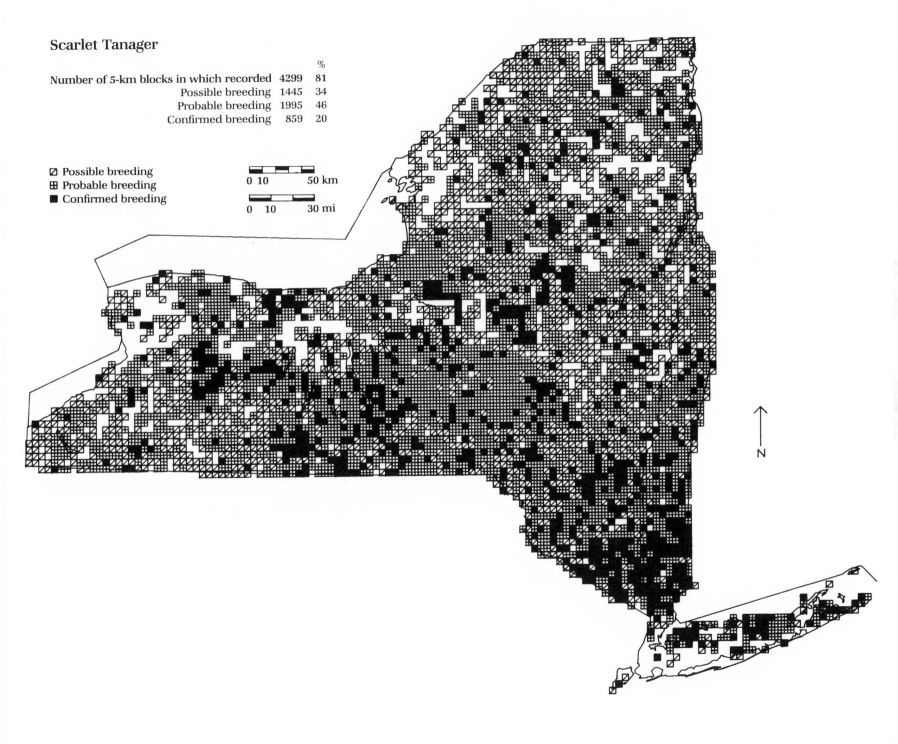

Scarlet Tanager

		%
Number of 5-km blocks in which recorded	4299	81
Possible breeding	1445	34
Probable breeding	1995	46
Confirmed breeding	859	20

▨ Possible breeding
⊞ Probable breeding
■ Confirmed breeding

0 10   50 km

0 10   30 mi

N

Northern Cardinal *Cardinalis cardinalis*

To a large majority of birders it will come as a surprise that the Northern Cardinal, an almost ubiquitous and striking part of our landscape, is a newcomer to New York's avifauna. This now abundant bird was a rarity only 40 years ago. In fact, it is a relatively recent arrival in most of the northeastern United States.

A picture of vast territorial expansion unfolds in successive editions of the AOU Check-list. In 1886 the Northern Cardinal was noted as "only casual north of the Ohio River"; in 1895 the range had extended "to the Great Lakes"; by 1895 it was north to the lower Hudson Valley and the Great Lakes; in both the 1910 and 1931 editions the range reached southern Iowa, northern Indiana and Ohio, southern Ontario, southeastern and southwestern Pennsylvania, and the southern Hudson Valley; in 1957 it was listed in western and southern Ontario, western New York, and southwestern Connecticut; and in 1983 its range included southwestern Quebec, northern New York, Massachusetts, and Nova Scotia. Thus, during the last decade of the 19th century and first half of the 20th, there was a fairly rapid expansion of range and increase in abundance of this formerly "southern" species. There were indications that much of the movement took place in the fall and winter seasons. For example, a marked incursion was noted throughout southern Ontario in the fall and winter of 1938–39 (Bent 1968a).

In the 1800s the Northern Cardinal was found only in the "Atlantic District," presumably Long Island and the area along Long Island Sound (De Kay 1844), and it apparently was a "very rare straggler" in western New York (Short 1896). Eaton (1914) said it was most commonly found in the lower Hudson Valley, Staten Island, and Long Island. Beardslee and Mitchell (1965), reporting on the Niagara Frontier region, said that before the 1920s the Northern Cardinal was merely a straggler. Later, in 1938, 2 nests were found, and by 1965 it was a regular visitor to feeders and an established breeder. Horsey (1938) said it was rare in the Rochester area, Monroe County, although a male spent the winter of 1917 there. A decade later it was seen during every month and bred successfully (Horsey 1949).

The Rochester CBCs show a steady rise in numbers, from 5 birds in 1913 and 1927, to 17 in 1946, 30 in 1949, 100 in 1950, 315 in 1971, and 510 in 1980. In the Syracuse area, Onondaga County, Burtt (1980) found that his annual winter feeder survey showed that the Northern Cardinal increased by 3% each year between 1959 and 1978. In the Adirondacks there were records in 1933 and 1935 along Lake Champlain, in the central mountains in 1961, and by 1965 it was reported frequently and is now fairly common (Beehler 1978; Carleton 1980).

According to Beddall (1963), there have been a number of factors producing the wide expansion of range and increases in numbers of the Northern Cardinal. First, in a period of several decades of moderating temperature, there was a high survival rate, increased success in raising young, and an increase in clutch size and number of broods, with resultant population increase. This led to strong population pressure, which pushed birds into previously unoccupied territory. The movement into new areas and establishment there was assisted by the species' inherent adaptability and motility, a tolerable climate, and a favorable change in habitat. Previously deforested land was growing up with new, young forest as farms were abandoned. Concurrently there was a rapid change of rural living to suburban and urban communities, to which the Northern Cardinal adapted well. Its expansion was abetted by the increase in winter bird feeding.

The Atlas map shows that this species is now widespread throughout most of southern New York. It was absent from much of the Adirondacks and other areas north of the Mohawk Valley, except along Lake Champlain, and at the eastern end of Lake Ontario. It is found sparsely in the eastern Appalachian Plateau. It seems to have largely avoided land above 305 m (1000 ft) except in well-settled communities. It is doubtful that the Northern Cardinal was overlooked by Atlas observers since its flaming plumage and bright, cheerful song are well known. In addition, according to Laskey (1944), during its lifetime the Northern Cardinal does not range far from its breeding territory.

The Northern Cardinal nests in a variety of habitats, including forest edges, open woods, swampland, and residential gardens, yards, and parks (DeGraaf et al. 1980). It builds its nest in shrubbery and vines around houses, in dense rose hedges, streamside thickets, shrubs in dry old fields, or in small trees, often some distance from its foraging area. In towns and villages it may choose lilac bushes and other dense shrubbery in parks and gardens (Bent 1968a).

Gordon M. Meade

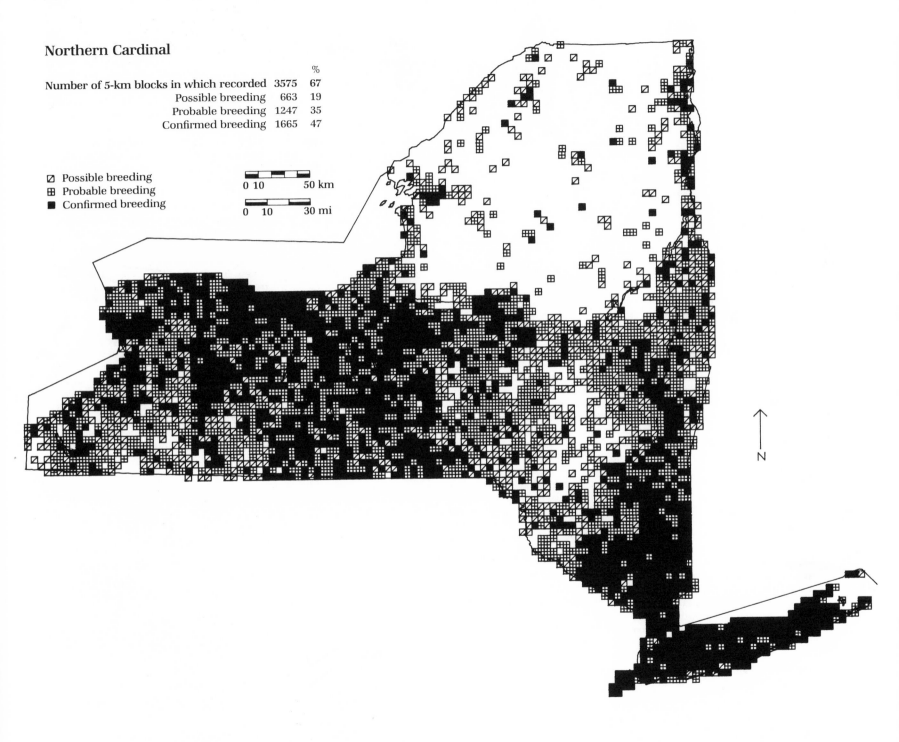

Northern Cardinal

Number of 5-km blocks in which recorded		3575	% 67
	Possible breeding	663	19
	Probable breeding	1247	35
	Confirmed breeding	1665	47

◨ Possible breeding
⊞ Probable breeding
■ Confirmed breeding

0 10   50 km

0 10   30 mi

N

Rose-breasted Grosbeak *Pheucticus ludovicianus*

The Rose-breasted Grosbeak is a large and handsome songbird that often whistles its loud, robinlike song from the very top of a tree. When not singing, it can sometimes be difficult to see, however, for it often stays hidden in the undergrowth of young deciduous forests. This bird is one of the most widely distributed species in New York, being fairly common to common everywhere in the state except on Long Island.

The Rose-breasted Grosbeak may be more widely distributed than it was earlier this century. Eaton (1914) stated that the species was fairly common throughout the Appalachian Plateau and a large portion of the Adirondacks but was rare or uncommon on Long Island and in the lower Hudson Valley. Bull (1974), however, stated that it was a widespread breeder everywhere north of the Coastal Lowlands, where it was rare and local and confined to the north shore. Finally, Atlas observers found the species to be well distributed throughout the entire state, including the lower Hudson Valley and even much of Long Island, except for southwestern Long Island, where it has never been present. In fact, though Bull (1974) noted only nine known breeding locations on Long Island, Atlas observers found more than 30 "confirmed" and several additional "possible" or "probable" locations. Robbins et al. (1986) noted a rapid increase in numbers of Rose-breasted Grosbeaks recorded on BBS routes in the eastern region, including New York, from 1965 to 1979.

This expanded distribution and perhaps increased abundance may have resulted from a change in habitat preference. Bent (1968b) noted that in the early 1900s the species moved into newly created suburban and farm habitats, and it appears to thrive in cut-over and disturbed habitats.

Although the Rose-breasted Grosbeak is widespread, it is not necessarily common everywhere. Stoner (1932) noted that it exhibits variations in its status that depend partly on local conditions, and Cruickshank (1942) stated that the species' numbers are subject to remarkable yearly variations. Atlas observers had little trouble recording this species. Both the rich song and the sharp *eek* call note are well known. The species also was fairly easy to "confirm," as adults were often seen with food for young, and fledglings were quite conspicuous.

The Rose-breasted Grosbeak is generally found in rich, moist, second-growth deciduous woodlands with dense undergrowth; in forested swamps; and along stream edges well grown with alder, maple, and birch. However, it also inhabits woodland with a mixture of hemlock, pine, or spruce; dense growths of small trees and shrubs along edges of woods and old pastures; shade trees along rural roads with extensive thickets and shrubbery; and gardens and parks of towns and villages (Eaton 1914; Bull 1974; Terres 1980). Pough (1946) described ideal habitat as the interface of tall forest trees with low shrubs. This type of habitat occurs most commonly along streams, ponds, and marshes. The bird's preference for edge habitat may explain why it is not common everywhere within its widespread range.

The Rose-breasted Grosbeak builds its nest in a bush or small tree, often an alder, maple, beech, or hemlock, usually from 1.2 to 9.1 m (4–30 ft) from the ground. Males sometimes select the nest site, may help the female build, and share in the incubation (Bent 1968b). The loosely constructed, flimsy-looking structure is typically placed in the fork or crotch of a limb and is made of fine twigs (especially beech and hemlock), small sticks, vine stems, and coarse straw; it is lined with grasses, fine twigs, rootlets, and sometimes hair (Eaton 1914; Saunders 1942; Bull 1974). Often, the eggs can be seen through the bottom of the nest (Bent 1968b).

Richard E. Bonney, Jr.

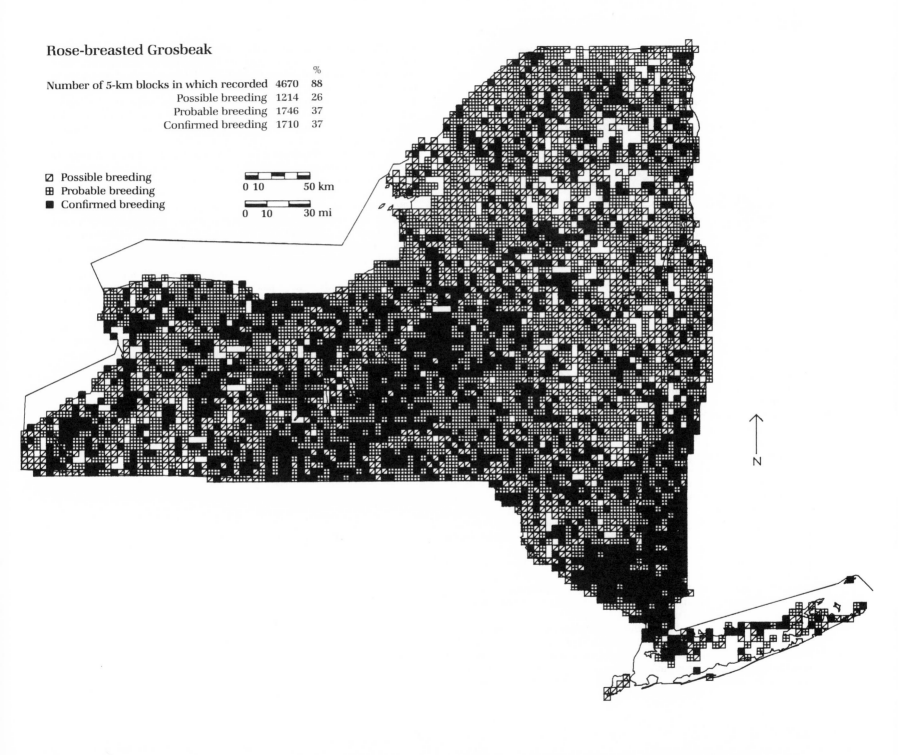

Rose-breasted Grosbeak

		%
Number of 5-km blocks in which recorded	4670	88
Possible breeding	1214	26
Probable breeding	1746	37
Confirmed breeding	1710	37

☒ Possible breeding
⊞ Probable breeding
■ Confirmed breeding

0 10   50 km

0 10   30 mi

N

Blue Grosbeak *Guiraca caerulea*

As Able (1983) and others predicted, the Blue Grosbeak has now been added to the list of breeding species of the state. Considering that this species has been successfully nesting at Hackensack Meadows in New Jersey since 1973, it should not be surprising that it made the rather short move to Staten Island. This bird is nonetheless a very rare breeder in New York, having been "confirmed" in only two Atlas blocks and recorded as "possible" in two others.

The Blue Grosbeak was considered a southern bird by Maynard (1881), and accidental in New England. Eaton (1914) described its range as the southeastern United States to southern Pennsylvania, Kentucky, and Kansas. Dwight and Griscom (1927) added New Jersey, West Virginia, Illinois, and Nebraska to the list of breeding locations. Bull (1974) reported it "breeding north to Maryland and Delaware, rarely to the southern portions of Pennsylvania and New Jersey; in recent years twice in central New Jersey." In 1973 a pair of Blue Grosbeaks raised 3 young at Hackensack Meadows (Buckley and Davis 1973), and in 1975 a group of 8–10 singing males was located at Assunpink Game Management Area north of Allentown, Pennsylvania (Buckley et al. 1975). Breeding was first recorded in Ohio in 1976 (Hall 1976).

The Blue Grosbeak has been observed, though rarely, in New York for many years, especially during the spring and fall. As far back as 1838, this species was observed in New York City (Cruickshank 1942). Thereafter it was found rarely in the southern areas of the state and very rarely farther north. A few summer observations were recorded during the 1960s and 1970s. The presence of a pair of Blue Grosbeaks on Staten Island in June

1981 was the first indication that breeding might have occurred. The following year nesting was documented at Heyerdahl's Hill, Staten Island (Siebenheller and Siebenheller 1982). Three eggs were laid and 2 young fledged. In 1983 a family group consisting of 4 birds was observed at Merritt's Island, Orange County (also near the New Jersey border), and in June 1984 in the same area, where a male Blue Grosbeak had previously been observed, a female was seen carrying food, "confirming" breeding again for the Atlas.

The call of this elusive and secretive species is similar to that of the Rose-breasted Grosbeak, although higher in pitch (F. and W. Abbott, v.r.). Observers reported that after initially locating the species, finding it again was difficult, if not impossible.

Bent (1968c) described its habitat, which is similar to that of the Indigo Bunting, as overgrown fields thick with brambles and dense shrubs along woods, streams, and roadsides. The Staten Island birds nested in a 6.1-ha (15-a) area of undeveloped land surrounded by deciduous woods. The area supported low growth, including primarily saplings and greenbrier (Siebenheller and Siebenheller 1982). The Orange County birds were found in a "weedy field adjacent to open woodland on one side with hedgerow between the field and alfalfa meadow on other side" (F. and W. Abbott, v.r.).

The Staten Island nest was "a neatly woven cup of dried grasses and weed stems" located approximately 81.3 cm (32 in) above the ground in a sweet-gum sapling (Siebenheller and Siebenheller 1982).

Janet R. Carroll

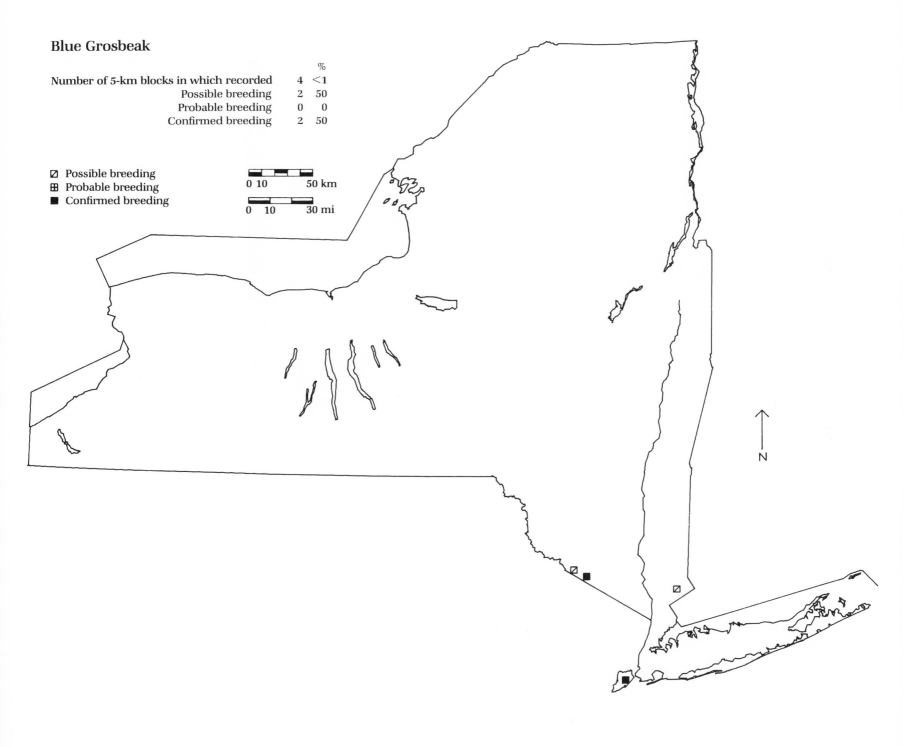

Blue Grosbeak

		%
Number of 5-km blocks in which recorded	4	<1
Possible breeding	2	50
Probable breeding	0	0
Confirmed breeding	2	50

☑ Possible breeding
⊞ Probable breeding
■ Confirmed breeding

0 10   50 km

0 10   30 mi

N

Indigo Bunting *Passerina cyanea*

The leisurely, finchlike, repetitive song of the Indigo Bunting is one of the most pleasant sounds of summer, and the singer is a tiny blue marvel. Fortunately, the Indigo Bunting is a widespread and common breeding species in New York.

Its early status is best summarized by Wells (1958), who wrote: "Perhaps originally a bird of successional vegetation within the Eastern Deciduous Forest . . . and of oak openings along the prairie ecotone, the Indigo Bunting was undoubtedly restricted in numbers by the relatively closed canopy of the climax forest. . . . In the East, the opening of the forest canopy by agriculture, logging and burning, and the western grasslands by planting of trees, coupled with the cessation of burning, converted great areas into potential Indigo Bunting habitat. This species has apparently responded . . . with a greatly increased population and extension of range." Bull (1974) noted the slow increase since Eaton's day, but by then the species was already far more abundant than in the historical past, and certainly more so than in precolonial centuries.

The distribution pattern today indicates that the Indigo Bunting is absent mainly from totally urban or industrial areas and at the higher elevations of the Catskills, Central Tug Hill, and Adirondacks. Even in those areas the Indigo Bunting is making inroads. Eaton (1914) stated that it was not found breeding in the Canadian Zone; Parkes (1952b) described its increase at higher elevations, apparently beginning no earlier than 1947. There have been environmental changes that negatively affect the Indigo Bunting, particularly on Long Island. Suburban sprawl, shopping malls, and road and highway expansion all reduce habitat. The Indigo Bunting was once a common summer resident on Long Island (Chapman 1906), but now it is rather sparsely represented there. Statewide it still prospers.

With its distinctive song and its habit of singing from exposed perches, as well as its startling plumage, the Indigo Bunting is a conspicuous bird. The male does not tolerate any intruders in his territory and makes a decidedly loud chip note and flies from perch to perch until the disturbing presence departs. "Confirming" the species was fairly difficult and most often done by observing adults carrying food and feeding young or by sighting fledglings.

Studies by numerous authors indicate that the Indigo Bunting has an almost infinite variety of favored habitats. It is found in forest clearing and edge locations, in very open second-growth woodland, and on brushy hillsides, but its preference for open landscapes with dense cover for nesting and tall trees for song perches is supported by studies summarized in Taber and Johnston (1968). In five of 14 Breeding Bird Censuses reported for New York (Van Velzen and Van Velzen 1983), the Indigo Bunting was breeding at an average density of 1 pair per 7.3 ha (18 a). In that same year a 6-year-old deciduous clear-cut in Virginia had 1 pair per 1 ha (2.5 a), and an old field with meadow and thickets in Pennsylvania 1 pair per 3.3 ha (8.3 a).

Nests are concealed in thickets, tangles, or shrubbery of a wide variety of plant species, from 0.6–4.6 m (2–15 ft) high. Construction of the typical cup nest is normally by the female; it is carefully woven of dried grasses, dead leaves, bark strips, and other available material. The cup is lined with finer grasses, feathers, and animal hairs. The nests are small and neat, about 9.7 cm (5.1 in) in outer diameter, 5.8 cm (2.3 in) less in inner dimension, and 5 cm (2 in) deep. Incubation is by the female. The female also does almost all of the feeding and nest sanitation. This is one species in which juveniles of previous broods (or possibly the previous year) have been observed as "helpers at the nest" (Skutch 1961). The species is a common victim of nest parasitism by the Brown-headed Cowbird (Friedmann 1929).

Robert Arbib

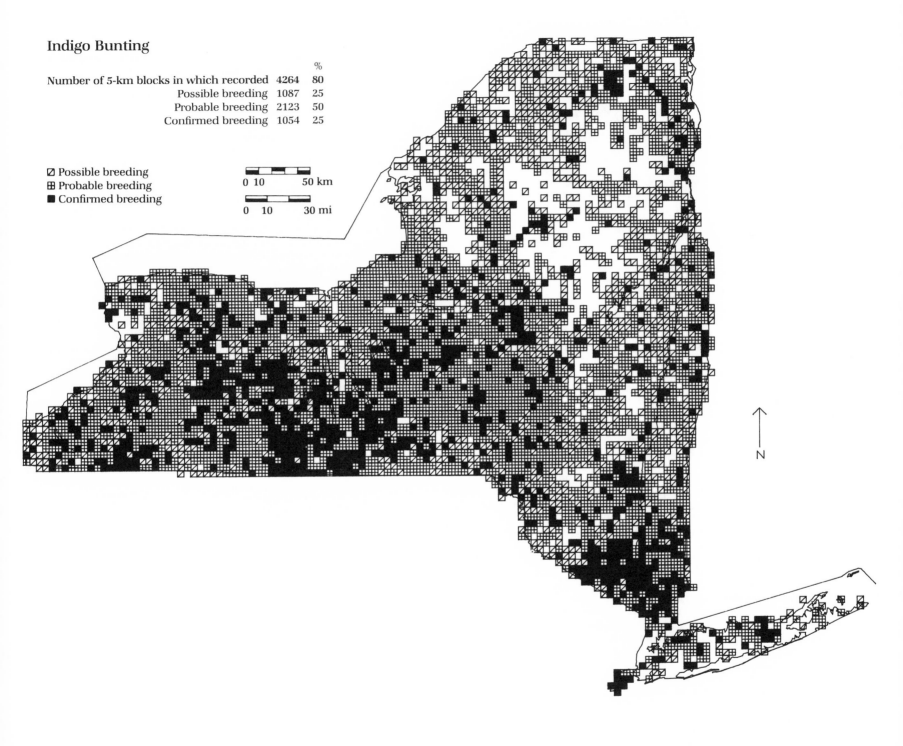

Indigo Bunting

		%
Number of 5-km blocks in which recorded	4264	80
Possible breeding	1087	25
Probable breeding	2123	50
Confirmed breeding	1054	25

☑ Possible breeding
⊞ Probable breeding
■ Confirmed breeding

0 10   50 km

0 10   30 mi

N

Rufous-sided Towhee *Pipilo erythrophthalmus*

There is nothing very secretive about the well-tailored Rufous-sided Towhee. A bustling presence, it is a common breeding species found in all but the mountainous areas of northern New York. Its signature call can be heard in brushy areas thoughout the state, but especially on Long Island in the pitch pine barrens and barrier beach tangles. Many Long Islanders claim to be presiding over the world's capitol of the "Towhee Confederacy." In two BBSs in pitch pine–scrub oak barrens and pitch pine–oak forest of Long Island, the average density of towhees was 1.9 and 2.3 pairs per 0.4 ha (1 a) (Kemnitzer et al. 1982; Clinton et al. 1982). There is evidence that this species has slowly but steadily spread northward in the state, while in the eastern United States its numbers appear to be declining (Robbins et al. 1986).

Roosevelt and Minot (1929) and Saunders (1929a) could not find this species breeding in the northern Adirondacks. Eaton (1914) reported that it did "not enter the mountain areas except in river valleys and settlements," and Hyde (1939) stated that there was no evidence that as a breeding species it reached the northern boundary of the state anywhere between Rochester and Lake Champlain. Eaton (1914) and earlier authors categorized the Rufous-sided Towhee as common in southern New York, abundant on Long Island, but scarce farther north. Later authors reiterated this appraisal, among them Griscom (1923), Cruickshank (1942), and Bull (1974), the last noting the spread upstate where "it has been increasing in recent years." The population on Long Island, though still optimal in ideal habitat, must be declining as this habitat disappears under the onslaught of development. Undoubtedly, part of the recent spread has been because of the abandonment of cropland and pastureland, with the resulting increase in shrubland and thickets. These afford welcome, if transitory habitat: when they grow back to forest, the towhee disappears.

As the Atlas map indicates, the Rufous-sided Towhee is widely distributed throughout New York. It is missing only from heavily urbanized areas, although it still hangs on in some of New York City's least manicured, least people-impacted parks. It is also absent in some areas of the Great Lakes Plain and Mohawk Valley where fields are plowed road-to-road, leaving no brushy borders. There are areas, such as along the Great Lakes Plain, where this species is scarce in some seemingly ideal locations (Spahn, pers. comm.) At higher elevations in the Adirondacks and Central Tug Hill area it is found infrequently, and although it is obviously not a high mountain denizen, it has nested at 823 m (2700 ft) in Essex County (Carleton 1980).

The Rufous-sided Towhee is found in pastures and hillsides grown to brush, roadside thickets, woodland edges, and dry open interiors and clearings, and in hedgerows. This habitat must have dense brushy cover to be suitable for this species (DeGraaf et al. 1980).

A Rufous-sided Towhee nest 3.7 m (12 ft) up in a shagbark hickory has been recorded (Dickinson 1968), but most nests are on or close to the ground under a bush or brush pile or clump of grass. Constructed by the female in about 5 days, it is a cup of leaves, stems, strips of bark and grass, lined with finer materials and is often partially roofed over and misshapen. It is so well camouflaged that the incubating female will often sit tight until almost stepped on. The female incubates, rarely assisted by the male, but he attends and feeds her at the nest (Forbush 1929; Dickinson 1968; Terres 1980). This species is a common victim of nest parasitism by the Brown-headed Cowbird (Bull 1974). Cowbirds are better nest-finders than Atlasers are; less than 3% of all Atlas records were of nests. The low score can be blamed on the beautiful concealment of the nest, the cryptic coloration of the female, and also the Atlaser's understandable distaste for bushwacking through greenbrier, bramble, and poison ivy. Even so, distraction displays were recorded in 57 blocks.

Robert Arbib

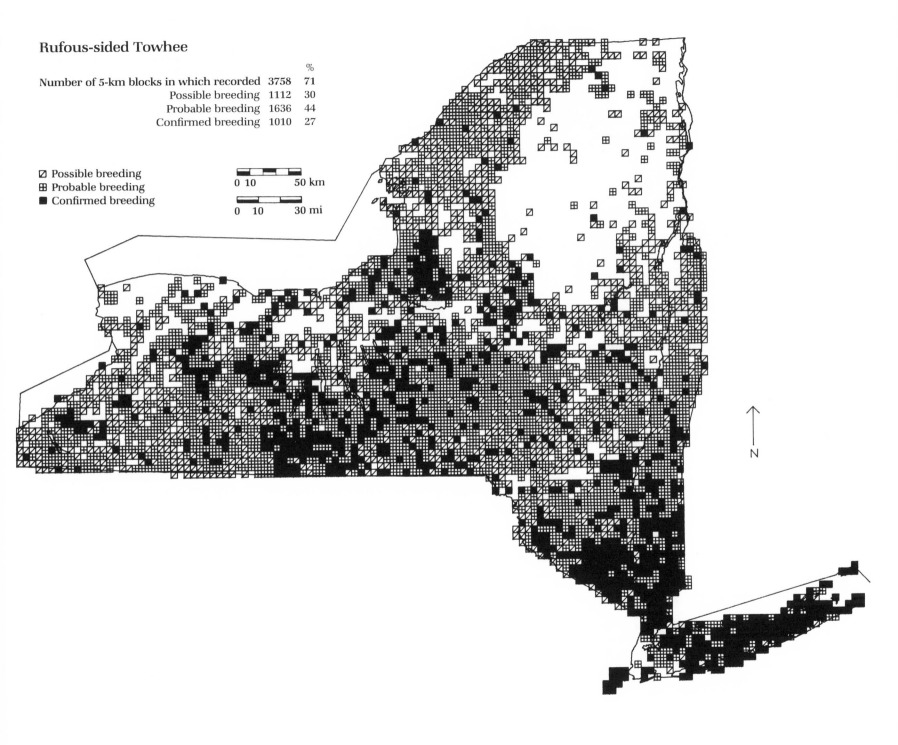

Rufous-sided Towhee

Number of 5-km blocks in which recorded	3758	71
Possible breeding	1112	30
Probable breeding	1636	44
Confirmed breeding	1010	27

%

☑ Possible breeding
⊞ Probable breeding
■ Confirmed breeding

0 10   50 km

0 10   30 mi

N

Chipping Sparrow *Spizella passerina*

A diminutive red-capped forager of lawn, garden, and field, the Chipping Sparrow is a familiar bird in New York. There is no doubt, however, that before the arrival of colonial settlers, the Chipping Sparrow was a much less common breeding species than it is today (Reilly 1964). "When man appeared on the scene and began to make clearings for his villages, he created open areas which the birds quickly occupied. The axes of European settlers made a hundred clearings where the aborigines had one, and undoubtedly the chipping sparrow population of today is many times that in pre-Columbian North America" (Stull 1968).

Audubon and Chevalier (1840–44) considered this sparrow to be one of the most common and widespread of breeding birds. Nearly all references in New York ornithological literature refer to it as common, very common, or abundant. However, early in the 20th century serious declines were noted in other states (Griscom 1949; Stull 1968) and blamed on Brown-headed Cowbird parasitism or House Sparrow competition.

There is evidence of breeding for the Chipping Sparrow in every county and in no less than 91% of the Atlas blocks. The Atlas map reveals an absence of records only in New York's northern forests, where the habitat is apparently not appropriate, and in New York City and southwestern Long Island, where the amount of open space is limited. Apparently there is a minimum size of open area required in its territory, which varies from 0.3 ha (0.7 a) or even less to about 0.6 ha (1.5 a) (Walkinshaw 1944; Sutton 1960). The smaller plots typical of recent suburban subdivisions are apparently not suitable for this sparrow.

The Chipping Sparrow's persistent trilled song makes it one of the most conspicuous birds of the summer scene, a species not often missed by Atlas observers. The noisy fledglings were an easy "confirmed" record for a block, especially since this sparrow tends to nest in areas of human habitation.

Although in earlier times its chosen habitats were forest clearings, open woodlands, and the edges of woodlands along watercourses, today it can be found around farms, gardens, suburban yards, orchards, city parks, estates, and settled rural areas. Golf courses, cemeteries, roadsides with grass margins, and forest clearings are also typical habitat (Stull 1968). The Chipping Sparrow was found atop Slide Mountain in Ulster County, at 1281 m (4204 ft) elevation, in 1965 and 1966 (Bull 1974).

The nest usually is in shrubs or trees in an open grassy area. The nest itself may actually be on the ground, but it is normally between 0.9 m (3 ft) and 6.1 m (20 ft) high. Nests have been found at heights reaching 17.7 m (56 ft) (Stull 1968). The nest is often placed at the end of a horizontal limb, with little regard for local disturbance. A nest found in Pelham Bay Park, Bronx County, in 1983 was out on a limb of a dogwood 3.1 m (10 ft) above a paved walk leading to the first tee of a golf course. Although hordes of noisy golfers passed under the tree every day, it was successful.

The future of this abundant and widespread breeding species seems assured in the state. It appears there will continue to be ample breeding habitat, and if increased timber cutting or recreational development takes place in the Adirondacks, the Chipping Sparrow may yet fill in those unmarked blocks on the map.

Robert Arbib

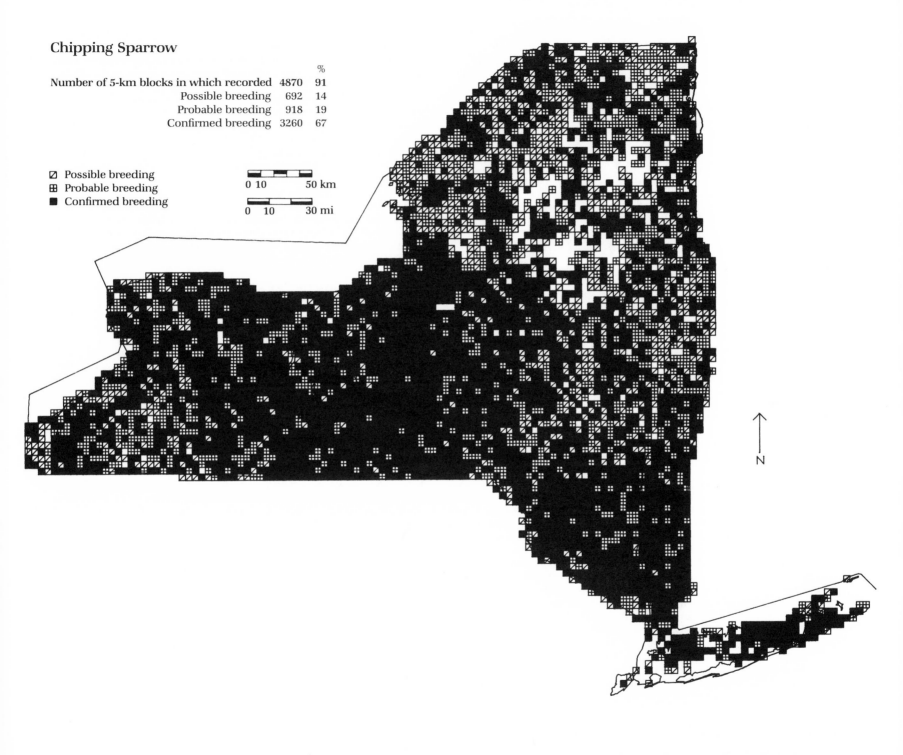

Chipping Sparrow

		%
Number of 5-km blocks in which recorded	4870	91
Possible breeding	692	14
Probable breeding	918	19
Confirmed breeding	3260	67

◨ Possible breeding
⊞ Probable breeding
■ Confirmed breeding

0 10   50 km

0 10   30 mi

N

© DAS 1987

Clay-colored Sparrow *Spizella pallida*

The Clay-colored Sparrow is a relative newcomer as a breeding species in New York. Since the 1970s it appears to have been increasing or, at least, occurring more regularly. Despite this apparent recent change, this sparrow remains a rare breeding species in New York, with a single "confirmed" record from a total of just 23 Atlas reports.

A specimen collected at Ithaca, Tompkins County, in 1935 represents the first record in New York. No other individuals were reported until a singing male was observed in Erie County in May 1943. Occasional spring and fall sightings were recorded during the 1950s; summer sightings were recorded in 1956 and 1959 (Bull 1974). In 1960 McIlroy documented an apparent mating between a male Clay-colored Sparrow and a female Chipping Sparrow at Ithaca, and a nest with eggs was found. Both birds were observed feeding nestlings for several days before the young disappeared, apparently taken by a predator (McIlroy 1961).

Although summer sightings were recorded at various locations in the next few years, the first documented successful breeding of this species in New York occurred in 1971 when a pair nested in a Scotch pine Christmas tree plantation in Alfred, Allegany County (Brooks 1971). A third breeding record was obtained the following year from the Rochester area, Monroe County, where a single adult was observed during July feeding 2 young birds (Claffey 1972).

The locations of the first state breeding records might suggest that the Clay-colored Sparrow arrived from the south, but the absence of Atlas records of this species from Ohio (Rice, pers. comm.) and Pennsylvania (Brauning, pers. comm.) undermines this theory. Instead, records from Ontario indicate that the species moved into New York on the heels of a range expansion into eastern Ontario. Breeding was documented in the Toronto area in 1950, near Hamilton, Ontario, in 1955, and in the Ottawa area in 1966 (Bull 1974).

The Atlas records in northern New York are from Jefferson, Lewis, and Franklin counties. These records probably reflect the expansion of birds from breeding populations in the Kingston and Ottawa, Ontario, areas. A cluster of 7 reports from eastern Allegany County indicates continued successful breeding in the Alfred area. The five remaining records, including the only "confirmed" record, come from five separate counties scattered across the state. One possible reason for these scattered reports is the species' ability to form pairs with Chipping Sparrows. In 1983 a male Clay-colored Sparrow was paired with a female Chipping Sparrow. A nest was located in July, but no eggs or fledglings were observed (N. Dill, v.r.). In 1984 a mixed pair was again present, and a nest with young was located (Dunham, pers. comm.).

The Clay-colored Sparrow's song is a series of 2 to 8 low, flat buzzes that have a remarkably nonbirdlike quality and may be easily mistaken for an insect or overlooked entirely. As more observers become familiar with this species, it may be found to be more widespread than Atlas records now indicate. Expansion from areas of current occupation is likely, and birders in northern and southwestern New York in particular should be on the lookout for this sparrow.

The Clay-colored Sparrow is found in uncultivated shrubby areas such as pastures, parks, cropland edges, woodland openings (Root 1968), grasslands, pine barrens, and conifer plantations (Harrison 1975). In New York it seems to require a site with scattered 1.5–6.1 m (5–20 ft) tall shrubs or trees interspersed with grassy, weedy openings. The majority of the Allegany County sightings are from an extensive area of young Christmas-tree plantations; several of the Jefferson County and Lewis County records are from broad sandy pine barrens on the Fort Drum Military Reservation. The remaining records are from inactive or lightly grazed pastureland with scattered low shrubs and trees, from successional old fields, and a single record from a Finger Lakes vineyard.

The nest of the Clay-colored Sparrow resembles that of its close relative the Chipping Sparrow but is not quite so compact. It is a cup-shaped structure of woven grasses lined with finer grasses, rootlets, or animal hair. The well-hidden nest is typically located in a low shrub or tree at a height from near ground level to 1.4 m (4.5 ft) (Root 1968). Both the 1983 and 1984 nests at Ghent, Columbia County, were typical Chipping Sparrow nests (N. Dill, v.r.).

Paul G. Novak

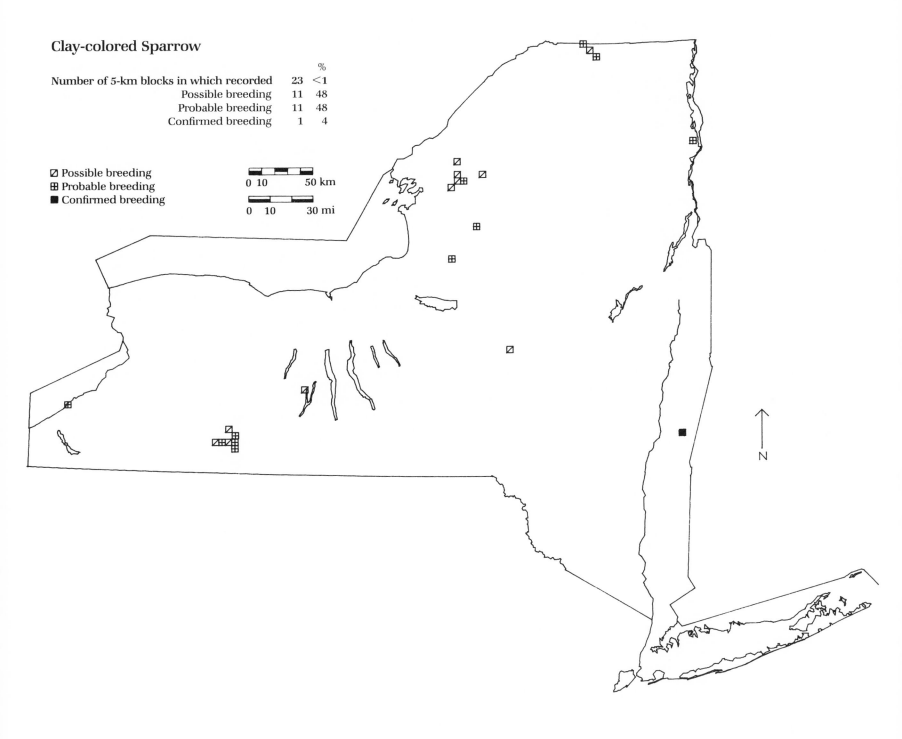

Clay-colored Sparrow

		%
Number of 5-km blocks in which recorded	**23**	**<1**
Possible breeding	11	48
Probable breeding	11	48
Confirmed breeding	1	4

◨ Possible breeding
⊞ Probable breeding
■ Confirmed breeding

0 10   50 km

0 10   30 mi

N

© DAS 1987

Field Sparrow *Spizella pusilla*

The clear, melodious, almost wistful song of the Field Sparrow floats across the brushy fields of summer. This sparrow is a very common and widespread breeding bird in New York. From the earliest commentaries on our native birds, it has been always been called common to abundant, and widespread in appropriate surroundings (De Kay 1844; Giraud 1844; Chapman 1906, 1927; Eaton 1914; Griscom 1923; Cruickshank 1942; Beardslee and Mitchell 1965; Bull 1974).

There seems little doubt, however, that it was not until the colonists arrived in the New World in the 17th and 18th centuries with axe, fire, and plow that this sparrow began to multiply and spread. The peak of abundance may have been in the late 19th century, since the 20th has seen brushy fields turn into shopping malls, rows of track houses, or woodland. Even so, this sparrow is still common.

As the map clearly shows, it is mainly at higher elevations and in heavily forested areas in the northern quadrant of the state where the Field Sparrow is absent. But even in this area and to the Canadian border it is found at lower elevations and occasionally in valleys where there are brush-dotted grassy slopes. The species turns up almost everywhere in southern New York except for Long Island, where its absence can be attributed to the annihilation of habitat by development, and metropolitan New York and scattered Catskill areas, where suitable habitat does not exist. There are other holes in its distribution, such as the areas east of Oneida Lake and along the Mohawk Valley, which are not easily explained, since the Field Sparrow is not difficult to locate by voice alone and can be readily "confirmed" by the sighting of adult birds carrying

food, or by the presence of fledglings being fed. It is "confirmed" in every county except New York (Manhattan) and Bronx.

The habitat of the Field Sparrow includes old fields, mowed orchards, overgrown pastures, and brushy roadsides. To attract this species, cleared areas must have grown back to brush, thickets, and tangles interspersed with an occasional clump of young trees. This bird is strongly territorial, and breeding densities have been calculated in various parts of the country from 3 pairs per 100 ha (7 pairs/100 a) in damp, deciduous scrub in Maryland to 32 pairs per 100 ha (79 pairs/100 a) in abandoned fields and 32 pairs per 100 ha (80 pairs/100 a) in orchards with unmowed ground cover (Stewart and Robbins 1958). Walkinshaw (1968) in upper Michigan found 1 pair per 1.2 ha (3 a).

The nest is placed in any one of a wide variety of small trees, shrubs, and even grasses; it is constructed by the female. First nests take 3–7 days to build; however, second nests are completed in only 2–3 days, suggesting that there is greater urgency later in the season. During the early part of the season the typical cup-shaped nest is found at any height, from on the ground to 31 cm (11.4 in), averaging 7.5 cm (2.9 in). The average is about 21.4 cm (8.4 in) in June and 30.5 cm (12 in) in July. This increase in height of nest placement may be a response to vegetational growth (Walkinshaw 1968).

The nest is woven of grass and weed roots and stems; it is lined with finer grasses and horsehairs when available. Females do the incubating, which begins the night before the last egg is laid. The male sometimes feeds its mate at the nest. Both sexes feed the young and attend to nest sanitation. The Field Sparrow is double- or triple-brooded and persistently rebuilds or builds a new nest when interrupted or disturbed by storms, predators, the Brown-headed Cowbird, or by death of the female and remating of the male. Walkinshaw (1968) cited a pair that built 7 nests and laid 17 eggs during one season, none of which fledged.

Like its companion of brushy hillsides, the Indigo Bunting, the Field Sparrow is a most welcome feature of the New York landscape, and Chapman's (1906) words still ring true: "To be convinced of its rare beauty one need only hear it as the sun goes down and the hush of early evening is quieting the earth."

Robert Arbib

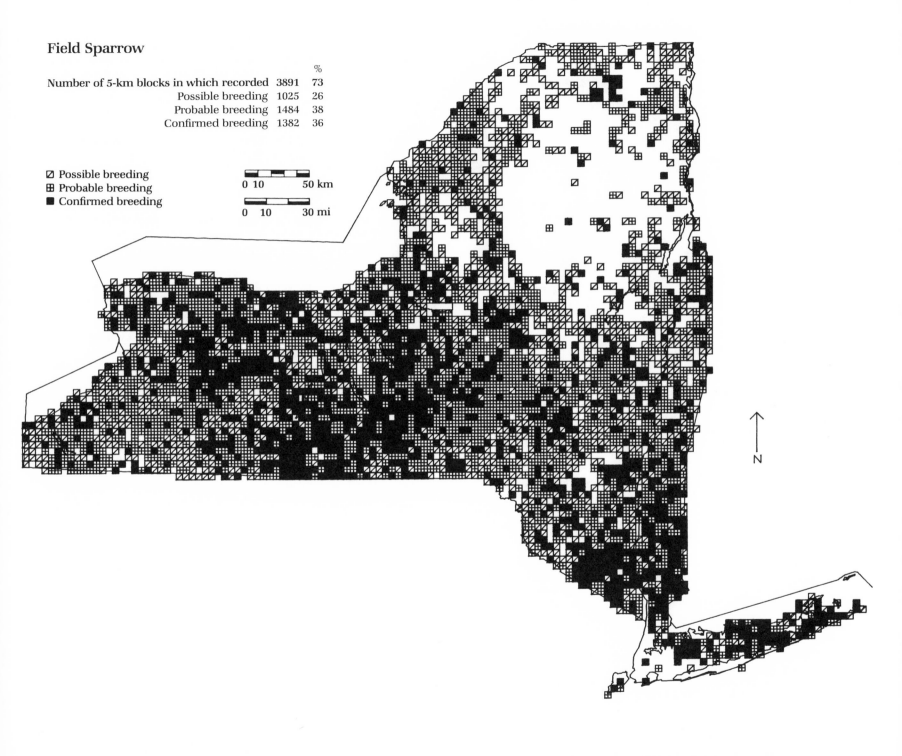

Field Sparrow

		%
Number of 5-km blocks in which recorded	3891	73
Possible breeding	1025	26
Probable breeding	1484	38
Confirmed breeding	1382	36

◩ Possible breeding
⊞ Probable breeding
■ Confirmed breeding

0 10   50 km

0 10   30 mi

N

©DAS 1987

Vesper Sparrow *Pooecetes gramineus*

The Vesper Sparrow, or bay-winged bunting as earlier ornithologists called it, is a typical bird of the farm country of central New York, where it is fairly common. It was listed as a Species of Special Concern by the NYSDEC in 1983 mainly because the loss of grassland habitat made its status uncertain. Robbins et al. (1986) found that the Vesper Sparrow's range appeared to be decreasing to a significant degree in the East and generally elsewhere.

Eaton (1901) called this sparrow abundant in western New York and in his 1914 publication stated that it was a common summer resident of all portions of the state. The Vesper Sparrow was much more common when small farms were in their heyday. Eaton (1914) stated that "on clear evenings in May and early June . . . the song of the Vesper Sparrow may often be heard in a dozen different directions at the same time." That is no longer true.

Bull (1964), writing of the birds of the New York City area, said, "Of all the breeding birds that have suffered because of rapid decline in agriculture in our area, the Vesper Sparrow has decreased the most as it is more dependent on farming than the other open country species are." He found it only in Orange County and on the extreme eastern end of Long Island. Although the once vast Hempstead Plains grassland in Nassau County had been a favored breeding ground in the early 1900's and before, Bull (1974) called it widespread in the state, but "rare and local on Long Island."

The decline of the Vesper Sparrow in New York is well documented. Clauser (1980) showed a rapid decline in numbers of Vesper Sparrows on BBS routes in New York from 1966 to 1978. The decline was steady and continuous, decreasing from 300 individuals reported in 1968 to 50 in 1974, when a leveling off occurred. Klabunde (1979), in a summary of 19 BBS routes in western New York, showed a decline from 122 individuals reported in 1967 to 15 in 1979. Scheider (1977) said that in the Oneida Lake area, grassland species, including the Vesper Sparrow, continued to decline as corn replaced hay fields. Treacy (1976) said that because of the species' scarcity, records were being kept in Dutchess County of every Vesper Sparrow; in July 1976 only four reports were made.

Atlas records continue to document a shrinking range, particularly in the Niagara Frontier region and the lower Hudson Valley. Although the species was once generally distributed in the Oneida Lake area, it has become more local. Atlas observers found it concentrated in the Central Appalachians, the Finger Lakes Highlands, and the Erie-Ontario Plain from Orleans County east to Oswego County. The Vesper Sparrow was reported in fewer blocks in the St. Lawrence and Champlain transitions and in the Lake Champlain Valley, the Mohawk Valley, the central Hudson Valley, Taconic and Hudson highlands, and eastern Long Island in Suffolk County. It was absent in the Allegany Hills and the Adirondack and Catskill ecozones, where the land is heavily forested. It did, however, extend into the Adirondacks along some of the larger river systems, including the Saranac and Black rivers.

Almost entirely a ground bird, the Vesper Sparrow inhabits open, grassy fields, preferring pastureland and cropland, either with row crops or field crops but with sparse cover of weeds and grasses. Usually found in drier sites than those inhabited by the Savannah Sparrow, it frequents the plowed fields and dusty roadsides preferred by the Horned Lark (Eaton 1914). The territory of this species is larger than that of most other grassland sparrows, ranging from 0.5 to 1.1 ha (1.2–2.7 a) (Berger 1968; Wiens 1969).

The nest is usually placed in a depression on the ground under cover of surrounding plants, or in a tussock of grass. It is built of dry grasses, weed stalks, and rootlets, and lined with finer grasses, rootlets, and occasionally hair (Harrison 1975). Atlas observers had difficulty locating nests; only 18 were found.

Stephen W. Eaton

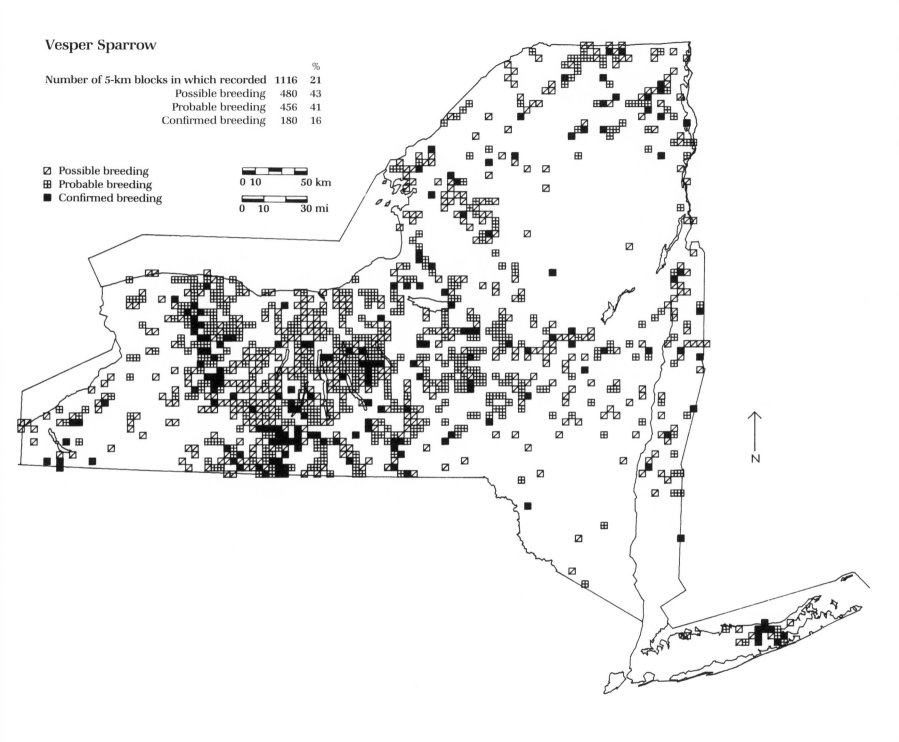

Vesper Sparrow

		%
Number of 5-km blocks in which recorded	1116	21
Possible breeding	480	43
Probable breeding	456	41
Confirmed breeding	180	16

◩ Possible breeding
⊞ Probable breeding
■ Confirmed breeding

0 10   50 km

0 10   30 mi

N

© DAS 1987

Savannah Sparrow *Passerculus sandwichensis*

A common breeder in open grasslands across most of the state, the Savannah Sparrow is rare in the forested mountains except where agriculture has provided suitable openings. It is a bird of mostly lower elevations, nesting from the seaside dunes to as high as 610 m (2000 ft) in open fields near the Olympic Bobsled Run at Lake Placid, Essex County.

The Savannah Sparrow, like many other grassland species, undoubtedly benefited from the growth of agriculture in the 18th and 19th centuries, especially from dairying, which requires large hay fields. As marginal farmland has been abandoned during this century and as more cropland is being used to produce corn rather than grasses, habitat suitable for the Savannah Sparrows has been slowly lost (Richmond and Nicholson 1985). A more rapid loss is caused by urban sprawl, with the big open fields favored by Savannah Sparrows converted to housing developments and industrial parks. At the turn of the century thousands of acres of grassland stretched unbroken across Nassau County. Of this area, near today's Garden City, Bull (1974) noted, "The once-large breeding population on the former Hempstead Plains was wiped out by destruction of that area after World War II." Caught between the pressures of a slowing farm economy and a growing demand for townhouses on outskirts of cities, the Savannah Sparrow may have seen the peak of its population in New York. Nevertheless, Eaton's (1914) statement that "throughout the interior of the State it is less common as a breeding species than the Vesper sparrow" is obviously no longer true. The two occupy similar ranges today, but the Savannah Sparrow is more abundant. Eaton (1914) is still correct, however, in stating that this bird is "decidedly more common and more generally distributed than the Grasshopper sparrow."

The Atlas map shows the main range of the Savannah Sparrow extending across the Great Lakes Plain and Appalachian Plateau eastward to the Mohawk Valley and northward onto the St. Lawrence and Malone plains. Savannah Sparrow records are especially scarce in the Central Adirondacks and Adirondack Foothills and Transition, although it can be found nesting locally on large bogs, sometimes close to Lincoln's or Vesper Sparrows. Areas such as the Central Tug Hill, Rensselaer Hills, Hudson Highlands, Manhattan Hills, Triassic Lowlands, Mongaup Hills, Delaware Hills, Catskill Peaks, and Allegany Hills are also largely devoid of this sparrow. In the Coastal Lowlands this bird is generally found near the shoreline and is absent inland.

The majority of Atlas records were provided by singing males; the distinctive song of the Savannah Sparrow, a lazy *tsit-tsit-tsit, tseeee-tsaaay,* wafts across the fields, dunes, and bogs until late summer. The alarm note, a sharp *tsip,* is ventriloquial, and when combined with the mouse-like tendency of the Savannah Sparrow to scurry through the grass, probably serves to confuse predators.

The Savannah Sparrow occupies strikingly different habitats in different parts of the state. It is usually considered a bird of low-lying open fields of grass with scattered forbs (Wiens 1969) but I have found it in fields of alfalfa, clover, and trefoil. On Long Island it favors maritime dunes and areas of sand fill covered with beach grass or weeds (Bull 1974). In boreal areas of the Adirondacks I have discovered it nesting on large, vast sphagnum bogs and singing from stunted black spruces. Such catholicity of nesting sites is suggested by its range, from northern Alaska to Labrador, from north of Hudson Bay south to Guatemala, and south into Maryland and West Virginia on the eastern seaboard (AOU 1983).

Wintering as far south as the Cayman Islands, Cuba, and Mexico, the Savannah Sparrow begins to return to its New York breeding grounds in late March and early April (Bull 1974; AOU 1983). It establishes its territory somewhat centrally within the grassland habitat and builds its nest in a natural depression concealed by low vegetation, either partially domed or placed under overhanging grasses (Wiens 1969). The nest is usually woven of coarse grasses and lined with finer grasses, hairs, or rootlets (Baird 1968). Over 60 nests were located by Atlas observers.

John M. C. Peterson

446

Savannah Sparrow

		%
Number of 5-km blocks in which recorded	3005	56
Possible breeding	836	28
Probable breeding	1233	41
Confirmed breeding	936	31

◪ Possible breeding
⊞ Probable breeding
■ Confirmed breeding

0 10   50 km

0 10   30 mi

N

©DAS 1987

Grasshopper Sparrow *Ammodramus savannarum*

If it didn't have wings it might be called the little mouse of the grass-roots, for the Grasshopper Sparrow must have grasslands to creep through and to nest in. Although its required habitat has appreciably decreased in recent years in areas of urban and suburban development, the Grasshopper Sparrow is still fairly common to uncommon in areas that provide grassland habitat. This sparrow was listed on the *American Birds'* Blue List of declining species from 1971 to 1986, with declines specifically reported in upstate New York and Vermont (Tate 1986). It is listed as a Species of Special Concern by the NYSDEC.

The Grasshopper Sparrow must have been a rare and local breeding bird in precolonial centuries when so much more of the state was forested, and it was probably confined to such natural grasslands as the Hempstead Plains in Nassau County. With colonization of New York and subsequent forest clearing, additional habitat was created. But its occupation of the new cropland was apparently slow, possibly because its original population was small and local. As late as 1870 it was considered rare in central New York (Benton 1951b), but in this century it has been considered fairly common throughout the state, until recently. This species seems to have been losing ground, as abandoned farms return to successional old field and woodland, agricultural land is developed for industry and housing, and row crops have replaced grain crops (Richmond and Nicholson 1985).

The map of the known breeding range of this species in 1906 published by Eaton (1914) showed that it was largely confined to southeastern New York and Long Island, the Great Lakes Plain and the southern Appala-

chian Plateau from Steuben and Broome to Tompkins and Madison counties. The Atlas map indicates a greatly reduced presence in the Hudson Valley, western Long Island, and the metropolitan area, but a more widely scattered presence elsewhere. The major areas of concentration are the Central Appalachians and the Eastern Ontario and western Erie-Ontario plains. On Long Island a disjunct population remains in Suffolk County. I noted the destruction of its habitat in Nassau County from 1930 to 1970.

Both Eaton (1914) and Forbush (1929) considered 305 m (1000 ft) close to the altitudinal limit for breeding, although the Grasshopper Sparrow has been found as high as 1311 m (4300 ft) (Brooks 1944). Atlas records indicate, however, that there are a number of scattered records at elevations approaching 610 m (2000 ft).

This sparrow is not hard to locate during the breeding season because the male is highly territorial, singing from available perches around and within its territory. The Grasshopper Sparrow is difficult to "confirm" and the nest is "extremely hard to find" (Smith 1968). Atlas observers found only 12. Females have been noted in distraction displays near occupied nests, and 12 Atlas records were of distraction displays.

Grain cropland and pastureland are primary habitats of the Grasshopper Sparrow. It favors orchard grass, alfalfa, red clover, and Lespedeza, poverty-grass in successional old fields, and many other planted grasses, but it will desert when shrub cover reaches 35% or more (Smith 1968). The nests, typical woven cups mostly of grass stems with finer linings, are roofed over with an entrance on one side, well concealed at the base of a grass clump, often with the nest depression below ground surface level. Both sexes build the nest and both incubate (Forbush 1929). Major threats to Grasshopper Sparrow population in New York are the mowing of fields during nesting season, heavy application of pesticides by farmers, but above all change of the habitat either by development or by plant succession.

Robert Arbib

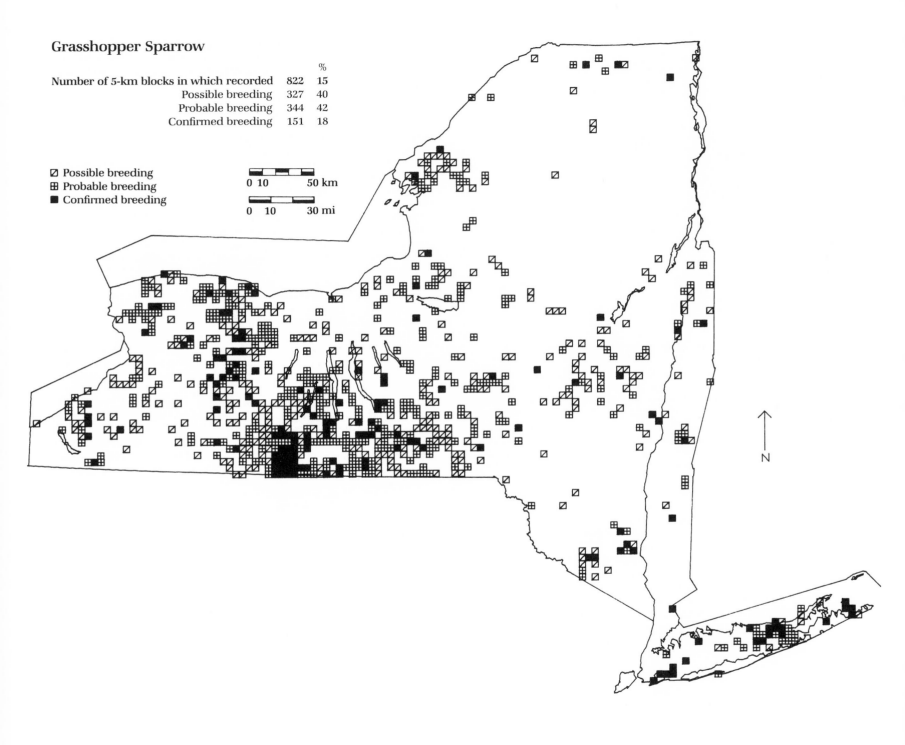

Grasshopper Sparrow

Number of 5-km blocks in which recorded	822	15 %
Possible breeding	327	40
Probable breeding	344	42
Confirmed breeding	151	18

Possible breeding
Probable breeding
Confirmed breeding

0 10   50 km

0 10   30 mi

N

© DAS 1987

Henslow's Sparrow *Ammodramus henslowii*

A specialist of fallow fields, the Henslow's Sparrow is more readily heard than seen and is one of the most inconspicuous land birds in the Northeast. It is rather uncommon, local in distribution, and appears to be decreasing in many areas. The Henslow's Sparrow is on the *American Birds'* list of Species of Local Concern (Tate 1986), and in New York it is listed as a Species of Special Concern.

In the early 1900s the Henslow's Sparrow was considered a local breeding species, rather uncommon or rare in all parts of the state (Eaton 1914). From the 1920s until the 1940s it appeared to be increasing. Two large colonies, each containing 20 pairs, were found in Albany County in 1925 and along the St. Lawrence River, St. Lawrence County, in 1939 (Bull 1974). Eaton (1953) wrote that by 1930 the Henslow's Sparrow was appearing at many new locations. In western New York increases were also documented (Stone 1933; Beardslee and Mitchell 1965). It appeared that this sparrow had moved into Long Island and the lower Hudson Valley from New Jersey and into central and western New York from Pennsylvania, apparently expanding along corridors of the Hudson, Delaware, and Susquehanna rivers (Graber 1968).

After 1950 this sparrow was reported less frequently and declines were noted. By 1952 there were no colonies found on the south shore of Long Island where it formerly bred, and decreases were noted in the lower Hudson Valley (Bull 1974). Several small colonies had been found in the Allegheny River Valley in the 1950s, but by the 1960s they were gone (Eaton 1981). In an examination of BBS records in New York for the period 1966–78, Clauser (1980) noted a rapid decline from a high count of 35 singing males in 1969 to a low of 5 in 1978.

Bull (1974) reported the Henslow's Sparrow to be local in central and western New York and in the highlands along the Pennsylvania border, and generally distributed along the Eastern Ontario Plains. The Atlas also documents an irregular and localized breeding distribution pattern. It was found in central and western New York along the Great Lakes Plain and Appalachian Plateau, especially in the Finger Lakes Highlands and Central Appalachians. It was reported rarely at the eastern end of the Appalachian Plateau and the Mohawk and Hudson valleys and was absent from the Coastal Lowlands. It had previously successfully colonized the St. Lawrence River Valley, especially in Jefferson County, but only a few records were reported in the Eastern Ontario and St. Lawrence plains by Atlas workers; these were directly across the St. Lawrence River from a population documented in Ontario (Ontario Breeding Bird Atlas, preliminary map). The Henslow's Sparrow is not too difficult to locate if the proper habitat and song are known, but many observers do not recognize the song, and the Atlas map probably is somewhat deficient.

Loss of grassland habitat through abandonment of farms and a change to monoculture production of corn and alfalfa (Richmond and Nicholson 1985) are undoubtedly factors in the species' decline. In addition, this sparrow often nests in wet meadows adjacent to wetlands that have been and continue to be drained mainly for development, further decreasing its habitat. Its localized distribution pattern is probably a result of its specific habitat requirements. The hilltop farms of the Appalachian Plateau, for example, have for the most part been abandoned, and many colonies occur in old fields mainly at a stage in succession in which goldenrod predominates. As more woody vegetation invades, the sparrow leaves.

The Henslow's Sparrow is a grassland species occupying meadows or marshy areas (Graber 1968). It is found in loose colonies occurring in scattered concentrations, while apparently identical habitat nearby is unoccupied (Wiens 1969). In Broome County, A. Peterson (1983) found it inhabiting mainly large ungrazed pastureland of tall herbaceous plants with virtually no woody invasion, in a variety of moisture regimes, and with long, unbroken, panoramic views to the horizon of more than 10.3 km (6.4 mi). In Cattaraugus County it was found in fallow fields of orchard grass, timothy, sweet vernalgrass, and goldenrod scattered throughout (Eaton 1981). In Wyoming County it regularly occurs near and at the tops of some ridges that border 610 m (2000 ft) in altitude (Rosche 1967). Novak (pers. comm.) observed it in extensive weedy fields at Fort Drum Military Reservation, Jefferson County, and in hilltop sites in Steuben County. The well-concealed nest is placed on or near the ground often at or near the base of a thick clump of grass with its bottom 5.1–7.6 cm (2–3 in) above the ground. The grass often arches over the nest to form a partial roof, but some nests are attached to vertical stems of grass and herbs 15.2–50.8 cm (6–20 in) above the ground. The nest is made of grasses and some hair and is formed into a deep cup (Harrison 1975). It is very difficult to find; Atlas observers found only 8.

Stephen W. Eaton

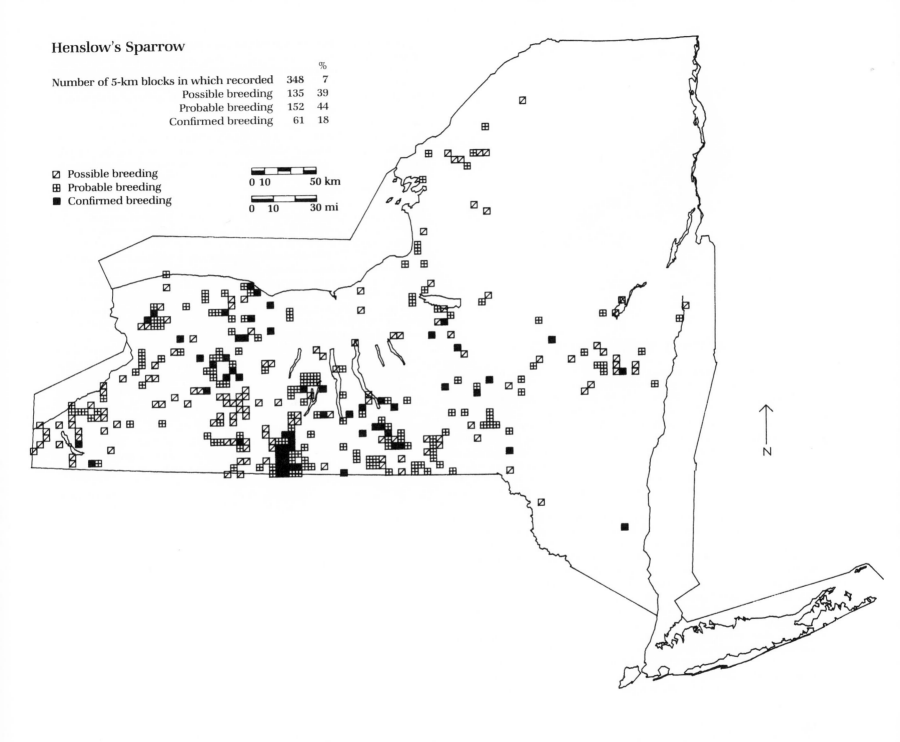

Henslow's Sparrow

		%
Number of 5-km blocks in which recorded	348	7
Possible breeding	135	39
Probable breeding	152	44
Confirmed breeding	61	18

◩ Possible breeding
⊞ Probable breeding
■ Confirmed breeding

0 10   50 km

0 10   30 mi

N

©DAS 1987

Sharp-tailed Sparrow *Ammodramus caudacutus*

A secretive, mouselike inhabitant of coastal salt marshes, the Sharp-tailed Sparrow is a common to locally abundant breeding species on Long Island (Bull 1974). Chapman (1927) noted that the Sharp-tailed Sparrow is "confined exclusively to the salt-water marshes of our coast, where it may be found in large numbers." It has a disjunct distribution along the Atlantic Coast as a result of the patchy distribution of its preferred salt-marsh habitat. Petersen (1983) stated that the Sharp-tailed Sparrow is a "semicolonial" inhabitant of coastal salt marshes. Inland populations occur in freshwater marshes. However, in New York the species is almost unknown away from salt marshes (Elliott 1962; Bull 1974).

The Sharp-tailed Sparrow was abundant only on the south shore of Long Island (Bull 1974), and it has been adversely affected by the destruction of marshes through drainage, filling, and development. Elliott (1962) stated that the loss of western Long Island salt-marsh habitat had an especially strong impact on the Sharp-tailed Sparrow, noting that a large colony was eliminated by marsh filling for what is now John F. Kennedy International Airport. This nesting colony, bordering Jamaica Bay, was estimated at over 200 individuals (Elliott 1962). Another notable historical breeding site was "the Piermont marshes in New York, 30 miles up the Hudson River from the Narrows, which formerly supported a colony of sharp-tailed sparrows that was apparently extirpated by pollution about 1930" (Hill 1968). Although Bull (1974) suggested that the species still nested there, Atlas observers did not find it at Piermont. Elliott (1962) stressed fragmentation of once continuous salt marshes as a factor in the

decrease of suitable nesting habitat for this species on western Long Island.

Atlasing has illustrated the restriction of the Sharp-tailed Sparrow's range to the Coastal Lowlands of Long Island; most "confirmed" breeding records were confined to the island's south shore. This agrees with L. Griscom's observation (Hill 1968) that "in a good marsh on the south shore of Long Island, for instance, Sharp-tails are ubiquitous and abundant." During Atlas work only a few nests were found; most "confirmed" breeding records were observations of fledglings or of adults feeding young. This species may have been missed because its song is not easy to hear and because it is often given while the bird perches near or on the ground (Elliott 1953; pers. obs.)

High salt marsh, characterized by salt-meadowgrass and spikegrass, is the typical breeding habitat of the Sharp-tailed Sparrow (Chapman 1927; Elliott 1962; Hill 1968; Bull 1974). It often occurs near the closely related Seaside Sparrow, with which it occasionally interbreeds (Hill 1968; Harrison 1975). Bull (1974) cited a breeding density of 13 pairs per 6.1 ha (15 a) of ditched salt marsh at Tobay Beach, Nassau County, noting that this density "appears to be the maximum recorded." Elliott (1953) suggested that patches of salt marsh 0.4–0.8 ha (1–2 a) in extent were too small to be used by breeding Sharp-tailed Sparrows. On Long Island nests are built in dryer areas on the surface of the marsh, sometimes in saltmarsh-elder and often in a clump of salt-meadowgrass (Elliott 1962).

The nest of the Sharp-tailed Sparrow is very difficult to locate because it is built in dense vegetation and because of the bird's habit of running along the ground, concealed by vegetation (Hill 1968; Harrison 1975). This behavior obscures the nest location and tends to lead the observer away from, rather than toward, the nest. Located either on or within a few inches of the surface of the marsh, nests are subject to destruction by unusually high tides. The nesting cycle of the Sharp-tailed Sparrow appears to be synchronized with spring tides, so that under normal conditions most broods fledge before monthly flooding of the marshes occurs (Hill 1968). The female builds the nest, which is constructed of loosely woven, dried grasses, or seaweed (Harrison 1975).

Richard A. Lent

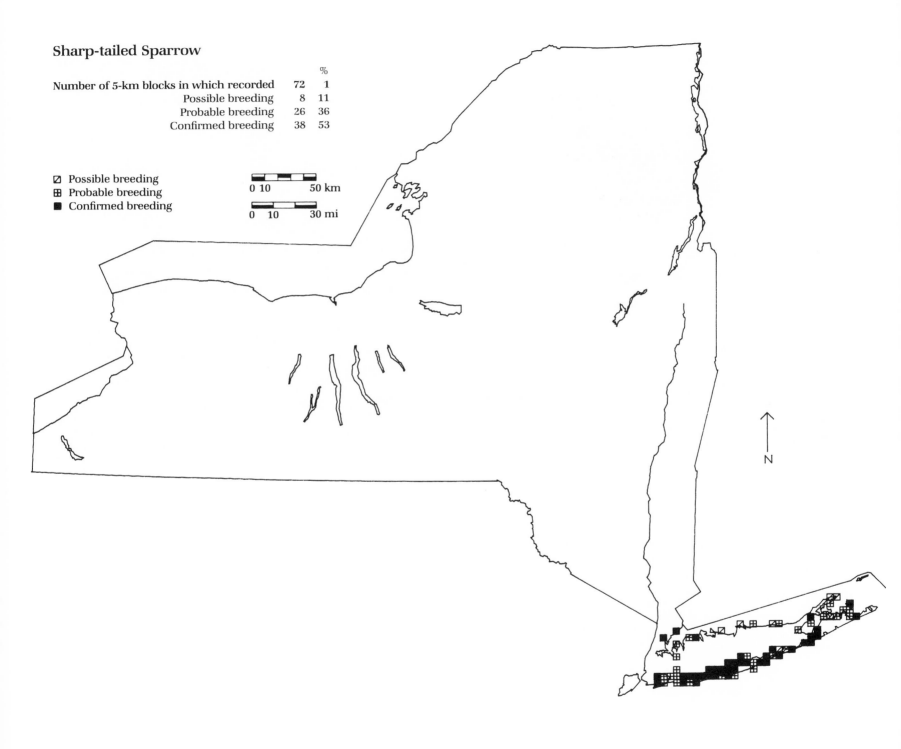

Sharp-tailed Sparrow

		%
Number of 5-km blocks in which recorded	72	1
Possible breeding	8	11
Probable breeding	26	36
Confirmed breeding	38	53

☑ Possible breeding
⊞ Probable breeding
■ Confirmed breeding

0 10   50 km

0 10   30 mi

N

©DAS 1987

Seaside Sparrow *Ammodramus maritimus*

Aptly named, the Seaside Sparrow inhabits tidal salt marshes, and its specific name *maritimus* coincidentally locates the species as a rare and extremely local breeder only in New York's maritime area. The Seaside Sparrow boasts one of the most restricted breeding ranges of any North American bird. It is found along the coastline where tidal salt marshes remain, from Massachusetts south to southeastern Texas, an area nearly 4827 km (3000 mi) long and normally only a few hundred meters wide (AOU 1983).

The species appears to have decreased in abundance in New York, and its distribution has become less widespread. Eaton (1914) spoke of it as abundant in the salt and brackish tidal marshes of Staten Island, Long Island, and the lower Hudson River north to Piermont, Rockland County. By the early 1940s this species was no longer found at Orient, Suffolk County, but still bred on the south shore of Staten Island, in the "few remaining tidal marshes on the eastern shore of Bronx and Westchester counties," and along the south shore of Long Island (Cruickshank 1942). Bull (1964) deemed it locally common to very common along the coast but rare and local at the extreme eastern end of Long Island and on the shore of Long Island Sound. He made no mention of breeding on Staten Island and indicated that there were no recent records at Piermont Marsh.

Presently the largest area of suitable habitat is along the north (Great South Bay) side of the outer barrier beaches on the south shore (Elliott 1962), which are still largely undisturbed. Of 48 Atlas records, only five are away from this area: one "confirmed" record on the Westchester County shore of Long Island Sound; two on the north shore of western Long Island; one on Fishers Island; and one on Gardiners Island. This species no longer breeds in the lower Hudson Valley or on Staten Island, where salt-marsh habitat is limited.

This is a dark, drab sparrow, painted with a somewhat softer brush, its somber mien lightened only by a touch of yellow in the lores. It could be difficult to locate. However, Atlas observers on Long Island have had experience with the Seaside Sparrow and recognize it by its rather "cheery, reedy performance" sung from a perch atop a shrub or stick, or on the wing during courtship (Elliot 1962). Within the marsh this sparrow seems to spend more time running than flying, using its large feet to scoot along the muddy substrate of the marsh. Normally the birder who flushes it will see it flutter away and then drop into the marsh grasses, never to be seen again. Because of this secretive behavior, diligent efforts are required to "confirm" this species. Over 60% of the records were "confirmed," most by observing fledglings or an adult with food for its young.

The preferred habitat is tidal wetlands with extensive stands of cordgrass and blackfoot rush, with saltmarsh-elder growing along the raised borders of mosquito-control ditches. The Seaside Sparrow prefers the wetter areas of the marsh (Elliott 1962), although it will nest, as it has at Jamaica Bay Wildlife Refuge, on dry ground adjacent to the marsh (Bull 1964). An apparent habitat anomaly occurred in New York, where, as early as 1906, a colony was noted in Piermont Marsh, where the Hudson River is tidal and brackish (Chapman 1906).

The Seaside Sparrow breeds in loose colonies with individual territories of 0.4–0.8 ha (1–2 a). The nest is well concealed in grass tussocks near the base of saltmarsh-elders or in the thicker stands of salt–meadow grass. Constructed by the female, it is a typical cup woven of grass stems, with finer grasses for the lining. Outside diameter of the cup is about 10.2 cm (4 in), with a depth of 3.8 cm (1.5 in). It is placed close to the ground but above the normal high-tide level, averaging 22.9–27.9 cm (9–11 in) above the muddy substrate (Terres 1980).

Robert Arbib

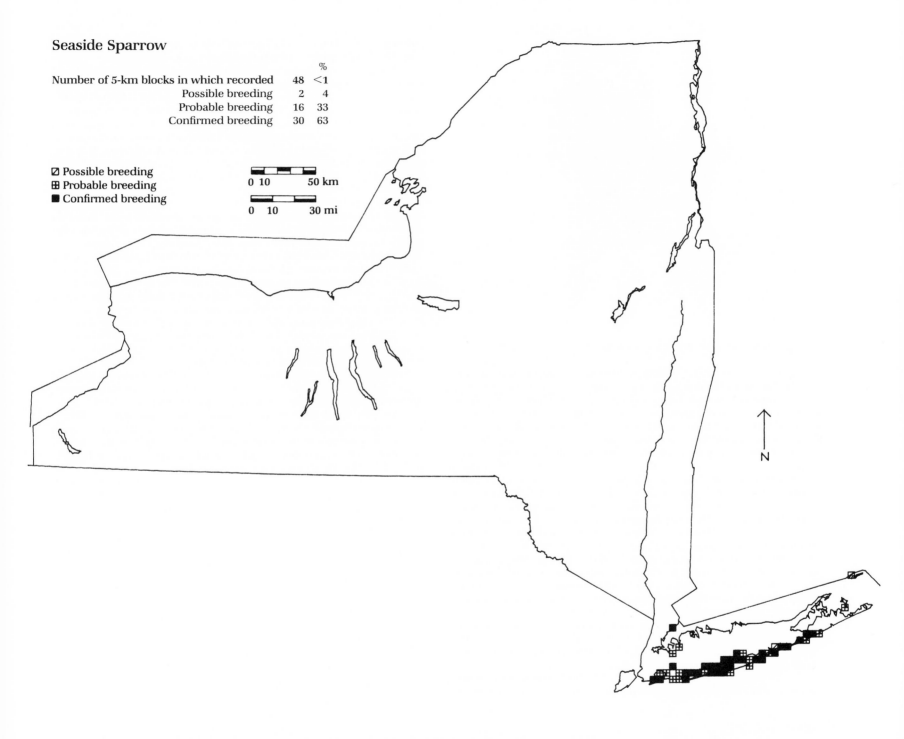

Seaside Sparrow

		%
Number of 5-km blocks in which recorded	48	<1
Possible breeding	2	4
Probable breeding	16	33
Confirmed breeding	30	63

☑ Possible breeding
⊞ Probable breeding
■ Confirmed breeding

0 10   50 km

0 10   30 mi

N

©DAS 1987

Song Sparrow *Melospiza melodia*

Nearly every birder knows the Song Sparrow, and no wonder. This thicket-loving bird is one of the most widespread breeding species in New York. It was recorded in more Atlas blocks than any other species.

Apparently the Song Sparrow always has been abundant and widespread in New York. Eaton (1914) stated that it was an abundant summer resident throughout the state and was fairly common even in the central portions of the Adirondack wilderness, except in the depths of the forest. Stoner (1932) said that it was an abundant summer resident throughout the Oneida Lake region; Saunders (1942) stated that it was a very common bird both in and near Allegany State Park, occurring abundantly into the mountain valleys; Cruickshank (1942) noted that it nested commonly throughout the New York City region in all but the very crowded sections and in dense woodlands; and Siebenheller (1981) said that the bird has been abundant on Staten Island for as long as records have been kept. The species' present abundance and widespread distribution results from its wide range of accepted habitats and its tolerance of civilization.

Examination of the Atlas map clearly shows its ubiquitous distribution. The Song Sparrow is a widespread breeder in every ecozone, and Atlas observers had no trouble locating it. It maintains its territory by singing during the entire breeding season; in a study done at Pymatuning Reservoir, Pennsylvania, the species continued to sing into mid-August (Mehner 1952). The Song Sparrow was "confirmed" most often by observations of fledglings. Also, distraction displays were recorded in 191 blocks. Typically, as an observer approached, the adults first gave a metallic warning call. Then, as the observer moved closer, one of the adults flew into the open and fluttered about, trying to lead the intruder away, while the other adult chipped continually from a nearby shrub.

The Song Sparrow breeds in a great variety of habitats from the interior mountains to the coastal beaches. It nests in gardens, city parks, suburban yards, and roadsides with numerous shrubs; in thickets, hedgerows, and fenceline vegetation in farm country; in woodland clearings; in thickets at edges of brooks, streams, lakes, marshes, and swamps; in brushy fields and bushy banks; and in thickets in coastal sand dunes. It has been found nesting in small woodland openings only a few meters in diameter and is absent only from deep forests, open fields, and marshes devoid of bushes (Eaton 1914; Bull 1974). The Song Sparrow requires an elevated perch for a song post in its breeding habitat (DeGraaf et al. 1980).

The nest is usually placed on the ground, securely hidden under a tussock of grass or a brush pile, or at the base of a bush. It also may nest off the ground in sedges, cat-tails, or low bushes, or even as high as 2.4 m (8 ft) or more in a small tree. Stoner (1932) suggested that elevated nest sites were more frequently chosen in the vicinity of human habitations than in wilderness situations, and Eaton (1914) noted that elevated nests tend to be those built for second or third broods. The bulky structure is built by the female and made of grass and weed stalks, rootlets, leaves, and bark strips, and lined with fine grasses, rootlets, and long hairs (Eaton 1914). Occasionally the nest is built in a cavity, especially hollows in old apple trees (Bent 1968e). When the young fledge, feeding of the brood is split between the male and the female, with each young being fed by only one adult (Smith and Merkt 1980).

The Song Sparrow is a frequent host of the Brown-headed Cowbird (Nice 1937), possibly because the Song Sparrow reacts aggressively when a cowbird approaches its territory, perhaps leading the cowbird to the nest location (Smith 1981). In their study area on Mandarte Island, British Columbia, Smith and Merkt (1980) found that cowbird fledging success was lower than that of the Song Sparrow host young. The low success was attributed to predation by the American Crow, which could easily find the boisterous, young food-begging cowbirds.

Richard E. Bonney, Jr.

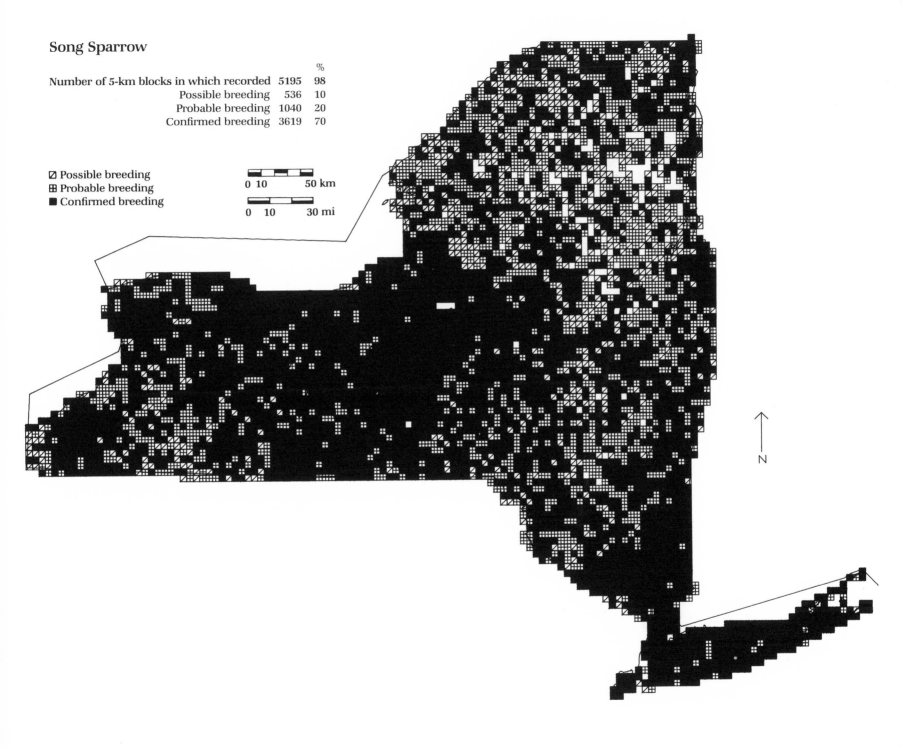

Song Sparrow

		%
Number of 5-km blocks in which recorded	5195	98
Possible breeding	536	10
Probable breeding	1040	20
Confirmed breeding	3619	70

◨ Possible breeding
⊞ Probable breeding
■ Confirmed breeding

0 10   50 km

0 10   30 mi

N

Lincoln's Sparrow *Melospiza lincolnii*

Following his Adirondack bird studies more than a half-century ago, Saunders (1929a) noted of the Lincoln's Sparrow: "This bird is rather rare and local in this region, but it is probably commoner than records would show, for it is little known and likely to be overlooked." Now found to be fairly common within the Adirondacks, this northern sparrow seems uncommon to rare in the outlying Adirondack and Champlain transitions and the Central Tug Hill; there it occurs only where suitable pockets of habitat are available.

Early ornithologists had varied success locating this rather secretive bird. Merriam collected the first state specimen of a breeding Lincoln's Sparrow at Locust Grove, Lewis County, 23 May 1873. He described this sparrow as a "regular summer resident, and apparently not very rare" a century ago (Merriam 1879, 1881). The first New York nest, containing 3 eggs, was found on 13 June 1878 at "Moose Lake"—now Little Moose Pond—in Hamilton County (Bagg 1878).

Following his visit in 1905 Eaton (1914) noted: "In the Adirondacks I found this bird very difficult to observe due to its shy, retiring habits. It was present, however, in the spruce and tamarack swamps of Essex County as well as in Hamilton and Herkimer Counties." At Cranberry Lake in 1916 Silloway (1923) noted that "a small colony of Lincoln's Sparrow inhabited the Bog, where perhaps four pairs were nesting, but it was not met with elsewhere," and he was the first to report seeing recently fledged young. Saunders (1929a) knew firsthand that the Lincoln's Sparrow could be overlooked: "I did not see it during 1925, but in 1926 found it first in the St. Regis region," later that summer finding others along the Heart Lake Road outside Lake Placid. Before the Atlas survey only 20 breeding localities in five Adirondack counties were known (Bull 1974).

The New York population constitutes an island—almost a peninsula tenuously extending southward along a thin neck from Quebec—that lies generally south of the main range, which stretches unbroken from western Alaska across boreal Canada to Newfoundland (AOU 1983). As the Atlas map indicates, within New York the range extends southward into the Champlain Transition and Malone Plain near the Canadian border, spreading south into the Central Adirondacks, Adirondack High Peaks, and Western Adirondack Foothills as far as southwestern Hamilton County. The Lincoln's Sparrow breeds as low as 259 m (850 ft) in bog habitat near The Gulf, a glacial ravine along the Canadian border in northern Clinton County, to as high as 853 m (2800 ft) along the shoreline of Flowed Land in the High Peaks wilderness area of Essex County. In hamlets bordered by forest, such as Onchiota, Franklin County, this sparrow may be found in openings with a scattered conifer edge (pers. obs.).

The Lincoln's Sparrow was not located in much of Warren County or the southeast section of the Central Adirondacks, where hardwoods prevail. Nineteenth-century clear-cuts have also removed much of the original softwood from large areas elsewhere in the Adirondacks, and broad-leaved trees have taken over. Forever wild Forest Preserve lands have continued to mature, and forest fires have been largely controlled. At the same time this sparrow has opportunistically moved into new niches along highways and on abandoned farms.

Atlasers' hearing the sweet song, much like that of the Purple Finch, accounted for the greatest number of records. Many of the other records were obtained by searching suitable habitat, where observers could elicit a response by "pishing."

The Lincoln's Sparrow nests in varied locations: bogs and fens, old clearings, and abandoned fields, roadsides, burns, and the banks of lakes, streams, and ponds (DeGraaf et al. 1980). It is a sparrow of fairly open transition zones marked by low and scattered conifers, generally black spruce and tamarack, often mixed. Erskine (1977) indicated that, in Canada, this must have originally been a bird largely restricted to bog habitats, burns, or other natural openings. When farms were abandoned, as in northern New York, scattered young conifers sometimes took over the early upland succession. Lincoln's Sparrow took advantage of this niche.

The ground nest of Lincoln's Sparrow is a well-hidden cup of dry grasses and dead leaves, lined with finer materials and measuring about 8.9 cm (3.5 in) outside diameter, 5.8 cm (2.3 in) inside, and 3.3 cm (1.3 in) deep (Speirs and Speirs 1968). Only 7 New York nests with eggs have been found since 1878 (Bull 1974), 2 of which were recorded by Atlas observers.

John M. C. Peterson

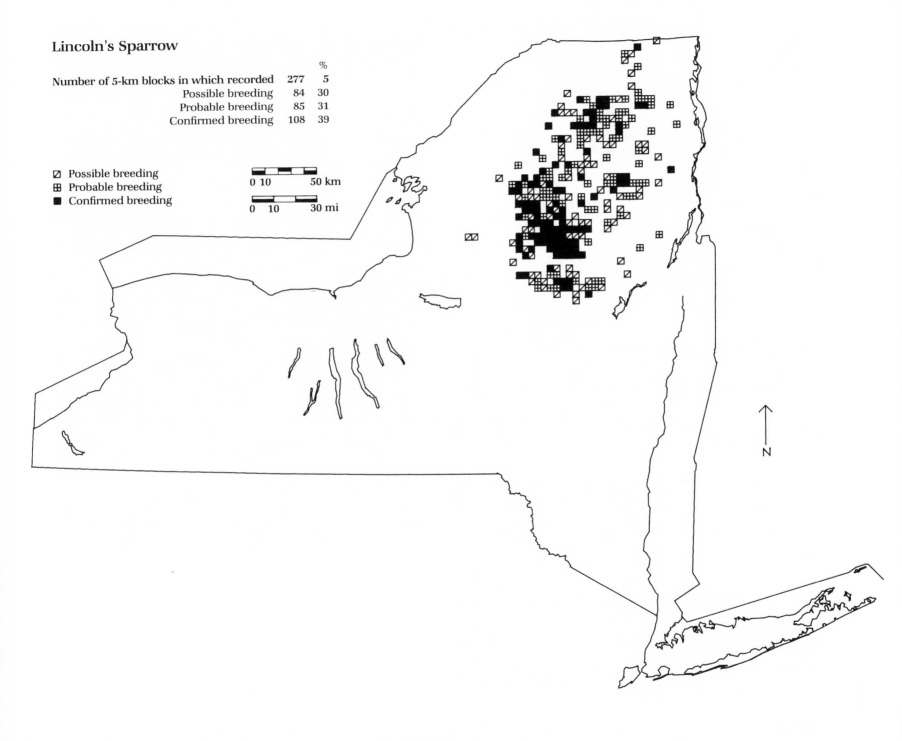

Lincoln's Sparrow

		%
Number of 5-km blocks in which recorded	277	5
Possible breeding	84	30
Probable breeding	85	31
Confirmed breeding	108	39

◪ Possible breeding
⊞ Probable breeding
◼ Confirmed breeding

0 10   50 km

0 10   30 mi

N

©DAS 1987

Swamp Sparrow *Melospiza georgiana*

The Swamp Sparrow has always been considered a common, if elusive and local, breeding bird in New York. Perhaps because of its secretive habits and often inaccessible breeding habitat, the literature is rather sparse on enthusiastic comment on this marshland species. Wilson and Bonaparte (1876) actually dismissed the Swamp Sparrow with these imperious words, "The history of this obscure and humble species is short and uninteresting."

Other ornithologists noted its presence in New York with varying opinions. De Kay (1844) called it abundant in the state, particularly in the western district. Chapman (1894) found it an abundant summer resident in the metropolitan areas. Eaton (1914) found it common in the Montezuma marshes, also around Lakes Erie and Ontario, where it was fairly abundant in the flooded portions of swamps in areas of dense vegetation. Eaton warned that the draining of swamps and marshes would restrict the habitat of the species and diminish its numbers. Historically, "while the creation of mill ponds by artificial damming during the 19th century probably increased the range and density of the Swamp Sparrow the draining of morasses for housing developments is currently reducing its habitat markedly" (Wetherbee 1968). Wetland destruction in more recent years is even greater. The species' decrease on Long Island, for example, is the end result of 20th-century marsh drainage, landfill, and stream channelization. Conversely, the slow return of the beaver, after centuries of exploitation and decline, must be providing new habitats.

As indicated on the map, this species is a widespread but somewhat local breeder in the state, absent where suitable wetland habitat is un-available. Strangely, its distribution does not seem to follow the major waterways of the state. It is most obviously absent from the wetland-deprived farmland, the higher mountainous regions, and urban areas. Breeding is "confirmed" in every county except New York and Nassau. Although the bird is easy to hear in appropriate habitat and responds readily to pishing, locating nests in the dense vegetation of a marsh is difficult; most "confirmed" records were of adults bringing food to young. Many of the "probable" records represent singing males in the breeding season somewhere out on an inaccessible marsh; the chances are excellent that they are breeding.

This sparrow's habitat includes the margins of freshwater marshes, grassy streamsides, reedbeds, large and small riverine areas, and less frequently the edges of salt marsh or salt meadow. A typical breeding site is the Basha Kill marsh, a flooded valley west of the Shawangunk ridge in Sullivan County, where hundreds of acres of dense freshwater marsh vegetation—grass, sedge, rush, flag—are home to such species as Pied-billed Grebe; both bitterns; American Black Duck and Mallard; Common Moorhen; Virginia Rail; and Sora. Here the trilled songs of the Swamp Sparrow ring out across the vegetation from early spring onward.

The Swamp Sparrow builds a loosely woven structure almost entirely of grasses that are finer in the inner lining. It is normally placed close to the ground or water substrate, in grass tussocks, or in reeds or cat-tails supported by stronger stalks. Sometimes it is partially or loosely roofed, with an entrance on the side. Nests are almost never on the ground, but if over water are at least 0.3 m (1 ft) above the substrate (Sutton 1928). The cup itself averages 10.2 cm (4 in) in outer diameter, with an inside diameter of 6.1 cm (2.4 in) and a depth of 3.8 cm (1.5 in). The female builds the nest, with very little or no help from her mate. The species is not usually colonial, but favorable habitat may result in local high densities of as many as 1 pair per acre (Bull 1964). Since nesting is normally just a few inches above the water level, nesting success is at risk from unexpected high water.

Where they coexist, the Swamp Sparrow is a frequent victim of nest parasitism by the Brown-headed Cowbird, with reports of from 10 to 80% of Swamp Sparrow nests parasitized (Berger 1951; Terrill 1961).

To this observer, the Swamp Sparrow is strikingly handsome: its browns redder, its grays richer, its markings unmistakable; its song is a lovely musical trill, often the only sound rising from an otherwise silent marsh. It will sing at night in spring, surely a welcome habit.

Robert Arbib

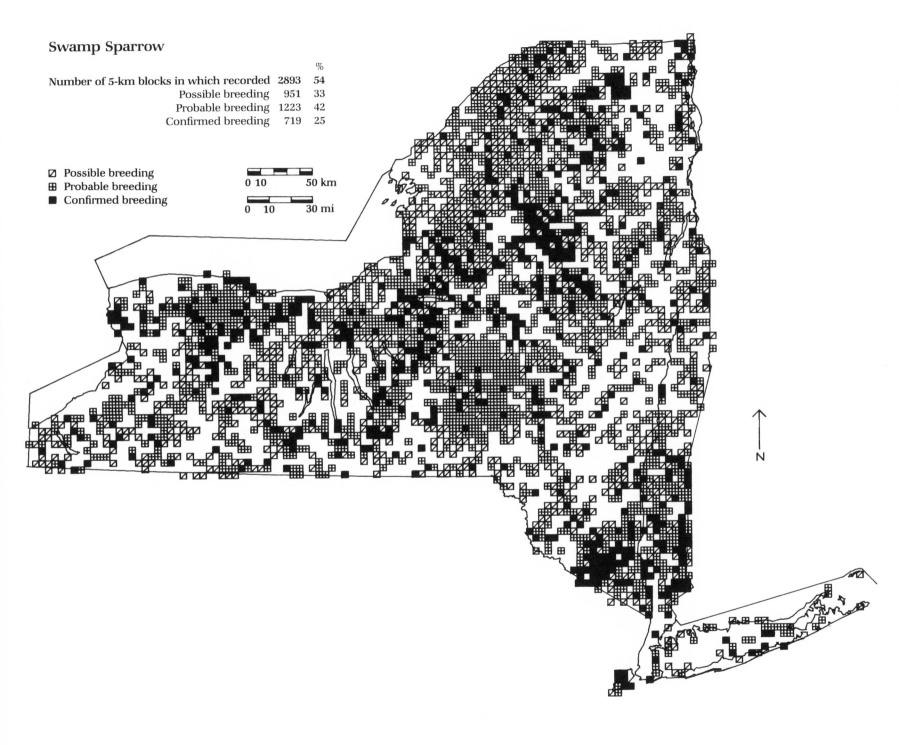

Swamp Sparrow

		%
Number of 5-km blocks in which recorded	2893	54
Possible breeding	951	33
Probable breeding	1223	42
Confirmed breeding	719	25

☒ Possible breeding
⊞ Probable breeding
■ Confirmed breeding

0 10   50 km

0 10   30 mi

N

©DAS 1987

White-throated Sparrow *Zonotrichia albicollis*

Familiar to many hikers and campers, the White-throated Sparrow is a common to abundant breeding bird in most of northern New York. Elsewhere it is most frequent at higher elevations in east-central New York. In recent decades its range in the state has expanded.

The White-throated Sparrow has been considered very numerous in the Adirondacks by authors from Merriam (1881) to Beehler (1978). Of the north central and High Peaks areas Saunders (1929a) wrote, "It is common throughout, from the lowest valleys almost to timberline." Eaton (1910, 1914) noted few breeding records for counties outside the Adirondack area but indicated that it was common in the Catskills. He stated that a few bred in southwestern New York, mentioned some central New York localities, but listed only Washington County east of the Hudson.

Bull (1974) documented a considerable range expansion for this species since 1950, especially into lower elevations. His range map showed records in central New York from western Jefferson County south to Broome County, and several others in eastern New York, with a scattering elsewhere. Robertson (1980) listed the White-throated Sparrow among 14 northern forest species recently expanding southward and to lower elevations in the East. Nevertheless, USFWS BBS data from 1966 to 1979 indicated a significant decrease in White-throated Sparrows in the northeastern states (Robbins et al. 1986), and low numbers have been reported locally in northern New York (Chamberlaine 1980).

The Atlas survey portrayed a broadly occupied zone from the Lake Champlain Valley west along the Champlain Transition and across the entire Adirondacks and Tug Hill plateau. This continuity is broken by open lowlands, especially in the Eastern Ontario Plains and Mohawk Valley southward, but favorable wooded habitat occurs in the Catskills and adjacent uplands west across the eastern Appalachian Plateau proper. The White-throated Sparrow's center of occurrence east of the Hudson River is in the thickly wooded, elevated Rensselaer Hills. Southernmost "confirmed" breeding is in southern Sullivan and Ulster counties. It was observed, but not "confirmed," on the Coastal Lowlands, mainly along the north shore of Long Island. The stronghold of this species continues to be in the Adirondacks, where only the Black-capped Chickadee was recorded in more blocks.

With its singular song and loud alarm note, the White-throated Sparrow was easily found by Atlas observers. The readily observed fledglings and adults carrying food permitted frequent "confirmed" records, but the inconspicuous nests were only occasionally found, requiring considerable time to locate. The White-throated Sparrow frequently sings in isolated and unlikely locations in summer, where breeding may or may not occur.

Essentially a bird of the vast boreal spruce-fir forest of Canada, in New York it is most numerous in mountain spruce–fir and spruce–fir–northern hardwood communities, but has increased in the widespread hemlock– and pine–northern hardwood forests. Bull (1974) listed extra-montane habitats, including northern white cedar, tamarack-alder, and white pine swamps; other New York habitats reported include assorted bogs, conifer plantations, lumber clearings, and overgrown pastureland. In general, the White-throated Sparrow prefers shrubbery, undergrowth, forest edge, and semi-open stands of trees. A most unusual habitat was urban downtown Buffalo, Erie County: in 1969 a pair fledged 2 young in dense shrubbery bordering the esplanade at the main library; in 1973, 2 adults, an empty nest on the ground in wood chips, and 3 fledged young were found in a shrubby area next to a cathedral (Andrle and Rew 1971; Andrle 1974).

The White-throated Sparrow is polymorphic, the two types or morphs differing most noticeably in the head striping. There are white-striped and tan-striped birds of both sexes, and almost all mated pairs consist of opposite color types, a form of selective mating apparently unique among birds (Lowther 1961). Remarkably, territories of white-striped males are in more open situations, whereas tan-striped males tend to be found in denser and more variable habitats (Knapton and Falls 1982).

The female builds the nest, which is almost always on the ground and well concealed in shrubs and ground vegetation. Many nests are under canopies—mats of dead bracken or other dried vegetation. Constructed of coarse grass, wood chips, twigs, pine needles, rootlets, and bark in the outer portion, it is lined mainly with fine grasses, rootlets, and deer hair (Lowther and Falls 1968; Harrison 1975).

Paul F. Connor

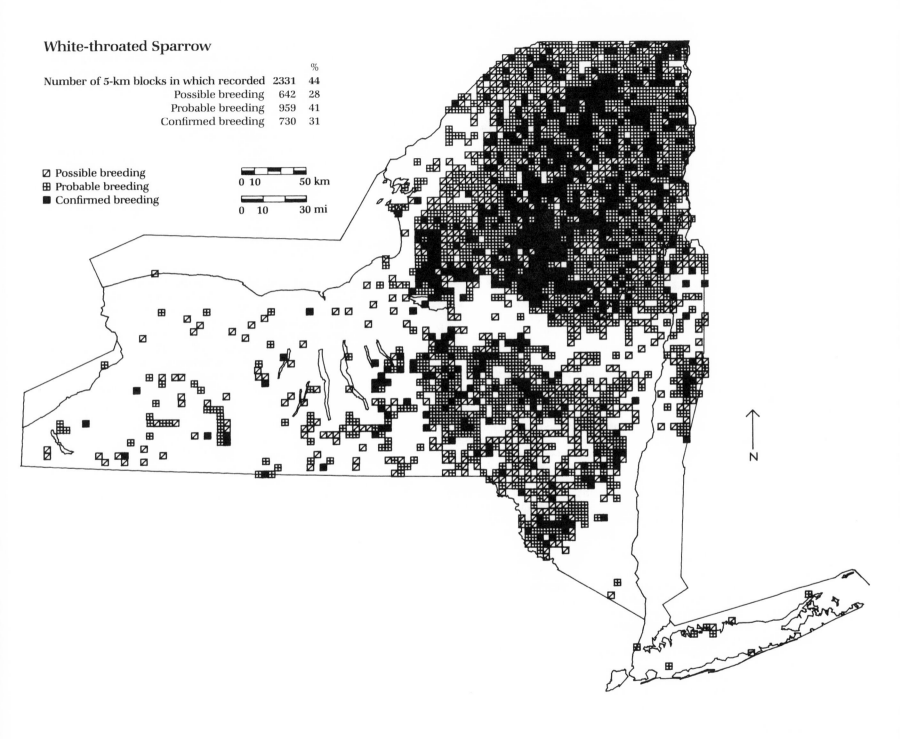

White-throated Sparrow

		%
Number of 5-km blocks in which recorded	2331	44
Possible breeding	642	28
Probable breeding	959	41
Confirmed breeding	730	31

◩ Possible breeding
⊞ Probable breeding
■ Confirmed breeding

0 10   50 km

0 10   30 mi

N

Dark-eyed Junco *Junco hyemalis*

The Dark-eyed Junco is one of the most arboreal of the sparrows, commonly found in New York wherever there is sufficient forest above the 305-m (1000-ft) contour and in cool, wooded ravines at lower elevations.

Eaton (1914) said that it was an abundant species, nesting throughout the Catskills and Adirondacks, and also fairly common as a breeding species in the highlands of western New York above 366 m (1200 ft). A few pairs were found during the nesting season in many of the colder swamps and gullies of central and western New York. Bull (1974) called it a widespread breeder at higher elevations, rare and local to absent elsewhere. Robbins et al. (1986) showed a slight but significant decline in this species' numbers in New York during the period 1965–79.

Atlas data show this species to be more generally distributed on the Appalachian Plateau, perhaps because of the recovery of much forest habitat there. The Dark-eyed Junco was not found in the heavily populated counties of southeastern New York, or in Niagara or Orleans counties. The distribution of this species is highly correlated with elevation. It was rarely found nesting on the Great Lakes Plain, the St. Lawrence Plains and Transition, the Mohawk Valley, and the Hudson Valley, and not at all on the Coastal Lowlands. Most of these either are agricultural areas, with only 50% forest cover, or are lower than 305 m (1000 ft). Exceptions are the cool glens of the Finger Lakes and seepage swamps, such as those in Genesee County.

The Dark-eyed Junco is one of the easier sparrows to "confirm," being a ground nester in wooded country. It becomes very agitated and conspicuous when its eggs or young are approached. It will sit tight while incubating, but when disturbed, the female may flutter away and then both parents will scold from nearby perches.

This species inhabits the more open northern woodlands and forest edges. In two Breeding Bird Surveys taken in New York 17 pairs per 100 ha (42 pairs/100 a) were recorded in a pine, spruce, and hardwood plantation in Allegany County (Brooks 1980); and 6 pairs per 100 ha (15 pairs/100 a) in a beech-spruce-maple forest in Essex County (Noon et al. 1982b).

The male usually arrives on its territory well in advance of nesting and sings from the tallest trees. The size of its territory is large for a sparrow, 0.8–1.2 ha (2–3 a), depending more on the availability of nest sites than on proper physiognomy of vegetation. Each territory seems to include an opening in the forest canopy surrounding a rock outcrop or an exposed soil bank (Eaton 1968).

The tendency for the species to build on or near a vertical wall probably explains many unusual nest sites. Eaton (1968) recorded nesting in a trellis overgrown with Virginia creeper, in a wind-vane feeder, and on a ledge beneath the gable of a house. Of 15 nests found in southwestern New York, 11 were along logging roads in the woods where cutbanks were topped by humus and leaf litter. The nests were located under the leaf litter at the top of cutbanks. Other nests have been found on the ledge of a vertical face of a large conglomerate boulder, in a small hemlock hedge, and on an upturned yellow birch root system (Eaton 1965). Later nests tend to be placed above ground level. The nest is a compact structure of grasses, rootlets, bark shreds, mosses, and twigs, lined with finer grasses, rootlets, and fur (Harrison 1975).

Stephen W. Eaton

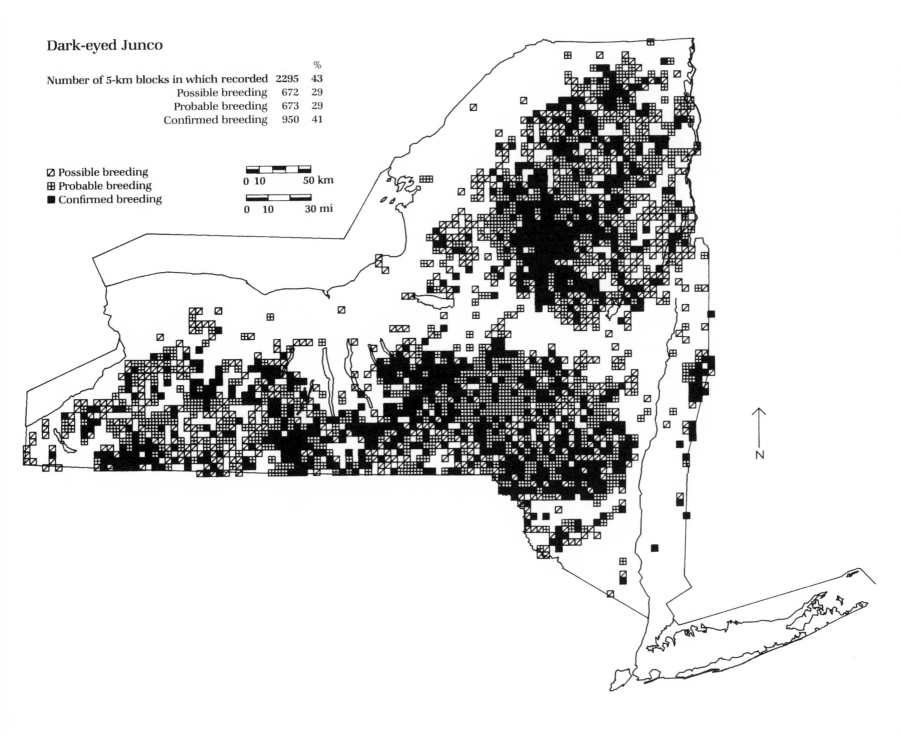

Dark-eyed Junco

		%
Number of 5-km blocks in which recorded	2295	43
Possible breeding	672	29
Probable breeding	673	29
Confirmed breeding	950	41

☑ Possible breeding
⊞ Probable breeding
■ Confirmed breeding

0 10   50 km

0 10   30 mi

N

Bobolink *Dolichonyx oryzivorus*

As a visual and aural adornment to the meadows and hay fields of New York, the Bobolink is unsurpassed. This widespread and fairly common breeding species is found in many areas of the state except the forested mountains, urban regions, and most of Long Island. It is surely one of the showiest native blackbirds, and its song, a bubbly, warbling repertoire, is everywhere welcomed.

Historically the Bobolink has been on a roller coaster typical of other meadow and hay field birds, but with even more severe ups and downs. Before colonial settlers drastically altered the landscape, the Bobolink was undoubtedly confined to river valleys with floodplain fields and marshy borders (Eaton 1914; Forbush 1929), but its distribution and population expanded rapidly as the inhospitable forests were cleared for pasture and agriculture. An enormous increase thus continued through the 17th and 18th centuries. But the mid-to-late 19th century brought a reversal. The hordes of Bobolinks that migrated down the eastern seaboard were in direct competition with humans for the rice grown in the southeastern states, and a great slaughter ensued. Each fall, thousands were harvested by shooting or jacklighting at night with torches, then sent to market. Even after rice was no longer a major crop in the Southeast, the "ricebirds" or "reedbirds" were harvested: delicious morsels served in expensive restaurants in northeastern cities (Bent 1958).

By 1910 the Bobolink populations of the Northeast, including New York, had reached their nadir. Forbush (1929) detailed the subsequent decline and then disappearance of the species as a breeder in New England. Griscom (1923) blamed the decrease on land development,

shooting, and its capture as a cagebird, and he called it completely extirpated in New York City, uncommon on Long Island. Although this species has long been protected, it has never recovered from its holocaust, and today's increased abandonment of farmland to plant succession or to development on rural land continues to reduce suitable habitat. Bull (1974) verified the scarcity downstate, judging it locally common upstate, particularly in the Great Lakes Plain. Robbins et al. (1986) could detect no increase in the eastern United States from 1965 to 1979.

Nonetheless, as the map shows, the Bobolink is still a widespread breeder, with "confirmed" records in every county except those making up New York City and in Nassau and Rockland counties. The sparsity of records in the Catskills, Adirondacks, Allegany Hills, and Central Tug Hill and Transition is probably more a result of lack of open areas than other altitude-related factors such as higher annual rainfall or lower July temperatures. The scattered "confirmed" records in these hilly and mountainous areas represent valleys under cultivation or in pasture.

Since the male is conspicuous in breeding season with aerial singing and courtship chases, and since the bird's habitat is visually unobstructed, records were often obtained from roadside observations. Singing males, territorial males, courtship, and pairs on territory, all easy to observe, were most frequently recorded. Nevertheless (barbed-wire fences notwithstanding), 38% of all records were "confirmed," with adults bringing food accounting for half of these; fledged young, one-third. Only 68 nests were found, confirming that they are extremely hard to find. Interestingly, 63 instances of distraction display were noted, far more than for the Eastern Meadowlark.

The habitat of the breeding Bobolink is tall grass meadows, hay fields, damp meadows near streams, drier portions of brackish marshes, and irrigated fields (Bent 1958). Grass and forb cover must be lush, and a mat of leaf litter that is not too deep is essential—where the litter is deep, the Bobolink is not present. The bird's territory is usually located closer to woods than that of other grassland species, and along fence rows (Wiens 1969).

The nest itself has been described and pictured many times: it is a cup of grasses in lush vegetation, loosely woven, lined with finer materials, on or near the ground beside or within a grass tussock. It is apparently entirely the work of the female (Bent 1958). Hazards are the same as for the Eastern Meadowlark: early mowing, pesticides, trampling by livestock, and ground predators.

Robert Arbib

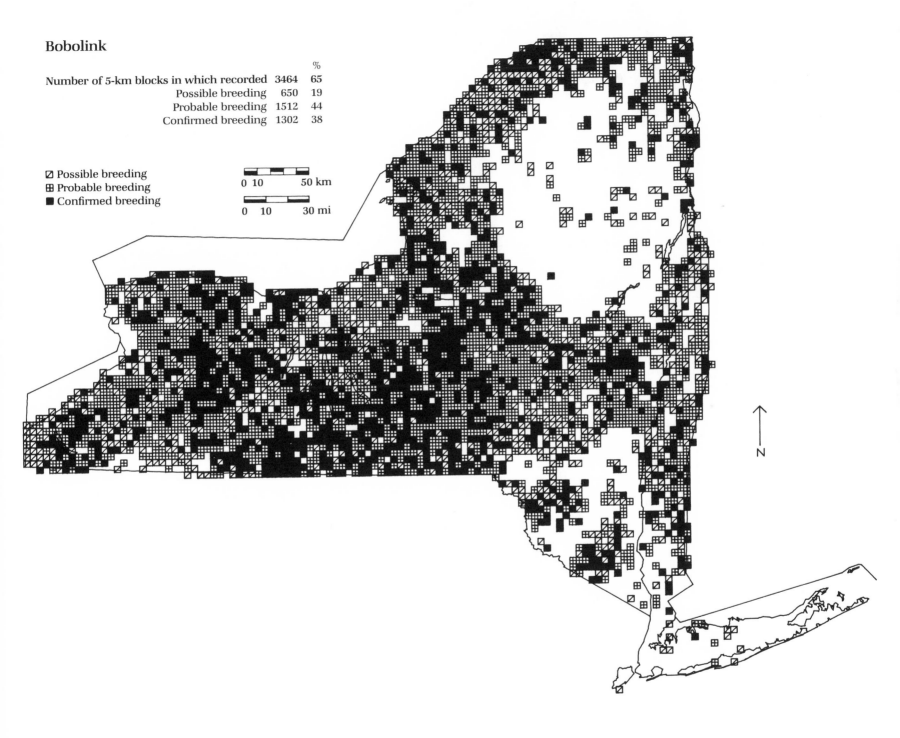

Bobolink

Number of 5-km blocks in which recorded	3464	%
		65
Possible breeding	650	19
Probable breeding	1512	44
Confirmed breeding	1302	38

◩ Possible breeding
⊞ Probable breeding
■ Confirmed breeding

0 10   50 km

0   10   30 mi

N

Red-winged Blackbird *Agelaius phoeniceus*

A noisy bird, the Red-winged Blackbird is one of the most abundant and widespread breeders in New York. It is found in virtually every swamp and marsh in the state. Eaton (1914) said that the species was a common summer resident of all districts, even the marshes of Staten Island and Long Island and the edges of the Flowed Land near Mount Marcy, high in the Adirondacks.

Wetlands are among the fastest disappearing habitats in New York because they are frequently drained for housing, industry, and agriculture, especially along the coast. A vast amount of wetland drainage has occurred during the past 50 years, so it is possible that the Red-winged Blackbird has declined in some areas. However, in recent years the species also has adapted to breed in upland locations, and in many areas Red-winged Blackbird numbers have increased dramatically (Case and Hewitt 1963). Now the species can be a major threat to New York's corn crops. In fact, in one day a flock can eat as much as 15% of the corn in a field (Giltz and Stockdale 1960).

Atlasing "confirmed" breeding in substantial portions of all ecozones. The species is absent from some heavily forested blocks in the Adirondack region and Catskill Peaks, probably because suitable swamps, marshes, and open upland fields are lacking. Where the Red-winged Blackbird does nest, it may be very abundant. For example, 70 pairs bred in 7.7 ha (19 a) of cat-tail marsh at Van Cortlandt Park, Bronx County, in 1965 (Bull 1974). The number of blocks in which this species was reported is exceeded only by that for the Song Sparrow, American Robin, Common Yellowthroat, Blue Jay, and Black-capped Chickadee.

The Red-winged Blackbird breeds not only in marshes and swamps and along their edges but also in the margins of ponds and streams. It is most commonly found in portions of wetlands where cat-tail, pickerelweed, arrowhead, sedge, and willow grow (Stoner 1932; Cruickshank 1942). The bird also breeds readily in upland grassy fields, often some distance from water (Case and Hewitt 1963; Bull 1974). Nests are more abundant in weedy fields than in clean ones, and in low lands than in hilly fields (Giltz and Stockdale 1960). The Red-winged Blackbird may have begun breeding in upland locations, including hay fields, abandoned pastures, and fallow fields, as recently as the 1930s (Ellison 1985h). In a study done near Ithaca, Tompkins County, nesting was generally more successful in marshes than in upland locations (Case and Hewitt 1963).

In wet areas the Red-winged Blackbird places its nest from a few centimeters to 3.7 m (12 ft) above water or ground, usually in cat-tail, bulrush, common reed, grass tussocks, buttonbush, willow, or alder, and cordgrasses in coastal salt marshes. Nests also are built in spruce, pine, elderberry, hawthorn, and other shrubs and small trees (Bull 1974). Often nests are in the middle of a tussock of sedge, or attached to cat-tails 0.3–0.6 m (1–2 ft) above water level (Saunders 1942). In dry areas the bird may nest on the ground.

The compact, strongly built nest is made by the female and constructed of grasses and sedges, firmly woven about the supporting structure. Usually it is lined with fine grasses, rushes, and sedges (Eaton 1914; Stoner 1932; Bent 1958). There are no dummy nests, although many nests are deserted before they are completed or before they receive eggs (Case and Hewitt 1963).

Richard E. Bonney, Jr.

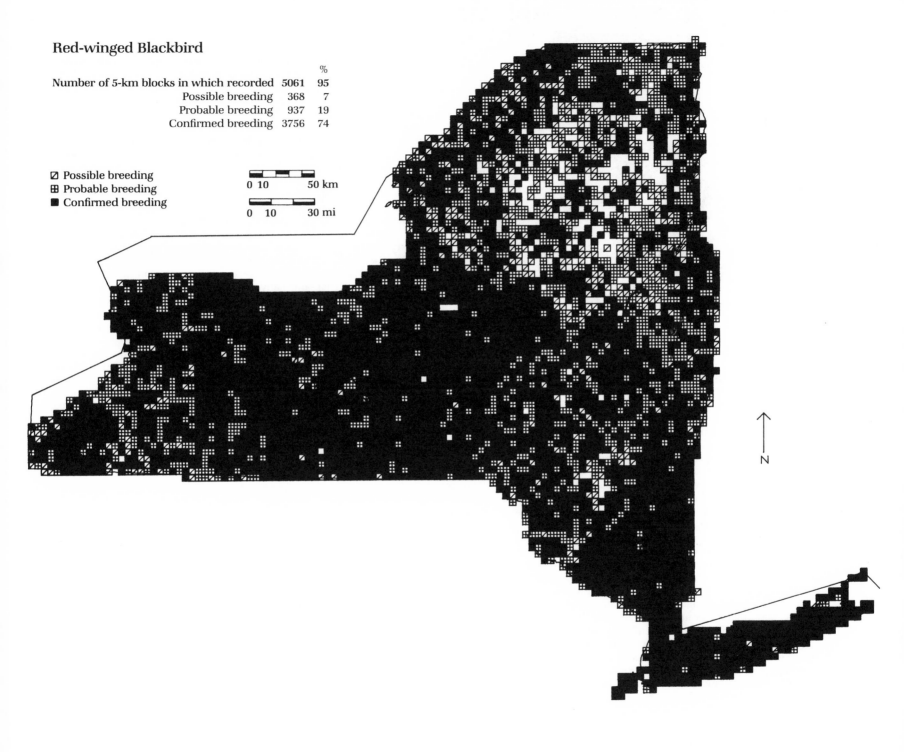

Red-winged Blackbird

		%
Number of 5-km blocks in which recorded	5061	95
Possible breeding	368	7
Probable breeding	937	19
Confirmed breeding	3756	74

☑ Possible breeding
⊞ Probable breeding
■ Confirmed breeding

0 10   50 km

0 10   30 mi

N

Eastern Meadowlark *Sturnella magna*

A singer of meadow and field, the Eastern Meadowlark is welcomed throughout its North American range for its song, which has been called plaintive, haunting, evocative, cheerful, and spirited. It is heard from early spring well into autumn. The Eastern Meadowlark is a common but declining breeding species in the state; it was on the *American Birds'* Blue List of declining species in 1986 (Tate 1986). Tate indicated that species numbers were down in the Northeast. Robbins et al. (1986) found that it had declined markedly in the East from 1965 to 1979.

Historically, the Eastern Meadowlark seems to have followed a familiar pattern; it was undoubtedly much scarcer in precolonial and early colonial times, when forest cover was far more extensive. Then it became increasingly abundant as forests and woodlands fell to axe and saw, and pastureland and cropland replaced forests on family farms (Eaton 1914). More recently, the species has declined, as abandoned farms turned to shrubland or second-growth woods, or were converted into highways, shopping malls, housing developments, or industrial parks (Bull 1964). In addition to habitat loss, the Eastern Meadowlark is a victim of nest-destroying early mowing, ground predators, trampling by livestock, and severe cold in its wintering area (Robbins et al. 1986).

The Atlas map portrays graphically the distribution described by Eaton (1914), who said it was "a common summer resident in all parts of the state except the forested portions of the Catskills, Adirondacks, and Allegany highland," to which we can now add the Central Tug Hill" Mongaup and Delaware hills, Hudson Highlands, Manhattan Hills, Tri-

assic Lowlands, the New York City metropolitan counties, and those areas of Long Island that are no longer prairies, cropland, or salt marsh.

This is a species of open areas and is fairly easy to locate because of its persistent singing. Roseberry and Klimstra (1970) found that in southern Illinois the Eastern Meadowlark favored pastures, both grazed and ungrazed, with hay fields (alfalfa, red clover, and mixed grains) and grassy soil banks, as well as idle and fallow fields. It has also been found nesting in corn fields, wheat fields, and other croplands, as long as there are dead grass stems at ground level and an absence of woody vegetation or numerous shrubs. The presence of elevated singing perches like fence posts, tall forbs, or isolated trees are essential parts of the habitat for this species (Wiens 1969). Most of the Atlas records were of singing males. Only about 4% of the "confirmed" records were of nests, suggesting the difficulty of locating this well-concealed structure.

The nest, often located in a beak-excavated depression, is a well-made grassy cup lined with finer grasses and loosely domed over with grasses, with an entrance, sometimes roofed, at one side (Gross 1958). Lanyon (1957) and Wiens (1969) observed that the orientation of the nest was either north or east, and Lanyon suggested that the prevailing winds, particularly in rain and sleet storms, would depress vegetation in those directions. Nest building and incubation are largely or entirely done by the female, as is much of the feeding of young in the nest, although the male may help (Gross 1958).

Being a ground nester, the Eastern Meadowlark is prey to skunks, snakes, house cats, dogs, and avian raptors and is the victim of nest parasitism of the Brown-headed Cowbird (Roseberry and Klimstra 1970). Undoubtedly the cause of the greatest mortality, however, is the early mowing of hay fields. According to Richmond and Nicholson (1985), "Recent developments in varieties and cultural practices . . . allow earlier and more frequent mowings, which may be extremely destructive to ground-nesting birds." They add that mortality could be significantly decreased if farmers would delay the first mowing until after the spring nesting season.

Robert Arbib

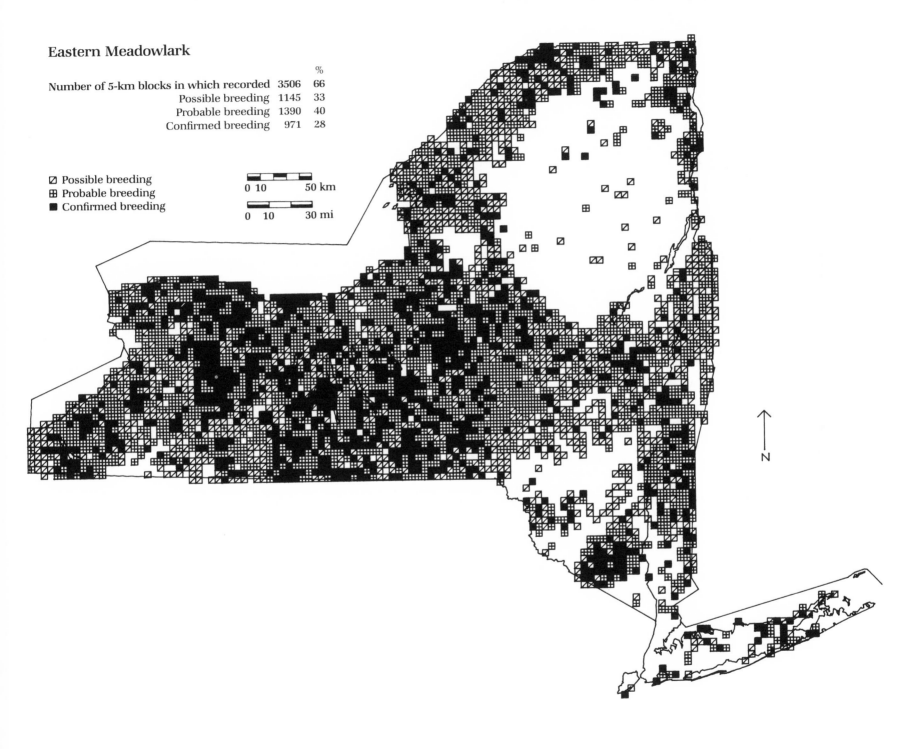

Eastern Meadowlark

		%
Number of 5-km blocks in which recorded	3506	66
Possible breeding	1145	33
Probable breeding	1390	40
Confirmed breeding	971	28

☑ Possible breeding
⊞ Probable breeding
■ Confirmed breeding

0 10   50 km

0 10   30 mi

N

Western Meadowlark *Sturnella neglecta*

A common resident of most of western North America, the Western Meadowlark is a very rare breeding species in New York. It is a relatively recent addition to the avifauna, having been first identified and collected in western Monroe County by A. S. Klonick in 1948 (Klonick 1951) and first found breeding near Braddock Bay, Monroe County, in 1957 (Miller 1958). Since the first sighting the Western Meadowlark has been observed almost every year, with reports from seven different localities in 1963 alone (Bull 1974). Thus far, there have been only two confirmed breeding records: the Braddock Bay record and another at Bangall, Dutchess County, in 1962, where a male Western Meadowlark and a female Eastern Meadowlark produced hybrid young (Bull 1974).

Most of the reports of this species in the state, including four of the six Atlas records, have been from the Great Lakes Plain and adjacent areas. The notable exceptions are several records from the lower Hudson Valley, including those noted by Bull (1974), more recent records from *The Kingbird* (Treacy 1982a, 1982b, 1983a, 1983b), and one Atlas record. This pattern of occurrence, as well as several earlier records of breeding in southern Ontario, suggests penetration from Ontario through the Niagara Frontier corridor, as stated by Bull (1974). The well-documented expansion of this species eastward in the northern part of its range was summarized and analyzed by Lanyon (1956).

In New York this species occupies the same open farm fields, meadows, and pastures that the Eastern Meadowlark inhabits. Almost all of the birds have been initially located by observers who heard its musical, bubbling song, so very different from that of its eastern relative. In some cases, observers were able to differentiate the Western Meadowlark from the Eastern Meadowlark by plumage or its distinctive call note.

Many of these birds reside for weeks or months in a given location with no indication of breeding. However, since the male does not visit the nest during building or incubation (Harrison 1975), evidence of breeding is hard to obtain. The association of a singing male with a nest may require careful detective work or, better, visits to the territory after the young have hatched when the male is helping with feeding. A classic case of long-term site fidelity, with no real evidence of breeding, is that of a male that returned to the same territory near Point Breeze, Orleans County, from 1979 to 1984. It arrived as early as mid-March and was heard singing as late as mid-September (Griffith 1983; Listman, pers. comm.).

Lanyon's long-term study (1966, 1979) of meadowlark hybridization, which utilized the captured adults and offspring of the Bangall nesting, suggested additional reasons for the lack of breeding evidence in the state. First, Western Meadowlark mates are scarce in such peripheral range locations, yet isolating mechanisms between the meadowlark species appear to remain strong and mixed pairings are rare. Should mixed pairings occur, hybrids are difficult to identify visually in the field, and the few male hybrids are likely to learn song repertoires with largely eastern variations. A colony will never result, as hybrid pairings are almost all infertile.

This species' nests and eggs are generally indistinguishable from those of the Eastern Meadowlark (Harrison 1975), although the Western Meadowlark may prefer nest sites on drier ground (Lanyon 1956; Terres 1980). The nest, which the female builds alone, is a scraped depression lined first with coarse grasses, then finer grasses and hair. Covered by a canopy of grasses interwoven with the surrounding vegetation, it has an entrance to one side. Relatively obvious trails often lead through the grass, providing observers with clues to the location of the nest (Harrison 1975).

Robert G. Spahn

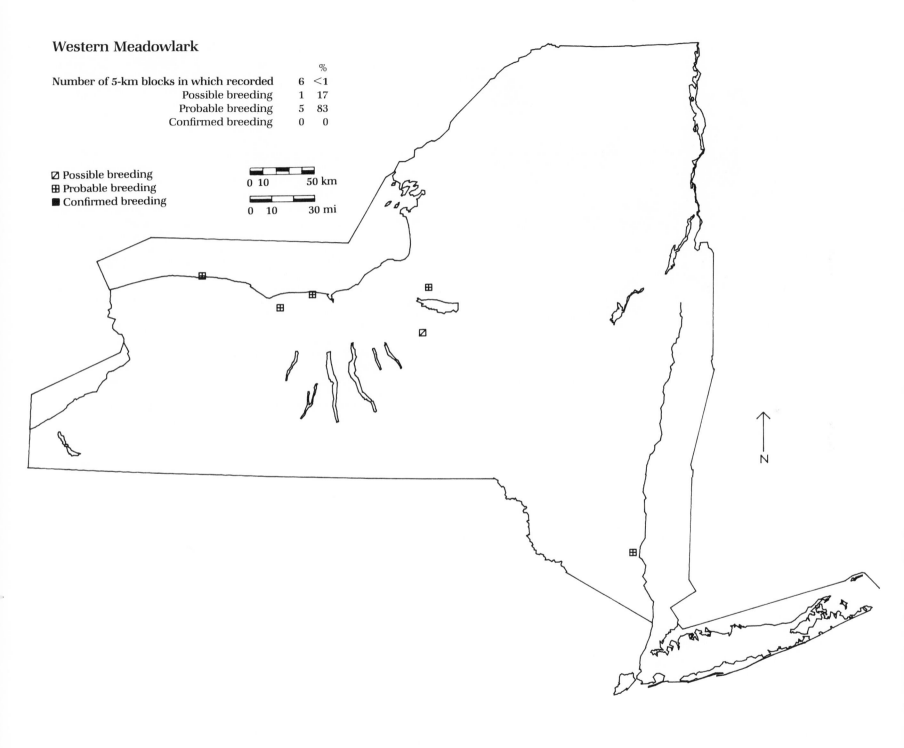

Western Meadowlark

		%
Number of 5-km blocks in which recorded	6	<1
Possible breeding	1	17
Probable breeding	5	83
Confirmed breeding	0	0

Possible breeding
Probable breeding
Confirmed breeding

0 10   50 km

0 10   30 mi

N

Rusty Blackbird *Euphagus carolinus*

A blackbird of northern swamps, the Rusty Blackbird is an uncommon local breeder over most of the Adirondacks. As Bendire (1895) observed at the end of the last century, the Rusty Blackbird is a forest-loving species: "Its favorite haunts in the Adirondacks are the swampy and heavily wooded shores of the many little mountain lakes and ponds found everywhere in this region, and here it spends the season of reproduction in comparative solitude."

De Kay (1844) believed the Rusty Blackbird probably bred "in the northern district, for according to Audubon it [bred] in Maine and further north." The first recorded sighting appears to be that of Henry Minot, who noted "two or three seen in June [1877]" near the St. Regis lakes of Franklin County (Roosevelt and Minot 1929). A nest was found on the Raquette River (Franklin or Hamilton County) by C. S. Pennock on 5 June 1878, according to Eaton (1914). Several years later Merriam (1881) declared this bird a common Adirondack region summer resident and collected several Herkimer County nests (Eaton 1914). Eaton, however, did not report seeing the Rusty Blackbird in Essex County during his 1905 visit, and he later described the species as a resident of the wilder portions of the Adirondacks, especially in Hamilton, Herkimer, and St. Lawrence counties. And Saunders (1929a) failed to find Rusty Blackbirds during his two summers in Essex and Franklin counties, concluding, "Evidently this bird is rare and local." More recently, Beehler (1978) described the bird as "uncommon and local in High Peaks, Lakes, and Northern districts."

The range of the Rusty Blackbird extends from Alaska to the Maritimes and into New York (AOU 1983), where the population forms an island south of the contiguous main range. Bull (1974) mapped just 20 breeding localities, whereas Atlas observers recorded the Rusty Blackbird in 151 blocks in the Adirondacks and Central Tug Hill, with breeding "confirmed" in 51 of these. Records of the New York range have been extended outward in all directions, probably in part because of improved observer coverage. A great deal of backwater habitat that appears suitable is still unoccupied by Rusty Blackbirds (pers. obs.)

Atlas observers of the 1980s could empathize with Bendire (1895): "The oologist who desires to study this species on its breeding grounds must make up his mind to endure all sorts of discomforts; millions of black flies, gnats, and mosquitoes make life a burden during his stay, while the bogs and swamps through which one is compelled to flounder in search of the nest render walking anything but pleasant." Moreover, there are problems of identification. The Rusty Blackbird can sometimes be mistaken for the Common Grackle. The rusty hinge "song" sounds much like the high whistle of a Broad-winged Hawk, and the *chack* call note can be confused with that of a grackle or Red-winged Blackbird.

Another factor has undoubtedly affected the distribution and may explain why Eaton and Saunders missed the Rusty Blackbird in their searches. Typical habitats were described by Atlas observers as boreal bogs, marshes, ponds, and swamps, often with standing dead or dying trees, surrounded by forest. The mention of a "half dried-up old beaver flow," "grassy swamp (old beaver pond)," and "mid-sized open water beaver pond" may be significant. Beavers were nearly extirpated in New York by 1894, when the Forest Commission estimated that only 10 were left in the entire state. Donaldson (1921) documented the subsequent protection and restocking efforts in the early years of this century, noting: "They seemed to take kindly to repatriation and increased rather rapidly." The Rusty Blackbird has undoubtedly benefited from the resurgence of the beaver in New York.

Rusty Blackbird nests are placed fairly low, from 0.3 to 6.1 m (1–20 ft) up in a conifer or shrub near water, usually where forest and swamp meet. The rather bulky nest is layered and quite distinctive, easily seen and identified during the following winter. The base is greenish *Usnea* lichen, the middle layer black duff, and the upper rim yellowish dried grasses, quite obvious in bare speckled alders bordering areas such as Chubb River Swamp, Essex County. Bendire (1895) noted that Adirondack nests last several years. The used nest code was used only once by Atlas observers, however, and only 1 nest with eggs and 2 nests with young were found.

John M. C. Peterson

474

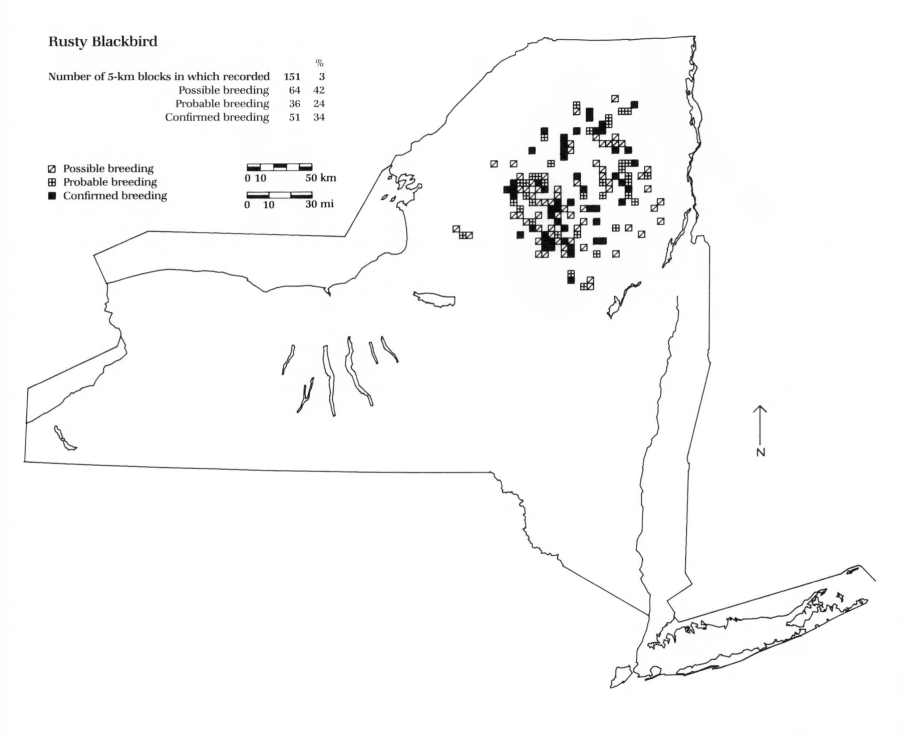

Rusty Blackbird

		%
Number of 5-km blocks in which recorded	151	3
Possible breeding	64	42
Probable breeding	36	24
Confirmed breeding	51	34

◨ Possible breeding
⊞ Probable breeding
■ Confirmed breeding

0 10   50 km

0 10   30 mi

N

Boat-tailed Grackle *Quiscalus major*

This huge grackle, much larger than the Common Grackle, is extending its range north along the coast and, except for Florida, is usually restricted to the vicinity of the shore. It is common along the Atlantic and Gulf coasts of southern states. Not even listed as accidental in New York before 1967, by the early 1980s the Boat-tailed Grackle had established a breeding presence on western Long Island. Although still rare and local, it may well become more numerous and widespread on the coast.

A report of an adult female at Brookhaven, Suffolk County, 1 September 1954 by Puleston (Bull 1964), following hurricane Carol on 31 August, placed the species on the state's list. First found breeding in New Jersey on Delaware Bay in 1952, it then rapidly increased in that area but was not noted again in New York until April 1967, when an adult male appeared at feeders in Far Rockaway, Queens County. It was photographed and became the first confirmed state record (Trimble and Post 1968; Bull 1974). It reappeared there annually for five years until 1971, arriving as early as 24 February with Common Grackles.

In 1972 and 1973 single Boat-tailed Grackles were reported along the south shore of Long Island at Jamaica Bay Wildlife Refuge, Jones Beach State Park, and Montauk Point, Suffolk County, on the eastern tip of Long Island (Bull 1976). But in 1979 the numbers of reports sharply increased, with possible nesting on Pearsall's Hassock, Hewlett Bay, Nassau County. Observations there included 2 males and 3 females (2 possibly immature) in early June and a full-grown immature in late June (Richards et al. 1979; Zarudsky and Miller 1983). Numbers were also up in New Jersey, with birds nesting as far north as Island Beach State Park.

In 1980 there were records for Jamaica Bay and also, remarkably, an individual in Monroe County in late October (DeBenedictis 1981). Evidence that the birds were indeed breeding came in 1981. On 23 June an adult female with 3 nearly full-grown young still being fed but capable of flight were observed on North Black Banks Hassock (south of Pearsall's Hassock) (Gochfeld and Burger 1981). Nesting was thought to have been either on Black Banks or nearby. On 18 and 19 July, 3, including 2 young that were flying but still begging and being fed, were at East Pond, Jamaica Bay (Davis 1981).

On 31 May 1982 breeding was "confirmed" on Pearsall's Hassock when 2 nests, one with 3 eggs, were located following the sighting of a male and 3 females; follow-up visits revealed a colony with 4 nests, 3 producing eggs and young (Zarudsky and Miller 1983). Nesting evidently continued through 1984 and 1985, with most reports, including winter, from Jamaica Bay, where sightings included a female with a juvenile 29 July 1984 and, to "confirm" breeding, a female carrying food 15 July 1985 (DiCostanzo 1984d, 1986). Also, a female was observed in the Black Banks Hassock complex, but in the Atlas block just south of the 1981 observations (J. Zarudsky, v.r.).

The Boat-tailed Grackle is conspicuous, and Sprunt (1958) wrote: "There may be more noisy birds, but I have yet to hear them!" The map approximates the present breeding distribution, except that the number of observations of young may indicate that there are unknown nesting sites on some of the many bay islands.

This species is rarely seen far from coastal lowland habitats. It frequents some coastal cities and towns in the South, but northward it is nearly confined to salt marshes and maritime shrubland, typical of bay and barrier islands. Zarudsky and Miller (1983) reported that on Pearsall's Hassock, which was covered by material dredged in the 1950s, the grackle nested in a plant association of bayberry and common reed on the edge of a heron colony. On Black Banks the family group (1981) was perched in common reed.

Nests are generally 0.6–3 m (2–10 ft) from the ground in bushes or canelike growth, occasionally high in trees (Pough 1946). On Pearsall's Hassock the 4 nests were 1.3–1.5 m (4.2–4.8 ft) from the ground (Zarudsky and Miller 1983). Built by the female, the nest is a bulky open cup on a foundation of grasses, stalks, and vines, coated inside with a layer of mud or decayed plants, plus a softer inner lining of fine grasses and other fibers (Harrison 1975).

Paul F. Connor

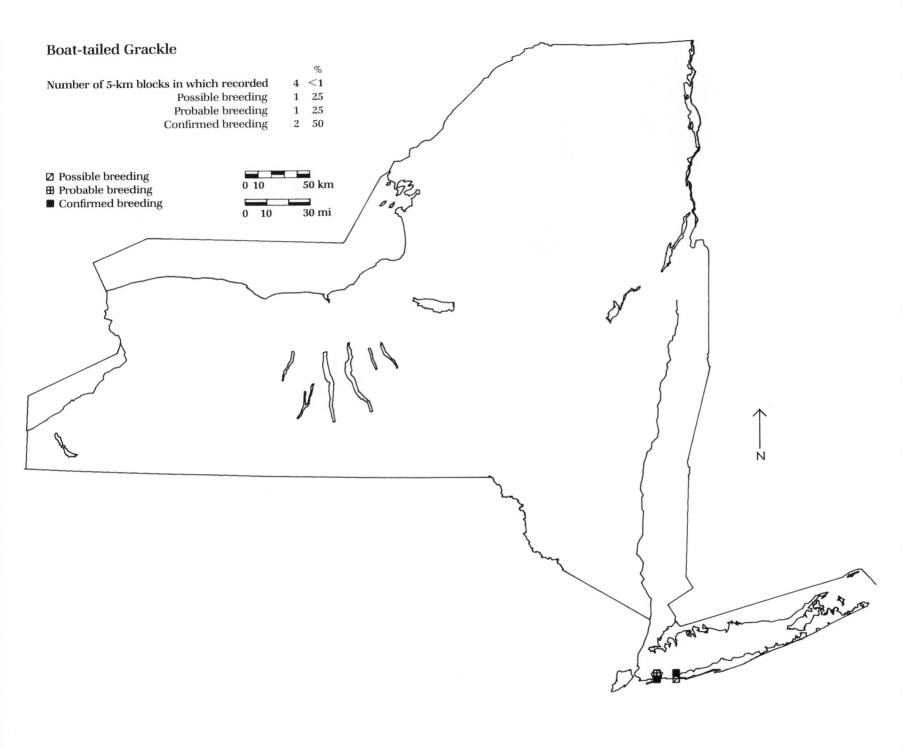

Boat-tailed Grackle

		%
Number of 5-km blocks in which recorded	4	<1
Possible breeding	1	25
Probable breeding	1	25
Confirmed breeding	2	50

◨ Possible breeding
⊞ Probable breeding
■ Confirmed breeding

0 10   50 km

0 10   30 mi

N

Common Grackle *Quiscalus quiscula*

The Common Grackle is one of the most common birds over much of New York—only six other species were found in more blocks. It is, however, rather uncommon at high elevations and is locally uncommon to rare in forested tracts.

Although no historical changes in distribution or abundance have been documented, this species has adapted to such a variety of urban and suburban habitats that increases may well have occurred over the past several centuries. The Atlas map indicates that the grackle was absent from the highest elevations of the Adirondacks and Catskills and observed infrequently in the heavily forested eastern Appalachian Plateau, Delaware and Schoharie hills, Central Tug Hill, and portions of the Adirondacks. If forest cover is a negative factor, population declines might be expected where old farms have been abandoned and fields have reverted to forest. At the same time, urban sprawl around metropolitan areas continues to provide new niches for this adaptable and opportunistic bird.

The Common Grackle is vocal, fairly tame, and easily identified. Surprisingly, among the "probable" records, few are in the category that includes courtship behavior. The grackle's noisy, raspy calls and ruffling of feathers should have attracted not only a mate but also more human attention than the records show. Most of the "confirmed" records were of adults carrying food or feeding young, or of fledglings. Although nests are not difficult to find, only about 10% of the "confirmed" records were of nests.

The Common Grackle prefers somewhat open areas, with scattered trees. In forested regions of the state it occupies open woodlands, bogs, and forest edges. Although found in wilderness areas, the grackle is also a familiar bird around human habitations (AOU 1983). In most parts of the state the Common Grackle has readily adapted to ornamental spruce and to conifer plantations, building conventional nests on branches. A number of less conventional New York nest sites have also been recorded. Bull (1974) documented nests built on the metalwork of a steel bridge, cat-tail stalks, grapevines, willows, the ledges of a brick house, in a Wood Duck nest box, and even against the side of an active Osprey nest. In wilder areas of the Adirondacks the Common Grackle frequently nests in cavities, occupying old nest holes of the Northern Flicker and other woodpeckers; it is most often found in dead snags common to swamps and beaver ponds and meadows. In such wild areas grackles may form loose colonies, as they do in settled areas. Saunders (1929a) observed grackles flying in and out of holes in stumps and stubs dotting an old burn in a marsh in the St. Regis lakes area, Franklin County. Atlas observers recorded the grackle at Marcy Dam, where the elevation is 721 m (2366 ft); earlier in the century Eaton (1914) located the species as high as Flowed Land at an elevation of 839 m (2753 ft). Clearly, the Common Grackle is well served by its adaptability.

For a month or more early in the breeding season the adults may do little more than drape a few grasses over the chosen nest site; then in a burst of activity the female will complete a nest in about 5 days. The nest is always a bulky cup of grasses, twigs, and other materials lined with mud, dry grass, rootlets, and other small stems. Once the eggs are laid, the male may remain to help guard the nest and rear the young; however, he frequently leaves his mate for a second female, and the abandoned mate rears the young alone. Pairing usually takes much less time with the second mate (Stokes 1979).

John M. C. Peterson

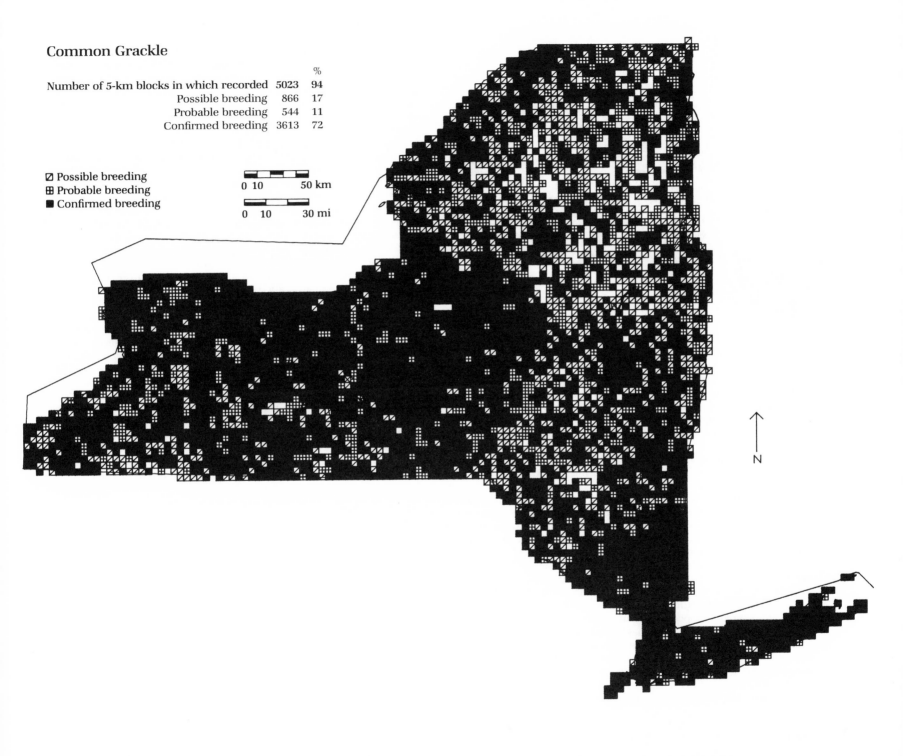

Common Grackle

		%
Number of 5-km blocks in which recorded	5023	94
Possible breeding	866	17
Probable breeding	544	11
Confirmed breeding	3613	72

◩ Possible breeding
⊞ Probable breeding
■ Confirmed breeding

0 10   50 km

0 10   30 mi

N

Brown-headed Cowbird *Molothrus ater*

The Brown-headed Cowbird is North America's only obligate brood parasite. It builds no nest of its own but lays its eggs in the nests of other species, which raise and care for the large and demanding cowbird young at the expense of their own. The cowbird breeds commonly throughout all of the low elevations of New York. Eaton (1914) said that the species was a common summer resident in all parts of the state except for the coniferous forests at higher elevations, but that it also invaded the valleys and cleared lands of the Adirondacks, as it does today.

The Brown-headed Cowbird is a relatively recent arrival in the eastern United States. Bent (1958) stated that this grassland species is supposed to have entered North America through Mexico, to have spread through the central prairies and plains with roving herds of wild cattle, and to have gradually extended its range westward and eastward to the coasts as forests disappeared, open lands became cultivated, and domestic cattle were introduced. However, the cowbird was common in New York as early as 1790 (Mayfield 1965) and continued to increase as more forests were cleared for agriculture.

A decline apparently occurred around the New York City area where both Griscom (1923) and Cruickshank (1942) said that it was decreasing. The species may be declining more recently as agricultural land reverts to forest. Robbins et al. (1986) found that it decreased on BBS routes in New York from 1965 to 1979. Nevertheless the cowbird remains abundant, to the detriment of other species. One reason for its success is the increasing fragmentation of large forests into small woodlots throughout much of New York. When forests are fragmented, the cowbird can penetrate them more easily as it looks for hosts. The Brown-headed Cowbird has been very successful in finding tolerant hosts in newly available fragmented forests, which may result in a surplus population that can move into peripheral areas (Mayfield 1965). Local changes in abundance can also be correlated with the presence of farm animals, including horses (Spencer 1983).

As seen from the Atlas map, the Brown-headed Cowbird is widespread in New York. It is absent from parts of the Adirondack High Peaks, Eastern and Western Adirondack Foothills, and Central Adirondacks. However, its distribution is probably less related to elevation than to the fact that the interior Adirondacks are too heavily forested. Saunders (1942) reported that it was a regular summer bird in the open farm country and the lower valleys of the mountainous Allegany State Park but rarely came to the higher parts of the park, and Atlas observers found that to be true as well. The cowbird is easy to "confirm" as a breeder, as fledglings are conspicuous because of their large size and noisy food-begging calls.

The Brown-headed Cowbird is found primarily in open country—fields, pastures, orchards, and residential areas—especially where there is short grass suitable for ground foraging. It often follows cattle and other pasture animals, alighting on their backs to capture insects that infest them or that are driven from the grass as the animals feed. The species avoids deep forests but does enter woodland to lay its eggs in the nests of woodland-dwelling birds. It also occasionally inhabits swamps, thickets, and semi-wooded conditions (Stoner and Stoner 1952), and city parks, suburban yards, farmland, open country, swamps, and forests.

The Brown-headed Cowbird parasitizes at least 77 species in New York. Most are passerines, especially flycatchers, thrushes, vireos, warblers, and icterids, including species that nest on the ground, in shrubs, high in trees, and even in tree cavities. The three most commonly parasitized species are the Yellow Warbler, Red-eyed Vireo, and Song Sparrow, in that order (Bull 1974; Bull 1976). In some cases birds parasitized by the cowbird never return to nest in that locality (Eaton 1914).

Richard E. Bonney, Jr.

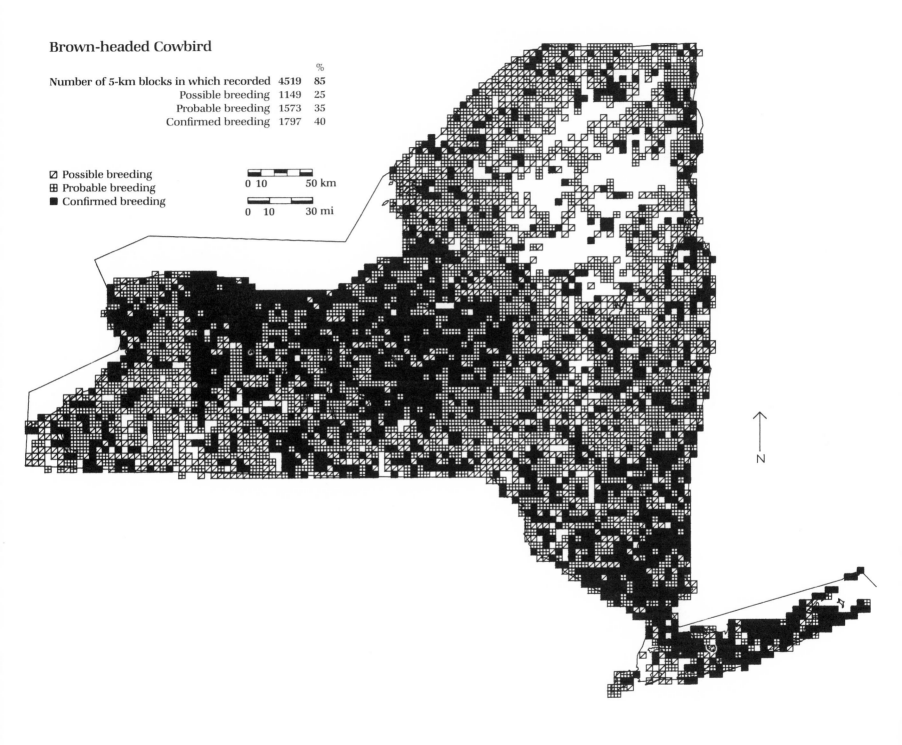

Brown-headed Cowbird

		%
Number of 5-km blocks in which recorded	4519	85
Possible breeding	1149	25
Probable breeding	1573	35
Confirmed breeding	1797	40

☒ Possible breeding
⊞ Probable breeding
■ Confirmed breeding

0 10   50 km

0 10   30 mi

N

Orchard Oriole *Icterus spurius*

At its northern nesting limit in New York, the Orchard Oriole is rather uncommon and is much sought after by upstate birders. This oriole is on *American Birds'* Special Concern list (Tate 1986) because of decreasing populations in several regions of the United States. However, after decades of decline in this century, a modest increase in numbers and a northward expansion has recently been reported in New York and some other states.

During the 1800s and early 1900s the Orchard Oriole was numerous and widespread in southeastern New York, including Long Island, but a decline began about 1920 (Bull 1974). During the 1800s it outnumbered the Northern Oriole on Long Island (Eaton 1910; Bull 1964). Eaton (1914) stated that it was common in the vicinity of New York City and the lower Hudson Valley but also fairly common north to Albany, and breeding was documented as far as Saratoga Springs, Saratoga County, and Granville, Washington County. Eaton also considered the species fairly common on Long Island as far east as Bellport, Suffolk County. Elsewhere it was uncommon and unknown at higher elevations, although it was known to breed in central and western New York.

Later, Cruickshank (1942) reported the Orchard Oriole as a very local nester on Long Island, with only 2–3 pairs remaining on Staten Island and but few pairs nesting in Rockland and Westchester counties. Bull (1964) stated that it was still fairly common locally along the Delaware River in northern New Jersey, an exception in the greater New York City region. By the late 1960s it had disappeared from its last site on Staten Island (Siebenheller 1981). The loss of agricultural land and orchards, as referred to by Bent (1958) and Bull (1964), may have contributed to the decline.

The increase in the Northeast has been evident for a decade or more. Bull (1976) reported an adult male Orchard Oriole seen by several observers in Essex County on Lake Champlain in 1974, an extension north of at least 20 miles, but without evidence of breeding. Ellison (1985i) described its reappearance in Vermont in the 1970s following an absence of more than 60 years, with breeding confirmed in 1977 and 1982. Buckley et al. (1976) reported an increase on Long Island, and in 1979 increases were reported in the lower Hudson Valley (Treacy 1979) and near Lake Ontario (Spahn 1979).

In 1982 Paxton et al. (1982) noted that the Orchard Oriole was locally abundant in coastal New Jersey and that in New York it was also doing well in Orange County and even in Queens County, New York City, where it bred for the first time in years. Kibbe and Boise (1983) stated that sightings upstate had increased. Also, from 1980 to 1985 it was reported nesting at sites near Lakes Erie and Ontario in southern Ontario (Goodwin 1981; Weir 1983; Ontario Breeding Bird Atlas, preliminary map). Thus populations appear to have attained a relatively high level recently in and near the state.

The Atlas map shows that the center of its distribution is found mainly in southeastern New York, although few were found in Rockland County and New York City, and none on Staten Island. Observers found the Orchard Oriole locally northward in the Hudson Valley, in the Finger Lakes and Oneida Lake regions, and along the south shore of Lake Ontario. Also scattered breeding localities were found in the Mohawk Valley, near Lake Erie, and along river systems on the Appalachian Plateau, but hardly any along the Delaware River system in the hill country of Delaware and Sullivan counties. "Confirmed" breeding was recorded as far north as Lake George and the Lake Champlain Valley. Less conspicuous than the Northern Oriole, this species may be overlooked high in the foliage of trees, although the musical, warbling song attracts attention. Recent increases seem gradual and may be partly due to increased observer activity during the breeding season.

The Orchard Oriole is mainly a species of estuarine, riverine, and lacustrine communities. It is commonly associated with farmyard, suburban, or roadside shade trees, nurseries, old fruit orchards, scattered trees in fields, and thin woods near water, including the edge of salt marshes. Observers have frequently found this oriole nesting in the same tree with the Eastern Kingbird. The nest is a suspended basketlike cup, woven of grasses, lined with fine grasses and plant down, and usually at a forked branch in a deciduous tree or occasionally a shrub 1.2–21.3 m (4–70 ft) above the ground (Harrison 1975). Occasionally ornamental spruces are chosen, and a Suffolk County pair nested 4.6 m (15 ft) above the ground in a privet hedge (Bull 1974).

Paul F. Connor

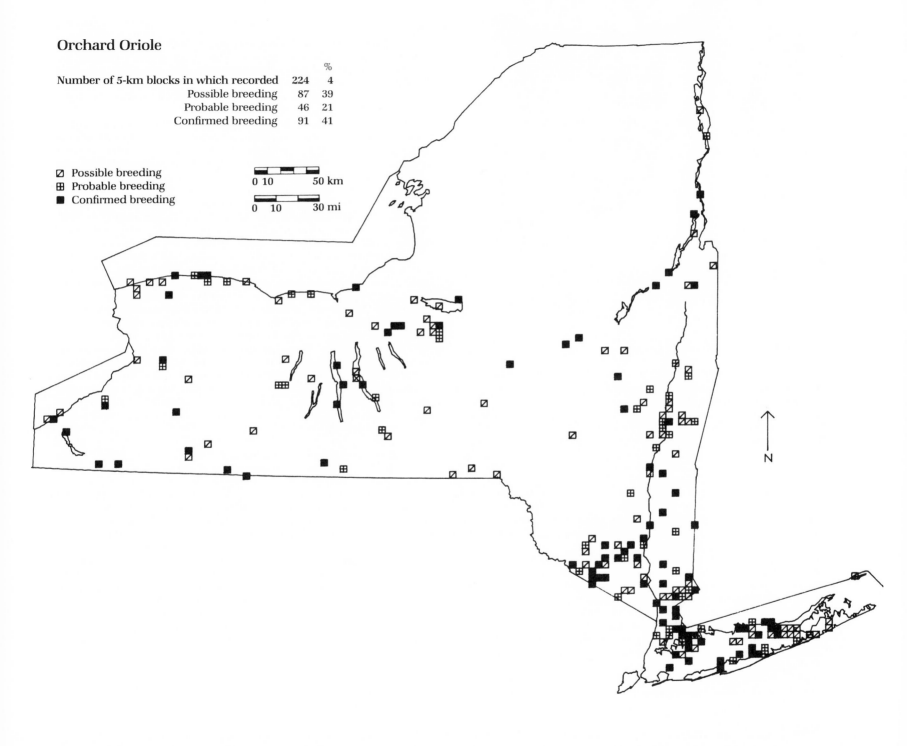

Orchard Oriole

		%
Number of 5-km blocks in which recorded	224	4
Possible breeding	87	39
Probable breeding	46	21
Confirmed breeding	91	41

◪ Possible breeding
⊞ Probable breeding
■ Confirmed breeding

0 10   50 km

0 10   30 mi

N

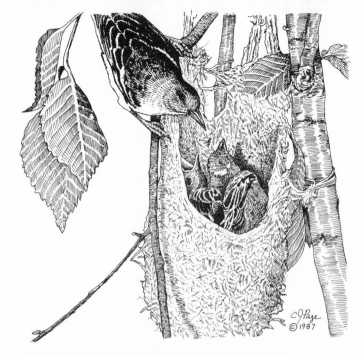

Northern Oriole *Icterus galbula*

The Northern Oriole is notable for its brilliant, unusual orange plumage, rich and clear song, and the remarkably strong and long-lasting suspended nest. Many casual observers are familiar with this bird, yet according to Stokes and Stokes (1983) it is, surprisingly, one of the least studied of our common species. This oriole is numerous and well distributed throughout most of the state, except for the Adirondack region.

There is little documentation of pronounced changes in the distribution pattern or abundance of the Northern Oriole in New York historically. It is not a bird of heavy forests, however, so it probably increased considerably in the state after the land was cleared and settled, as mentioned by Eaton (1914). This species was apparently formerly outnumbered by the Orchard Oriole on Long Island. Eaton (1914) stated that it was common in summer in all of the state except the wooded portion of the Catskills and Adirondacks, but it entered river valleys and cleared lands in various localities of the Adirondacks and was most common in the warmer parts of the state. More recently, Cruickshank (1942) stated that breeding pairs are few and steadily decreasing in heavily populated areas in the immediate vicinity of New York City. In the period 1965–79 Robbins et al. (1986) reported a widespread increase in the eastern states, despite loss of elms to Dutch elm disease.

The Atlas map portrays the extensive, nearly continuous distribution of the Northern Oriole in New York. However, it was not found in many blocks in the Central Adirondacks and adjacent portions of the Adirondack High Peaks, Western Adirondack Foothills, and nearby Central Tug Hill. These areas are heavily forested with spruce–fir–northern hardwood forest and successional northern hardwoods (Stout 1958; Will et al. 1982). Otherwise, it is generally found throughout and is especially numerous in lowland ecozones and valleys. It is absent locally, primarily in unbroken forests, high elevations, and downtown urban areas. But it still breeds in several sections of New York City, especially Staten Island and Bronx and Queens counties.

This is one of the easiest breeding birds to find. Although often invisible high in tree foliage, the conspicuous males sing their distinctive song from a prominent perch, and the durable nests are among the most obvious. Also, this species may attain a high nesting density, since it appears to defend only a small area around the nest and areas of activity of pairs overlap (Sealy 1980). This oriole may be missed in midsummer, however, as it is among the first of northeastern birds to cease singing, generally very early in July, although it is heard again singing its "fall" song during the latter part of August (Saunders 1929b). Family groups are conspicuous in summer.

The Northern Oriole nests in tall trees, usually deciduous, along roads and waterways, in fruit orchards or in scattered trees in fields, in towns and parks, and occasionally at the edge of woods (Cruickshank 1942; Saunders 1942). Eaton (1914) observed that in different villages in western New York, tree preference seemed to be in the following order: American elm, silver maple, and apple; but also it may nest in any tree, even Norway spruce, hemlock, or horse-chestnut. Along the Allegheny River, Eaton (1981) noted nests in silver maples and American elms, and also observed nest building in willow and wild cherry. In the upper Hudson Valley nests are frequently in cottonwood, silver maple, and small elm trees on roadsides and along the river, and also beside roads in sugar maples (frequent), and wild cherry and ash trees (occasional) (pers. obs.). In a northern New Jersey survey most nests were found in elm and maple and placed over or near roads or, less often, in fields, but none in woodlands (Bull 1964).

Northern Oriole nests were found by Eaton (1914) to be placed 2.1–18.3 m (7–60 ft) above ground, but they averaged about 7.6–9.1 m (25–30 ft). The placement and structure of the nests differ geographically for various reasons; in the Northeast they are built on thinner branches, probably to make them less susceptible to squirrel predation (Schaefer 1976). Eaton (1914) reported that red and gray squirrels, the Eastern Screech-Owl, and the American Crow removed young from nests (the squirrels also take eggs). The beautiful nests, hung by the rim, are gourd-shaped, larger at the bottom to provide room for the young. The nest is almost always grayish, as gray plant fibers such as the outside of milkweed stalks and often pieces of string are used in construction. Lining consists of soft materials and also horsehairs where available (Eaton 1914; Saunders 1942). Atlas observers used the UN (used nest) code to "confirm" the Northern Oriole in 515 cases, more than for any other species.

Paul F. Connor

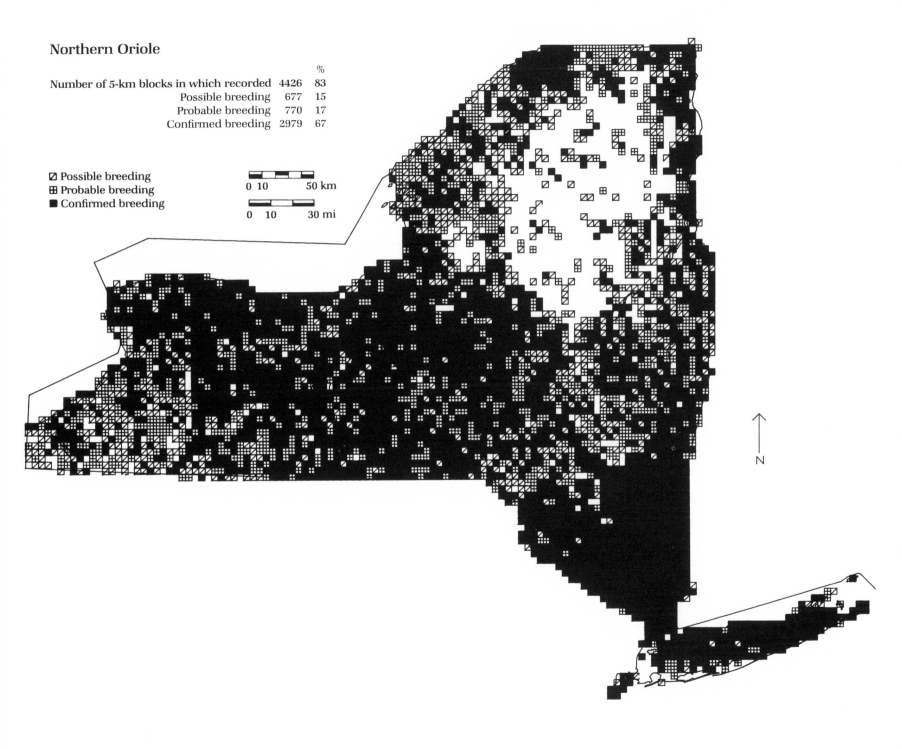

Northern Oriole

		%
Number of 5-km blocks in which recorded	4426	83
Possible breeding	677	15
Probable breeding	770	17
Confirmed breeding	2979	67

◩ Possible breeding
⊞ Probable breeding
■ Confirmed breeding

0 10   50 km

0 10   30 mi

N

Purple Finch *Carpodacus purpureus*

The Purple Finch, a handsome bird with a lovely song and nomadic winter habits, has an uncertain history as a breeding bird in the Northeast and in New York. Some writers in the first half of the 19th century spoke of the species as being uncommon or rare. De Kay (1844), for example, said it nested only in the northern part of the state, and Giraud (1844) never found a nest on Long Island. Allen (1869), however, noted a great increase in southern New England, and Bent (1968d) remembered it as "common enough" in the 1870s.

During the 20th century it appears to have become more and more common, except from New York City and Long Island, where it remains rare, and in western New York, where it was apparently once quite common but is now uncommon (Beardslee and Mitchell 1965). Eaton (1914) described it as one of the most common species in western New York and the spruce and balsam belt in the Adirondacks. He declared that it was "one of the characteristic summer residents in the greater portion of the state, found in open evergreen forests and swamps where cedars and firs are scattered about." Chapman (1906) noted that the Purple Finch was rather rare in the New York City area, and Griscom (1923) found it breeding in the city but only rarely on Long Island. Cruickshank (1942) reported a few pairs scattered along the north shore of Long Island. Bull (1974) considered this species common and widespread statewide, except for the New York City and Long Island areas.

The distribution map for the Purple Finch should be a pleasant surprise to those worried that this species was suffering from competition from the House Finch for winter food supplies and nest sites. It seems that the Purple Finch has not suffered greatly during the last 40 years, a period when the House Finch population burgeoned. The Atlas map indicates that the Purple Finch is widely and generally distributed in New York. Where there are gaps, the closest correlation seems to be mean July temperatures, with fewer records on Long Island, metropolitan New York City, the Hudson-Mohawk drainage, the St. Lawrence Plains, and western Great Lakes Plain, where the means are above 21°C (70°F). This species was also found in areas of heavy forest cover. There is a further correlation with elevation: lands below 152 m (500 ft) are least favored, but this is also related to mean temperatures and forest cover. Apparently, the Purple Finch in New York favors cool, forested, high elevations.

The favored nesting habitat of the Purple Finch varies. DeGraaf et al. (1980) reported that this bird could be found at "edges of coniferous forests, evergreen plantations, ornamental conifers in residential areas, parks, open mixed woodlands" and that the key habitat requirement was the presence of coniferous trees. The planting of conifers during the 1930s in New York most certainly provided additional habitat for this species.

The nest, usually placed in a conifer at a height of from 1.5 to 18.3 m (5–60 ft) (Forbush 1929), is most often on a horizontal limb close to the stem. It is woven of rootlets and weed stems and smoothly lined with finer vegetable matter and animal hair. Both adults build the nest; however, only the female incubates, attended by and often fed by the male. Both feed the young (Bent 1968d).

Robert Arbib

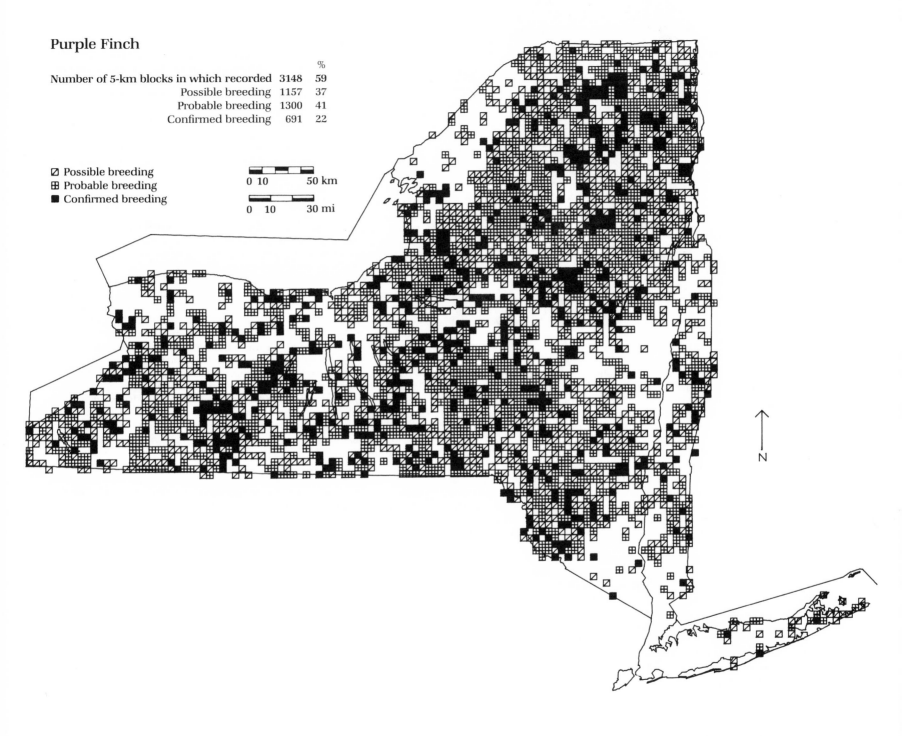

Purple Finch

Number of 5-km blocks in which recorded	3148	59
Possible breeding	1157	37
Probable breeding	1300	41
Confirmed breeding	691	22

%

☑ Possible breeding
⊞ Probable breeding
■ Confirmed breeding

0 10 50 km

0 10 30 mi

N

House Finch *Carpodacus mexicanus*

Perhaps no bird has so often been misidentified in New York as has the House Finch. Until recently, the House Finch was not found in the state, nor in the East; it was not illustrated in Peterson's widely used field guide before the 1980 edition; and it closely resembles the Purple Finch and suggests the Common Redpoll. Today, it is common in much of New York.

The House Finch invasion of New York began in the late 1930s when pet shop owners in the New York City area were importing thousands from California to sell as "red-headed or California linnets, or Hollywood finches." The National Audubon Society and the Bureau of Biological Survey were alerted to this illegal practice by a Dr. Fleisher, a Brooklyn birder, and ultimately the trapping and selling of these birds was banned. Apparently one dealer released his unsaleable birds in 1940 (Elliott and Arbib 1953).

A male House Finch was seen at Jones Beach, Nassau County, from 11 to 20 April 1941, and a year later, in March 1942, 7 birds were observed at Babylon, western Suffolk County. These birds were observed throughout the winter and into the following spring, and a nest with young was found in the summer of 1943. By 1946 the number at Babylon had increased to 50. The birds became established at several locations in Nassau County, and by 1949 the Long Island population increased to several dozen. By 1951 and 1952 there were frequent records outside of Long Island, in Westchester County and at Ridgewood, New Jersey (Elliott and Arbib 1953).

From 1948 onward the House Finch continued to expand farther into New York. CBC records between 1947 and 1985 reveal the pattern of its spread. Its first appearance was on the southern Nassau County CBC in 1947; then on the western Long Island CBC in 1949. In 1953 it was counted on the Brooklyn and Bronx CBCs. By the early 1970s it was observed during both the Troy and Schenectady CBCs in the upper Hudson Valley, as well as along the Appalachian Plateau on the Binghamton and Elmira CBCs. Along the Great Lakes Plain in 1976 the House Finch was reported on the Geneva and Rochester CBCs, and also in western New York on the Olean CBC. In 1979 it was reported on the Jamestown and Hamburg counts in far western New York. The House Finch was reported in 1985 on all New York CBCs but one, the Thousand Islands Ontario/New York count (*Audubon Field Notes* 1948–70; *American Birds* 1971–86).

While it was extending its range north and west, its numbers were also steadily increasing. For example, on the Binghamton CBC 2 birds were reported in 1971 and 610 in 1985; 2 birds were reported on the 1977 Buffalo CBC and 132 in 1985; and 34 birds were counted on the Oswego CBC in 1982, and in 1985 there were 283. There were similar increases throughout the state (*American Birds* 1972, 1978, 1983, 1986).

Mundinger and Hope (1982) stated that the House Finch had increased its range in the Northeast by two processes—diffusion, the gradual moving of a population across hospitable terrain over many generations, and jump-dispersal, the movement of individuals across great distances over inhospitable terrain. Its avoidance of such terrain as areas of the New Jersey pine barrens and the pitch pine–scrub oak forests of east central Long Island are examples of this latter process.

The Atlas map shows that the House Finch has moved up the Hudson Valley, spreading into the bordering highlands and as far north as the Canadian border through the Lake Champlain Valley. It also moved to central New York, from where it spread north along the Eastern Ontario Plains and along the Black River Valley and into western New York, where it is now absent mainly from the Cattaraugus Highlands and the Allegany Hills. In the Adirondacks it is now found in several villages.

Possibly a desirable result of the House Finch invasion was suggested by Kricher (1983), who found that where both the House Finch and the House Sparrow occur, the former species increased and the latter declined. It should be noted that there is also a decline of House Sparrow populations in parts of the country where there are, as yet, no House Finches.

The amazing expansion of the House Finch range and increase of its numbers may derive from several interacting factors—its fecundity, its ability to disperse, and its preference for towns and cultivated lands. The House Finch normally produces 2 but often 3 broods, with 4 or 5 eggs each, each year adding potentially 8–15 young (Woods 1968).

About towns and farmlands it uses an available tree, wall, tin can, old hat, stovepipe, hedge, flower basket, ledge, eaves, or mail box for a nest site (Woods 1968).

Gordon M. Meade

House Finch

Number of 5-km blocks in which recorded	2871	54 %
Possible breeding	518	18
Probable breeding	724	25
Confirmed breeding	1629	57

◩ Possible breeding
⊞ Probable breeding
■ Confirmed breeding

0 10   50 km

0 10   30 mi

N

Red Crossbill *Loxia curvirostra*

The nomadic Red Crossbill is rare to casual over New York in most years, although a few are found annually in suitable habitat, with most sightings made in the conifers of the Adirondacks. During years when a bumper crop of cones is available, crossbills of both species appear in great numbers, and the Red Crossbill becomes fairly common to locally abundant.

The first Red Crossbill nest recorded in the state, and apparently in North America, was discovered in late April 1875 at Riverdale, New York City, by Eugene Bicknell (1880). He found a female building a nest into which were laid 3 eggs. The first Long Island nest was found by Arthur H. Helme (1883) near Miller Place, Suffolk County, on 10 April 1883. Merriam (1881) wrote of this crossbill in the Adirondacks: "Abundant resident, rather scarce and irregular in summer, but the commonest bird in winter and early spring. Breeds in February and March while the snow is still four or five feet deep on the level and the temperature below zero [F]," and he collected fully fledged young as early as April. The first documented nest in northern New York was apparently not found until late May 1890 near Brandreth Lake, Hamilton County (Kennard 1895). Bull (1974) presented a map showing 23 known breeding locations: in the Adirondacks; in Schenectady, Schoharie, and Sullivan counties and on Long Island in eastern New York; in Schuyler and Tompkins counties in central New York; and in Niagara, Erie, and Allegany counties in western New York.

Between 1980 and 1984 the Red Crossbill was recorded in just 48 Atlas blocks, with the majority of records north of the Mohawk Valley, but by the end of those 5 years an invasion was commencing. Early in 1984 a nest was found on Tongue Mountain, west of Lake George in Warren County (Benkman, pers. comm.), and during the fall and winter of that year, reports became more and more numerous across the state. By January and February 1985 nesting was in progress, not only in the Adirondacks (Peterson 1985a), but also on the Appalachian Plateau (Crumb 1985; Messineo 1985). By late April nesting was ending, and many Red Crossbills were moving northward (Peterson 1985b). The Atlas map suggests the extent of the 1984 and 1985 invasion and nesting by Red Crossbills, thanks to a special concerted effort by many persons.

The Red Crossbill is generally found in coniferous or mixed conifer-hardwood forest. During the Atlas invasion year the Red Crossbill exploited various spruces, tamarack, and hemlock, both planted and naturally occurring. The century-old Adirondack and Catskill Forest Preserves contain a considerable amount of both spruce-fir and hemlock-hardwood forest (Considine 1984). On the Appalachian Plateau, New York State purchased almost 283,290 ha (700,000 a) of worn-out abandoned farmland during the Great Depression and reforested about half with spruce, pine, and European larch (Messineo 1985). All these areas had bumper cone crops in 1984 and 1985, which drew the Red Crossbill.

There are four bill size classes of the Red Crossbill, the differences in bill structure allowing birds in a given class to feed efficiently on particular conifer species (Phillips 1981). These classes may have evolved during the Pleistocene era in four isolated refugia of conifers composed of differing tree species and thus cone sizes (Benkman, pers. comm.) At least two forms were present during the recent invasion.

Dickerman (1986a) presented a map showing the North American "core ranges" of the four size classes of Red Crossbill, noting, "In an irruptive species such as the Red Crossbill, which during a year of cone-crop failure may move hundreds or thousands of miles and, upon reaching a region with abundant food, may settle and even nest within the range of another subspecies, the usual term 'breeding range' is meaningless." At least two, and probably three, subspecies have bred in New York, according to Dickerman (1986b). The most numerous nesting form in New York may well be the small subspecies whose core range is the coastal region of the Pacific Northwest. New York specimens from the 1984–85 invasion were of this small northwestern and the large Rocky Mountain forms; this was the first invasion of the state in which two "exotic" subspecies were proven to occur (Dickerman 1986b).

Three nests of the Red Crossbill found in February 1985 were bulky structures placed in red spruce trees at heights of 7.6, 10.7, and 12.2 m (25, 35, and 40 ft), in a forest of red spruce, hemlock, white pine, and northern hardwoods in Hamilton County (C. Benkman, v.r.). Other New York nests have been found in pine and cedar (Bull 1974).

John M. C. Peterson

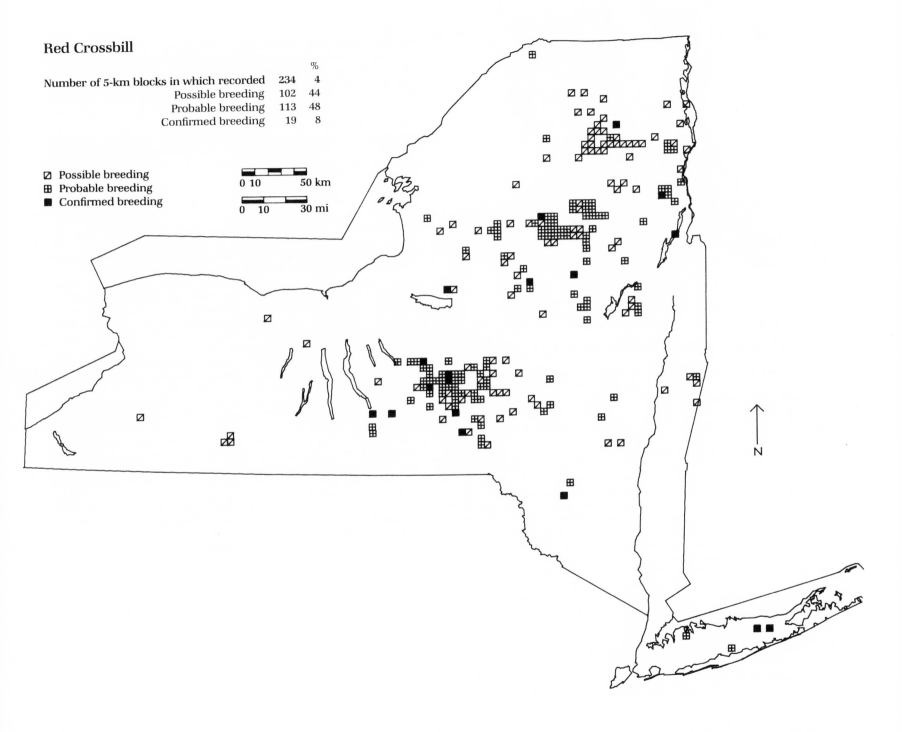

Red Crossbill

		%
Number of 5-km blocks in which recorded	234	4
Possible breeding	102	44
Probable breeding	113	48
Confirmed breeding	19	8

☒ Possible breeding
⊞ Probable breeding
■ Confirmed breeding

0 10   50 km

0 10   30 mi

N

White-winged Crossbill *Loxia leucoptera*

During years when the spruce cone crop is abundant, the White-winged Crossbill is locally common to abundant in parts of the Adirondacks and elsewhere in New York. Usually, however, it is a rare breeder, restricted to the Adirondacks.

Merriam (1881) called the White-winged Crossbill a resident of the Adirondacks, but not nearly so common as the Red Crossbill. Eaton (1914) noted that in years when food was abundant it had wintered in western New York, the Finger Lakes area, in New York City, and on Long Island to Montauk Point. In the fifth edition of the AOU Check-list the White-winged Crossbill was said to breed in northeastern New York's "eastern Lewis County, Long Lake [Hamilton County]" (AOU 1957). Up to then, however, no breeding had been documented. But on 19 July 1958 a pair was observed in Essex County by Hugh Fosburgh, who reported: "They were obviously searching for something—food I thought—but then the female flew onto the Hereford-skin rug that I had out drying in the sun, and proceeded to tweak a beakful of hair from it. The male sat on the edge of the wheelbarrow and waited quietly until she had accomplished this purpose, then they flew down in the swamp together" (Fosburgh 1960). No nest was found (J. Peterson 1980).

Bull (1974), unaware of this record ("probable" by Atlas criteria), noted that although the White-winged Crossbill very likely nested in the Adirondacks, he was unable to find a single positive breeding record for the state. Finally, on 22 February 1975, a search by members of High Peaks Audubon Society was rewarded when a female White-winged Crossbill was found building a nest at Chubb River Swamp, Essex County, while the male sang atop a nearby spruce. The nest bough in a black spruce was covered with heavy snow and ice by a snowstorm 3–5 April, but up to 20 singing males along the river between 20 January and 17 May suggested that successful nesting might have occurred (Peterson 1975). On 30 December 1979, just 2 days before Atlas fieldwork began, a pair was again found at Chubb River, the female carrying nesting material (Nickerson, pers. comm.).

During the next 4 years spruce cones were poor to nonexistent in the state, and the White-winged Crossbill was recorded in only nine Atlas blocks. A bumper cone crop, the best since 1974 and 1975, began to form in 1984, and this crossbill began to arrive in July. The first "confirmed" record of nesting came on 11 September 1984, at Bloomingdale Bog, Franklin County, when I saw a young fledgling accompanying an adult male in a cone-laden white spruce.

By the winter of 1984–85 the White-winged Crossbill was reported to be in almost every stand of spruce in the Adirondacks; by early 1985 it was clearly breeding outside the Adirondacks in conifer plantations on the Appalachian Plateau (Crumb 1985; Messineo 1985), areas acquired by the state and planted in conifers during the 1930s. Although the White-winged Crossbill had previously occurred in numbers outside the Adirondacks and past breeding is a possibility, the first "confirmed" record came on 1 April 1985, when a fledgling White-winged Crossbill was found in the Town of Otselic, Chenango County (Messineo 1985). The accompanying map reflects this invasion. The species is certainly underrecorded, given the attendant difficulties of obtaining winter coverage in remote areas. For the first time, however, the extent of a breeding invasion by this highly nomadic species can be seen.

Breeding White-winged Crossbills occur in coniferous forest (especially spruce, balsam fir, and tamarack), mixed conifer-deciduous woodland, and forest edge (AOU 1983). It nests during three periods, which are marked by diet shifts to different conifer cones as seeds become available (Benkman, pers. comm.). Plantations of European larch and Norway spruce have provided a new food source and nesting habitat on the Appalachian Plateau. The Norway spruce holds seeds for a long period, and European larch matures late. This may be one reason nesting of this crossbill in this area was later. Road salt lures countless White-winged Crossbills—together with Pine Grosbeaks, Pine Siskins, and Red Crossbills—to winter highways, where thousands are struck and killed by motor vehicles during flight years (Meade 1942).

The nest found at Chubb River Swamp in 1975 was located deep within a thick black spruce bough that extended over a trail at a height of approximately 7.6 m (25 ft), and the principal material of the rounded nest appeared to be *Usnea* lichen (Peterson 1975). No nest with eggs or young of White-winged Crossbill has apparently been found in New York.

John M. C. Peterson

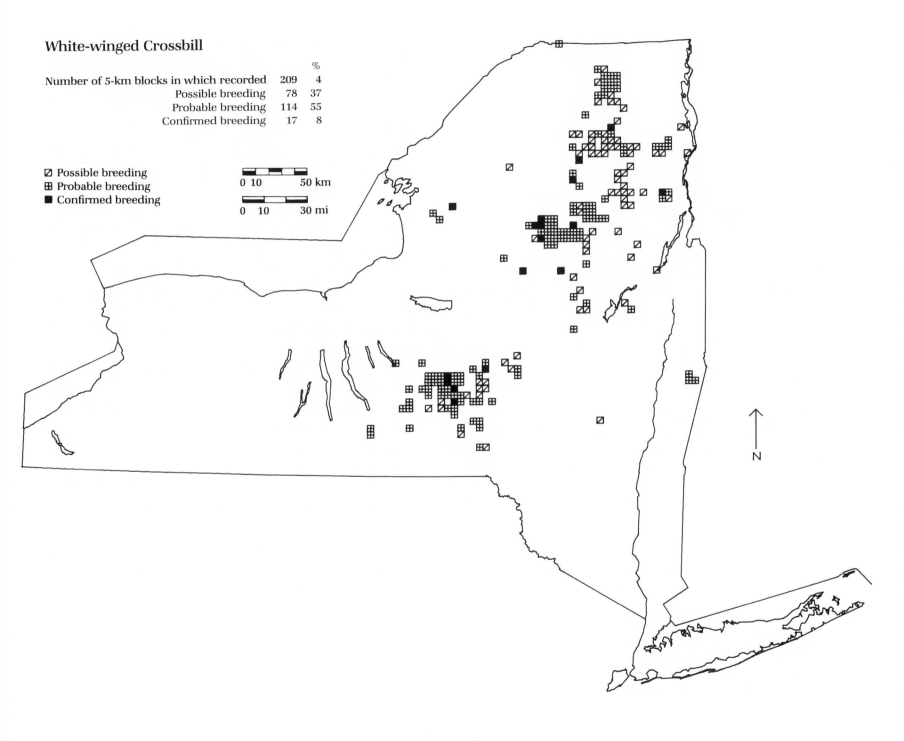

White-winged Crossbill

		%
Number of 5-km blocks in which recorded	209	4
Possible breeding	78	37
Probable breeding	114	55
Confirmed breeding	17	8

◩ Possible breeding
⊞ Probable breeding
■ Confirmed breeding

0 10   50 km

0 10   30 mi

N

Pine Siskin *Carduelis pinus*

Except during major breeding years, the Pine Siskin, a small finch, is a rare nester over most of New York, and even rather uncommon in the Adirondacks. Yet in the scattered years when the large, wandering flocks arrive in the state and subsequently break up into smaller nesting bands, the Pine Siskin can suddenly appear common to locally abundant wherever suitable habitat is available. During the period of the Atlas fieldwork there were big incursions in parts of New York during the winter and spring of 1981, and again in 1982 (Mutchler 1983), but with little subsequent nesting. Finally, in the winter of 1985 enormous numbers of siskins settled down within the forests of the Adirondacks and elsewhere around the state to breed.

These somewhat erratic appearances have been commented upon for over a century, but there may be certain long-term changes in distribution or abundance. Merriam (1881) described the pine linnet or siskin within the Adirondack region as "an irregular visitor; sometimes breeding in vast multitudes, and during other seasons not seen at all." Several decades later Eaton (1914) commented, "It must not be assumed that it does not occur nearly every season, especially in the eastern and northern portions of the State, but certainly is rather uncommon except at intervals of a few years." Eaton described breeding records outside the Adirondacks at Ossining, Westchester County, and Cornwall, Orange County, as "certainly very exceptional." Six decades later Bull (1974) observed, "Like the Red Crossbill, the Pine Siskin is little known as a breeder in the state and, if possible, even more erratic and unpredictable in its time of nesting than that species." Bull's accompanying map of breeding distribution showed only eight additional sites away from the Adirondacks.

The Atlas map shows a rather dramatic change in distribution, although this can be attributed, in part, to the concerted effort of observers, together with the good fortune that 1985 was an exceptional year for siskin nesting in New York. The Pine Siskin was found in the Catskill region and many of the reforestation and conifer plantations along the eastern Appalachian Plateau that have helped increase the ranges of other state birds. In addition, it was found along the Great Lakes Plain and in the Cattaraugus Highlands, where previous records were scarce or lacking. At least some of the "possible" records may reflect blocks where siskins were present in suitable habitat but no breeding took place. Yet, had it been possible to revisit every Adirondack block during the winter and spring of 1985, the number of blocks with siskin records in northern New York would probably be far greater. The present Atlas map provides a generally accurate picture of an incursive species, which includes data from a major nesting year.

If there have been few historical changes in its irregular abundance, the seasonal movement of the Pine Siskin suggests that it may have once been entirely absent from New York. Within the interior of North America many siskins show a southeasterly movement in fall and a return flight to the Northwest in spring (Palmer 1968). Palmer has also suggested that the Pine Siskin "may have spread eastward, as the evening grosbeak did at a later period, but before the arrival could be chronicled."

Once wandering flocks locate an adequate food supply, they break up into smaller social flocks of perhaps a half-dozen adults to initiate nesting. During this period, which may begin as early as February, siskins engage in territorial chases, song flights, singing, and mate feeding. Nesting may be loosely colonial, or a single pair nesting alone may be accompanied at times by other pairs.

The nest is a rather flat structure of twigs, rootlets, and grasses, lined with feathers, fur, or hair; it is usually placed on a dense outer limb in a conifer. Before the Atlas survey, only 5 New York nests apparently had been found. Those were placed in cedar, hemlock, pine, or spruce trees (Bull 1974). One at Chubb River Swamp, Essex County, in 1975 was being constructed on 29 March (Peterson 1975); another near Elmira, Chemung County, in 1893 held a single egg on 20 July (Bull 1974). Heights of these nests were 7.6–10.7 m (25–35 ft) above the ground. Some young may be fully fledged just as neighboring pairs begin nest building (pers. obs.).

On the basis of his banding studies in Saratoga and Schenectady counties, Yunick (1976) suggested, "It is not unrealistic to expect that some individuals breed very late in the course of the season and give rise to a group of young that possess molting traits well into winter." Breeding Pine Siskins may leave during late spring, or remain into summer, before resuming their wanderings.

John M. C. Peterson

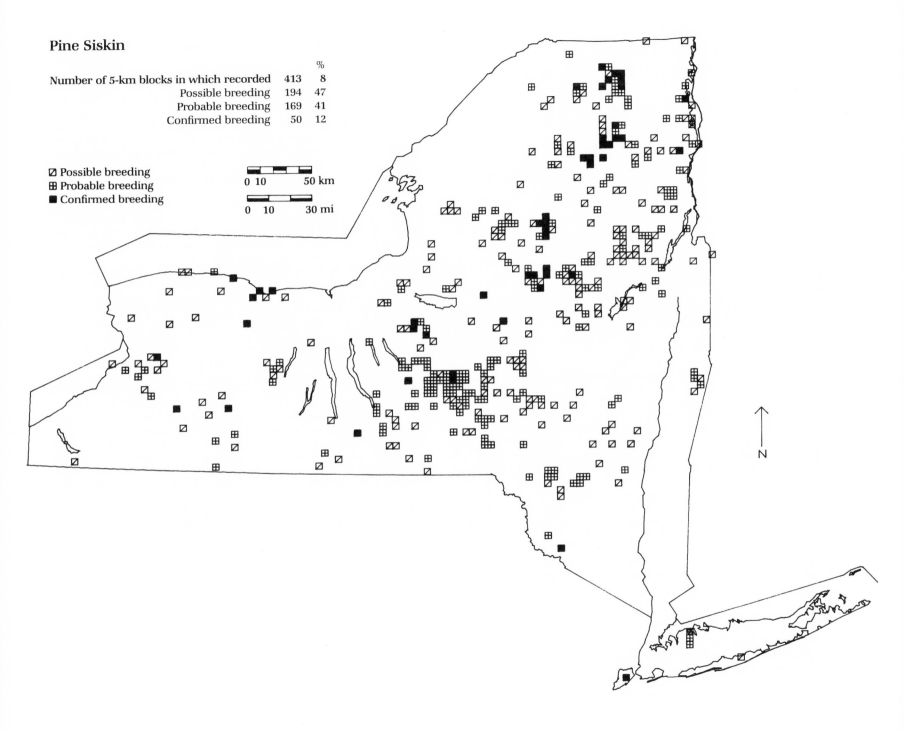

Pine Siskin

		%
Number of 5-km blocks in which recorded	413	8
Possible breeding	194	47
Probable breeding	169	41
Confirmed breeding	50	12

☑ Possible breeding
⊞ Probable breeding
■ Confirmed breeding

0 10   50 km

0 10   30 mi

N

American Goldfinch *Carduelis tristis*

The American Goldfinch is a common breeding species in New York and apparently has been for some time. Eaton (1914) considered it to be a "fairly abundant" breeding species throughout the state except in densely forested areas, and Bull (1974) described it as a "widespread breeder." Its status, however, may be changing. Using BBS data, Robbins et al. (1986) described a significant continental population decline for the American Goldfinch. A similar trend was reported for New York, even though New York and Ontario represent the geographic region of greatest abundance for the species.

The Atlas map shows a breeding distribution for the American Goldfinch like that described by Eaton and Bull, with absences of breeding evidence only from parts of the Adirondacks and from the densely populated areas of New York City and Long Island. Only 12 species were reported in more Atlas blocks.

The American Goldfinch begins nesting later than any other breeding species in the state, rarely beginning nest building before the last week of June (Eaton 1914). This late nesting was a disadvantage to Atlas observers, who usually finished fieldwork by August, when nestlings and fledglings normally appeared (Bull 1974). "Confirmed" breeding activity was consequently observed less frequently than is suggested by the substantial number of reports of "probable" breeding. Late nesting has at least two advantages to the bird. First, it makes the species less susceptible to brood parasitism by the Brown-headed Cowbird (Friedmann 1963). Second, it assures the availability of seeds for feeding young, since the adult goldfinch typically feeds its young a regurgitate of partially digested seeds (Stokes 1979).

Nickell (1951) commented that the American Goldfinch is predominantly a bird of open country and believed that both its abundance and geographic range had increased since precolonial times. Bull (1974) reported that the goldfinch favors croplands and pastureland, successional old fields, hedgerows, orchards, and woodland borders, all of which result from human activities. Before the arrival of western European settlers, the American Goldfinch probably depended largely upon openings associated with swamps and marshes, beaver meadows, lake shores, river banks, and areas cleared by forest fires. Nickell (1951) suggested that in such areas the goldfinch placed its nest near the tops of swamp shrubs or near the ends of horizontal branches at the edges of forests, much as it does in similar situations today. Nickell also remarked that the current widespread distribution of two important introduced food plants, the thistle and the dandelion, may also have contributed to the goldfinch's success as a species.

In his studies of songbirds' use of upland clear-cuts on the Connecticut Hill WMA in Tompkins and Schuyler counties, Keller (1980) found the goldfinch nesting in successional old fields, grassy areas interspersed with saplings, and on clear-cuts 2 to 3 years after cutting had occurred. None were found nesting on clear-cuts older than 3 years after cutting. Keller reported nesting densities of from 3 to 21 pairs per 40.5 ha (100 a).

The nest of the American Goldfinch is a remarkable example of avian architecture. It typically is lined with thistle or cat-tail down, with an outer supporting layer woven from fibers stripped from the stems of milkweed, sometimes held in place with spider or tent caterpillar webs (Nickell 1951). Stokes and Stokes (1986) offered a fascinating account of a female American Goldfinch collecting webbing from a nest of fall webworms: "She fluttered her wings in the webbing, then flew to her nest where she carefully pulled pieces of it off her wings with her bill and applied them to the nest." The nest of the American Goldfinch typically is cradled in an upright fork, but occasionally it can be found saddled on a horizontal branch, at an average height of about 1.8 m (6 ft) above the ground (Nickell 1951).

The goldfinch nest is an extremely durable structure. It is so tightly constructed that it can hold water after a heavy rain, and Allen (1935) reported that young goldfinches have been found drowned in water-filled nests. Used nests usually persist for two or three seasons, but their similarity in placement and construction to nests of the Yellow Warbler made it difficult for Atlas observers to distinguish used nests of the two species in the field. Sometimes birds dismantle old nests and recycle materials to build the nests of a new season.

Charles R. Smith

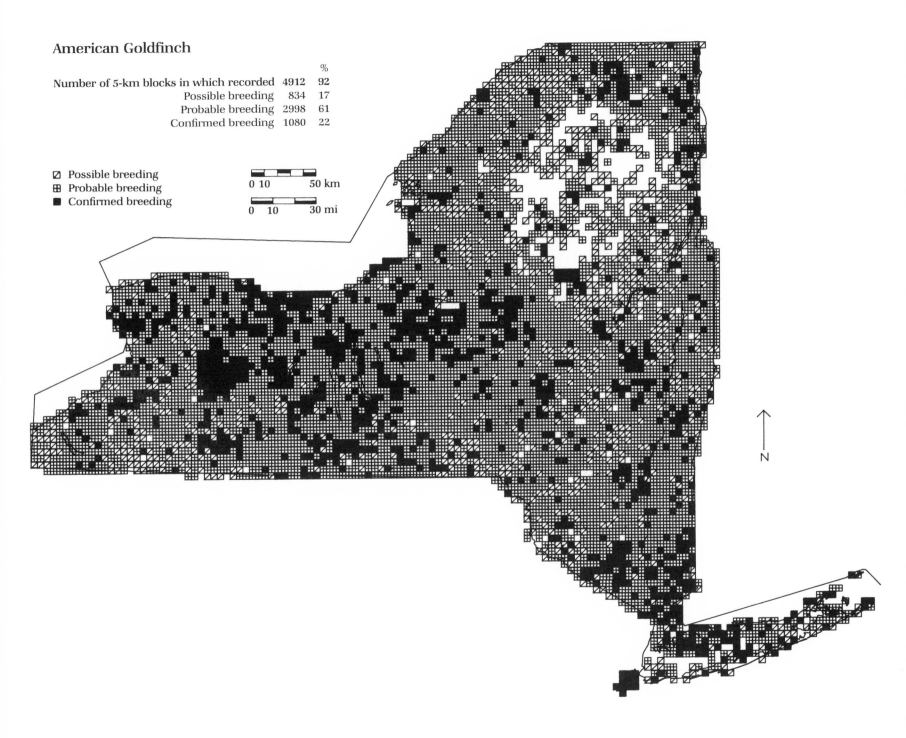

American Goldfinch

		%
Number of 5-km blocks in which recorded	4912	92
Possible breeding	834	17
Probable breeding	2998	61
Confirmed breeding	1080	22

◨ Possible breeding
⊞ Probable breeding
■ Confirmed breeding

0 10   50 km

0 10   30 mi

N

Evening Grosbeak *Coccothraustes vespertinus*

A relatively recent addition to the list of breeding birds in New York, the Evening Grosbeak is now a fairly common, but at times localized, nester in the Adirondacks. In recent years new populations have appeared in other parts of the state, where this chunky finch is still a rare to uncommon breeder.

The range of the Evening Grosbeak began its eastward growth over a century ago. Although the first Evening Grosbeak may have been seen near New York City as early as 1866 (Lawrence 1866), the first documented observation was made by the Rev. Dr. Sewall Sylvester Cutting at Elizabethtown, Essex County, in the winter of 1875 (Brewer 1875). The first summer record was at Marcellus, Onondaga County, on 8 July 1882, but this was exceptional (Eaton 1914). The first major invasion of the state began with sightings at Ithaca, Tompkins County, at Lockport, Niagara County, and at Brockport, Monroe County, in December 1889, and by April of 1890 reports came from at least 15 communities across New York (Eaton 1914).

Thereafter, observers were eager to document their first encounters with this striking bird, and for a half-century or more notes appeared in ornithological journals, tracing the grosbeak invasion eastward across Canada and the United States, with at least 30 new locations described in New York. These were all winter sightings, however, and birds disappeared before mid-May. Yet by 1940 the Evening Grosbeak occupied "a rather narrow but probably unbroken transcontinental belt" in summer, extending from western Alberta to eastern Massachusetts, but New York was represented by only a single sighting on Grindstone Island in the St. Lawrence River on 2 August 1939 (Baillie 1940).

More suggestive breeding evidence was obtained in July 1942 at Elk Lake and Clear Pond, Essex County, where Edward Fleisher saw several birds, including a pair (Fleisher 1943). In St. Lawrence County a pair was observed on 31 May, and by June there were at least 3 pairs on the campus of the New York State Ranger School, with another pair visiting a probable nest site in the crown of a red spruce on 27 June at Cranberry Lake. Breeding was finally confirmed with the discovery of young birds near Bay Pond, Franklin County, on 3 August 1946 (Beehler 1978). Thereafter the spread was rapid, with several observations of young accompanying adults to feeders (Schaub 1951). In 1962, following a major flight, the Evening Grosbeak moved outside the Adirondacks to nest: "It was the first and *only* time that the species bred far outside the usual boreal evergreen forest zone," according to Bull (1974).

The map of distribution presented by Bull (1974) showed the Evening Grosbeak breeding in 13 counties in what he called the "greater" Adirondack area, with extralimital 1962 records in Dutchess, Monroe, Schenectady, and Tompkins counties. The Atlas map presents a similar pattern of distribution, with a strong concentration in the Adirondack High Peaks, Central Adirondacks, and Western Adirondack Foothills, with breeding extending outward. Elsewhere in the state there are scattered records from a number of areas. Records were added in some 17 counties, yet there were no records in Dutchess, Fulton, or Jefferson counties where noted earlier by Bull. The Evening Grosbeak's fondness for the seeds of the "ash-leaved maple," or box-elder, was noted by Verdi Burtch as early as 1911 (Eaton 1914). This tree is credited with luring the Evening Grosbeak eastward, P. A. Taverner describing a "baited highway" along which the birds were able to pass (Baillie 1940). Since then, Belknap (1973) noted, "In more recent years the very extensive availability of sunflower seed at feeding trays has been an influence."

Nesting habitat has been described as "coniferous (primarily spruce and fir) and mixed coniferous-deciduous woodland, second growth, and occasionally parks" (AOU 1983). The Evening Grosbeak is more of a generalist than many other northern species, tolerating a wider variety of habitats; even within the Adirondacks this grosbeak is more common around hamlets than in wilderness areas. Many "confirmed" records (22 FL [fledglings], 14 FY [adults carrying food or feeding young]) came from family groups at feeders in villages or rural yards, the young sometimes hardly capable of sustained flight.

Nests appear difficult to locate, only 4 nests with eggs having been found prior to the Atlas effort. Two were in spruce, 1 in pine, and 1 in a maple, ranging from 12.2 to 15.2 m (40–50 ft) above the ground at the forest edge (Bull 1974). Only 1 nest was found during the Atlas period; it was the first nest with young ever found in New York.

John M. C. Peterson

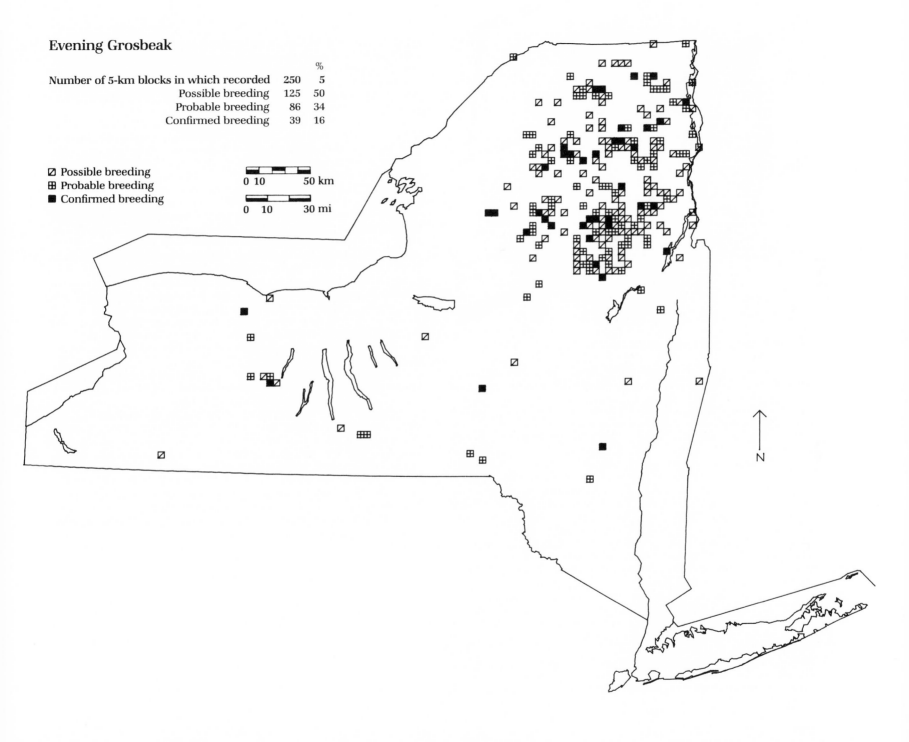

Evening Grosbeak

		%
Number of 5-km blocks in which recorded	250	5
Possible breeding	125	50
Probable breeding	86	34
Confirmed breeding	39	16

◩ Possible breeding
⊞ Probable breeding
■ Confirmed breeding

0 10   50 km

0 10   30 mi

N

CJ Page
©1987

House Sparrow *Passer domesticus*

The ubiquitous House Sparrow, now classified as a member of the family Passeridae, the Old World Sparrows (AOU 1983), is one of the most familiar and least loved breeding birds. It is common to abundant everywhere in close association with humans—in cities, towns, villages, around farms, feedlots, and garbage dumps.

It was first unsuccessfully introduced in 1850 from England (hence the earlier name, English sparrow) into Brooklyn by a persistent and misguided Nicholas Pike. He released it again at The Narrows in 1852 (fate unknown), and successfully in Greenwood Cemetery in 1853. The hardy, adaptable species spread rapidly. Other introductions followed: Portland, Maine, in 1854 and 1858, and then Boston, Rochester, New Haven, Galveston, Cleveland, Philadelphia, Salt Lake City, and many other sites as far west as San Francisco were victims of this ill-advised fad (Barrows 1889). By 1910, spreading from community to community, especially along railway lines, it had become, in the opinion of some observers, the most abundant species in the continental United States, except in forest, alpine, and desert regions (Bent 1958). The great increase of the House Sparrow and European Starling populations are two outstanding examples of the unforeseen and disastrous effects of foreign introductions.

The peak of abundance apparently came about 1910–15 (Bent 1958); then a sharp diminution, especially in the cities, was recorded. It is accepted dogma that the replacement of the urban horse by the automobile, which produced no edible by-product, was a major cause. The enormous flocks of city sparrows, which had fed around the crushed and dried, oat-laden droppings, diminished. Similar decreases occurred when the rural horse was supplanted by truck and tractor. A factor not often mentioned is the evolution of both urban and suburban architecture, from Victorian decoration with its myriad nesting places to modern simplicity that eliminated nooks and crannies where sparrows could nest. Nonetheless, the House Sparrow must still be classified as an abundant and widespread breeding bird today.

Some authors have detected a continued decline in the numbers of the House Sparrow, especially since the introduction of the House Finch in the Northeast (Kricher 1983), but analysis of CBC data from four areas with long, continuous records reveals no positive trends. Considering only Bronx-Westchester, Buffalo (Erie County), Rochester (Monroe County), and Lower Hudson (Manhattan), birds per party hour averaged 7.47 in 1947, 13.82 in 1957, 14.93 in 1967, 13.10 in 1977, and 11.31 in 1985.

As the Atlas map clearly shows, this bird is absent only from forested or sparsely populated areas at higher elevations, especially in the Adirondacks, Central Tug Hill, Catskills, and Allegany Hills. Even in these areas it can be seen in and around human habitation in the narrow valleys. It gleans sustenance from garden and lawn seeds, farm fields, bird feeders (where it is a wasteful glutton), and barnyard waste. The species has been blamed for severe negative effect on the nesting of bluebirds, swallows, and other hole or box nesters by usurping their nest sites. It is said, with exaggeration, that every martin house has one pair of House Sparrows acting as landlords! Numerous authors earlier in this century deplored the reduction in numbers of native species attributed to the aggressive ways of this interloper.

The House Sparrow is not a species that could easily be missed by Atlas workers, nor was "confirming" the bird difficult. About half of the "confirmed" records were of nests, half of fledglings.

The House Sparrow is an exceedingly sloppy homemaker. Its nests are commonly a loose aggregation of fibers, grass stems, leaves, feathers, string, or other adventitious materials generally roofed over and lined. It is stuffed into any available sheltered cavity: a bluebird box, martin house, swallow nest, eaves under buildings, attic louvres, behind shutters, or anywhere it can be fitted. The nest may be as large as 30.5 by 50.8 cm (12 × 20 in) with a 7.6 cm (3 in) inside diameter "pocket." It may be as high as 15.2 m (50 ft), but it is usually lower. Both sexes construct the nest. This highly fecund species is double- or perhaps even triple-brooded (Bent 1958).

Robert Arbib

House Sparrow

		%
Number of 5-km blocks in which recorded	4196	79
Possible breeding	377	9
Probable breeding	407	10
Confirmed breeding	3412	81

▨ Possible breeding
⊞ Probable breeding
■ Confirmed breeding

0 10   50 km

0 10   30 mi

N

THE ECOZONES OF NEW YORK STATE

The ecological zones portrayed on Map 3 in this book were derived from somewhat differing bases in the northern part of the state than in the southern and western parts. However, this map shows the physiography of New York State very well. In many cases these ecozones relate to the distribution of breeding birds. Some natural and cultural factors such as land forms and topography, bedrock geology, climate, vegetation, and land use interact to delineate these zones. Some can be considered subzones of major zones, and others transitions relating to forest and nonforest cover between zones. More detailed discussions of physical and cultural features can be found in Dickinson (1979, 1983) and Will et al. (1982), from which much of this section is derived. Following are brief descriptions of the characteristics of each area.

Lowlands

Great Lakes Plain

The Great Lakes Plain, covering approximately 18,664 sq km (7206 sq mi), is a low terrain ecozone with horizontal rock formations. It is essentially a flat plain having little local relief, with the exception of the drumlin area between Rochester and Syracuse. The plain has a simple erosional topography of glacial till, modified by moraines, shoreline deposits and drumlins. The natural vegetation consists of elm–red maple–northern hardwoods, with beech, white ash, basswood, sugar maple, hickory, hemlock, and tulip tree predominating on better drained sites. Farms and orchards predominate, with only about one-fifth of the land forested, mainly in a disrupted pattern. Climate is equable, modified by Lakes Erie and Ontario. Elevation ranges from 75 m (245 ft) at the Lake Ontario shoreline to about 305 m (1000 ft), but it is mostly under 244 m (800 ft).

Erie-Ontario Plain

A large subzone of the Great Lakes Plain, the Erie-Ontario Plain borders Lakes Erie and Ontario and consists of relatively level terrain that is about

15% wooded in disrupted sections. Forests are composed of species similar to those of the Great Lakes Plain generally. Much of the area is farmland. It covers 13,080 sq km (5050 sq mi). Elevation along the Lake Ontario shore begins at 75 m (245 ft) and along the Lake Erie shore at about 174 m (527 ft). Almost the entire subzone is under 244 m (800 ft) in elevation.

Oswego Lowlands

Rolling plains of about 777 sq km (300 sq mi) with elevation varying from 91 to 152 m (300–500 ft) are typical of the Oswego Lowlands, a small lowland subzone. Lake Ontario exerts a moderating effect conducive to agriculture. Forest types of elm and red maple exist but are not extensive. Moderately productive muck soils overlie sandstone.

Drumlins

Prominent oval, elongated drumlins, generally aligned northeast-southwest and composed of glacial deposits, cover 2849 sq km (1100 sq mi) in the subzone known as the Drumlins. The area is about 25% forested and has some farming. Elevation ranges from 91 m (300 ft) near Lake Ontario to a maximum of about 244 m (800 ft) above sea level at the summits of drumlins.

Eastern Ontario Plains

The Eastern Ontario Plains, a nearly level area, ranges in elevation from 76 to 152 m (250–500 ft). Lake Ontario moderates the climate for agriculture and dairying. Elm–red maple and northern hardwoods are the dominant forest types. The subzone covers about 1958 sq km (756 sq mi). Soils are mostly lake sediments over limestone bedrock.

St. Lawrence Plains

The St. Lawrence Plains extends over about 2222 sq km (858 sq mi) and generally is a flat and rolling plain ranging in elevation from 76 to 122 m

(250–400 ft). Northern hardwoods are dominant in small woodlots, often in low, swampy areas. Land abandonment has resulted in considerable shrubland, and it possesses more hills and swamp forest with elm and red maple than does the Eastern Ontario Plains. Soils of medium productivity overlie limestone and sandstone. Higher agricultural production distinguishes this area from the transition zones to the north and east.

Malone Plain

Topography in the Malone Plain varies from flat to rolling plains. The area covers about 453 sq km (175 sq mi) with elevation from 122 to 305 m (400–1000 ft). Aspen, gray birch, and paper birch make up the principal forest type. Agriculture is quite extensive but declining as land is abandoned. Clay soils cover sandstone.

Lake Champlain Valley

A lowland that covers about 746 sq km (288 sq mi), the Lake Champlain Valley contains natural forests of white pine, northern hardwoods, and pioneer species. Elevation ranges from 30 to 213 m (100–700 ft). Gentle relief and a rather mild climate owing to its proximity to Lake Champlain are characteristic, with farming prominent. It is bounded on the west and south by the Precambrian Shield and has clay soils derived from sediments of glacial lakes.

Indian River Lakes

Encompassed by the St. Lawrence Plains and Eastern Ontario Plains, the Indian River Lakes ecozone, a lowland of 624 sq km (241 sq mi), consists primarily of rolling hills and granite outcrops ranging in elevation from 107 to 152 m (350–500 ft). Precambrian granite and intruding Potsdam sedimentary sandstones underlie the shallow, poorly drained soils. Forests are a transition between northern hardwoods and oak-hickory of more southern affinity. Glacial lakes, outcrops, and rough terrain prevail in an area of chiefly recreational use.

Black River Valley

A largely agricultural area with some woodlots of northern hardwoods, the Black River Valley occupies about 907 sq km (350 sq mi) and has an average elevation of about 305 m (1000 ft). It lies between two large second-growth forest regions and old fields with the higher Tug Hill Plateau to the west, the Precambrian Shield on the east, the divide between the Mohawk and Black River drainages on the south, and the escarpment of the Tug Hill Plateau and margin of the flat Eastern Ontario Plains on the north. Rich loam soils cover the limestone bedrock.

Oneida Lake Plain

The Oneida Lake Plain surrounds Oneida Lake but is not named on the ecozone map. It is largely flat plain with elevation from 91 to 183 m (300–600 ft) and occupying about 798 sq km (308 sq mi). Farming is widespread. Forests are mainly elm and red maple with swamps prominent. Northern hardwoods occur in small woodlots on the higher portions. Highly productive soils cover the sandstone and limestone bedrock. Oneida Lake helps moderate the climate.

Mohawk Valley

The Mohawk Valley, a zone of over 4765 sq km (1840 sq mi), is characterized by a valley of variable terrain with soft sedimentary rock overlaid by glacial till. Rolling plains with gentle slopes and low local relief occur, as well as some hills with moderate slopes and higher relief. Elevation is from 152 m (500 ft) to over 518 m (1700 ft). The narrow inner river valley has elevations 305 m (1000 ft) below the country to the north and south. This zone generally has cooler temperature than the Great Lakes Plain. Northern hardwoods and associated species predominate, with hemlock stands in some ravines, white pine and cedar in a few swamps, and oaks on the shale slopes. Forested areas are mostly on farms and occupy less than 20% of the land. Manufacturing industries, intensive dairy farming, and a variety of crop farms are major uses of land in this area.

Hudson Valley

Rolling plains and hills bound the Hudson Valley, a 7485 sq km (2890 sq mi) depression, interlaced with long, narrow stream bottomlands. Most elevations are below 152 m (500 ft) and near sea level to the south. Hills and terraces overlie highly folded sedimentary rock. Hills exceeding 305 m (1000 ft) above sea level rise toward the south. The climate is relatively mild and the vegetation, oak–northern hardwoods and white pine, and the Albany sand plains support pitch pines and scrub oaks. Oaks are common on south slopes and red cedar in abandoned fields. In many places up to 50% of the land is wooded. Several types of agriculture are practiced, there are major transportation centers, and population density is high, particularly in the north.

Central Hudson

The Central Hudson subzone, 5672 sq km (2190 sq mi) in area, is generally flat to rolling land below 152 m (500 ft) but with a number of hilltops reaching over 305 m (1000 ft). Northern and pioneer hardwoods are the most extensive forest types, with some white pine and red cedar. There is a mixture of industry, residential centers, and a variety of farms in the area.

Triassic Lowlands

A small 311 sq km (120 sq mi) zone of low relief and gently rolling plains, the Triassic Lowlands includes on the east the Palisades, a large igneous escarpment adjacent to the Hudson River. Soils of glacial till are deep and loamy from the underlying sandstone, limestone, and shale, with intervening lava flows protruding in some places. About 75% of the land is covered by oak–northern hardwood forest, despite the acceleration of residential and commercial development. Climate is warm and humid in summer and mild and wet in winter. At the base of the Palisades altitude is near sea level, but most of the area is above 61 m (200 ft) with maximum local elevation about 183 m (600 ft).

Coastal Lowlands

The Coastal Lowlands ecozone, a part of the Atlantic Coastal Plain, has very low relief, mostly below 61 m (200 ft) but reaching a maximum of 122 m (400 ft). It totals 3445 sq km (1330 sq mi). Covered by glacial drift, it is underlain by sands and clays. The climate is moderated by the ocean. Because of poor soils, scrub oaks dominate the zone, with pitch pine the main conifer often in mixture. In addition, tulip tree, sweet birch, sugar and red maples, and elm occur. Much of the vegetation of the zone, as well as farmland, is being lost by rapid urban and suburban expansion, although this development is less pronounced toward the eastern parts of Long Island.

Highlands and Transitions

The Highlands and Transitions are composed of mountains and hills (highlands), as well as ecotones between cleared land and forest (transitions) typically having gently rolling plains and low hills.

Appalachian Plateau

A 43,434 sq km (16,770 sq mi) zone, the Appalachian Plateau is mainly hill country with deeply dissected valleys, but it retains a plateau aspect with more or less flat-topped hills and with a skyline about 610 m (2000 ft). Elevation ranges from 152 m (500 ft) near the Hudson Valley to over 1219 m (4000 ft) in the Catskills. Water erosion largely determined the topography, with glaciers later modifying it. Escarpments exist in some places. Most of the plateau has cold, snowy winters and cool, wet summers. Northern hardwoods cover much of the zone, with oaks abundant especially on south-facing slopes. Beech, sugar maple, basswood, white ash, and black cherry predominate; red oak is the most frequent where oaks occur. Hemlock and white pine are also found, along with many plantations of spruce and pine. About one-third of the total area is forested. Land abandonment is widespread as farming decreases.

There are 10 subzones in this largest physiographic area of the state. These are described below, generally from west to east and ending at the western edge of the Shawangunk Hills of the Hudson Valley zone.

Cattaraugus Highlands

With the exception of the valleys along the boundary with the Great Lakes Plain, the Cattaraugus Highlands subzone ranges in elevation from 305 to 549 m (1000–1800 ft), with a few altitudes in the southeast exceeding 610 m (2000 ft). Covering 6242 sq km (2410 sq mi), and about 30% wooded, northern hardwoods predominate, with oaks frequently found

on the south slopes. Deep valleys dissect the rather flat-topped uplands. Agricultural activity is primarily dairying.

Allegany Hills

The average town in the Allegany Hills is two-thirds wooded, with northern hardwoods covering the ridges and with more oaks on the south slopes than farther to the north. Allegany State Park covers much of this area, which totals 1943 sq km (750 sq mi). Altitude ranges from 427 m to above 732 m (1400 to above 2400 ft). It contains one of the very few areas in the state that was not glaciated. There is less land in agriculture than in the subzones to the north and west.

Central Appalachians

Most of the Central Appalachians, the largest subzone, lies above 457 m (1500 ft) and has a few heights reaching 701 m (2300 ft) and more. It encompasses 22,870 sq km (8830 sq mi). The economy is dominated by several large urban industrial areas, with the average town 36% wooded with mixtures of hardwoods, hemlocks in the ravines, and oaks on the south slopes. Agriculture is predominant in areas outside of urban communities, with much of the woodland on farms.

Finger Lakes Highlands

The Finger Lakes Highlands subzone possesses relatively level uplands dissected by deep valleys. It extends over about 3160 sq km (1220 sq mi). Oak forests are most widespread, but there are sections of pure northern hardwoods, hemlock–northern hardwoods, and white pine. About 27% wooded, this subzone has agricultural activity of several kinds on the hills, slopes, and in the valleys. Except for the deeper valleys, elevation ranges generally from 305 to 518 m (1000–1700 ft), with a few points over 610 m (2000 ft).

Delaware Hills

Extensive, solid blocks of woodlands characterize the Delaware Hills, a subzone of about 1735 sq km (670 sq mi). Forests are mainly young, even-aged hardwoods of sugar maple, white ash, black cherry, and basswood, often with hemlock. Dairying predominates. Elevation is quite variable, ranging from a minimum of 274 m (900 ft) to some peak elevations up to 914 m (3000 ft).

Schoharie Hills

The Schoharie Hills subzone, covering 1943 sq km (750 sq mi), is characterized by rolling uplands cut by deep ravines. The area is 47% wooded by oak, pine, hemlock, and northern hardwoods. Minimum altitude is 274 m (900 ft), most elevations are above 610 m (2000 ft), and some peaks reach 1219 m (4000 ft). Dairy farming is the principal activity in the area.

Helderberg Highlands

Flat hilltops intermixed with steep valleys occur throughout the Helderberg Highlands, a rather small subzone of about 1347 sq km (520 sq mi). About one-third of the land is woodland, which in many places follows the steep, shallow-soil valleys; higher lands are often in fields or shrubs. Oak and pine are found on the slopes, with mixtures of northern hardwoods elsewhere.

Dairy farming is common. Elevation in the subzone, except for the lower east edge bordering the Hudson Valley, ranges from 274 to 488 m (900–1600 ft), although there are a few higher peaks, including one near 640 m (2100 ft).

Catskill Peaks

A rugged area of 1632 sq km (630 sq mi) dominated by erosion-resistant sandstones and few valleys and streams, the Catskill Peaks subzone is about 90% forested, mainly with northern hardwoods and some spruce and fir. Recreation and forestry are prominent, along with a few farms. In the eastern portion along the Hudson Valley zone, altitudes begin at about 152 m (500 ft). In the western part of this subzone, however, elevations are generally over 610 m (2000 ft), with many peaks topping 914 m (3000 ft), and with the highest 1280 m (4200 ft). Climate is generally cool and winters severe.

Neversink Highlands

The Neversink Highlands subzone contains numerous well-known Catskills resorts. The average town is 55% wooded, primarily with northern hardwoods, black cherry, and ash, but with hemlock and white pine in the ravines. Some farming contributes to the economy. The area covers about 1606 sq km (620 sq mi), with altitude starting at 183 m (600 ft). Most of the highlands are over 366 m (1200 ft); maximum local height ranges from 457 to 610 m (1500–2000 ft); and a few peaks exceed this. Relief is low compared with subzones to the north.

Mongaup Hills

Relief in the Mongaup Hills subzone, an area of 958 sq km (370 sq mi), is quite low except in the steep southern ravines. Minimum altitude is 183 m (600 ft), but most of the area is above 305 m (1000 ft) with a few heights reaching 488 m (1600 ft). The average town is 81% wooded, with mixtures of oak and red maple on the original oak-chestnut sites and with hemlock and white pine in pure stands or in mixture with hardwoods. Isolated by mountains and the Delaware Valley, and with poor soil conditions precluding extensive agricultural activities, this subzone has large amounts of land that are held for recreational purposes.

Hudson Valley

Shawangunk Hills

A part of the Hudson Valley zone, the Shawangunk Hills subzone is about 90% wooded, with generally high local relief ranging from 122 to 698 m (400–2289 ft). The area covers only 337 sq km (130 sq mi). Oaks are abundant on upper slopes and ridge tops, while tulip tree, white ash, hard maple, and hemlock, mixed with oak, occur on the lower slopes and better sites.

Manhattan Hills

Covering 1295 sq km (500 sq mi) with elevation ranging from near sea level close to the Hudson River to over 213 m (700 ft) on many hilltops, the Manhattan Hills ecozone has mild wet winters and warm humid summers. Oak and oak–northern hardwoods predominate, with pioneer trees most common as in the Hudson Highlands. Despite much recent residential and industrial development, all towns remain more than 50% wooded. The country is rolling; there are many rock exposures; and soils vary in type and depth.

Hudson Highlands

The Hudson Highlands ecozone covers 1347 sq km (520 sq mi), and the highlands themselves are steep, rough, and stony from water erosion. Underlying igneous and metamorphic rocks are complex, with many folds and faults. Soils are shallow and acid. Altitude ranges from 61 m (200 ft) to a peak of 488 m (1600 ft), with much of the zone over 213 m (700 ft); maximum local elevations are near 305 m (1000 ft), but many altitudes are as high as 427 m (1400 ft). Oak is the natural vegetation, with northern hardwoods much less abundant. About 75% of the land is forested; industrialization has spread to many small communities; and there is considerable demand for homes by commuters.

Taconic Highlands

Terrain in the Taconic Highlands, a zone 3750 sq km (1448 sq mi), is rolling near the Hudson Valley and more hilly toward the eastern border of the state. Geologically it is a very complex area, with intensely folded and faulted rocks covered with acid till. Woodlands on good soils contain white ash, tulip tree, basswood, sugar maple, black cherry, hemlock, white pine, and red oak. Those on the poorer, shallower soils are scrubby pioneer forests with gray birch, black birch, maple, and red cedar. The northern part of this zone contains successional-growth northern hardwoods, oak, and hickory with large areas of shrubland. Elevation starts at 122 m (400 ft) in the west and gradually trends higher to about 610 m (2000 ft) along the Massachusetts state line. Maximum height reaches about 853 m (2800 ft). Agriculture is the chief land use in this ecozone.

Taconic Foothills

A 40% wooded area of rolling terrain, the Taconic Foothills subzone contains forests that range from scrub pioneer types to old-growth stands of oaks and other hardwoods of a variety of species. It covers an area of 3913 sq km (1511 sq mi), and altitude varies from 122 to over 366 m (400–1200 ft) on many hilltops on the east. Agriculture is also the main occupation, but, as with the Taconic Highlands, it has decreased and been replaced by old fields and woodlands in various stages of succession.

Rensselaer Hills

Over 67% of the Rensselaer Hills, a subzone of 466 sq km (180 sq mi), is wooded, with spruce and balsam fir common. Elevation begins at 213 m

506

(700 ft), but most of the varied landscape is well above 305 m (1000 ft). Some hilltops reach 579 m (1900 ft). Predominately stony soils of low agricultural value have led to a considerable decrease in farming.

Taconic Mountains

The Taconic Mountains subzone is a very small area of 337 sq km (130 sq mi) in two sections, with rock outcrops and shallow soils of low agricultural quality covering 67% of it. Extensive forests of northern hardwoods are mixed with white birch and oak; spruce and fir occur at higher altitudes. Elevation starts near 305 m (1000 ft), with peaks averaging over 579 m (1900 ft) and the highest reaching 853 m (2798 ft).

Tug Hill Plateau

The Tug Hill Plateau is an outlier of the Appalachian Plateau. It comprises the hilly Tug Hill Transition and the largely flat and undulating Central Tug Hill, and descends to lowlands on all sides.

Tug Hill Transition

The forest type in the low hills of the Tug Hill Transition is mainly northern hardwoods, with hemlock in the western section. Elevation ranges between 305 m and 518 m (1000–1700 ft). Soils are of low productivity over Hudson River shale. Marginal dairy farms are located mostly in the southern part and many have been abandoned. Pioneer species and the introduction of state reforestation softwood plantations have replaced many farms in this ecozone of 2883 sq km (1113 sq mi).

Central Tug Hill

Poorly drained soils are the reason for the large area of wetlands in Central Tug Hill, a zone covering 774 sq km (299 sq mi). It is mainly flat and rolling terrain, with elevation ranging from 457 to 579 m (1500–1900 ft). Large accumulations of snow occur on this isolated plateau from moisture-laden air of Lake Ontario. Dense forests of cut-over northern hardwoods, spruce, and fir are to be found. Severe climate and poor soils have prevented agricultural development, and much privately owned land is used for logging and outdoor recreation.

St. Lawrence Transition

About 1199 sq km (463 sq mi) in area and averaging elevations of 122 m (400 ft) the St. Lawrence Transition has soil of low to medium productivity, with a resulting decrease in farming and increase in state ownership. The area is 65% wooded, mainly with aspen, birch, and shrubland.

Champlain Transition

Lower elevations and gentle topography combined with better soil productivity make the Champlain Transition attractive for agriculture. It occupies about 694 sq km (268 sq mi), with elevation varying from 91 to 366 m (300–1200 ft) and averaging 213 m (700 ft). Forests consist primarily of aspen, birch, and northern hardwoods with some white pine, red spruce, and balsam fir. The greater amount of land use for agriculture here contrasts with ecozones to the west and south, where much land has reverted to second-growth forest.

Western Adirondack Transition

Poor soils over Precambrian bedrock, rougher topography, and more severe climate than in the other transition areas characterize the Western Adirondack Transition. A mixture of old fields, successional forests, and farms occurs in this northern area of 3204 sq km (1237 sq mi).

Eastern Adirondack Transition

The rolling plains of the Eastern Adirondack Transition have a more significant proportion of oak and a higher human population density than do the other transition areas. There is a high degree of land abandonment and many former agricultural clearings. The zone covers about 886 sq km (342 sq mi), with altitude extending from 91 to 610 m (300–2000 ft) and averaging 213 m (700 ft). The bedrock of granites, sandstones, limestones, and acid glacial till is covered by soils of medium to low productivity.

Western Adirondack Foothills

A large area of 8785 sq km (3392 sq mi), the Western Adirondack Foothills ecozone is physically similar to the Central Adirondacks and Eastern Adirondack Foothills but topography is more gentle here. Average elevation is 457 m (1500 ft). Climax forest types are spruce, balsam fir, and northern hardwoods occupying, with shrubland, 85% of the area. Wetlands are characteristic of the floodplains adjacent to many rivers and streams. Logging here has created more variety in forest composition than in any other forest zone. Paper companies, hunting clubs, and large private estates own much of the land. Human settlement is rather low, but density is double that of the Central Adirondacks.

Eastern Adirondack Foothills

Hills and rounded mountains cover some 3983 sq km (1538 sq mi) of the Eastern Adirondack Foothills, with elevations ranging from 152 to 914 m (500–3000 ft) and averaging 427 m (1400 ft). White pine, oak, and northern hardwoods typify woodlands. The zone lacks agriculture. The abrupt eastern margin of the Precambrian Shield marks a change in typical vegetation from the spruce-fir to the west, and it possesses a somewhat milder climate than the Western Adirondack Foothills.

Sable Highlands

An isolated ecozone in the Western Adirondack Foothills, the Sable Highlands region consists of hills and rounded mountains covered by pioneer spruce, balsam fir, and northern hardwoods. About half of the land is state owned. The area is about 588 sq km (277 sq mi), with altitude varying from 396 to 1036 m (1300–3400 ft) and averaging 610 m (2000 ft). Topography and poor soils have determined land use, which is largely recreational.

Central Adirondacks

The Central Adirondacks ecozone, a large area of 8283 sq km (3198 sq mi), is composed of hills and rounded mountains ranging from 396 to 1219 m (1300–4000 ft). Spruce, balsam fir, and northern hardwoods are the main forest components. Over 75% of the land is State Forest Preserve; this fact largely determines land use and human population den-

sity. The eastern boundary of this zone represents the separation of the spruce-fir associations from the white pine of the Eastern Adirondack Foothills.

Adirondack High Peaks

Climax forest in the Adirondack High Peaks ecozone is spruce, balsam fir, and northern hardwoods. Topography in this rugged terrain, which includes the alpine biome, largely determines it use—mostly recreational. About 90% of the total area is included in the State Forest Preserve. The ecozone encompasses about 2033 sq km (785 sq mi), with altitude ranging from 305 to 1629 m (1000–5344 ft) and averaging 762 m (2500 ft).

Appendix B

NATURAL AND CULTURAL ECOLOGICAL
COMMUNITIES OF NEW YORK STATE

Breeding birds in the state occupy a great variety of both natural and man-made habitats for nest sites, territories, and home ranges. The species accounts in the Atlas illustrate this well. The ecological communities below are from a draft (dated March 1987) compiled by Carol Reschke, Community Ecologist, New York Natural Heritage Program, Wildlife Resources Center, Delmar, NY 12054.

The primary objective of this classification is to identify all the natural and cultural ecological communities that constitute the full ecological and biotic array in the state. It is intended to serve a variety of needs, among them a consistent means of describing wildlife habitats. A community, as used herein, is defined as a reoccuring assemblage of populations of all the resident organisms that share a common environment. Each community is conceived as being based on its dominant and characteristic species, as well as on its most apparent and significant environmental features.

This classification is organized into seven systems, each of which is divided into subsystems. Each system and subsystem is defined, and the most apparent features used to identify and name the communities are listed. Brief descriptions are in preparation. The classification is based on a literature review, recent field surveys, and discussions with ecologists, naturalists, and wildlife biologists. Although currently it is considered a working hypothesis that requires further definition, this listing nonetheless provides an unprecedented means of identifying the many ecological communities utilized by New York's breeding birds.

Marine System—consists of the open ocean overlying the continental shelf, its associated high-energy coastline, and shallow coastal indentations or bays without appreciable freshwater inflow. The limits extend from mean high water to the seaward limits of rooted vascular vegetation.

Subtidal—substrate is continuously submerged.
  Marine deepwater
  Marine subtidal aquatic bed

Intertidal—subtrate is exposed and flooded by tides.
  Marine intertidal mudflat
  Marine intertidal gravel/sand beach
  Marine rocky intertidal

Cultural—communities that are either created and maintained by human activities or modified by human influence to such a degree that the physical conformation of the land or substrate, and/or the biological composition of the resident community, is substantially different from the character of the land, substrate, or community prior to human influence.
  Marine artificial impoundment/bay
  Marine submerged artifical structure or reef
  Marine dredge spoil shore
  Riprap shore/artificial marine shore

Estuarine System—consists of deepwater tidal habitats and adjacent tidal wetlands that are usually semi-enclosed but have open, partly obstructed, or sporadic access to open ocean, and in which ocean water is at least occasionally diluted by freshwater runoff. The limits extend from the upstream limit of tidal influence seaward to an imaginary line closing the mouth of a river or bay.

Subtidal—substrate is continuously submerged.
  Estuarine deepwater/tidal river

Brackish subtidal aquatic bed
Freshwater subtidal aquatic bed

Intertidal—substrate is exposed and flooded by tides.
Coastal salt pond
Salt marsh
Salt flat
Salt shrub
Brackish tidal marsh
Brackish intertidal mudflats
Brackish intertidal shore
Freshwater tidal swamp
Freshwater tidal marsh
Freshwater intertidal mudflats
Freshwater intertidal shore

Cultural
Estuarine channel/artificial impoundment
Mosquito ditch
Estuarine artificial structure or reef
Estuarine impoundment marsh
Estuarine diked marsh
Riprap shore/artificial estuarine shore
Estuarine dredge spoil shore

Riverine System—consists of natural, flowing waters from the source of origin downstream to the limits of tidal influence, and bounded by the channel bank, but not including seasonally or permanently vegetated banks or shores.

Natural—ecological communities that are not substantially modified by human activities.
Major river
Backwater slough
Low-gradient stream
Medium-gradient stream
High-gradient stream
Calcareous waterfall and plunge pool
Noncalcareous waterfall and plunge pool
Calcareous spring and spring run
Noncalcareous spring and spring run
Intermittent stream

Cultural
Canal/artificial impoundment
Ditch/artificial intermittent stream
Riverine submerged artificial structure
Industrial effluent stream

Lacustrine System—consists of waters situated in topographic depressions or dammed river channels, lacking persistent emergent vegetation but including areas with submerged and/or floating-leaved vegetation.

Natural
Great Lakes deepwater community
Great Lakes aquatic bed
Inland softwater lake/pond
Inland moderately alkaline lake/pond
Inland highly alkaline lake
Marl pond
Inland salt pond
Oxbow lake
Coastal plain pond

Cultural
Inland acidified lake/pond
Farm pond/artificial pond
Reservoir/artificial impoundment
Lacustrine submerged artificial structure
Quarry pond
Swimming pool
Highway drainage pond (engineered feature)
Commercial/residential drainage pond (engineered feature)
Industrial cooling pond
Sewage treatment pond

Palustrine System—consists of nontidal perennial wetlands characterized by emergent vegetation. The system ranges from wetlands permanently saturated by seepage to permanently flooded wetlands and may include seasonally or intermittently flooded wetlands if the vegetative cover is predominantly hydrophytic.

Open Canopy Wetlands—wetlands with less than 50% canopy cover of trees (see definition of "tree," below)
Deep emergent marsh
Shallow marsh
Sedge meadow
Beaver pond and meadow
Inland salt marsh
Inland calcareous lake/pond shore
Marl pond shore
Inland noncalcareous lake/pond shore
Coastal plain pond shore
Sinkhole wetland
Marl fen
Rich graminoid fen (strongly minerotrophic)
Rich shrub fen (strongly minerotrophic)
Poor fen (weakly minerotrophic)
Coastal plain poor fen (weakly minerotrophic)

Patterned peatland
Rich hillside fen (strongly minerotrophic)
Poor hillside fen (weakly minerotrophic)
Boreal acid bog
Appalachian acid bog
Maritime interdunal swale
Pine barrens vernal pond
Coastal plain groundwater seep
Pine barrens shrub swamp
Shrub swamp

Forested Wetlands—wetlands with at least 50% canopy cover of trees (woody plants usually having one principal stem or trunk, and characteristically reaching a mature height of at least 4.9 m [16 ft]).
Inland Atlantic white cedar swamp
Coastal plain Atlantic white cedar swamp
Floodplain forest
Rich red maple–tamarack swamp
Northern white cedar swamp
Red maple–hardwood swamp
Silver maple–ash swamp
Elm–white pine swamp
Black spruce–tamarack swamp
Rich hemlock–hardwoods swamp
Poor hemlock–hardwoods swamp
Red spruce–balsam fir swamp

Cultural
Reverted drained muckland
Waterfowl management marsh
Waterfowl management swamp
Phragmites/purple loosestrife marsh
Dredge spoil wetland
Mine spoil wetland
Palustrine impoundment marsh
Palustrine diked marsh
Flooded unpaved road
Water recharge basin

Terrestrial System—consists of habitats that are often termed uplands; these habitats have vegetative cover that is never predominantly hydrophytic and soil that is never hydromorphic, even if the surface is occasionally or seasonally flooded or saturated.

Open Canopy Uplands—upland communities with less than 40% canopy cover of trees, usually dominated by shrubs, herbs, or cryptogamic plants (lichens, mosses, etc.).
Great Lakes dune system
Maritime dune system
Maritime shrubland

Maritime heathland
Maritime grassland
Hempstead Plains grassland
Riverside ice meadow
Riverside sand/gravel bar
Acidic shoreline outcrop
Calcareous shoreline outcrop
Acidic cobble shore
Calcareous cobble shore
Alvar grassland
Alpine meadow
Acidic cliff community
Calcareous cliff community
Shale cliff community
Successional fern meadow
Successional blueberry heath
Successional old field
Successional shrubland

Barrens and Woodlands—upland communities with approximately 40–60% canopy cover of trees, or with stunted or dwarf trees (less than 4.9 m [16 ft tall]); includes wooded communities occurring on thin soils over bedrock with numerous bedrock outcrops.
Serpentine barrens
Dwarf pine plains
Dwarf pine ridges
Pitch pine–scrub oak barrens
Pitch pine–heath barrens
Sandstone pavement barrens
Oak opening
Calcareous pavement barrens
Alpine krummholz
Talus/ice cave community
Calcareous talus slope woodland
Acidic talus slope woodland
Shale talus slope woodland
Appalachian acidic rocky summit
Boreal acidic rocky summit
Appalachian calcareous rocky summit
Boreal calcareous rocky summit
Limestone woodland
Successional red cedar woodland

Forested Uplands—upland communities with more than 60% canopy cover of trees, occurring on substrates with less than 50% rock outcrop or thin soil over bedrock.
Maritime oak-holly forest
Maritime red cedar forest
Pitch pine–oak forest
Appalachian oak-hickory forest

Allegheny oak forest
Oak–tulip tree forest
Appalachian oak-pine forest
Mixed mesophytic forest
Beech-maple mesic forest
Maple-basswood rich mesic forest
Hemlock–northern hardwood forest
Pine–northern hardwood forest
Spruce flats
Spruce–fir–northern hardwood forest
Mountain spruce–fir forest
Successional northern hardwoods
Successional southern hardwoods

Cultural
Cropland/row crops
Cropland/field crops
Pastureland
Flower/herb garden
Orchard
Vineyard
Hardwood plantation
Pine plantation
Spruce/fir plantation
Other conifer plantation
Mowed lawn with trees (e.g., park, shady residential yard)
Mowed lawn (e.g., golf course, sunny residential yard)
Mowed roadside/pathway
Herbicide-sprayed roadside/pathway
Unpaved road/path
Paved road/path
Roadcut cliff/slope
Riprap/erosion control roadside
Rock quarry
Gravel mine

Sand mine
Brushy cleared land/woody debris
Artificial imported sand beach
Riprap/erosion control lake/pond shore
Dredge spoil lake/pond shore
Construction/road maintenance spoils
Dredge spoils
Mine spoils
Landfill/dump (primarily organic waste disposal)
Junkyard (primarily nonorganic waste disposal)
Urban vacant lot
Metal structure exterior
Concrete/rock structure exterior
Wood structure exterior
Interior of barn/agricultural building
Interior of nonagricultural building

Subterranean System—consists of both aquatic and nonaquatic habitats beneath the earth's surface, including air-filled cavities with openings to the surface (caves), water-filled cavities and aquifers, and interstitial habitats in small crevices.

Natural
Solution cave aquatic community
Solution cave terrestrial community
Tectonic cave aquatic community
Tectonic cave terrestrial community
Talus cave community

Cultural
Calcareous mine/artificial cave community
Noncalcareous mine/artificial cave community
Sewer
Tunnel
Basement/building foundation

Appendix C

BREEDING SEASON TABLE

This table was compiled by Gordon M. Meade as an aid to Atlasers in their field surveying. The data on which it is based were derived from Forbush (1929), Bull (1974), and Harrison (1978). Additional data submitted by surveyors and regional coordinators have been incorporated into it. Information on the Canvasback and Brewer's Blackbird is also added, but the two exotic parakeets are omitted as are the hybrids. This table is still incomplete, however, because data on breeding in New York are minimal or lacking for many species.

The "Egg dates" are the earliest and latest dates within which eggs have been found for each species. The "Incubation period" refers to the period during which each species incubates and hatches its clutch of eggs.

The "Nestling period" is the time during which the young bird is dependent on its parents for survival. Its length varies depending on several factors, including whether the species is altricial or precocial. The young of some species may remain with their parents after fledging and achieving independence. Because severing contact from the parents is a gradual process with many species, the times given for this period are necessarily approximations.

The dates given for "Unfledged juveniles" are those within which young have been found in the nest (altricial), and both in the nest and after they have left it (precocial) but before they are able to fly. Those dates in the table for "Fledglings" are the periods within which young have been found that are able to fly. Dates for "Unfledged juveniles" can be earlier than those for "Egg dates" because some data are incomplete, certain species may have more than one brood during the season, some single-brooded species replace broods if they are lost, and there is often a differential in time within a species as to when it commences egg laying. For some species only single dates rather than a period are known.

Species	Egg dates	No. of broods	Incubation period (days)	Nestling period (days)	Unfledged juveniles	Fledglings
Common Loon	5/15–7/17	1, occ. 2	29–30	Lv. @ 10–17 (usual 12), near nest 2–3	6/5–8/22	6/20–9/15
Pied-billed Grebe	4/21–7/2	1, poss. 2	23–24	**	5/14–8/20	6/30–9/23
Double-crested Cormorant	6/2	1	25–29	Yg. wander @ 3–4 wks, fly @ 5–6 wks, indep. @ 10 wks	8/31; 9/19	*
American Bittern	5/10–6/29	1	24–29	Lv. nest at 14	5/23–7/24	6/14–8/3
Least Bittern	5/15–7/29	1 or 2	15–19	Lv. nest @ 5–14, flight age ?	6/10–7/20	7/2–9/4
Great Blue Heron	4/15–6/9	1	25–29	Yg. fly @ 60, Lv. nest @ 64–90	5/19–7/17	from 7/17
Great Egret	5/23–6/4	1	25–28	Yg. fly @ 35–42	6/25–7/25	7/25
Snowy Egret	4/16–6/25	1	21–23	Yg. leave nest for branches @ 21–28	5/16–7/14	7/31–9/17
Little Blue Heron	6/18	1	21–24	Lv. nest @ 12, fly @ 28, indep. @ 35–40	7/7	7/4–7/18
Tricolored Heron	begin mid-May	1(?)	21	Yg. climb @ 11–17, fed away from nest @ 24	July	July
Cattle Egret	6/7	1	21–25	Yg. fly @ 40, indep. @ 60	6/9–7/7	*

Species	Egg dates	No. of broods	Incubation period (days)	Nestling period (days)	Unfledged juveniles	Fledglings
Green-backed Heron	4/29–8/4	1, occ. 2	17–21	Yg. fly @ 21–23, indep. @ 35–40	5/22–8/24	7/4–9/19
Black-crowned Night-Heron	4/1–7/23	1	24–26	Lv. nest @ 14–21, fly @ 6 wks	5/21–7/26	6/30–8/25
Yellow-crowned Night-Heron	4/30–6/10	1 or 2	24	**	5/30–6/24	6/22–7/4
Glossy Ibis	5/3–7/27	1	21	On branches @ 14, fly by 42	6/24–8/25	7/1–9/14
Mute Swan	3/26–5/26	1	34–38	Independent @ about 4 months	5/16–6/21	*
Canada Goose	3/28–5/14	1	25–30	Fly @ 9 weeks	4/28–6/27	from 5/18
Wood Duck	3/28–7/15	1	28–32	Lv. nest in 24–30 hours	5/15–8/7	5/22–9/23
Green-winged Teal	5/25–7/15	1	21–24	Fly @ 6 weeks	6/16–7/28	7/5–8/11
American Black Duck	4/2–6/22	1	26–28	Fly @ 7–8 weeks	4/28–7/14	*
Mallard	3/25–7/9	1–2	23–29	Fly @ 7–8 weeks	4/24–8/16	*
Northern Pintail	May–early June	1	22–26	Fly @ 7 weeks	*	*
Blue-winged Teal	5/3–7/4	1	23–24	Fly @ 7 weeks	5/17–8/7	*
Northern Shoveler	5/29–6/11	1	21–26	Indep. @ 6–7 weeks	6/12–7/18	7/18
Gadwall	5/30–7/25	1	25–28	Fly @ 7 weeks	5/26–8/25	6/29–9/19
American Wigeon	late May–mid-June	1	24–25	Indep. @ 6–7 weeks	6/24–8/6	*
Canvasback	*	1	24–27	Fly @ 10–12 weeks	7/3; 7/7	*
Redhead	mid-May–early June	1	22–24	**	6/4–7/27	August
Ring-necked Duck	5/20–6/30	1	26	**	5/29–7/11	7/25–8/22
Lesser Scaup	mid-May–June	1	21–28	**	6/1	*
Common Goldeneye	mid-Apr–mid-June	1	27–32	Fly @ 51–60	*	7/21
Hooded Merganser	4/25–6/2	1†	31	**	5/11–7/17	6/21–8/18
Common Merganser	5/5–7/10	1	28–32	Indep. @ 5 weeks	5/15–8/18	7/12–8/25
Red-breasted Merganser	early June	1	26–35	Fly by 59	*	*
Ruddy Duck	June–early July (Ont.)	1	24–30	**	5/30–9/1	into Sept.
Turkey Vulture	5/4–6/20	1	38–41	Fly @ 11 weeks	6/15–8/27	7/14–9/24
Osprey	4/27–6/21	1	32–33	Fly @ 51–59	6/18–7/25	7/10–8/22
Bald Eagle	3/16–5/14	1	28–46	Lv. nest at 10–11 weeks	4/11–6/30	from 5/20
Northern Harrier	4/20–6/25	1	21–36	Fly @ 37	5/30–7/18	7/4–8/11
Sharp-shinned Hawk	4/16–6/21	1	21–35	Fly @ 23	6/8–7/23	7/3–7/25
Cooper's Hawk	4/20–6/16	1†	21–36	Lv. nest—male @ 30, female @ 34; indep. @ 8 weeks	6/2–7/2	7/2–8/3
Northern Goshawk	4/20–5/15	1	28–41	Fly @ 45, hunt @ 50, indep. @ 70	5/18–7/1	6/14–7/27
Red-shouldered Hawk	3/25–5/26	1†	23–25	Lv. nest @ 5–6 weeks	5/5–7/5	early as 6/6
Broad-winged Hawk	4/27–6/26	1	23–28	Lv. nest @ 29–30	5/30–7/27	7/4–8/16
Red-tailed Hawk	3/8–5/16	1	23–28	Fly @ 45	4/17–6/20	6/1–7/8
Golden Eagle	Mar.–June (U.S.)	1	27–45	Fly @ 9–10 weeks	7/10	7/24
American Kestrel	4/5–6/29	1	29–30	Fly @ 30	5/19–8/2	6/12–8/10
Peregrine Falcon	3/2–5/31	1	28–29	Fly @ 35–42	4/19–7/10	5/21–7/27
Gray Partridge	late Mar.–early June	1–2	21–26	Fly @ 16	*	*
Ring-necked Pheasant	4/14–8/16	1–2	23–27	Fly @ 12–14	6/22–8/16	8/14
Spruce Grouse	Mid–early June	1	17–24	Fly @ 10–12	6/19–7/16	8/17–8/22
Ruffed Grouse	4/1–6/22	1†	23–24	Fly @ 10–12	5/27–7/5	6/15–9/4
Wild Turkey	4/26–7/9	1	28	Fly @ 14	5/13–8/13	6/1–9/7
Northern Bobwhite	5/25–9/14	1	23–24	Fly @ 14; full grown @ 60	6/11–9/27	7/5–10/11
Black Rail	6/20–7/12	*	*	**	*	*
Clapper Rail	4/11–8/4	1	20–24	Lv. nest soon after hatching; swim @ 1	6/6–8/20	*
King Rail	5/20–7/3	1	21–24	Lv. nest soon after hatching	6/16–8/6	8/2–8/31
Virginia Rail	5/5–7/13	1	20	Lv. nest soon after hatching	5/11–8/14	7/23–9/8
Sora	4/30–7/17	1	14–20	Lv. nest @ 1–2, fly @ 36	5/19–8/8	6/9–9/15
Common Moorhen	5/14–7/25	1	19–25	Indep. @ 5 weeks	6/3–8/27	7/9–9/17
American Coot	4/25–7/14	1–2	21–24	Indep. @ 8 weeks	5/17–8/12	6/29–8/21

Species	Egg dates	No. of broods	Incubation period (days)	Nestling period (days)	Unfledged juveniles	Fledglings
Piping Plover	4/18–7/23	1†	26–30	Fly @ 30–35	5/21–7/24	6/2–8/18
Killdeer	4/3–7/4	1–2	24–28	Fly @ 40	5/3–7/30	5/21–8/12
American Oystercatcher	5/25–7/22	1†	24–27	Indep. @ 34–37	5/30–7/28	6/7–8/19
Willet	5/19–6/30	1	22	**	6/15	*
Spotted Sandpiper	5/6–7/26	1	18–24	Fly @ 16–18	6/2–8/19	*
Upland Sandpiper	4/23–6/15	1	17–21	Full grown @ 30	5/28–7/18	6/15–8/11
Common Snipe	4/20–6/16	1	18–20	Fly @ 19–20	5/19–6/20	7/5
American Woodcock	3/24–6/17	1	20–21	Fly @ 14–15	4/17–6/29	4/29–8/2
Laughing Gull	late May; 6/14, 6/28	1	21–23	Fly @ 4–6 weeks		
Ring-billed Gull	5/3–7/10	1	21–23	Fly @ 35	5/16–7/10	6/25–7/24
Herring Gull	4/27–6/26	1	24–28	Fly @ 6 weeks	5/17–7/24	7/5–8/31
Great Black-backed Gull	4/25–6/19	1	26–30	Fed for 7 weeks, then begin to fly	5/30–6/27	7/10–7/26
Gull-billed Tern	6/2–7/8 (Va.)	1	22–23	Fly @ 4–5 weeks	*	*
Caspian Tern	6/23, 7/6	1†	20–22	Fly @ 25–30	6/23, 7/6	*
Roseate Tern	5/20–7/27	1	21–26	**	6/13–8/31	7/11–9/9
Common Tern	5/12–8/15	1†	20–23	Fly @ 4 weeks	6/11–9/3	7/10–9/9
Forster's Tern	6/8	1	23–25	**	6/16, 6/23	*
Least Tern	5/9–7/27	1†	14–22	Fly @ 15–17	6/4–8/11	7/15–8/29
Black Tern	5/27–7/23	1	20–22	Fly @ 3–4 weeks	6/13–8/5	7/3–8/25
Black Skimmer	5/31–9/3	1	*	**	6/20–9/24	7/17–10/11
Rock Dove	every month	2–3	14–19	Indep. @ 30–35		
Mourning Dove	3/9–9/28	2–3	12–15	Fly @ 13–15	4/6–10/5	4/24–10/26
Black-billed Cuckoo	5/20–8/28	1	14	Fly @ 21–24	6/1–9/10	6/20–9/27
Yellow-billed Cuckoo	5/26–8/19	1	14	**	6/21–9/17	6/23–9/23
Common Barn-Owl	all mos. usually Apr.–June	1–2	32–34	Fly @ 60, indep. @ 70	all months	all months
Eastern Screech-Owl	3/23–5/11	1	21–26	Lv. nest @ 35	4/24–6/25	5/5–8/17
Great Horned Owl	1/28–5/8	1†	30–35	Lv. nest @ 31–35	3/8–6/12	4/9–6/9
Barred Owl	3/23–5/3	1†	21–28	Fly @ 6 weeks	4/14–6/11	5/13–7/1
Long-eared Owl	3/21–5/23	1, occ. 2	21–30	Lv. nest @ 23–24	5/5–6/24	6/1–8/8
Short-eared Owl	4/2–5/19	1, occ. 2	24–28	Lv. nest @ 12–17, fly @ 22–27	5/7–6/19	6/11–7/13
Northern Saw-whet Owl	3/31–6/11	1	26–28	Lv. nest @ 36, occ. longer	4/21–7/16	5/28–8/22
Common Nighthawk	5/25–7/25	1	16–19	Fly @ 23, indep. @ 30	6/14–8/14	7/7–8/30
Chuck-will's-widow	5/23 (Va.)	1†	20	**	6/22–6/28	6/29
Whip-poor-will	5/6–6/30	1	14–20	**	6/2–7/14	6/16–8/8
Chimney Swift	5/30–7/27	1	18–22	Fly @ 24–26	6/25–8/12	7/18–9/1
Ruby-throated Hummingbird	5/21–8/16	1–2	14–16	Lv. nest @ 19	6/24–9/6	7/12–9/30
Belted Kingfisher	4/28–6/10	1†	17–24	Lv. nest @ 30–35	6/8–7/14	7/29–8/9
Red-headed Woodpecker	5/16–6/19	1–2	14	Lv. nest @ 27	5/31–8/26	7/5–9/15
Red-bellied Woodpecker	4/26–6/28	1†	12–14	Lv. nest @ 26	5/18–8/29	6/23–8/13
Yellow-bellied Sapsucker	4/29–6/19	1†	12–14	Lv. nest @ 25–29, depend. 1–2 wks more	5/29–7/8	6/12–8/15
Downy Woodpecker	5/6–6/30	1	12	Lv. nest @ 20–22, depend. 3 wks more	5/31–7/3	6/9–7/16
Hairy Woodpecker	4/23–5/19	1†	11–14	Lv. nest @ 28–30, depend. 2 wks more	5/5–6/14	6/13–8/1
Three-toed Woodpecker	5/14–6/14	1	14	**	7/2, 7/31	7/9–7/24
Black-backed Woodpecker	5/18–6/12	1	14	**	5/30–6/20	6/20–7/23
Northern Flicker	4/20–6/19	1–2†	11–16	Lv. nest @ 25–28	5/18–7/26	6/19–8/15
Pileated Woodpecker	4/22–5/20	1†	18	Lv. nest @ 22–26	5/10–6/21	6/9–7/15
Olive-sided Flycatcher	6/9–6/27	1	14–17	Lv. nest @ 15–19	6/22	7/10–7/24
Eastern Wood-Pewee	5/30–8/15	1	12–13	Lv. nest @ 15–18	6/22–8/13	8/3, 9/16

Species	Egg dates	No. of broods	Incubation period (days)	Nestling period (days)	Unfledged juveniles	Fledglings
Yellow-bellied Flycatcher	6/10–6/27	1	12–15	Lv. nest @ 13	*	7/25
Acadian Flycatcher	5/28–7/4	1	12–14	Lv. nest @ 13, fed by parents 12 more	6/19	*
Alder Flycatcher	6/2–7/29	1	12	Lv. nest @ about 14	6/21–8/14	7/11–8/24
Willow Flycatcher	6/11–7/29	1	13–15	Lv. nest @ 12–15	6/21–8/14	7/11–8/24
Least Flycatcher	5/16–6/28	1–2	12–16	Lv. nest @ 13–16	6/22–8/6	7/8–8/16
Eastern Phoebe	4/20–8/4	1–3	12–16	Lv. nest @ 15–17, fed by parents 2–3 wks more	5/13–8/10	6/9–8/24
Great Crested Flycatcher	5/22–7/11	1	13–15	Lv. nest @ 14–15	6/10–7/26	6/27–9/14
Eastern Kingbird	5/20–7/18	1–2	12–16	Lv. nest @ 13–14, fed by parents 5 wks more	6/3–8/5	6/21–8/21
Horned Lark	2/28–7/31	1–3	11–14	Lv. nest @ 9–10, fly well @ 20	3/11–8/4	3/31–9/13
Purple Martin	5/21–7/13	1, occ. 2	12–20	Lv. nest @ 24–28, roost in nest after leaving	6/22–8/15	7/30–8/22
Tree Swallow	5/5–7/18	1–2	13–16	Lv. nest @ 16–24	5/22–8/10	6/10–8/2
Northern Rough-winged Swallow	5/12–7/5	1	15–16	Lv. nest @ 18–21	6/14–7/11	7/6–7/28
Bank Swallow	5/15–7/13	1–2	12–16	Fly @ 17–18, lv. nest 1–2 days later	5/31–8/12	6/28–9/1
Cliff Swallow	5/9–7/14	1–2	12–16	Fly @ 23, return to nest for 2–3 more	5/29–8/19	6/23–8/23
Barn Swallow	5/15–8/4	2–3	13–16	Lv. nest @ 17–24	5/24–8/28	6/25–9/22
Gray Jay	3/10–4/10	1	16–18	Lv. nest @ about 15	*	5/19–8/12
Blue Jay	4/15–6/17	1†	15–18	Lv. nest @ 17–21, indep. in 3 wks more but may be fed longer	5/18–7/5	6/1–7/31
American Crow	3/30–6/14	1	15–18	Lv. nest @ about 5 wks	5/1–7/28	*
Fish Crow	3/20–6/5	1†	16–18	**	*	*
Common Raven	3/26–4/14	1	19–21	Lv. nest @ 5–6 wks	3/21, 4/12	4/17, 5/30, 6/14
Black-capped Chickadee	4/29–7/15	1	11–14	Lv. nest @ 16	5/21–7/20	5/21–8/3
Boreal Chickadee	6/11–7/17	*	**	**	6/27–7/26	7/2–8/27
Tufted Titmouse	4/29–5/27	1	12–13	Lv. nest @ 15–16	5/13–6/30	5/20–8/4
Red-breasted Nuthatch	4/30–6/17	1	12	Lv. nest @ 18–21	5/15–7/1	6/6–8/18
White-breasted Nuthatch	4/13–6/6	1	12 (?)	Fed for 2 wks after leaving nest	5/8–6/11	6/3–6/22
Brown Creeper	4/24–6/30	1	14–15	Lv. nest @ 14–16	5/27–7/28	6/24–8/20
Carolina Wren	4/1–8/5	2–3	12–14	Lv. nest @ 12–14	4/21–10/2	5/8–8/29
Bewick's Wren	late Mar.–early Apr.	2–3	about 14	Lv. nest @ about 14, fed for 2 wks more	*	*
House Wren	5/15–7/31	1–2	13–15	Lv. nest @ 12–18, feed selves @ 13	5/22–8/28	6/26–9/11
Winter Wren	5/22–7/29	1–2	14–17	Lv. nest @ 15–20	6/3–8/4	6/15–8/16
Sedge Wren	5/28–7/30	1–2	12–14	Lv. nest @ 12–14	6/30–8/22	8/4–9/15
Marsh Wren	5/22–8/7	2–3	10–14	Lv. nest @ 13–15, fed for 7 more	6/21–8/12	7/2–8/31
Golden-crowned Kinglet	5/28–7/26	1–2	12–17	**	6/11–7/25	6/17–8/30
Ruby-crowned Kinglet	May–6/29	1–2	14–15	**	7/2	7/24
Blue-gray Gnatcatcher	5/14–6/17	1	15	Lv. nest @ 12–13, fed for up to 19 more	6/1–7/11	6/28–7/25
Eastern Bluebird	4/1–8/18	2–3	12	Lv. nest @ 15–18, male may continue to feed yg.	4/28–9/6	5/10–9/17
Veery	5/15–6/25	1–2	10–12	Lv. nest @ 10–12	6/14–7/22	6/20–7/31
Gray-cheeked Thrush	6/12–6/27	1	13–14	Lv. nest @ 11–13	7/1–7/25	7/12–8/7
Swainson's Thrush	5/31–7/11	1	10–13	Lv. nest @ 10–12	6/30–7/22	7/10–8/10
Hermit Thrush	5/10–8/24	2–3	12–13	Lv. nest @ 10	5/30–8/31	6/9–9/23
Wood Thrush	5/14–7/7	1–2	12–14	Feed selves @ 10, lv. nest @ 12–13	5/22–8/1	6/9–8/31
American Robin	3/23–7/19	2–3	11–14	Lv. nest @ 14–16	4/21–8/30	5/25–9/10
Gray Catbird	5/5–8/12	2–3	10–14	Lv. nest @ about 10	5/29–8/20	6/6–9/21
Northern Mockingbird	4/27–7/21	2–3	10–14	Lv. nest @ 12–14	5/5–8/11	5/25–8/29

Species	Egg dates	No. of broods	Incubation period (days)	Nestling period (days)	Unfledged juveniles	Fledglings
Brown Thrasher	5/6–6/26	1–2	11–14	Lv. nest @ 9–12	5/19–7/29	6/19–7/26
Cedar Waxwing	6/5–9/23	1–2	12–16	Lv. nest @ 16–18	6/12–10/1	6/16–10/8
Loggerhead Shrike	4/18–6/28	1–2	13–16	Lv. nest @ 20, indep. @ 26–35	5/18–6/25	5/25–7/26
European Starling	4/10–6/15	1–2	12–16	Fed by parents for 20–22	5/1–7/30	5/19–8/30
White-eyed Vireo	5/17–7/17	1	12–15	**	6/18	6/30
Solitary Vireo	5/10–8/9	1	10–11	**	6/7–8/13	6/28–8/31
Yellow-throated Vireo	5/17–6/18	1	12–14	**	6/16–7/30	7/1–8/14
Warbling Vireo	5/16–6/16	1	12	Lv. nest @ 16	5/31–6/29	6/21–7/24
Philadelphia Vireo	June–July	*	13–14	Lv. nest @ 13–14	*	*
Red-eyed Vireo	5/13–8/1	1–2	12–14	Lv. nest @ 12	6/8–8/17 (2nd brood: 9/4)	8/6–9/13
Blue-winged Warbler	5/18–6/17	1	10–11	Lv. nest @ 8–10	6/4–7/11	6/8–8/12
Golden-winged Warbler	5/18–6/16	1	10–11	Lv. nest @ 10	6/8–7/6	6/27–8/6
Tennessee Warbler	June–July	1	*	**	*	*
Nashville Warbler	5/19–6/10	1	11	Lv. nest @ 11	5/30–6/22	6/15–8/17
Northern Parula	5/17–6/27	1–2	12–14	**	6/6–7/4	7/4–8/5
Yellow Warbler	5/15–7/3	1–2	9–15	Lv. nest @ 9–12	6/4–7/23	6/12–8/1
Chestnut-sided Warbler	5/20–7/25	1, occ. 2	10–13	Lv. nest @ 10–12	6/15–8/6	6/22–8/20
Magnolia Warbler	5/25–7/11	1–2	11–13	Lv. nest @ 8–10	6/5–7/24	6/15–8/26
Cape May Warbler	6/6–6/16	1	*	**	*	6/23–7/4
Black-throated Blue Warbler	5/29–7/17	*	12	Lv. nest @ 10	6/14–7/29	6/22–8/14
Yellow-rumped Warbler	5/19–7/10	1	12–13	Lv. nest @ 12–14	6/2–7/22	6/8–8/17
Black-throated Green Warbler	5/24–7/2	1–2	12	Lv. nest @ 8–10	6/11–7/29	6/23–8/15
Blackburnian Warbler	6/1–6/24	*	*	**	6/17–7/1	7/13–8/4
Yellow-throated Warbler	*	*	*	**	7/21	*
Pine Warbler	5/4–6/6	1–2	*	**	5/19–6/17	5/30–8/8
Prairie Warbler	5/25–6/29	1	12–14	Lv. nest @ 8–10	6/19–7/15	6/30–7/14
Palm Warbler	7/8	1–2(?)	12	Lv. nest @ 12	*	*
Bay-breasted Warbler	mid-June	1	12–13	Lv. nest @ 11	6/25–7/6	7/23
Blackpoll Warbler	6/5–7/10	*	11	Lv. nest @ 10–11	*	6/30
Cerulean Warbler	5/19–6/23	1	**	**	6/12–7/6	6/22–7/22
Black-and-white Warbler	5/10–6/30	1	11–13	Lv. nest @ 8–12	6/5–7/23	6/19–7/31
American Redstart	5/14–7/16	1	12	Lv. nest @ 9	6/4–8/5	6/26–8/19
Prothonotary Warbler	5/17–6/29	1–2	10–14	Lv. nest @ 10–11	6/8–7/6	7/10–8/6
Worm-eating Warbler	5/24–6/18	*	13	Lv. nest @ 10	6/6–7/15	6/16–7/29
Ovenbird	5/17–7/22	1–2	12–14	Lv. nest @ 8–10	6/8–8/8	6/18–9/10
Northern Waterthrush	5/10–6/28	1	14	**	5/24–7/5	6/4–7/20
Louisiana Waterthrush	4/25–6/20	1	12–14	Lv. nest @ 10, fly @ 16	5/20–7/6	6/9–7/25
Kentucky Warbler	6/1–6/27	1	12–13	Lv. nest @ 8–10, fed for 17 more	6/20	6/29
Mourning Warbler	5/28–7/7	1	12–13	Lv. nest @ 7–9, fly 2nd wk	6/17–7/28	6/27–8/16
Common Yellowthroat	5/15–7/12	1–2	12	Lv. nest @ 9–10	6/2–8/22	6/15–9/11
Hooded Warbler	5/25–7/10	1–2	12	Lv. nest @ 8–9	6/14–8/12	7/8–9/10
Wilson's Warbler	8/1	1	11–13	Lv. nest @ 10–11	*	*
Canada Warbler	5/31–7/24	1	*	**	6/14–7/29	6/20–8/15
Yellow-breasted Chat	5/25–7/13	1	11–15	Lv. nest @ 8–11	6/8–7/17	6/22
Scarlet Tanager	5/20–7/23	1	13–14	Lv. nest @ 15	6/9–Aug.	7/4–9/19
Northern Cardinal	4/10–9/9	2–3	12	Lv. nest @ 9–11, fly well @ 19, indep. @ 38–45	4/23–9/23	4/30–9/23
Rose-breasted Grosbeak	5/6–7/19	1–2	12–14	Lv. nest @ 9–12, depend. 3 wks more	5/30–7/26	6/11–8/15
Blue Grosbeak	6/17	1–2	11	Lv. nest @ 9–13	*	7/1
Indigo Bunting	5/20–8/3	1–2	12–13	Lv. nest @ 9–13	6/18–8/14	6/21–9/20

Species	Egg dates	No. of broods	Incubation period (days)	Nestling period (days)	Unfledged juveniles	Fledglings
Dickcissel	May–6/29	1–2	11–13	Lv. nest @ 7–10, fly @ 11–12	*	*
Rufous-sided Towhee	5/15–8/4	1–2	12–13	Lv. nest @ 8–10	5/18–8/15	6/2–8/31
Chipping Sparrow	5/2–7/19	1–2	10–14	Lv. nest @ 9–12, fly @ 14	5/23–9/3	6/4–9/21
Clay-colored Sparrow	May–June	1–2	10–11	Lv. nest @ 7–9, fed for 8 more	6/15	6/20–7/15
Field Sparrow	5/16–8/17	2–3	10–13	Lv. nest @ 7–8, fly @ 12, indep. 18–20 later	5/26–8/20	6/17–6/20
Vesper Sparrow	5/5–8/16	1–3	11–13	Lv. nest @ 9–13, depend. 21 more	6/11–7/16	7/11–7/31
Savannah Sparrow	5/11–6/16	1–2	12	**	5/30–7/23	6/12–8/30
Grasshopper Sparrow	5/17–8/2	1–3	11–12	Lv. nest @ 9	6/29–8/19	7/21–9/5
Henslow's Sparrow	5/17–7/5	1–2	11	Lv. nest @ 9–10	6/1–7/22	6/19–7/30
Sharp-tailed Sparrow	5/30–7/21	1	11	Lv. nest @ 10, depend. 20 more	6/11–8/5	8/1
Seaside Sparrow	5/25–7/10	1–2	11–12	Lv. nest @ 9, depend. 21 more	6/8–7/23	*
Song Sparrow	4/17–8/13	1–3	12–14	Lv. nest @ 10, fly @ 17, depend. 18–20 more	5/5–9/3	5/18–9/23
Lincoln's Sparrow	6/10–6/28	1–2	13–14	Lv. nest @ 10–12	6/18	7/21
Swamp Sparrow	5/5–7/22	1–2	12–15	Lv. nest @ 9–10	5/21–7/30	6/28–8/3
White-throated Sparrow	5/30–7/21	1–2†	11–14	Lv. nest @ 7–12, fly 3 later	6/14–8/16	6/27–8/31
Dark-eyed Junco	4/28–8/13	1–3	11–13	Lv. nest @ 10–13, depend. 21 more	5/16–8/17	6/7–8/27
Bobolink	5/18–6/20	1	10–13	Lv. nest @ 10–14, fly a few days later	5/30–7/20	*
Red-winged Blackbird	4/26–7/9	1–2, occ. 3	10–15	Lv. nest @ 10–11, stay near nest 10 more	5/29–7/19	6/20–7/30
Eastern Meadowlark	5/9–8/1	1–2	13–17	Lv. nest @ 11–12	5/24–8/12	6/5–8/24
Western Meadowlark	May–July	1–2	13–15	Lv. nest @ 12, fed for a few days more	6/23	6/26
Rusty Blackbird	5/17–6/15	1	14	Lv. nest @ 13	5/30–7/8	7/7–7/24
Brewer's Blackbird	*	2	12–13	Lv. nest @ 13, fed for further 12–13	*	*
Boat-tailed Grackle	5/31–6/15	1–2–3	13	Lv. nest @ 20–23	6/9–7/11	6/23, 7/29
Common Grackle	4/12–6/4	1–2	12–14	Lv. nest @ 10–17, near nest only 2–3	5/3–6/28	5/18–7/29
Brown-headed Cowbird	4/23–7/31	*	10–12	Lv. nest @ 10, usually before host yg., fed for 2 wks	5/19–8/2	5/30–8/19
Orchard Oriole	5/18–6/22	1	12–15	Lv. nest @ 11–14	5/28–7/26	6/19–8/21
Northern Oriole	5/15–6/13	1	14	**	6/6–7/9	6/15–7/14
Purple Finch	5/13–7/16	1	13	Lv. nest @ 14	6/2–7/24	6/10–9/3
House Finch	4/11–8/6	2–3	12–14	Lv. nest @ 14–16	4/24–8/23	5/18–8/11
Red Crossbill	3/30–4/30	1–2	12–16	Lv. nest @ 17–23, depend. 3–4 wks more	4/24–5/27	3/29–6/19
White-winged Crossbill	mid-Jan–Aug	*	*	**	*	2/4, 6/15, 9/11, 10/10, 11/25
Pine Siskin	4/25–5/25	1–2	13	Lv. nest @ 14–15	4/13–6/10	6/11–7/16
American Goldfinch	6/25–9/16	1	12–14	Lv. nest @ 11–17	7/24–9/30	8/17–10/10
Evening Grosbeak	5/19–6/4	*	12–14	Lv. nest @ 13–14	5/31–6/17	6/15–9/5
House Sparrow	3/23–7/16	2–3	11–14	Lv. nest @ 15	4/15–8/4	6/24–9/6

* No New York data available.
** No information from references checked.
(?) Probable.
† If brood is lost, it usually will be replaced.

Able, K.P. 1983. Trends in the state list of New York birds. Kingbird 33:6–11.

Able, K.P., and B.R. Noon. 1976. Avian community structure along elevational gradients in the northeastern United States. Oecologia 26:275–294.

Achilles, L. 1906. Nesting of the Arctic Three-toed Woodpecker in the Adirondacks. Bird-Lore 8:158–160.

Aldrich, J.W. 1953. Habits and habitat differences in two races of Traill's Flycatcher. Wilson Bulletin 65:9.

Allen, A.A. 1911. A note on the Prothonotary Warbler. Auk 28:115.

——. 1935. American bird biographies. Comstock Publishing Co., Ithaca, NY.

Allen, C.S. 1892. Breeding habits of the fish hawk on Plum Island, New York. Auk 9:313–321.

Allen, E.G. 1951. The history of American ornithology before Audubon. Transactions of the American Philosophical Society 44: 465–466.

Allen, J.A. 1869. Notes on some of the rarer birds of Massachusetts. American Naturalist 3:505–519, 568–585, 631–648.

——. 1900. The Little Black Rail. Auk 17:1–8.

Allen, R.P. 1938. Black-crowned Night-Heron colonies on Long Island. Proceedings of the Linnaean Society of New York 49:43–54.

Allen, R.P., and M.M. Nice. 1952. A study of the breeding biology of the Purple Martin (*Progne subis*). American Midland Naturalist 47:606–665.

Ambuel, B., and S.A. Temple. 1982. Songbird populations in southern Wisconsin forests: 1954 and 1979. Journal of Field Ornithology 53:149–158.

American Birds. 1971–1986. Vols. 25–40. Christmas bird count issue (no. 4) in each volume.

American Ornithologists' Union. 1957. Check-list of North American birds. 5th ed. Lord Baltimore Press, Baltimore.

——. 1983. Check-list of North American birds. 6th ed. American Ornithologists' Union, Washington, DC.

Anderson, D.W., and J.J. Hickey. 1972. Eggshell changes in certain North American birds. *In* Proceedings of the XV International Ornithological Congress, ed. K.H. Voous. E.J. Brill, Leiden, Netherlands. P. 517.

Andrle, R.F. 1969. Red-tailed Hawks nesting on cliffs in Ontario. Canadian Field-Naturalist 83:165.

——. 1971a. Range extension of the Golden-crowned Kinglet in New York. Wilson Bulletin 83:313–316.

——. 1971b. The birds of McCarty Hill, Cattaraugus County, NY. Prothonotary 37:90–93.

——. 1974. White-throated Sparrow nesting again in downtown Buffalo, New York. Auk 91:837–839.

——. 1976. Herring Gulls breeding on cliff at Niagara Falls, New York. Canadian Field-Naturalist 90:480–481.

——. 1978. Ruby-crowned Kinglet breeding in Cattaraugus County. Kingbird 28:29–30.

——. 1985. Personal communication. Buffalo Museum of Science, Humboldt Parkway, Buffalo, NY 14211.

Andrle, R.F., and F.M. Rew. 1971. White-throated Sparrow breeding in downtown Buffalo, New York. Auk 88:172–173.

Anonymous. 1912. These gulls have a private guardian. New York Herald, May 19.

Anonymous. 1985. Annual report 1984. Incorporated Orange County Chapter, New York State Archeological Association. Middletown, NY. Mimeo.

Apfelbaum, S.I., and P. Seelbach. 1983. Nest tree, habitat selection, and productivity of seven North American raptor species based on the Cornell University Nest Record Card Program. Raptor Research 17:97–113.

Arbib, R.S. 1963. The Common Loon in New York State. Kingbird 13:132–140.

——. 1975. Blue List for 1976. American Birds 29:1067–1072.

——. 1977. The Blue List for 1978. American Birds 31:1087–1096.

Armistead, H.T. 1983. Purple Martin. *In* The Audubon Society master guide to birding, ed. J. Farrand, Jr. Vol. 2, Gulls to dippers. Alfred A. Knopf, New York. P. 298.

Armstrong, W.H. 1958. Nesting and food habits of the Long-eared Owl in Michigan. Michigan State Univ. Biology Series 1:61–96.

Askildsen, J., ed. 1986. Loons-herons. Upper Westchester-Putnam Field Notes 3:2.

Audubon Field Notes. 1948–1970. Vols. 2–24. Christmas bird count issue (no. 2) in each volume.

Audubon, J.J., and J.B. Chevalier. 1840–1844. The Birds of America. Vols. 1–7. Reprint. Dover Publications, New York, 1967.

Austin, D.E. 1964. Status of the Wild Turkey in New York, 1963–1964. New York Federal Aid in Wildlife Restoration Project W-81-R-11:I-C. New York State Conservation Dept. Albany. Mimeo.

——. 1980. The distribution and density of Gray Partridge in northern New York. *In* Proceedings of *Perdix* II Gray Partridge workshop, ed. S.R. Peterson, and L. Nelson, Jr. Forest Wildlife and Range Experiment Station, Univ. of Idaho, Moscow. Pp. 1–7.

——. 1987. Personal communication. NYSDEC, Wildlife Resources Center, Delmar, NY 12054.

Austin, D.V. 1975. Bird flowers in the eastern United States. Florida Science 38:1–12.

Bagg, A.M., and H.M. Parker. 1951. The Turkey Vulture in New England and eastern Canada up to 1950. Auk 68:218.

Bagg, E. 1878. Lincoln's finch (*Melospiza lincolnii*) breeding in Hamilton County, New York. Bulletin of the Nuttall Ornithological Club 3:197–198.

——. 1911. Annotated list of the birds of Oneida County, New York, and of the West Canada Creek Valley. Transactions of the Oneida Historical Society 12:17–86.

Baillie, J.L., Jr. 1940. The summer distribution of the Eastern Evening Grosbeak. Canadian Field-Naturalist 54:15–25.

Baird, J. 1968. Eastern Savannah Sparrow. *In* Life histories of North American cardinals, grosbeaks, buntings, towhees, finches, sparrows, and allies, by A.C. Bent and collaborators, ed. O.L. Austin, Jr. U.S. National Museum Bulletin no. 237, pt. 2. Washington, DC. Pp. 678–696.

Baird, S.F., T.M. Brewer, and R. Ridgway. 1905. A history of North American birds. Vol. 1. Little, Brown and Co., Boston.

Baird, T. 1983. Ecology of the birds of the Quaker Run Valley, Allegany State Park, New York. Report to the Wild Wings/ Underhill foundations. New York.

——. 1984. A first record of nesting Yellow-throated Warblers in New York State. Kingbird 343:221–223.

——. 1986a. Ecology of the birds of the Quaker Run Valley, Allegany State Park, New York. Report to the Wild Wings/ Underhill foundations. New York.

——. 1986b. A comparison of breeding bird populations in the Quaker Run Valley of Allegany State Park, NY, 1930–31 and 1983–85. Report to the Wild Wings/Underhill foundations. New York.

Banks, R.C. 1976. Wildlife importation into the United States, 1900–1972. U.S. Dept. of the Interior Special Scientific Report, Wildlife no. 200. Washington, DC.

Barlow, J.C., and J.C. Rice. 1977. Aspects of the comparative behavior of Red-eyed and Philadelphia vireos. Canadian Journal of Zoology 55:528–542.

Barnett, T. 1985. Personal communication. The Nature Conservancy, Elizabethtown, NY 12932.

Barnum, M.K. 1886. A list of the birds of Onondaga County. Syracuse Univ. Press, Syracuse.

Barrows, W.B. 1889. The English sparrow (*Passer domesticus*) in North America. U.S. Dept. of Agriculture Bulletin no. 1. Washington, DC.

——. 1912. Michigan bird life. Michigan Agricultural College, Lansing.

Bart, C. 1985. Stream and lake edges reward diligent atlasers. Ohio Breeding Bird Atlas Newsletter 2:3.

Bateman, H.A., Jr. 1977. King Rail. *In* Management of migratory shore and upland game birds in North America, ed. G.C. Sanderson. International Association of Fish and Wildlife Agencies, Washington, DC. Pp. 93–104.

Baumgartner, F.M. 1939. Territory and population in the Great Horned Owl. Auk 56:274–282.

Beardslee, C.S. 1932. Prothonotary Warblers nesting near Buffalo, N.Y. Auk 49:91.

Beardslee, C.S., and H.D. Mitchell. 1965. Birds of the Niagara Frontier region. Bulletin of the Buffalo Society of Natural Sciences. Vol. 22.

Beddall, B.G. 1963. Range expansion of the Cardinal and other birds in the northeastern states. Wilson Bulletin 75:140–156.

Bednarz, J.C., and J.J. Dinsmore. 1981. Status, habitat use, and management of Red-shouldered Hawks in Iowa. Journal of Wildlife Management 45:236–241.

——. 1982. Nest sites and habitat for Red-shouldered Hawks and Red-tailed Hawks in Iowa. Wilson Bulletin 94:31–45.

Beehler, B.M. 1978. Birdlife of the Adirondack Park. Adirondack Mountain Club, Glens Falls, NY.

Belknap. J.B. 1955. The expanding range of the Ring-billed Gull. Kingbird 5:63–64.

——. 1967. Turkey Vulture in northern New York. Kingbird 18:25.

——. 1968. Little Galloo Island—a twenty-year summary. Kingbird 18:80–81.

——. 1973. The Evening Grosbeak in New York State. Kingbird 23:122–123.

Bellrose, F.C. 1978. Ducks, geese, and swans of North America. Stackpole Books, Harrisburg.

Bendire, C.E. 1895. Life histories of North American birds. U.S. National Museum Special Bulletin no. 3. Washington, DC.

Benkman, C. 1985. Personal communication. Dept. of Zoology, Univ. of Georgia, Athens, GA 30602.

Benning, W.E. 1969. Survey of Great Blue Heron heronries, 1964–68. Kingbird 19:85–90.

——. 1979. Regional report (Region 3—Finger Lakes). Kingbird 24:219.

——. 1983. Regional report (Region 3—Finger Lakes). Kingbird 33:271.

Benson, D. 1966. What's happening to Black Ducks? New York State Conservationist 21(2):14–15, 37.

——. 1986. Personal communication. 47 Forest Hill Ave., Saranac Lake, NY 12983.

Benson, D., and S.D. Browne. 1972. Establishing breeding colonies of Redheads in New York by releasing hand-reared birds. New York Fish and Game Journal 19:59–72.

Benson, M.H. 1939. A study of the American Redstart (*Setophaga ruticilla* Swainson). Master's thesis, Cornell Univ., Ithaca, NY.

Bent, A.C. 1919. Life histories of North American diving birds. U.S. National Museum Bulletin no. 107. Washington, DC.

——. 1921. Life histories of North American gulls and terns. U.S. National Museum Bulletin no. 113. Washington, DC.

——. 1923. Life histories of North American wild fowl, pt. 1. U.S. National Museum Bulletin no. 126. Washington, DC.

——. 1926. Life histories of North American marsh birds. U.S. National Museum Bulletin no. 135. Washington, DC.

——. 1927. Life histories of North American shore birds, pt. 1. U.S. National Museum Bulletin no. 142. Washington, DC.

——. 1929. Life histories of North American shore birds, pt. 2. U.S. National Museum Bulletin no. 146. Washington, DC.

——. 1932. Life histories of North American gallinaceous birds. U.S. National Museum Bulletin no. 162. Washington, DC.

——. 1937. Life histories of North American birds of prey, pt. 1. U.S. National Museum Bulletin no. 167. Washington, DC.

——. 1938. Life histories of North American birds of prey, pt. 2. U.S. National Museum Bulletin no. 170. Washington, DC.

——. 1939. Life histories of North American woodpeckers. U.S. National Museum Bulletin no. 174. Washington, DC.

——. 1940. Life histories of North American cuckoos, goatsuckers, hummingbirds, and their allies. U.S. National Museum Bulletin no. 176. Washington, DC.

——. 1942. Life histories of North American flycatchers, larks, swallows, and their allies. U.S. National Museum Bulletin no. 179. Washington, DC.

——. 1946. Life histories of North American jays, crows, and titmice. U.S. National Museum Bulletin no. 191. Washington, DC.

——. 1948. Life histories of North American nuthatches, wrens, thrashers, and their allies. U.S. National Museum Bulletin no. 195. Washington DC.

——. 1949. Life histories of North American thrushes, kinglets, and their allies. U.S. National Museum Bulletin no. 196. Washington, DC.

——. 1950. Life histories of North American wagtails, shrikes, vireos, and their allies. U.S. National Museum Bulletin no. 197. Washington, DC.

——. 1953. Life histories of North American wood warblers. U.S. National Museum Bulletin no. 203. Washington, DC.

——. 1958. Life histories of North American blackbirds, orioles, tanagers, and allies. U.S. National Museum Bulletin no. 211. Washington, DC.

——. 1968a. Eastern Cardinal. *In* Life histories of North American cardinals, grosbeaks, buntings, towhees, finches, sparrows, and allies, by A.C. Bent and collaborators, ed. O.L. Austin, Jr. U.S. National Museum Bulletin no. 237, pt. 1. Washington, DC. Pp. 1–8

——. 1968b. Rose-breasted Grosbeak. *In* Life histories of North American cardinals, grosbeaks, buntings, towhees, finches, sparrows, and allies, by A.C. Bent and collaborators, ed. O.L. Austin, Jr. U.S. National Museum Bulletin no. 237, pt. 1. Washington, DC. Pp. 36, 39.

——. 1968c. Eastern Blue Grosbeak. *In* Life histories of North American cardinals, grosbeaks, buntings, towhees, finches, sparrows, and allies, by A.C. Bent and collaborators, ed. O.L. Austin, Jr. U.S. National Museum Bulletin no. 237, pt. 1. Washington, DC. Pp. 67–75.

——. 1968d. Purple Finch. *In* Life histories of North American cardinals, grosbeaks, buntings, towhees, finches, sparrows, and allies, by A.C. Bent and collaborators, ed. O.L. Austin, Jr. U.S. National Museum Bulletin no. 237, pt. 1. Washington, DC. Pp 264–278.

——. 1968e. Song Sparrow. *In* Life histories of North American cardinals, grosbeaks, buntings, towhees, finches, sparrows, and allies, by A.C. Bent and collaborators, ed. O.L. Austin, Jr. U.S. National Museum Bulletin no. 237, pt. 1. Washington, DC. P. 1495.

Bentley, A., ed. 1981. The Annual Report of the Province of Quebec Society for the Protection of Birds. Tchebec 11:14–15.

Benton, A.H. 1949. The breeding birds of Cayuga County, New York. Ph.D. dissertation, Cornell Univ., Ithaca, NY.

——. 1950. Notes on the breeding birds of Cayuga County, New York. Kingbird 1(1):8–10.

——. 1951a. Bird population changes in a central New York county since 1870. Kingbird 1(2):2–11.

——. 1951b. Yellow-billed Cuckoo at Lake Placid. Kingbird 1(3):59.

——. 1960. Southern warblers in central New York. Kingbird 10:137–141.

——. 1961a. Nest sharing by Robin and Catbird. Kingbird 11:81–82, 137.

——. 1961b. Notes on some unusual nesting sites. Kingbird 11:201.

——. 1986. Personal communication. 292 Water St., Fredonia, NY 14063.

Benton, A.H., and H. Tucker. 1968. Weather and Purple Martin mortality in western New York. Kingbird 18:71–75.

Berger, A.J. 1951. The Cowbird and certain host species in Michigan. Wilson Bulletin 63:28.

——. 1968. Eastern Vesper Sparrow. *In* Life histories of North American cardinals, grosbeaks, buntings, towhees, finches, sparrows, and allies, ed. O.L. Austin, Jr. U.S. National Museum Bulletin no. 237, pt. 2. Washington, DC. Pp. 868–882.

Bergman, R.D., P. Swain, and M.W. Weller. 1970. A comparative study of nesting Forster's and Black terns. Wilson Bulletin 82:435–444.

Bertin, R.I. 1977. Breeding habitats of the Wood Thrush and Veery. Condor 79:303–311.

Bicknell, E.P. 1880. Remarks on the nidification of *Loxia curvirostra americana*, with a description of its nest and eggs. Bulletin of the Nuttall Ornithological Club 5:7–11.

Birmingham, M. 1987. Personal communication. NYSDEC, 50 Wolf Rd., Albany, NY 12233.

Bishop, P.G., Jr. 1980. The history and recent breeding of the Common Raven in New York State. Master's thesis, SUNY College of Environmental Science and Forestry, Syracuse.

Black, C.P. 1975. The ecology and bioenergetics of the northern Black-throated Blue Warbler (*Dendroica caerulescens caerulescens*). Ph.D. dissertation, Dartmouth College, Hanover, NH.

Blancher, P.J., and R.J. Robertson. 1985. A comparison of Eastern Kingbird breeding biology in lakeshore and upland habitats. Canadian Journal of Zoology 63:2305–2312.

Blokpoel, H. 1986. Personal communication. 1725 Woodward Dr., Ottawa, ON K1A 0E7.

Blokpoel, H., and G.D. Tessier. 1986. The Ring-billed Gull in Ontario: a review of a new problem species. Canadian Wildlife Service Occasional Paper no. 57. Ottawa, ON.

Blokpoel, H., and D.V. Weseloh. 1982. Status of colonial nesting birds on Little Galloo Island, Lake Ontario. Kingbird 32:149–157.

Bollinger, P.B. 1985. Status of the Common Tern population breeding on Oneida Lake in 1985. New York Cooperative Wildlife Research Unit, Cornell Univ., Ithaca, NY. Mimeo.

Borror, D.J., and W.W.H. Gunn. 1985. Songs of the warblers of North America. Cornell Univ. Laboratory of Ornithology, Ithaca, NY.

Bouta, R.P. 1987. Population status and habitat characteristics of Adirondack Spruce Grouse. SUNY College of Environmental Science and Forestry, Syracuse. Mimeo.

Boyd, E.M. 1962. A half-century's changes in the bird-life around Springfield, Massachusetts. Bird-Banding 33:137–148.

Brand, A.R. 1938. Vibration frequencies of passerine bird song. Auk 55:263–268.

Brauning, D.L. 1986. Personal communication. Pennsylvania Breeding Bird Atlas, Academy of Natural Sciences, 19th and The Parkway, Philadelphia, PA 19103.

Breckenridge, W.J. 1956. Measurements of the habitat niche of the Least Flycatcher. Wilson Bulletin 68:50–51.

Brewer, T.M. 1875. Catalogue of the birds of New England. Proceedings of the Boston Society of Natural History 17:451.

Brewster, W. 1938. The birds of the Lake Umbagog region of Maine, pt. 4. Comp. L. Griscom. Bulletin of the Museum of Comparative Zoology 66:525–620.

Briggs, J.N., and J.R. Haugh. 1973. Habitat selection in birds, with consideration of the potential establishment of the parakeet (*Myiopsitta monachus*) in North America. Kingbird 23:3–13.

Brittingham, M.C., and S.A. Temple. 1983. Have cowbirds caused forest songbirds to decline? BioScience 33:31–35.

Brooks, E.W. 1971. A nesting record of the Clay-colored Sparrow in Allegany County. Prothonotary 37:99.

——. 1980. Breeding Bird Survey no. 50: upland mixed pine–spruce–hardwood plantation. American Birds 34:57.

Brooks, M.G. 1944. A check-list of West Virginia birds. West Virginia Univ. Bulletin no. 316. Charleston, WV.

Brown, C., J. Kelley, J. Penman, and J. Sparling. 1973. Faunal analysis of the Cole Quarry Archaic Site. Bulletin of the New York State Archeological Association 58:25–40.

Brown, C.P. 1954. Distribution of the Hungarian Partridge in New York. New York Fish and Game Journal 1:119–129.

Brown, C.P., and S.B. Robeson. 1959. The Ring-necked Pheasant in New York. Bulletin of the Division of Conservation Education. New York State Conservation Dept. Albany.

Brown, L., and D. Amadon. 1968. Eagles, hawks, and falcons of the world. Vol. 2. Country Life Books, Feltham, England.

Browne, S.D. 1975. Hooded Mergansers breeding in New York. New York Fish and Game Journal 22:68–70.

Buckley, P.A., and F.G. Buckley. 1980. Population and colony-site trends of Long Island waterbirds for five years in the mid 1970's. Transactions of the Linnaean Society of New York 9:23–56.

——. 1981. The endangered status of North American Roseate Terns. Colonial Waterbirds 4:166–173.

——. 1984a. Expanding Double-crested Cormorant and Laughing Gull populations on Long Island, New York. Kingbird 34:146–155.

——. 1984b. Seabirds of the north and middle Atlantic Coast of the United States: their status and conservation. *In* Status and conservation of the world's seabirds, ed. J.P. Croxall, P.G.H. Evans, and R.W. Schreiber. International Council for Bird Preservation Technical Publication no. 2. Paston Press, Norwich, England. Pp. 101–133.

Buckley, P.A., and T.H. Davis. 1973. Regional report: Hudson-Delaware region. American Birds 27:852.

Buckley, P.A., F.G. Buckley, and M. Gochfeld. 1975. Gull-billed Tern: New York State's newest breeding species. Kingbird 25:178–183.

Buckley, P.A., R.O. Paxton, and D.A. Cutler. 1975. Regional report: Hudson-Delaware region. American Birds 29:954.

——. 1976. Regional report: Hudson-Delaware region. American Birds 30:937.

Buckley, F.G., M. Gochfeld, and P.A. Buckley. 1978. Breeding Laughing Gulls return to Long Island. Kingbird 28:203–207.

Bull, J. 1964. Birds of the New York area. Harper & Row, New York.

521

——. 1970. Supplement to "Birds of the New York area." Proceedings of the Linnaean Society of New York 71:16.

——. 1971. Monk Parakeets in the New York City region. Linnaean News-Letter 25:1–2.

——. 1974. Birds of New York State. Doubleday/Natural History Press, Garden City, NY. Reprint. Cornell Univ. Press, Ithaca, NY, 1985.

——. 1975. Introduction to the United States of the Monk Parakeet—a species with pest potential. Bulletin of the International Council for Bird Preservation 12:98.

——. 1976. Supplement to "Birds of New York State." Federation of New York State Bird Clubs, Cortland, NY. Reprint. Cornell Univ. Press, Ithaca, NY, 1985.

——. 1981. Double-crested Cormorants breeding at Fishers Island. Kingbird 32:83.

Bump, G. 1941. The introduction and transplantation of game birds and mammals into the state of New York. Transactions of the 5th North American Wildlife Conference 5:409–420.

Bump, G., R.W. Darrow, F.C. Edminster, and W.F. Crissey. 1947. The Ruffed Grouse: life history, propagation, management. New York State Dept. of Conservation, Albany.

Bureau of Census. 1977. 1974 census of agriculture: New York state and county data. Vol. 1, pt. 32. U.S. Dept. of Commerce, Washington, DC.

Bureau of Wildlife. 1979. Management plan for Ring-necked Pheasant in New York. NYSDEC, Albany. Mimeo.

Burger, J. 1978. Competition between Cattle Egrets and native North American herons, egrets, and ibises. Condor 80:15–23.

——. 1979a. Competition and predation: Herring Gulls versus Laughing Gulls. Condor 81:269–277.

——. 1979b. Resource partitioning: nest site selection in mixed species colonies of herons, egrets, and ibises. American Midland Naturalist 101:191–210.

Burger, J., and M. Gochfeld. 1985. Nest site selection by Laughing Gulls: comparison of tropical colonies (Culebra, Puerto Rico) with temperate colonies (New Jersey). Condor 87:364–373.

Burns, F.L. 1900. Monograph of the Flicker. Wilson Bulletin 7:1–82.

Burtch, V. 1910. Turkey Vulture in northern Steuben County. Auk 27:208.

Burtt, B. 1960. Mockingbird nesting in Onondaga County. Kingbird 10:95.

——. 1980. The feeder survey and trends in central New York Cardinal populations. Kingbird 30:138–139.

Buss, I.O. 1942. A managed Cliff Swallow colony in southern Wisconsin. Wilson Bulletin 54:153–161.

Bystrak, D. 1975. Data print-out of breeding bird surveys taken from 112 routes run in New York, 1966–1974, of 32 species of warblers. USFWS Migratory Bird Population Station, Laurel, MD.

——. 1986. Personal communication. USFWS, Patuxent Research Center, Laurel, MD 20708.

Cade, T.J., and P.R. Dague, eds. 1985. New York and California provide key support. Peregrine Fund Newsletter no. 13. Ithaca, NY. Pp. 1–2.

Cadman, M.D. 1985. Status report on the Loggerhead Shrike in Canada to Committee on the Status of Endangered Wildlife in Canada (draft). Federation of Ontario Naturalists, Don Mills, ON.

Cairns, W.E., and I.A. McLaren. 1980. Status of the Piping Plover on the East Coast of North America. American Birds 32:206–208.

Cameron, E.S. 1907. The birds of Custer and Dawson counties, Montana. Auk 24:259.

Capen, D.E. 1979. Management of northeastern pine forests for nongame birds. In Management of north central and northeastern forests for nongame birds. U.S. Dept. of Agriculture, Forest Service General Technical Report NC-51. St. Paul, MN. Pp. 90–109.

Carleton, G.C. 1935. Notes from Essex County, NY. Auk 52:197.

——. 1958. The birds of Central and Prospect parks. In Proceedings of the Linnaean Society of New York, ed. L.S. Pearl. Nos. 66–70. New York. Pp. 1–60.

——. 1963. Regional report: Hudson-St. Lawrence region. Audubon Field Notes 17:456.

——. 1971. Wood Duck presumed nesting in cliff. Kingbird 21:212.

——. 1980. Birds of Essex County, New York. 2d ed. High Peaks Audubon Society, Elizabethtown, NY.

——. 1986. Personal communication. Elizabethtown, NY 12932.

Carleton, G., H.H. Poor, and O.K. Scott. 1948. Cape May Warbler breeding in New York State. Auk 65:607.

Carney, S.M., M.F. Sorensen, and E.M. Martin. 1986. Waterfowl harvest and hunter activity in the United States during the 1985 hunting season. USFWS administrative report. Washington, DC.

Carroll, D. 1986. Personal communication. NYSDEC, 6274 East Avon-Lima Rd., Avon, NY 14414.

Carroll, J. 1982. Occurrence of the Nanday Conure in Westchester County, New York. New York Fish and Game Journal 29:217.

——. 1985. Personal communication. NYSDEC, Wildlife Resources Center, Delmar, NY. 12054.

Carroll, J., and J.M.C. Peterson, eds. 1982. Confirming the Belted Kingfisher. New York State Breeding Bird Atlas Newsletter no. 5. Delmar, NY. P. 2.

——. 1983. Northern Saw-whet Owl. New York State Breeding Bird Atlas Newsletter no. 7. Delmar, NY. P. 3.

——. 1984. An atlas glossary. New York State Breeding Bird Atlas Newsletter no. 10. Delmar, NY. P. 1.

Case, N.A., and O.H. Hewitt. 1963. Nesting and productivity of the Red-winged Blackbird in relation to habitat. Living Bird 2:7–20.

Caslick, J.W. 1975. Measuring revegetation rates and patterns on abandoned agricultural land. Search 5:1–27.

Chabreck, R.H. 1963. Breeding habits of the Pied-billed Grebe in an impounded coastal marsh in Louisiana. Auk 80:447–452.

Chadwick, N. 1987a. The nest box network 1986 report. National Audubon Society, Delmar, NY.

——. 1987b. Personal communication. Box 162, West Berne, NY 12023.

Chamberlaine, L.B. 1978. Regional report (Region 6—St. Lawrence). Kingbird 28:251.

——. 1980. Regional report (Region 6—St. Lawrence). Kingbird 30:244–245.

——. 1986. The Double-crested Cormorant on Lake Ontario. NYSDEC, Watertown, NY. Mimeo.

Chambers, R.E. 1980a. Distribution and abundance of Spruce Grouse in the Adirondack region of New York: final report to NYSDEC. SUNY College of Environmental Science and Forestry, Syracuse. Mimeo.

——. 1980b. Report on 1980 census of Adirondack Spruce Grouse. SUNY College of Environmental Science and Forestry, Syracuse. Mimeo.

Chapman, A., and W. Chapman. 1985. Personal communication. 32 Marcy Lane, Newcomb, NY 12852.

Chapman, F.M. 1889. Notes on birds observed in the vicinity of Englewood, New Jersey. Auk 6:302–305.

——. 1894. Visitors' guide to the collection of birds found within 50 miles of New York City. American Museum of Natural History, New York.

——. 1906. Birds in the vicinity of New York City. American Museum of Natural History Guide Leaflet no. 22. New York.

——. 1907. The warblers of North America. D. Appleton and Co., New York.

——. 1927. Handbook of birds of eastern North America. D. Appleton and Co., New York.

Chase, G.T. 1986. Personal communication. Ampersand Bay, Saranac Lake, NY 12983.

Christy, B.H. 1939. Pileated Woodpecker. In Life histories of North American woodpeckers, by A.C. Bent. U.S. National Museum Bulletin no. 174. Washington, DC. Pp. 172–177.

——. 1942. Acadian Flycatcher. *In* Life histories of North American flycatchers, larks, swallows, and their allies, by A.C. Bent. U.S. National Museum Bulletin no. 179. Washington, DC. Pp. 183–197.

Claffey, J. 1972. Clay-colored Sparrow, *Spizella pallida*. Goshawk 28(7):50–52.

Clark, D.B. 1985. Blue-winged Warbler. *In* The atlas of breeding birds of Vermont, ed. S.B. Laughlin, and D.P. Kibbe. Univ. Press of New England, Hanover, NH. Pp. 274–275.

Clark, J.N. 1884. Nesting of Little Black Rail in Connecticut. Auk 1:393–396.

Clark, R.C. 1975. A field study of the Short-eared Owl, *Asio flammeus*, in North America. Wildlife Monographs no. 47.

Clark, R.C., and S.W. Eaton. 1977. The breeding biology of the Eastern Phoebe (abstract). American Ornithologists' Union Annual Meeting, Berkeley.

Clauser, M. 1980. The Grasshopper, Henslow's, and Vesper sparrows: their declining numbers in New York State. Research report. St. Bonaventure Univ., St. Bonaventure, NY.

Claypoole, K. 1986. Status of the Common Tern population breeding on Oneida Lake, New York, in 1986. New York Cooperative Fish and Wildlife Research Unit, Cornell Univ., Ithaca, NY. Mimeo.

Clinton, J., D. Larsen, G. Raynor, J. Ruscica, and K. Tuohy. 1980. Breeding Bird Survey no. 49: young red maple–gray birch. American Birds 34:57.

Clinton, J., C. Dodge, G. Raynor, and K. Tuohy. 1982. Breeding Bird Survey no. 56: second growth oak-pine forest. American Birds 36:66.

Clum, N.J. 1986. The effects of prey quantity on reproductive success of Osprey (*Pandion haliaetus*) in the Adirondack Mountains. Master's thesis, Cornell Univ., Ithaca, NY.

Cohen, R. 1966. Saw-whet Owls in Atlantic Beach. Kingbird 16:90.

Colvin, B.A. 1985. Common Barn-Owl population decline in Ohio and the relationship to agricultural trends. Journal of Field Ornithology 56:224–235.

——. 1986. Barn-owls: their secrets and habits. Illinois Audubon no. 216. Wayne, IL. Pp. 9–13.

Comar, M.C. 1974. Fish Crow in Ithaca, New York. Kingbird 24:124.

Confer, J.L., and K. Knapp. 1979. The changing proportions of Blue-winged and Golden-winged warblers in Tompkins County and their habitat selection. Kingbird 29:8–14.

——. 1981. Golden-winged Warblers and Blue-winged Warblers: the relative success of a habitat specialist and a generalist. Auk 98:108–114.

Confer, J.L., and P. Paicos. 1985. Downy Woodpecker predation at goldenrod galls. Journal of Field Ornithology 56:56–64.

Conner, R.N. 1975. Orientation of entrances to woodpecker nest cavities. Auk 92:371–374.

——. 1981. Seasonal changes in woodpecker foraging patterns. Auk 98: 562–570.

Conner, R.N., and C.S. Adkisson. 1977. Principal component analysis of woodpecker nesting habitat. Wilson Bulletin 89:122–129.

Conner, R.N., R.G. Hooper, H.S. Crawford, and H.S. Mosby. 1975. Woodpecker nesting habitat in cut and uncut woodlands in Virginia. Journal of Wildlife Management 39:148.

Conner, R.N., O.K. Miller, Jr., and C.S. Adkisson. 1976. Woodpecker dependence on trees infected by fungal heart rots. Wilson Bulletin 88:575–581.

Connor, P.F. 1986. Personal communication. 1506 Sunset Rd., Castleton, NY 12033.

Considine, T.J., Jr. 1984. An analysis of New York's timber resources. U.S. Dept. of Agriculture, Forest Service Resource Bulletin NE-80. Broomall, PA.

Cooch, F.G. 1964. A preliminary study of the survival value of a functional salt gland in prairie Anatidae. Auk 81: 380–393.

Cook, A.H. 1957. Man, marshes, and—ducks. New York State Conservationist 11(5):12–13.

Cook, B. 1983. Distraction display in Northern Waterthrush. New York State Breeding Bird Atlas Newsletter no. 9. Delmar, NY. P. 2.

Cooke, W.W. 1914. Distribution and migration of North American rails and their allies. U.S. Dept. of Agriculture Bulletin no. 128. Washington, DC. Pp. 14–17.

Cooper, R.J. 1981. Relative abundance of Georgia Caprimulgids based on call-counts. Wilson Bulletin 93:363–371.

Cornwell, G.W. 1963. Observations of the breeding biology and behavior of a nesting population of Belted Kingfishers. Condor 65:426–431.

Coulter, M.W., and W.R. Miller. 1968. Nesting biology of Black Ducks and Mallards in northern New England. Vermont Fish and Game Department Bulletin 68:1–74.

Courtney, P.A., and H. Blokpoel. 1983. Distribution and numbers of Common Terns on the lower Great Lakes during 1900–1980: a review. Colonial Waterbirds 6:107–120.

Cowardin, L.M., G.E. Cummings, and P.B. Reed, Jr. 1967. Stump and tree nesting by Mallards and Black Ducks. Journal of Wildlife Management 31:229–235.

Craighead, J.J., and F.C. Craighead. 1956. Hawks, owls, and wildlife. Stackpole Co., Harrisburg.

Crawford, R.D. 1977. Polygynous breeding of Short-billed Marsh Wrens. Auk 94:359–361.

Crocoll, S.T. 1984. Breeding biology of Broad-winged and Red-shouldered hawks in western New York. Master's thesis, State Univ. College, Fredonia, NY.

Crombie, J.M., and V.H. Blackman. 1911. Lichens. Encyclopedia Britannica. 11th ed. Vol. 16. Encyclopedia Britannica, Inc., New York. P. 583.

Crowell, K.L. 1982. Regional report (Region 6—St.Lawrence). Kingbird 32:971.

Crowell, K.L., and G.A. Smith. 1985. Regional report (Region 6—St. Lawrence). Kingbird 35:277–281.

——. 1986. Regional report (Region 6—St. Lawrence). Kingbird 36:230.

Cruickshank, A.D. 1942. Birds around New York City. American Museum of Natural History Handbook Series no. 13. New York.

Crumb, D. 1985. Nesting finches in the highlands of southern Onondaga and Madison counties. Kingbird 35:238–240.

——. 1986. Personal communication. 3983 Gates Rd., Jamesville, NY 13078.

Davids, M. 1986. Personal communication. 121 Hillside Dr., Hilton, NY 14468.

Davis, C.M. 1978. A nesting study of the Brown Creeper. Living Bird 17:237–267.

Davis, G.L. 1971. Cliff Swallow nest in active Bank Swallow colony. Kingbird 21:67.

Davis, L.R. 1974. The Monk Parakeet: a potential threat to agriculture. *In* Proceedings of the 6th Vertebrate Pest Control Conference, ed. W.V. Johnson and R.E. Marsh. Anaheim. Pp. 253–256.

Davis, M.B. 1983. Holocene vegetational history of the eastern United States. *In* Late Quaternary environments of the United States, ed. H.E. Wright, Jr. Vol. 2, The Holocene, Univ. of Minnesota Press, Minneapolis. Pp. 166–181.

Davis, T.H. 1968. Willet nesting on Long Island, NY. Wilson Bulletin 80:330.

——. 1972. Photographs of New York rarities—22: Chuck-will's-widow. Kingbird 22:157.

——. 1975. Notes concerning the first New York nesting of Chuck-will's-widow. Kingbird 25:132.

——. 1981. Boat-tailed Grackles breeding at Jamaica Bay Wildlife Refuge. Kingbird 31:214.

——. 1982. The 1981 fall shorebird season at Jamaica Bay Wildlife Refuge. Kingbird 32:85–96.

Davis, T.H., and L. Morgan. 1968. Regional report (Region 10—Marine). Kingbird 17:244.

Davis, T.H., and L. Morgan. 1968. Regional report (Region 10— Marine). Kingbird 18:229.

——. 1971. Regional report (Region 10—Marine). Kingbird 21:245.

——. 1972. Regional report (Region 10—Marine). Kingbird 22:192.

Dawson, D.K. 1979. Bird communities associated with successional and management of lowland conifer forests. *In* Management of north central and northeastern forests for nongame birds, comp. R.M. DeGraaf and K.E. Evans. U.S. Dept.

of Agriculture, Forest Service General Technical Report NC-51. St. Paul. Pp. 120–131.

DeBenedictis, P.A. 1981. Report of the New York State Avian Records Committee. Kingbird 31:209.

———. 1982. Regional report (Region 5—Oneida Lake Basin). Kingbird 32:284.

———. 1983. Regional report (Region 5—Oneida Lake Basin). Kingbird 33:278.

———. 1984. Regional report (Region 5—Oneida Lake Basin). Kingbird 34:253.

Decker, D.J. 1985. More Mourning Doves in New York? New York State Conservationist 39(4):34–38.

Deed, R. 1984. Cerulean Warbler. New York State Breeding Bird Atlas Newsletter no. 11. Delmar, NY. Pp. 2, 5.

DeGraaf, R.M., and D.D. Rudis. 1986. New England wildlife: habitat, natural history, and distribution. U.S. Dept. of Agriculture, Forest Service General Technical Report NE-108. Broomall, PA.

DeGraaf, R.M., J.R. Pywell, and J.W. Thomas. 1975. Relationships between nest height, vegetation, and housing density in New England suburbs. In Transactions of the Northeast Section, the Wildlife Society, ed. D. DeCarli. Northeast Fish and Wildlife Conference. New Haven. Pp. 135–136.

DeGraaf, R.M., G.M. Witman, J.W. Lanier, B.J. Hill, and J.M. Keniston. 1980. Forest habitat for birds of the Northeast. U.S. Dept. of Agriculture, Forest Service, Washington, DC.

DeGraff, L.W. 1973. Return of the Wild Turkey. New York State Conservationist 28(2):24–27, 47.

———. 1975. The Ring-necked Pheasant. New York State Conservationist 30(2):12–13, 46.

———. 1985. Managing Mourning Doves in New York. New York State Conservationist 39(4):39–40.

De Kay, J.E. 1844. Zoology of New York or the New York fauna. Pt. 2, Birds. Carroll and Cook, Albany.

Delacour, J. 1954. Waterfowl of the world. Vol. 1. Country Life, London.

Delafield, H.L. 1968. Common Raven observations in Essex County, northern Adirondacks, spring and summer, 1968. Kingbird 18:198–199.

Dennis, J.V. 1969. The Yellow-shafted Flicker (Colaptes auratus) on Nantucket Island, Massachusetts. Bird-Banding 40:290–308.

Derrickson, S.R. 1978. The mobility of breeding Pintails. Auk 95:104–114.

Dewey, V. 1987. Personal communication. 45 Owasco St., Auburn, NY 13201.

Dexter, R.W. 1962. Attempted reuse of old nest by Chimney Swift. Wilson Bulletin 74:284–285.

———. 1981a. Chimney Swifts reuse ten-year-old nest. North American Bird Bander 6:136–137.

———. 1981b. Nesting success of Chimney Swifts related to age and number of adults at the nest, and the subsequent fate of the visitors. Journal of Field Ornithology 52:228–232.

Dickerman, R.W. 1986a. A review of the Red Crossbill in New York State. Pt. 1, Historical and nomenclatural background. Kingbird 36:73–78.

———. 1986b. A review of the Red Crossbill in New York State. Pt. 2, Identification of specimens from New York. Kingbird 36:127–134.

Dickinson, J.C., Jr. 1968. Rufous-sided Towhee. In Life histories of North American cardinals, grosbeaks, buntings, towhees, finches, sparrows, and allies, by A.C. Bent and collaborators, ed. O.L. Austin, Jr. U.S. National Museum Bulletin no. 237, pt. 1. Washington, DC. Pp. 562–579.

Dickinson, N.R. 1979. A division of southern and western New York State into ecological zones. NYSDEC, Albany. Mimeo.

———. 1983. Physiographic zones of southern and western New York. NYSDEC, Albany. Mimeo.

DiCostanzo, J.A. 1983a. Regional report (Region 10—Marine). Kingbird 33:226.

———. 1983b. Regional report (Region 10—Marine). Kingbird 33:298–299.

———. 1984a. Regional report (Region 10—Marine). Kingbird 34:71.

———. 1984b. Regional report (Region 10—Marine). Kingbird 34:142.

———. 1984c. Regional report (Region 10—Marine). Kingbird 34:210.

———. 1984d. Regional report (Region 10—Marine). Kingbird 34:270–272.

———. 1985. Regional report (Region 10—Marine). Kingbird 35:71.

———. 1986. Regional report (Region 10—Marine). Kingbird 36:51.

Dilger, W.C. 1956. Adaptive modifications and ecological isolating mechanisms in the thrush genera Catharus and Hylocichla. Wilson Bulletin 68:171–199.

———. 1987. Personal communication. 107 Niemi Rd., R.D. 1, Freeville, NY 13068.

Dingle, E. von S. 1942. Rough-winged Swallow. In Life histories of North American flycatchers, larks, swallows, and their allies, by A.C. Bent. U.S. National Museum Bulletin No. 179. Washington, DC. Pp. 424–433.

DiNunzio, M.G. 1984. Adirondack wild guide: a natural history of the Adirondack Park. Adirondack Conservancy Committee and Adirondack Council, Elizabethtown, NY. Pp. 103–122.

Dolton, D.D. 1986. Mourning Dove: 1986 breeding population status. Administrative report. USFWS, Office of Migratory Bird Management, Laurel, MD.

Donaldson, A.L. 1921. A history of the Adirondacks. Vols. 1 and 2. Century Co., New York.

Drumm, J. 1963. Mastodons and mammoths, Ice Age elephants of New York. New York State Museum and Science Service Educational Leaflet no. 13. Albany, NY. Pp. 1–31.

Dudones, T. 1983. Redhead with ducklings on the Saranac River, Franklin County. Kingbird 33:106.

Duffy, D. 1977. Breeding populations of terns and skimmers on Long Island Sound and eastern Long Island: 1972–1975. In Proceedings of the Linnaean Society of New York, ed. C. Pessino. No. 73. New York. Pp. 1–41.

Dunham, K. 1985. Personal communication. Old Chatham, NY 12136.

Dunne, P. 1984. Northern Harrier breeding survey in coastal New Jersey. Records of New Jersey Birds 10:2–5.

Dwight, J. 1900. The sequence of plumages and moults of the passerine birds of New York. Annals of the New York Academy of Science 13:73–360.

Dwight, J., and L. Griscom. 1927. A revision of the geographical races of the Blue Grosbeak. American Museum Novitates no. 257.

Dye, H.G. 1924. Ruby-crowned Kinglet summering at Rochester, N.Y. Auk 41:349.

Eastern Peregrine Falcon Recovery Team. 1979. Eastern Peregrine Falcon recovery plan. USFWS, Washington, DC.

Eaton, E.H. 1901. Birds of western New York. Proceedings of the Rochester Academy of Sciences 4:45.

———. 1910. Birds of New York, pt. 1. Univ. of the State of New York, Albany.

———. 1914. Birds of New York, pt. 2. Univ. of the State of New York, Albany.

———. 1953. Birds of New York: 1910 to 1930. Kingbird 3:53.

Eaton, R.J. 1931. Great Black-backed Gull (Larus marinus) breeding in Essex County, Massachusetts. Auk 48:588–589.

Eaton, S.W. 1957. A life history study of Seiurus noveboracensis. St. Bonaventure Univ. Science Studies 19:7–36.

———. 1958. A life history study of the Louisiana Waterthrush. Wilson Bulletin 70:211–236.

———. 1959. The Tufted Titmouse invades New York. Kingbird 9:59–62.

———. 1965. Juncos of the high plateaus. Kingbird 15:141–146.

———. 1968. Northern Slate-colored Junco. In Life histories of North American cardinals, grosbeaks, buntings, towhees, finches, sparrows, and allies, by A.C. Bent and collaborators, ed. O.L. Austin, Jr. U.S. National Museum Bulletin no. 237, pt. 2. Washington, DC. Pp. 1029–1043.

———. 1979. Notes on the reproductive behavior of the Yellow-billed Cuckoo. Wilson Bulletin 91:154–155.

———. 1981. Birds of Cattaraugus County, New York. Bulletin of the Buffalo Society of Natural Sciences vol. 29.

———. 1985. Regional report (Region 1—Niagara Frontier). Kingbird 35:262.

Eaton, S.W., P.D. O'Connor, M.B. Osterhaus, and B.Z. Anicete. 1963. Some osteological adaptations in Parulidae. In Proceedings of the XIII International Ornithological Congress, ed. C.G. Sibley. Vol. 1. American Ornithologists' Union, Baton Rouge. Pp. 71–83.

Eaton, S.W., T.L. Moore and E.N. Saylor. 1970. A ten-year study of the food habits of a northern population of Wild Turkeys. St. Bonaventure Univ. Science Studies 26:43–64.

Edminster, F.C. 1954. American game birds of field and forest. Charles Schribner's Sons, New York.

Elliott, J.J. 1953. The nesting sparrows of Long Island. Long Island Naturalist 2:15–24.

——. 1958. Regional report (Region 10—Marine). Kingbird 7:104.

——. 1961. Regional report (Region 10—Marine). Kingbird 11:176.

——. 1962. Sharp-tailed and Seaside sparrows on Long Island, New York. Kingbird 12:115–123.

Elliott, J.J., and R. Arbib. 1953. Origin and status of the House Finch in the eastern United States. Auk 70:31–37.

Ellison, W.G. 1985a. Cattle Egret. In The atlas of breeding birds of Vermont, ed. S.B. Laughlin and D.P. Kibbe. Univ. Press of New England, Hanover, NH. Pp. 40–41.

——. 1985b. Common Goldeneye. In The atlas of breeding birds of Vermont, ed. S.B. Laughlin and D.P. Kibbe. Univ. Press of New England, Hanover, NH. Pp. 62–63.

——. 1985c. Eastern Wood-Pewee. In The atlas of breeding birds of Vermont, ed. S.B. Laughlin and D.P. Kibbe. Univ. Press of New England, Hanover, NH. Pp. 172–173.

——. 1985d. White-breasted Nuthatch. In The atlas of breeding birds of Vermont, ed. S.B. Laughlin and D.P. Kibbe. Univ. Press of New England, Hanover, NH. Pp. 218–219.

——. 1985e. House Wren. In The atlas of breeding birds of Vermont, ed.S.B. Laughlin and D.P. Kibbe. Univ. Press of New England, Hanover, NH. Pp. 224–225.

——. 1985f. Cerulean Warbler. In The atlas of breeding birds of Vermont, ed. S.B. Laughlin and D.P. Kibbe. Univ. Press of New England, Hanover, NH. Pp. 308–309.

——. 1985g. Canada Warbler. In The atlas of breeding birds of Vermont, ed. S.B. Laughlin and D.P. Kibbe. Univ. Press of New England, Hanover, NH. Pp. 326–327.

——. 1985h. Red-winged Blackbird. In The atlas of breeding birds of Vermont, ed. S.B. Laughlin and D.P. Kibbe. Univ. Press of New England, Hanover, NH. Pp. 360–361.

——. 1985i. Orchard Oriole. In The atlas of breeding birds of Vermont, ed. S.B. Laughlin and D.P. Kibbe. Univ. Press of New England, Hanover, NH. Pp. 370–371.

Emlen, J.T., Jr. 1954. Social behavior in nesting Cliff Swallows. Condor 54:177–199.

England, M.E. 1985. 1985 Northern Harrier productivity, Long Island, New York. Report to NYSDEC. Delmar, NY. Mimeo.

Ermer, E.M. 1984. The status of beaver management in New York. Report to the Annual Meeting of the New York Chapter of the Wildlife Society. NYSDEC, Utica, NY. Mimeo.

Erskine, A.J. 1977. Birds in boreal Canada: communities, densities, and adaptations. Canadian Wildlife Service Report Series no. 41. Minister of Supply and Services, Ottawa, ON.

Erwin, R.M. 1979. Coastal waterbird colonies: Cape Elizabeth, Maine to Virginia. FWS/OBS-79/10. USFWS, Office of Biological Services, Washington, DC.

Erwin, R.M., and C.E. Korschgen. 1979. Coastal waterbird colonies: Maine to Virginia, 1977. FWS/OBS-79/08 USFWS, Office of Biological Services, Washington, DC.

Evans, K.E., and R.N. Conner. 1979. Snag management. In Management of north central and northeastern forests for nongame birds, comp. R.M. DeGraaf and K.E. Evans. U.S. Dept. of Agriculture, Forest Service General Technical Report NC-51. St. Paul. Pp. 214–225.

Ewert, D. 1974. First Long Island nesting record of the Kentucky Warbler. Proceedings of the Linnaean Society of New York 72:77–79.

Faxon, W. 1911. Brewster's Warbler. Museum of Comparative Zoology Memoirs 40:57–78.

Fetterolf, P.M., and H. Blokpoel. 1983. Reproductive performance of Caspian Terns at a new colony on Lake Ontario, 1979–1981. Journal of Field Ornithology 54:170–186.

Fichtel, C. 1985. Great Egret. In The atlas of breeding birds of Vermont, ed. S.B. Laughlin and D.P. Kibbe. Univ. Press of New England, Hanover, NH. Pp. 392–393.

Ficken, M.S. 1962. Agonistic behavior and territory in the American Redstart. Auk 79:607–632.

Ficken, M.S., and R.W. Ficken. 1962. The comparative ethology of the wood warblers. Living Bird 1:113.

——. 1967. Singing behavior of Blue-winged and Golden-winged warblers and their hybrids. Behavior 28:149–181.

——. 1968. Reproductive isolating mechanisms in the Blue-winged and Golden-winged Warbler complex. Evolution 22:166–179.

——. 1969. Responses of Blue-winged Warblers and Golden-winged Warblers to their own and the other species' song. Wilson Bulletin 81:69–74.

Fimreite, N., R.W. Fyfe, and J.A. Keith. 1970. Mercury contamination of Canadian prairie seed eaters and their avian predators. Canadian Field-Naturalist 84:269–276.

Fischer, R.B. 1958. The breeding biology of the Chimney Swift Chaetura pelagica (Linnaeus). New York State Museum Bulletin no. 368.

Fisher, A.K. 1878. Kentucky Warbler. Bulletin of the Nuttall Ornithological Club 3:191–192.

Fisher, D.W. 1955. Prehistoric mammals of New York. New York State Conservationist 9(4):18–22.

Fisk, L.H., and D.M. Crabtree. 1974. Black-hooded Parakeet: a new feral breeding species in California. American Birds 28:11–13.

Fleisher, E. 1943. Evening Grosbeak in summer in the Adirondack Mountains. Auk 60:107.

Fogarty, M.J., K.A. Arnold, L. McKibben, L.B. Pospichal, and R.J. Tully. 1977. Common Snipe (Gallinago gallinago). In Management of migratory shore and upland game birds in North America, ed. G.C. Sanderson. International Association of Fish and Wildlife Agencies, Washington, DC. Pp. 188–209.

Foley, D.D. 1954. Survival and establishment of waterfowl released as ducklings. New York Fish and Game Journal 1:206–213.

——. 1956. Primary waterfowl of New York. New York State Conservationist 11(2):22–23.

——. 1960. Recent changes in waterfowl populations in New York. Kingbird 10:86.

——. 1963. Our wary, Wild Turkey. New York State Conservationist 14(2):2–3, 30.

Forbes, J.E., and D.W. Warner. 1974. Behavior of a radio-tagged Saw-whet Owl. Auk 91:783–795.

Forbush, E.H. 1912. A history of the game birds, wildfowl, and shore birds of Massachusetts and adjacent states. Massachusetts State Board of Agriculture, Boston.

——. 1929. Birds of Massachusetts and other New England states, pts. 2 and 3. Norwood Press, Norwood, MA.

Forbush, E.H., and J.B. May. 1936. A natural history of American birds of eastern and central North America. Bramhall House, New York.

Forness, M. 1985. Personal communication. Haskell Rd., Cuba, NY 14727.

Forshaw, J.W. 1978. Parrots of the world. 2d ed. Landsdowne Press, Melbourne, Australia.

Fosburgh, H. 1960. One man's pleasure: a journal of the wilderness world. William Morrow and Co., New York.

Frederickson, L.H. 1971. Common gallinule breeding biology and development. Auk 88:914–919.

Frederickson, L.H., J.M. Anderson, F.M. Kozlik, and R.A. Ryder. 1977. American Coot. In Management of migratory shore and upland game birds in North America, ed. G.C. Sanderson. International Association of Fish and Wildlife Agencies, Washington, DC. Pp. 122–147.

Freer, V.M. 1979. Factors affecting site tenacity in New York Bank Swallows. Bird-Banding 50:349–357.

Friedmann, H. 1929. The cowbirds: a study in the biology of social parasitism. Charles C Thomas, Springfield, IL.

——. 1963. Host relations of the parasitic cowbirds. U.S. National Museum Bulletin no. 233. Washington, DC.

Fritz, R.S. 1977. The distribution and population status of the Spruce Grouse in the Adirondacks. Master's thesis, SUNY College of Environmental Science and Forestry, Syracuse.

——. 1979. Consequences of insular population structure: distribution and extension of Spruce Grouse populations. Oecologia 42:57–65.

Frohling, R.C. 1965. American Oystercatcher and Black Skimmer nesting on salt marsh. Wilson Bulletin 77:193–194.

Funk, R.E. 1973a. The Scaccia Site (Cda 17–3). In Aboriginal settlement patterns in the Northeast, by W.A. Ritchie and R.E. Funk. New York State Museum and Science Service Memoir 20:99–116.

——. 1973b. The Garoga Site (Las. 7). In Aboriginal settlement patterns in the Northeast, by W.A. Ritchie and R.E. Funk. New York State Museum and Science Service Memoir 20:313–332.

——. 1976. Recent contributions to Hudson Valley prehistory. New York State Museum and Science Service Memoir 22.

Funk, R.E., G.R. Walters, and W.F. Ehlers, Jr., eds. 1969. The archeology of Dutchess Quarry Cave, Orange County, New York. Pennsylvania Archeologist 39:7–22.

Gawalt, J. 1986. Personal communication. NYSDEC, Wildlife Resources Center, Delmar, NY 12054.

Gibson, S.J. 1968. The Oran-Barnes Site. Bulletin of the Chenango Chapter of the New York State Archeological Association 10:1–22.

Gill, F.B. 1980. Historical aspects of hybridization between Blue-winged and Golden-winged warblers. Auk 97:1–18.

——. 1985. Whither two warblers? Living Bird Quarterly 4(4):4–7.

Gill, F.B., and B.G. Murray, Jr. 1972. Discrimination behavior and hybridization of the Blue-winged and Golden-winged warblers. Evolution 26: 282–293.

Gilroy, M.J. 1985. Report of the 1985 Peregrine Falcon survey in New York City and Adirondack regions. Peregrine Fund, Ithaca, NY. Mimeo.

Giltz, M., and T. Stockdale. 1960. The Red-winged Blackbird story. Ohio Agricultural Experiment Station Special Circular no. 95. Wooster.

Gingrich, T.A. 1986. Personal communication. Montezuma NWR, Seneca Falls, NY 13148.

Giraud, J.P., Jr. 1844. Birds of Long Island. Wiley and Putnam, New York.

Glidden, J.W. 1977. Net productivity of a Wild Turkey population in southwestern New York. Transactions of the Northeast Fish and Wildlife Conference 34:13–21.

——. 1986. Personal communication. NYSDEC, Wildlife Resources Center, Delmar, NY 12054.

Glover, F.A. 1956. Nesting and production of the Blue-winged Teal (Anas discors Linnaeus) in northwest Iowa. Journal of Wildlife Management 20:28–46.

Gochfeld, M. 1976. Waterbird colonies of Long Island, New York. 3. Cedar Beach ternery. Kingbird 26:62–80.

Gochfeld, M., and J. Burger. 1981. Boat-tailed Grackles in Hewlett Bay, Long Island. Kingbird 31:214.

Godfrey, W.E. 1966. The birds of Canada. National Museum of Canada Bulletin 203. Ottawa, ON.

Goodwin, C.E. 1976. Regional report: Ontario region. American Birds 30:950.

——. 1981. Regional report: Ontario region. American Birds 35:936.

Goodwin, R.E. 1953. A study of the breeding behavior of the Black Tern, Chlidonias niger surinamensis. Master's thesis, Cornell Univ., Ithaca, NY.

Gordon, D.C. 1959. Owls of New York State. Kingbird 9:103–107.

Gorenzel, W.P., R.A. Ryder, and C.E. Braun. 1982. Reproduction and nest site characteristics of American Coots at different altitudes in Colorado. Condor 84:59–65.

Graber, J.W. 1968. Western Henslow's Sparrow. In Life histories of North American cardinals, grosbeaks, buntings, towhees, finches, sparrows, and allies, by A.C. Bent and collaborators, ed. O.L. Austin, Jr. U.S. National Museum Bulletin no. 237, pt. 2. Washington, DC. Pp. 779–788.

Gradoni, P.B., and D.L. Bishop. 1981. Distribution and abundance of the Spruce Grouse in the Altamont Farms area. Final report for Adirondack Park Agency Project 81–51. SUNY College of Environmental Science and Forestry, Syracuse. Mimeo.

Graham, F., Jr. 1976. Will the Bald Eagle survive to 2076? Audubon 78:99.

Grandy, J.W. 1983. The North American Black Duck (Anas rubripes): a case study of 28 years of failure in American wildlife management. International Journal of the Study of Animal Problems, Supplement 4(4):1–35.

Granger, J.E., Jr. 1978. Meadowood phase settlement pattern in the Niagara Frontier region of western New York State. Museum of Anthropology Paper no. 65. Univ. of Michigan, Ann Arbor.

Greenlaw, J.S. 1986. Personal communication. Dept. of Biology, C.W. Post Center, Greenvale, NY 11548.

Greenlaw, J.S., and R.F. Miller. 1982. Breeding Soras on a Long Island salt marsh. Kingbird 32:78–84.

Gretch, M. 1984. Sedge Wren. New York State Breeding Bird Atlas Newsletter no. 11. Delmar, NY. P. 2.

——. 1986. Personal communication. Box 748, Champlain, NY 12919.

Gretch, M., T. O'Connell, and L.O. O'Connell. 1981. Another breeding locality for the Redhead. Kingbird 31:226.

Gretch, M., H. Booth, and R. Booth. 1982. Red-headed Woodpecker breeding in Clinton County. Kingbird 32:176.

Griffith, K.C. 1983. Regional report (Region 2—Genesee). Kingbird 33:191.

Griscom, L. 1923. Birds of the New York City region. The American Museum of Natural History Handbook Series no. 9. New York.

——. 1949. The birds of Concord. Harvard Univ. Press, Cambridge.

Griscom, L., and D.E. Snyder. 1955. The birds of Massachusetts. Peabody Museum, Salem, MA.

Griscom, L., and A. Sprunt, Jr., eds. 1979. The warblers of America. Doubleday and Co., Garden City, NY.

Gross, A.O. 1923. The Black-crowned Night Heron (Nycticorax nycticorax naevius of Sandy Neck. Auk 40:1–30, 191–214.

——. 1940. Eastern Nighthawk. In Life histories of North American cuckoos, goatsuckers, hummingbirds, and their allies, by A.C. Bent. U.S. National Museum Bulletin no. 176. Washington, DC. Pp. 209–212.

——. 1942. Cliff Swallow. In Life histories of North American flycatchers, larks, swallows, and their allies, by A.C. Bent. U.S. National Museum Bulletin no. 179. Washington, DC. Pp. 463–484.

——. 1945. The present status of the Great Black-backed Gull on the coast of Maine. Auk 62:241–265.

——. 1948a. House Wren. In Life histories of North American nuthatches, wrens, thrashers, and their allies, by A.C. Bent. U.S. National Museum Bulletin no. 195. Washington, DC. Pp. 118–121.

——. 1948b. Gray Catbird. In Life histories of North American nuthatches, wrens, thrashers, and their allies, by A.C. Bent. U.S. National Museum Bulletin no. 195. Washington, DC. Pp. 322–323.

——. 1949. Hermit Thrush. In Life histories of North American thrushes, kinglets, and their allies, by A.C. Bent. U.S. National Museum Bulletin no. 196. Washington, DC. Pp. 143–162.

——. 1953a. Blackpoll Warbler. In Life histories of North American wood warblers, by A.C. Bent. U.S. National Museum Bulletin no. 203. Washington, DC. Pp. 389–408.

——. 1953b. Common and Maryland Yellowthroat. In Life histories of North American wood warblers, by A.C. Bent. U.S. National Museum Bulletin no. 203. Washington, DC. Pp. 542–565.

——. 1958. Eastern Meadowlark. *In* Life histories of North American blackbirds, orioles, tanagers, and allies, by A.C. Bent. U.S. National Museum Bulletin no. 211. Washington, DC. Pp. 53–80.

Grzybowski, J.A. 1974. Nest site selection in the Wild Turkey (*Meleagris gallopavo silvestris*) of western New York. Master's thesis, St. Bonaventure Univ., St. Bonaventure, NY.

Guilday, J.E. 1969. Faunal remains from Dutchess Quarry Cave no. 1. *In* The archeology of Dutchess Quarry Cave, Orange County, New York, by R.E. Funk, G.R. Walters, and W.F. Ehlers, Jr. Pennsylvania Archeologist 39:17–19.

——. 1973a. Faunal remains from the Nahrwold No. 1 Site. *In* Aboriginal settlement patterns in the Northeast, by W.A. Ritchie and R.E. Funk. New York State Museum and Science Service Memoir 20:288–290.

——. 1973b. Vertebrate remains from the Garoga Site, Fulton County, New York. *In* Aboriginal settlement patterns in the Northeast, by W.A. Ritchie and R.E. Funk. New York State Museum and Science Service Memoir 20:329–330.

——. 1980. Bone refuse from the Lamoka Lake Site. *In* The archeology of New York State, by W.A. Ritchie. Harbor Hill Books, Harrison, NY. Pp. 54–59.

Gullion, G.W. 1952. The displays and calls of the American Coot. Wilson Bulletin 64:83–97.

——. 1984. Managing northern forests for wildlife. Minnesota Agricultural Experiment Station, St. Paul.

Gullion, G.W., and T. Martinson. 1984. Grouse of the North Shore. Willow Creek Press, Oshkosh, WI.

Hagar, D.C. 1957. Nesting populations of Red-tailed Hawks and Horned Owls in central New York State. Wilson Bulletin 69:263–272.

Hall, A. 1976. Regional report: Appalachian region. American Birds 30:7.

Hamilton, W.J., and M.E. Hamilton. 1965. Breeding characteristics of Yellow-billed Cuckoos in Arizona. Proceedings of the California Academy of Science 32:405–432.

Hansen, P.W. 1987. Acid rain and waterfowl: the case for concern in North America. Izaak Walton League of America, Arlington, VA.

Hardy, J.W. 1961. Studies in behavior and phylogeny of certain new world jays (*Garrulinae*). Univ. of Kansas Science Bulletin 42:28.

Hargrave, L.L. 1933. The Western Gnatcatcher also moves its nest. Wilson Bulletin 45:30–31.

Harris, R. 1986. Personal communication. New York Botanical Gardens, Brooklyn, NY 10400.

Harrison, C. 1978. A field guide to the nests, eggs, and nestlings of North American birds. William Collins Sons and Co. New York.

Harrison, H.H. 1975. A field guide to birds' nests of 285 species found breeding in the United States east of the Mississippi River. Houghton Mifflin Co., Boston.

——. 1984. Wood warblers' world. Simon and Schuster, New York.

Hays, H. 1967. The adaptive ibis. Natural History 76(7):32–33.

——. 1969. Second nesting record of the Louisiana Heron in New York State. Kingbird 19:93–94.

——. 1972. Polyandry in the Spotted Sandpiper. Living Bird 11:43–57.

——. 1984. The vole that soared. Natural History 93(5):7–16.

Heintzelman, D.S., and A.C. Nagy. 1968. Clutch sizes, hatchability rates, and sex ratios of Sparrow Hawks in eastern Pennsylvania. Wilson Bulletin 80:306–311.

Helm, R.M., D.M. Pashley, and P.J. Zwank. 1987. Notes on the nesting of the Common Moorhen and Purple Gallinule in southwest Louisiana. Journal of Field Ornithology 58:55–61.

Helme, A.H. 1883. Red Crossbills. Ornithologist and Oologist 8:68–69.

Herbert, R.A., and K.G.S. Herbert. 1965. Behavior of Peregrine Falcons in the New York City region. Auk 82:62–94.

Hespenheide, H.A. 1971. Flycatcher habitat selection in the eastern deciduous forest. Auk 88:61–74.

Heusmann, H.W. 1974. Mallard–Black Duck relationships in the Northeast. Wildlife Society Bulletin 2:171–177.

Hickey, J.J. 1940. Territorial aspects of the American Redstart. Auk 57:255–256.

Hill, N.P. 1968. Eastern Sharp-tailed Sparrow. *In* Life histories of North American cardinals, grosbeaks, buntings, towhees, finches, sparrows, and allies, by A.C. Bent and collaborators, ed. O.L. Austin, pt. 2. Washington, DC. Pp. 795–812. U.S. National Museum Bulletin no. 237.

Holmes, R.T., and T.W. Sherry. 1986. Bird community dynamics in a temperate deciduous forest: long-term trends at Hubbard Brook. Ecological Monographs 56:201–220.

Holmes, R.T., C.P. Black, and T.W. Sherry. 1979. Comparative population bioenergetics of three insectivorous passerines in a deciduous forest. Condor 81:9–20.

Horning, E. 1986. Personal communication. Fishers Island, NY 06390.

Horsey, R.E. 1938. Birds of Rochester and Monroe County—1913–1936. Privately printed, Rochester, NY.

——. 1949. Supplement to "Birds of Rochester and Monroe County—1938–1948." Privately printed, Rochester, NY.

Hotopp, K.P. 1986. Status of Common Terns on the Buffalo Harbor and Upper Niagara River. NYSDEC, Buffalo, NY. Mimeo.

Hough, F.N. 1964. The thrushes (*Turdidae*): their occurrence in Ulster County, New York. Burroughs Natural History Society Bulletin 7:17–19.

Howard, D.V. 1967. Variation in the breeding season and clutch-size of the Robin in northeastern United States and the Maritime Provinces of Canada. Wilson Bulletin 79:432–440.

Howell, J., B. Smith, J.B. Holt, Jr., and D.R. Osborne. 1978. Habitat structure and productivity in Red-tailed Hawks. Bird-Banding 49:162–171.

Howell, J.C. 1942. Notes on the nesting habits of the American Robin (*Turdus migratorius* L.). American Midland Naturalist 28:529–603.

Howes, P. 1926. Turkey Vulture nest in New York. Bird-Lore 28:175.

Hoyt, S.F. 1957. The ecology of the Pileated Woodpecker. Ecology 38:247–256.

——. 1959. Regional report (Region 3—Finger Lakes). Kingbird 9:81.

——. 1962. Regional report (Region 3—Finger Lakes). Kingbird 12:149.

Hurley, R.J., and E.C. Franks. 1976. Changes in the breeding ranges of two grassland birds. Auk 93:108–115.

Hyde, A.S. 1939. The ecology and economics of the birds along the northern boundary of New York State. Roosevelt Wild Life Bulletin 7:61–215.

Jackson, C.F., and P.F. Allen. 1932. Additional notes on the breeding of the Great Black-backed Gull in Maine. Auk 49:349–350.

Jackson, J.A. 1970. A quantitative study of the foraging ecology of Downy Woodpeckers. Ecology 51:318–323.

——. 1983. Nesting phenology, nest site selection, and reproductive success of Black and Turkey vultures. *In* Vulture biology and management, ed. S.R. Wilbur and J.A. Jackson. Univ. of California Press, Berkeley. Pp. 245–270.

James, R.D. 1976. Foraging behavior and habitat selection of three species of vireos in southern Ontario. Wilson Bulletin 88:62–75.

Johnson, C.E. 1927. The Bay-breasted Warbler in the Adirondacks of New York. Auk 44:255–256.

Johnson, R.W. 1973. The ecology of the northern Clapper Rail, *Rallus longirostris crepitans*. Ph.D. dissertation, Cornell Univ., Ithaca, NY.

Johnsgard, P.A. 1961. Evolutionary relationships among North American Mallards. Auk 78:3–43.

——. 1967. Sympatry changes and hybridization in Mallards and Black Ducks. American Midland Naturalist 77:51–63.

——. 1973. Grouse and quails of North America. Univ. of Nebraska Press, Lincoln.

——. 1975. Waterfowl of North America. Indiana Univ. Press, Bloomington.

——. 1978. Ducks, geese, and swans of the world. Univ. of Nebraska Press, Lincoln.

——. 1979. Birds of the Great Plains. Univ. of Nebraska Press, Lincoln.

——. 1983. The hummingbirds of North America. Smithsonian Institution Press, Washington, DC.

Johnston, D.W. 1971. Niche relationships among some deciduous forest flycatchers. Auk 88:796–804.

Jones, M. 1975. Brown-headed Cowbird brood parasitism on the Eastern Phoebe. Kingbird 25:119–123.

——. 1980. The New York State waterfowl count—a quarter century report. Kingbird 30:210–216.

Jones, P.D., T.M.L. Wigley, and P.B. Wright. 1986. Global temperature variations between 1861 and 1984. Nature 322:430–434.

Jones, S. 1979. The accipiters—Goshawk, Cooper's Hawk, Sharp-shinned Hawk. Habitat management series for unique or endangered species report no. 17. U.S. Dept. of Agriculture, Bureau of Land Management, Denver, CO.

Jordan, A.H.B. 1888. A visit to the Four Brothers, Lake Champlain. Ornithologist and Oologist 13:138–139.

Judd, W.W. 1907. The birds of Albany County. Brandow Printing Co., Albany.

Kaufman, K. 1984. The changing seasons. American Birds 38:994.

Keeler, J.E., C.C. Allin, J.M. Anderson, S. Gallizioli, K.E. Gamble, D.W. Hayne, W.H. Kiel, Jr., F.W. Martin, J.L. Ruos, K.C. Sadler, L.D. Soileau, C.E. Braun, and H.D. Funk. 1977. Mourning Dove (*Zenaida macroura*). *In* Management of migratory shore and upland game birds in North America, ed. G.C. Sanderson. International Association of Fish and Wildlife Agencies, Washington, DC. Pp. 274–298.

Keller, J.K. 1980. Species composition and density of breeding birds in several habitat types on the Connecticut Hill Wildlife Management Area. Master's thesis, Cornell Univ., Ithaca, NY.

——. 1986. Personal communication. 611 Pennbrook Ave., Lansdale, PA 19446.

Kelly, S. 1986. American Woodcock—1986 breeding population status. USFWS administrative report. Laurel, MD.

Kemnitzer, E., R. Adamo, G. Raynor, A. Scherzer, and B. Scherzer. 1982. Breeding Bird Survey no. 57: pitch pine–scrub oak barrens. American Birds 36:66.

Kendeigh, S.C. 1945. Community selection by birds on the Helderberg Plateau of New York. Auk 62:418–436.

——. 1963. Regulation of nesting time and distribution in the House Wren. Wilson Bulletin 75:418–427.

Kennard, F.H. 1895. Notes on the breeding of the American Crossbill in Hamilton County, New York. Auk 12:304–305.

Ketchledge, E. 1986. Personal communication. Bloomingdale, NY 12913.

Kibbe, D.P. 1976. Regional report: Niagara-Champlain region. American Birds 30:952.

——. 1985a. Pied-billed Grebe. *In* The atlas of breeding birds of Vermont, ed. S.B. Laughlin and D.P. Kibbe. Univ. Press of New England, Hanover, NH. P. 32.

——. 1985b. Upland Sandpiper. *In* The atlas of breeding birds of Vermont, ed. S.B. Laughlin and D.P. Kibbe. Univ. Press of New England, Hanover, NH. Pp. 108–109.

——. 1985c. Sedge Wren. *In* The atlas of breeding birds of Vermont, ed. S.B. Laughlin and D.P. Kibbe. Univ. Press of New England, Hanover, NH. Pp. 228–229.

——. 1985d. American Robin. *In* The atlas of breeding birds of Vermont, ed. S.B. Laughlin and D.P. Kibbe. Univ. Press of New England, Hanover, NH. Pp. 250–251.

——. 1987. Personal communication. 312 Bloomfield St., Hoboken, NJ 07030.

Kibbe, D.P., and C.M. Boise. 1983. Regional report: Niagara-Champlain region. American Birds 37:987.

Kilgore, D.L., and K.L. Knudsen. 1977. Analysis of materials in Cliff and Barn Swallow nests: relationship between mud selection and nest architecture. Wilson Bulletin 89:562–571.

Kilham, L. 1971a. Reproductive behavior of Yellow-bellied Sapsuckers. 1. Preference for nesting in Fomes-infected aspens and nest hole interrelations with flying squirrels, raccoons, and other animals. Wilson Bulletin 83:159–171.

——. 1971b. Roosting habits of White-breasted Nuthatches. Condor 73: 113–114.

——. 1972. Death of Red-breasted Nuthatch from pitch around nest hole. Auk 89:451–452

——. 1977. Nesting behavior of Yellow-bellied Sapsuckers. Wilson Bulletin 89:310–324.

——. 1983. Life history studies of woodpeckers of eastern North America. Publications of the Nuttall Ornithological Club no. 20. Cambridge, MA.

Kisiel, D.S. 1972. Foraging behavior of *Dendrocopos villosus* and *D. pubescens* in eastern New York State. Condor 74:393–398.

Kittredge, J., Jr. 1925. Ruby-crowned Kinglet in summer in the Adirondack Mountains, N.Y. Auk 42:144.

Kiviat, E., R. Schmidt, and N. Zeising. 1985. Bank Swallow and Belted Kingfisher nest in dredge spoil on the tidal Hudson River. Kingbird 35:3.

Klabunde, W. 1979. U.S. Fish and Wildlife Service Cooperative breeding bird survey—1979. Prothonotary 45:131–135.

Klimstra, W.D., and W.O. Stieglitz. 1957. Notes on reproductive activities of Robins in Iowa and Illinois. Wilson Bulletin 69:333–337.

Klingensmith, C.W. 1970. Breeding bird census no. 21: maturing upland red pine–white pine plantation. Audubon Field Notes 24:754–755.

Klonick, A.S. 1951. Western Meadowlark, *Sturnella neglecta*, in New York State. Auk 68:107.

Knapton, R.W., and J.B. Falls. 1982. Polymorphism in the White-throated Sparrow: habitat occupancy and nest-site selection. Canadian Journal of Zoology 60:452–458.

Knight, C.W.R. 1932. Photographing the nest life of the Osprey. National Geographic 62:247–260.

Knight, O.W. 1908. The birds of Maine. C.H. Glass and Co., Bangor, ME.

Knight, R.L., and M.W. Call. 1980. The Common Raven. U.S. Dept. of Interior, Bureau of Land Management Technical Note no. 344. Denver, CO.

Kogut, K.L. 1979. An investigation of the breeding status and population distribution of the Northern Harrier (*Circus cyaneus hudsonius* L.) in New York State. NYSDEC, Delmar, NY. Mimeo.

Kortright, F.H. 1942. The ducks, geese, and swans of North America. Stackpole Co./Wildlife Management Institute, Harrisburg.

Krapu, G.L. 1974. Feeding ecology of Pintail hens during reproduction. Auk 91:278–290.

Kress, S.W. 1983. The use of decoys, sound recordings, and gull control for reestablishing a tern colony in Maine. Colonial Waterbirds 6:185–196.

Kricher, J.C. 1983. Correlation between House Finch increase and House Sparrow decline. American Birds 37:358–360.

Kroodsma, D.E. 1980. Winter Wren singing behavior: a pinnacle of song complexity. Condor 82:357–365.

Kroodsma, D.E., and J. Verner. 1978. Complex singing behaviors among *Cistothorus* wrens. Auk 95:703–716.

Krug, H.H. 1956. Great Black-backed Gull nesting on Haystack Island, Lake Huron. Auk 73:559.

Kuerzi, R.G. 1941. Life history studies of the Tree Swallow. Proceedings of the Linnaean Society of New York nos. 52–53. New York. Pp. 1–52.

Kutz, H.L. 1946. Breeding of the Ring-billed Gull in New York. Auk 63:591.

Kutz, H.L., and D.G. Allen. 1946. The American Pintail breeding in New York. Auk 63:596.

——. 1947. Double-crested Cormorant nesting in New York. Auk 64:137.

Lacombe, I. 1986. Personal communication. Old Lake Shore Rd., Peru, NY 12972.

LaFrance, F. 1973. Observations of the Three-toed Woodpeckers in an Adirondack bog with notes on plumage. Kingbird 23:190–191.

——. 1975. Ferd's Bog. Kingbird 25:184–190.

Lamendola, J. 1986. Personal communication. NYSDEC, 317 Washington Ave., Watertown, NY 13601.

Lanyon, W.E. 1956. Ecological aspects of the sympatric distribution of meadowlarks in the north-central states. Ecology 37:98–108.

——. 1957. The comparative biology of the meadowlarks, (*Sturnella*) in Wisconsin. Publications of the Nuttall Ornithological Club no. 1. Cambridge, MA.

——. 1966. Hybridization in meadowlarks. Bulletin of the American Museum of Natural History 134:1–25.

——. 1979. Hybrid sterility in meadowlarks. Nature 279:557–558.

Larson, E. 1982. Ducks in the chimney. Allegany County Bird Club News 16:71.

Laskey, A.R. 1944. A study of the Cardinal in Tennessee. Wilson Bulletin 52:183–190.

——. 1962. Breeding biology of Mockingbirds. Auk 79:596–597.

Laub, R.S., M.F. De Remer, C.A. Dufort, and W.L. Parsons. In press. The Hiscock Site: a rich late Quaternary locality in western New York State. In Late Pleistocene and early Holocene paleoecology and archeology of the eastern Great Lakes region, ed. R.S. Laub and D.W. Steadman. Bulletin of the Buffalo Society of Natural Sciences vol. 33.

Laughlin, S.B. 1985. Common Barn-Owl. In The atlas of breeding birds of Vermont, ed. S.B. Laughlin and D.P. Kibbe. Univ. Press of New England, Hanover, NH. Pp. 130–131.

Lauro, A.J. 1977. Regional report (Region 10—Marine). Kingbird 27:238.

Lauro, A.J., and B.J. Spencer. 1975. Regional report (Region 10—marine). Kingbird 25:234.

Lawrence, G.N. 1866. Catalogue of birds observed on New York, Long, and Staten islands, and adjacent parts of New Jersey. Annals of the New York Lyceum of Natural History 8:279–300.

Lawrence, L. deK. 1967. A comparative life history study of four species of woodpeckers. Ornithological Monographs no. 5. American Ornithologists' Union, Lawrence, KS.

Lea, R.B. 1942. A study of the nesting habits of the Cedar Waxwing. Wilson Bulletin 54:225–237.

Leck, C.F. 1975. The birds of New Jersey—their habits and habitats. Rutgers Univ. Press, New Brunswick, NJ.

——. 1984. The status and distribution of New Jersey birds. Rutgers Univ. Press, New Brunswick, NJ.

Lehman, J.G. 1986. Personal communication. RD 2, Box 68C, Norwich, NY 13815.

Lehman, J.G., and J.M.C. Peterson. 1983. Hints on haunts. NYSDEC, Delmar, N.Y. Mimeo.

Leopold, A. 1949. A Sand County almanac. Oxford Univ. Press, New York.

Lesperance, T.A. 1960. A ground nesting of Mourning Doves. Kingbird 10:166–167.

Leukering, A. 1985. Personal communication. American Birds, 950 Third Ave., New York, NY 10022.

Lima, S.L. 1984. Downy Woodpecker foraging behavior: efficient sampling in simple stochastic environments. Ecology 65:166–174.

Lincoln, F.C. 1979. Migration of Birds. Rev. by S.R. Peterson. USFWS, Washington, DC.

Lintner, G.A. 1884. Duck Hawks breeding in the Helderberg Mountains, New York. Auk 1:391.

Listman, W. 1986. Personal communication. 14775 Lakeshore Rd., Kent, NY 14477.

Lombardo, M.P. 1986. Attendants at Tree Swallow nests. 1. Are attendants helpers at the nest? Condor 88:297–303.

——. 1987. Attendants at Tree Swallow nests. 2. The exploratory-dispersal hypothesis. Condor 89:138–149.

Long, J.L. 1981. Introduced birds of the world. Universe Books, New York.

Loucks, B.A. 1986a. Status, research, and management of Osprey in New York. New York Federal Aid in Wildlife Restoration Project W-166-E. NYSDEC, Albany. Mimeo.

——. 1986b. Status, research, and management of the Peregrine Falcon in New York. New York Federal Aid in Wildlife Restoration Project W-166-E. NYSDEC, Albany. Mimeo.

Lowther, J.K. 1961. Polymorphism in the White-throated Sparrow, Zonotrichia albicollis (Gmelin). Canadian Journal of Zoology 39:281, 289–291.

Lowther, J.K., and J.B. Falls. 1968. White-throated Sparrow. In Life histories of North American cardinals, grosbeaks, buntings, towhees, finches, sparrows, and

allies, by A.C. Bent and collaborators, ed. O.L. Austin, Jr. U.S. National Museum Bulletin no. 237, pt. 3. Washington, DC. Pp. 1368–1371.

Ludwig, J.P. 1965. Biology and structure of the Caspian Tern (Hydroprogne caspia) population of the Great Lakes from 1896–1964. Bird Banding 36:217–233.

——. 1974. Recent changes in the Ring-billed Gull population and biology in the Laurentian Great Lakes. Auk 91:575–594.

Lundelius, E.L., Jr., R.W. Graham, E. Anderson, J. Guilday, J.A. Holman, D.W. Steadman, and S.D. Webb. 1983. Terrestrial vertebrate faunas. In Late Quaternary environments of the United States, ed. H.E. Wright, Jr. Vol. 1, The Pleistocene, ed. S.C. Porter. Univ. of Minnesota Press, Minneapolis. Pp. 311–353.

Lunk, W.A. 1962. The Rough-winged Swallow: a study based on its breeding biology in Michigan. Publications of the Nuttall Ornithological Club no. 4. Cambridge, MA.

MacArthur, R.H. 1958. Population ecology of some warblers of northeastern coniferous forests. Ecology 39:599–619.

MacClintock, L., R.F. Whitcomb, and B.L. Whitcomb. 1977. Evidence for the value of corridors and minimization of isolation in preservation of biotic diversity. American Birds 31:6–16.

Mack, T.D. 1974. Regional report (Region 7—Adirondack-Champlain). Kingbird 24:204.

——. 1975. Regional report (Region 7—Adirondack-Champlain). Kingbird 25:226.

——. 1979. Regional report (Region 7—Adirondack-Champlain). Kingbird 29:230.

——. 1980. Regional report (Region 7—Adirondack-Champlain). Kingbird 30:247–248.

——. 1982. Regional report (Region 7—Adirondack-Champlain). Kingbird 32:291.

MacLean, D.C. 1985. Personal communication. Seatuck Research Program, P.O. Box 31, Islip, NY 11751.

MacNamara, E.E., and H.F. Udell. 1970. Clapper Rail investigations on the south shore of Long Island. Proceedings of the Linnaean Society of New York 71:120–131.

Marquiss, M., I. Newton, and D.A. Ratcliffe. 1978. The decline of the raven, Corvus corax, in relation to afforestation in southern Scotland and northern England. Journal of Applied Ecology 15:129–144.

Marti, C.D. 1976. A review of prey selection by the Long-eared Owl. Condor 78:331–336.

Martin, M. 1978. Status report on endangered wildlife in Canada: Caspian Tern. Committee on the Status of Endangered Wildlife in Canada, Museum of Natural Sciences, Ottawa, ON.

Martin, N. 1987. Personal communication. Vermont Institute of Natural Science, Woodstock, VT 05091.

Matray, P.F. 1974. Broad-winged Hawk nesting and ecology. Auk 91:307–324.

Matthiessen, P. 1967. Chapter 1. In The shorebirds of North America, ed. G.D. Stout. Viking Press, New York.

Maxwell, G.R. 1986. Personal communication. Rice Creek Biological Station, Oswego, NY 13126.

Maxwell, G.R., and G.A. Smith. 1980. Breeding Bird Survey no. 6: young white ash–basswood forest. American Birds 34:46.

——. 1983. Nest site competition and population estimates of island-nesting terns, Ring-billed Gulls, and Herring Gulls in the St. Lawrence River. Kingbird 33:19–25.

Mayfield, H. 1965. The Brown-headed Cowbird, with old and new hosts. Living Bird 4:13–28.

Maynard, C.J. 1881. The birds of eastern North America with original descriptions of all species which occur east of the Mississippi River. C.J. Maynard and Co., Newtonville, MA.

McAtee, W.L. 1911. Our vanishing shorebirds. U.S. Dept. of Agriculture, Bureau of Biological Survey Circular no. 79. Washington, DC.

——. 1926. Relation of birds to woodlots. Roosevelt Wild Life Bulletin 4:7–152.

McCrimmon, D.A., Jr. 1981. The status and distribution of the Great Blue Heron

(*Ardea herodias*) in New York State: results of a two-year census effort. Colonial Waterbirds 4:85–90.

———. 1982. Populations of the Great Blue Heron (*Ardea herodias*) in New York State from 1964 to 1981. Colonial Waterbirds 5:87–95.

McIlroy, D. 1986. Personal communication. 419 Triphammer Rd., Ithaca, NY 14850.

McIlroy, M. 1961. Possible hybridization between a Clay-colored Sparrow and a Chipping Sparrow at Ithaca. Kingbird 11:7–10.

McKee, R. 1984. Swan shift. Michigan Natural Resources Magazine 53(4):10–15.

McKinney, R. 1985. Personal communication. 198 Parkview Dr., Rochester, NY 14625.

McLaren, M.A. 1975. Breeding biology of the Boreal Chickadee. Wilson Bulletin 87:344–354.

McNeil, J.P. 1987. Personal communication. 168 Lexington Rd., Shirley, NY 11967.

Meade, G.M. 1942. Calcium chloride—a death lure for crossbills. Auk 59:439–440.

Meanley, B. 1969. Natural history of the King Rail. North American Fauna no. 67. U.S. Bureau of Sport Fisheries and Wildlife, Washington, DC.

———. 1985. The Marsh Hen. A natural history of the Clapper Rail of the Atlantic Coast salt marsh. Tidewater Publishers, Centreville, MD.

Mehner, J.F. 1952. Notes on song cessation. Auk 69:467.

Melvin, S.M. 1986. Personal communication. Natural Heritage Program, Division of Fish and Wildlife, 100 Cambridge St., Boston, MA 02202.

Mendall, H.L. 1936. The home-life and economic status of the Double-crested Cormorant *Phalacrocorax auritus auritus*. Univ. of Maine Studies, 2d ser., no. 38. Maine Bulletin 39(3):1–159.

———. 1958. The Ring-necked Duck in the Northeast. Univ. of Maine Bulletin 60(16):20.

Meng, H. 1951. The Cooper's Hawk. Ph.D. dissertation, Cornell Univ., Ithaca, NY.

Mengel, R.M. 1964. The probable history of species formation in some northern wood warblers (Parulidae). Living Bird 3:27.

Merriam, C.H. 1878a. Remarks on some birds of Lewis County, northern New York. Bulletin of the Nuttall Ornithological Club 3:52–56.

———. 1878b. Nesting of the banded Three-toed Woodpecker (*Picoides americanus*) in northern New York. Bulletin of the Nuttall Ornithological Club 3:200.

———. 1879. Remarks on some of the birds of Lewis County, northern New York, with remarks by A.J. Dayan. Bulletin of the Nuttall Ornithological Club 4:1–7.

———. 1881. Preliminary list of birds ascertained to occur in the Adirondack region, northeastern New York. Bulletin of the Nuttall Ornithological Club 6:225–235.

———. 1884. Third addendum to the preliminary list of birds ascertained to occur in the Adirondack region, northeastern New York. Auk 1:58–59.

Messineo, D.J. 1985. The 1985 nesting of Pine Siskin, Red Crossbill, and White-winged Crossbill in Chenango County, NY. Kingbird 35:233–237.

Meyerriecks, A.J. 1957. Louisiana Heron breeds in New York City. Wilson Bulletin 69:184–185.

———. 1960. Comparative breeding behavior of four species of North American herons. Publications of the Nuttall Ornithological Club no. 2. Cambridge, MA.

Milburn, E., and T. Cade. 1977. Summary report of the 1976 Bald Eagle reintroduction project at the Montezuma National Wildlife Refuge, New York. Cornell Univ., Laboratory of Ornithology, Ithaca, NY. Mimeo.

Miller, H.S. 1958. The Western Meadowlark in Monroe County. Kingbird 7:115.

———. 1959. Regional report (Region 2—Genesee). Kingbird 9:78.

———. 1961. Chuck-will's-widow in Monroe County. Kingbird 11:149.

Miller, N.G. 1973a. Late-glacial and postglacial vegetation change in southwestern New York State. New York State Museum and Science Service Bulletin 420:1–102.

———. 1973b. Late glacial plants and plant communities in northwestern New York State. Journal of the Arnold Arboretum 54:123–159.

———. In press. The late Quaternary Hiscock Site, Genesee County, New York: paleoecological studies based on pollen and plant macrofossils. Bulletin of the Buffalo Society of Natural Sciences vol. 33.

Miller, R.L. 1967. A field trial of Sudan Black B as an antifertility agent on a

population of breeding starlings. Master's thesis, Cornell Univ., Ithaca, NY.

———. 1981. The atlas grid: mapping scales. *In* Proceedings of the Northeastern Breeding Bird Atlas Conference, ed. S.B. Laughlin. Vermont Institute of Natural Science, Woodstock. Pp. 58–59.

Miller, R.S., and R.E. Miller. 1971. Feeding activity and color preference of Ruby-throated Hummingbirds. Condor 73:309–313.

Miller, R.S., and R.W. Nero. 1983. Hummingbird-sapsucker associations in northern climates. Canadian Journal of Zoology 61:1540–1546.

Mills, A.M. 1986. The influence of moonlight on the behavior of goatsuckers (Caprimulgidae). Auk 103:370–378.

Minor, W.F., and M.L. Minor. 1981. Nesting of Red-tailed Hawks and Great Horned Owls in central New York suburban areas. Kingbird 31:68–76.

Minton, C.D.T. 1968. Pairing and breeding of Mute Swans. Wildfowl 19:41–60.

Mitchell, C.W. 1986. Personal communication. 93 Park Ave. West, Plattsburgh, NY 12901.

Mitchell, H.D. 1954. An unusual Killdeer nesting. Kingbird 4:66–68.

Mitchell, R.S. 1986. A checklist of New York State plants. New York State Museum Bulletin no. 458, Albany, NY.

Montagna, W. 1942. The Sharp-tailed Sparrows of the Atlantic Coast. Wilson Bulletin 54:107–120.

Moore, G.C., and A.M. Pearson. 1941. The Mourning Dove in Alabama. Alabama Dept. of Conservation Bulletin. Montgomery.

Morris, M.M.J., and R.E. Lemon. 1983. Characteristics of vegetation and topography near Red-shouldered Hawk nests in southwestern Quebec. Journal of Wildlife Management 47:138–145.

Morse, D.H. 1979. Habitat use by the Blackpoll Warbler. Wilson Bulletin 91:234–241.

Moser, J. 1982. Waterfowl population study. New York Federal Aid in Wildlife Restoration Project W-39-R (final report). NYSDEC, Albany. Mimeo.

———. 1986. Personal communication. NYSDEC, Wildlife Resources Center, Delmar, NY 12054.

Mudge, E.T. 1962. A collection of chats. Kingbird 12:142–143.

Muldal, A., H.L. Gibbs and R.J. Robertson. 1985. Preferred nest spacing of an obligate cavity-nesting bird, the Tree Swallow. Condor 87:356–364.

Mundinger, P.C., and S. Hope. 1982. Expansion of the winter range of the House Finch, 1949–1979. American Birds 36:347–353.

Munoff, J.A. 1963. Food habits, growth, and mortality in Marsh Hawks. Kingbird 13:67–74.

Murray, W.H.H. 1869. Loon-shooting in a thunder-storm. *In* Adventures in the wilderness; or, camp-life in the Adirondacks. Fields, Osgood, and Co., Boston. Pp. 101–113.

Mutchler, N.M. 1983. Pine Siskins: spring of 1982. North American Bird Bander 8:14–16.

New York State Department of Environmental Conservation (NYSDEC), Nongame Unit. 1985. Checklist of the amphibians, reptiles, birds and mammals of New York State. NYSDEC, Delmar, NY.

New York State Department of Transportation (NYSDOT). 1983. New York State atlas. NYSDOT, Albany, NY.

Nice, M.M. 1937. Studies in the life history of the Song Sparrow, pt. 1. Transactions of the Linnaean Society of New York 4:1–247.

Nickell, W.P. 1951. Studies of habitats, territory, and nests of the Eastern Goldfinch. Auk 68:447–470.

Nickerson, D. 1978. Wilson's Warbler nests in New York State. Kingbird 28:215–220.

———. 1985. Personal communication. 9 Guptill Ave., Freeport, ME 04032.

Niedermyer, W.J., and J.J. Hickey. 1977. The Monk Parakeet in the United States, 1970–1975. American Birds 31:273–278.

Nilsson, G. 1981. The bird business and a study of the commercial cage bird trade. Animal Welfare Institute, Washington, DC.

Nisbet, I.C.T. 1971. The Laughing Gull in the Northeast. American Birds 25:677–683.

Nolan, V., Jr., and C.F. Thompson. 1975. The occurrence and significance of anomalous reproductive activities in two North American non-parasitic cuckoos *Coccyzus* spp. Ibis 117:496–503.

Noon, B.R., and K.P. Able. 1976. Avian community structure along elevational gradients in the northeastern United States. Oecologia 26:275–294.

———. 1978. A comparison of avian community structure in the northern and southern Appalachian Mountains. *In* Management of southern forests for nongame birds. U.S. Dept. of Agriculture, Forest Service General Technical Report SE-14. Atlanta. Pp. 98–117.

Noon, B.R., V.P. Bingman, and J.P. Noon. 1979. The effects of changes in habitat on northern hardwood forest bird communities. *In* Management of north central and northeastern forests for nongame birds, comp. R.M. DeGraaf and K.E. Evans. U.S. Dept. of Agriculture, Forest Service General Technical Report NC-51. St. Paul. Pp. 33–48.

Noon, B.R., S. Droege, R.A. Bingham, and D.K. Dawson. 1982a. Breeding Bird Survey no. 54: beech-spruce-hemlock forest. American Birds 36:66.

———. 1982b. Breeding Bird Survey no. 55: beech-spruce-hemlock forest. American Birds 36:66.

Norse, W.J. 1985a. Pileated Woodpecker. *In* The atlas of breeding birds of Vermont, ed. S.B. Laughlin and D.P. Kibbe. Univ. Press of New England, Hanover, NH. Pp. 168–169.

———. 1985b. Willow Flycatcher. *In* The atlas of breeding birds of Vermont, ed. S.B. Laughlin and D.P. Kibbe. Univ. Press of New England, Hanover, NH. P. 178.

Norse, W.J., and D.P. Kibbe. 1985. Chimney Swift. *In* The atlas of breeding birds of Vermont, ed. S.B. Laughlin and D.P. Kibbe. Univ. Press of New England, Hanover, NH. Pp. 148–149.

Norton, A.H., and R.P. Allen. 1931. Breeding of Great Black-backed Gull and Double-crested Cormorant in Maine. Auk 48:589–592.

Novak, P. 1986a. Possible factors influencing the distribution, status, and abundance of the Loggerhead Shrike (*Lanius ludovicianus*) in New York State. Kingbird 36:176–181.

———. 1986b. Personal communication. 1097 Warren Rd., Apt. 2, Ithaca, NY 14850.

Nye, P.E. 1979. Survey of historical Bald Eagle nesting, migration, and wintering areas. New York Federal Aid in Wildlife Restoration Project E-1-3 (final report). NYSDEC, Albany. Mimeo.

———. 1986. Personal communication. NYSDEC, Wildlife Resources Center, Delmar, NY 12054.

Nye, P.E., and M.L. Allen. 1983. Return of the natives. Living Bird Quarterly 2(1):16–19.

Nye, P.E., and A. Peterson. 1980. New York State potential Bald Eagle nesting and wintering areas (unpublished report). NYSDEC, Delmar, NY. Mimeo.

Oatman, G.F. 1985. Common Raven. *In* The atlas of breeding birds of Vermont, ed. S.B. Laughlin and D.P. Kibbe. Univ. Press of New England, Hanover, NH. Pp. 208–209.

O'Brien, K. 1986. Personal communication. NYSDEC, Wildlife Resources Center, Delmar, NY 12054.

O'Brien, M., and R.A. Askins. 1985. The effects of Mute Swans on native waterfowl. Connecticut Warbler 5:27–31.

Odom, R. 1977. Sora (*Porzana carolina*). *In* Management of migratory shore and upland game birds in North America, ed. G.C. Sanderson. International Association of Fish and Wildlife Agencies, Washington, DC. Pp. 57–65.

Odum, E.P. 1941a. Annual cycle of the Black-capped Chickadee—1. Auk 58:314–333.

———. 1941b. Annual cycle of the Black-capped Chickadee—2. Auk 58:518–535.

Office of Migratory Bird Management. 1985. National species of special emphasis: management plan for Ring-necked Ducks. USFWS, Laurel, MD.

O'Hara, R.T. 1974. Regional report (Region 2—Genesee). Kingbird 24:189.

Olds, B. 1985. Personal communication. 40 Fall Brook Pk., Canandaigua, NY 14424.

Oring, L.W., and M.L. Knudsen. 1972. Monogamy and polyandry in the Spotted Sandpiper. Living Bird 11:59–73.

Oring, L.W., D.B. Lank, and S.J. Maxson. 1983. Population studies of the polyandrous Spotted Sandpiper. Auk 100:272–285.

Ozard, J. 1987. Personal communication. NYSDEC, Wildlife Resources Center, Delmar, NY 12054.

Palmer, R.S. 1949. Maine birds. Bulletin of the Museum of Comparative Zoology at Harvard College 102:481–484.

———, ed. 1962. Handbook of North American birds. Vol. 1, Loons through flamingos. Yale Univ. Press, New Haven.

———. 1967. Killdeer. *In* The shorebirds of North America, ed. G.D. Stout. Viking Press, New York. Pp. 167–168.

———. 1968. Pine Siskin. *In* Life histories of North American cardinals, grosbeaks, buntings, towhees, finches, sparrows, and allies, by A.C. Bent and collaborators, ed. O.L. Austin, Jr. U.S. National Museum Bulletin no. 237, pt. 1. Washington, DC. Pp. 424–447.

———, ed. 1976a. Handbook of North American birds. Vol. 2, Waterfowl (pt. 1). Yale Univ. Press, New Haven.

———, ed. 1976b. Handbook of North American birds. Vol. 3, Waterfowl (pt. 2). Yale Univ. Press, New Haven.

Palmer, T.S. 1912. Chronology and index of the more important events in American game protection, 1776–1911. U.S. Dept. of Agriculture Biological Survey Bulletin 41:20.

Parker, K.E. 1985. Foraging and reproduction of the Common Loon (*Gavia immer*) on acidified lakes in the Adirondack Park, New York. Master's thesis, SUNY College of Environmental Science and Forestry, Syracuse.

Parker, K.E., and R.H. Brocke. 1985. Common Loon chick feeding on acidified lakes in the Adirondack Park, New York. Report to the North American Loon Fund. Meredith, NH.

Parker, K.E., R.L. Miller, and S. Isil. 1986. Status of the Common Loon in New York State. NYSDEC, Wildlife Resources Center, Delmar, NY 12054. Mimeo.

Parkes, K.C. 1951. The genetics of the Golden-winged × Blue-winged Warbler complex. Wilson Bulletin 63:4–15.

———. 1952a. Notes on some birds of the Cayuga Lake Basin. Kingbird 2:56–59.

———. 1952b. The birds of New York State and their taxonomy, pt. 2. Ph.D. dissertation, Cornell Univ., Ithaca, NY.

Paxton, R.O., W.J. Boyle, Jr., and D.A. Cutler. 1982. Regional report: Hudson-Delaware region. American Birds 36:961.

———. 1985. Regional report: Hudson-Delaware region. American Birds 39:893.

Peakall, D.B. 1963. Highlights of the winter season. Kingbird 13:99.

———. 1967. Recent changes in status of the Great Black-backed Gull. Kingbird 17:69–73.

Peakall, D.B., and M.S. Rusk. 1963. Regional report (Region 5—Oneida Lake Basin). Kingbird 13:218.

Pearson, T.G., ed. 1916. A ground-nesting Flicker. Bird-Lore 18:399–400.

———. 1917. Birds of America, pt. 2. University Society, New York.

———, ed. 1936. Birds of America. Garden City Publishing Co., Garden City, NY.

Peck, G.K., and R.D. James. 1983. Breeding birds of Ontario: nidiology and distribution. Vol. 1, Non-passerines. Royal Ontario Museum, Toronto, ON.

Penrod, B., and D.E. Austin. 1985. Ringneck—the road to recovery. New York State Conservationist 40(2):8–11.

Perrins, C.M., and A.L.A. Middleton. 1985. The Encyclopedia of birds. Facts on File Publications, New York.

Petersen, W.R. 1983. Sharp-tailed Sparrow. *In* The Audubon Society master guide to birding, ed. J. Farrand, Jr. Vol. 3, Old World Warblers to sparrows. Alfred A. Knopf, New York. Pp. 258–260.

Peterson, A. 1983. Observations on habitat selection by Henslow's Sparrow in Broome County, New York. Kingbird 33:155–164.

Peterson, D.M. 1984. A survey of colonial waterbirds on the southwestern shore of Long Island, N.Y.—1982–1984. Paper presented at the 8th Colonial Waterbird Group meeting, Ithaca, NY.

Peterson, D.M., T.S. Litwin, D.C. MacLean, and R.A. Lent. 1985. 1985 Long Island colonial waterbird and Piping Plover survey. Cornell Univ. Laboratory of Ornithology, Seatuck Research Program, Islip, NY.

Peterson, J.M.C. 1975. Attempted nesting of the White-winged Crossbill in New York State. Kingbird 25:191–193.

——. 1980. An earlier breeding record for White-winged Crossbill in New York State. High Peaks Audubon Newsletter 8:44.

——. 1983. Regional report (Region 7—Adirondack-Champlain). Kingbird 33:283–287.

——. 1984a. First record of the Palm Warbler nesting in New York State. Kingbird 34:3–7.

——. 1984b. Regional report (Region 7—Adirondack-Champlain). Kingbird 34:259–263.

——. 1985a. Regional report (Region 7—Adirondack-Champlain). Kingbird 35:139–142.

——. 1985b. Regional report (Region 7—Adirondack-Champlain). Kingbird 35:206–209.

——. 1985c. Regional report (Region 7—Adirondack-Champlain). Kingbird 35:282.

——. 1986a. Four Brothers Island Report no. 4. High Peaks Audubon Newsletter 14:7–14.

——. 1986b. Personal communication. Discovery Farm, Elizabethtown, NY 12932.

Peterson, R.T. 1980. A field guide to the birds. 4th ed. Houghton Mifflin Co., Boston.

Pettingill, O.S., Jr. 1936. The American Woodcock, *Philohela minor* (Gmelin). Memoirs of the Boston Society of Natural History 9:182–313.

Phillips, A.R. 1981. The races of Red Crossbill, *Loxia curvirostra*, in Arizona. *In* Annotated checklist of the birds of Arizona, ed. G. Monson and A.R. Phillips. Univ. of Arizona Press, Tucson. Pp. 223–230.

Pink, E., and O. Waterman. 1967. Birds of Dutchess County: 1933–1964. Unpublished report.

Pitzrick, V. 1977. Regional report (Region 1—Niagara Frontier). Kingbird 27:215.

Poole, E.L. 1964. Pennsylvania birds/an annotated list. Livingston Publishing Co., Narberth, PA.

Porter, W.F. 1980. An evaluation of Wild Turkey brood habitat in southeastern Minnesota. *In* Proceedings of the 4th National Wild Turkey Symposium, ed. J.M. Sweeney. National Wild Turkey Federation, Edgefield, SC. Pp. 203–211.

Portnoy, J.W., and W.E. Dodge. 1979. Red-shouldered Hawk nesting ecology and behavior. Wilson Bulletin 91:104–117.

Pospichal, L.B., and W.H. Marshall. 1954. A field study of Sora rail and Virginia Rail in central Minnesota. Flicker 26:2–32.

Post, P.W. 1961. The American Oystercatcher in New York. Kingbird 11:3–6.

Post, P.W., and G.S. Raynor. 1964. Recent range expansion of the American Oystercatcher into New York. Wilson Bulletin 76:339–346.

Post, P.W., and D. Riepe. 1980. Laughing Gulls colonize Jamaica Bay. Kingbird 30:11–13.

Post, W., and F. Enders. 1969. Reappearance of the Black Rail on Long Island. Kingbird 19:189–191.

——. 1970. Notes on a salt marsh Virginia Rail population. Kingbird 20:61–67.

Potter, E.F. 1980. Notes on nesting Yellow-billed Cuckoos. Journal of Field Ornithology 51:17–28.

Pough, R.H. 1946. Audubon land bird guide. Doubleday and Co., Garden City, NY.

——. 1951. Audubon water bird guide. Doubleday and Co., Garden City, NY.

Prescott, K.M. 1965. The Scarlet Tanager. New Jersey State Museum Investigations no. 2. Trenton.

Probst, J.R. 1979. Oak forest bird communities. *In* Management of north central and northeastern forests for nongame birds, comp. R.M. DeGraaf and K.E. Evans. U.S. Dept. of Agriculture, Forest Service General Technical Report NC-51. St. Paul. Pp. 80–89.

Puleston, D. 1977. Osprey population studies on Gardiners Island. *In* Transactions of the North American Osprey Research Conference, ed. J.C. Ogden. U.S. Dept. of Interior, National Park Service Transactions and Proceedings Series no. 2. Washington, DC. Pp. 95–99.

Purdy, K., and R. Malecki. 1984. Canada Goose. New York's wildlife resources. New York State College of Agriculture and Life Sciences Extension Publication no. 20. Cornell Univ., Ithaca, NY.

Quilliam, H.R. 1973. History of the birds of Kingston, Ontario. Kingston Field Naturalists, Kingston, ON.

Rauber, T.J. 1976. Notes on a New York nest of the Bald Eagle. Kingbird 26:122–135.

Raynor, G.S. 1941. The nesting habits of the Whip-poor-will. Bird Banding 12:98–104.

——. 1959. Recent range extension of the Veery on Long Island. Kingbird 9:68–69.

——. 1983. A method for evaluating quality of coverage in breeding bird atlas projects. American Birds 37:9–13.

——. 1984. Personal communication. Schultz Rd., Manorville, NY 11949.

Reed, H.D., and A.H. Wright. 1909. The vertebrates of the Cayuga Lake Basin, NY. Proceedings of the American Philosophical Society 48:370–459.

Reese, J.G. 1975. Productivity and management of feral Mute Swans in Chesapeake Bay. Journal of Wildlife Management 39:80–286.

Reilly, E.M., Jr. 1964. Birds and geography in New York State. Kingbird 14:197–200.

——. 1986. Personal communication. P.O. Box 21, Old Chatham, NY 12136.

Reynolds, R. 1985. Report to Atlantic Waterfowl Council. Technical Section Minutes. Atlanta.

Reynolds, R.T., E.C. Meslow, and H.M. Wight. 1982. Nesting habitat of coexisting *Accipiter* in Oregon. Journal of Wildlife Management 46:124–137.

Rice, D.L. 1986. Personal communication. Ohio Breeding Bird Atlas, 4147 Leather Stocking Trail, Gahanna, Ohio 43224.

Richards, K.C., R.O. Paxton, and D.A. Cutler. 1979. Regional report: Hudson-Delaware region. American Birds 33:850.

Richardson, F. 1967. Black Tern nest and egg-moving experiments. Murrelet 48:52–55.

Richmond, M.E., and A.G. Nicholson. 1985. New York's farming history: implications for wildlife abundance and habitat. Conservation Circular no. 23. New York State College of Agriculture and Life Sciences, Cornell Univ., Ithaca, NY. P. 8.

Riegner, M. 1982. Foraging behavior of Yellow-crowned Night-Herons in relation to behavior, distribution, and abundance of prey. Paper presented at the 6th Colonial Waterbird Group meeting. Chevy Chase, MD.

Riehlman, D. 1987. Personal communication. NYSDEC, Stony Brook, NY 11790.

Riepe, D. 1986. Personal communication. Jamaica Bay Wildlife Refuge, Gateway National Recreation Area, Floyd Bennett Field, Brooklyn, NY 11234.

Riexinger, P., W. Corbett, and W. Sharick. 1978. Hooded Merganser breeding in Schoharie County. Kingbird 28:30.

Ritchie, W.A. 1928. An Algonkian village site near Levanna, New York. Research Records of the Rochester Museum of Arts and Sciences no. 1.

——. 1946. A stratified prehistoric site at Brewerton, New York. Research Records of the Rochester Museum of Arts and Sciences no. 8.

——. 1947. Archaeological evidence for ceremonialism in the Owasco culture. Research Transactions of the New York State Archeological Association 11(2).

——. 1973a. The Sackett or Canandaigua Site (Can. 1). *In* Aboriginal settlement patterns in the Northeast, by W.A. Ritchie and R.E. Funk. New York State Museum and Science Service Memoir 20:213–225.

——. 1973b. The Bates Site (Grn. 1). *In* Aboriginal settlement patterns in the Northeast, by W.A. Ritchie and R.E. Funk. New York State Museum and Science Service Memoir 20:226–252.

——. 1973c. The Nahrwold no. 1 Site (Shr. 51–4). *In* Aboriginal settlement patterns in the Northeast, by W.A. Ritchie and R.E. Funk. New York State Museum and Science Service Memoir 20:276–290.

——, ed. 1980. The archaeology of New York State. Harbor Hill Books, Harrison, NY.

Ritchie, W.A., and R.E. Funk, eds. 1973a. Aboriginal settlement patterns in the Northeast. New York State Museum and Science Service Memoir 20.

——. 1973b. The Kelso Site (Bwv. 12). *In* Aboriginal settlement patterns in the Northeast, by W.A. Ritchie and R.E. Funk. New York State Museum and Science Service Memoir 20:253–275.

Ritchie, W.A., D. Lenig, and P.S. Miller. 1953. An early Owasco sequence in eastern New York. New York State Museum Circular no. 32.

Ritter, L.V. 1983. Growth, development, and behavior of nestling Turkey Vultures in central California. *In* Vulture biology and management, ed. S.R. Wilbur and J.A. Jackson. Univ. of California Press, Berkeley. Pp. 287–302.

Robbins, C.S. 1984. Management to conserve forest ecosystems. *In* Proceedings of the workshop on management of nongame species and ecological communities, ed. W.C. McComb. Univ. of Kentucky, Lexington. Pp. 101–107.

Robbins, C.S., D. Bystrak, and P.H. Geissler. 1986. The Breeding Bird Survey: its first fifteen years, 1965–1979. USFWS Resource Publication no. 157. Washington, DC.

Robertson, W.B., Jr. 1980. The changing seasons. American Birds 34:873.

Roosevelt, T., Jr., and H.D. Minot. 1929. The summer birds of the Adirondacks in Franklin County, New York. Roosevelt Wild Life Bulletin 5:501–504.

Root, O.M. 1968. Clay-colored Sparrow. *In* Life histories of North American cardinals, grosbeaks, buntings, towhees, finches, sparrows, and allies, by A.C. Bent and collaborators, ed. O.L. Austin, Jr. U.S. National Museum Bulletin no. 237, pt. 2. Washington, DC. Pp. 1190–1191.

Root, R.B. 1969. The behavior and reproductive success of the Blue-gray Gnatcatcher. Condor 71:16–31.

Rosche, R.C. 1966. An unusual Blue Jay nest. Kingbird 16:212.

——. 1967. Birds of Wyoming County, New York. Bulletin of the Buffalo Society of Natural Sciences Vol. 23.

Roseberry, J.W., and W.D. Klimstra. 1970. The nesting ecology and reproductive performance of the Eastern Meadowlark. Wilson Bulletin 82:243–267.

Rothstein, S.I. 1971. High nest density and non-random nest placement in the Cedar Waxwing. Condor 73:483–485.

Rusk, M.S. 1968. Birds of the Rome Sand Plains. Kingbird 18:124–127.

Rusk, M.S., and F.G. Scheider. 1966. Regional report (Region 5—Oneida Lake Basin). Kingbird 16:228.

——. 1969. Regional report (Region 5—Oneida Lake Basin). Kingbird 19:222–229.

——. 1970. Regional report (Region 5—Oneida Lake Basin). Kingbird 20:194–197.

Rusk, M.S., and C.G. Spies. 1971. Regional report (Region 5—Oneida Lake Basin). Kingbird 21:231.

Russell, R.P., Jr. 1983. The Piping Plover in the Great Lakes region. American Birds 37:951–955.

Sabo, S.R. 1980. Niche and habitat relations in subalpine bird communities of the White Mountains of New Hampshire. Ecological Monographs 50:241–259.

Safina, C. 1985a. A roseate by any other name would still be *Sterna dougallii*. Bird Watcher's Digest 7(6):66–68.

——. 1985b. Tern conservation and research—1985 report National Audubon Society. Scully Sanctuary, Islip, NY.

Salzman, E. 1983. Cerulean Warbler breeding in Suffolk County. Kingbird 33:105.

——. 1985a. A ground-nesting mixed heronry in Suffolk County. Kingbird 35:253.

——. 1985b. Breeding birds of Shinnecock Bay—birds of the barrier beach, salt marshes, and islands of the bay. Report to South Fork Chapter of The Nature Conservancy. East Hampton, NY. Mimeo.

——. 1985c. Personal communication. Box 775, East Quogue, NY 11942.

Samuel, D.E. 1971. The breeding biology of Barn and Cliff swallows in West Virginia. Wilson Bulletin 83:284–301.

Saunders, A.A. 1919. Geographical variation in the song of the Ruby-crowned Kinglet. Auk 36:525–528.

——. 1923. The summer birds of the Allegany State Park. Roosevelt Wild Life Bulletin 1:239–354.

——. 1926. The summer birds of central New York marshes. Roosevelt Wild Life Bulletin 3:324–475.

——. 1929a. The summer birds of the northern Adirondack mountains. Roosevelt Wild Life Bulletin 5:327–499.

——. 1929b. Bird song. New York State Museum Handbook no. 7. Albany.

——. 1936. Ecology of the birds of Quaker Run Valley, Allegany State Park, New York. New York State Museum Bulletin no. 16. Albany.

——. 1938. Studies of breeding birds in the Allegany State Park. New York State Museum Bulletin no. 318. Albany.

——. 1942. Summer birds of the Allegany State Park. New York State Museum Handbook no. 18. Albany.

Schaefer, V.H. 1976. Geographic variation in the placement and structure of oriole nests. Condor 78:447.

Schaeffer, F.S. 1968. Saw-whet Owl nesting in Tobay Sanctuary, Long Island. Kingbird 18:143–144.

Scharf, W.C., and G.W. Shugart. 1983. New Caspian Tern colonies in Lake Huron. Jack-Pine Warbler 61:13–15.

Schaub, B.M. 1951. Young Evening Grosbeaks, *Hesperiphona vespertina*, at Saranac Lake, New York, 1949. Auk 68:517–519.

Scheibel, M.S. 1981. Attempted breeding by Hooded Mergansers in Suffolk County. Kingbird 31:77.

——. 1982. Survey of Long Island area tern colonies. NYSDEC, Stony Brook, NY. Mimeo.

Scheibel, M.S., and I. Morrow. 1979. Survey of Long Island area tern colonies. NYSDEC, Stony Brook, NY. Mimeo.

Scheider, F.G. 1957. Regional report (Region 5—Oneida Lake Basin). Kingbird 7:97.

——. 1958. Regional report (Region 5—Oneida Lake Basin). Kingbird 8:86–87.

——. 1959. Warblers in southern New York. Kingbird 9:13–19.

——. 1971a. Highlights of the winter season. Kingbird 21:71.

——. 1971b. Highlights of the summer season. Kingbird 21:215.

——. 1975. Regional report (Region 5—Oneida Lake Basin). Kingbird 25:222.

——. 1977. Regional report (Region 5—Oneida Lake Basin). Kingbird 27:228.

——. 1986. Personal communication. 4304 Belmont Dr., Liverpool, NY 13088.

Schorger, A.W. 1952. Introduction of the Domestic Pigeon. Auk 69:462–463.

Schriver, E.C., Jr. 1969. Brief reports: the status of certain other raptors. The status of Cooper's Hawks in western Pennsylvania. *In* Peregrine Falcon populations: their biology and decline, ed. J.J. Hickey. Univ. of Wisconsin Press, Madison. Pp. 356–359.

Scott, S.L., ed. 1983. Field guide to the birds of North America. National Geographic Society, Washington, DC.

Sealy, S.G. 1980. Reproductive responses of Northern Orioles to a changing food supply. Canadian Journal of Zoology 58:224–225.

Sedwitz, W., I. Alperin, and M. Jacobson. 1948. Gadwall breeding on Long Island, New York. Auk 65:610–612.

Seeber, E.L. 1963. Red-bellied Woodpecker in western New York. Kingbird 13:188–190.

Sepik, G.F., R.B. Owen, Jr., and M.W. Coulter. 1981. A landowner's guide to woodcock management in the Northeast. USFWS Miscellaneous Report no. 253. Washington, DC.

Severinghaus, C.W., and D. Benson. 1947. Ring-necked Duck broods in New York State. Auk 64:626–627.

Seymour, M. 1986. Personal communication. NYSDEC, Wildlife Resources Center, Delmar, NY 12054.

Sharrock, J.T.R. 1976. The atlas of breeding birds in Britain and Ireland. T. and A.D. Poyser, Calton, England.

Sheffield, R., and M. Sheffield. 1963. Nesting of the Philadelphia Vireo in the Adirondacks. Kingbird 13:204–205.

Sheldon, W.G. 1971. The book of the American Woodcock. Univ. of Massachusetts Press, Amherst.

Sherry, T.W. 1979. Competitive interactions and adaptive strategies of American Redstarts and Least Flycatchers in northern hardwoods forest. Auk 96:265–275.

Shields, W.M. 1984. Factors affecting nest and site fidelity in Adirondack Barn Swallows (*Hirundo rustica*). Auk 101:780–789.

Short, E.H. 1896. Birds of western New York. F.H. Lattin, Albion, NY.

Short, L.L. 1962. The Blue-winged Warbler and the Golden-winged Warbler in central New York. Kingbird 12:59–76.

———. 1963. Hybridization in the wood warblers *Vermivora pinus* and *V. chrysoptera. In* Proceedings of the XIII International Ornithological Congress, ed. C.G. Sieley. Vol. 1. American Ornithologists' Union, Baton Rouge. Pp. 147–160.

———. 1969. Taxonomic aspects of avian hybridization. Auk 86:84–105.

———. 1982. Woodpeckers of the world. Delaware Museum of Natural History Monograph Series no. 4. Greenville.

Shugart, G.W., W.C. Scharf, and F.J. Cuthbert. 1978. Status and reproductive success of the Caspian Tern (*Sterna caspia*) in the U.S. Great Lakes. *In* Proceedings of the Colonial Waterbird Group, comp. W.E. Southern. Available from Cornell Laboratory of Ornithology, Ithaca, N.Y. Pp. 146–156.

Sidle, R.D. 1985. Endangered and threatened wildlife and plants; determination of endangered and threatened species status for the Piping Plover. Federal Register 50:50726–50733.

Siebenheller, N. 1981. Breeding birds of Staten Island, 1881–1981. Staten Island Institute of Arts and Sciences, Staten Island, NY.

Siebenheller, N., and B. Siebenheller. 1982. Blue Grosbeak nesting in New York State: a first record. Kingbird 32:234–238.

Silloway, P.M. 1920. Guide to the summer birds of the Bear Mountain and Harriman Park sections of the Palisades Interstate Park. Bulletin of the New York State College of Forestry at Syracuse 11:44, 71.

———. 1923. Relation of summer birds to the western Adirondack forest. Roosevelt Wild Life Bulletin 1:397–486.

Singer, F.J. 1974. Status of the Osprey, Bald Eagle, and Golden Eagle in the Adirondacks. New York Fish and Game Journal 21:18–30.

Skutch, A.F. 1961. Helpers among birds. Condor 63:198–226.

———. 1985. Life of the woodpecker. Ibis Publishing Co., Santa Monica.

Slingerland, D. 1983. The Mourning Dove in New York. NYSDEC, Albany. Mimeo.

———. 1986. Personal communication. NYSDEC, Wildlife Resources Center, Delmar, NY 12054.

Sloss, R. 1986. Personal communication. American Museum of Natural History, Central Park West at 79th St., New York, NY 10024.

Smiley, D. 1964. Thirty-three years of bird observations at Mohonk Lake, New York. Kingbird 14:205–208.

Smith, C.G. 1927. The Mockingbird nesting in New York. Oologist 44:44.

Smith, C.R. 1982. What constitutes adequate coverage? New York State Breeding Bird Atlas Newsletter no. 5. Delmar, NY. P. 6.

———. 1986. Personal communication. Cornell Univ. Laboratory of Ornithology, Sapsucker Woods Rd., Ithaca, NY 14850.

Smith, G.A. 1986. Personal communication. RR 1, Box 498, Mexico, NY 13114.

Smith, G.A., and J.M. Ryan. 1978. Annotated checklist of the birds of Oswego County and northern Cayuga County, New York. Rice Creek Biological Field Station Bulletin no. 5. Oswego, NY.

Smith, G.A., K. Karwowski, and G. Maxwell. 1984. The current breeding status of the Common Tern in the international sector of the St. Lawrence River, eastern Lake Ontario, and Oneida Lake. Kingbird 34:18–31.

Smith, J.L. 1978. Northward expansion of the Yellow-throated Warbler. Redstart 45:84–86.

Smith, J.N.M. 1981. Cowbird parasitism, host fitness, and age of the host female in an island Song Sparrow population. Condor 83:152–161.

Smith, J.N.M., and J.R. Merkt. 1980. Development and stability of single-parent family units in the Song Sparrow. Canadian Journal of Zoology 58:1869–1875.

Smith, P.W., R.O. Paxton, and D.A. Cutler. 1978. Regional report: Hudson-Delaware region. American Birds 32:1143.

Smith, R.L. 1968. Grasshopper Sparrow. *In* Life histories of North American cardinals, grosbeaks, buntings, towhees, finches, sparrows, and allies, by A.C. Bent and collaborators, ed. O.L. Austin, Jr. U.S. National Museum Bulletin no. 237, pt. 2. Washington, DC. Pp. 725–744.

Snapp, B.D. 1976. Colonial breeding in the Barn Swallow (*Hirundo rustica*) and its adaptive significance. Condor 78:471–480.

Southwood, T.R.E., and D.J. Cross. 1969. The ecology of the partridge. 3. Breeding success. Journal of Animal Ecology 38:497–509.

Spahn, R. 1979. Regional report (Region 2—Genesee). Kingbird 29:217.

———. 1984. Highlights of the summer season. Kingbird 34:234.

———. 1985. Personal communication. 716 High Tower Way, Webster, NY 14580.

Speier, O.T. 1965. A preliminary study of the mobbing behavior of swallows during the nesting season. Master's thesis, St. Bonaventure Univ., St. Bonaventure, NY.

Speirs, J.M., and D.H. Speirs. 1968. Lincoln's Sparrow. *In* Life histories of North American cardinals, grosbeaks, buntings, towhees, finches, sparrows, and allies, by A.C. Bent and collaborators, ed. O.L. Austin, Jr. U.S. National Museum Bulletin no. 237, pt. 3. Washington, DC. Pp. 1434–1467.

Speiser, R. 1982. Recent observations on breeding birds in the Hudson Highlands. Kingbird 32:97–101.

Spencer, B.J. 1979. Regional report (Region 10—Marine). Kingbird 29:240.

———. 1981. Regional report (Region 10—Marine). Kingbird 31:263–268.

Spencer, S.J. 1983. Of cowbirds and horses: nineteen-year summary. North American Bird Bander 8:123.

Spendelow, J.A. 1982. An analysis of temporal variation in, and the effects of habitat modification on, the reproductive success of Roseate Terns. Colonial Waterbirds 5:19–31.

Sperry, C.C. 1940. Food habits of a group of shorebirds: woodcock, snipe, knot, and dowitcher. U.S. Dept. of the Interior, Bureau of Biological Survey Wildlife Research Bulletin no. 1. Washington, DC.

Spitzer, P.R. 1978. Osprey egg and nestling transfers: their value as ecological experiments and as management procedures. *In* Endangered birds: management techniques for preserving threatened species, ed. S.A. Temple. Univ. of Wisconsin Press, Madison. Pp. 171–182.

Spofford, S.H. 1985. Some sticky solutions. Living Bird Quarterly 4(2):4–9.

Spofford, W.R. 1959. The white-headed eagle survey. Kingbird 9:22.

———. 1969. Hawk Mountain counts as population indices in northeastern America. *In* Peregrine Falcon populations: their biology and decline, ed. J.J. Hickey. Univ. of Wisconsin Press, Madison. Pp. 323–332.

———. 1971. The breeding status of the Golden Eagle in the Appalachians. American Birds 25:3–7.

Sprunt, A., Jr. 1940. Chuck-will's-widow. *In* Life histories of North American cuckoos, goatsuckers, hummingbirds, and their allies, by A.C. Bent. U.S. National Museum Bulletin no. 176. Washington, DC. Pp. 147–162.

———. 1948. Eastern Mockingbird. *In* Life histories of North American nuthatches, wrens, thrashers, and their allies, by A.C. Bent. U.S. National Museum Bulletin no. 195. Washington, DC. Pp. 299–308.

———. 1958. Eastern Boat-tailed Grackle. *In* Life histories of North American blackbirds, orioles, tanagers, and allies, by A.C. Bent. U.S. National Museum Bulletin no. 211. Washington, DC. P. 371.

Stapleton, J. 1980. Breeding Bird Survey no. 4: maple-oak forest. American Birds 34:45.

Stauffer, D.F., and L.B. Best. 1980. Habitat selection by birds of riparian communities: evaluating effects of habitat alterations. Journal of Wildlife Management 44:1–15.

Steadman, D.W. 1986. Personal communication. New York State Museum, 3140 Cultural Education Center, Albany, NY 12230.

———. In press. Vertebrates of the late Quaternary Hiscock Site, Genesee County, New York. *In* Late Pleistocene and early Holocene paleoecology and archeology of the eastern Great Lakes region, ed. R.S. Laub and D.W. Steadman. Bulletin of the Buffalo Society of Natural Sciences vol. 33.

Steadman, D.W., and N.G. Miller. In press A. Paleoecology of the late Quaternary Hiscock Site, Genesee County, New York: an initial report (abstract). New York State Museum Bulletin no. 462.

534

——. In press B. California Condor associated with spruce–jack pine woodland in the late Pleistocene of New York. Quaternary Research.

Steadman, D.W., R.S. Laub, and N.G. Miller. 1986. The late Quaternary Hiscock Site, Genesee County, New York: progress report. Current Research Pleistocene 3:22–23.

Stein, R.C. 1958. The behavioral, ecological, and morphological characteristics of two populations of the Alder Flycatcher, *Empidonax traillii.* New York State Museum Bulletin no. 371. Albany, NY. Pp. 26–27, 37–41.

——. 1963. Isolating mechanisms between populations of Traill's Flycatchers. Proceedings of the American Philosophical Society 107:25–27.

Stevens, W.K. 1985. Lush northeastern forests are at a 150–year peak. New York Times, July 7. P. 24.

Stewart, R.E. 1953. A life history study of the Yellow-throat. Wilson Bulletin 65:99–115.

——. 1958. Distribution of the Black Duck. USFWS Circular no. 51. Washington, DC. Pp. 2–4.

——. 1962. Waterfowl populations in the upper Chesapeake region. USFWS Special Scientific Report: wildlife—no. 65. Washington, DC.

Stewart, R.E., and C.S. Robbins. 1958. Birds of Maryland and the District of Columbia. North American Fauna no. 62. U.S. Bureau of Sport Fisheries and Wildlife, Washington, DC.

Stokes, D.W. 1979. A guide to the behavior of common birds. Little, Brown and Co., Boston.

Stokes, D.W., and L.Q. Stokes. 1983. A guide to bird behavior. Vol. 2. Little, Brown and Co., Boston.

——. 1986. Watching: American Goldfinches. Living Bird Quarterly 5(3):22.

Stone, C.F. 1933. Notes from northern Steuben County, New York. Auk 50:227–229.

Stone, W. 1937. Bird studies at Old Cape May. Vol. 1. The Delaware Valley Ornithological Club, Lancaster, PA.

Stone, W.B., and P. Gradoni. 1985. Recent poisonings of wild birds by diazinon and carbofuran. Northeastern Environmental Science 4:160–164.

——. 1986. Poisoning of birds by cholinestrase inhibitor pesticides. *In* Wildlife rehabilitation, vol. 5, ed. P. Beaver and D.J. Mackey. D.J. Mackey, Coconut Creek, FL. Pp. 12–28.

Stoner, D. 1932. Ornithology of the Oneida Lake region, with reference to the late spring and summer seasons. Roosevelt Wild Life Annals 2:277–764.

——. 1936. Studies on the Bank Swallow in the Oneida Lake Region. Roosevelt Wild Life Annals 4:129–233.

Stoner, D., and L.C. Stoner. 1952. Birds of Washington Park, Albany, New York. New York State Museum Bulletin no. 344.

Stout, N.J. 1958. Atlas of forestry in New York. State Univ. College of Forestry at Syracuse Univ., Syracuse.

Stoutenburgh, P. 1986. Personal communication. Skunk Land, Cutchogue, NY 11935.

Strohmeyer, D.L. 1977. Common gallinule. *In* Management of migratory shore and upland game birds in North America, ed. G.C. Sanderson. International Association of Fish and Wildlife Agencies, Washington, DC. Pp. 110–116.

Stull, W.DeM. 1968. Eastern and Canadian Chipping Sparrows. *In* Life histories of North American cardinals, grosbeaks, buntings, towhees, finches, sparrows, and allies, by A.C. Bent and collaborators, ed. O.L. Austin, Jr. U.S. National Museum Bulletin no. 237, pt. 2. Washington, DC. Pp. 1166–1184.

Sundell, R.A. 1967. Regional report (Region 1—Niagara Frontier). Kingbird 17:217.

——. 1969. Regional report (Region 1—Niagara Frontier). Kingbird 19:213.

——. 1973. Regional report (Region 1—Niagara Frontier). Kingbird 23:109.

Sutcliffe, S.M. 1986. Personal communication. 679 Waterburg Rd., RD 1, Box 37, Trumansburg, NY 14886.

Sutton, G.M. 1928. The birds of Pymatuning Swamp and Conneaut Lake, Crawford County, Pennsylvania. Annals of the Carnegie Museum 18:19–39.

——. 1960. Nesting fringillids of the George Reserve, southeastern Michigan. Jack-pine Warbler 38:46–65, 125–139.

Swift, B. 1986. Personal communication. NYSDEC, Wildlife Resources Center, Delmar, NY 12054.

Symonds, W. 1986. Personal communication. 9022 Ridge Rd. West, Brockport, NY 14420.

Taber, W., and D.W. Johnston. 1968. Indigo Bunting. *In* Life histories of North American cardinals, grosbeaks, buntings, towhees, finches, sparrows, and allies, by A.C. Bent and collaborators, ed. O.L. Austin, Jr. U.S. National Museum Bulletin no. 237, pt. 1. Washington, DC. Pp. 80–111.

Tate, J., Jr. 1981. The Blue List for 1981. American Birds 35:3–10.

——. 1986. The Blue List for 1986. American Birds 40:234–235.

Tate, J., Jr., and D.J. Tate. 1982. The Blue List for 1982. American Birds 36:132.

Temple, S.A., and B.L. Temple. 1976. Avian population trends in central New York State, 1935–1972. Bird-Banding 47:238–257.

Temple, S.A., M.J. Mossman, and B. Ambuel. 1979. The ecology and management of avian communities in mixed hardwood-coniferous forests. *In* Management of north central and northeastern forests for nongame birds, comp. R.M. DeGraaf and K.E. Evans. U.S. Dept. of Agriculture, Forest Service General Technical Report NC-51. St. Paul. Pp. 132–153.

Terres, J., Jr. 1941. Short-billed Marsh Wren in the western Adirondacks. Auk 58:263–264.

Terres, J.K. 1980. Audubon Society encyclopedia of North American birds. Alfred A. Knopf, New York.

Terrill, L.M. 1961. Cowbird hosts in southern Quebec. Canadian Field-Naturalist 75:10.

Thomas, J.W., R.G. Anderson, C. Maser, and E.L. Bull. 1979. Wildlife habitats in managed forests in the Blue Mountains of Oregon and Washington—Snags. U.S. Dept. of Agriculture Handbook no. 553. Washington, DC.

Thompson, Z. 1853. Natural history of Vermont. Published by the author, Burlington, VT.

Tilghman, N.G. 1979. The Black Tern survey, 1979. Passenger Pigeon 42:1–8.

Todd, R.L. 1977. Black Rail, Little Black Rail, Black Crake, Farallon Rail. *In* Management of migratory shore and upland game birds in North America, ed. G.C. Sanderson. International Association of Fish and Wildlife Agencies, Washington, DC. Pp. 71–83.

Todd, W.E.C. 1940. Birds of western Pennsylvania. Univ. of Pittsburgh Press, Pittsburgh.

Tomkins, I.R. 1947. The oystercatcher of the Atlantic Coast of North America and its relation to oysters. Wilson Bulletin 59:204–208.

——. 1954. Life history notes on the American Oystercatcher. Oriole 19:37–45.

Townsend, C.W. 1926. Green Heron. *In* Life histories of North American marsh birds, by A.C. Bent. U.S. National Museum Bulletin no. 135. Washington, DC. Pp. 186–187.

——. 1932. Hudsonian Spruce Grouse. *In* Life histories of North American gallinaceous birds, by A.C. Bent. U.S. National Museum Bulletin no. 162. Washington, DC. Pp. 120–129.

Trautman, M.B. 1940. The birds of Buckeye Lake, Ohio. Museum of Zoology Miscellaneous Publication no. 44. Univ. of Michigan, Ann Arbor.

Treacy, E.D. 1968. Regional report (Region 9—Delaware-Hudson). Kingbird 16:240.

——. 1976. Regional report (Region 9—Delaware-Hudson). Kingbird 26: 234.

——. 1979. Regional report (Region 9—Delaware-Hudson). Kingbird 29: 237.

——. 1982a. Regional report (Region 9—Delaware-Hudson). Kingbird 32: 216–221.

——. 1982b. Regional report (Region 9—Delaware-Hudson). Kingbird 32:296–300.

——. 1983a. Regional report (Region 9—Delaware-Hudson). Kingbird 33: 220–224.

——. 1983b. Regional report (Region 9—Delaware-Hudson). Kingbird 33:292–296.

——. 1985. Regional report (Region 9—Delaware-Hudson). Kingbird 35: 67.

Trimble, B., and P.W. Post. 1968. Photographs of New York State rarities—15. Boat-tailed Grackle. Kingbird 18:182–183.

Tripp, N. 1983. Acid rain: now it's threatening our forests. Country Journal 10:63–70.

Trivelpiece, W., S. Brown, A. Hicks, R. Fekete, and N.J. Volkman. 1979. An analysis of the distribution and reproductive success of the Common Loon in the Adirondacks. *In* The Common Loon: Proceedings of the 2d North American Conference on Common Loon Restoration and Management, ed. S.A. Sutcliffe. National Audubon Society, New York.

Tuck, J.A. 1971. Onondaga Iroquois prehistory. Syracuse Univ. Press, Syracuse.

Tuck, L.M. 1965. On the life history of the Common Snipe, *Capella gallinago* (L.). *In* Transactions of the 22d Northeast Fish and Wildlife Conference, January 1965, Harrisburg.

Tyler, W.M. 1929. Spotted Sandpiper. *In* Life histories of North American shore birds, pt. 2, by A.C. Bent. U.S. National Museum Bulletin no. 146. Washington, DC. Pp. 81–82.

———. 1937. Turkey Vulture. *In* Life histories of North American birds of prey, pt. 1, by A.C. Bent. U.S. National Museum Bulletin no. 167. Washington, DC. Pp. 12–28.

———. 1939. Yellow-bellied Sapsucker. *In* Life histories of North American woodpeckers, by A.C. Bent. U.S. National Museum Bulletin no. 170. Washington, DC. Pp. 126–141.

———. 1940a. Whip-poor-will. *In* Life histories of North American cuckoos, goatsuckers, hummingbirds, and their allies, by A.C. Bent. U.S. National Museum Bulletin no. 176. Washington, DC. Pp. 337–339.

———. 1940b. Chimney Swift. *In* Life histories of North American cuckoos, goatsuckers, hummingbirds, and their allies, by A.C. Bent. U.S. National Museum Bulletin no. 176. Washington, DC. Pp. 273–275.

———. 1942. Tree Swallow. *In* Life histories of North American flycatchers, larks, swallows, and their allies, by A.C. Bent. U.S. National Museum Bulletin no. 179. Washington, DC. Pp. 387–389.

———. 1946a. Blue Jay. *In* Life histories of North American jays, crows, and titmice, by A.C. Bent. U.S. National Museum Bulletin no. 191. Washington, DC. Pp. 35–37.

———. 1946b. Black-capped Chickadee. *In* Life histories of North American jays, crows, and titmice, by A.C. Bent. U.S. National Museum Bulletin no. 191. Washington, DC. Pp. 324–325.

———. 1948a. Red-breasted Nuthatch. *In* Life histories of North American nuthatches, wrens, thrashers, and their allies, by A.C. Bent. U.S. National Museum Bulletin no. 195. Washington, DC. Pp. 22–35.

———. 1948b. Brown Creeper. *In* Life histories of North American nuthatches, wrens, thrashers, and their allies, by A.C. Bent. U.S. National Museum Bulletin no. 195. Washington, DC. Pp. 56–70.

———. 1949. Veery. *In* Life histories of North American thrushes, kinglets, and their allies, by A.C. Bent. U.S. National Museum Bulletin no. 196. Washington, DC. Pp. 217–231.

———. 1950. Cedar Waxwing. *In* Life histories of North American wagtails, shrikes, vireos, and their allies, by A.C. Bent. U.S. National Museum Bulletin no. 197. Washington, DC. Pp. 79–102.

———. 1953. Golden-winged Warbler. *In* Life histories of North American wood warblers, by A.C. Bent. U.S. National Museum Bulletin no. 203, pt. 1. Washington, DC. Pp. 47–57.

U.S. Fish and Wildlife Service (USFWS). 1976. Nest boxes for Wood Ducks. U.S. Dept. of the Interior, Wildlife Leaflet 510. Washington, DC.

Van Camp, L.R., and C.J. Henny. 1975. The Screech Owl: its life history and population ecology in northern Ohio. North American Fauna no. 71. USFWS, Washington, DC.

Van Riet, J. 1983. Breeding Bird Survey no. 4: young red maple–gray birch forest. American Birds 37:53.

———. 1986. Personal communication. 26 Williams St., Massena, NY 13662.

Van Tyne, J. 1926. An unusual flight of Arctic Three-Toed woodpeckers. Auk 43:469–474.

Van Velzen, W.T., and C.S. Robbins. 1969. The Breeding Bird Survey, 1969. Migratory Bird Populations Station administrative report. Laurel, MD.

Van Velzen, W.T., and A.C. Van Velzen, eds. 1983. Forty-sixth Breeding Bird Census. American Birds 37:49–108.

Vargo, J., and D. Vargo. 1983. The Rabuilt Cave Site-Pke 4-1 site report. Bulletin of the New York State Archeological Association 87:13–39.

Verner, J. 1963. Song rates and polygamy in the Long-billed Marsh Wren. *In* Proceedings of the XIII International Ornithological Congress, ed. C.G. Sibley. Vol. 1. American Ornithologists' Union, Baton Rouge. P. 300.

Versaggi, N.M., J.M. McDonald, A.T. Mair II, and S.C. Prezzano. 1982. The Southside Treatment Plant Site, SuBi-672 Village of Owego. Cultural Resource Survey Report to Village of Owego, NY.

Walker, R.C. 1986. Personal communication. RD 3, Dry Hill Rd., Watertown, NY 13601.

Walkinshaw, L.H. 1935. Studies of the Short-billed Marsh Wren (*Cistothorus stellaris*) in Michigan. Auk 52:362–364.

———. 1940. Summer life of the Sora Rail. Auk 57:153–168.

———. 1944. The eastern Chipping Sparrow in Michigan. Wilson Bulletin 56:50–59.

———. 1968. Field Sparrow. *In* Life histories of North American cardinals, grosbeaks, buntings, towhees, finches, sparrows, and allies, by A.C. Bent and collaborators, ed. O.L. Austin, Jr. U.S. National Museum Bulletin no. 237, pt. 2. Washington, DC. Pp. 1217–1235.

Wallace, G.J. 1949. Bicknell's Thrush. *In* Life histories of North American thrushes, kinglets, and their allies, by A.C. Bent. U.S. National Museum Bulletin no. 196. Washington, DC. Pp. 199–204.

Ward, P. 1953. The American Coot as a game bird. *In* Transactions of the 18th North American Wildlife Conference, ed. J.B. Trefethen. Wildlife Management Institute, Washington, DC. Pp. 322–327.

Warren, P.H. 1979. Birds of Clinton County. Northern Adirondack Chapter, National Audubon Society, Plattsburgh, NY.

Watts, W.A. 1983. Vegetational history of the eastern United States 25,000 to 10,000 years ago. *In* Late Quaternary environments of the United States, ed. H.E. Wright, Jr. Vol. 1, The Pleistocene, ed. S.C. Porter. Univ. of Minnesota Press, Minneapolis. Pp. 294–310.

Weaver, F.G. 1949. Wood Thrush. *In* Life histories of North American thrushes, kinglets, and their allies, by A.C. Bent. U.S. National Museum Bulletin no. 196. Washington, DC. Pp. 105–106.

Weber, J.A. 1927. Bay-breasted Warbler breeding in the Adirondacks, New York. Auk 44:111.

Weeks, H.P., Jr. 1984. Importance and management of riparian bridges and culverts for nesting passerines. *In* Proceedings of the workshop on management of nongame species and ecological communities, ed. W.C. McComb. Univ. of Kentucky, Lexington. Pp. 163–175.

Weinman, P.L., and T.P. Weinman. 1969. The Moonshine Rockshelter. Bulletin of the New York State Archeological Association 46:11–15.

Weir, R. 1982. Regional report: Ontario region. American Birds 36:971.

———. 1983. Regional report: Ontario region. American Birds 37:985.

Weissman, B. 1985. Personal communication. 15 Laurel Hill Rd., Dobbs Ferry, NY 10522.

Weller, M.W. 1959. Parasitic egg laying in the Redhead (*Aythya americana*) and other North American Anatidae. Ecological Monographs 29:333–365.

———. 1961. Breeding biology of the Least Bittern. Wilson Bulletin 73:11–35.

———. 1964. Distribution and migration of the Redhead. Journal of Wildlife Management 28:64–103.

Wells, P.V. 1958. Indigo Bunting in Lazuli Bunting habitat in southwestern Utah. Auk 75:223–224.

Wells, R.A. 1951. More Huns than you think. New York State Conservationist 6(1):24–25.

Welter, W.A. 1935. The natural history of the Long-billed Marsh Wren. Wilson Bulletin 47:3–34.

Weseloh, D.V. 1986. Personal communication. Canada Centre for Inland Waters, Box 5050, Burlington, ON L7R 4A6.

Weseloh, D.V., P. Mineau, S.M. Temple, H. Blokpoel, and B. Ratcliff. 1986. Colonial waterbirds nesting in Canadian Lake Huron in 1980. Canadian Wildlife Service Progress Notes no. 164.

Weston, F.M. 1949. Blue-gray Gnatcatcher. *In* Life histories of North American thrushes, kinglets, and their allies, by A.C. Bent. U.S. National Museum Bulletin no. 196. Washington, DC. Pp. 345–355.

Wetherbee, D.K. 1968. Swamp Sparrow. *In* Life histories of North American cardinals, grosbeaks, buntings, towhees, finches, sparrows, and allies, by A.C. Bent and collaborators, ed. O.L. Austin, Jr. U.S. National Museum Bulletin no. 237, pt. 3. Washington, DC. Pp. 1474–1490.

Wetmore, A. 1930. Turkey Vulture in western New York. Auk 47:81.

Weyl, E. 1927a. Notes from the Mt. Marcy region, New York. Auk 44:112–114.

——. 1927b. Philadelphia Vireo and Bay-breasted Warbler in the Adirondacks. Auk 44:570–571.

Wheat, M.C., Jr. 1979. Prothonotary Warbler breeding on Long Island. Kingbird 29:190–191.

Whipple, G. 1935. A history of the management of the natural resources of the Empire State, 1885–1935. New York State Conservation Dept. and New York State College of Forestry, Albany.

Whitcomb, R.F., C.S. Robbins, J.F. Lynch, B.L. Whitcomb, M.K. Klimkiewicz, and D. Bystrak. 1981. Effects of forest fragmentation on avifauna of the eastern deciduous forests. *In* Forest island dynamics in man-dominated landscapes, ed. R.L. Burgess and D.M. Sharpe. Springer-Verlag, New York. Pp. 125–205.

White, R.P. 1983. Distribution and habitat preference of the Upland Sandpiper (*Bartramia longicauda*) in Wisconsin. American Birds 37:16–22.

Wickham, P.P. 1962. Regional report (Region 8—Hudson-Mohawk). Kingbird 12:160.

Wiens, J.A. 1969. An approach to the study of ecological relationships among grassland birds. Ornithological Monographs no. 8. American Ornithologists' Union, Lawrence, KS.

Wilcove, D.S. 1985. Nest predation in forest tracts and the decline of migratory songbirds. Ecology 66:1211–1214.

Wilcox, L. 1939. Notes on the life history of the Piping Plover. Birds of Long Island 1:3–13.

——. 1944. Great Black-backed Gull breeding in New York. Auk 61:653–654.

——. 1959. A 20-year banding study of the Piping Plover. Auk 76:129–152.

——. 1980. Observations on the life history of Willets on Long Island, NY. Wilson Bulletin 92:253–258.

Will, G.B., R.D. Stumvoll, R.F. Gotie, and E.S. Smith. 1982. The ecological zones of northern New York. New York Fish and Game Journal 29:20–23.

Wilmore, S.B. 1977. Crows, jays, ravens, and their relatives. David and Charles, London.

Wilson, A. 1811. American ornithology. Vol. 3. Bradford and Inskeep, Philadelphia.

——. 1812. American ornithology. Vol. 5. Bradford and Inskeep, Philadelphia.

Wilson, A., and C.L. Bonaparte. 1870. American ornithology. Vol. 2. Porter and Coates, Philadelphia.

——. 1876. American ornithology. Vol. 1. Chatto and Windus, Piccadilly (London).

Wolfe, L.R. 1923. The Herring Gulls of Lake Champlain. Auk 40:621–625.

Woodford, J. 1962. Regional report: Ontario–western New York region. Audubon Field Notes 16:474.

Woods, R.S. 1968. House Finch. *In* Life histories of North American cardinals, grosbeaks, buntings, towhees, finches, sparrows, and their allies, by A.C. Bent and collaborators, ed. O.L. Austin, Jr. U.S. National Museum Bulletin no. 237, pt. 1. Washington, DC. Pp. 290–308.

Wray, C.F. 1964. The bird in Seneca archeology. Proceedings of the Rochester Academy of Sciences 11:1–56.

Wright, A.H., and A.A. Allen. 1910. The increase of austral birds in Ithaca. Auk 27:63.

Young, A.D., and R.D. Titman. 1986. Costs and benefits to Red-breasted Mergansers nesting in tern and gull colonies. Canadian Journal of Zoology 64:2339–2343.

Yunick, R.P. 1976. Delayed molt in the Pine Siskin. Bird-Banding 47:306–309.

——. 1984. An assessment of the irruptive status of the Boreal Chickadee in New York State. Journal of Field Ornithology 55:31–37.

——. 1985. A review of recent irruptions of the Black-backed Woodpecker and Three-toed Woodpecker in eastern North America. Journal of Field Ornithology 56:138–152.

Zarudsky, J.D. 1980. Town of Hempstead colonial bird nesting survey, 1980. Town of Hempstead Dept. of Conservation and Waterways, Point Lookout, NY. Mimeo.

——. 1981. Forster's Tern breeding on Long Island. Kingbird 31:212–213.

——. 1985a. Breeding status of the American Oystercatcher in the Town of Hempstead. Kingbird 35:105–113.

——. 1985b. Town of Hempstead colonial bird nesting survey, 1985. Town of Hempstead Dept. of Conservation and Waterways, Point Lookout, NY. Mimeo.

——. 1986. Personal communication. Town of Hempstead Dept. of Conservation and Waterways, Point Lookout, NY 11569.

Zarudsky, J.D., and R. Miller. 1983. Nesting Boat-tailed Grackles on Pearsall's Hassock. Kingbird 33:2–5.

Zeleny, L. 1976. The bluebird. Indiana Univ. Press, Bloomington.

Zimmerman, J.L. 1977. Virginia Rail (*Rallus limicola*). *In* Management of migratory shore and upland game birds in North America, ed. G.C. Sanderson. International Association of Fish and Wildlife Agencies, Washington, DC. Pp. 46–56.

Zumeta, D.C., and R.T. Holmes. 1978. Habitat shift and roadside mortality of Scarlet Tanagers during a cold wet New England spring. Wilson Bulletin 90:575–586.

ATLAS PARTICIPANTS

In recognition of their contribution to New York ornithology, the individuals who collected the data for this book are listed below. These members of the birding community contributed more than 200,000 hours to this project.

Members of the
NYSDEC Block-busting Teams

James Clinton, 1983	Diane Emord, 1984	Malcolm Hodges, 1985	Douglas Judell, 1983	Paul Martin, 1984	Daniel Niven, 1984–85	Amy Peterson, 1985
Martha Dunham, 1982	David Gagne, 1984–85	Jack Holloway, 1984	Anthony Leukering, 1985	David Medd, 1983	Paul Novak, 1983–85	Towne Peterson, 1985
Ellen Ehrhardt, 1985	Frank Gardiner, 1985	Peter Hunt, 1983–85	Wendy Malpass, 1985	Michael Milligan, 1982–84	Barbara Olds, 1984	Douglas Zitzmann, 1984

Volunteers

Richard A'Brook	John R. Alexander	Linda Anderson	James Ash	Barbara Bailey	Roxanne Baran	Lillian Bassett
Fred Abbott	Linda Alexander	Ted Anderson	John Askildsen	Phyllis Bailey	Maxine Barber	Mrs. Frank W. Bassett
Jim Abbott	Karen Allaben-Confer	Tom Anderson	Catherine Atchinson	Jack Baird	Thomas Barber	Gordon Batcheller
Wava Abbott	Richard Allaire	Betty Andrews	Natasha Atkins	James Baird	Wavel Barber	Mary Batcheller
Kenneth Able	Dick Allen	Joseph Andrews	Lorraine L. Aust	Lois Baird	Spider Barbour	Christopher Bates
Viola Able	Douglass Allen	Ronald Andrews	Virginia Austen	Timothy Baird	Jack Barclay	Michael Bates
William Able	Elizabeth Allen	William B. Andrews	Don Avener	Berry Baker	Michele Barclay	Timothy Bates
Linda Abraham	G. Allen	Christopher Andrle	Alice Avouris	Betty Baker	Helen Bard	William Bates
Roger Abraham	L. Allen	Patricia Andrle	Phaedon Avouris	Georgia Baker	Robert Bard	Doug Batt
Roger Abramson	Marty Allen	Robert F. Andrle	Harvey Ayers	Merry Baker	Barbara Barker	Karen Batt
Dorette Abruzzi	Mary Jane Allen	Lana Andrus	Jane Ayers	Naomi B. Baker	John M. Barna	Trudy Battaly
R.B. Acker	Michael Allen	Richard Andrus	Marty Ayers	Ron Baker	B. Barnard	Shirley Baty
Gwen Ackermann	Paul Allen	M. Annsen	Sam Ayers	Cutler Baldwin	Robert Barnard	Ann Bauer
Joanne Ackermann	Louise Allexanderson	Teresa Anson	Douglas Ayres, Jr.	Jeannette Baldwin	B. Barnes	Lisa Bauer
Norbert Ackermann	T. Allworth	Rich Anthony	Barry Babcock	Joseph Baldwin	Tim Barnett	Sharon Baumgardner
Thomas Ackermann	Charles Alsheimer	Steve Antonini	H. Babcock	Justin O. Baldwin	Elsa Barnouw	Alan Bayless
Dorothy Ackley	Lawrence Alson	Robert Arbib	Marian Bach	Karle Baldwin	Betty Barnum	Alison Beal
Otto Adamec	Dean Amadon	Margaret Arinsen	Robert Bach	Kay Ballard	Jane Barry	Robert Beal
Robert Adamo	Tavi Amadon	S. Aris	Gladys Bachelet	Ron Ballard	Elinore Barth	Karl Beard
Ann Adams	P. Ambrose	Betty Armbruster	Janet Bachman	Anne D. Ballman	J. Barthelme	Nancy Beard
C. Adams	Kathy Amitrani	Kurt Armstrong	Jennifer Back	Kenneth Balmas	John Barthelmess	John Beattie
Elliott Adams	Arvid Anderson	Mrs. Ralph Arrandale	Anna Mae Bacon	S. Balzato	Richard Bartlett	Laura Beattie
Robert Adams	Avis Anderson	Ralph Arrandale	Betty Bacon	William A. Banaszewski	William Bartlett	Karen Becher
Terri Adams	Bill Anderson	Jane Arsenau	Bob Bacon	John Bandfield	Don Barton	Peter Becher
F. Albrecht	Harold Anderson	Jim Artale	C. Bader	Norma Bandy	Jennie Barton	Bob Beck
Mary Aldrich	Kaye Anderson	Bruce Artz	Marilyn Badger	Tom Baptist	Douglas Bassett	Charles C. Beck
Gladys Alexander	Kristen Anderson	Mrs. Bruce Artz	Paul Baglia	Joseph Baran	Frank W. Bassett	Gwen Beck

538

Clayton Becker
D. Becker
Dennis Becker
Esther Becker
Joseph Becker
K. Becker
Raymond Becker
Peg Beckman
Rick Bedicini
Alta Bedner
Marion Bee
Roger Beebe
Cecil Beers
Pat Beetle
Andy Behrend
Donald Behrend
Barbara Belanger
Elisabeth Belfer
Nathan Belfer
Betty Belge
John Belknap
Albert Bell
Don Bell
Judith Bell
Sara Bell
Susan Bell
Wayne Bell
Leslie Bemont
Cathy Bender
Michael Bender
Ivy Benedykt
W. Benedykt
R. W. Bengston
Craig Benkman
Bob Bennett
E. Bennett
Jean Bennett
Judd Bennett
Michael Bennett
Reynolds Bennett
Sharon Bennett
Walter E. Benning
Dirck Benson
Mary Benson
Allen H. Benton
Francis Benton
Seth Benz
J. Benzinger
Angie Berchielli
Lou Berchielli
Gil Bergen
Jeanne Bergen
Byron Berger
Myrtle Bergman
Bob Bergmann
Andrea Bergstrom
Barry Bermudez
Paul Bernath
Roberta Bernstein
Irene Bertok
Bob Bessette
Jerry Bessette
Bill Betts
Charles Beyah
Herda Bhrvil
R. Bialick

Dean Bianco
Jo Ann Bicchele
Lance Bicchele
Larry Biegel
Norman Bigelow
Charlotte Bigger
Irving Biggers
Virginia H. Billings
Lois Bingley
Georgia Binns
Rick Bircheler
David Birdsall
Nancy Birdsall
Betty J. Birdsey
Jack Bisgrove
Paul Bishop
Alice Bisk
J. Bissell
C. Blackburn
G. Blackburn
Anthony Blackett
Barbara Blackie
Ruth Blackmer
Lee Blades
Mrs. Lee Blades
Jean Blair
Patty Blair
Agnes Blake
Roger Blanchard
Gregory E. Blanchet
Charles J. Blanchet, Jr.
Thomas Bleasdale
John Bleiler
Gary Bletsch
Margery Blew
Richard Blinn
Hans Blokpoel
Clifford Blood
Bertha E. Bloom
Eugene Bloom
Mortimer L. Bloom
Karen Blumer
Robert Boardman
Frank Bobson
Sue Bock
Alan Boczkiewicz
Barbara Bodi
John Bodi
J. Bodor
Mrs. A.J. Bodurtha
Susan D Boettger
William Bogacki
Ray Bogart
Fran Bogausch
Lee Ann Bogert
Barbara Bohn
Cheryl Boise
Leslie Boise
Marilyn Bollino
Beryl W. Bond
Eric Bond
Rosann Bongey
K. Bonnlander
Joseph Bookalam
Pete Bookalam
Anna Boolukus

Helen Booth
Patricia Booth
Robert T. Booth
Debbie Boots
Dorothy Borg
Arlene Borko
Judith Borko
Martin Borko
Laura Borland
Leslie Borland
Melissa Borys
Pat Bosco
Mel Boskin
Nancy Boudrie
Paul B. Bough
Jim Bougill
Barbara Bouguio
George Bouguio
Jean Bourque
Ronald Bourque
Rob Bouta
Jeff Bouton
Gerald Bouvay
Joan Bowden
Barbara Bower
Noah Bower
George Bown
J. Bowser
Nancy M. Bowser
Lee Boyd
Alexa Boyes
Mary Boyes
Mary F. Boynton
Bruce Bozdos
Lynn Braband
Fran Brabazon
Pauline Bracy
Noel Bradbear
Diane Bradley
Joseph Bradley
Naomi Bradshaw
J. Brady
Keith Brady
S. Bramer
Ann Branch
Cynthia Brandreth
Doris Brann
Al Brayton
Marty Brazeau
Martin Brech
Abraham Bregman
Nettie Bregman
William Breidinger
Al Breisch
Sharon Breisch
Pierce Brennan
William Breslingame
Tim Brew
Lorna G. Brewer
Melissa Brewer
Naomi Brewer
Richard Brewer
Robert C. Brewer
Beverly Brickford
Jim Bridges
Martin Bridgham

B. Briggs
M. Briggs
Virginia P. Briggs
Frank Brignoli
E. Brill
Theodora Brimmer
Thomas Brink
Lysle Brinker
Eileen Bristol
Gerry Britton
Robert W. Brock
Rachel Brody
Charles Brogan
John Bronson
Marjorie Bronson
Elizabeth Brooks
R. Brooks
Charles L. Brosnan
Chris Brothers
Donald Brott
Marion Brouse
Richard Brouse
Louis Brousseau
Ray Brousseau
Arlene Brown
Art Brown
Barb Brown
Bea Brown
Betty Ann Brown
Beverly Brown
C. Brown
D. Brown
Diana N. Brown
Frank Brown
George Brown
J. Brown
Joe Brown
John W. Brown
Leonard Brown
Lynn Brown
Mark K. Brown
Myrt Brown
Rae Brown
Robert D. Brown
Rory Brown
Ruth Brown
Seth Brown
Sharon Brown
Stewart Brown
Thomas Brown
Vera Brown
Wilfred Brown
Jim Browne
Stephen Browne
Jean Browning
Jack Brubaker
Richard Brubaker
John Bruce
Lyda Bruce
Rick Bruce
Rex Brugul
Joanne Brule
Nancy D. Brundage
Susan Brunell
Mary Brusgul
John Brush

Bill Brust
Timothy Bryant
Leonard F. Bryniarski
Christopher Bubel
Frank Bucella
Lois Buck
A.R. Buckelew, Jr.
J.A. Buckelew, Jr.
Francine G. Buckley
Paul A. Buckley
Carol Budd
Ada Budda
William Budda
Robert Budliger
Greg Budney
Dan Buehler
Burrell Buffington
John Bull
Elsa Bumstead
Norma Bundy
Richard Bunting
Grant Bunzey
Ann Burch
Virginia Burch
William Burch
Cretta Burchard
Betty Burdick
Carl Burdick
Emmanuelle Burger
John S. Burgess
Thomas W. Burke
William S. Burke
Jay Burney
James Burns
Marian Burns
Jeanne Burr
Hal Burrell
Margo Burrell
Chuck Burt
Don Burt
Doris Burton
Hazel Burton
Lou Burton
Evelyn Bussman
Gregory Butcher
Diane Butcinno
Ron Butcinno
Paul Butkereit
Barbara Butler
Bryce Butler
Karen Butler
Lewis Butler
John Buttner
Bobbie Byron
Richard Byron
Danny Bystrak
Christa Cacciotti
Thomas Caggiano
C. Cahill
Marie T. Cain
Gary Calabrese
Andrea Calderone
J. Callagee
C. Calson
Bruce Campbell
Irving Campbell

Joan Campbell
Parker Cane
B.C. Cannon
John Cannon
Nancy Cannon
B. Cannon, Jr.
Irving Cantor
Peter Capainolo
Richard Capaldo
Anthony Caparolio
Estelle Capelin
Evelyn Capell
Victor Capelli
Roberta Capers
Mary Capkanis
Gregory L. Capobianco
Katherine Caputo
Peggy Caraher
Joanne Cardinali
Fred Carey
Laura Carey
Martin Carille
George Carle
Geoffrey C. Carleton
Penelope Carlin
Carolyn Carlson
Douglas Carlson
Joe Carlson
Mrs. Otto Carlson
Otto Carlson
Bernard Carman
David S. Carman
James H. Carman
Judy Carman
W.H. Carman
Gary Carmen
Mrs. Gary Carmen
Joseph Caron
Mrs. Joseph Caron
John Carpenter
Ruby Carpenter
Victor Carpenter
Tom Carrolan
Janet Carroll
Thomas D. Carroll
Tim Carroll
Bruce Carter
John B. Caruso
F. Carver
M. Carver
Carol Casazza
Gerald Cashion
Carolyn Cass
Roger Cass
Anna Casselberry
Bruce Caswell
Phil Caswell
Carolyn Catalano
D. Catalano
G. Catalano
S. Catalano
T. Catalano
Ian Cattell
Stuart Cattell
Kathy Cawrse
Karl Cerasoli

D.G. Cerretani
Robert Cerwonka
Candis Cesari
David J. Cesari
Nan Chadwick
W. Chaisson
Betty Chamberlaine
Lee Chamberlaine
Kenneth A. Chambers
Robert E. Chambers
Roger Chambers
Stephen Chang
Susan Chang
Ann Chapman
Donna Chapman
Fred Chapman
Glen Chapman
Jay Chapman
Lois Chapman
Lorry Chapman
Melinda Chapman
Mrs. James Chapman
Walter Chapman
Gerald Chapple
R. Charif
Greenleaf T. Chase
Dawn Chatfield
Dean Chatfield
Don Chatfield
Frank Chestnut
Sammy Chevalier
Chris Chiappa
Marietta Chiappa
Gene Chichester
Richard Chicoine
Margaret Childers
Roberta Childers
Wesley Childers
Emily R. Childs
Mark Chipkin
Isabel Chiquoine
Steven Chorvas
Richard Christensen
D. Christenson
L. Christenson
M. Christoff
Philip Chu
Gerald Church
M. Church
Sylvia Churgin
David Cilley
Kathy Ann Cilley
Ralph Cioffi
Julie Claffey
Faye Clancy
Charles Clapp
Arthur R. Clark
Dewey Clark
Donald F. Clark
Ed Clark
Jan Clark
Linda Clark
M. E. Clark
Margaret Clark
Mildred Clark
Mrs. Dewey Clark

Mrs. Donald F. Clark
Richard J. Clark
Roberta Clark
Alice Clarke
Anne Clarridge
Jim Class
Andrew Clauson
Terry Clavs
June Clement
Phill Clement
Walter Clement
Richard Clements
Robert Clermont
Don Cleveland
R.S. Cleveland
S. Clifford
Frank Clinch
Mildred Clinch
James Clinton
James Clinton, Sr.
Lee Clukey
Beverly Clum
Kenneth Clum
Linda Clum
Nancy Clum
C. Coats
K. Coats
Donald G. Cobb
Jean Cochrane
Jim Coe
T. Coe
Iris Cohen
Lee Cohen
S. Cohen
Roger Cohn
Barbara Cole
Doris Cole
John Cole
Leila Cole
Louis Cole
Nancy Cole
Pat Cole
M.C. Coleman
Don Colemno
R. T. Colesante
P. Collier
Mrs. Herbert Collins
Peter Collins
Ray Collins
Ruth Collins
Tim Collins
Rhea Colsman
William Colsman
Helen Colver
Nina Combes
Nina Comins
Alina Commerford
Spencer Commerford
Betty Compton
John Compton
Rosa Conbeels
Norma Coney
John Confer
Elmira Conklin
Paul Conklin
Jean F. Conley

Mae Conner
Daniel Connor
Martha Connor
Paul Connor
Barbara Conolly
Frances Converse
C. Coogan
Brad Cook
Dean Cook
Donald Cook
Eleanor Cook
Jack Cook
Juanita Cook
Kathy Cook
Leon Cook
Ray Cook
Robert Cook
Sharon Cook
Shirley Cook
Susan Cook
William Cook
Arthur Cooley
Meredith Coombs
Shannon Coombs
Bruce Coons
David R. Cooper
Janet Cooper
John Cooper
Joyce Cooper
Michael Cooper
Jim Copenheaver
Joe Corbett
Laurie Corbett
William Corbett
Bob Corbo
John Corcoran
Lester Corliss
Renee Corliss
Bart Cormier
Dick Cornell
Marvin Cornell
Robert Cornell
David Corse
Michael Corse
Alan Cortright
Linda Cortright
Steve M. Corvas
Bridget Coullon
Dominic Couzens
Joyce Covert
Wilson Covert
James C. Covert, III
Chad Covey
Bill Cox
Phil Cox
Edward Coyle
Kate Coyle
Sal Cozzolino
Ruth W. Craig
Mrs. C. Cramer
Alice Crandall
Matt Crandall
Kay Crane
P. Crane
Liza Crawford
Michael Crevier

Linda Cristman
Donn Critchell
Scott Crocoll
Edward Crohn
Estelle Cronauer
Jan Crook
John W. Crosby
Margaret D. Cross
Mrs. Kenneth Cross
Sue Cross
Lindsey M. Crouch
J. R. Crouse
Bruce Crowder
Pat Crowder
Kenneth L. Crowell
T.L. Crowell
Tom Crowell
Janice Crowfoot
Tom Crowley
Dorothy Crumb
Lloyd Crumb
Jean Crump
C. Cruz
John Cryan
Elek Csont
Wilma Csont
Joseph Cullen
Cynthia Culley
Emily Culley
Paul T. Culley
Sharon Cumm
Terrance Cumm
Jack Cunningham
Jim Cunningham
Elizabeth Curley
Hazel Curtis
Karl E. Curtis
Nancy Curtis
Van Curtis
John Curtiss
J. Cushing
Sandra Custer
Jerry H. Czech
Eric Czirr
Teresa Czirr
Brian Czora
Douglas Czora
Stephen D'Amato
William D'Anna
George Dadone
Willard T. Daetsch
M. Daigle
Paul Daigle
Roger Daigle
Ruth Dairdson
Lois Dake
Helen Dakin
Bob Dale
Matthew Dale
Charlotte Daley
Jeff Daley
Judy Damkoehler
Steven Daniel
Janee Daniele
Peggy Daniels
Roland Daniels

Rosemary Daniels
Taddy Dann
Bill Darling
Thomas Darling
Betsy Darlington
Carol Daugherty
Rod Daugherty
Peggy Daulton
Tom Daulton
Karl David
Mike Davids
Carole Davies
Elmer Davies
Angie Davis
Carolyn Davis
Dan Davis
F. Davis
Fred Davis
George Davis
Kirsten Davis
Marie Davis
N. Davis
Nelson Davis
Ruth Davis
Thomas H. Davis
Georgia Davison
Marjorie S. Dawes
D. Dawson
Deanna Dawson
Mary Dawson
Odeal Day
Olita Day
Ann Daye
John De Bell
J. De Benedictis
Paul A. De Benedictis
Laura De Chene
Peggy De Chene
Robert De Chene
Margret De Francisco
J. De George
Frances De Groff
Harrison De Groff
Ben De Heus
J. B. De Heus
Alice De Jaeger
Joan De Orsey
Stan De Orsey
S. De Silva
Bruno De Simone
Chad Deal
Louise Dean
Pete Dean
Ellie Dearn
Peter Debes
Agnes Decker
Alfred Decker
Beverly Decker
Bob Decker
Janet Decker
Robert Decocker
Donald Deed
Louise Deed
Robert Deed
Jerome Degen
Louise Cason Del Savio

James Delaney
David Delaparra
Charlcie Delehanty
John E. Delehanty
Julio Deltorro
Mrs. S. Delucia
Marion Deming
Stephen B. Dempsey
Ruth Denholm
Jack Denning
Chuck Dente
Lynn Dente
Sara Derane
Joan Derose
Ellen Derven
Peter Derven
Jane Desotelle
Charles Devan
Harriet Deverell
Larry Deverell
P. Devine
Lucinda Devlin
Susan Devlin
Alice Dewell
Vernon Dewey
Bob Dewire
Barbara Dewitt
David Deyo
Joseph Di Costanzo
Michael Di Giorgio
Mike Di Nunzio
Nola Di Simone
Dean Di Tommaso
David W. Diamond
Kenneth Dickens
Robert Dicker
Robert Dickerman
Barbara Diebeler
Dora Diesewroth
Tom Diesewroth, Jr.
Tom Diesewroth, Sr.
Robert Dieterich
Betty Dietert
Roy Dietert
Adrian J. Dignan
Helen Dildine
Walter Dildine
William C. Dilger
Nancy B. Dill
Dale Dillavon
M. Dimick
Jean Dingerson
William Dingerson
Elsie Dingman
Barb Dinse
Charles R. Dishaw
Ernest J. Dishaw, Jr.
Kathy Disney
David Dister
Courtland Dixon
Joan Dobert
Mary Dobinsky
Robert Dobson
Rena Dodd
Barbara Dodge
Cleveland Dodge

Jeff Dodge
Martin Dodge
Kelly Dodson
Ron Dodson
Lucy Doering
Eric Donahue
M. Donald
Earl Donaldson
Margaret Donaldson
William Donaldson
Carl Donath
Nancy Donath
Wilma Dondero
Harold Donner
Charles Donohue
Eric C. Donohue
Lynn Donohue
Vilda Dora
Jeanette Doran
Sadie Dorber
Jim Doris
Kay Doris
Jean Dorman
Marian Dornhaffer
James Dorr
L. Dorr
Gertrude Dorschide
Nancy Dorwart
Robert Doty
Judith Doubleday
Carol Douglas
Thelma Douglas
Mrs. W. Dousharm
Aline Dove
Harriet Dowdall
George Doyle
Terry J. Doyle
Ruth Draffen
Caroline K. Drake
Milo Drake
Fran Drakert
William Drakert
Georges Dremeaux
Myra Dremeaux
Fred Drews
Sam Droege
Trina Droisen
Hermine Drossos
Jon Drossos
Eunice Dudley
James Dudley
David Dudones
Janet Dudones
Jenny Dudones
Tom Dudones
Werner Duerr
Ruth Dufault
Keith Duffy
Sean Duffy
Debra Dufrese
Ward Dukelow
Francis Dumas
W. Dumas
Elizabeth M. Dumont
Lewis A. Dumont
Alex Dunbar

Mrs. Duane Dunbar
Kate Dunham
David Dunn
Steven A. Dunne
Carl Durdleman
George Durner
S. Dutcher
David Dygert
Bill Dyrkacz
Mary Dyrkacz
Mary Ann Dzamba
Theodore Earl
Ellen E. Eaton
Stephen W. Eaton
Blain B. Eaves
Robert Ebel
Amy Ebersbach
Richard Ebersbach
Patricia Eckel
Robert Eckhardt
Ruth Eckhardt
Jim Eckler
Evan Edelbaum
Beth Edmonds
Elaine Edmonds
Lynn Edmonds
K. Edwards
Jane Egan
Chris Egeland
Douglas Egeland
Rebecca Egeland
Tabitha Egeland
Thor Egeland
Tina Egeland
Mike Ehlers
Sally Ehrman
Bill Eichenlaub
Flora Elderkin
Frank Eldridge
Sam Eliot
Marge Elitharp
Ken Ellinger
Lang Elliot
Lynda Elliot
Sarah Elliot
M. Ellis
Marian Ellis
Walter Ellison
Susan Ells
Sally Ellyson
Mary Elmer
Virginia Elson
Al Emerton
Eleanor Emery
Helen Emery
Kurt Emmanuelle
Diane Emord
M. Emrick
Jody Enck
Mary Enck
Mary Ende
Ed Endres
Kay Endres
Marilyn England
Juanita Engle
Herbert Engman

Ronda Engman
William Ephraim
Claude Epstein
Linda Epstein
Kay Erwood
Wayne Estabrook
Mary Estey
Jean Eustis
Alan G. Evans
Alice Evans
Fred Evans
Marsha Evans
Richard Evans
Robert Evans
Ruth Evans
Michael Evanyke
Harold Evarts
Dexter Eves
Mary Eves
Thelma Eves
L. Ewert
Charles J. Fallon
Jeff Fallon
M. Fallorino
Norman Fancher
Tom Fanton
Laurie Farber
Chuck Fargione
M. Fargione
Joseph Farkas, Jr.
Evelynn Farley
G. Farley
Jane Farley
Terry Farnham
Todd Farnham
Peter Farnsworth
William Farren
Cindy Fates
M. Faust
P. Fauth
John J. Fay, Jr.
Ruth Fayerweather
Dawn Fazio
Jim Fazio
Jon Fazio
Debbi Featherly
June Feder
Leslie Federoff
Steve Feeney
Peter Feinberg
Larry Feldman
Lucille Feldman
Angel Feliciano
Henry Felle
Mary Felle
Mrs. Henry Felle
Bethany Felt
Carol Felt
Clifford Felt
Lloyd Felt
Vincent Felt
Charles Ferguson
F. Ferguson
Christine Ferrand
Warren Ferris
Margaret Ferronti

Connie Fessler
Edward C. Fessler
Ken Feustel
Sue Feustel
Jim Fiedler
Norm Fields
Ed Fiesinger
Pat Figary
Eugene File
Fran File
Audrey Finley
Nyanda R. Finley
A. Finney
Bernard Finney
Heidi Firstencel
Grace Fischer
Howie Fischer
Frank Fish
Gen Fish
James Fish
Glenn Fisher
Marian Fisher
Michael Fisher
Michael Fishman
Alice Fisk
Norma Fissler
Ethel Fitzgerald
Robert J. Fitzgerald
Connie Fitzmaurice
Francis Fitzmaurice
Janet Fitzpatrick
Christy Fitzsimmons
Mark Fitzsimmons
Henry F. Flamm
C. Flemming
Dustin Flemming
E. Fletcher
Dorothy Fleury
Gloria Flick
Larry Flick
Jeanine Flory
William Floyd
John Flurschutz
John Flynn
Jack Focht
Jack Foehrenbach
D. Fog
Chris Foltman
Dwight Folts
Nancy Fontana
Michael Foor-Pessin
Willard Foote
Ann Ford
Darrell Ford
Mariam Fordham
Stephen C. Fordham, Jr.
John Forsyth
Arleen Foster
Clarence Foster
John Foster
Marcya N. Foster
Sylvia Foster
Lew Foulke
Walter Foulke
J. Fowler
Marion G. Fox

Wyn Frahn
Elizabeth Francis
James L. Francis
Virginia Francis
Carolyn Frank
Marie Frappier
Dorothy Frederick
Ron Fredericks
Barry Freelove
Marguerite Freer
Valerie Freer
Bea Freiburg
Mrs. John Freitas
Anne French
Padraic French
Susie Frenette
Anne Frey
Elaine Frost
Robert Frost
Roger Frost
Wilma Fryer
Greg Fuerst
David Fuller
Paul Fuller
Thomas Fuller
Vivian Fulton
Clarice Furness
Greg Furness
David Gagne
Norma Galan
Joana Galdore
Barbara Gale
Wayne Gall
Pete Gallagher
Red Gallarneau
R. Gallo
Diane Galusha
Lynn Galusha
Frances Galvin
Lee Gammage
Natalia Garcia
Barbara Gardina
Hugo Gardina
Mrs. M.M. Gardner
Raylene Gardner
Chandler Gardyne
Frances Gardyne
Martin Garilla
P. Garille
Arthur Garland
Mrs. S. Garlick
Delight Gartlein
Jay Gartner
Anne Garvey
Edward Gates
Phyllis Gates
Jean Gawalt
Paul Gebhard
Lois Gebhardt
Emily Geddes
John P. Gee
Peggy A. Gee
Karen Geiger
Ralph Geiser
Ralph Genito
Sid Genoux

Gloria Gentile
Marion George
Robert Gerdts
Ludlow Gere
E. Gerhards
Ann German
Florence Germond
David Gersak
Ritz Geuv
Sandy Giandana
Frieda Gibbons
Barbara Gibson
Bruce Gibson
Dave Gibson
Doris J. Gibson
Elizabeth Gibson
Jay Gibson
K. Gibson
Louise Gibson
Bob Giffen
Neil Giffen
Donna Gilbert
Sandy L. Gilbert
Sibyll Gilbert
T. Gilbert
Violet Gilbert
Paul Gildersleeve
David P. Gillen
David T. Gillen
Thomas Gillen
Carlton Gillette
J. Gilligan
Louis Gillmeister
Jean Gilmore
Art Gingert
Arthur Ginter
T. Giovanelli
Susan Girard
Chris Given
K. Given
Jeffrey Glassberg
Vince Glasser
John Gleason
D. Gleick
Lois Glenn
Robert Gloria
Michael Gochfeld
Joanne Goetz
Sandy Goetz
Aden Gokay
Bella Golden
Mary Golino
D. Gomez
D. Goodfellow
Peter Goodfriend
David A. Gooding
Synnova H. Gooding
Lindsey Goodloe
Robert J. Goodrich
Bob Goodsell
R. Goodwin
David C. Gordon
Richard Gordon
Dianne Gorman
William G. Gorman
Ed Gormley

Pera Gorson
T. Goss
Richard Gostic
Bob Gotie
Mildred Gould
H.J. Goulding
Marrack I. Goulding
S.R. Goulding
L. Gouthreau
Austin Gowan
Myrtle Gowan
Ike Goyette
Scott Graber
Elisabeth Grace
Peter Gradoni
Alma Graef
Andy Graef
Florence Graham
Rich Graham
Edward Grant
George Grant
Lucile Grant
Marilyn Grant
Norma Grant
Lena A. Graton
Clarence Gravelding
Esther Graves
Helen Graves
Daniel Gray
Douglas J. Gray
Mal Gray
Mary Gray
Susan Gray
Tim Gray
Julie Greco
Brian Green
Charles Green
John I. Green
Mrs. Charles Green
Pat Green
Brian Greene
D. Greene
Don Greene
Evelyn Greene
Philip Greene
Warren Greene
Jon S. Greenlaw
Thomas Gregg
Steve Gregor
David Gregory
James Gregory
Norma Gregory
Amy Greher
Warren Greher
Jennifer Gretch
Mark Gretch
Theodore Gretch
Carrie Grey
Gregory Griffen
Harold Griffin
James Griffin
Judy C. Griffith
Kevin C. Griffith
Vincent Grilli
Raymonde Grimaldi
Cardelion T. Grimm

Bob Grindrod
L. Grindrod
Ted Grisez
Marian Griswold
Thomas Griswold
Charles Groescup
Madelene Groescup
Walter Gronski
Tim Grotheer
Pat Grover
William T. Gruenbaum
Sophie Gryzlec
Joseph Grzybowski
Dotty Guard
Howard Guard
John Guarnaccia
Tom Guba
Dick Guck
Marguerite Guernsey
Annette Gula
Joseph Gula
Jean Gunderson
Emily A. Guse
William Guse
Laura Gussman
Dorothy Gustafson
John Gustafson
Andrew Guthrie
Isabel Guthrie
Richard P. Guthrie
Tom Guthrie
Bea Guyette
Cassius Guyette
Lois Gyr
Rudy Gyr
Elizabeth Haas
Marybeth Haer
Robert Hagar
Tom Hagar
Richard Haggerty
Harold Haglund
Paul Haight
Thelma Haight
Judy Haines
Sol Hait
Henry Halama
Rita Halbeisen
M. Halce
Harland Hale
Richard Hale
J.L. Haley
B. Hall
Kirsten Hall
Michael Hall
Peter Hall
Barbara Halliley
Jerry Halliley
James Halling
Emma Hallopeter
Everett Hallopeter
Kathy Halloran
June Hamel
David Hamelin
D. Hamilton
Harriet Hamilton
Jill Hamilton

M. Hamilton
Ken Hamm
Sarah Hamm
Donald Hammel
Christina Hammond
Dorothy Hammond
Julia Hammond
Wesley Hammond
Kay Hampshire
Thomas M. Hampson
Jong U. Han
M. L. Hanaka
S. Hanaka
Ann Hance
Clifton Hance
Wayne Hance
D. Hancock
Fran Hanes
Polly Hanford
Mary Hannan
Vincent Hannan
Evan Hannay
Ives Hannay
Doris Hansen
Dorothy L. Hansen
Lawrence Hansen
John Hanyak
Douglas Happ
Bob Harcleroad
Sue Harden
Mrs. William H. Harder
Betty Harms
Bonnie Harnish
James Harnish
Thomas Harper
Genevieve Harrington
Jonathan Harrington
C. Harris
L. Harris
Scott Harris
Constance Hart
Shirley Hartman
W. Hartman
William Hartranft
Reg Hartwell
Doris Hartz
Jackie Harvey
Jason Harvey
Marjorie Harwitz
Hazel Haselton
Celia Hastings
Julius Hastings
Phyllis J. Hatch
Gertrude Hauck
Galvin Hausen
Mrs. Galvin Hausen
Joanne Hauser
Josef Hauser
P. Havemeyer
Barrington S. Havens
Elva Hawken
Dolly Hawkins
George Hawkins
Sandra Rae Hawkins
Helen Hays
Lois Head

Diana Heath
John Heath
Paul A. Hebert, Jr.
John Hecklau
Sue Heckman
Elizabeth Hedges
Helen Hedlund
Dorothy Heidler
Richard S. Heil
Douglas Heilbrun
Scott Heim
Beth Heimlich
Ralph Heimlich
Jane Hein
Raymond C. Hein
Andrew Heineman
William Heineman
Paul Heinz
H. Heinzerling
Teresa Heise
Jean Held
Charles Helgeson
Dolores Helgeson
Ton Hellman
Carl Helms
Lynn D. Hemink
Myrna Hemmerick
S. Hendler
J. Hendrason
Judy Hendrix
D. Heneka
Kurt R. Henning
Robert Henrickson
M. Henry
Richard Henry
D. Herder
Barbara Herlich
Pete Herlich
C. Herman
Virginia Herman
Joel Hermes
Lee Hershon
Jeff Herter
Jan Hess
Peter Hess
Jane Heyer
Cathy Hibbard
Alan Hicks
Betty Hicks
Charlotte Higbee
George Higbee
Tom Higgins
Seward Highley
Dale Hildebrand
Bernice Hilfiker
Armas Hill
Cheryl Hill
Cora Hill
Maurice Hill
Harlan Hine
John Hines
Lisle T. Hines
Barbara Hipp
Helen Hirschbein
Bruce Hiscock
Fred Hiscock

Nancy Hiscock
T. Hisko
Ethel Hoag
Al Hobart
Seldon Hochschartner
June Hodges
Russel Hoeflich
Mildred Hoffman
Charles Holbrook
Michael Holbrook
Janet Holmes
Robert Holmes
Anne Holt
Fenton Holt
Patricia Holt
John Holtzapple
Joseph Holveck
J. Homburger
J. Hood
Chris Hoogendyk
Betty Hooker
Dean Hoover
Roy Hopkins
John Hor
Audrey Horbett
Blythe Horman
Andy Horn
Bob Horn
Rhoda Horne
Edwin Horning
Katherine Horning
Martha Horning
Steve Horowitz
Stuart Hosler
Mrs. Thomas Hoster
Mathew Hotchkiss
Esther Houghtaling
Mrs. Isaac Houghtaling
Gertrude Houghton
David Houle
Peter Houlihan
Pauline Hovemeyer
Barbara Howard
Billy Howard
E. Howard
Frieda Howard
Gordon Howard
Hank Howard
M. Howard
Mrs. E. Douglas Howard
Robert Howard
Wilifred I. Howard
Claude Howard, Jr.
Gretchen Howe
Harlan Howe
Shirley Howe
Sam Hoye
Harold Hoyt
Linda Hoyt
D. Hubbard
Margaret Hubbard
S. Hubbard
Hans Huber
Gene Huck
Herbert Hudnut
T. Hudson

Gene Huggins
Lee Hughes
William Hughes
John Hull
O. Hull
Rose Humegay
Anne Hungerford
David Hungerflch
Pam Hungerford
Bea Hunt
C. Hunt
Jim Hunt
Michelle Hunt
Peter Hunt
John Hunter
Robert Hunter
C. William Huntley
Peter Hurd
Robert Hurd
Alan Hurst
Joan Hurst
R.E. Hurst
Alice Husar
Betty Huth
Alvin Hyde
Larry Hymes
Sarah Jane Hymes
Tomia Hysko
Stephen Iachetta
Johan Inborr
Philip Ingalls
John Ingham
Isabel Ingraham
Tony Ingraham
Veronica Insel
Charlie Intres
Rick Intres
Irene Irwin
Selma Isil
Nurak Israsena
Robert Italiano
Mary Ellen Ivers
Norman Ives
J. Ivett
Martha B. Jablonski
Carol Jack
James Jackson
Lawrence W. Jackson
Hal Jacobi
Mrs. Hal Jacobi
Constance M. Jacobs
Jack Jaenike
B. James
Catherine Jansen
J.R. Jaquillard
Richard D. Jarvis
Joe Java
Nellie Jeanblanc
Paul Jeheber
Harry Jenkins
Jerry Jenkins
Scotty Jenkins
Larry Jenks
Stanton Jenks
James R. Jennings
Oivind E. Jensen

Kira Jenssen
Peter J. Jerkowicz
Frank S. Jewett
Wayne A. Jock
Edward Johann
Arthur Johnsen
A.J. Johnson
Alene Johnson
Carol Johnson
Catherine Johnson
Dave Johnson
Eleanor Johnson
Elsbeth Johnson
Glenn Johnson
P. Johnson
Richard Johnson
Suzy L. Johnson
Alice Jones
Anita Jones
Carolyn Jones
Dorothy A. Jones
Dovie Jones
Euclid Jones
Gilbert Jones
Hannah Jones
John Jones
Marcia Jones
Morgan Jones
Nancy Jones
Norman Jones
Sheila Jones
Wardell Jones
Morgan Jones, Jr.
Joan Jordan
Mary Jordan
Peter Joust
Bea Joy
Lois Juckett
Cec Judd
Marjorie Julian
Marion Junes
David Junkin, II
David Junkin, III
David Junkins
Edward Just
Scott Just
Ben Kagen
Leroy Kahl
Steven Kahl
David Kain
A. Kaiser
I. Kaiser
Gail Kalison
Ed Kanze
George D. Kappus, Jr.
Marian Karl
Mike Karp
Henry Karsch
Betty Karuth
Kenneth Karwowski
M. Kashiwa
Judy Kauffeld
Karen Kauffeld
Donald Kauffman
Jeanne Kauffman
Doris Kaufman

Jeanne T. Kaufmann
Dean Kear
Herbert Keating
Polly Keating
Jean Keck
David Keefer
Delight Keefer
Brian Keelan
Steven Keene
George Kehr
Jeddu Keil
Karen Keil
Caroline Keinath
David Keller
Henry Keller
Jeffrey Keller
John Keller
Robert K. Keller, Jr.
Harry Kelley
Joan Kelley
Genevieve Kellock
Barbara Kellogg
Dorothy Kelly
Karen Kelly
Larry Kelly
Richard Kelly
Homer Kelsey
Paul Kelsey
Edward Kemnitzer
Mary Kemnitzer
John Kemnitzer, Jr.
Norm Kendall
Bev Kennison
D. Kent
Robert J. Kent
Ruth Kent
Brendon Keogh
Jeanne Keplinger
Joseph Keplinger
Donald Kern
Marion Kern
Nancy Kern
Roger Kern
Chris Ketcham
Thelma Kett
Jim Key
Mary Key
Richard Keyel
C.L. Keyworth
Douglas P. Kibbe
Andrew Kibler
Lewis Kibler
Jane Kidney
Mrs. Edward Kieb
Jennifer Kiely
Charles Kiewig
Douglas Kiewig
Ellen Kiewig
Heather Kiewig
Margo Kilburn
Carolyn Kilby
L.J. Kimmel
Irma Kimpton
Lynn Kindinger
Mark Kindinger
Alice King

Blanche King
Emma King
Floyd King
Laura King
Marie King
Paul King
Sharri King
Stanley King
Wallace G. King, Jr.
Hugh E. Kingery
Urling Kingery
Marguerite Kingsbury
M. Kingsway
Anne Kinner-Johnson
Dave Kintzer
Hazel Kinzly
Gail Kirch
Robert Kirker
Jane Kirkpatrick
Nancy Kirst
K. Kirwan
John Kiseda
Joan Kissling
Francis Kittner
Erik Kiviat
Jo Kixmiller
Robert Kixmiller
Harriette Klabunde
Walter Klabunde
Bert Klein
Bob Klein
Harold Klein
Georges Kleinbaum
Christel Kleinhuber
Gina Klick
Tom Kligerman
Clarence Klingensmith
Margaret Klingensmith
Dawn Klinko
Robert Klips
Allan Klonick
Helen D. Klopf
John Knapik
Emery Knapp
Roberta Knapp
Ann Knight
Bill Knight
Fritz Knight
Laura Knight
Marilyn Knight
Ruth Knight
Alan Koechlein
Margaret Koechlein
Paul Koehler
Ruth Koehler
Joan Kogut
Ken Kogut
Lorraine Kolb
Barbara Koll
Helen Kolo
William Kolodnicki
Alexei Kondratiev
Gregory Konesicy
John Koopmans
Art Kopp
Lillian Korner

George Kosmaler
Mrs. Thomas Koster
Bob Kousi
Sue Krafchak
Ed Krautz
Thomas C. Kressley
Daniel Kriesberg
Robert Krinsky
Helen Krog
Kimberly N. Krog
Norman Krog
Rodney Kromer
Dorothy Krotje
William Krueger
Robert Krug
M. Kruger
Derrick Krym
Mrs. Joseph Kucher
Boon Kuhn
Marvin Kuhn
William Kulinek
Judy Kumler
Philip Kumler
Dorothea Kunz
Hans Kunze
Allen Kurtz
Jeff Kurtz
Elliot Kutner
Ray Kutzman
Hazel La Barge
Louis H. La Barge
Mrs. Louis H. La Barge
Mary La Belt
Mary La Croy
Ferdinand La France
Harmon T. La Mar
Helen La Mar
Jo Anne La Mar
Norman La Mar
Tom La Mont
Fred La Pann
B. La Roer
Dorothy Lacombe
Ignatius Lacombe
Anne Lacy
Marge Lafayette
J. Lafever
Arthur Laflamme
James Lafley
William Lafley
Carolyn Laforce
Ronald Laforce
Phil Laino
Jeanne Lally
Eric Lamont
Mary Laura Lamont
William Lamoureaux
Julie Lamy
Danny Lancor
Judy Land
Laurie Landau
Bruce Landon
Barb Lane
Dave Lane
Dorothy Lang
Mrs. David Langdon

Betty Langendorfer
John Langendorfer
M. Lango
Clarice Lanigan
B. Lansing
Helen Lapham
Frances Lapinski
Betty Laros
David Larsen
Gail Larsen
Nels Larsen
Erma Larson
Lucy Laser
Margot Lash
Sonny Lasher
Bob Laskowski
Michael J. Laspia
Wilma Lass
Robert Lastine
Fuat Latif
Carol Latone
Mrs. Ralph Latten
Ralph Latten
Sarah B. Laughlin
B. Laundry
Anthony Lauro
Brook Lauro
Alton Lavack
David Lavack
June Lavack
Mary Lavender
Mark Lawler
Bob Lawrence
John Lawrence
Joyce Lawrence
Mrs. Burgess Lawrence
Richard Lawrence
Anne Lay
Margaret Layton
Jun Ming Le
Joe Le Clair
M. Le Marchant
Terence Leahy
Alex Leal
John Leal
Meg Leal
Susan Leal
Kent Lechner
Sandra G. Lednor
Tony Lednor
Brian Lee
E. Lee
Gary Lee
William J. Lee
D. Leete
Claudia Leff
Mrs. Elton Lefure
Harold Legg
Chris Lehman
Jay Lehman
Paul Lehman
Lewis Leidwinger
Joanne Leisenring
Gerry Lemmo
Richard Lent
Alice Lentz

L. Leombruno
B. Leonard
Kris Leonard
Lucille Leonard
Natalie Leonard
Robert Leonard
William Leonard
Lorinda Leonardi
Nick Leone
Lynn Leopold
Ixora Lerch
Malcolm Lerch
Eugene Lesser
George Lesson
Al Lester
Tony Leukering
John Leverett
Carl Levickas
Dawn Levickas
C. Levin
Emanuel Levine
Rob Levine
L. Levitan
Anne Levy
David Levy
Ira Levy
Olive Levy
Bob Lewis
Ken Lewis
Michael Lewis
Mildred Lewis
Norm Lewis
Robert H. Lewis
Lu Li
Barbara Lide
Mary Lienke
Lou Lieto
J. Lilliock
Florence Linaberry
Harold Linaberry
Stanley Lincoln
Tom Linda
Allan Lindberg
Lois Lindberg
Larry Linder
Barb Lindholm
A. Lindsey
Mark Lindvall
Rachel Lindvall
Helen Link
Robert Link
Clayton Linstead
Gary Lintner
Helen Lippert
John Lippert
John Liptak
Matt Liptak
George D. List, Jr.
Walter Listman
Phil Littler
Thomas S. Litwin
Barbara Litz
Valerie Lloyd
Warren Lloyd
Mary Lo Guidice
Patricia Lo Guidice

Bob Lobou
Jim Loe
Audrey Loener
Gladys Loeven
Janice Loftus
Arthur Long
Dorothy Long
Robert Long
John M. Longyear
Gladys Loomis
Bud Lorch
Charles M. Lord
William Lord
E. Lorette
K. Lorette
Christopher Loscalzo
Eleanor Loucks
D. Louden
Maggie Lovass
Tom Lovass
Lydia Love
Ronald Love
Louis Loveland
James Lowe
Donald E. Lown
Bob Lubeck
Ellen C. Luce
Vince Lucid
Marianne Ludwigsen
Eva Lukert
Edward Lutz
Jean Lutz
Madeline Lutz
Richard Lyle
Clellie Lynch
Danny Lynch
Joanne Lytle
Elizabeth Mac Arthur
Joyce Mac Arthur
Travis Mac Clendon
Esther Ann Mac Cready
Beryl Mac Donald
Carol Mac Donald
Jill S. Mac Donald
Joseph Mac Donald
Willis Mac Donald
Kristi Mac Dougal
M. Mac Gibbon
William C. Mac Gregor
Margaret Mac Intyre
Donald Mac Kenzie
Cynthia Mac Lean
David Mac Lean
Diana Mac Lean
Barbara Mac Turk
William Mac Turk
Michael Machette
Theodore Mack
John Macomber
Marjorie Madajewicz
Emily Madden
Harry Madden
J. Hayward Madden
Sheila Madden
Samuel R. Madison
Gertrude Magaverin

Derek Mahon
Jean Mahon
Terry Mahoney
Barbara Mai
William Mai
Jean Mailey
Josephine Main
Marjorie Main
Earl Mainville
Ken Malcolm
Pam Malcolm
Richard Malecki
Andrea Malik
Marlene Mallen
Mary Mallory
Lorraine Mancroni
K. Mann
Marc Manske
Helen Manson
Harold Manzer
Dave Manzo
Alan A. Mapes
Kathy Mapes
Barb Maple
William T. Maple
Bienvenido Mardones
David Markham
Peg Markham
Ted Markham
Judy Markowsky
Marcia Marks
B. Marleau
Joe Marlo
Harry Maroncelli
Frances Marrus
Richard Marrus
Gordon H. Marsh
Harriet Marsi
Rick Marsi
Alan Martel
G. Martens
Audrey Martin
Betty Martin
Janet Martin
Nancy Martin
J. Marx
Robert Marx
Brad Masem
Andrew Mason
Jane Mason
Pat Mason
Sterling Mason
T. Norman Mason
Julie Masowich
Kurt Massey
Larry Master
Ray Masters
Alan Masterson
J. Mastini
Dana Mather
Dian Mathews
Nancy Mathews
Marge Mathis
V. Mathis
Tom Matterfis
Dianne Matthews

Michael Matthews
Eleanor Mattusch
M. Maurer
George Maxwell
Susan Maxwell
Al May
Bob May
Larry Mayer
Douglas Maynard
Bob Mays
Patricia Mazzarella
Richard Mazzarella
Lynn Mc Allister
Ed Mc Bride
Laura Mc Bride
Eugene R. Mc Caffrey
M. Mc Cann
Gary Mc Chesney
Alexis Mc Clean
Helen Mc Clure
Jerry Mc Conville
Glen Mc Cormick
Judy Mc Cormick
Cindy Mc Cranels
Joe Mc Cranels
Linda Mc Cullough
Russell Mc Cullough
Enoch Mc Cune
Joan Mc Dermott
Judy Mc Dermott
Kenneth Mc Dermott
Douglas Mc Donald
Lisa Mc Donald
Florence Mc Donough
Leona Mc Dowell
K. Mc Garigal
Robert Mc Grath
Thomas A. Mc Grath
Vincent Mc Grath
James Mc Graw
Dave Mc Gregor
Ginny Mc Gregor
Gail Mc Guire
Laura Mc Guire
Dorothy Mc Ilroy
John Mc Ilwaine
Leo Mc Intosh
Sue Mc Intosh
Carol Mc Intyre
Don Mc Intyre
Jo Ann Mc Intyre
Judith Mc Intyre
Kay Mc Intyre
Alice Mc Kale
Willard Mc Kale
Lisa Mc Kasty
Alex Mc Kay
May Mc Kay
Stanley Mc Kay
K. Mc Keaver
Christopher Mc Keever
Kay Mc Keever
David Mc Keith
Roger D. Mc Keon
J. Mc Kernan
Margaret Mc Kinney

Robert Mc Kinney
Lerena Mc Lead
Melville Mc Lear
Lorena Mc Leod
Sally L. Mc Leod
Ed Mc Mahon
Sheila Mc Mahon
Dorothy Mc Michael
Gloria Mc Murtry
John Mc Neil
Pamela Mc Neil
Charles Mc Nichols
David Mc Nicolas
Don Mc Nin
William Mc Niven
Ora Mc Nulty
Jim Mc Vey
Eric Mc Williams
Susan Mc Williams
John Mead
Mrs. George Mead
Gordon M. Meade
Hildred Meade
Laura S. Meade
Robert Meade
Oliver Meddaugh
Robert B. Meech
Doug Meillhant
Phil Meisner
Claudia K. Melin
Diane Meltzer
Ferne Merrill
Sheldon Merritt, Sr.
Larry Merryman
James Mershon
Bob Merweck
Mrs. Alfred Meschinelli
Jack Mesick, Jr.
Wendy Mesnikoff
David Messineo
Edith Messinger
Stan Metcalf
Ellie Metzler
Sarah Meurs
John Meyer
Kay Meyer
Barbara Michelin
Jim Micklas
Mary Mickle
Philip Mickle, Jr.
B. Mihalyi
Louis Mihalyi
Peter Milburn
Marilyn Miles
Carl Miller
Dave Miller
Ellen Miller
George P. Miller
Gretchen Miller
J. Robert Miller
Jean Miller
John S. Miller
Laurie Miller
Marian Miller
Marianne Miller
Melinda Miller

Mrs. George P. Miller
Mrs. J. Robert Miller
Raymond A. Miller
Renee E. Miller
Robert L. Miller
Rosemary Miller
Ruth Miller
Sandra Z. Miller
Rose Millerolte
Michael Milligan
Bob Milliken
Kathy Milliken
Ron Milliken
Greg Millman
M. Millman
Bernard Mills
Herb Mills
Louise Mills
Melissa Mills
Mrs. Kenneth D. Mills
Ted Mills
G.J. Milmoe
Jim Miner
Ron Miner
William F. Minor
Michael Miranda
Anna Miranian
A. Mishkit
Elizabeth Miskis
Mrs. Peter Miskis
Helen Misner
Karl Misner
Bradley Mitchell
Harold D. Mitchell
J. V. Mitchell
Rita Mitchell
Robert Mitchell
Charles Mitchell, Jr.
Dianne Mitchelson
John Mitton
Kathy Moberg
Tom Moberg
Ginny Moede
Peter Mohr
Barbara Molyneaux
Erwin Molyneaux
Dennis Money
Sheila Money
John Monje
David Monk
Michelle Monnier
Bobbi Monroe
Laura Moon
Neil Moon
Betty Moore
Charles Moore
David Moore
Jane Moore
Kevin Moore
Louise Moore
Lucille Moore
Sharon Moore
Carrol Moores
Ray Morantz
Jerry Morea
Donald R. Morey

Victoria Morey
Janice Morgan
Lee Morgan
Nancy Morgan
R. Morgan
Tom Morgan
Celeste Morien
Thomas E. Morien, Jr.
Thomas E. Morien, Sr.
Arthur Morris
Louisa Morris
Robert H. Morris
Beth Morrison
Frank Morrison
L. Morrison
Vaughn Morrison
Ian Morrow
Helen Morse
John S. Morse
Karen Mort
Linda Morzillo
Ken Mosher
Terence D. Mosher
Bette Moss
Bob Moss
Steve Moss
George Mosseau
Richard Mostent
John Moyle
Joanne Mrstik
Richard Mrstik
Helga Mueller
D. Muir
John Mulholland
Andrew Mullen
Henry Muller
Judy Muller
James Munves
Bryce Murphy
Dennis Murphy
Diana Murphy
Frank Murphy
Keith Murphy
Darrell Murray
Ellen Murray
Michael Murray
Sally Murray
Mark Mushkat
Nadia Mutchler
Tom Mutchler
Valerie J. Myers
Eric Mynter
Dorothy Nagel
Pat Nagle
John Nagy
Raymond Nava
Wendover Neefus
Milton Nehrke
Carole Neidich-Ryder
Arden Nelson
David Nelson
Donald Nelson
Eileen Nelson
Mary Nelson
Robert Nelson
Jeff Nerp

K. Delphena Nessle
Susie Nettleton
Mrs. Peter Nevaldine
Peter Nevaldine
Kathleen Nevins
John G. New
Dan Newby
Ruth Newman
William Floyd Nichols
Dan Nicholson
Daniel Nickerson
Frank Nicoletti
Charles Nicosia
Robert Nightman
Leona Nilhelmsen
Rose Nitschke
Daniel Niven
Kenneth Niven
Norma Niven
Charles Noble
Francesca Noble
Calvin Nobles
Mrs. Calvin Nobles
Helen Nodecker
Barry Noon
Steve Nord
Irvin Norman
Phil Norman
Connie Norte
Daniel V. Norte
Barbara North
Elaine H. Norton
Mrs. Yale Norton
Winifred Norton
Yale Norton
Elizabeth Notman
Will Notman
Betty Novak
Paul Novak
Jeff Nugent
B. Nunn
Lincoln Nutting
Peter E. Nye
Bill O'Brien
Diane O'Brien
Harold O'Brien
Linda O'Brien
Mrs. William O'Brien
William O'Brien
Toni O'Bryan
Alan O'Connell
Louise O'Connell
Terry O'Connell
C. O'Connor
Mary O'Connor
Suzanne O'Dea
Thomas O'Dea
Ann O'Dell
Richard O'Hara
Tim O'Hara
Katherine O'Keeffe
Dorothy O'Leary
Francis O'Leary
Russ O'Malley
Terry O'Neill
John O'Pezio

Peter O'Shea
Bob O'Sullivan
Frank O'Sullivan
James O'Toole
David Odell
Ralph W. Odell, Jr.
Douglas Odermatt
Rabbit Oehlbeck
Joan Ogrodnik
Joseph Okoniewski
Susan Olcott
Barbara Olds
Mark Oliver
Alice Oliveri
Gene Olnyk
Peggy Olnyk
Odmund Olsen
Rodney Olsen
Gus Olson
John Olson
Sylvia Olson
Julius Olszewski
Florence Olyphant
David Oneil
Susan Opotow
S. D. Oppenheimer
Dan Orcutt
Emerson C. Orford
Robert Orkovitz
Mildred Orlow
Chuck Orne
Sherrie Orne
Jack Orth
Nancy Osborn
Murray Osmond
William Ostrander
Joyce Ostrowski
Dawn Otello
Jesse Ottesen
Dianne Outlaw
Frank Overton
William N. Overton
Bob Packard
A. Paddock
Ford Paddock
Helen Paddock
Kay Paddock
Cindy Page
Randy Page
Ruth Paige
Patty Paine
Peter Paine, Sr.
Diane Palaszynski
Edward Palaszynski
J. Palmer
Katherine Palmer
Peter Palmer
Stanley Palmer
Thomas R. Palmer
Jack Pangburn
Drew Panko
Irene Panko
Laura Panko
David Pannone
Robert Pantle
Nola Paquette

Sylvia Paquette
Olive Paquin
Francis S. Park
Nelson Park
William J. Park
William T. Park
John Parke
Renate Parke
Cherry Parker
Dan Parker
Jack Parker
Rosemary Parker
Pat Parkinson
Irene Parks
Kitty Parks
Barbara Parnass
Norma Parson
Jack H. Parsons
Norman Parsons
Arthur Partridge
Mrs. Tom Passamonte
Vince Passamonte
Bea Pastel
Robert Pastel
Sandie Paston
Bruce Pataky
Bill Paterson
Debbie Patrick
James E. Patterson
David Patton
Bernard Paul, Jr.
Bernard Paul, Sr.
Bob Pavelka
J. Pawlak
Robert O. Paxton
John Paye
Maxine Payne
Ron Payson
Ruth Peacor
C. Pear
R. Pearlmutter
Carol Pearsall
Edmund Pease
John Peasley
L.L. Pechuman
Belle Peebles
Manuel Peebles
Owen Peet
Clark Pell
Howard Pellet
Patricia G. Pendell
Larry Penny
Leonard Pepkowitz
Philip Pepper
Abby Perelman
Gerald S. Peretz
Terry Perkins
B. Perretti
Claudia Perretti
George Perrin
Lucinda Perrin
William Pesold
Brooky C. Peters
Chris Peters
Al Peterson
Allen Peterson

David Peterson
Janet Peterson
John M. C. Peterson
Nancy Peterson
Robert Peterson
William D. Peterson
Pete Petokas
Mary Pettit
Ted Pettit
Marie Petuh
Audrey Pfund
Mrs. Alex Phillips
Ruth Phillips
Mona Philson
Cynthia Phinney
Helen Phinney
D. Pica
Catherine Pierce
Halliday Pierce
Dorothy Pihlblad
Thelma Pikett
Cay Pillsbury
Elizabeth Pillsbury
D. Pinckney
Norman C. Pinckney
Eleanor Pink
Clive Pinnock
M. James Pion
Dan Pisanello
Alan Pistorius
Raymond Pitzrick
Vivian Pitzrick
Rick Plage
Marge Plant
Jerry Platt
Rachel Pleuthner
Peter Pliniski
Doris Plunkett
Floyd Plunkett
Carrie Poelvoorde
David Pohle
Gunther Pohle
G. Pohlig
Mrs. G. Pohlig
Lambert Pohner
J. Lloyd Poland
Susan Pollard-Walker
Hilda Pollock
Ned Pollock
Peter Polshek
Geoff Pond
John G. Poor
A. Poore
Robert G. Popp
Walter Popp
Grace Porter
Bea Post
Brian Post
Harold Post
Paul Post
Peter W. Post
Robert Post
Betsy Potter
Dave Potter
Lynn Potter
Morris Potter

Mrs. Ted Potter
Ted Potter
Tim Potter
Mark Potter, Jr.
Barry Powell
Jon Powell
Sue Powell
Elsie K. Powell, Jr.
Anne Powers
David B. Powers
Steve Pradon
Joseph Praino
Charles Pratt
Delia Pratt
Rosemary Pratt
Mrs. Sam Prentiss
Sam Prentiss
Richard Prestopnik
D. Preziosi
Steven Prievo
Zu Proly
Jean W. Propst
Ken Propst
Alec Proskine
Louise Proskine
S. Prossner
T. Prusa
Dorothy Ptak
Louis Pugliani
Clement Puleston
Dennis Puleston
Jennifer Puleston
Eunice H. Purcell
Carl Purple
Barbara Putnam
Bob Putnam
Vic Putnam
Linda Quackenbush
P. Quillan
Charles Quinlan
Adeline Quinn
Candace Quinn
Glenn Quinn
Willa Rabeler
Mary Radlubowski
Gary Randorf
Mike Randsford
Joe Ranski
Craig Ransom
Mary Ransom
Bob Rasmussen
Gerald P. Rasmussen
Joan Rasmussen
Susan Rasmussen
Dorothy E. Rauch
Pat Raup
Frank Rawls
Katherine A. Ray
Michael Ray
David Raymond
Gilbert S. Raynor
Marie P. Read
Jennifer Reaser
Sue Reaser
Cindy Rechlin
Alan Reckhow

Ruth Redjives
David Reed
E. Reed
Susan Reed
David Reedy
Gary Reeves
William Reeves
Hilo Reich
H. James Reiche
Fiona Reid
Ruth Reid
Marge Reilly
Edgar M. Reilly, Jr.
David Reimers
Bob Reiner
Cindy Reiner
Bernd Reinhardt
Bernice Reinhardt
Martha Reinhardt
Patty Reister
John Remias
Beryl Remington
J. Remy
Ronnie Renoni
Karen Rentz
Lee Rentz
S. Resler
Douglas Resme
Mike Restuccia
Stephanie Restuccia
Debbie Retzlaff
James Retzlaff
Frances M. Rew
Dolly Rewinski
M. Reynolds
Sara Reynolds
Jean Rezelman
John Rezelman
Meredith Rhindress
Richard Rhindress
Agnes Rhodes
Everett Rhodes
Harold Rhodes
Lori Riccardo
Dawn Rice
Doris Rice
Helen Rice
Jim Rice
Vesta Rice
Arthur Richard
Hanna Richard
Mary Richards
Sue Richards
Holly Richardson
Jane Richardson
Jim Richardson
June Richardson
Scott Richardson
John Ricks
Doug Riedman
Alfred Rieker
David Riemers
Don Riepe
Patricia Riexinger
Bob Rifenburg
Evelyn Rifenburg

Brandon Riggins
Leslie Rigley
Mary Rigley
Frank Riley
Mike Riley
Susan Riley
Terry Riley
Thomas Riley
Atea Ring
B. Rist
Betty Rist
Tom Ritchie
Betty Robb
Roger Robb
Charles Robbins
Edythe Robbins
C. Roberts
Richard Roberts
T. Roberts
Hammond Robertson
Joan A. Robertson
Alice Robinson
Dorothy Robinson
H. Robinson
Ralph Robinson
Jonathan Robson
Jim Rod
John Rodgers
D. Roeske
Mark Roffe
Ed Rogers
George Rogers
Ivan Rogers
M. Rogers
Nancy Rogers
R. Rogers
S. Rogers
B. Rollfinke
James Romansky
Aline Romero
Forrest Romero
M. Ronins
Darwin Roosa
Pat Root
Arthur Rosche
Lee Rosche
Mrs. Richard Rosche
Olga Rosche
Richard Rosche
Mrs. Donald P. Rose
Barry Rosen
Diana Rosen
Edith Rosen
Jerry Rosenband
Richard Rosenbloom
Chuck Rosenburg
Martin Rosencranz
Bob Rosenheck
Robert Rosenswie, Jr.
Jonathan Rosenthal
Mike Rosenthal
Robert Rosenzwig
Sheila Rosenzwig
David Rosgen
Alice Ross
David Ross

Dave Rossie
Jane Rossman
Ray Roswell
Barbara Roth
Herbert Roth
Mrs. Peter Roth
Polly Rothstein
Lee Roversi
Bill Rowe
Karen Rowley
Charles Royal
Richard Royal
Carl Ruff
Brian Ruffe
A. Rugerson
Gertrude M. Ruggles
Allan Ruhl
H. Ruhl
K. Ruhl
Mae Ruhl
Sterling Ruhl
Lorrie Rumble
Marguerite Rumsey
Peg Rumsey
Roger Rumsey
Kenneth Rundell
Carl Runyon
Art Runzo
Barbara Rupp
Bill Ruppert
Charles Ruppert
Sarah Ruppert
John Ruscica
Margaret S. Rusk
Beverly Ruska
John Ruska
Eleanor M. Rusling
Arlene Russ
S. Russel
Robert Russell
Robert P. Russell, Jr.
Paul Russo
Frank Rutkowski
Jean Rutkowski
Kathy Rutkowski
Maria Rutkowski
Sylvia Rutkowski
David Rutkowski, Jr.
David Rutkowski, Sr.
Steve Ryan
Robert Rybzynski
Ronald T. Rycroft
Rich Ryder
Mona Rynearson
Walton B. Sabin
Ann Saeli
Carl Safina
Marie Sage
Sam Sage
Ed Sallie
Judy Sallie
Herbert Saltford
Barbara Salthe
Rebecca Salthe
Stanley N. Salthe
Eric Salzman

Lorna Salzman
Gorden Samer
Helen Samson
John Sandwick
John F. Sandwick, Jr.
Steven Jay Sanford
Nelson Santiago
P. Santora
Starr Saphir
Ingeborg B. Sapp
Rick Saracelli
Paul Saraceni
Sarah Saul
J. Saunders
Richard Sautkulis
Mary Saville
Dorothy Sawyer
Mark Sayre
James Scaff
P. Scannell
Al Scarpinato
Kathi Schachinger
Alfred J. Schadow
John H. Schadow
Shirley M. Schadow
S. Schaffer
Arthur Schaffner
Henrietta Scharran
Melinda Schatz
Billy Scheffel, Jr.
L.C. Scheibel
Michael Scheibel
Pat Scheible
Fritz Scheider
Erma Schell
Betty Schenk
George Schermerhorn
Alfred Scherzer
Barbara Scherzer
Albert Schiavone
Helen Schiel
Adam Schiff
Seymour Schiff
Susan Schirger
Violet Schirone
Martin Schlabach
Onolee Schlageter
Fred Schlauch
Marge Schmale
Diane Schmalz
D. Schmidt
Kathy Schmidt
Patricia Schmidt
Robert Schmidt
Carol Schmitt
Ron Schmitt
Arnie Schmollinger
Bruce S. Schneider
Elinor Schneider
Susan Schneider
Art Schneier
Lil Schneier
Paul Schnell
M. Schnitt
Diane Schoder
Nancy Schoen

Magda Schoenwetter
Syd Schoenwetter
David Schonfeld
Dave Schonfield
Lore Schore
Robert Schrader
Albert Schreck
Joel Schreck
L. Schreiber
Terry Schreiner
Tootie Schrieber
Earl Schriver
Alan B. Schroeder
Edith Schrot
J. Rudolph Schrot
Kathy Schwab
Steve Schwab
Emil Schwarting
Jeff Schwartz
Mark Schwartz
Pauline Schwartz
Robert Schwartz
Victor Schwartz
David Schwarz
Wally Schweinsberg
Jolene Scofield
Robert Scofield
Ann Scott
Esther R. Scott
Janet V. Scott
Kenneth Scott
Morris Scott
Richard Scott
Jean Scrocarelli
Lynn Seabrook
Dolores Seaman
Gail Seaman
Joseph Sedlacek
Susan Sedlacek
Mary Sedler
Walter Sedwitz
Bonnie Seegmiller
Steve Seegmiller
Beth Seery
John Sefcik
Benton Seguin
Jim Seibert
J. Seifts
Diane Semanski
Gail Senesac
Pixie Senesac
Judy Senio
Caryl Seplow
Lucia Severinghaus
Eric Seyferth
A. Seymour
David Seymour
Sue Seymour
Vicki Seymour
Joseph Shafer
William Sharick

Millie Sharp
Eric Shattuck
Mary Jeanne Shaughnessy
Debbie Shaw
Ellen Shaw
Lee Shaw
Mary Shea
Dick Shear
Robert I. Shear
A. Sheets
E. Sheets
Mary Sheffield
Robert Sheffield
Blair Sheilds
Tony Shelley
Frank Shellhorn
Sandy Shellhorn
Don Shelling
Margaret L. Sheperd
Julian Shepherd
Margaret Sheppard
Marianne Sherman
Tim Sherrat
E. Shettine
William Shields
Ruth Shipman
William Shipman
Katherine Sholtz
S. Shore
Judy Short
Avery Shuey
Betty Shuey
Jane Shumsky
Nancy Shuster
William Shuster
David Sibley
Steve Sibley
Tom Siccama
Richard Sichel
R. Sick
Glenn Sickels
Jane Sickels
Sue Sickels
Grace Sicles
Bill Siebenheller
Norma Siebenheller
Edward Siemon
Nancy Siemon
Emily Sillars
Earl Silvernail
Jane M. Silvernail
Janet Silvernail
James Sime
Peter Sime
Dorothy Simmons
Grant Simmons
Thomas Simmons
Michael Singer
Vera Sinning
Jean Siracuse
Grace Sisler
Kay Sisson
Phyllis Skalwold
Robert Skalwold
Jeanne Skelly
Durward C. Skiff

Ellen Skiff
Florence Skiff
David Skinner
Norman Skinner
Cathie Slack
H. Slack
Michele Slack
Roy Slack
Carole Slatkin
Robert Slechta
Elanor Sleeth
Don Slingerland
Richard Sloss
Karen Slotnick
C. Walter Smallman
Mary H. Smallman
Sarah Smeltzer
Keith Smiley
Ruth Smiley
Alberta Smith
Alice Smith
Ann Smith
Anne Smith
Arnold H. Smith
C. Lester Smith
Charles Smith
Charles R. Smith
David H.A. Smith
Ed Smith
Edyth Smith
Emmette Smith
Ethel Smith
Evelyn Smith
George Smith
Gerald A. Smith
Greg Smith
Harold Smith
Helen Smith
James Smith
Jeff Smith
John Smith
Lester Smith
Marion Smith
Mary Smith
Meg Smith
Mrs. Donald F. Smith
Mrs. Henry Smith
Nicholas Smith
P. Wayland Smith
Robert Smith
Royal Smith
Ruth Smith
Stephen W. Smith
Thelma Smith
Tom Smith
Warren Smith
Arthur C. Smith, Jr.
Charles J. Smith, Jr.
Harold H. Smith, Jr.
Sally Smithson
Skip Smithson
Robert Smolker
Dot Smrtic
Natalie C. Snare
Glen Snell
Brad Snider

Pat Snyder
Alex Soda
Dorothy Soda
John Sojda, III
Vivian Sokol
Olivia Solomon
Ken Soltesz
Marilyn Sorrell
Bettye Southwick
Robert Spahn
Susan Spahn
Gigi Spates
Dean Spaulding
Evan Speck
Stan Speck
Carol Speegle
Barbara Spencer
Charlotte Spencer
James Spencer
Robert Spencer
Selden Spencer
C. Speno
Michael Speno
Ray Spenser
Burt Sperling
Jon Sperling
Linda Sperling
Christopher Spies
Lee Spinning
Lorie Sposato
John Sprague
M. Sprague
Elizabeth Spraker
Sherm Squires
James Stacey
Jim Stack
B. Stahlberg
Charles Staloff
Maureen Staloff
Joyce Stammer
Ann Stanbury
Birgette Stanford
Ora Stanford
Rex Stanford
Steve Stanne
Susan Stappers
Betty Starr
Harold Stasch
Mrs. Harold Stasch
David States
Rosemary States
Rich Stavdal
Anne Stear
Bruce Stebbins
Paul F. Steblein
George Steele
Mrs. Frank Steele
Nancy Steele
Joseph Steeley
Ruth Steese
Mary Ann Stegmeier
Rose Steidl
Burnadean Stein
Alma K. Steinbacher
Judith Stelboum
Martha Stenta

Amy Stephens
Connor Stephens
Trudy Stephens
Greg Sterling
Frank Sterrett
Jamie Sterrett
Mae J. Sterrett
Marianne Sterrett
Nat Sterrett
Tim Sterrett
Craig Stevens
Sally Stevens
Langdon Stevenson
Mac Kenzie Stewart
Margaret Stewart
Mike Stickles
Sherrie Stieberitz
William Stieberitz
Carey Stiffler
Steve Stiffler
Chris Stiles
Sandy Stiles
Timothy Stiles
Betsy Stille
John Stock
Tom Stock
Charles Stoffel
Chris Stokes
Mark Stolzenburg
Abby Stone
Garry Stone
Gary Stone
Udell B. Stone
John Stonick
Magdalena Stooks
Norman Stotz
Deborah Stout
Luella Stout
Randy Stout
William Stout
Paul Stoutenburgh
Peter Stoutenburgh
Steve Stowe
Carlos H. Straight
Anne Strain
Susan Strange
Barrie Strath
Betty Strath
Gary Strathearn
Beatrix Strauss
Ed Street
Jerry Strein
Harry Strickling
Ruth Strickling
Louis Strndtka
Mildred Sturken
Connie M. Sturm
Wesley Suhr
A. Sulger
Connie Sulivan
M. Sullivan
Dorie Summers
Drew Summers
June Summers
Robert Sundell
Roger Sundell

Ann Sunderland
Carolyn Sunderland
Lincoln Sunderland
Vincent Sunderland
Mary Ann Sunderlin
Dennis Sustare
Scott Sutcliff
Reed Sutherland
Gen Sutter
Claudia E. Swain
Clarence Swanson
Francis Swanson
Norma Swanson
Irene W. Swanton
Douglas Swartout
Elaine Swartout
Lois Swarts
William Swarts
Howard Sweet
Louise Sweet
James Swiencicki
Bryan Swift
Byron Swift
Morrisen Swift
Gary Sylvester
William Symonds
Michael Symons
Jerry Tabbia
Kay Taber
Patricia A. Taber
Walt Taber
Stan Taft
Mrs. Bruce Taggert
Mrs. Gordon Taggert
D. Takacs
Shirley Talbert
Tammy Talbot
Sarah Talpey
Abbie Tamber
John E. Tanck
Barbra Tancredi
Helen E. Tank
Mrs. M.L. Tanner
Pat Tarkowski
Clay Taylor
Helen Taylor
John Taylor
Joseph Taylor
Norma Taylor
Robert E. Taylor
Russell Taylor
Wendell Taylor
Czecher Terhune
D. Terry
Ellen Terry
Jim Terry
Malcome W. Terry
Tim R. Terry
Robert Terwilliger
Steven Tetley
David Tetlow
Joan Thiele
John Thill
Joseph F. Thill
Constance T. Thomas
Elizabeth Thomas

Evelyn Thomas
Jack Thomas
Jane Thomas
Lester Thomas
M. Thomas
Marguerite Thomas
Michael J. Thomas
Olive Thomas
Ruth Thomas
Wilford J. Thomas
Emmy Thomee
Bev Thompon
Agnes Thompson
Craig Thompson
Don Thompson
Edna Thompson
Elizabeth Thompson
Frances Thompson
Ginny Thompson
John Thompson
Judy Thompson
Martin Thompson
Mrs. George Thompson
William Thompson
Grace Thomson
Jerry Thomson
G. Thorne
Kay Thorne
Melvin Thorne
Judy Thurber
Grace Thurston
John Thurston
E. B. Tierson
William Tierson
Chris Tiesl
L. Tillinghast
W. Tillinghast
J. Timmerman
Donald Timmons
Dorothy Timmons
Janice Timmons
Jeanne Tinnsman
Betsy Tisdale
Russell Titus
Barbara Tomasi
Bill Toner
Melissa Toner
Nick Torino
Ann Townsend
Anthony Townsend
Sarah L. Townsend
William Townsend
Peter Tozzi
Matthew Trail
Paris Trail
Pepper Trail
Harriet Trainer
Lee Traines
John Tramontano
Mary Trautwein
Donald Traver
Donna Traver
James F. Traynor
Edward Treacy
Joanne Treffs
Karl Trever

Lynn Trevett
G. Trieller
Nathan Tripp
Marguerite Trocme
Dorothy Troike
David Truscott
Nancy Truscott
Jan Trzeciak
Alfred Tucker
Arvis Tucker
Louis Tucker
Mavis Tucker
Guy Tudor
Kathleen Tuohy
B. Turin
Michael Turisk
Celia Turk
Helen Turnbull
Henry Turner
Joan Turner
John Turner
Scott Turner
Jean Tuthill
Marion Tuthill
Ralph Tuthill, Jr.
Josephine Tuttle
Merrile Tuttle
Joanna Twinstrom
Bud Tyler
Betty Tyrell
Kim Tyrell
M. Tyrell
R. Tyrell
Debbie Tyrrell
Jeffrey Ulmer
Marion Ulmer
Ed Ulrich
Marge Ulrich
Bob Unnasel
Illiam E. Uook
John Updyke
Tom Utley
Jim Utter
Lorraine Utter
Elaine Vadnais
Stephen Vahk
Armand Vaillancourt
Scott Vakay
Shelley Vakay
J. Valente
Mrs. J. Valin
Andrew Vallely
Mary Van Buren
David Van Etten
Martha Van Loan
Muriel Van Orden
Emily Van Riper
Millicent Van Schaick
James Van Scoy
Regina Van Scoy
Ann Van Sweringen
Philip Van Valen
Marion Van Wagner
Jean Vanbencoten
Barb Vanderberg
Carolyn Vanderwerf

Gary Vanderwyst
Elba Vandresar
Jim Vandresar
J. Vandrie
Dolores Vanetti
Joannes Vanriet
Tony Vanriet
Kenneth Vansickle
Greg Vanslyke
Marty Vanslyke
Doris C. Vaughan
William C. Vaughan
Thomas Vawter
Francis A. Velasquez
Rosario Venuto
Alphonse Vercek
Hazel Verde
Bob Vetter
Margaret Vetter
Bonnie Vicki
William Vierhile
R. Viet
Jason Viglietta
Joe Viglietta
Mercedes Villamil
Maurice Villeneuve
Pauline Villeneuve
Marion Vince
Chester Vincent
Elizabeth Vincent
Barry Virgilio
K. Voegtle
Douglas Vogt
Barbara Voight
John Voight
Chris Von Schilgen
Alfred Voorhies
Fran Vosburgh
George Wade
Arthur Wagner
John Wagner
Louise Wagner
Millie Wagner
Robert M. Wagner
Betty Wahl
Chris Wainwright
John Wainwright
Pat Wainwright
Esther Waite
A. Wakefield
J. Wakefield
Robert Wakefield
R. Wakeman
John Walcott
Steven Waldron
Tom Waldron
Ellen Waldstein
Robert Waldstein
Henry Walike
Barbara Walker
John Walker
June I. Walker
Philip Walker
Robert C. Walker
Jeffrey Walters
Robert Walters

Brian Walton
Richard Walton
Linda Waltos
S. Wang
Ann Wanser
Clyde Ward
Cornelius Ward
Gladys Ward
Harold R. Ward
James A. Ward
Jeanne L. Ward
Ken Ward
Pat Ward
Charlotte Warden
Robyn Warn
Sandy Warn
William Warne
Allen Warner
Dale Warner
Harvey F. Warner
Jane Warner
M. Warner
Steve Warner
Peter Warny
Charles R. Warren
Glen Waruch
Anna Washburn
Elizabeth Washburn
Loren Washburn
Margaret Washburn
Lillian Washington
Otis Waterman
Irene Waters
Jane Waters
Bruce Watson
Dorothy Watson
Louis Watson
William Watson
Hyla Watters
W. Wayne
Marjorie Weatherly
Jim Weaver
Lesley Weaver
Elsie Webb
Gary Webb
Joseph Webb
Alice L. Weber
Dick Weber
Hans W. Weber
Loni Weber
Dan Webster
Edna Webster
Beverly Weeks
Harold Weeks
John A. Weeks
Mrs. Charles Weiczochowski
Steve Weiczochowski
Randy Weidner
A. Weigand
Helen S. Weil
Susan Weil
E. Weinberger
Marc Weinberger
Mathilde P. Weingartner
Kenneth Weinstein
Joe Weise

Kurt Weiskotten
Carol Weiss
Berna Weissman
Thomas D. Welch
Eleanor Weld
Paul Weld
Floyd Welker
George Welles
Mary Welles
William Welles
Cora Wellman
Richard Wells
Chris Welsh
Marie Wendling
Herman Wenz
Susan Wernert
Robert J. Werrlein
Bill Wertenbaker
Judy Wertenbaker
D.V. Weseloh
Patti West
Richard L. West
Maxwell Corydon Wheat, Jr
Cheryl Wheaton
Diane Wheelock
G. Whitbeck
Clarence G. Whitcomb
Connie Whitcomb
Florence Whitcomb
Blake White
Christopher White
Hollis White
James White
John H. White
Linda White
Mary White
Mrs. Hollis White
Randall White
Robert E. White
Ruth White
Kathryn Whitehorse
Donna Whiteman
John Whiting
Polly Whitney
Jan Whittaker
Guy Whittall
Edna Whyman
Richard Wiesenthal
Sandra T. Wiesenthal
Debbie Wiggens
Robert Wightman
Sue Wijeyesekera
Suril Wijeyesekera
Linda Wilcox
Barbara Wild
Irene Wilder
M. J. Wildermuth
Harold Wiley
Joan Wilgus
B. Wilhelm
Fred Wilhelm
Llona Wilhelm
Bunny Wilkening
Gary Will
Munro Will
Robert Will

Norma Willard
Cleon J. Williams
Clifford Williams
E.A. Williams
Gertrude Williams
Henry J. Williams
Jim Williams
Liz Williams
Mrs. Tom Williams
Tom Williams
Lila Willis
T. Wills
Betsy Wilmarth
Jay Wilmarth
Aubrey Wilson
Betty Wilson
Emma Wilson
Glenn Wilson
Grace Wilson
J. Wilson

Kathryn Wilson
Larry R. Wilson
Lorna Wilson
Margaret Wilson
Roger Wilson
Ruth Wilson
Doris Wilton
Mary Windels
Rosemary Windheim
Don Windsor
Joan Windsor
Leila Windsor
B. Wineman
Mildred Wingert
Elizabeth H. Winn
Arnold Wise
K. Wisniewski
Lawrence F. Withington
James Woehr
Tim Woehr

Mrs. William Woernley
Dorothy Wohlbach
Cecelia Wojciukiewicz
Karen Wolf
D. Wolfe
Gerald Wolfe
Martina Wolfe
Susan Wolfe
Pat Wolff
Peter Wolff
Lynn Wolfgram
Cyril T. Wolfling
Marguerite Wolfsson
Alvin Wollin
Alice Wood
Carl Wood
Carol Wood
S. Wood
B. Woodall
L. Woodard

Robert Woodard
Rose Ann Woodard
Olwen Woodier
Nick Woodin
Helen Woodmancee
N. Woodrey
Don Wooster
Tim Wootin
Tim Wooton
Tom Wooton
Florence R. Worden
Richard Wormwood
Julie Worth
Betty Worthington
Peter Worthington
Sarah Worthington
Richard Worthley
L. Wovner
Alice Wray
Crystal Wright

David Wright
John Wright
Margaret Wright
Robert Wright
Stacey Wright
Ed Wroblewski
John Wunderlich
Robert Wyatt
K.A. Yager
David Yando
Mrs. James Yando
Pat Yando
Miriam Yarnell
Roy Yarnell
M. Yarnes
Clifford Yates
Gregory Yates
Lori Yates
Joseph Ycas
Mary A. Ycas

C. Yeager
Sam Yeaton
Mary Yegella
Raymond Yelle
George Yerdon
Peter Yoerg
David Young
Kathy Young
Kim Young
Kristine Young
Lucille Young
Steve Young
John Yrizarry
Elizabeth Yule
Robert P. Yunick
Richard Zaineldeen
Richard Zander
Katherine Zarik
John Zarudsky
Michael Zebehazy

Joseph Zelinski
Harvey Zendt
Mary Zendt
Rita Zendt
Steve Zendt
Edward Zero
Julie Zickerfoose
Eileen Zimmer
M. Zimmer
A. Zimmerman
Fred Zirega
C. Kendall Zoller
Frank Zoller
Marilyn Zoller
Fred Zolna
Joe Zoranski

PHOTOGRAPHERS

We are grateful to the following for supplying photographs that proved a valuable resource to the artists: R. and D. Aitkenhead, Tom Barden, Richard E. Bird, Hans Blokpoel, Wilson C. Bloomer, Robin Bouta, L. Page Brown, Alvin Carpenter, H. R. Comstock, John L. Confer, Betty D. Cottrille, Christopher Crowley, Allan D. Cruickshank, H. Cruickshank/VIREO, Dorothy W. Crumb, J. H. Dick, John Dunning, Bill Duyck, Bill Dyer, Lang Elliott, Diane Emord, John Gavin, B. B. Hall, Jean R. Harris, Mike Hopiak, Ule F. Hublitz, Isador Jeklin, Elaine Kibbe, Rick Kline, Steve Kress, John Krimmel, Dorothy Levy, Judith McIntyre, Denny Mallory, H. Mayfield, W. A. Paff, O. S. Pettingill, Esther and Don Phillips, Berthold Seeholzer, F. Shanholtzer, C. Singletary, Walter Spofford, J. Surman, Mary Tremaine, Fred K. Truslow, Y. R. Tymstra, Donald Waite, Lawrence Wales, L. H. Walkinshaw, P. H. Watson, James Weaver, J. Weeks, John Wiessinger, J. R. Woodward, Zoo-America ®.

548

INDEX

549